Selected Papers from PRES 2018

Selected Papers from PRES 2018

The 21st Conference on Process Integration, Modelling and Optimisation for Energy Saving and Pollution Reduction

Special Issue Editors

Jiří Jaromír Klemeš
Petar Sabev Varbanov
Paweł Ocłoń
Hon Huin Chin

MDPI • Basel • Beijing • Wuhan • Barcelona • Belgrade

Special Issue Editors

Jiří Jaromír Klemeš
SPIL, NETME Centre, FME,
Brno University of
Technology—VUT Brno
Czech Republic

Petar Sabev Varbanov
SPIL, NETME Centre, FME,
Brno University of
Technology—VUT Brno
Czech Republic

Paweł Ocłoń
Institute of Thermal Power
Engineering, Faculty of
Mechanical Engineering,
Cracow University
of Technology,
Poland

Hon Huin Chin
SPIL, NETME Centre, FME,
Brno University of
Technology—VUT Brno
Czech Republic

Editorial Office
MDPI
St. Alban-Anlage 66
4052 Basel, Switzerland

This is a reprint of articles from the Special Issue published online in the open access journal *Energies* (ISSN 1996-1073) from 2018 to 2019 (available at: https://www.mdpi.com/journal/energies/special_issues/PRES_2018).

For citation purposes, cite each article independently as indicated on the article page online and as indicated below:

LastName, A.A.; LastName, B.B.; LastName, C.C. Article Title. *Journal Name* **Year**, *Article Number*, Page Range.

ISBN 978-3-03928-134-3 (Pbk)
ISBN 978-3-03928-135-0 (PDF)

© 2020 by the authors. Articles in this book are Open Access and distributed under the Creative Commons Attribution (CC BY) license, which allows users to download, copy and build upon published articles, as long as the author and publisher are properly credited, which ensures maximum dissemination and a wider impact of our publications.

The book as a whole is distributed by MDPI under the terms and conditions of the Creative Commons license CC BY-NC-ND.

Contents

About the Special Issue Editors . . . . **vii**

Preface to "Selected Papers from PRES 2018" . . . . **ix**

Jiří Jaromír Klemeš, Petar Sabev Varbanov, Paweł Ocłoń and Hon Huin Chin
Towards Efficient and Clean Process Integration: Utilisation of Renewable Resources and Energy-Saving Technologies
Reprinted from: *Energies* **2019**, *12*, 4092, doi:10.3390/en12214092 . . . . **1**

Jayne Lois G. San Juan, Kathleen B. Aviso, Raymond R. Tan and Charlle L. Sy
A Multi-Objective Optimization Model for the Design of Biomass Co-Firing Networks Integrating Feedstock Quality Considerations
Reprinted from: *Energies* **2019**, *12*, 2252, doi:10.3390/en12122252 . . . . **33**

Ron-Hendrik Peesel, Florian Schlosser, Henning Meschede, Heiko Dunkelberg and Timothy G. Walmsley
Optimization of Cooling Utility System with Continuous Self-Learning Performance Models
Reprinted from: *Energies* **2019**, *12*, 1926, doi:10.3390/en12101926 . . . . **57**

Miyer Valdes, Juan G. Ardila, Dario Colorado and B.A. Escobedo-Trujillo
Computational Model to Evaluate the Effect of Passive Techniques in Tube-In-Tube Helical Heat Exchanger
Reprinted from: *Energies* **2019**, *12*, 1912, doi:10.3390/en12101912 . . . . **74**

Inesa Barmina, Antons Kolmickovs, Raimonds Valdmanis, Maija Zake, Sergejs Vostrikovs, Harijs Kalis and Uldis Strautins
Electric Field Effect on the Thermal Decomposition and Co-combustion of Straw with Solid Fuel Pellets
Reprinted from: *Energies* **2019**, *12*, 1522, doi:10.3390/en12081522 . . . . **86**

András Éles, László Halász, István Heckl and Heriberto Cabezas
Evaluation of the Energy Supply Options of a Manufacturing Plant by the Application of the P-Graph Framework
Reprinted from: *Energies* **2019**, *12*, 1484, doi:10.3390/en12081484 . . . . **106**

Jara Laso, Isabel García-Herrero, María Margallo, Alba Bala, Pere Fullana-i-Palmer, Angel Irabien and Rubén Aldaco
LCA-Based Comparison of Two Organic Fraction Municipal Solid Waste Collection Systems in Historical Centres in Spain
Reprinted from: *Energies* **2019**, *12*, 1407, doi:10.3390/en12071407 . . . . **130**

Samir Meramo-Hurtado, Adriana Herrera-Barros and Ángel González-Delgado
Evaluation of Large-Scale Production of Chitosan Microbeads Modified with Nanoparticles Based on Exergy Analysis
Reprinted from: *Energies* **2019**, *12*, 1200, doi:10.3390/en12071200 . . . . **148**

Pavel Charvát, Lubomír Klimeš and Martin Zálešák
Utilization of an Air-PCM Heat Exchanger in Passive Cooling of Buildings: A Simulation Study on the Energy Saving Potential in Different European Climates
Reprinted from: *Energies* **2019**, *12*, 1133, doi:10.3390/en12061133 . . . . **164**

Khairulnadzmi Jamaluddin, Sharifah Rafidah Wan Alwi, Zainuddin Abdul Manan, Khaidzir Hamzah and Jiří Jaromír Klemeš
A Process Integration Method for Total Site Cooling, Heating and Power Optimisation with Trigeneration Systems
Reprinted from: *Energies* **2019**, *12*, 1030, doi:10.3390/en12061030 181

Jan Najser, Petr Buryan, Sergej Skoblia, Jaroslav Frantik, Jan Kielar and Vaclav Peer
Problems Related to Gasification of Biomass—Properties of Solid Pollutants in Raw Gas
Reprinted from: *Energies* **2019**, *12*, 963, doi:10.3390/en12060963 215

Florian Schlosser, Ron-Hendrik Peesel, Henning Meschede, Matthias Philipp, Timothy G. Walmsley, Michael R.W. Walmsley and Martin J. Atkins
Design of Robust Total Site Heat Recovery Loops via Monte Carlo Simulation
Reprinted from: *Energies* **2019**, *12*, 930, doi:10.3390/en12050930 229

Zhi-chuan Sun, Xiang Ma, Lian-xiang Ma, Wei Li and David J. Kukulka
Flow Boiling Heat Transfer Characteristics in Horizontal, Three-Dimensional Enhanced Tubes
Reprinted from: *Energies* **2019**, *12*, 927, doi:10.3390/en12050927 246

Parkpoom Sriromreun and Paranee Sriromreun
A Numerical and Experimental Investigation of Dimple Effects on Heat Transfer Enhancement with Impinging Jets
Reprinted from: *Energies* **2019**, *12*, 813, doi:10.3390/en12050813 271

Matthias Rathjens and Georg Fieg
Cost-Optimal Heat Exchanger Network Synthesis Based on a Flexible Cost Functions Framework
Reprinted from: *Energies* **2019**, *12*, 784, doi:10.3390/en12050784 287

Aristotle T. Ubando, Isidro Antonio V. Marfori III, Kathleen B. Aviso and Raymond R. Tan
Optimal Operational Adjustment of a Community-Based Off-Grid Polygeneration Plant using a Fuzzy Mixed Integer Linear Programming Model
Reprinted from: *Energies* **2019**, *12*, 636, doi:10.3390/en12040636 305

Libor Kudela, Radomir Chylek and Jiri Pospisil
Performant and Simple Numerical Modeling of District Heating Pipes with Heat Accumulation
Reprinted from: *Energies* **2019**, *12*, 633, doi:10.3390/en12040633 322

Guillermo Martínez-Rodríguez, Amanda L. Fuentes-Silva, Juan R. Lizárraga-Morazán and Martín Picón-Núñez
Incorporating the Concept of Flexible Operation in the Design of Solar Collector Fields for Industrial Applications
Reprinted from: *Energies* **2019**, *12*, 570, doi:10.3390/en12030570 345

Dominika Fialová and Zdeněk Jegla
Analysis of Fired Equipment within the Framework of Low-Cost Modelling Systems
Reprinted from: *Energies* **2019**, *12*, 520, doi:10.3390/en12030520 365

Daniel Leitold, Agnes Vathy-Fogarassy and Janos Abonyi
Evaluation of the Complexity, Controllability and Observability of Heat Exchanger Networks Based on Structural Analysis of Network Representations
Reprinted from: *Energies* **2019**, *12*, 513, doi:10.3390/en12030513 382

Sarah Hamdy, Francisco Moser, Tatiana Morosuk and George Tsatsaronis
Exergy-Based and Economic Evaluation of Liquefaction Processes for Cryogenics Energy Storage
Reprinted from: *Energies* **2019**, *12*, 493, doi:10.3390/en12030493 . . . . **405**

Xuexiu Jia, Jiří Jaromír Klemeš, Petar Sabev Varbanov and Sharifah Rafidah Wan Alwi
Analyzing the Energy Consumption, GHG Emission, and Cost of Seawater Desalination in China
Reprinted from: *Energies* **2019**, *12*, 463, doi:10.3390/en12030463 . . . . **424**

Mato Perić, Ivica Garašić, Sandro Nižetić and Hrvoje Dedić-Jandrek
Numerical Analysis of Longitudinal Residual Stresses and Deflections in a T-joint Welded Structure Using a Local Preheating Technique
Reprinted from: *Energies* **2018**, *11*, 3487, doi:10.3390/en11123487 . . . . **440**

Jara Laso, Daniel Hoehn, María Margallo, Isabel García-Herrero, Laura Batlle-Bayer, Alba Bala, Pere Fullana-i-Palmer, Ian Vázquez-Rowe, Angel Irabien and Rubén Aldaco
Assessing Energy and Environmental Efficiency of the Spanish Agri-Food System Using the LCA/DEA Methodology
Reprinted from: *Energies* **2018**, *11*, 3395, doi:10.3390/en11123395 . . . . **453**

Jan Poláčik, Ladislav Šnajdárek, Michal Špiláček, Jiří Pospíšil and Tomáš Sitek
Particulate Matter Produced by Micro-Scale Biomass Combustion in an Oxygen-Lean Atmosphere
Reprinted from: *Energies* **2018**, *11*, 3359, doi:10.3390/en11123359 . . . . **471**

About the Special Issue Editors

Jiří Jaromír Klemeš is now Head of the Laboratory and Key Foreign Scientist at the Sustainable Process Integration Laboratory (SPIL). Previously, he was Project Director, Senior Project Officer and Hon. Reader at the Department of Process Integration at UMIST, the University of Manchester and the University of Edinburgh, UK. He was awarded with the Marie Curie Chair of Excellence (EXC) by the EC and has a track record of managing and coordinating 96 major EC, NATO, bilateral and UK Know-How projects, with research that has attracted over 33 M in funding. He is Co-Editor-in-Chief of top journals Journal of Cleaner Production (IF 6.395) and Chemical Engineering Transactions (Scopus), Subject Editor of Energy (IF 5.537) and Emeritus Executive Editor of Applied Thermal Engineering (IF 4.026). He is the founder of the Process Integration for Energy Saving and Pollution Reduction (PRES) conference and has been President for 23 years. He has been Chair of the CAPE-WP of the European Federation of Chemical Engineering (EFCE) for 7 years and is a member of the Sustainability Platform and Process Intensification WP. In 2015, he was awarded with the EFCE Life-Time Achievements Award.

Petar Sabev Varbanov is Senior Researcher and Associated Professor at the Sustainable Process Integration Laboratory (SPIL), NETME Centre, Faculty of Mechanical Engineering, Brno University of Technology (VUT Brno), Czech Republic. His main fields of activity are total site and regional integration for energy and water, including industry interaction interfaces, retrofit, waste to energy and wastewater minimisation. Since February 2016, Dr. Varbanov has been affiliated with the Centre for Process Systems Engineering and Sustainability at Pázmány Péter Catholic University in Budapest. Dr. Varbanov acts as Subject Editor of ENERGY: The International Journal, Scientific Secretary of the PRES series of conferences and was Executive Co-Chair of PRES 2018 and PRES 2019. He is also a member of the International Scientific Committee of the SDEWES series of conferences, a member of the Editorial Board of Applied Thermal Engineering, Guest Editor for the Journal of Cleaner Production, Cleaner Technologies and Environmental Policy and Theoretical Foundations of Chemical Engineering.

Paweł Ocłoń currently works at the Institute of Thermal Power Engineering, Cracow University of Technology as Associate Professor. His research topics cover (1) energy systems analysis, (2) underground energy systems, (3) optimisation of thermal systems, such as heat exchangers, heating networks and underground power cable systems, (4) experimental investigation of high-performance heat exchangers, (5) energy storage systems, (6) photovoltaic cooling, (7) finite element method and (8) CFD simulation of energy devices.

Hon Huin Chin is Junior Researcher at the Sustainable Process Integration Laboratory (SPIL), NETME Centre, Faculty of Mechanical Engineering, Brno University of Technology (VUT Brno), Czech Republic. In 2018, he graduated from the Faculty of Chemical and Environmental Engineering at the University of Nottingham Malaysia (UNMC). His field of study focuses on chemical process design, with the option to extend to environmental/sustainability assessment. His current research interests are reliability and safety for asset optimisation, total site process integration, heat exchanger network synthesis, and retrofit.

Preface to "Selected Papers from PRES 2018"

In a rapid developing country, the economic growth and rising global population cause the reliance on energy to surge continuously, provided the resources are available and the government is capable of providing them. Almost a one-third increase in energy consumption is expected in the next 20 years. The overall carbon emissions are also expected to increase, due to the heavy dependence on natural gas. This in turn requires intensive research and discovery of cleaner energy sources and proper industrial practices to advance toward the goals of sustainability. The Special Issues (SI) from the Process Integration for Energy Saving and Pollution Reduction (PRES) 2018 conference aim to address the issues of boosting energy and environmental performances for processes. The collected articles focus on the process analysis, modelling and optimisation as well as design modifications to minimise the energy loss or exergy destructions. The problem domain size ranges from process level to total site. Studies on the allocation of renewable energy resources in a regional supply chain are also included. In this SI, the strategies used by researchers are divided into two major groups: (a) optimisation of process network topology and resources utilisation in a total site and (b) energy and environmental efficiency analysis of process technologies.

The thermochemical conversion of biomass appears to be a promising way to generate process heat. The energy generated can potentially be used in a gas turbine or in steam generation. It could aid in reducing natural energy resources, but the pollutants produced from biomass are difficult to identify due to their complex nature. Flue gas generation from biomass gasification contains a complex distribution of solid pollutants with a range of different sizes. These pollutants could cause not only severe air pollution but also be detrimental to human health—especially the ultrafine particles. Solar energy is also exploited as a clean energy source. The inflexibility of the operation of solar collectors hinders their large-scale industrial application. This issue is also addressed in this SI through the designing of solar collector arrays to improve operational flexibility.

The process improvements through bottlenecks analysis and technologies shifting are also alternative strategies to consume less energy. The additional insights from this SI also determine the energy efficiency, environmental performance and economic evaluation of various energy-saving and clean production technologies. For example, separation units and drying are the main contributors in high exergy destructions for the large-scale production of bio-adsorbent chitosan microbeads for wastewater purification. The recycling of organic waste in the existing municipal solid waste collection systems through pneumatic collection could be an energy-saving technology, as could reusing organic waste as energy providers. Another article compares the seawater desalination technologies in China, evaluating their economic, environmental and energy performance and provides insights on the potential in-process improvements through regional water–energy integration. The technologies shifting through modifying the structures of heat exchanger, piping and building supports can enhance the heat transfer area and facilitate energy saving as well. The design modifications are validated through empirical and numerical analysis, e.g., by computational fluid dynamics (CFD).

This Special Issue also deals with the enhancement of process integration with more robust approaches. The cost-optimal synthesis of heat exchanger network (HEN) synthesis is addressed, considering the 2-D and 3-D layout representation of the process. The controllability and complexity of HEN are also determined through network modelling to locate the sensor installation. The operation of a cooling utility system is also optimised through the self-adaptive model with non-steady data. Another interesting study is the integration of energy usage in cryogenic energy storage with the liquefaction process. The cold and heat generated during the charging and discharging processes can be integrated and reused. The studies also extend to the regional boundary through total site heat and power integration. This SI also addresses total site heating and cooling with tri-generation systems and the robust design of total site heat recovery loops with transient data. The site-wide optimisation of the renewable energy sources' supply chain, mainly biomass, is also highlighted in this SI. The targeted processes that consist of polygeneration plant and manufacturing plant are optimised through cost-effective operational adjustments with various energy supply options.

Jiří Jaromír Klemeš, Petar Sabev Varbanov, Paweł Ocłoń, Hon Huin Chin
Special Issue Editors

Review

Towards Efficient and Clean Process Integration: Utilisation of Renewable Resources and Energy-Saving Technologies

Jiří Jaromír Klemeš [1,*], Petar Sabev Varbanov [1], Paweł Ocłoń [2] and Hon Huin Chin [1]

[1] Sustainable Process Integration Laboratory—SPIL, NETME Centre, Brno University of Technology—VUT Brno, Faculty of Mechanical Engineering, Technická 2896/2, 616 69 Brno, Czech Republic; varbanov@fme.vutbr.cz (P.S.V.); chin@fme.vutbr.cz (H.H.C.)

[2] Institute of Thermal Power Engineering, Faculty of Mechanical Engineering, Cracow University of Technology, Al. Jana Pawła II 37, 31-435 Cracow, Poland; poclon@mech.pk.edu.pl

* Correspondence: jiri.klemes@vutbr.cz

Received: 2 August 2019; Accepted: 23 October 2019; Published: 26 October 2019

Abstract: The strong demand for sustainable energy supplies had escalated the discovery, and intensive research into cleaner energy sources, as well as efficient energy management practices. In the context of the circular economy, the efforts target not only the optimisation of resource utilisation at various stages, but the products' eco-design is also emphasized to extend their life spans. Based on the concept of comprehensive circular integration, this review discusses the roles of Process Integration approaches, renewable energy sources utilisation and design modifications in addressing the process of energy and exergy efficiency improvement. The primary focus is to enhance the economic and environmental performance through process analysis, modelling and optimisation. The paper is categorised into sections to show the contribution of each aspect clearly, namely: (a) Design and numerical study for innovative energy-efficient technologies; (b) Process Integration—heat and power; (c) Process energy efficiency or emissions analysis; (d) Optimisation of renewable energy resources supply chain. Each section is assessed based on the latest contribution of this journal's Special Issue from the 21st conference on Process Integration, Modelling and Optimisation for Energy Saving and Pollution Reduction (PRES 2018). The key results are highlighted and summarised within the broader context of the state of the art development.

Keywords: process integration; renewable energy sources; energy-saving technologies

1. An Overview and Introduction

Wang et al. [1] have stated that the economic growth of countries features a strong correlation to their energy consumption, especially for rapidly developing countries. The reliance on energy will only continue to surge as long as the resources are available, and governments are capable of providing them. As the economies continue growing by the utilisation of natural resources, society has started to take their abundance for granted and created an excessive amount of waste that could otherwise be reusable. Fossil fuels, natural gas and coal have been serving humanity as the primary sources of power generation for decades. Inadequate management of the used resources has led to numerous environmental issues, of which climate change (global warming) is one of the manifestations. Melorose et al. [2] have predicted that the Earth's population is expected to grow by more than 1×10^9 inhabitants by 2035. The rapid rise of the global population results in almost one-third of the expected increase in energy consumption—see Figure 1. The global oil and natural gas reserves are expected to be exhausted within 60 years if they are exploited at the current rate as predicted by British Petroleum Company (BP) [3], which raises the issues of energy security. As reported by Mah et al. [4], since the

Paris Agreement of 2015, 195 countries are committed to investing efforts to reduce the global average temperature rise to below 2 °C and limit the warming threshold to 1.5 °C. Greenhouse gas (GHG) emissions in a business-as-usual scenario are expected to cause the planet to heat up by 4 °C, creating imbalanced disruptions to the ecological systems. Considering the severe environmental impacts and quick diminishing of natural resources due to exploitation, a cleaner and more sustainable energy resource alternative is needed.

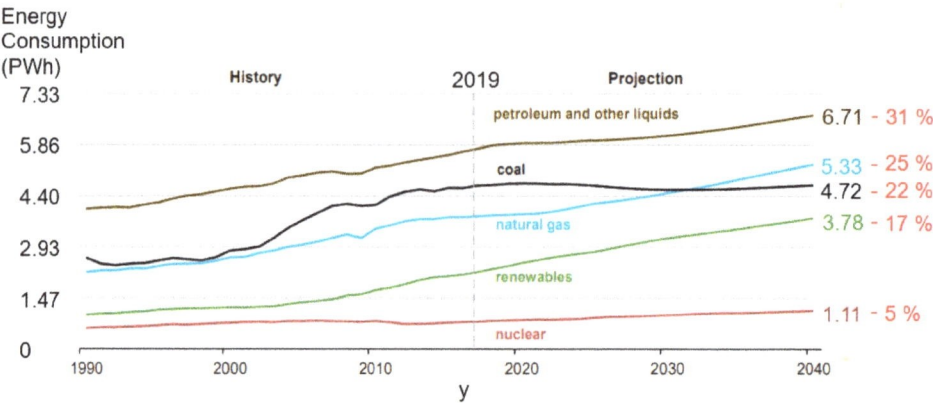

Figure 1. Global energy consumption for different fuels adapted from Capuano [5].

Some renewable energies resources (e.g., nuclear energy and virtually carbon-free energy sources) have emerged as promising solutions for ecological degradation and energy security problems. Many countries have started investing in nuclear and renewable energies to reduce over-dependency on the crude oil trade, stemming from the desires for sustainable energy [6]. Based on the data from Figure 1, conventional energy sources (petroleum and liquids, coal and natural gas) are still forecast to account for about 78% of the global energy consumption in 2040. There might be several reasons for the continuous reliance on non-renewable sources. The scalability of renewable technology to the industrial application is the primary challenge. Most of the successful energy harvesting from renewable sources, such as thin-film solar based on Dharmadasa [7], algae-based biofuel according to Vo et al. [8] and cellulosic ethanol in Zabed et al. [9] are developed at a small scale due to the nature of the resource supply—distribution over the harvesting area [10].

Power generation technologies, including renewable energy technologies and innovative low carbon emissions technologies, take a lot of time and efforts for full commercialisation. Gross et al. [11] mentioned that it could take from 20 to almost 70 years for a full technology and product to emerge from the invention, diffuse into market and reach widespread commercial deployment. According to their study, electricity generation technologies, such as combined cycle gas turbine (CCGT), nuclear power, wind electricity and solar photovoltaic (PV), can take approximately 43 years from invention to widespread commercialisation. No clear reasoning has been provided, but it can be deduced that the asset lifespans of electricity generation technologies are usually very long– several decades in the case of power stations. Longer time duration is often necessary to completely replace the existing facilities. Mature technologies often require years of development and inventions so that they can be stable enough to replace existing technologies. In the case of nuclear power, social acceptance is one of the most significant obstacles to achieving the goal of cleaner energy production.

The requirement of large land space is another obstacle to the implementation of renewable energy technology. At present, the world requires a continuous and consistent energy supply, which is the second critical hurdle for solar and wind technology, because the resources for these energy options are available at rates that vary in time. The need for extensive integration with the grid or installation of battery storage to become the chief power generating sources generates a penalty cost of the large land footprint. Brook and Bradshaw [12] estimated that around 50% of the total energy demand in the United States could be satisfied with renewable energy technologies, but it would require at least 17% of the land. This is evident from the data in Table 1, which shows the land use for renewables is significantly larger than for conventional sources.

These clean energy sources are claimed to mitigate climate change or improve air quality. Pablo-Romero et al. [13] conducted a Life Cycle Analysis (LCA) on various renewable technologies for electricity generation. They showed that the carbon emission intensities of renewable technologies (including solar power, geothermal power, hydropower, nuclear energy and wind energy) are negligible as compared to fossil fuels and coal. Mathiesen et al. [14] conducted a scenario study on health costs estimation based on the case of Denmark, in the year 2050 when 100% renewable energy systems are used. The health costs are estimated based on the basis of various emissions: SO_2, CO_2, NO_x, $PM_{2.5}$, mercury and lead, and the results show significant health savings can be achieved. Likewise, Patridge and Gamkhar [15] quantified the health benefits of replacing coal-fired generation with 100% wind or small hydro in China. The scenario analysis shows a significant reduction number of hospital stays due to reductions in SO_2, NO_x and particulate emissions. Heat generation from renewables might not be the ideal case for air pollutant reduction. For example, the gasification of biomass can produce various ultrafine particles that are detrimental to human health. Poláčik et al. [16] conducted the analysis and concluded that the oxygen contents in the atmosphere could result in higher particulate matter production from biomass combustion. This requires post-treatment for biomass conversion before discharging the flue gas into the atmosphere. To compare between energy supply options, Table 1 shows the economic-environmental impact indicators for the power generating options, adopted from Brook and Bradshaw [12].

Table 1. Relative ranking of the power generation options, data adapted from Brook and Bradshaw [12]. The values in brackets are the relative ranking between energy options.

Indicator Category	GHG Emissions (kt CO_2/TWh)	Electricity Cost ($/TWh)	Land Use (km²/TWh)	Safety (Fatality/TWh)	Solid Waste (kt/TWh)	Capacity Factors * (%)	Toxic Waste Amount
Coal	1,001 (7)	100.1 (4)	2.1 (3)	161 (7)	58.6 (7)	70–90 (2)	Mid (6)
Natural gas	469 (6)	65.6 (1)	1.1 (2)	4 (5)	NA (1)	60–90 (3)	Low (3)
Nuclear	16 (3)	108.4 (5)	0.1 (1)	0.04 (1)	NA (1)	60–100 (1)	High (7)
Biomass	18 (4)	111 (6)	95 (7)	12 (6)	9.17 (6)	50–60 (4)	Low (3)
Hydro	4 (1)	90.3 (3)	50 (6)	1.4 (4)	NA (1)	30–80 (5)	Trace (1)
Wind (onshore)	12 (2)	86.6 (2)	46 (5)	0.15 (2)	NA (1)	30–50 (6)	Trace (1)
Solar PV	46 (5)	144.3 (7)	5.7 (4)	0.44 (3)	NA (1)	12–19 (7)	Trace (1)

* The capacity factors represent the percentage of time that the power generation plant operates at full rated capacity. Data adapted from IRENA [17].

Brook and Bradshaw [12] ranked the preferences of different energy options with various criteria. The readers are referred to them for more information. Table 1 shows that the emissions for coal are the highest among all options. From the estimations in Figure 1, the utilisation of coal is expected to increase in the next few years and around 2030, the natural gas is expected to exceed the consumption of coal. This might be due to the fact that emissions and waste produced from coal usage are higher compared to natural gas. Rapid progress in strengthening of project financing by government policies is enabling more cost-effective installation of dynamic renewable technologies like solar photovoltaics, solar-thermal plants, hydrothermal plants and onshore wind worldwide [18]. The surges in renewable energy consumptions estimated from Figure 1 show the global acceptance and evolution of renewable technologies. Although nuclear power generation is virtually carbon-free and the most efficient option, high amount of radioactive waste produced still creates ethical and ecological issues. The slow expansion rate of the nuclear energy shown in Figure 1 is probably due to its hazardous level, which causes low social acceptance. Figure 2 shows that the traditional biofuel use is preferred by consumers compared to other renewable options. The reason behind is probably because of the capacity factor for biofuel is also comparatively consistent than the non-nuclear renewable options, as shown in Table 1. Biomass-related fuel requires several stages of energy-intensive pre- and post-treatment. This might be the cause of large land requirements, mainly for biomass upgrading. The electricity cost is relatively higher due to the high operating temperature and heating medium. The generation of an excessive amount of low-quality waste creates a significant obstacle for the green policy. The enforced government policies to reduce environmental emissions and waste decrease the reliance on biofuels slowly, emphasising the use of alternative cleaner renewable options such as hydropower and solar power.

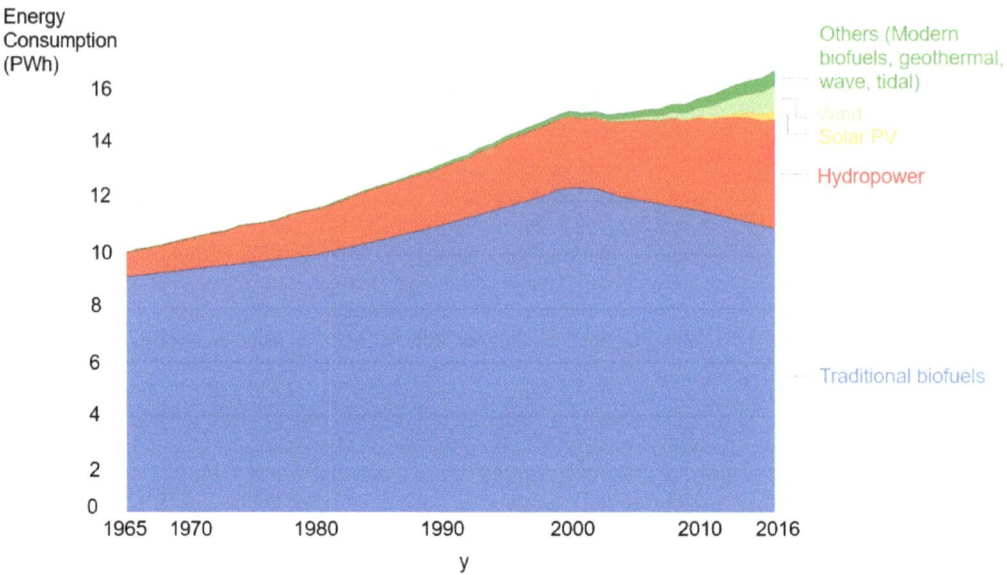

Figure 2. Global renewable energy consumption, adapted from Ritchie and Roser [19].

The Renewables 2018 Global Status Report [20] mentioned that the power sector is driving a rapid change towards a renewable energy future, and others are not advancing with the needed rate. The power sector is progressing with positive momentum, but the emissions reductions targets were not being met. In Figure 3, it is clearly shown that the overall carbon emissions are expected to increase, due to the increased dependence on natural gas if compared to other fuels—see also Figure 1. The primary energy demand for natural gas continues to rise in the next 20 years. The World Energy Outlook (WEO) from the International Energy Agency [21] has predicted that the global CO_2 emissions would continue to rise with existing energy policies in the next 20 years. The analysis from WEO also predicted that future electrification in transportation, buildings and industry would lead to a peak oil demand before 2030 and reduce air pollutants. However, the impacts on GHG emissions are negligible after the year 2030. Stronger efforts are needed to increase the utilisation of renewables with low carbon contents so that the predicted CO_2 emissions could decrease (Figure 3). With agreed international objectives, government from all countries could cooperate to tackle the issues of air quality, climate change and universal access to modern energy, ideally results in a significant reduction of global CO_2 emissions. The policies integration around the world is the key to achieve common targets of sustainability, global Circular Economy and the objectives of the Paris Agreement on climate change.

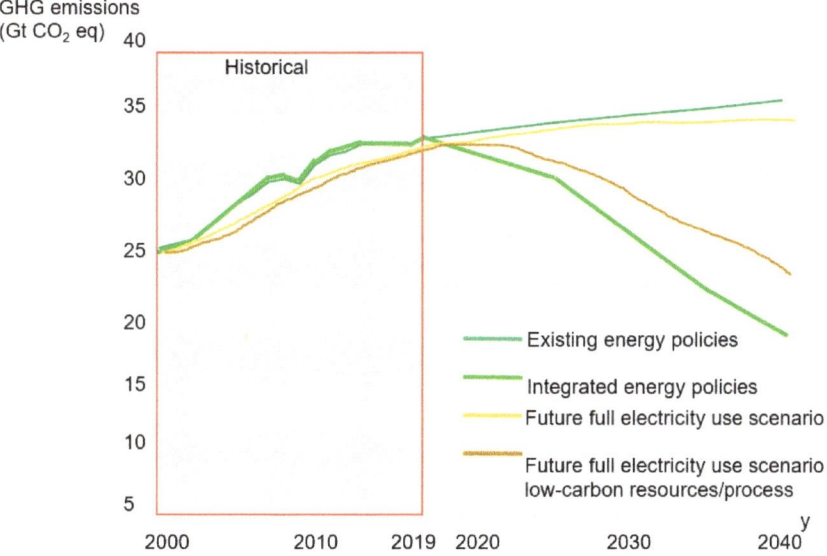

Figure 3. Predicted global GHG emissions with different energy policies, adapted from WEO [21].

For better and consistent use of renewable energy, energy storage plays a crucial role. The energy surplus from solar or wind technologies can be stored in a storage unit during their peak seasons and off-peak demand season. The stored energy can then be used to satisfy the energy's demand during peak season, reducing the outsourced electricity from the grid—see Figure 4. As electricity prices are often the highest during peak demand season, the utilisation of energy-efficient storage technologies could aid in electricity saving. This topic is discussed more in Section 2. Main topics in this Special Volume.

One of the typical methods in assessing the energy efficiency of a particular fuel is the so-called Energy Return on Investment (EROI). It measures the quality of fuels by calculating the ratio of energy that can be delivered to the society to the energy invested in the harvesting, based on Murphy and Hall [22]. Hall et al. [23] mentioned that the majority of current EROI analyses tend to focus on the 'energy break-even' point of EROI for different fuels, i.e., whether it is greater than 1:1. The metric is a straightforward analysis. However, the variations of the findings can be wide depending on the selection of study boundaries. The possible boundaries that are often studied are illustrated in Figure 5. EROI analyses are typically categorised into three levels: (a) Standard EROI ($EROI_{STD}$), a conventional analysis that focuses on the energy input for the extraction and the energy needed to generate the desired output; (b) Point of Use EROI ($EROI_{POU}$), the boundary is extended to the additional spent energy to refining and transporting the fuel; (c) Extended EROI ($EROI_{EXT}$), this EROI analysis consider the further the use of that energy in specific applications, for example, to run a boiler. Lambert et al. [24] further emphasised that the EROI boundaries should sum up the entire gains from the fuels and entire energy spent on the fuels. They stress that the vision for a modern energy system should extend to any non-energy cost for setting up the energy system. The temporal boundaries also should cover from the starts of the project until the end of the project. Figure 6 shows the temporal boundaries for determining the comprehensive net energy requirement of thermal technology.

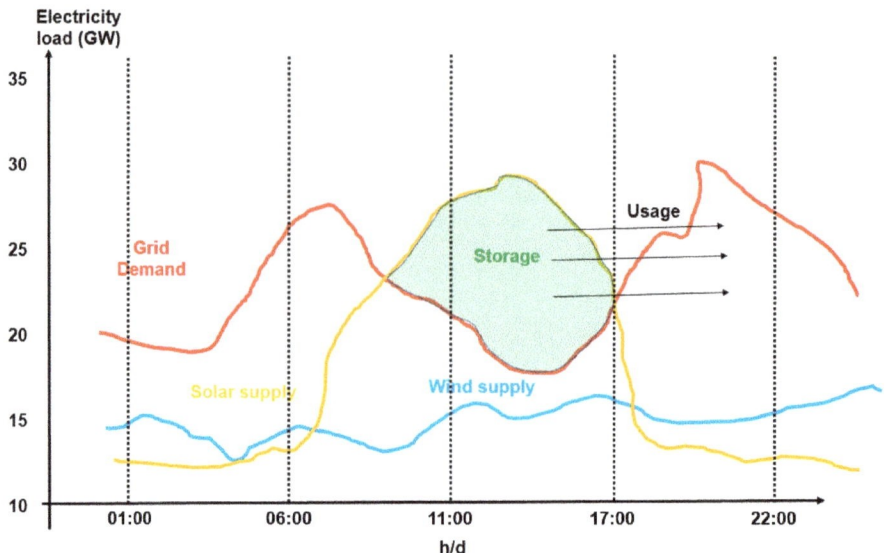

Figure 4. Utilising stored renewable energy surplus to satisfy the energy demand, adapted from Movallen [25].

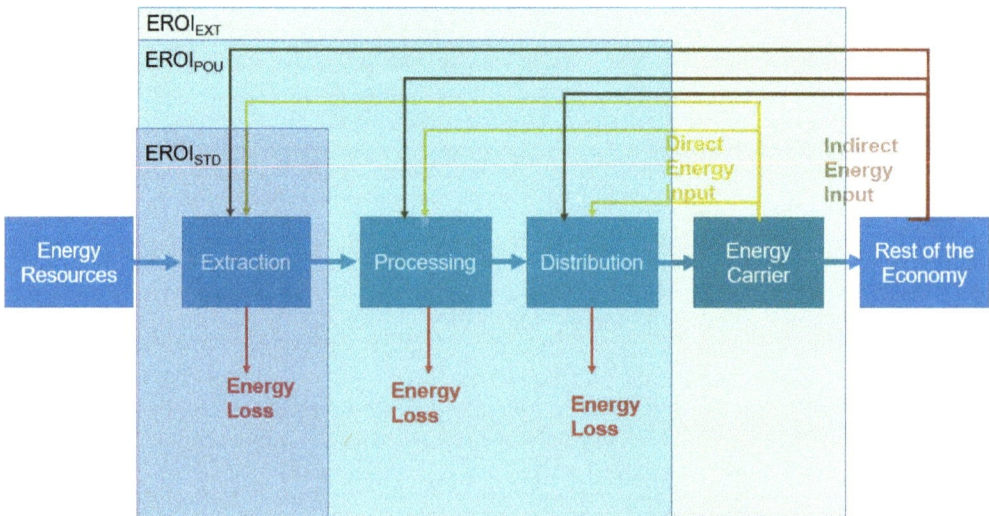

Figure 5. Possible boundaries of a net energy assessment, adapted from Hall et al. [23].

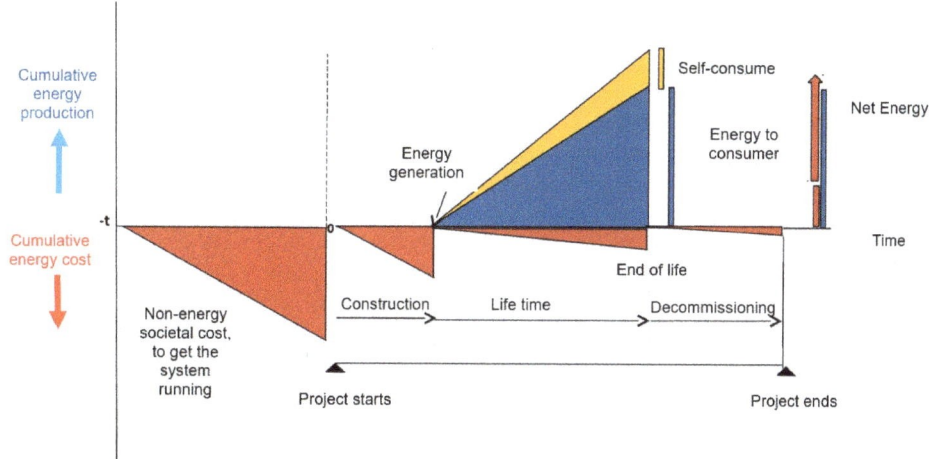

Figure 6. Net energy requirements for technology, adapted from Priesto and Hall [26].

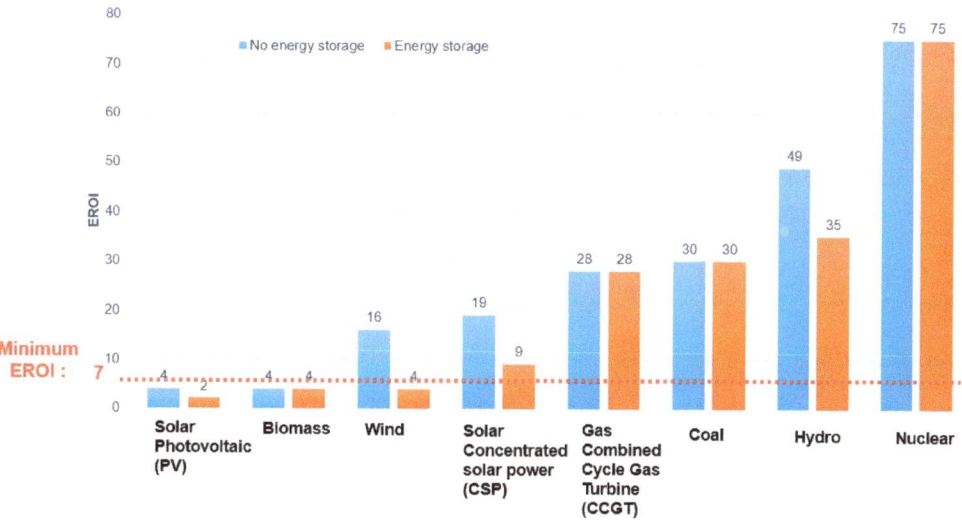

Figure 7. EROI of different fuels, data adapted from Weißbach et al. [27]. The minimum EROI is taken as the average minimum EROI value from Hall et al. [28].

To illustrate the EROI values for different energy options, the facilities in Germany, which one of the most advanced economies in the EU, are used as an example. Figure 7 shows the EROI values for different energy options in Germany country presented by Weißbach et al. [27]. The EROI boundary is contained within the project implementation period. It should be noted that pump storage is used as the energy storage option. The requirements of additional energy for storage reduce the EROI values for the fuels which decrease their economic preferences. Solar PV in Germany has an EROI far below the economic limit, even with the most effective roof installation. A study from Hall et al. [28] indicated that the minimum EROI limit is about 5–10 for any energy supply option, which is just enough for civilisation. Wind energy has a preferable EROI, but falls below an economic threshold, even when combined with pump storage. Installation in the German coast to enhance the EROI values is also futile. Biogas-fired plants have the problem of requiring enormous fuel provisioning efforts, which brings them clearly below the economic limit with no potential for achievable improvements contemplated. Solar CSP has better performance among the new solar/wind technologies. However, pump storage is often not available in regions with high solar irradiation. Less effective storage techniques like molten salt thermal storage and the connection to the European grid probably brings the EROI again below the economic limit. Further information on energy storage is briefly discussed in Section 2.3.

Even though EROI is the most critical parameter to measure the energy effectiveness of power technology, it is neither fixed nor the only parameter for such an assessment. EROI slowly changes with time due to fossil fuels extraction—when they become harder to access, but also when processes are improved as it happened with the steel production and the uranium enrichment [27]. The land consumption, the impact on nature, and the scope of the stockpiles have to be taken into account separately. This is where the term 'exergy' comes into play. The central definition of exergy is bounded to a physical process which usually measures the utilisation of energy. It is defined as the maximum attainable work inside a system. The exergy of a heat flow measures theoretically the amount of work that can be generated if it is to be discharged into a reference environment (usually ambient), which mimics the heat engine configuration. Exergy accounts for both the energy in the system and the condition of the system relative to the environment. The high energy efficiency of the system does not necessarily mean that the energy is used to its full potential. Consider two energy storage systems (ESS1 and ESS2) as an example; both ESS1 and ESS2 are charged with the same amount of input energy and produced the same amount of output. The energy efficiency for both systems can be similar, but the exergy efficiency can be different. If ESS1 produces a thermal energy flow with a higher temperature than ESS2, the exergy efficiency of ESS1 would be higher than ESS2. This is due to the output from ESS1 has a higher quality of heat (higher temperature), which theoretically can be used to generate more work. The output from ESS2 has lower quality as exergy is lost in the system, which stems the term 'exergy efficiency'.

Exergy efficiency is a useful indicator to pinpoint the technical inefficiency of any process. For the example of energy storage systems above, if both have over 90% efficiency, one can actually satisfy with the performance of the system without modifications. However, if there are secondary processes after the primary process that utilised heat from the storage systems, the 'higher' quality heat can further be used as a heating source. The lower quality heat has limited usage. The strategies to reduce the exergy destruction for ESS2 can be developed to improve the performance. Exergy can be applicable in any physical transformation process, not limited to only thermodynamic or chemical processes. It can be used for resource accounting by evaluating the resource consumption, depletion and degradation in a spatial and temporal boundary. This concept is also widely used in a few Life-Cycle Analysis (LCA) studies of energy processes. The thermal and electrical energy is converted into equivalent 'works' which provide the 'weighting' for various forms of energy. The exergy concept can be implemented into EROI calculations to determine the technical limits for each thermal processes, see Figure 8 for the illustration of the potential study boundary. The evaluation should be performed through comprehensive top-down analysis, ranging from raw materials, extraction, consumptions, recycling, energy usage to environmental emissions. The exergy can also be used to represent the degradation of the resources over time, which energy is not able to do so. To pinpoint the technical bottlenecks, the exergy efficiency should be evaluated starting from the resource input exergy until it is used up and released to ground state. The temporal effect can also be incorporated into the analysis to evaluate the overall efficiency affected by resource degradations. The resource upgrading units vary depending on the type of resource, for example, heat pumps for energy resources and desalination for water resources. The upgraded resources can be utilised not only to the original consumption process but can be used in a secondary process. The maximisation of resource utilisation is in-line with the concept of the circular economy.

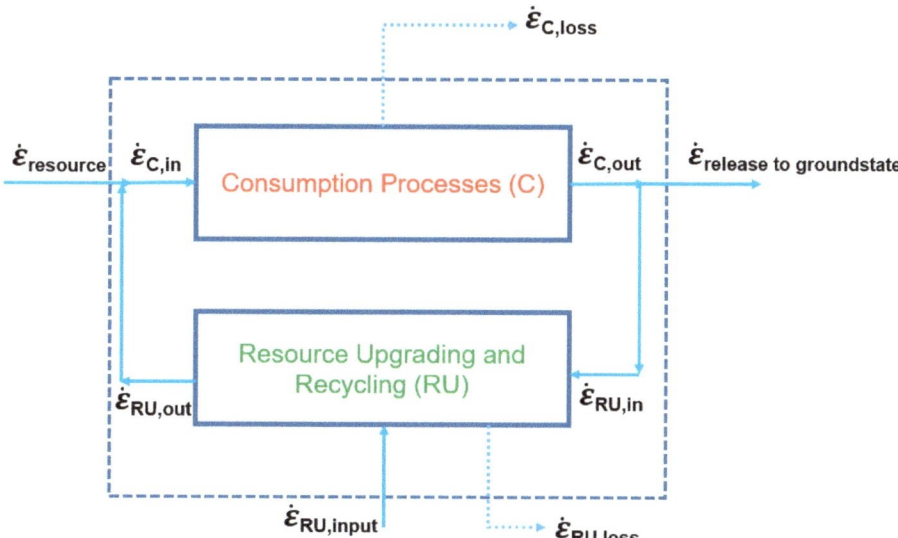

Figure 8. Proposed exergetic efficiency calculation boundary to access resource consumption and depletion, adapted from Connelly and Koshland [29].

In the light of the mentioned issues, the Special Issue (SI) from the conference PRES 2018 aims to address the issues of boosting energy and environmental performances for processes. Based on Figure 5, this study is targeted to maximise the energy production and saving within the system lifetime period. The collected Special Issues (SI) focus on the process analysis, modelling and optimisation as well as design modifications to minimise the energy loss or exergy destructions. The problem domain size ranges from process level to total site. The studies on the allocation of renewable energy resources in a regional supply chain are collected as well. This SV aims to provide high-end researches in dealing with better energy resource management and product designs, advancing towards the goal of the global circular economy.

2. Main Topics in this Special Volume

In this work, the state-of-the-art Process Integration and Intensification strategies through the utilisation of renewable resources towards sustainable production are collected and analysed. More emphasis is given on the efficiency improvements in heat and power usage for various technologies. In this special issue, the strategies used by the researchers are divided into two major groups: (a) Optimisation of process network topology and resources utilisation in a total site; (b) Energy and environmental efficiency analysis of process technologies.

This review paper is supported by the published works in the same journal, within the Special Volume (SV) presented in the conference series "Process Integration, Modelling and Optimisation for Energy Saving and Pollution Reduction" in 2018 (PRES 2018). This conference has served as a knowledge-sharing platform among international experts from multi-disciplinary domain since 1998, producing high-quality researches through intellectual integration. A total of 40 selected papers invited to be submitted for this SV, 24 of them have been accepted. The contributors in this special issue has a wide range of country distribution, which involve six from the Czech Republic, four from Germany, two from Mexico, two from Spain, one from Colombia, one from Croatia, one from Hungary, one from Latvia, one from Malaysia, two from The Philippines, one from Thailand and one from the USA) The statistics shows the active participation of researchers with a wide variety of nationalities. Upon close examination of the keywords in this special issue, the articles are further categorised into four main topics that are overviewed to achieve the review's target:

(a) Design and numerical study for innovative energy-efficient technologies
(b) Process Integration—heat and power
(c) Process energy efficiency or emissions analysis
(d) Optimisation of renewable energy resources supply chain

The biomass-related studies commonly refer to the biomass sources as abundantly found in various agriculture process. The thermochemical conversion of biomass appears to be a promising way to generate process heat. The energy can be potentially used in a gas turbine or steam generation. It could aid in reducing the natural energy resources through upgrading, but the pollutants produced from biomass are hardly identified due to their complex nature. The flue gas generation from biomass gasification contains a complex distribution of solid pollutants with different size ranges. They could cause not only severe air pollutions but also detrimental to human health, specifically the ultrafine particles. Solar energy is also exploited as a clean energy source. The inflexibility of the operation of solar collectors hinders their large-scale industrial application. This issue is also addressed in this SV through designing solar collector arrays to improve operational flexibility.

The process improvements through bottlenecks analysis and technologies shifting are also alternative strategies to consume less energy. The additional insights from the special issue also determine the energy efficiency, environmental performance and economic evaluation of various energy-saving and clean production technologies. For example, the large-scale production of bio-adsorbent chitosan microbeads for wastewater purification is limited by the low exergy efficiencies. The separation units and drying are the main contributors in high exergy destructions. One of the articles determined that the recycling of organic wastes in the existing municipal solid waste collection systems through pneumatic collection could be an energy-saving technology, also reusing organic waste as energy providers. The economic, environmental and energy performance the aspects of the current practice of seawater desalination in China is assessed. By comparing desalination technologies, it provides insights on the potential in process improvements through regional water and energy integration. The technologies shifting through modifying the structures of heat exchanger fired equipment, piping and building supports can enhance the effective heat transfer area and facilitate energy saving as well. The design modifications are validated through empirical and numerical analysis, e.g., Computational Fluid Dynamics (CFD).

This special issue also deals with the enhancement of Process Integration with more robust approaches in driving towards sustainable operation. The cost-optimal synthesis of heat exchanger network (HEN) synthesis is addressed, considering the 2-D and 3-D layout representation of the process. The controllability and complexity of HEN are also determined through network modelling to provide a solid foundation for the location of sensor installation. The operation of cooling utility system is also optimised through the self-adaptive model with non-steady data. Another interesting study is the integration of energy usage in the cryogenic energy storage with the liquefaction process. A large amount of cold and heat generated during the charging and the discharging process can be integrated and reused. The studies also extend to the regional boundary through Total Site Heat and Power Integration. This special issue particularly addresses for total site heating and cooling with tri-generation systems, and robust design of total site heat recovery loops with transient data. The site-wide optimisation of the renewable energy sources supply chain, mainly biomass, is also highlighted in this Special Issue. The targeted processes that consist of polygeneration plant and manufacturing plant are optimised through cost-effective operational adjustments with various energy supply options.

2.1. Design and Numerical Study for Innovative Energy-Efficient Technologies

In recent years the share of Renewable Energy Sources (RES) in the total energy budget has increased significantly. Especially in the EU, where it is planned to cover at least 20% of electrical energy production by the year 2020, according to WEO [21]. The major problem related to RES is low efficiency, and electrical energy production depending on the climatic conditions. The solutions to improve the energy-efficiency of RES are highly welcomed. The use of RES requires the advanced energy storage systems, since the operation of solar energy-driven RES is efficient mostly in the spring and summer period, while the efficiency is very low in the heating season. Also, due to the increased electricity production from RES, the conventional power plants, need to improve their flexibility.

2.1.1. Core Developments

The primary issue is related to more frequent start-up and shut-down of conventional power units. For this reason, the extensive fundamental research on both renewable energy sources and conventional power plants are carried out. Both related to design optimisation as well as to the improvement of energy efficiency. The following topics are nowadays of high-importance:

(a) Research on electric energy storage from Khodadoost et al. [30], including: battery energy storage systems (BESSs), flywheel energy storage systems (FESS), supercapacitors (SC) or ultracapacitors, superconducting magnetic energy storage (SMES), and compressed air energy storage (CAES), among others. To minimise the total costs of hybrid power systems (HPS), Jiang et al. [31] proposed a mathematical model for the configuration of BESSs with multiple types of battery. The authors studied the effect of battery types and capacity degradation characteristics on the optimal capacity configuration of the BEES alongside with power scheduling schemes of the hybrid power systems. The performance of the proposed model was verified through the case study of HPS with photovoltaic-wind-biomass-batteries. The authors found the BESS with multiple types of battery is superior to the one with a single battery type. Duan et al. [32] studied a hybrid generation system consisting of a micro gas turbine (MGT) generator system coupled with a supercapacitor (SC) energy storage. The authors proposed two cooperative control methods for the hybrid generation system. The first one was a PI-based control algorithm, and the other is the electric power coordinated control method through MGT output power forecast. The authors found that the electric power dynamic response of SC energy storage can compensate for the low dynamic responses of MGT, which allows achieving a transient power equilibrium state in real-time. Santos et al. [33] studied the possibility of adapting superconducting magnetic energy storage (SMES) in smart grids since the characteristics of smart cities enhance the use of high power density storage systems such as SMES. The authors simulated the effects of an energy storage system with the high power density and designed an electrical and control adaptation circuit for storing energy. The simulation results show the possibility of controlling the energy supply as the storage. The authors also discussed the drawbacks of SMES, such as the high cost of construction and operation compared with other EES, i.e., superconductors. Compressed air energy storage (CAES) is an up-and-coming large capacity energy storage technology, primarily due to the increased share of renewable energy sources. Venkataramani et al. [34] performed a comprehensive thermodynamic analysis for conventional and modified configurations of CAES, with increased round-trip efficiency. The results showed that when the compressed air is kept isothermal at atmospheric conditions, the mass of air stored in the tank will be high, so the size of the storage tank can be reduced. The authors also studied the possibility of cooling energy generation along with power generation during the expansion of compressed air from the atmospheric temperature. The results showed that even the round-trip efficiency is weak, in the case when the heat of compression and cold energy generated during expansion are utilized for other applications, the overall polygeneration efficiency is very high.

(b) Research on thermal energy storage, including short-term and long-term storages, based on Guelpa and Verda [35]. Also depending on the physical phenomenon used for storing heat TES: sensible, latent and chemical storages, and depending on the size: small TES with a low capacity, and large capacity TES systems. TES can also be classified depending on the mobility, i.e., mobile and stationary TES. Bhagat et al. [36] proposed the finned multi-tube latent heat thermal energy storage system (LHTES) for medium temperature (approximate 200 °C) solar thermal power plant in reducing the fluctuations in heat transfer fluid (HTF) temperature caused by the intermittent solar radiation. The authors used phase change materials (PCMs) as the storage material in the shell of LHTES while a thermal oil-based HTF is flowing through the tubes. The authors applied thermal conductivity enhances (TCE)—fins to improve the heat transfer in PCM. The fluid flow and heat transfer were studied numerically. The coupling with the enthalpy technique to account for the phase change process in the PCM was also performed. The developed model was validated experimentally. The results showed that the number of fins and fin thickness considerably affect the thermal performance of the storage system, whereas the enhancement in heat transfer for high thermal conductivity material fin is low. Silakhori et al. [37] investigated the potential of copper oxide for both thermal energy storage and oxygen production in a liquid chemical looping thermal energy storage system. Thermogravimetric analysis was used as the assessment method. The significant advantage of liquid chemical looping thermal energy storage is the availability of stored thermal energy (through sensible heating, phase change, and thermochemical reactions) and oxygen production. The authors achieved isothermal reduction and oxidation reactions by varying the partial pressure of oxygen, through the change in concentrations of oxygen and nitrogen. The experimental confirmed that copper oxide could be reduced in the liquid state. However, thermochemical storage mainly occurred in the solid phase. Heat storage plays a crucial role in the buildings. Taler at al. [38] studied the thermal performance of the heat storage unit made of repeatable modules. The heat accumulator proposed by the authors that are used in solar installations may be a separate unit, or it may be a building wall insulated on the inner and outer surfaces. The accumulator works as a heat storage unit with electric discharge using forced airflow through the channels. The authors determined the transient temperature field in the walls of the channel. Three various methods were used: finite volume method (FVM), control volume-based finite element method (CVFEM), and finite element method (FEM). The preferred method, due to simplicity in the discretization of the governing equation, is CVFEM. Therefore, it was chosen for the construction of a full model of the heat storage to model a solid filling of a heat storage unit with a complex shape. The authors also developed a numerical model of the heat storage unit with the airflow through the channel and CFD simulation was employed. The airflow from the laminar, transitional, or turbulent flow regimes was considered. The airflow was modelled using the finite volume method with integral averaging of air temperature over the finite volume length, and the accurate air temperature distribution can be determined even with a coarse finite volume mesh. The performance of the heat accumulator model was validated experimentally in an experimental study [39] and further evaluated using numerical models [40]. Thermal energy storage techniques are highly required during the operation of solar collector networks. Martínez-Rodríguez et al. [41] proposed a stepwise design approach for solar collectors' networks. The approach allows the assessment of the effect that design variables have on the size of the solar collector. Also, a design strategy is proposed to obtain the network of solar collectors with the smallest surface and the most extended operation during the day.

(c) Research on photovoltaic, and photovoltaic-thermal systems, including the improvement of PV efficiency by PV panels cooling. The efficiency of converting solar energy into electricity is still relatively low, which causes a significant amount of solar energy is not utilised. The excess of solar energy that remains unused is still absorbed, in some part, by a PV module and may cause a significant increase in the PV panel temperature. According to Kalogirou and Tripanagnostopoulos [42], PV efficiency decreases by 0.45% per each 1 °C temperature increase above 25 °C. In order to increase the energy efficiency of photovoltaic modules by using the effect of PV panel heating, and to increase the efficiency of solar to electricity conversion, cooling systems for the PV modules are used. Few PV cooling techniques may be distinguished, including active and passive techniques. For active cooling, a forced flow of cooling fluid (e.g., water or air) or water spraying may be used, among others. Passive cooling uses natural convection and heat conduction to dissipate and remove heat from the PV cell. Passive cooling techniques increase energy efficiency and cost-effectiveness of the system, but still, active cooling removes more efficient, due to the higher heat transfer coefficient. The analysis of passive cooling for the photovoltaic modules using selective spectral cooling and radiative cooling was performed by Li et al. [43]. The cooling processes are based on the principle of suppressing heating by the PV module itself. The investigation proved that PV modules with selective spectral cooling, passive radiative cooling, and combined cooling could increase the efficiency by 0.98%, 2.40%, and 4.55%. Alizadeh et al. [44] studied the use of a single turn pulsating heat pipe (PHP) for PV cooling. A two-phase heat transfer mechanism ensures high thermal efficiency of PHP. The corresponding 3D numerical models were developed, and PV cooling by applying a single turn PHP was analysed. Moreover, a copper fin with the same dimensions as the PHP for cooling the PV panel was simulated to compare the performance of the PHP with a solid metal like copper. The performance investigation of the PV panel has proved that PHP cooling ensures the reduction of the PV panel surface temperature by 16.1 ^{0}C while the use of copper PHP only by 4.9 °C.

(d) Research on energy systems components monitoring and design. Thick-wall boiler components are limiting the maximum heating and cooling rates during start-up or shut-down of the boiler. Taler et al. [45] presented a method for thermal monitoring stresses in the thick-walled pressure components of steam boilers. The allowable heating rates of the critical pressure components of the boiler shall be determined, alongside with the temperature of the fluid. The rate of change of the wall temperature of the pressure component and the thermal stress on the inner surface are controlled online and compared with the allowable values. The boiler's manufacturers designate thermal stresses on the inner surface of the pressure component on the edge of the hole based on the measurement of the wall temperature at two points located inside it. However, the accuracy of the method used by boiler manufacturers is low. The authors proposed a new method for thermal stress determination. The method is based only on the internal temperature measurement point to determine the stresses on the inner surface of the component. The method employs the inverse heat conduction algorithms to find the internal surface temperature, and then the stresses are calculated. The authors also performed computational tests for cylindrical and spherical elements. The thermal stresses on the inner surface were also determined using the actual temperature data. The significant advantage of the proposed method is the high accuracy even at rapid changes in the fluid temperature. Trzcinski and Markowski [46] proposed a data-driven framework of diagnosing fouling effects on shell and tube heat exchangers using an artificial neural network. The data are continuously sampled and collected to estimate the pressure drop increment or heat transfer drop in the outlet. The fouling effect can thus be predicted using the model automatically and provides a base for tubes cleaning scheduling. Oravec et al. [47] proposed a closed-loop model predictive control using the novel soft-constrained based strategy for plate heat exchanger. The strategy keeps the control inputs and outputs within the required operation ranges. The experimental results show the improved control performance with such a strategy, and future application in laboratory implementation is undertaking. Taler et al. [48] proposed two methods for monitoring of thermal stresses in pressure components of thermal power plants. The first method determines the transient temperature distribution by measuring the transient wall temperature distribution at several points located at the outer insulated surface of the pressure component. Taking the outer surface temperature measurements as the input, the inverse heat conduction algorithm calculates the temperature distribution in the pressure component. Based on the temperature field determined, it is possible to calculate the thermal stresses. The second method proposed by the authors involves the finite element method (FEM) calculation of thermal stresses, taking as the input the measured fluid temperature and heat transfer coefficient. The method is suitable for pressure components with complex shape. Other applications of inverse heat conduction algorithms in the monitoring and optimisation of the heating rate of pressure components of steam boilers are presented in [49] dealing with the thermal stresses in the pipes and in [50] focusing on thick-wall components. Perić et al. [51] performed a numerical analysis of longitudinal residual stresses and deflections in a T-joint filled welded structure using a local preheating technique. FEM calculations were performed. The authors found that by applying a preheating temperature prior to starting of welding, the post-welding deformations of welded structures can be considerably reduced. The authors also studied the effect of inter-pass time (i.e., 60 s and 120 s) between two weld passes on the longitudinal residual thermal stress state and plate deflection. The results showed that with the increase of inter-pass time, the plate deflections significantly increase, while the effect of the inter-pass time on the longitudinal residual stress field is marginal. Fialová and Jegla [52] proposed a novel framework for the efficient design of fired equipment. The authors supplemented the traditional thermal-hydraulic calculation of the radiant and convective section by the low-cost modelling systems taking into account the real distribution of heat flux and process fluids. The application of the low-cost models was demonstrated in the industrial steam boiler case. The significant advantages of the proposed approach are that the presented framework links calculations of radiant and convective sections in the combustion chamber, and offers a fast rating calculation of complex fired equipment. The proposed approach can successfully supplement CFD simulation, that should be used for critical components of power boilers. Sriromreun and Sriromreun [53] studied the numerical and experimental characteristics of the airflow impinging on a dimpled surface for air at Re numbers varied from 1500 to 14,600. The authors compared the heat transfer coefficient between the jet impingement on the dimpled surface and the flat plate. The CFD simulations results showed the different airflow characteristics for the dimpled surface and the flat plate. For a particular case, it was shown that a thermal enhancement of up to 5.5 could be achieved by using the dimpled surface. Flow boiling heat transfer is characterized by high heat transfer coefficient. Sun et al. [54] performed an experimental investigation to explore the flow boiling characteristics of R134A and R410A refrigerants flowing inside enhanced tubes. The experimental conditions included saturation temperatures of 6 °C and 10 °C, mass velocities from 70 to 200 kg/(m^2 s) and heat fluxes from 10 to 35 kW/m^2. The inlet and outlet vapour quality was equal to 0.2 and 0.8. The results showed that the dimples/protrusions and petal arrays are the effective surface structures for enhancing the tube-side evaporation. Moreover, the Re-EHT tube has the largest potential for boiling heat transfer enhancement. García-Castillo et al. [55] also discovered new opportunities to utilise plate-fin surfaces as a secondary surface in a multi-stream heat exchanger. They considered the theoretical design study of such new heat exchanger design, emphasising on the surface design to improve heat transfer coefficients. However, since the design is at the conceptual stage, reliable and accurate thermal-hydraulic correlations are needed. Heat transfer enhancement is highly required in energy equipment. Valdes et al. [56] studied the effect of twists in the internal tube of tube-in-tube helical heat exchanger keeping constant one type of ridges. The CFD simulations were performed to study the effect of the fluid flow rate on heat transfer in the internal and annular flow. The counter-current flow mode operation with hot fluid in the internal tube and cold fluid in the annular domain was considered. The flow and thermal development in a tube-in-tube helical heat exchanger were predicted. The double passive technique was provided within the internal tube to improve the turbulence in the outer region. The results showed that the addition of four ridges in the inner tube increases the heat transfer up to 28.8% when compared to the smooth tube. Kukulka et al. [57] also studied the flow characteristics for condensing and evaporating streams inside Vipertex stainless steel enhanced heat transfer tubes using R410A refrigerants. They proposed that using the Vipertex enhanced tubes are more energy-efficient than using old technology for phase change streams. As condensation and evaporation processes increase the interfacial turbulence, the proposed technology produces flow separation, secondary flows and higher heat flux from the wall to the working fluid.

2.1.2. Possible Future Development

From the performed literature survey, it is evident that efficient energy technologies are needed to improve the energy efficiency of energy systems. This can be achieved by improvement of the unit processes occurring in energy devices, such as enhancement of heat transfer, improvement of energy storage techniques, or improvement of the flexibility of power systems. A very good example here is the improvement of the electrical efficiency of PV panels by using active or passive cooling. Gaining heat from PV allows one to use it as low-temperature waste heat, and couple the PVT systems with heat pumps or underground energy storage systems. This kind of energy systems may attract widespread attention in the near future, due to the high efficiency, and utilization of waste heat. In the author's opinion, the major improvement may be made in the field of energy storage, which is crucial for electrical and thermal energy storage from renewable energy sources. A very important and challenging topic is also increasing the flexibility of hard-coal fired power units. This topic is important due to the large fluctuations in the power of wind farms, and due to the significant fluctuation in PV electricity production. In every moment, the generated power and demand should be equal. In case of rapid decrease of renewable energy production by wind farms and photovoltaics, the rapid start-up of steam boilers is needed. Thus, it is very important to improve the flexibility of thermal power units to shorten their start-up. Therefore, the new method on online calculation and monitoring of thermal stresses occurring in boiler's pressure components are highly needed by the industry, to allow safe start-up and shut-down of power units.

2.2. Process Integration—Heat and Power

The PRES conferences have been traditionally providing momentum to Process Integration research and development, not for more than 20 years and been analysed in detail at a jubilee PRES'17 conference [58]. The initial idea was based on the Heat Integration pioneered by the Centre for Process Integration at UMIST (Manchester, UK) started from 1998 hosted in Prague (Czech Republic). Process integration development continues to increase in scope and coverage. This has been in the recent period overviewed and analysed by several review papers. The methodology has been consistently extended to the Water Integration, combined energy and water, hydrogen network synthesis, regional resources planning and power system planning. The illustration of using graphical Pinch Methodology for Heat Integration is shown in Figure 9. In the aspects of heat and power, the methodology is consistently extended from process level to total sites, see Figure 10. The previous period was assessed by Klemeš et al. [59] in 2013 and more recently by Klemeš et al. [60] in 2018. It is worth to remind the contribution from Bandyopadhyay [61] who provided a detailed mathematical formulation of Pinch Methodology.

2.2.1. Core Developments

Pereira et al. [62] created a web-based Pinch Analysis tool for heat management called FI^{2}EPI. The tool can handle several energy management scenarios automatically, saving a significant amount of time for tedious routine calculations. It not only features energy or cost targeting, but it also identifies the optimal heat exchanger network design opportunities based on the heat exchanger loops and utility paths. The trade-off between design for minimum total annual cost or minimum temperature difference can also be identified, based on the preferences of the users. It is especially useful for Process Integration practitioners, saving extensive time or efforts in performing targeted calculations or network optimisation.

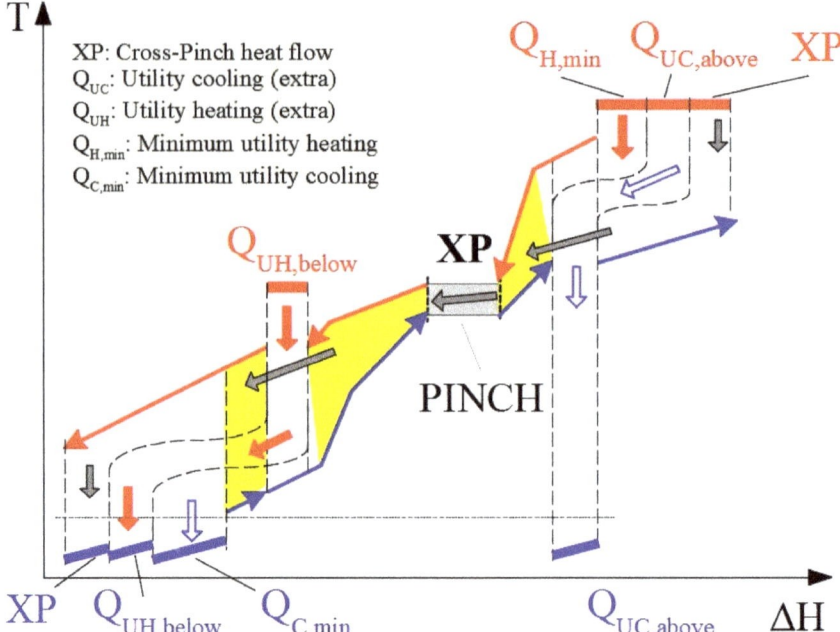

Figure 9. Pinch Methodology in energy targeting (adapted from Klemeš et al. [59]).

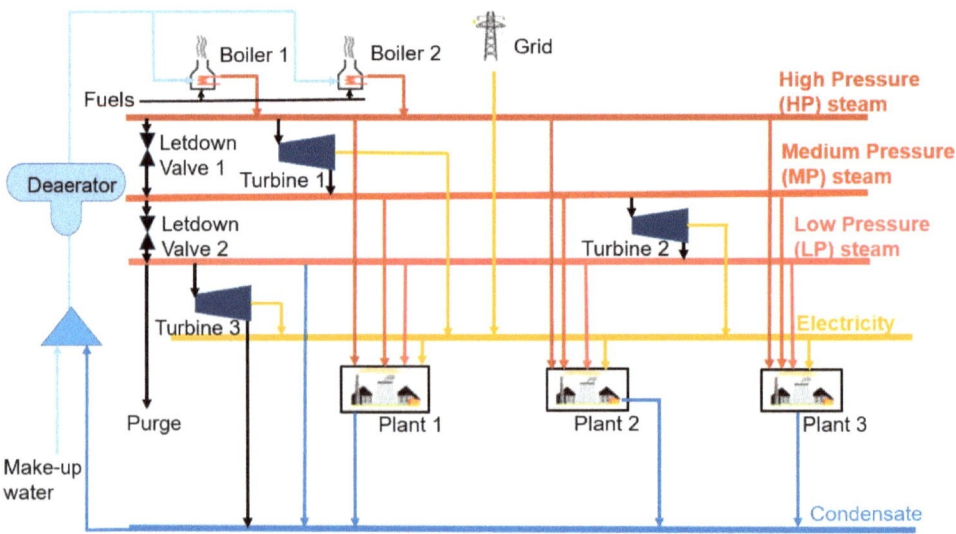

Figure 10. Illustration of industrial Total Site processes with central utility system (adapted from Klemeš et al. [60]).

Esfahani et al. [63] extended Power Pinch Analysis (EPoPA), which was developed by Wan Alwi et al. [64] in 2013 and extended by Rozali et al. [65] in 2017 for the integration of renewable energy systems with battery/hydrogen storage systems. This concept is based on the usage of hydrogen as an energy storage medium for the wasted electricity, which cannot be stored by the battery bank in the conventional PoPA [66], see Figure 11 for the original concept representation. This graphical tool not only provides visualisation by targeting the minimum required external electricity source and wasted electricity, but the appropriate hydrogen storage system capacity can also be identified during first and regular operation. They showed that integration of renewable hydrogen storage with a diesel generator is cost-effective. More renewable energy storage systems can be considered for in future work, to provide a more sustainable supply of electricity.

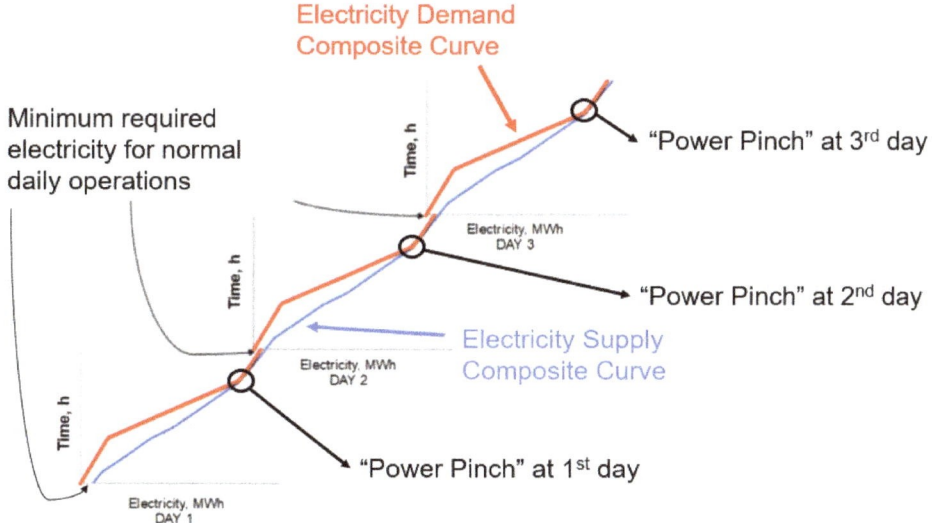

Figure 11. Illustration of Power Pinch Analysis (PoPA), adapted from Wan Alwi et al. [66].

The relation and contribution of process integration to cleaner production were studied by Fan et al. [67]. This paper indicated a very considerable contribution in process sustainability improvement, by reviewing recent progress in waste-to-energy, pollution prevention and remediation. The relation to CO_2 and GHG generally was highlighted by Manan et al. [68]. Another attempt to extend the Pinch Analysis was presented by Li et al. [69] focusing on the retrofitting of heat exchanger networks. The graphical approach provides interfaces to the users to get insights into the system bottlenecks. Iterative Pinch Analysis to address non-linearity in a stochastic Pinch problem was very recently studied by Arya and Bandyopadhyay [70]. Jain and Bandyopadhyay [71] developed multi-objective optimisation for segregated targeting problems using Pinch Analysis, which again extends the scope of Process Integration.

Interesting work was presented by Martinez-Hernandez et al. [72] dealing with the conceptual design of integrated production of arabinoxylan products using bioethanol Pinch Analysis. The Mass Integration allows significant advances in biorefineries achieved by retrofitting existing biorefineries. High value-added integrated production could be achieved through this method, which in turn provides potential in heat and power saving.

Walmsley et al. [73] developed another significant enhancement by analysing the possible contribution of Process Integration to the circular economy. The extensions of Pinch Analysis to other fields were reached by Roychaudhuri et al. [74] energy conservation projects through financial Pinch Analysis, and Ekvall et al. [75] presented a serious of works developing and applying material Pinch Analysis.

From the heat and power field come a couple of useful analyses. Jamaluddin et al. [76] presented an enhanced targeting tool for trigeneration problems. Chauhan and Khanam [77] reported enhancement of efficiency for the steam cycle of thermal power plants using applying Pinch Analysis—identifying and eliminating Cross-Pinch heat transfer in the steam system of the power plant.

Tie et al. [78] studied a specific impact of Process Integration on a classical chemical engineering issue—the production of glycol ether. Bandyopadhyay et al. [79] applied a combined pinch and exergy analysis for the energy-efficient design of the diesel hydrotreating unit. Malham et al. [80] contributed with hybrid exergy/pinch process integration methodology. Chen et al. [81] succeeded with another extension of process integration optimal heat rejection pressure of co_2 heat pump water heaters based on Pinch Analysis.

The PRES 2018 Special Issue also contributed very significantly to the following topics:

Jankowski et al. [82] applied process integration methodology of ORC plant using a multiobjective approach to recover low-potential heat. It has shown some new possible implementation of Process integration.

Schlosser et al. [83] paid attention to robust total site heat recovery and applied a Monte Carlo simulation successfully. Jamaluddin et al. [84] attempt to deal with the Heat and power at a total site by a trigeneration system and reached an extension of the methodology. Another contribution to a most studied issue dealing with this topic a cost-optimal heat exchanger network synthesis enhanced with flexible cost function network was presented by Rathjens et al. [85]. The utilisation of an air-PCM heat exchanger in passive cooling of buildings was presented by a team from the VUT Brno Energy Institute [86]. From the same research group came the contribution presented by Kudela et al. [87] stressing heat accumulation, an essential issue in district heating pipes. Leithoid et al. [88] dealt with controllability and observability of heat exchanger networks, the key equipment used and studied by Heat Integration. Kamat et al. [89] studied the heat integrated water regeneration networks, considering variable regeneration temperature. They formulated a linear model to optimise the freshwater use, utilities requirement and regenerated water. Sequential and simultaneous optimisation are also considered. Sensitivity analysis between water regeneration unit vs the total operating cost is also performed. The higher the temperature, the higher the operating cost due to more expensive heating utility required.

Outside the Special Issue, the thematic support provided by the work of Ong et al. [90] dealing with the total site mass, heat and power integration using process integration published in 2017 and Kim et al. [91] with clean and energy-efficient mass production of biochar by process integration should be considered.

2.2.2. Possible Future Developments

From the review of the area, one can note a clear trend of expansion of the scope of integration problems, which increases the complexity of the obtained models. One example to be given includes the addition of the power management domain to the family of Process Integration areas. Other examples are the combination of Heat Integration with power generation (leading to the CHP domain) and with Mass Integration.

Another noteworthy development is the attempt at developing tools for further improvement and application of the Process Integration methods—as in the web-based tool by Pereira et al. [62]. This is by no means the only tool on the market. One can mention the flagship products—SuperTarget (version 7.0.15) that comes bundled with the PetroSim software (version 7.0) by KBC in London, UK [92] as well as the software suite for process integration by the company of the same name [93]. While these tools are quite suitable for final use by industrial and consultancy companies, their use for research is inherently limited by the fixed context and procedures built into the software. To support new research and further improvement of the Process Integration methods, including their interactions and combinations, the development of an integrated software platform tailored to the Process Integration thinking would be very beneficial. This should allow researchers and users to define new methods and algorithms while reusing a unified code base of already established core methods like Heat Integration and Water Integration and the well-known resource cascades.

2.3. Process Energy Efficiency/Emissions Analysis

As has been reasoned previously by Varbanov et al. [94], one of the core problems with achieving sustainable development lies in the reduction of wasted energy. This is often referred to as increasing energy efficiency. It can be achieved by heat recovery and reuse [60], as well as by heat upgrade and recycling—widely known as heat pumping in the energy engineering community [95].

2.3.1. Recent Developments

There has been the argument that increased energy efficiency is bound to lead to decrease in the prices for the energy services, in turn inducing increased demand and finally—compensating or even overshooting the prior consumption of primary energy resources, which is referred to as the "rebound effect". An empirical study from Greening et al. [96] based on USA data sources corroborates this argument to a moderate extent. A more recent study for China [97] also supports the existence of such an effect, varying between 50% and a 2-3-fold increase in energy demands as a result of energy efficiency improvement.

While the reasons for these trends are under investigation, one has also to pay attention to a subtle difference in the argument to reduce energy waste. While Varbanov et al. [94] put forward the argument that energy waste has to be reduced, the official statistics detect only the energy waste within the supply chain of delivering energy services. This implicitly excludes the waste of the energy-based services themselves. This is the core of the problem. If one analyses the extended onion diagram for integrating user demands to production processes and supply chains based on Walmsley et al. [73], the picture becomes much clearer. The use of food, lighting and direct energy use all involve waste. Further emissions reduction studies should consider all opportunities for wasting energy and minimise them.

Biomass is viewed as one of the ways of increased use of renewable energy sources and decrease GHG footprints, but it is also associated with other issues—such as increased nitrogen footprint in the study by Čuček et al. [98] and increased release of fine solids by Bartington et al. [99]. While it is common knowledge that coal combustion causes significant release of particulate matter (PM), it is less-known that burning biomass causes similar problems, but on a smaller scale. Al-Naiema et al. [100] evaluated the PM emissions from the co-firing coal and biomass, reporting reduced PM levels, compared to burning coal alone, including solids (PM) by 90%, polycyclic aromatic hydrocarbons by 40% and metals by 65%.

Poláčik et al. [16] presented a parametrical study that assessed the influence of the composition of the atmosphere and the temperature on the formation and release of ultra-fine solids (PM) by micro-scale combustion of biomass. The described laboratory procedure employing thermogravimetric analysis (TGA) and a detailed assessment of the size distribution of the produced fine particles. The authors concluded that the particle sizes feature a strong correlation to the concentration of light volatiles released from the heated wood sample. They have also established a trend of increased formation of PM with the reduction of oxygen content in-stream fed to the test chamber, featuring twice more PM particle count for pyrolysis (zero oxygen) than for regular combustion in normal air.

Biomass gasification is also associated with PM formation and release, as discussed in [101]. The authors investigated experimentally the content and properties of polluting solid particles present in the synthesis gas, resulting from a test gasifier of the "Imbert" type. The analysis also included an evaluation of the particle size distribution. The author's reason that it is more efficient to clean the synthesis gas before burning than to leave the cleaning to the flue gas stage. The analysis of the filtration cake established significant amounts of aluminium, calcium and silicon oxides, as well as SO_3—all in the range of 12-16 mass%. There was also non-negligible and dangerous content of metal oxides—including MnO (7.6%) and Na_2O (2.4%), as well as heavy metal oxides (Cr_2O_3, CdO, TiO_2, SrO) and even P_2O_5. These results indicate the need to deeper investigate the process of biomass gasification, for establishing the true extent of the resulting pollution and footprints and the means of their minimisation, for providing sustainable solutions.

Agricultural activities are associated with a number of environmental impacts and risks [102]—including business risks of varying prices and regulatory uncertainties, as well as environmental impacts and risks such as unforeseen emissions when the activities are conducted inappropriately. Such issues can be tackled by comprehensive optimisation methods—such as the maximisation of Sustainability Net Present Value [103].

There have been increasing concerns in providing sufficient food at acceptable quality to the continuously rising human population worldwide, addressed in [104] for the case of Spain. The study applies a combination of life-cycle assessment and data envelopment analysis to assess the energy efficiency of the Spanish agri-food system. Potential improvement actions, aimed at reducing energy usage and GHG emissions, were also proposed. Energy Return on Energy Invested (EROI) is used as a criterion. For more complex food types (meat, eggs, seafood) the primary energy and GHG footprint contributing stage have been the core product, while for vegetables, this was found to be the energy use for cooking and cooling. The authors report that, for sufficient efficiency of the system, it is necessary to implement energy-saving measures, resulting in approximately 70% energy savings.

Water-related issues involve the need for reliable supply. Evermore frequently, it becomes necessary to generate freshwater by desalination. Due to the high cost of the product, its reliable distribution is also a vital issue [105]. Water desalination industry is a good illustration of water-energy linkages. Some technologies rely on membranes—for instance, this is the case with some installations in Jordan [106]. The strength of the links and their correlations to energy consumption and GHG emissions have been evaluated by Jia et al. [107]. The authors presented an overview of the seawater desalination developments in China and evaluated the annual energy consumption, GHG emissions, and cost of seawater desalination plants from 2006 to 2016. The results indicate that the energy consumption increased from 81 MWh/y to 1,561 MWh/y during the 11 y period, and the GHG emissions increase from 85 Mt CO_2eq/y to 1,628 Mt CO_2eq/y, representing an increased rate of 180%. The authors concluded that the current unit product cost of seawater desalination in China is still higher than other water alternatives, but it has good potential for reduction with the improvement of desalination technologies. The unit product cost shows a decreasing trend with increasing the processing capacity.

It has been well-known that the intermittence of supply is the critical barrier before the effective integration of renewable energy sources, which is made even more dynamic by the variability of the user energy demands based on Varbanov and Klemeš [108]. Many energy storage technologies are available—both for thermal as presented by Alva et al. [109] and electrical forms as discussed by Cheng et al. [110]. The popularity of storing electrical energy grows further, fueled by the increased attention to electric cars, as can be traced by the search for novel materials for electrochemical energy storage according to Chen et al. [111]. The availability of storage technologies fulfils only part of the task. To use them, they need to be integrated with the energy supply, delivery and use systems. Rozali et al. [112] mentioned that the efforts include the optimisation of electrical storage size, based on the dynamics of power generation and use. Jamaluddin et al. [76] provided a further extension has been the optimisation of a combination of electricity and heat storage facilities, for catering for heating, cooling and power flows.

A hybrid energy storage method, based on cryogenics, has been investigated by Hamdy et al. [113] and reported in this Special Issue. Cryogenics-based energy storage (CES) is a technology for thermoelectric energy storage at a larger scale. Using this method, electricity is stored in the form of liquefied gas at cryogenic temperatures. The charging process consists of the gas liquefaction process. That represents the limiting factor to the round-trip efficiency (RTE) of the storage method. During discharge, the liquefied gas is pressurised, evaporated and then super-heated to drive a gas turbine. The cold released during evaporation can be stored and supplied to the subsequent charging process. In the research by Hamdy et al. [113], several liquefaction processes are evaluated to identify the most cost-efficient one, using exergy analysis. The authors have concluded that the integration of cold storage enhances the liquid yield, in this way, reducing the specific power requirement by 50–70%.

Besides the evaluation of the energy conversion networks, exergy analysis is starting to play an essential role in also evaluating production processes and their energy relationships. For instance, Ghannadzadeh and Sadeqzadeh [114] have used exergy analysis as a scoping tool for optimising an ethylene production process. This is an energy-intensive process, in which the authors have successfully identified significant efficiency improvement options. The same type of analysis can also be used for decision-making in comparing process design alternatives [115].

Meramo-Hurtado et al. [116] presented an exergy analysis of a bio-adsorbent production process, aiming at the identification of opportunities and measures for reducing the energy demand of the process. Three bio-adsorbent production process networks for large-scale production of chitosan microbeads have been evaluated. Exergy efficiencies, total process irreversibilities, energy consumption, and exergy destruction were calculated for the analysed alternatives. While the authors could not find crucial differences among the evaluated processes, they did identify process improvement opportunities in the product drying and washing water recovery stages of all investigated processes.

Municipal solid waste (MSW) collection is an essential activity in modern cities, which, when combined with appropriate separation, materials recovery and waste-to-energy recovery based on Tomić and Schneider [117], can bring about several synergies. This is practised world-wide for simultaneous reduction of fresh resource intake, fuel use and for GHG emissions reduction. Since door-to-door collection generates significant direct greenhouse gas emissions from trucks, pneumatic collection emerges as an alternative to the trucking system. While this technology apparently reduces local direct air emissions, it has a large energy demand caused by the need for generating vacuum for waste suction. Laso et al. [118] presented an analysis that compares conventional door-to-door and pneumatic waste collection systems using Life Cycle Analysis. The considered system boundary includes accounting for the creation, installation, maintenance, and decommissioning of the waste collection system, as well as for the waste transfer, sorting and waste processing sub-systems. The focus is on the biodegradable fraction of the collected waste. The authors report that the energy savings from the recycling of the organic fraction outweigh the energy requirements for the operation. Based on that, they suggest that pneumatic collection could be an environmentally-friendly option for MSW management under a circular economy, pointing out that waste could be a valuable source of materials and energy.

2.3.2. Possible Future Developments

It has been shown in the previous section that the optimisation of energy sourcing, conversion and use has to be considered holistically. This involves two key dimensions of the problem. One is the consideration of the complete product chains—from "cradle" to "cradle", as advocated by some Life Cycle Analysis branches [119]. The other dimension consists of accounting for the trade-offs in emissions and other footprints when substituting currently used fossil fuels by renewable alternatives.

Important directions for further research include providing additional degrees of freedom in the energy-related networks that would allow the increased use of renewables, compensating for or eliminating some of the problems associated with their exploitation. Such directions certainly include energy storage—using all forms: thermal, electrical, chemical, mechanical, to name a few.

Another key topic is innovation for separation and neutralisation of harmful emissions of sulphur and nitrogen compounds resulting from biomass use. In this regard, especially in the utilisation of biomass waste, it is important to minimise the CO_2 emission overhead of logistics. A good step in this direction has been the framework by How et al. [120], which needs, however, further development of technology solutions that are closer to practical implementation.

The development of more durable and efficient energy conversion technologies should also be kept on the front burner. In this regard, fuel cells and the microbial fuel cell variety, for generating power from organic waste, are a good example. The development of lower-cost materials [121] is a good step that should be followed by similar studies in related areas.

2.4. Optimisation of Renewable Energy Resources Supply Chain

Renewable energy is clean and can be efficient in energy supply if adequately developed, based on Kong et al. [122]. The significant advantage is no fuel consumption and relatively low costs when compared to conventional power plants. Therefore, green electricity can be competitive with conventional electrical energy in a long time period. Kong et al. [122] built a renewable energy electricity supply chain collaboration model by employing the revenue-sharing contract to achieve the green power grid-connection and consumption optimisation. The authors used continuous random variables to describe the intermittency of green power output intensity and the fluctuations of power market demand. Afterwards, the authors coordinated the profit distribution between the power generator and the grid company by adjusting the revenue-sharing contract and analysed the optimal decisions taken by the companies for different power market demand price. Through the numerical simulation analysis, the authors obtained the equilibrium solutions of contract satisfied various conditions and investigated the relationship among the optimal variables and profits, obtaining the management suggestions. The critical point is, the authors also investigated the influences of market demand price elasticity and power output efficiency.

2.4.1. Core Developments

A sustainable supply chain should involve coordination among resources, flows and stocks with a well-defined sustainability concept. Saavedra et al. [123] presented a literature overview on system dynamics modelling applied in the renewable energy supply chain, considering works published between 2007 and 2017. The review provides new insight into the analysis of the supply chain in renewable energy using systems dynamics. The authors showed that the system dynamics approach provides harmony between its subsystems and processes, understanding the system behaviour, testing policies for improvement, and assessing impacts over time. The system dynamics approach was presented for the Biomass Scenario Model (BSM). The authors discussed the scenario analysis process to determine the most significant factors affecting the overall performance of the supply chain model. Beside this was discussed the application for the Hybrid Modelling Framework that integrates multiple tools to study complex system problems, such as different actors in the supply chain with various needs, objectives and decision-making behaviours.

Fernando et al. [124] studied the effect of energy management practices on renewable energy supply chain initiatives in 151 certified manufacturing companies in Malaysia. The results showed three dimensions of energy management practices, i.e., top management commitment, energy awareness and energy auditing. Those practices were positively linked with the development of renewable energy supply chain initiatives. The authors found that insufficient knowledge of energy efficiency means does not allow to manage energy effectively, constraining opportunities such as converting waste into energy to support business' targets. The authors suggested transferring the energy efficiency management knowledge and technology from multinationals to local companies. Local companies would be able to generate renewable energy through supply chain networks.

Nugroho and Zhu [125] developed a biofuel platform for planning and optimisation. The platform unifies biofuel product, production process and networks design. The authors considered the design of the biofuel supply chain network under various production paths. The authors studied the optimum region of the composition ratio between rice straws and waste cooking oils and found its value between 0% to 50%. The results showed that the combined raw materials increase the supply flexibility and supply chain responsiveness. The hydrocarbon biofuels are favoured over ethanol in minimizing the overall carbon emissions.

Sarker et al. [126] controlled the supply chain costs of biomethane gas (BMG) production systems, optimised the location of BMG plants and determined the routing network for transporting the feedstock and the work-in-process materials. The authors proposed an efficient mixed-integer nonlinear model to optimise biogas a plant location problem. The algorithm was used to find a solution to locate hubs, reactors and condensers to minimise the total costs.

Li et al. [127] stated that renewable energy systems are constantly affected by weather or climatic conditions (i.e., solar irradiation, wind speed, external temperature). In order to handle the effect of external disturbances on RES systems performance, the dynamic forecasting, as well as energy storage, shall be provided. The advanced prediction algorithms like coupled autoregressive and dynamic system (CARDS) from Huang and Boland [128] or artificial neural networks (ANN) from Gupta et al. [129] can be applied for solar radiation and wind speed, wind power and solar power prediction among others. Even more challenging is the long-term forecasting for future power output, according to Zhao et al. [130], with multiobjective optimisation from Behzadi et al. [131]. Luo et al. [132] mentioned that deep learning techniques are also commonly used in long term prediction of wind speed.

Azevedo et al. [133] performed a comprehensive bibliometric analysis of studies in the field of supply chain performance and renewable energies. The review was focused on the most productive authors and institutions, as well as the most cited articles from the field. According to Azevedo et al. [133], most articles in the field focus on the design optimisation of renewable energies supply chain. Among the analysed methods, the mixed-integer linear programming (MILP) model is the most popular. Nevertheless, other methods, like case studies, surveys, simulations, modelling, genetic algorithms, multi-scale modelling, and optimisation, are being used successfully.

Zakaria et al. [134] reviewed stochastic optimisation techniques in renewable energy applications. The authors found that the stochastic optimisation exhibit enhanced performances and can deliver accurate representations in capturing the uncertainties of renewable systems. The authors also found that with a rapid increment of data and size of renewables' problem, the model-driven approaches alone could not adequately address and handle with the underlying complexity in vast multivariate and expanding renewable systems. The data-driven scenario generations could be an excellent future choice. Chen et al. [135] presented a method for scenario generation, which has been complemented with complete scenario-based forecasting by Chen et al. [136]. Also, in the field of renewable energy integration, when the problems are of higher dimension, there is a need to hybridize the existing optimisation methods with intelligent search. Those methods can reduce the computational time with proper accuracy, as can be seen from the work of Dufo-López et al. [137] on the optimisation of stand-alone energy systems on the optimisation of stand-alone energy systems on the optimisation of stand-alone energy systems, comprising photovoltaics, wind and diesel generators, combined with batteries for electricity storage. Another work by Rahmani-Andebili [138] focuses on the management of power storage systems with the goal of minimising the power losses. Sharafi et al. [139] discussed the stochastic optimisation of renewable energy systems. Zakaria et al. [134] also identified further research areas in the field of stochastic renewable energy problems such as (a) plug-in electric vehicles integration—for example. Thompson [140] considered charging and scheduling of plug-in EV, renewable energy integration via vehicle to grid operation (b) demand-side management (c) multi-scale and multi-time-scale distributed renewable energy systems.

Ubando et al. [109] proposed a fuzzy mixed-integer linear programming model to achieve an optimal operational adjustment of an off-grid micro-hydropower-based polygeneration plant and maximize the satisfaction levels of the community utility demands, which are represented as fuzzy constraints. The authors considered three case studies to demonstrate the developed model. The results showed that the use of a diesel generator for back-up power is considered as an option to mitigate inoperability during extreme drought conditions.

Éles et al. [141] developed a new P-graph model to study the synthesis of the energy supply options of a manufacturing plant in Hungary. The authors applied a multi-periodic scheme for heating and electricity demands. The modified P-graph was applied to model the pelletizer and biogas plant investments. The authors found the best solution in terms of total costs. The results showed that a long-term investment horizon is needed in order to make incorporation of sustainable energy sources into the system economically beneficial.

San Juan et al. [142] developed a MILP model for optimising a biomass co-firing supply chain network. The model considers feedstock properties while minimising economic cost and environmental emissions through goal programming. The effect of feedstock, transportation and pre-treatment requirements was incorporated in the model. The authors found that minimising either the financial or environmental objective individually emphasised the conflicting nature of the two objectives. Simultaneously optimising both objectives created a network which balanced performance on both objectives. The results showed that without considerations for feedstock properties, costs and emissions were artificially decreased, leading to the purchase of insufficient fuel and combustion of inappropriate fuel. This situation may lead to damage or loss of inefficiency of the equipment. The model proposed by the authors is a better fit to design and manage a biomass co-firing network.

Peesel et al. [143] proposed a predictive optimisation algorithm to calculate the optimal operating conditions of multiple chillers. The authors applied a sprinkler tank that allows storing cold-water for later utilization. The load shifting potential of the cooling system was demonstrated by using a variable electricity price as an input variable to the optimisation. The dynamic simulation was used to adjust the setpoints from the optimisation continuously. The results showed that by applying an optimal chiller sequencing and charging strategy of a sprinkler, tank leads to electrical energy savings of up to 43%. The purchasing electricity on the EPEX SPOT market leads to additional costs savings of up to 17%. It was shown that the total energy savings highly depend on the weather conditions and the prediction horizon.

Barmina et al. [144] studied the electric field effect on the thermal decomposition and co-combustion of straw with solid fuel pellets. The fixed bed experimental setup with a heat output of 4 kW was used. The authors found that the co-firing of straw with wood or with peat pellets provide the enhanced decomposition of the mixture, with the best performance when straw mass share in the mixture is about 20–30%. The authors performed extensive experimental research and found that the field-induced ion current in the space between the electrodes is responsible for the field-enhanced reverse axial heat/mass transfer of the flame species, that provides the enhanced heating and thermal decomposition of biomass pellets. The results also showed that the electric field-induced processes of heat and mass transfer allow to control and improve the main combustion characteristics so enhancing the fuel burnout and increasing the heat output.

Recently due to the increased share of electricity from renewable energy sources, many studies are performed by integrating them with the electrical energy network. Fichera et al. [145] studied the energy mapping of the urban flows using the implementation of the network theory. The scenarios analysis for the elaboration of the energy strategies for the promotion and installation of cogeneration systems using the RES was performed. The authors developed a tool that characterises the energy profile of an urban area, and the model was tested with the data of Catania municipality. What is important, the developed model is able to define the interaction between the nodes and enables to formulate the urban energy trajectory relatively to the energy demand of each district.

Gonzalez de Durana et al. [146] presented a generalised energy networks modelling approach using the agent-based method. The model proposed by the authors can be used to represent integrated utility infrastructures, including the systems in which not only one but different carriers are managed together by a multi-energy utility. The application of the proposed method can be performed from small, rural or microgrid systems up to large energy infrastructures in an urban context. What is also important, the model can be used to perform exploratory simulations to better understand those systems behaviour, and further to test and develop operation management strategies. In other work, Gonzalez de Durana et al. [147] developed a complete and self-contained model of a simple microgrid. The procedure based on the system dynamics not making use of any technological parts is adopted. The model can be applied to simulate the energy performance of a number households in a neighbourhood, at high time resolution, including energy generation and consumption, allowing the user for trying and designing particular generation and storage methods or demand-side management procedures.

Another study that considers microgrid modelling was performed by Kremers et al. [148]. The authors created a systemic modular model for a microgrid with a load flow calculation. The communication layer was also included in the model. The applied agent-based approach enables to include the intelligent strategies on every node of the studied system. Classical tools struggle with the implementation of dynamic interaction and message passing among individual devices. The agent-based approach, which is used alongside a classical load flow treats the simulation from a different side. The case study performed by the authors shows the interest of being able to reproduce both effects on the power grid and the communication network and observe the complex system behaviour as a whole.

One of the promising methods of electrical energy storage involves the use of stationary battery energy storage systems (BESS), Zeh et al. [149] studied the application of battery energy storage systems to provide primary control reserves (PCR) in the Union for the Coordination of the Transmission of Electricity (UCTE) area. The authors discussed the technical requirements for BESS operation as PCR provision systems, provided explanations of the PCR market and regulation; and demonstrated the approach for operating BESS as PCR storage systems, showing potential outcome for such a system.

Tran and Smith [150] investigated residential rooftop photovoltaic (PV) systems for long-term thermos-economic benefits from PV homeowners' perspectives and for impacts on the electrical distribution network from grid operators' perspectives. The authors studied the costs of generating electricity from grid-connected PV systems with and without Energy storage. The case study was performed for three different scenarios, including net metering, wholesale pricing, and no payback. PV systems in Utah. The simulation results showed that the addition of PV systems reduces the annual electricity bill up to 75%. A net metering policy offers PV homeowners with the most benefit in terms of annual electricity bills. However, the addition of energy storage under the net metering and wholesale pricing policies increases the annual electricity bills compared to similar systems without energy storage. The reason is related to losses associated with charging and discharging the battery.

2.4.2. Possible Future Development

Due to the rapid increase in renewable energy electricity demands, there is a need to develop an efficient method for the integration of renewable energy sources and coal-fired power plants. A first important research area here is the efficient energy demand prediction and optimization of RES electricity usage, to address the customers' needs. The second important research area is the electrical and thermal energy storage, which can assure the proper usage of renewable energy sources in case of high energy demands. The third research area of high interest is the optimization of RES operation, including photovoltaics, wind farms, and biomass plants, to allow those energy systems operate in the most efficient mode. In authors opinion those two research directions will be on the highest priority in the following years since the EU demands a significant contribution of RES in the energy supply chain.

3. Conclusions

This review editorial was initiated by the Special Issue for carefully selected papers from the 2018 Process Integration, Modelling and Optimisation for Energy Saving and Pollution Reduction (PRES 2018) conference. The conference, which attracted more than 550 leading researchers worldwide, boosted the interest in a number of highly appealing scientific issues: Design and numerical study for innovative energy-efficient technologies, Process Integration—Heat and Power, Process energy efficiency or emissions analysis, Optimisation of renewable energy resources supply chain. These issues are having a strong influence on sustainability and the circular economy. The authors endeavoured to overview this field by adding relevant and recent references and suggested some conclusions for future research directions. The feedback from the readers and especially researchers in the field is most welcome and appreciated, and it should provide a ground for the next PRES'19 panel discussion devoted to the future Special Issues of the journal Energies.

Author Contributions: All authors contributed to the overview. J.J.K. has managed the overall process of the article creation., supervising and directing the work of the other authors. He has also been responsible for writing the abstract and the conclusions section, as well as most of Section 2.3. H.H.C. has contributed with Sections 1 and 2.4 and been responsible for the overall arrangement and graphics. P.O. has contributed with Section 2.1., P.S.V.—for Section 2.2 and partly Section 2.3 and provided the overall advise.

Funding: The EU supported project Sustainable Process Integration Laboratory—SPIL funded as project No.CZ.02.1.01/0.0/0.0/15_003/0000456, by Czech Republic Operational Programme Research and Development, Education, Priority 1: Strengthening capacity for quality research in collaboration with Cracow University of Technology, Poland.

Conflicts of Interest: The authors declare no conflict of interest.

References

1. Wang, S.; Li, Q.; Fang, C.; Zhou, C. The relationship between economic growth, energy consumption, and CO_2 emissions: Empirical evidence from China. *Sci. Total Environ.* **2016**, *542*, 360–371. [CrossRef] [PubMed]
2. Melorose, J.; Perroy, R.; Careas, S. *World population prospects*; United Nations: New York, NY, USA, 2015; Volume 1, pp. 587–592.
3. BP Statistical Review of World Energy. Available online: https://www.bp.com/content/dam/bp/business-sites/en/global/corporate/pdfs/energy-economics/statistical-review/bp-stats-review-2018-full-report.pdf (accessed on 28 July 2019).
4. Mah, A.X.Y.; Ho, W.S.; Bong, C.P.C.; Hassim, M.H.; Liew, P.Y.; Asli, U.A.; Kamaruddin, M.J.; Chemmangattuvalappil, N.G. Review of hydrogen economy in Malaysia and its way forward. *Int. J. Hydrogen Energy* **2019**, *44*, 5661–5675. [CrossRef]
5. Capuano, D.L. International Energy Outlook 2018 (IEO2018). Available online: https://www.eia.gov/pressroom/presentations/capuano_07242018.pdf (accessed on 28 July 2019).
6. Suman, S. Hybrid nuclear-renewable energy systems: A review. *J. Clean. Prod.* **2018**, *181*, 166–177. [CrossRef]
7. Dharmadasa, I.M. *Advances in Thin-Film Solar Cells*; Pan Stanford Publishing Pte. Ltd.: Singapore, 2018; ISBN 978-981-4800-12-9.
8. Vo Hoang Nhat, P.; Ngo, H.H.; Guo, W.S.; Chang, S.W.; Nguyen, D.D.; Nguyen, P.D.; Bui, X.T.; Zhang, X.B.; Guo, J.B. Can algae-based technologies be an affordable green process for biofuel production and wastewater remediation? *Bioresour. Technol.* **2018**, *256*, 491–501. [CrossRef]
9. Zabed, H.; Sahu, J.N.; Boyce, A.N.; Faruq, G. Fuel ethanol production from lignocellulosic biomass: An overview on feedstocks and technological approaches. *Renew. Sustain. Energy Rev.* **2016**, *66*, 751–774. [CrossRef]
10. Lam, H.L.; Varbanov, P.; Klemeš, J. Minimising carbon footprint of regional biomass supply chains. *Resour. Conserv. Recycl.* **2010**, *54*, 303–309. [CrossRef]
11. Gross, R.; Hanna, R.; Gambhir, A.; Heptonstall, P.; Speirs, J. How long does innovation and commercialisation in the energy sectors take? Historical case studies of the timescale from invention to widespread commercialisation in energy supply and end use technology. *Energy Policy* **2018**, *123*, 682–699. [CrossRef]
12. Brook, B.W.; Bradshaw, C.J.A. Key role for nuclear energy in global biodiversity conservation. *Conserv. Biol.* **2015**, *29*, 702–712. [CrossRef]
13. Pablo-Romero, M.D.P.; Román, R.; Sánchez-Braza, A.; Yñiguez, R. Renewable Energy, Emissions, and Health. In *Renewable Energy—Utilisation and System Integration*; IntechOpen: London, UK, 2016.
14. Mathiesen, B.V.; Lund, H.; Karlsson, K. 100% Renewable energy systems, climate mitigation and economic growth. *Appl. Energy* **2011**, *88*, 488–501. [CrossRef]
15. Partridge, I.; Gamkhar, S. A methodology for estimating health benefits of electricity generation using renewable technologies. *Environ. Int.* **2012**, *39*, 103–110. [CrossRef] [PubMed]
16. Poláčik, J.; Šnajdárek, L.; Špiláček, M.; Pospíšil, J.; Sitek, T. Particulate Matter Produced by Micro-Scale Biomass Combustion in an Oxygen-Lean Atmosphere. *Energies* **2018**, *11*, 3359. [CrossRef]
17. IRENA Renewable Power Generation Costs in 2017. Available online: https://www.irena.org/-/media/Files/IRENA/Agency/Publication/2018/Jan/IRENA_2017_Power_Costs_2018.pdf (accessed on 29 September 2019).
18. Twidell, J.; Weir, T. *Renewable Energy Resources*; Routledge: New York, NY, USA, 2015; ISBN 978-1-317-66037-8.
19. Ritchie, H.; Roser, M. Renewable Energy. Available online: https://ourworldindata.org/renewable-energy (accessed on 28 July 2019).
20. REN21 Renewables 2018 Global Status Report. Available online: http://www.ren21.net/gsr-2018 (accessed on 28 July 2019).

21. WEO 2018. Available online: https://www.iea.org/weo2018/ (accessed on 28 July 2019).
22. Murphy, D.J.; Hall, C.A.S. Year in review—EROI or energy return on (energy) invested. *Ann. N. Y. Acad. Sci.* **2010**, *1185*, 102–118. [CrossRef]
23. Hall, C.A.S.; Lambert, J.G.; Balogh, S.B. EROI of different fuels and the implications for society. *Energy Policy* **2014**, *64*, 141–152. [CrossRef]
24. Lambert, J.G.; Hall, C.A.S.; Balogh, S.; Gupta, A.; Arnold, M. Energy, EROI and quality of life. *Energy Policy* **2014**, *64*, 153–167. [CrossRef]
25. Movellan, J. Fighting Blackouts: Japan Residential PV and Energy Storage Market Flourishing. Available online: https://www.renewableenergyworld.com/articles/2013/05/fighting-blackouts-japan-residential-pv-and-energy-storage-market-flourishing.html (accessed on 30 July 2019).
26. Prieto, P.A.; Hall, C.A.S. Spain's Photovoltaic Revolution the Energy Return on Investment. Available online: science-and-energy.org/wp-content/uploads/2016/03/20160307-Des-Houches-Case-Study-for-Solar-PV.pdf (accessed on 28 July 2019).
27. Weißbach, D.; Ruprecht, G.; Huke, A.; Czerski, K.; Gottlieb, S.; Hussein, A. Energy intensities, EROIs (energy returned on invested), and energy payback times of electricity generating power plants. *Energy* **2013**, *52*, 210–221. [CrossRef]
28. Hall, C.; Balogh, S.; Murphy, D. What is the Minimum EROI that a Sustainable Society Must Have? *Energies* **2009**, *2*, 25–47. [CrossRef]
29. Connelly, L.; Koshland, C.P. Exergy and industrial ecology. Part 2: A non-dimensional analysis of means to reduce resource depletion. *Exergy Int. J.* **2001**, *1*, 234–255. [CrossRef]
30. Khodadoost Arani, A.A.B.; Gharehpetian, G.; Abedi, M. Review on Energy Storage Systems Control Methods in Microgrids. *Int. J. Electr. Power Energy Syst.* **2019**, *107*, 745–757. [CrossRef]
31. Jiang, Y.; Kang, L.; Liu, Y. A unified model to optimize configuration of battery energy storage systems with multiple types of batteries. *Energy* **2019**, *176*, 552–560. [CrossRef]
32. Duan, J.; Liu, J.; Xiao, Q.; Fan, S.; Sun, L.; Wang, G. Cooperative controls of micro gas turbine and super capacitor hybrid power generation system for pulsed power load. *Energy* **2019**, *169*, 1242–1258. [CrossRef]
33. Colmenar-Santos, A.; Molina-Ibáñez, E.L.; Rosales-Asensio, E.; López-Rey, Á. Technical approach for the inclusion of superconducting magnetic energy storage in a smart city. *Energy* **2018**, *158*, 1080–1091. [CrossRef]
34. Venkataramani, G.; Vijayamithran, P.; Li, Y.; Ding, Y.; Chen, H.; Ramalingam, V. Thermodynamic analysis on compressed air energy storage augmenting power/polygeneration for roundtrip efficiency enhancement. *Energy* **2019**, *180*, 107–120. [CrossRef]
35. Guelpa, E.; Verda, V. Thermal energy storage in district heating and cooling systems: A review. *Appl. Energy* **2019**, *252*, 113474. [CrossRef]
36. Bhagat, K.; Prabhakar, M.; Saha, S.K. Estimation of thermal performance and design optimization of finned multitube latent heat thermal energy storage. *J. Energy Storage* **2018**, *19*, 135–144. [CrossRef]
37. Silakhori, M.; Jafarian, M.; Arjomandi, M.; Nathan, G.J. Experimental assessment of copper oxide for liquid chemical looping for thermal energy storage. *J. Energy Storage* **2019**, *21*, 216–221. [CrossRef]
38. Taler, D.; Dzierwa, P.; Trojan, M.; Sacharczuk, J.; Kaczmarski, K.; Taler, J. Mathematical modeling of heat storage unit for air heating of the building. *Renew. Energy* **2019**, *141*, 988–1004. [CrossRef]
39. Sacharczuk, J.; Taler, D. Numerical and experimental study on the thermal performance of the concrete accumulator for solar heating systems. *Energy* **2019**, *170*, 967–977. [CrossRef]
40. Taler, D.; Dzierwa, P.; Trojan, M.; Sacharczuk, J.; Kaczmarski, K.; Taler, J. Numerical modeling of transient heat transfer in heat storage unit with channel structure. *Appl. Therm. Eng.* **2019**, *149*, 841–853. [CrossRef]
41. Martínez-Rodríguez, G.; Fuentes-Silva, A.L.; Lizárraga-Morazán, J.R.; Picón-Núñez, M. Incorporating the Concept of Flexible Operation in the Design of Solar Collector Fields for Industrial Applications. *Energies* **2019**, *12*, 570. [CrossRef]
42. Kalogirou, S.A.; Tripanagnostopoulos, Y. Hybrid PV/T solar systems for domestic hot water and electricity production. *Energy Convers. Manag.* **2006**, *47*, 3368–3382. [CrossRef]
43. Li, H.; Zhao, J.; Li, M.; Deng, S.; An, Q.; Wang, F. Performance analysis of passive cooling for photovoltaic modules and estimation of energy-saving potential. *Sol. Energy* **2019**, *181*, 70–82. [CrossRef]
44. Alizadeh, H.; Ghasempour, R.; Shafii, M.B.; Ahmadi, M.H.; Yan, W.M.; Nazari, M.A. Numerical simulation of PV cooling by using single turn pulsating heat pipe. *Int. J. Heat Mass Transf.* **2018**, *127*, 203–208. [CrossRef]

45. Taler, J.; Dzierwa, P.; Jaremkiewicz, M.; Taler, D.; Kaczmarski, K.; Trojan, M.; Sobota, T. Thermal stress monitoring in thick walled pressure components of steam boilers. *Energy* **2019**, *175*, 645–666. [CrossRef]
46. Trzcinski, P.; Markowski, M. Diagnosis of the fouling effects in a shell and tube heat exchanger using artificial neural network. *Chem. Eng. Trans.* **2018**, *70*, 355–360.
47. Oravec, J.; Bakošová, M.; Vašičkaninová, A.; Meszaros, A. Robust model predictive control of a plate heat exchanger. *Chem. Eng. Trans.* **2018**, *70*, 25–30.
48. Taler, J.; Taler, D.; Kaczmarski, K.; Dzierwa, P.; Trojan, M.; Sobota, T. Monitoring of thermal stresses in pressure components based on the wall temperature measurement. *Energy* **2018**, *160*, 500–519. [CrossRef]
49. Taler, J.; Zima, W.; Jaremkiewicz, M. Simple method for monitoring transient thermal stresses in pipelines. *J. Therm. Stress.* **2016**, *39*, 386–397. [CrossRef]
50. Dzierwa, P.; Trojan, M.; Taler, D.; Kamińska, K.; Taler, J. Optimum heating of thick-walled pressure components assuming a quasi-steady state of temperature distribution. *J. Therm. Sci.* **2016**, *25*, 380–388. [CrossRef]
51. Perić, M.; Garašić, I.; Nižetić, S.; Dedić-Jandrek, H. Numerical Analysis of Longitudinal Residual Stresses and Deflections in a T-joint Welded Structure Using a Local Preheating Technique. *Energies* **2018**, *11*, 3487. [CrossRef]
52. Fialová, D.; Jegla, Z. Analysis of Fired Equipment within the Framework of Low-Cost Modelling Systems. *Energies* **2019**, *12*, 520. [CrossRef]
53. Sriromreun, P.; Sriromreun, P. A Numerical and Experimental Investigation of Dimple Effects on Heat Transfer Enhancement with Impinging Jets. *Energies* **2019**, *12*, 813. [CrossRef]
54. Sun, Z.C.; Ma, X.; Ma, L.X.; Li, W.; Kukulka, D.J. Flow Boiling Heat Transfer Characteristics in Horizontal, Three-Dimensional Enhanced Tubes. *Energies* **2019**, *12*, 927. [CrossRef]
55. Garcia-Castillo Jorge, L. Picon-Nunez Martin Design and operability of multi-stream heat exchangers for use in LNG liquefaction processes. *Chem. Eng. Trans.* **2018**, *70*, 31–36.
56. Valdes, M.; Ardila, J.G.; Colorado, D.; Escobedo-Trujillo, B.A. Computational Model to Evaluate the Effect of Passive Techniques in Tube-In-Tube Helical Heat Exchanger. *Energies* **2019**, *12*, 1912. [CrossRef]
57. Kukulka, D.J.; Smith, R.; Li, W.; Zhang, A.F.; Yan, H. Condensation and evaporation characteristics of flows inside Vipertex 1EHT and 4EHT small diameter enhanced heat transfer tubes. *Chem. Eng. Trans.* **2018**, *70*, 13–18.
58. Klemeš, J.J.; Varbanov, P.S.; Fan, Y.V.; Lam, H.L. Twenty Years of PRES: Past, Present and Future—Process Integration Towards Sustainability. *Chem. Eng. Trans.* **2017**, *61*, 1–24.
59. Klemeš, J.J.; Varbanov, P.S.; Kravanja, Z. Recent developments in Process Integration. *Chem. Eng. Res. Des.* **2013**, *91*, 2037–2053. [CrossRef]
60. Klemeš, J.J.; Varbanov, P.S.; Walmsley, T.G.; Jia, X. New directions in the implementation of Pinch Methodology (PM). *Renew. Sustain. Energy Rev.* **2018**, *98*, 439–468. [CrossRef]
61. Bandyopadhyay, S. Mathematical Foundation of Pinch Analysis. *Chem. Eng. Trans.* **2015**, *45*, 1753–1758.
62. Pereira, P.M.; Fernandes, M.C.; Matos, H.A.; Nunes, C.P. FI^{2}EPI: A heat management tool for process integration. *Appl. Therm. Eng.* **2017**, *114*, 523–536. [CrossRef]
63. Janghorban Esfahani, I.; Lee, S.; Yoo, C. Extended-power pinch analysis (EPoPA) for integration of renewable energy systems with battery/hydrogen storages. *Renew. Energy* **2015**, *80*, 1–14. [CrossRef]
64. Wan Alwi, S.R.; Tin, O.S.; Rozali, N.E.M.; Manan, Z.A.; Klemeš, J.J. New graphical tools for process changes via load shifting for hybrid power systems based on Power Pinch Analysis. *Clean Technol. Environ. Policy* **2013**, *15*, 459–472. [CrossRef]
65. Rozali, N.E.M.; Alwi, S.R.W.; Ho, W.S.; Manan, Z.A.; Klemeš, J.J. PoPA—SHARPS: A New Framework for Cost-Effective Design of Hybrid Power Systems. *Chem. Eng. Trans.* **2017**, *56*, 559–564.
66. Wan Alwi, S.R.; Mohammad Rozali, N.E.; Abdul-Manan, Z.; Klemeš, J.J. A process integration targeting method for hybrid power systems. *Energy* **2012**, *44*, 6–10. [CrossRef]
67. Fan, Y.V.; Varbanov, P.S.; Klemeš, J.J.; Nemet, A. Process efficiency optimisation and integration for cleaner production. *J. Clean. Prod.* **2018**, *174*, 177–183. [CrossRef]
68. Manan, Z.A.; Mohd Nawi, W.N.R.; Wan Alwi, S.R.; Klemeš, J.J. Advances in Process Integration research for CO_2 emission reduction—A review. *J. Clean. Prod.* **2017**, *167*, 1–13. [CrossRef]
69. Li, B.H.; Chota Castillo, Y.E.; Chang, C.T. An improved design method for retrofitting industrial heat exchanger networks based on Pinch Analysis. *Chem. Eng. Res. Des.* **2019**, *148*, 260–270. [CrossRef]

70. Arya, D.; Bandyopadhyay, S. Iterative Pinch Analysis to address non-linearity in a stochastic Pinch problem. *J. Clean. Prod.* **2019**, *227*, 543–553. [CrossRef]
71. Jain, S.; Bandyopadhyay, S. Multi-objective optimisation for segregated targeting problems using Pinch Analysis. *J. Clean. Prod.* **2019**, *221*, 339–352. [CrossRef]
72. Martinez-Hernandez, E.; Tibessart, A.; Campbell, G.M. Conceptual design of integrated production of arabinoxylan products using bioethanol pinch analysis. *Food Bioprod. Process.* **2018**, *112*, 1–8. [CrossRef]
73. Walmsley, T.G.; Ong, B.H.Y.; Klemeš, J.J.; Tan, R.R.; Varbanov, P.S. Circular Integration of processes, industries, and economies. *Renew. Sustain. Energy Rev.* **2019**, *107*, 507–515. [CrossRef]
74. Roychaudhuri, P.S.; Kazantzi, V.; Foo, D.C.Y.; Tan, R.R.; Bandyopadhyay, S. Selection of energy conservation projects through Financial Pinch Analysis. *Energy* **2017**, *138*, 602–615. [CrossRef]
75. Ekvall, T.; Fråne, A.; Hallgren, F.; Holmgren, K. Material pinch analysis: A pilot study on global steel flows. *Rev. Métall.* **2014**, *111*, 359–367. [CrossRef]
76. Jamaluddin, K.; Alwi, S.R.W.; Manan, Z.A.; Klemeš, J.J. Pinch Analysis Methodology for Trigeneration with Energy Storage System Design. *Chem. Eng. Trans.* **2018**, *70*, 1885–1890.
77. Chauhan, S.S.; Khanam, S. Enhancement of efficiency for steam cycle of thermal power plants using process integration. *Energy* **2019**, *173*, 364–373. [CrossRef]
78. Tie, S.; Sreedhar, B.; Donaldson, M.; Frank, T.; Schultz, A.K.; Bommarius, A.; Kawajiri, Y. Process integration for simulated moving bed reactor for the production of glycol ether acetate. *Chem. Eng. Process.* **2019**, *140*, 1–10. [CrossRef]
79. Bandyopadhyay, R.; Alkilde, O.F.; Upadhyayula, S. Applying pinch and exergy analysis for energy efficient design of diesel hydrotreating unit. *J. Clean. Prod.* **2019**, *232*, 337–349. [CrossRef]
80. Malham, C.B.; Tinoco, R.R.; Zoughaib, A.; Chretien, D.; Riche, M.; Guintrand, N. A novel hybrid exergy/pinch process integration methodology. *Energy* **2018**, *156*, 586–596. [CrossRef]
81. Chen, Y.G. Optimal heat rejection pressure of CO_2 heat pump water heaters based on pinch point analysis. *Int. J. Refrig.* **2019**, *106*, 592–603. [CrossRef]
82. Jankowski, M.; Borsukiewicz, A.; Szopik-Depczyńska, K.; Ioppolo, G. Determination of an optimal pinch point temperature difference interval in ORC power plant using multi-objective approach. *J. Clean. Prod.* **2019**, *217*, 798–807. [CrossRef]
83. Schlosser, F.; Peesel, R.H.; Meschede, H.; Philipp, M.; Walmsley, T.G.; Walmsley, M.R.W.; Atkins, M.J. Design of Robust Total Site Heat Recovery Loops via Monte Carlo Simulation. *Energies* **2019**, *12*, 930. [CrossRef]
84. Jamaluddin, K.; Wan Alwi, S.R.; Abdul Manan, Z.; Hamzah, K.; Klemeš, J.J. A Process Integration Method for Total Site Cooling, Heating and Power Optimisation with Trigeneration Systems. *Energies* **2019**, *12*, 1030. [CrossRef]
85. Rathjens, M.; Fieg, G. Cost-Optimal Heat Exchanger Network Synthesis Based on a Flexible Cost Functions Framework. *Energies* **2019**, *12*, 784. [CrossRef]
86. Charvát, P.; Klimeš, L.; Zálešák, M. Utilization of an Air-PCM Heat Exchanger in Passive Cooling of Buildings: A Simulation Study on the Energy Saving Potential in Different European Climates. *Energies* **2019**, *12*, 1133. [CrossRef]
87. Kůdela, L.; Chýlek, R.; Pospíšil, J. Performant and Simple Numerical Modeling of District Heating Pipes with Heat Accumulation. *Energies* **2019**, *12*, 633. [CrossRef]
88. Leitold, D.; Vathy-Fogarassy, A.; Abonyi, J. Evaluation of the Complexity, Controllability and Observability of Heat Exchanger Networks Based on Structural Analysis of Network Representations. *Energies* **2019**, *12*, 513. [CrossRef]
89. Kamat, S.; Bandyopadhyay, S.; Garg, A.; Foo, D.C.Y.; Sahu, G.C. Heat integrated water regeneratin networks with variable regeneration temperature. *Chem. Eng. Trans.* **2018**, *70*, 307–312.
90. Ong, B.H.Y.; Walmsley, T.G.; Atkins, M.J.; Walmsley, M.R.W. Total site mass, heat and power integration using process integration and process graph. *J. Clean. Prod.* **2017**, *167*, 32–43. [CrossRef]
91. Kim, M.; Park, J.; Yu, S.; Ryu, C.; Park, J. Clean and energy-efficient mass production of biochar by process integration: Evaluation of process concept. *Chem. Eng. J.* **2019**, *355*, 840–849. [CrossRef]
92. KBC Petro-SIM. Available online: https://www.kbc.global/software/process-simulation-software (accessed on 19 August 2019).
93. Process Integration Limited Chemical Engineering Consultancy 2019. Available online: https://www.processint.com/software/ (accessed on 29 September 2019).

94. Varbanov, P.S.; Sikdar, S.; Lee, C.T. Contributing to sustainability: Addressing the core problems. *Clean Technol. Environ. Policy* **2018**, *20*, 1121–1122. [CrossRef]
95. Hamsani, M.N.; Liew, P.Y.; Walmsley, T.G.; Alwi, S.R.W. Compressor Shaft Work Targeting using New Numerical Exergy Problem Table Algorithm (Ex-PTA) in Sub-Ambient Processes. *Chem. Eng. Trans.* **2018**, *63*, 283–288.
96. Greening, A.L.; Greene, D.L.; Difiglio, C. Energy efficiency and consumption—The rebound effect—A survey. *Energy Policy* **2000**, *28*, 389–401. [CrossRef]
97. Li, J.; Lin, B. Rebound effect by incorporating endogenous energy efficiency: A comparison between heavy industry and light industry. *Appl. Energy* **2017**, *200*, 347–357. [CrossRef]
98. Čuček, L.; Klemeš, J.J.; Kravanja, Z. Carbon and nitrogen trade-offs in biomass energy production. *Clean Technol. Environ. Policy* **2012**, *14*, 389–397. [CrossRef]
99. Bartington, S.E.; Bakolis, I.; Devakumar, D.; Kurmi, O.P.; Gulliver, J.; Chaube, G.; Manandhar, D.S.; Saville, N.M.; Costello, A.; Osrin, D.; et al. Patterns of domestic exposure to carbon monoxide and particulate matter in households using biomass fuel in Janakpur, Nepal. *Environ. Pollut.* **2017**, *220*, 38–45. [CrossRef]
100. Al-Naiema, I.; Estillore, A.D.; Mudunkotuwa, I.A.; Grassian, V.H.; Stone, E.A. Impacts of co-firing biomass on emissions of particulate matter to the atmosphere. *Fuel* **2015**, *162*, 111–120. [CrossRef]
101. Najser, J.; Buryan, P.; Skoblia, S.; Frantik, J.; Kielar, J.; Peer, V. Problems Related to Gasification of Biomass—Properties of Solid Pollutants in Raw Gas. *Energies* **2019**, *12*, 963. [CrossRef]
102. Yatim, P.; Lin, N.S.; Lam, H.L.; Choy, E.A. Overview of the key risks in the pioneering stage of the Malaysian biomass industry. *Clean Technol. Environ. Policy* **2017**, *19*, 1825–1839. [CrossRef]
103. Zore, Ž.; Čuček, L.; Širovnik, D.; Novak Pintarič, Z.; Kravanja, Z. Maximizing the sustainability net present value of renewable energy supply networks. *Chem. Eng. Res. Des.* **2018**, *131*, 245–265. [CrossRef]
104. Laso, J.; Hoehn, D.; Margallo, M.; García-Herrero, I.; Batlle-Bayer, L.; Bala, A.; Fullana-i-Palmer, P.; Vázquez-Rowe, I.; Irabien, A.; Aldaco, R. Assessing Energy and Environmental Efficiency of the Spanish Agri-Food System Using the LCA/DEA Methodology. *Energies* **2018**, *11*, 3395. [CrossRef]
105. Ubando, A.T.; Marfori, I.A.V.; Aviso, K.B.; Tan, R.R. Optimal Operational Adjustment of a Community-Based Off-Grid Polygeneration Plant using a Fuzzy Mixed Integer Linear Programming Model. *Energies* **2019**, *12*, 636. [CrossRef]
106. Novosel, T.; Ćosić, B.; Pukšec, T.; Krajačić, G.; Duić, N.; Mathiesen, B.V.; Lund, H.; Mustafa, M. Integration of renewables and reverse osmosis desalination—Case study for the Jordanian energy system with a high share of wind and photovoltaics. *Energy* **2015**, *92*, 270–278. [CrossRef]
107. Jia, X.; Klemeš, J.J.; Varbanov, P.S.; Wan Alwi, S.R. Analyzing the Energy Consumption, GHG Emission, and Cost of Seawater Desalination in China. *Energies* **2019**, *12*, 463. [CrossRef]
108. Varbanov, P.S.; Klemeš, J.J. Integration and management of renewables into Total Sites with variable supply and demand. *Comput. Chem. Eng.* **2011**, *35*, 1815–1826. [CrossRef]
109. Alva, G.; Lin, Y.; Fang, G. An overview of thermal energy storage systems. *Energy* **2018**, *144*, 341–378. [CrossRef]
110. Cheng, X.; Pan, J.; Zhao, Y.; Liao, M.; Peng, H. Gel Polymer Electrolytes for Electrochemical Energy Storage. *Adv. Energy Mater.* **2018**, *8*, 1702184. [CrossRef]
111. Chen, W.; Yu, H.; Lee, S.Y.; Wei, T.; Li, J.; Fan, Z. Nanocellulose: A promising nanomaterial for advanced electrochemical energy storage. *Chem. Soc. Rev.* **2018**, *47*, 2837–2872. [CrossRef]
112. Mohammad Rozali, N.E.; Ho, W.S.; Wan Alwi, S.R.; Manan, Z.A.; Klemeš, J.J.; Mohd Yunus, M.N.S.; Syed Mohd Zaki, S.A.A. Peak-off-peak load shifting for optimal storage sizing in hybrid power systems using Power Pinch Analysis considering energy losses. *Energy* **2018**, *156*, 299–310. [CrossRef]
113. Hamdy, S.; Moser, F.; Morosuk, T.; Tsatsaronis, G. Exergy-Based and Economic Evaluation of Liquefaction Processes for Cryogenics Energy Storage. *Energies* **2019**, *12*, 493. [CrossRef]
114. Ghannadzadeh, A.; Sadeqzadeh, M. Exergy analysis as a scoping tool for cleaner production of chemicals: A case study of an ethylene production process. *J. Clean. Prod.* **2016**, *129*, 508–520. [CrossRef]
115. Zhu, L.; Zhou, M.; Shao, C.; He, J. Comparative exergy analysis between liquid fuels production through carbon dioxide reforming and conventional steam reforming. *J. Clean. Prod.* **2018**, *192*, 88–98. [CrossRef]
116. Meramo-Hurtado, S.; Herrera-Barros, A.; González-Delgado, Á. Evaluation of Large-Scale Production of Chitosan Microbeads Modified with Nanoparticles Based on Exergy Analysis. *Energies* **2019**, *12*, 1200. [CrossRef]

117. Tomić, T.; Schneider, D.R. The role of energy from waste in circular economy and closing the loop concept—Energy analysis approach. *Renew. Sustain. Energy Rev.* **2018**, *98*, 268–287. [CrossRef]
118. Laso, J.; García-Herrero, I.; Margallo, M.; Bala, A.; Fullana-i-Palmer, P.; Irabien, A.; Aldaco, R. LCA-Based Comparison of Two Organic Fraction Municipal Solid Waste Collection Systems in Historical Centres in Spain. *Energies* **2019**, *12*, 1407. [CrossRef]
119. McDonough, W.; Braungart, M. *Cradle to Cradle—Remaking the Way We Make Things*, 1st ed.; North Point Press: New York, NY, USA, 2002; ISBN 978-0-86547-587-8.
120. How, B.S.; Yeoh, T.T.; Tan, T.K.; Chong, K.H.; Ganga, D.; Lam, H.L. Debottlenecking of sustainability performance for integrated biomass supply chain: P-graph approach. *J. Clean. Prod.* **2018**, *193*, 720–733. [CrossRef]
121. Sonawane, J.M.; Al-Saadi, S.; Singh Raman, R.K.; Ghosh, P.C.; Adeloju, S.B. Exploring the use of polyaniline-modified stainless steel plates as low-cost, high-performance anodes for microbial fuel cells. *Electrochim. Acta* **2018**, *268*, 484–493. [CrossRef]
122. Kong, L.C.; Zhu, Z.N.; Xie, J.P.; Li, J.; Chen, Y.P. Multilateral agreement contract optimization of renewable energy power grid-connecting under uncertain supply and market demand. *Comput. Ind. Eng.* **2019**, *135*, 689–701.
123. Fontes, C.H.O.; Freires, F.G.M. Sustainable and renewable energy supply chain: A system dynamics overview. *Renew. Sustain. Energy Rev.* **2018**, *82*, 247–259.
124. Fernando, Y.; Bee, P.S.; Jabbour, C.J.C.; Thomé, A.M.T. Understanding the effects of energy management practices on renewable energy supply chains: Implications for energy policy in emerging economies. *Energy Policy* **2018**, *118*, 418–428. [CrossRef]
125. Nugroho, Y.K.; Zhu, L. Platforms planning and process optimization for biofuels supply chain. *Renew. Energy* **2019**, *140*, 563–579. [CrossRef]
126. Sarker, B.R.; Wu, B.; Paudel, K.P. Modeling and optimization of a supply chain of renewable biomass and biogas: Processing plant location. *Appl. Energy* **2019**, *239*, 343–355. [CrossRef]
127. Li, Q.; Loy-Benitez, J.; Nam, K.; Hwangbo, S.; Rashidi, J.; Yoo, C. Sustainable and reliable design of reverse osmosis desalination with hybrid renewable energy systems through supply chain forecasting using recurrent neural networks. *Energy* **2019**, *178*, 277–292. [CrossRef]
128. Huang, J.; Boland, J. Performance Analysis for One-Step-Ahead Forecasting of Hybrid Solar and Wind Energy on Short Time Scales. *Energies* **2018**, *11*, 1119. [CrossRef]
129. Gupta, R.A.; Kumar, R.; Bansal, A.K. BBO-based small autonomous hybrid power system optimization incorporating wind speed and solar radiation forecasting. *Renew. Sustain. Energy Rev.* **2015**, *41*, 1366–1375. [CrossRef]
130. Zhao, J.; Guo, Z.H.; Su, Z.Y.; Zhao, Z.Y.; Xiao, X.; Liu, F. An improved multi-step forecasting model based on WRF ensembles and creative fuzzy systems for wind speed. *Appl. Energy* **2016**, *162*, 808–826. [CrossRef]
131. Behzadi Forough, A.; Roshandel, R. Multi objective receding horizon optimization for optimal scheduling of hybrid renewable energy system. *Energy Build.* **2017**, *150*, 583–597. [CrossRef]
132. Liu, H.; Mi, X.; Li, Y. Wind speed forecasting method based on deep learning strategy using empirical wavelet transform, long short term memory neural network and Elman neural network. *Energy Convers. Manag.* **2018**, *156*, 498–514. [CrossRef]
133. Azevedo, S.G.; Santos, M.; Antón, J.R. Supply chain of renewable energy: A bibliometric review approach. *Biomass Bioenergy* **2019**, *126*, 70–83. [CrossRef]
134. Zakaria, A.; Ismail, F.B.; Lipu, M.S.H.; Hannan, M.A. Uncertainty models for stochastic optimization in renewable energy applications. *Renew. Energy* **2019**, *145*, 1543–1571. [CrossRef]
135. Chen, Y.; Wang, Y.; Kirschen, D.; Zhang, B. Model-Free Renewable Scenario Generation Using Generative Adversarial Networks. *IEEE Trans. Power Syst.* **2018**, *33*, 3265–3275. [CrossRef]
136. Chen, Y.; Wang, X.; Zhang, B. An Unsupervised Deep Learning Approach for Scenario Forecasts. In Proceedings of the 2018 Power Systems Computation Conference (PSCC), Dublin, Ireland, 11–15 June 2018; pp. 1–7.
137. Dufo-López, R.; Cristóbal-Monreal, I.R.; Yusta, J.M. Stochastic-heuristic methodology for the optimisation of components and control variables of PV-wind-diesel-battery stand-alone systems. *Renew. Energy* **2016**, *99*, 919–935. [CrossRef]

138. Rahmani-Andebili, M. Stochastic, adaptive, and dynamic control of energy storage systems integrated with renewable energy sources for power loss minimization. *Renew. Energy* **2017**, *113*, 1462–1471. [CrossRef]
139. Sharafi, M.; Elmekkawy, T.Y. Stochastic optimization of hybrid renewable energy systems using sampling average method. *Renew. Sustain. Energy Rev.* **2015**, *52*, 1668–1679. [CrossRef]
140. Thompson, A.W. Economic implications of lithium ion battery degradation for Vehicle-to-Grid (V2X) services. *J. Power Sources* **2018**, *396*, 691–709. [CrossRef]
141. Éles, A.; Halász, L.; Heckl, I.; Cabezas, H. Evaluation of the Energy Supply Options of a Manufacturing Plant by the Application of the P-Graph Framework. *Energies* **2019**, *12*, 1484. [CrossRef]
142. San Juan, J.L.G.; Aviso, K.B.; Tan, R.R.; Sy, C.L. A Multi-Objective Optimization Model for the Design of Biomass Co-Firing Networks Integrating Feedstock Quality Considerations. *Energies* **2019**, *12*, 2252. [CrossRef]
143. Peesel, R.H.; Schlosser, F.; Meschede, H.; Dunkelberg, H.; Walmsley, T.G. Optimization of Cooling Utility System with Continuous Self-Learning Performance Models. *Energies* **2019**, *12*, 1926. [CrossRef]
144. Barmina, I.; Kolmickovs, A.; Valdmanis, R.; Zake, M.; Vostrikovs, S.; Kalis, H.; Strautins, U. Electric Field Effect on the Thermal Decomposition and Co-combustion of Straw with Solid Fuel Pellets. *Energies* **2019**, *12*, 1522. [CrossRef]
145. Fichera, A.; Fortuna, L.; Frasca, M.; Volpe, R. Integration Of Complex Networks For Urban Energy Mapping. *Int. J. Heat Technol.* **2015**, *33*, 181–184. [CrossRef]
146. Gonzalez de Durana, J.M.; Barambones, O.; Kremers, E.; Varga, L. Agent based modeling of energy networks. *Energy Convers. Manag.* **2014**, *82*, 308–319. [CrossRef]
147. Gonzalez de Durana, J.; Barambones, O. Technology-free microgrid modeling with application to demand side management. *Appl. Energy* **2018**, *219*, 165–178. [CrossRef]
148. Kremers, E.; Gonzalez de Durana, J.; Barambones, O. Multi-agent modeling for the simulation of a simple smart microgrid. *Energy Convers. Manag.* **2013**, *75*, 643–650. [CrossRef]
149. Zeh, A.; Müller, M.; Naumann, M.; Hesse, H.C.; Jossen, A.; Witzmann, R. Fundamentals of Using Battery Energy Storage Systems to Provide Primary Control Reserves in Germany. *Batteries* **2016**, *2*, 29. [CrossRef]
150. Tran, T.T.D.; Smith, A.D. Thermoeconomic analysis of residential rooftop photovoltaic systems with integrated energy storage and resulting impacts on electrical distribution networks. *Sustain. Energy Technol. Assess.* **2018**, *29*, 92–105. [CrossRef]

© 2019 by the authors. Licensee MDPI, Basel, Switzerland. This article is an open access article distributed under the terms and conditions of the Creative Commons Attribution (CC BY) license (http://creativecommons.org/licenses/by/4.0/).

Article

A Multi-Objective Optimization Model for the Design of Biomass Co-Firing Networks Integrating Feedstock Quality Considerations

Jayne Lois G. San Juan [1,*], Kathleen B. Aviso [2,3], Raymond R. Tan [2,3] and Charlle L. Sy [1,3]

1. Industrial Engineering Department, De La Salle University, Manila 0922, Philippines; charlle.sy@dlsu.edu.ph
2. Chemical Engineering Department, De La Salle University, Manila 0922, Philippines; kathleen.aviso@dlsu.edu.ph (K.B.A.); raymond.tan@dlsu.edu.ph (R.R.T.)
3. Center for Engineering and Sustainable Development Research, De La Salle University, Manila 0922, Philippines
* Correspondence: jayne.sanjuan@dlsu.edu.ph

Received: 17 March 2019; Accepted: 8 June 2019; Published: 12 June 2019

Abstract: The growth in energy demand, coupled with declining fossil fuel resources and the onset of climate change, has resulted in increased interest in renewable energy, particularly from biomass. Co-firing, which is the joint use of coal and biomass to generate electricity, is seen to be a practical immediate solution for reducing coal use and the associated emissions. However, biomass is difficult to manage because of its seasonal availability and variable quality. This study proposes a biomass co-firing supply chain optimization model that simultaneously minimizes costs and environmental emissions through goal programming. The economic costs considered include retrofitting investment costs, together with fuel, transport, and processing costs, while environmental emissions may come from transport, treatment, and combustion activities. This model incorporates the consideration of feedstock quality and its impact on storage, transportation, and pre-treatment requirements, as well as conversion yield and equipment efficiency. These considerations are shown to be important drivers of network decisions, emphasizing the importance of managing biomass and coal blend ratios to ensure that acceptable fuel properties are obtained.

Keywords: biomass co-firing; biomass quality; network optimization; goal programming; mixed integer nonlinear programming

---

1. Introduction

Society depends heavily on energy to support nearly all of its activities. Not surprisingly, the world is currently facing an unrivaled colossal energy threat. Alongside with the increase in global population, energy demand and consumption are projected to increase by 30% by 2040 [1]. However, fossil fuel resources such as oil, gas and coal, are unevenly distributed among nations and are now rapidly depleting. This thus raises issues on energy security and sustainability [2]. Furthermore, the continued use of fossil fuels has resulted in both health and environmental problems due to hazardous air emissions [3]. Recent studies assert that disastrous environmental problems will occur if the world does not reduce the emission of greenhouse gases (GHGs), making global warming a crucial issue. Hence, governments and policy makers are trying to take steps towards minimizing causes of global warming and climate change [4].

Because of these, the development of more sustainable and renewable sources of energy (e.g., solar, biomass, hydro, geothermal, and wind) as well as innovative strategies for cleaner production, and efficient utilization of products is a necessity. Energy derived from biomass plays an important role in this. It is a clean, natural, renewable energy source. If one considers its entire life cycle, burning biomass results in net zero carbon emissions since CO_2 was initially sequestered from the atmosphere during its growth. Furthermore, countries may utilize indigenous resources to replace current coal demand, thereby reducing dependency on conventional fossil fuels.

Even though the use of biomass for energy production has risen in the past few years, dedicated biomass-fired power plants remain to have small capacities (e.g., typically only 100 MW [5]) because of difficulties associated with seasonal availability, inherent quality variations, and the wide geographical distribution of feedstock supply. To deal with varying biomass quality, advanced technologies for pre-treatment are used, such as drying, pelletization, torrefaction, and pyrolysis. These technologies can help reduce the moisture and ash contents, and bulk density of biomass feedstock without compromising their energy content significantly. Pre-treatment can thus improve the durability of biomass thereby reducing the costs associated to their storage and transport [6]. Nonetheless, performing pre-treatment entails additional costs and can result in additional environmental impacts, which may not be necessarily favorable for the system as a whole [7].

Co-firing of biomass with coal is a more practical interim approach towards increasing the utilization of renewable energy sources. This strategy requires minimal modification of existing power plants and allows for the continued use of high capacity coal power plants. Furthermore, biomass can easily be integrated into the energy supply chain by utilizing the existing infrastructure for fuel storage, transport and handling [4]. Biomass co-firing also improves the net energy and emissions balance of energy generation because it will require less coal to meet energy demands and thus less emissions associated with the mining and transportation of coal [8]. Biomass co-firing also provides an alternative to open field burning where the latter results in the generation of pollutants such as dioxins and furans because of uncontrolled burning conditions [9].

According to Ba et al. [10], the planning and management of biomass and biomass co-firing supply chains have generally been modelled numerically using two main approaches: (1) simulation and (2) optimization models. Although, simulation modelling has the advantage of being highly flexible with the capability of handling stochastic events in complex supply chains, it is critiqued because of its inability to design large-scale optimal supply chains considering multiple objectives, which is usually the case in biomass co-firing supply chains.

Zandi Atashbar et al. [7] in a recent review, provided a critical analysis of various mathematical modelling approaches which have been used for biomass supply chains. Zandi Atashbar et al. [7] identified that the predominant objective among existing studies focused on minimizing overall costs, while some researchers define their objective function as to maximize overall profits or net present value (NPV). Most studies optimize based only on a single objective which may either be economic, environmental, or social. In fact, Shang [11] comments that existing epidemic modelling studies are similarly limited to considering single objectives. Environmental impact has usually been measured based on emissions, while the number of local jobs created has been used to measure the performance of the social objective. More recently, there have been limited studies focused on multi-objective optimization of biomass co-firing supply chains [7]. Pérez-Fortes et al. [12] asserts that the consideration of multiple objectives in the optimization of biomass co-firing supply chains is crucial because the design of such systems necessitates the satisfaction of conflicting goals, particularly those associated with economic and environmental factors. Multi-objective optimization models allow for the consideration of varied priorities of several stakeholders and balance the tradeoffs that exist between the objectives.

Only three studies dealing with multiple objectives in biomass co-firing were presented in literature. Mohd Idris et al. [13] and Griffin et al. [14] proposed a biomass co-firing supply chain optimization model that minimized the cost and emissions of the system, while Pérez-Fortes et al. [12] formulated a model in which decisions were assessed based on minimum NPV losses and maximum environmental impact annual savings. All three studies approached the problem by solving the economic and environmental objective functions separately.

Thus far, no studies have been able to optimize both the economic and environmental objectives simultaneously. Savic [15] explains that single objective optimization is only useful as a tool to allow decision makers to understand the nature of a problem. However, it cannot yield a set of alternative solutions that account for the trade-offs between conflicting objectives. Furthermore, considering only economic costs in optimizing a supply chain may result in a design which fails to consider critical processes and options to achieve the lowest cost at the expense of environmental sustainability. Alternatively, when a system is optimized in terms of environmental benefits, costs may be dramatically inflated making the solution impractical for implementation. Goal programming is an appropriate approach to simultaneously account for two or more conflicting objectives. It has been applied to several multi-objective optimization problems, demonstrating its efficiency and effectiveness as an approach for tackling such problems [16]. The goal programming optimal solution for an industrial water network design problem was compared against other approaches used to solve multi-objective optimization problems, such as M-TOPSIS, LMS-TOPSIS, reference point method. Their results showed that although all approaches were able to obtain points on the Pareto front, the goal programming approach consistently obtained the best Pareto optimal solution with minimal computational effort [17].

Furthermore, there have been limited studies on the impact of feedstock quality [18] on the design of an optimal supply chain network. Most models overlook considering quality related issues, lowering logistics costs and emissions artificially. Scale-up scenarios become an important consideration when technologies expand from laboratory to commercial use. For instance, consider the implications of a conversion technology, which was rated to work with feedstock having a moisture content of about 10%, which in reality needs to work with fuels with moisture content of more than 25%. Moreover, significant financial losses will ensue when two batches of feedstock yield considerably different amounts of energy. Several case studies establish that both scenarios are highly likely to take place in practice [19].

Pérez-Fortes et al. [12] has identified the following critical fuel properties: bulk density, moisture content, lower heating value, and ash content. For biomass, the bulk density and lower heating values are typically low while the moisture content is typically high. These properties are interdependent and affect different phases of the supply chain. Low heating values for example, will require more biomass to satisfy the needed energy while the low bulk density will need higher capacity vehicles or storage units [7]. High moisture and ash content will decrease the lower heating value [20]. Furthermore, with the ash in biomass feedstock being more alkaline that those from coal, fouling problems can potentially decrease the efficiency of boilers [21]. Biomass use must be managed carefully to avoid these effects [22].

Mohd Idris et al. [13] and Dundar et al. [4] identified the optimal blending ratios for fuels to satisfy a minimum biomass percentage regulation. Pérez-Fortes et al. [12] attempted to address the impact of biomass properties on the supply chain by integrating the pretreatment options into the optimization model. Required quality levels for the feedstock were considered but the impact of fuel properties during combustion were not captured. However, conversion technology usually ends up working with feedstock that do not follow the rated requirements, causing a corresponding decrease in yield or in the life of the equipment. In particular, fouling of heat transfer surfaces can become problematic. The impact of storage on the quality of biomass was also neglected in the study.

To address these gaps, a mathematical optimization model focused on a biomass co-firing network that simultaneously optimizes the economic and environmental objectives of the system is developed. Costs associated with retrofitting, storage, transport and pre-treatment are considered and the impact of biomass properties on blending ratio decisions, conversion efficiencies and equipment life are also taken into account. Capturing these parameters increase the complexity of the model, but the solutions obtained provide more realistic insights into the behavior of the system and can be more reliable for decision-making.

The rest of the paper is organized as follows: Section 2 gives the formal problem statement. Section 3 gives a description of the system considered, while the MINLP model formulation is described in Section 4. The model capabilities are illustrated with a case study and scenario analysis in Sections 5 and 6. Finally, conclusions and prospects for future work are given in Section 7.

2. Problem Definition

The formal problem statement can be stated as follows:

- planning horizon which consists of time intervals $t \in T$;
- A set of biomass sources $i \in I$, with a maximum available amount of s_{it}^b in period t with bulk density, ρ_{it}^r, moisture content, m_{it}^r, ash content, a_{it}^r, and higher heating value g;
- The biomass has an associated cost of p_{it}^b;
- The biomass has to be transported from source i to pre-treatment facility j a distance of d_{ij}^r and that the biomass quality degrades with a damage factor of b_{ij}^{tr};
- The transport of biomass from source i to pretreatment facility j has a weight capacity of u_{ijt}^r and volume capacity of v_{ijt}^r in period t, an associated cost of tc_{ijt}^r and emission of te_{ijt}^r;
- A set of biomass pre-treatment facilities $j \in J$ which can process with a capacity of c_{jt}^p and store a capacity of c_{jt}^s in period t;
- Pre-treatment facility j can improve the biomass properties based on the facility's ash improvement efficiency, ae_j, and moisture content improvement efficiency, me_j, and improve bulk density to ρ_{jt}^p;
- There are associated costs for the operation of the pretreatment facility (oc_{jt}^p), the processing of biomass (pc_{jt});
- The biomass stored in the pre-treatment facility degrades with a damage factor of b_j^s and increases in moisture content by z_{jt} in period t;
- The processed biomass has to be transported from pre-treatment facility j to coal power plant l a distance of d_{jl}^p and that the processed biomass degrades by a factor b_{jl}^{tp} during transport;
- Transport from pretreatment facility j to coal power plant l has a weight capacity of u_{jlt}^p and volume capacity v_{jlt}^p in period t;
- A set of coal sources $k \in K$ which can provide a maximum s_{kt}^c amount of coal in period t with bulk density, ρ^c, moisture content, m^c, ash content, a^c, and lower heating value, q^c;
- The coal has an associated cost of p_{kt}^c;
- The coal should be transported from coal source k to coal power plant l a distance of d_{kl}^c;
- The transport of coal from source k to powerplant l has a transport weight capacity of u_{klt}^c and transport volume capacity of v_{klt}^c, an associated cost of tc_{klt}^c and emission of te_{klt}^c;
- A set of coal-fired power plants $l \in L$ with combustion capacity c_{lt}^c which can be retrofitted for biomass co-firing to meet the total demand of energy D_t for period t;
- The coal power plant l will have an efficiency of λ_{lt} in period t;
- The coal power plant l will have upper (L_l^u) and lower (L_l^l) coal displacement limits if retrofitted, maximum allowable ash content (a_l^U), and upper (m_l^U) and lower (m_l^L) moisture content limits;

The problem may be visualized using the superstructure in Figure 1 where biomass and coal are obtained from their respective supply locations, biomass is pre-treated and then co-fired with coal in the identified powerplants. The objective is to determine the optimal allocation of biomass from the sources to the pretreatment facilities (w_{ijt}), allocation of processed biomass from the pretreatment facilities to the coal power plant (x_{jlt}), the amount of coal that should be transported from the coal source to the power plant (y_{klt}), the choice of which power plant should be retrofitted (R_l), when the pretreatment facilities (F_{jt}) and coal power plants (A_{lt}) should be operating, and when the biomass option is implemented in the power plant (O_{lt}) to achieve the simultaneous reduction in costs and environmental emissions. The solution should also indicate if the biomass should be stored in a pretreatment facility during period t (S_{jt}) and how much to keep in inventory (I_{jt}), if a power plant (C_{lt}) or pretreatment facility (P_{jt}) should increase its capacity in another time period and by how much the capacities of the power plants (f_{lt}^c) and pretreatment facilities (f_{jt}^p) should be increased.

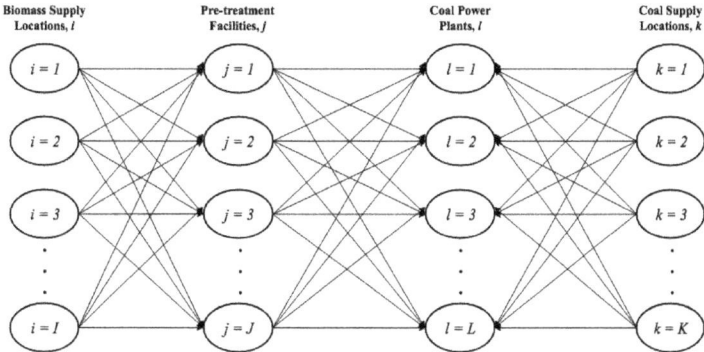

Figure 1. Network superstructure.

3. System Definition

Biomass waste must be allocated from a set of source locations $i \in I$ to a set of existing coal power plants $l \in L$ to partially displace coal consumption. Coal is supplied to these plants by $k \in K$ coal supply locations. Coal power plants generate electricity by co-firing biomass with coal to satisfy certain demands. Each biomass supply location provides biomass with certain properties, which is improved through pre-treatment in a given set of facilities $j \in J$ before they are brought to the coal power plants for combustion. Biomass may be stored in pre-treatment facilities prior to transport.

3.1. Biomass Co-Firing Network

The biomass and coal sources have predefined supply capacities which vary between periods. Different biomass source localities also experience variations in the biomass properties due to climate differences. The carbon dioxide emissions from the cultivation and harvesting of biomass residues may be neglected since baseline conditions still require these operations and focus is given to emissions generated as a consequence of using the residues for co-firing.

Biomass are transported to pre-treatment facilities to improve their quality. Each facility performs a specific pre-treatment process, but can only process a certain amount of biomass each period. In addition, each process addresses only a set of properties and improves them only to certain extents. After pre-treatment, the biomass may be: (1) kept in storage until the succeeding period or (2) transported to coal power plants for co-firing. Storing biomass may result in deterioration and additional holding costs.

The model will also decide whether each existing coal power plant must be retrofitted for co-firing. Retrofitting a power plant will require capital investments. For plants that will be retrofitted, the model decides when the retrofit is implemented, when co-firing is activated, and where the biomass and coal will be sourced from. Each power plant has processing capacities, fuel property limits, and electricity demands that need to be satisfied. Equipment degradation is expressed as a function of the volume and quality of the feedstock they process. Higher usage rates or processing of feedstock with unsuitable quality levels will accelerate the degradation of the equipment. Unlike previous studies which model fuel property limits of power plants as hard constraints, this model allows for flexible feedstock properties and instead accounts for the impact of feedstock quality on equipment efficiency and environmental emissions.

3.2. Economic Considerations

The system considers six cost components: (1) feedstock costs, (2) capital costs, (3) transportation costs, (4) fixed operating costs, (5) variable operating costs, and (6) holding costs. Feedstock costs represent the cost of purchasing biomass waste (p_{it}^b) and coal (p_{kt}^c) from their respective sources during a specific period, expressed in cost per kiloton. Capital costs include the investment costs for retrofitting the handling systems of existing coal power plants (ic_l) to accept diverse feedstocks and costs to expand the capacities of the pretreatment facility (ec_{jt}^p) and the coal power plant (ec_{lt}^c). Expansion costs are also based on the increase in capacity of pretreatment facilities (e_{jt}^p) and coal power plants (e_{lt}^c). Transportation costs refer to the costs incurred in transporting biomass from the source (tc_{ijt}^r) and pre-treatment facility (tc_{jlt}^p); and in transporting coal (tc_{klt}^c). Fixed operating costs represent the costs brought about by operating the coal power plants (oc_{lt}^c), using the co-firing option (bc_{lt}), operating the pre-treatment facilities (oc_{jt}^p), and storing biomass (sc_{jt}). Variable operating costs include two types of costs, namely (1) pre-treatment costs (pc_{jt}) and (2) combustion costs for coal (r_{lt}^c) and biomass (r_{lt}^b) expressed as a cost per feedstock kiloton. Pre-treatment cost represents the cost of treating biomass to improve the properties of the biomass, while combustion cost is the cost to process and generate power from biomass and coal. Lastly, holding cost (h_{jt}) refers to the cost to store biomass across each period.

3.3. Environmental Considerations

CO_2 emissions from transportation (e.g., of biomass (te_{jlt}^p) and coal (te_{klt}^c)), combustion (e.g., of biomass (ce_{lt}^b) and coal (ce_{lt}^c)), and biomass pre-treatment are the environmental emissions considered in the system. The use of biomass is preferred from an environmental standpoint because emissions released are significantly less when burning biomass compared to coal. Pre-treating biomass to improve its properties also result in greenhouse gas emissions (pe_{jt}).

4. Model Formulation

The following section presents the model formulation for the biomass co-firing network described. A MINLP model is developed for the network, which aims to make investment and operational decisions that simultaneously minimizes the costs and environmental emissions while satisfying energy demand and capacity constraints. The model also considers the impact of fuel properties on the efficiency and life of the conversion equipment. Table 1 shows the indices, as well as the relevant parameters and variables used in the model.

Table 1. Notations.

Indices	Definition	
i	Biomass source locations	
j	Pretreatment facilities	
k	Coal source locations	
l	Coal power plants	
t	Time period	
Parameters	**Definition**	**Units**
$Cost_{max}$	Cost value when environmental emissions is minimized	Million US$
$Cost_{pot}$	Minimum achievable cost	Million US$
Env_{max}	Environmental emissions value when cost is minimized	kt CO_2
Env_{pot}	Minimum achievable environmental emissions	kt CO_2
D_t	Amount of energy demanded on period t	MJ
s_{it}^b	Amount of biomass available at biomass source location i in period t	kt
s_{kt}^c	Amount of coal available a coal source location k in period t	kt
L_l^u	Upper coal displacement limit of coal power plant l	%
L_l^l	Lower coal displacement limit of coal power plant l	%
c_{jt}^s	Storage capacity in pretreatment facility j in period t	kt
ρ_{it}^r	Bulk density of raw biomass from source i in period t	kg/m^3
ρ_{jt}^p	Bulk density of pretreated biomass in pretreatment facility j in period t	kg/m^3
ρ^c	Bulk density of coal	kg/m^3

Table 1. *Cont.*

Symbol	Description	Unit
m_{it}^r	Moisture content of raw biomass from source i in period t	% wt.
m^c	Moisture content of coal	% wt.
q^c	Lower heating value of coal	MJ/kg
g	Higher heating value of biomass	MJ/kg
a_{it}^r	Ash content of raw biomass from source i in period t	% wt.
a^c	Ash content of coal	% wt.
a_l^U	Maximum allowable ash content in coal power plant l	% wt.
m_l^U	Maximum allowable moisture content in coal power plant l	% wt.
m_l^L	Minimum allowable moisture content in coal power plant l	% wt.
ae_j	Ash content improvement efficiency in pretreatment facility j	%
me_j	Moisture content improvement efficiency in pretreatment facility j	%
b_j^s	Biomass damage factor from storing in pretreatment facility j	%
b_{ij}^{tr}	Biomass damage factor from transporting raw biomass from source i to pretreatment facility j	%
b_{jl}^{tp}	Biomass damage factor from transporting pretreated biomass from pretreatment facility j to coal power plant l	%
z_{jt}	Increase in moisture content due to storage in pretreatment facility j in period t	%
d_{ij}^r	Distance from biomass source i to pretreatment facility j	km
d_{jl}^p	Distance from pretreatment facility j to coal power plant l	km
d_{kl}^c	Distance from coal source k to coal power plant l	km
u_{ijt}^r	Transport weight capacity from biomass source i to pretreatment facility j in period t	kt
u_{jlt}^p	Transport weight capacity from pretreatment facility j to power plant l in period t	kt
u_{klt}^c	Transport weight capacity from coal source k to coal power plant l in period t	kt
v_{ijt}^r	Transport volume capacity from biomass source i to pretreatment facility j in period t	m^3
v_{jlt}^p	Transport volume capacity from pretreatment facility j to power plant l in period t	m^3
v_{klt}^c	Transport volume capacity from coal source k to coal power plant l in period t	m^3
ic_l	Cost to retrofit coal power plant l	Million US$
oc_{lt}^c	Fixed cost to operate coal power plant l on period t	Million US$
bc_{lt}	Fixed cost to use biomass option in coal power plant l on period t	Million US$
r_{lt}^b	Biomass combustion cost in coal power plant l on period t	US$/kg
r_{lt}^c	Coal combustion cost in coal power plant l in period t	US$/kg
oc_{jt}^p	Fixed cost to operate pretreatment facility j in period t	Million US$
pc_{jt}	Biomass pretreatment cost in facility j in period t	US$/kg
ec_{jt}^p	Fixed cost to expand the capacity of pretreatment facility j in period t	Million US$
e_{jt}^p	Unit capacity expansion cost of pretreatment facility j in period t	US$/kg
ec_{lt}^c	Fixed cost to expand the capacity of power plant l in period t	Million US$
e_{lt}^c	Unit capacity expansion cost of coal power plant l in period t	US$/kg
sc_{jt}	Fixed cost to store in pretreatment facility j in period t	Million US$
h_{jt}	Unit holding cost in pretreatment facility j in period t	US$/kg
tc_{ijt}^r	Cost to transport raw biomass from source i to pretreatment facility j in period t per trip	US$/kg-km
tc_{jlt}^p	Cost to transport pretreated biomass from pretreatment facility j to coal power plant l in period t per trip	US$/kg-km
tc_{klt}^c	Cost to transport coal from source k to coal power plant l in period t per trip	US$/kg-km
p_{it}^b	Cost of biomass from source i in period t	US$/kg
p_{kt}^c	Cost of coal from source k in period t	US$/kg
pe_{jt}	Emissions due to biomass pretreatment in facility j in period t	kg CO$_2$/kg
ce_{lt}^b	Emissions due to biomass combustion in coal power plant l in period t	kg CO$_2$/kg
ce_{lt}^c	Emissions due to coal combustion in power plant l in period t	kg CO$_2$/kg
te_{ijt}^r	Emissions due to transporting raw biomass from source i to pretreatment facility j in period t	kg CO$_2$/kg-km
te_{jlt}^p	Emissions due to transporting pretreated biomass from pretreatment facility j to coal power plant l in period t	kg CO$_2$/kg-km
te_{klt}^c	Emissions due to transporting coal from source k to power plant l in period t	kg CO$_2$/kg-km

Table 1. Cont.

System Variables	Definition	Units
I_{jt}	Ending biomass inventory in pretreatment facility j in period t	kt
n_{jt}	Weight of biomass received and pretreated in pretreatment facility j in period t	kt
q^b_{lt}	Lower heating value of biomass in coal power plant l in period t	MJ/kg
q_{lt}	Lower heating value of the mixed feedstock in coal power plant l in period t	MJ/kg
m^t_{jt}	Moisture content of pretreated biomass from source j in period t	% wt.
m^p_{jt}	Moisture content of all biomass in pretreatment facility j in period t	% wt.
m^{ppb}_{lt}	Moisture content of biomass in coal power plant l in period t	% wt.
m^{bp}_{lt}	Moisture content of mixed feedstock in power plant l in period t	% wt.
a^t_{jt}	Ash content of pretreated biomass from source j in period t	% wt.
a^p_{jt}	Ash content of all biomass in pretreatment facility j in period t	% wt.
a^{ppb}_{lt}	Ash content of biomass in coal power plant l in period t	% wt.
a^{bp}_{lt}	Ash content of mixed feedstock in coal power plant l in period t	% wt.
m^+_{lt}	Accumulated excess moisture content of feedstock in power plant l in period t	% wt.
m^-_{lt}	Accumulated lack in moisture content of feedstock in power plant l in period t	% wt.
a^+_{lt}	Accumulated excess ash content of feedstock in coal power plant l in period t	% wt.
Q_{lt}	Accumulated feedstock processed in power plant l in period t	kt
λ_{lt}	Efficiency loss of equipment in coal power plant l in period t	-
t^r_{ijt}	Number of trips to transport raw biomass from source i to pretreatment facility j in period t	Trips
t^p_{jlt}	Number of trips to transport pretreated biomass from pretreatment facility j to coal power plant l in period t	Trips
t^c_{klt}	Number of trips to transport coal from source k to coal power plant l in period t	Trips
c^c_{lt}	Combustion capacity of coal power plant l in period t	kt
c^p_{jt}	Pretreatment capacity in facility j in period t	kt

Decision Variables	Definition	Units
w_{ijt}	Amount of biomass transported from biomass source locations i to pretreatment facilities j in period t	kt
x_{jlt}	Amount of biomass transported from pretreatment facilities j to power plant l in period t	kt
y_{klt}	Amount of coal transported from coal source location k to power plant l in period t	kt
f^p_{jt}	Capacity expansion for pretreatment facility j in period t	kt
f^c_{lt}	Capacity expansion for coal power plant l in period t	kt
R_l	Binary variable, 1 if coal power plant l is retrofitted	-
O_{lt}	Binary variable, 1 if biomass option of coal power plant l is used in period t	-
S_{jt}	Binary variable, 1 if storage in pretreatment facility j is used in period t	-
F_{jt}	Binary variable, 1 if pretreatment facility j is operating in period t	-
A_{lt}	Binary variable, 1 if coal power plant l is operating in period t	-
P_{jt}	Binary variable, 1 if pretreatment facility j undergoes capacity expansion in period t	-
C_{lt}	Binary variable, 1 if coal power plant l undergoes capacity expansion in period t	-

4.1. Constraints

The demand for energy is applied in Equation (1). The amount of biomass and coal that undergo combustion multiplied by the combustion efficiency and the lower heating value of the feedstock must be greater than or equal to demand. Equation (2) enforces the processing capacity of the power plants. Equations (3) and (4) ensure that the amount of biomass and coal delivered from their corresponding source locations are limited by the amount that is available in each period:

$$\sum_l \lambda_{lt} q_{lt} \left(\sum_j x_{jlt} \left(1 - b_{jl}^{tp}\right) + \sum_k y_{klt} \right) \geq D_t \quad \forall t \tag{1}$$

$$\sum_j x_{jlt} \left(1 - b_{jl}^{tp}\right) + \sum_k y_{klt} \leq c_{lt}^c A_{lt} \quad \forall lt \tag{2}$$

$$\sum_j w_{ijt} \leq s_{it}^b \quad \forall it \tag{3}$$

$$\sum_l y_{klt} \leq s_{kt}^c \quad \forall kt \tag{4}$$

The amount of biomass brought to the pre-treatment facilities should be less than or equal to the processing capacity of each facility, shown in Equation (5). The inventory of biomass kept in the pre-treatment facilities is defined by Equation (6). This amount is equal to the amount of biomass carried from the previous period, plus the amount of biomass that were delivered from sources and have undergone pre-treatment, less the biomass transported to the coal power plants in the current period. Equation (7) makes sure that the amount of biomass held each period is restricted by the facility's storage capacity:

$$\sum_i w_{ijt} \left(1 - b_{ij}^{tr}\right) \leq c_{jt}^p F_{jt} \quad \forall jt \tag{5}$$

$$I_{jt+1} = I_{jt} \left(1 - b_j^s\right) + n_{jt+1} - \sum_l x_{jlt+1} \quad \forall jt \tag{6}$$

$$I_{jt} \leq c_{jt}^s S_{jt} \quad \forall jt \tag{7}$$

The capacity of the pre-treatment facilities and the coal power plants may be expanded. Equations (8) and (9) show how the capacities of each facility and power plant in a given period are increased according to the expansion in the previous period. Binary variables for expanding the capacities of the pre-treatment facilities and coal power plants are switched on in Equations (10) and (11):

$$f_{jt}^p P_{jt} + c_{jt}^p = c_{jt+1}^p \quad \forall jt \tag{8}$$

$$f_{lt}^c C_{lt} + c_{lt}^c = c_{lt+1}^c \quad \forall lt \tag{9}$$

$$f_{jt}^p \leq MP_{jt} \quad \forall jt \tag{10}$$

$$f_{lt}^c \leq MC_{lt} \quad \forall lt \tag{11}$$

Equation (12) requires an existing power plant to first undergo retrofitting before the biomass co-firing option can be used. Meanwhile, Equation (13) sets upper and lower limits to the amount of biomass to displace coal in the power plants if the biomass co-firing option is activated:

$$R_l \geq O_{lt} \quad \forall lt \tag{12}$$

$$L_l^l O_{lt} \leq \frac{\sum_j x_{jlt}\left(1 - b_{jl}^{tp}\right)}{\sum_j x_{jlt}\left(1 - b_{jl}^{tp}\right) + \sum_k y_{klt}} \leq L_l^u O_{lt} \quad \forall lt \tag{13}$$

The weight of the biomass in a pre-treatment facility after pre-treatment is given in Equation (14). This is computed for by adding the dry (moisture and ash-free) biomass weight (first term) and the remaining amount of moisture and ash in mass units after completing treatment (second and third term, respectively). This is dependent on the effectiveness of the pretreatment process:

$$\sum_i w_{ijt}\left(1 - b_{ij}^{tr}\right)\left[\left(1 - m_{it}^r - a_{it}^r\right) + \left(m_{it}^r\right)\left(1 - me_j\right) + \left(a_{it}^r\right)\left(1 - ae_j\right)\right] = n_{jt} \quad \forall jt \tag{14}$$

Equation (15) computes for the moisture content of the biomass that has just completed pre-treatment. This biomass is mixed with the existing biomass in stock. The moisture content of the biomass from inventory was determined in the previous period; however, this is increased by a certain factor as an effect of storage. The moisture content of all the biomass in each pre-treatment facility is shown in Equation (16). Equation (17) defined the average moisture content of all the biomass received by a coal power plant in each period, while Equation (18) computes for the moisture content of the feedstock mix received by each power plant each period. Equations (19)–(22) compute for the ash content of the feedstock as it flows through the supply chain in the same manner. Equations (15)–(22) are conceptually similar to the generating function methodology, which are intensively discussed and applied by Shang on his works on the robustness of complex networks against failure [23,24]:

$$m_{jt}^{t} = \frac{\sum_i w_{ijt}\left(1 - b_{ij}^{tr}\right)\left(m_{it}^{r}\right)\left(1 - me_j\right)}{n_{jt}} \quad \forall jt \tag{15}$$

$$m_{jt+1}^{p} = \frac{m_{jt+1}^{t} n_{jt+1} + I_{jt}m_{jt}^{p}\left(1 + z_{jt}\right)}{n_{jt+1} + I_{jt}\left(1 - m_{jt}^{p}\right) + I_{jt}m_{jt}^{p}\left(1 + z_{jt}\right)} \quad \forall jt \tag{16}$$

$$m_{lt}^{ppb} = \frac{\sum_j m_{jt}^{p} x_{jlt}\left(1 - b_{jl}^{tp}\right)}{\sum_j x_{jlt}\left(1 - b_{jl}^{tp}\right)} \quad \forall lt \tag{17}$$

$$m_{lt}^{pp} = \frac{\sum_j m_{jt}^{p} x_{jlt}\left(1 - b_{jl}^{tp}\right) + \sum_k m^c y_{klt}}{\sum_j x_{jlt}\left(1 - b_{jl}^{tp}\right) + \sum_k y_{klt}} \quad \forall lt \tag{18}$$

$$a_{jt}^{t} = \frac{\sum_i w_{ijt}\left(1 - b_{ij}^{tr}\right)\left(a_{it}^{r}\right)\left(1 - ae_j\right)}{n_{jt}} \quad \forall jt \tag{19}$$

$$a_{jt+1}^{p} = \frac{a_{jt+1}^{t} n_{jt+1} + I_{jt}a_{jt}^{p}}{n_{jt+1} + I_{jt}} \quad \forall jt \tag{20}$$

$$a_{lt}^{ppb} = \frac{\sum_j a_{jt}^{p} x_{jlt}\left(1 - b_{jl}^{tp}\right)}{\sum_j x_{jlt}\left(1 - b_{jl}^{tp}\right)} \quad \forall lt \tag{21}$$

$$a_{lt}^{pp} = \frac{\sum_j a_{jt}^{p} x_{jlt}\left(1 - b_{jl}^{tp}\right) + \sum_k a^c y_{klt}}{\sum_j x_{jlt}\left(1 - b_{jl}^{tp}\right) + \sum_k y_{klt}} \quad \forall lt \tag{22}$$

Equation (23) determines the lower heating value of the biomass in each power plant. This equation is adapted from Hernández et al. [25]. The average lower heating value of the feedstock mix considering the biomass and coal blend is given in Equation (24):

$$q_{lt}^{b} = g\left(1 - m_{lt}^{ppb}\right)\left(1 - a_{lt}^{ppb}\right) - 2.443 m_{lt}^{ppb} \quad \forall lt \tag{23}$$

$$q_{lt} = \frac{\sum_j q_{lt}^{b} x_{jlt} + \sum_k q^c y_{klt}}{\sum_j x_{jlt} + \sum_k y_{klt}} \quad \forall lt \tag{24}$$

The efficiency in each power plant is defined in Equation (25). This is a function of the excess and lacking moisture content of the feedstock, excess ash content, and the total amount of feedstock processed by the plant, as these values increase, the efficiency will decrease. Equation (26) describes excess moisture content to be equal to the maximum between zero and the difference between the actual moisture content of the feedstock and the upper limit, while Equation (27) defines shortage in moisture. Similarly, Equation (28) computes for the excess ash content based on its maximum allowable amount. With this approach, there will be no amount stored if the difference returned is negative. Equation (29) sums up the biomass and coal processed in a coal power plant each period to get the total feedstock handled by the equipment:

$$\lambda_{lt} = f(m_{lt}^+, m_{lt}^-, a_{lt}^+, Q_{lt}) \quad \forall lt \quad (25)$$

$$m_{lt+1}^+ = \max(m_{lt+1}^{pp} - m_l^U, 0) + m_{lt}^+ \quad \forall lt \quad (26)$$

$$m_{lt+1}^- = \max(m_l^L - m_{lt+1}^{pp}, 0) + m_{lt}^- \quad \forall lt \quad (27)$$

$$a_{lt+1}^+ = \max(a_{lt+1}^{pp} - a_l^U, 0) + a_{lt}^+ \quad \forall lt \quad (28)$$

$$Q_{lt+1} = \sum_j x_{jlt+1} + \sum_k y_{klt+1} + Q_{lt} \quad \forall lt \quad (29)$$

Equations (30) and (31) compute for the number of trips needed to transport biomass from source to pre-treatment facilities and from pre-treatment facilities to coal power plants based on weight and volume capacities. Likewise, Equation (32) defines the number of trips required to deliver coal from source locations to coal power plants. Lastly, non-negativity, binary, and integer constraints apply to relevant variables:

$$t_{ijt}^r \geq \max\left\{\frac{w_{ijt}}{u_{ijt}^r}, \frac{w_{ijt}}{\rho_{it}^r v_{ijt}^r}\right\} \quad \forall ijt \quad (30)$$

$$t_{jlt}^p \geq \max\left\{\frac{x_{jlt}}{u_{jlt}^p}, \frac{x_{jlt}}{\rho_{jt}^p v_{jlt}^p}\right\} \quad \forall jlt \quad (31)$$

$$t_{klt}^c \geq \max\left\{\frac{y_{klt}}{u_{klt}^c}, \frac{y_{klt}}{\rho^c v_{klt}^c}\right\} \quad \forall klt \quad (32)$$

4.2. Objective Function

The model seeks to maximize the performance of both objectives, which are to minimize total cost and emissions; a balance is achieved by maximizing the smaller desirability value to prevent optimizing one objective at the expense of the other as shown in Equation (33). Dimensionless efficiency values (not to be confused with thermodynamic efficiency) are obtained by dividing the improvement achieved (difference between worst and actual values) and the potential improvement (difference between worst and potential values). Potential objective values are obtained by minimizing each corresponding objective as single objective optimization models. The worst value that the cost objective may take is its value when the environmental objective is optimized, and vice-versa. Note that this max-min aggregation approach always optimizes the less satisfied objective:

$$\text{Max } Z = \min\left[\left(\frac{Cost_{max} - Cost}{Cost_{max} - Cost_{pot}}\right), \left(\frac{Env_{max} - Env}{Env_{max} - Env_{pot}}\right)\right] \quad (33)$$

4.2.1. Cost Component

The first sub-objective of the model is to minimize total costs incurred by the system as indicated in Equations (34)–(36). The total fixed cost (Equation (35)), includes costs obtained from retrofitting existing coal power plants, from operating and expanding coal power plants and pre-treatment facilities, from using the biomass option in modified power plants and using storage areas in pre-treatment facilities. Variable costs, shown in Equation (36), include costs to purchase feedstock, convert biomass and coal to energy, pretreat biomass, keep biomass in inventory, transport biomass and coal, and expand the capacities of power plants and pretreatment facilities. Transportation costs are based on the average cost (which may include fuel and labor costs, loading and unloading costs, insurance, taxes, etc.) per distance travelled, which is the applied convention in industry [26]:

$$Cost = \sum_t Fixed\ Cost_t + \sum_t Variable\ Cost_t \qquad (34)$$

$$Fixed\ Cost_t = \sum_l ic_l R_l + \sum_l oc_{lt} A_{lt} + \sum_l bc_{lt} O_{lt} + \sum_j oc^p_{jt} F_{jt} + \sum_j sc_{jt} S_{jt} + \sum_l rc_{lt} N_{lt} + \\ \sum_j ec^p_{jt} P_{jt} + \sum_l ec^c_{lt} C_{lt} \qquad \forall t \qquad (35)$$

$$Variable\ Cost_t = \sum_j \sum_l r^b_{lt} x_{jlt} + \sum_k \sum_l (r^c_{lt} + p^c_{kt}) y_{klt} + \sum_i \sum_j (pc_{jt} + p^b_{jt}) w_{ijt} + \sum_j h_{jt} I_{jt} + \\ \sum_i \sum_j t^r_{ijt} tc^r_{ijt} d^r_{ij} + \sum_j \sum_l t^p_{jlt} tc^p_{jlt} d^p_{jl} + \sum_k \sum_l t^c_{klt} tc^c_{klt} d^c_{kl} + \sum_j e^p_{jt} f^p_{jt} + \sum_l e^c_{lt} f^c_{lt} \qquad \forall t \qquad (36)$$

4.2.2. Emissions Component

Another sub-objective of the model is to minimize the system's emissions, which includes emissions from pre-treatment, combustion, and transport processes. The environmental objective is shown in Equation (37). Similarly, the amount of carbon footprint attributed to transportation is based on emissions per unit distance and total distance travelled:

$$Env = \sum_i \sum_j \sum_t pe_{jt} w_{ijt} + \sum_k \sum_l \sum_t ce_{lt} y_{klt} + \sum_i \sum_j \sum_t t^r_{ijt} te^r_{ijt} d^r_{ij} + \sum_j \sum_l \sum_t t^p_{jlt} te^p_{jlt} d^p_{jl} + \\ \sum_k \sum_l \sum_t t^c_{klt} te^c_{klt} d^c_{kl} \qquad (37)$$

5. Model Implementation

The model was implemented in General Algebraic Modelling System (GAMS) and solved using the nonlinear solver Convex Over and Under Envelopes for Nonlinear Estimation (COUENNE), with a solution time of 285.66 s and integer gap of 0.0523 on a MacBook Pro with a 3.1 GHz Intel Core i5 processor and 8 GB 2133 MHz LPDDR3 RAM. The case study considers three potential locations each for biomass sources, coal sources, pre-treatment/storage facilities, and coal power plants considered over three time periods. The biomass considered is rice straw. The resulting model has 656 continuous variables, 164 integer variables, and 425 constraints. Figure 2 illustrates the exponential increase in solution times (expressed in seconds) as the number of potential locations in each echelon is increased from 2 to 5. As the number of nodes in each echelon is increased even further, it is expected that the computation times will also increase following the same trend. Although the model seems complex, it can be easily implemented with most commercially available solvers. The data inputs required to run the model are also typically available to the user. Parameter values were used based on various literature sources.

The electricity demand and supply of biomass and coal are given in Table 2. The bulk density, moisture content, and ash content of raw biomass, particularly rice straw, from each source are summarized in Table 3. The higher heating value of rice straw is 18 MJ/kg. Data on rice straw properties were adapted from Liu et al. [27] and Kargbo et al. [28]. The improvement effectiveness for ash and moisture content, and resulting bulk density of each pretreatment facility are shown in Table 4. Table 4 also gives the amount of damage in biomass when it is stored, and the storage capacity in each pretreatment facility. The damage factor for transporting raw and pretreated biomass are 0.10 and 0.05, respectively, while moisture content increases by 0.10 due to storage. Additionally, the initial processing capacity of the pretreatment facilities is 500 kt. The displacement, ash content, and moisture content limits of each power plant are given in Table 5. Upper and lower coal displacement limits are strictly enforced in the power plants. On the other hand, the feedstock may violate the maximum preferred ash content and allowable range for moisture content, but this would lead to negative consequences on equipment efficiency; thus, they act only as soft constraints. Input parameters for coal composition are summarized in Table 6 and were adapted from Bains et al. [29]. The distances between biomass sources and pretreatment facilities, pretreatment facilities to power plants, and coal sources to power plants are shown in Tables A1–A3 of the Appendix A. The weight and volume capacities are 450 kt and 75 m^3 respectively. Costs to purchase biomass and coal, and to retrofit each coal power plant for co-firing are given in Table A4. Power plant associated costs are summarized in Table A5, while costs associated with processing and storing biomass in pretreatment facilities are given in Table A6. Transportation costs are assumed to be US$ 18/km-kg. Lastly, Table A7 summarizes the emissions from biomass pretreatment, transporting biomass, and the combustion of biomass and coal. Cost and emissions parameters were adapted from the studies of Griffin et al. [14] and Mohd Idris et al. [13] respectively.

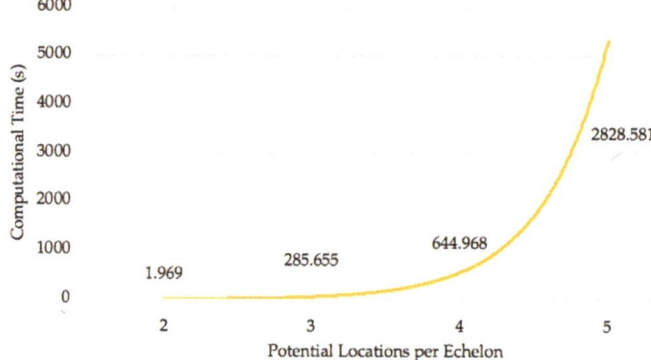

Figure 2. Solution Time of Varying Model Sizes.

Table 2. Supply and Demand Parameters.

	Period 1	Period 2	Period 3
Demand (MJ)	47,000	48,500	49,000
Biomass Supply (kt)			
B 1	800	900	1050
B 2	1000	750	1200
B 3	1200	1050	850
Coal Supply (kt)			
C 1	1550	1350	1000
C 2	800	1000	900
C 3	3000	2250	2500

Table 3. Biomass Quality Parameters.

	Bulk Density (kg/m^3)			Moisture Content (% wt.)			Ash Content (% wt.)		
	Period 1	Period 2	Period 3	Period 1	Period 2	Period 3	Period 1	Period 2	Period 3
B 1	50	30	55	17	18	20	17	18	20
B 2	35	40	40	25	20	19	25	20	19
B 3	60	70	55	18	23	22	18	23	22

Table 4. Pretreatment Facility Characteristics.

	Ash	Moisture	Bulk Density (kg/m^3)	Storage Damage Factor	Storage Capacity (kt)
PF 1	65%	35%	25	2	1350
PF 2	76%	52%	20	7	1000
PF 3	54%	67%	30	5	2250

Table 5. Power Plant Facility Characteristics.

	Displacement Limits (%)	Ash Content (% wt.)	Moisture Content (% wt.)
PP 1	[0, 40]	0	[10, 12]
PP 2	[0, 45]	0	[10, 12]
PP 3	[0, 40]	0	[10, 12]

Table 6. Coal Quality Parameters.

Bulk Density (kg/m^3)	78.5
Lower Heating Value (MJ/kg)	30
Ash Content (% wt.)	5
Moisture Content (% wt.)	9

The relationship between conversion equipment efficiency, feedstock property violations and total feedstock processed is modeled with an arbitrary function for the model validation. An exponential decay function is used as in Equation (38):

$$\lambda_{lt} = constant^{-(m_{lt}^{+} + m_{lt}^{-} + a_{lt}^{+} + Q_{lt})} \tag{38}$$

Statistical experiments support that the negative exponential function may be used to describe the performance degradation of equipment. The specific behavior of degradation differs between each unit of equipment and, as a result, would have a unique combination of input parameters to accurately predict behavior. These parameters may be obtained through exponential regression [30]. An exponential decay function with a base that is between 0 to 1 will return decreasing values as the exponent variables increase. Efficiency will begin at 1 when the exponent is 0 or when no feedstock property violations have been made and/or no feedstock has been handled by the power plant yet. The efficiency value will then decrease, approaching 0 as the exponent increases. The constant dictates the rate of decrease per unit increase in the exponent variables. The higher the constant is, the faster the rate of decrease will be. As shown in Figure 3, the decrease in efficiency is more significant when the constant used is 4 compared to when the constant is equal to 2. For the purpose of validating the model, a hypothetical system is captured, and the constant for the negative exponential function was set to 2.

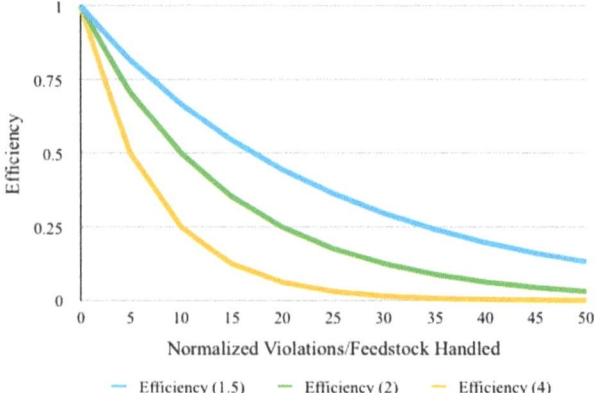

Figure 3. Conversion Efficiency Function Graphs.

The results are presented in three parts, namely where each sub-objective is optimized separately, followed by the complete model run. Running the model wherein each objective is minimized individually is necessary to obtain the potential and worst values for cost and emissions needed in the full model run.

5.1. Base Case

5.1.1. Minimizing Cost

In minimizing the cost component separately, model results show a bias towards using only coal to satisfy the demand for energy as presented in Figure 4. This is because coal is relatively cheap compared to biomass, especially when considering transportation and storage requirements, pre-treatment costs, and investment costs for retrofitting existing coal power plants. Furthermore, using biomass, which are not within machine specifications, decreases conversion efficiency which then requires more feedstock to satisfy demand. However, optimizing the network based solely on minimizing costs sacrifices the environmental objective as shown in Table 7.

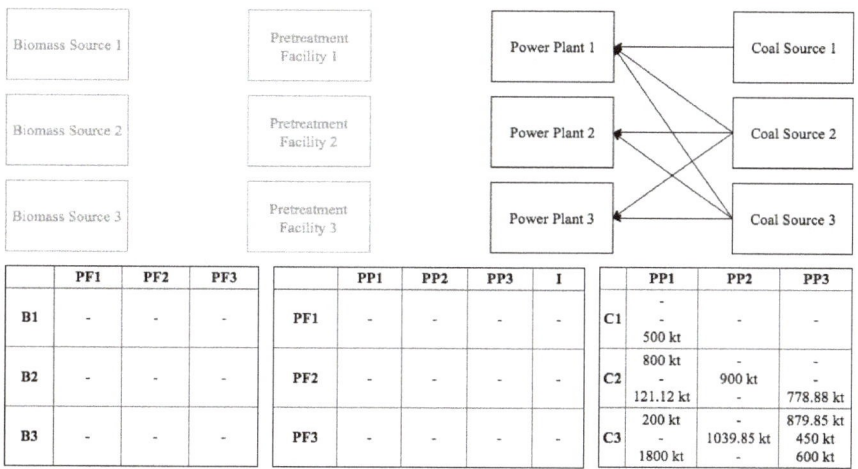

Figure 4. Minimum Cost Network.

Table 7. Comparison of costs and emissions objective performances.

	Potential	Minimizing Cost	Minimizing Emissions	Complete Model Run	
					Efficiency
Cost (Million US$)	45,959.97	45,959.97	360,920.00	80,859.36	0.8892
Emissions (kt CO_2)	3491.29	4858.40	3491.29	3642.78	0.8892

5.1.2. Minimizing Emissions

On the other hand, when the environmental objective is optimized solely, more biomass is purchased and used in all existing coal power plants to prevent incurring the much larger emissions from coal firing as shown in Figure 5. However, cost inflates significantly (Table 7) because of several reasons. Transporting biomass is relatively more expensive because of its inherent properties, in addition pre-treatment will also have associated fixed and operating costs. Without regard for costs, all the pre-treatment facilities are opened depending on the pre-treatment process and effectiveness of each facility most suited to the initial quality of the biomass. In addition, the use of more biomass results in efficiency loss in the equipment leading to the purchase of more fuel to reach demand. This is why significant increases are seen in fuel use in the second and third periods.

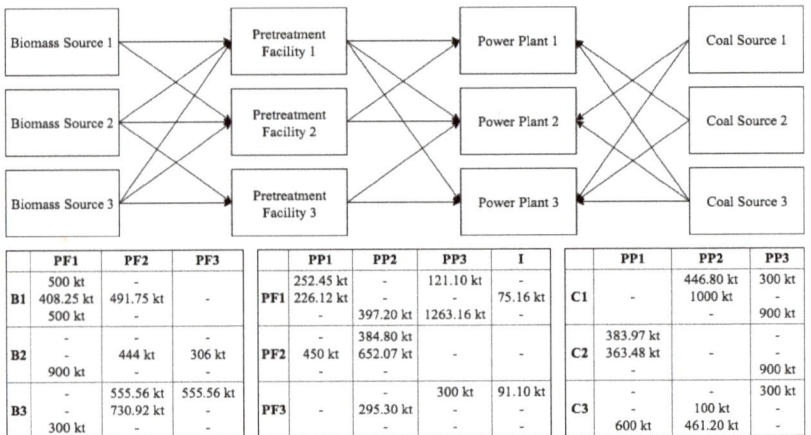

	PF1	PF2	PF3		PP1	PP2	PP3	I		PP1	PP2	PP3
	500 kt	-	-		252.45 kt	-	121.10 kt	-		-	446.80 kt	300 kt
B1	408.25 kt	491.75 kt	-	PF1	226.12 kt	-	-	75.16 kt	C1	-	1000 kt	-
	500 kt	-	-		-	-	397.20 kt	1263.16 kt		-	-	900 kt
	-	-	-		-	-	384.80 kt	-		383.97 kt	-	-
B2	-	444 kt	306 kt	PF2	450 kt	652.07 kt	-	-	C2	363.48 kt	-	-
	900 kt	-	-		-	-	-	-		-	-	900 kt
	-	555.56 kt	555.56 kt		-	-	300 kt	91.10 kt		-	-	300 kt
B3	-	730.92 kt	-	PF3	-	295.30 kt	-	-	C3	-	100 kt	-
	300 kt	-	-		-	-	-	-		600 kt	461.20 kt	-

Figure 5. Minimum Environmental Emissions Network.

Optimizing each objective as single optimization models reiterate that a compromise must be found between the two conflicting objectives. One objective should not be minimized too much that no attention is given to the other. Considering only economic costs in optimizing the network result to a scheme where crucial investments and processes are disregarded to reduce costs, significantly compromising environmental sustainability. Similarly, when the system is optimized solely on environmental performance, costs are dramatically increased which may make the solution impractically expensive.

A goal programming approach is used to address this and achieve a solution that balances the two objectives. Each sub-objective–the economic and environmental–are optimized as single objective models and the results are recorded. The desirability levels of the objectives are computed for by dividing the actual improvement achieved by the potential improvement. In this case, the potential improvements for cost and environmental emissions are Million US$ 314,960.03 and 1367.1058 kt CO_2, respectively, based on the values shown in Table 7. The minimum between two efficiencies are maximized to obtain a solution that balances the minimization of costs and emissions.

5.1.3. Full Model Run

As shown in Table 7, simultaneously optimizing both objectives allow the system to reach efficiency ratings closer to each other, both objectives getting 0.8892 desirability levels. Figure 6 also show a more manageable network configuration. In an effort to control both costs and emissions, biomass is used by the system and only two pre-treatment facilities are opened. Less biomass is used compared to when only the emissions were minimized. Only two of the three coal power plants are retrofitted for co-firing. Because less biomass is used, the decrease in boiler efficiency is slower. As a result, when comparing the fuel usage of the optimal network and the network which optimized the environmental aspect only, the increase across periods is not as dramatic. Aside from this, overall fuel usage is also less because a lower biomass-to-coal blend ratio means that the lower heating value of the feedstock is higher, resulting in a higher electricity yield. In addition, only two pre-treatment facilities are chosen to avoid the additional costs needed to operate more pre-treatment facilities. This required the model to choose the facility which costs the least to operate but resulted in the best improvements in biomass properties.

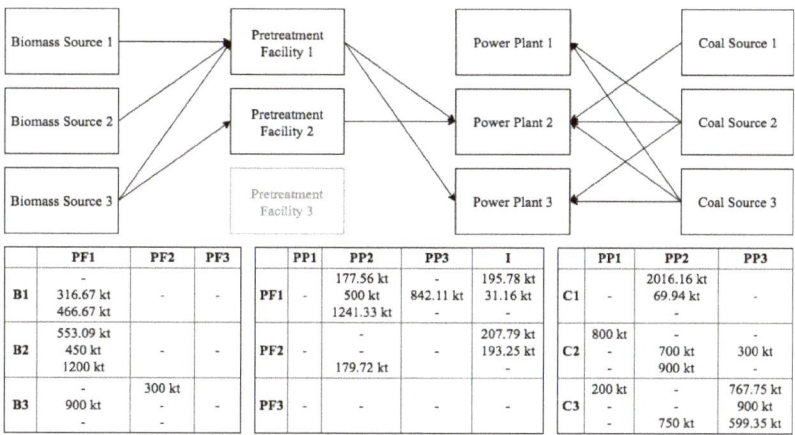

Figure 6. Optimal Biomass Co-firing Network.

6. Scenario Analysis

6.1. Impact of Feedstock Properties Consideration

The model was optimized without considering feedstock properties and compared to the results of the proposed model to demonstrate the impact of these considerations on a biomass co-firing supply chain, particularly on storage, transportation, and pretreatment decisions, and conversion yield. The resulting network is presented in Figure 6.

The optimized network without quality considerations (Figure 7) is compared to the optimal network obtained from the model proposed in this work (Figure 6). Without the consideration of biomass and coal properties in the network, biomass is sourced only from two locations and only two coal power plants are activated. Less fuel is used by the network because no damage occurs to the fuel during storage and transport and coal power plant conversion equipment does not experience any changes in yield or capacity. Thus, the supply and capacity of two biomass sources and two coal power plants are already enough to satisfy the demand for power in each period. Selecting where to source biomass is based only on distance and costs, unlike in the proposed model where the properties of the biomass from each source is also a factor in this decision. Similarly, pretreatment facilities/processes are chosen based on distance and pretreatment costs instead of their effectiveness in improving the qualities of the biomass. The amount of inventory held across periods is reduced also because the cost of purchasing and pretreating biomass remains relatively stable, so there is no need to keep inventory. On the other hand, the proposed model chooses to hold inventory on certain periods to avoid periods where the quality of the biomass is worse. In addition, only one of the two active coal power plants are retrofitted for co-firing. The system becomes less careful with distributing the amount of biomass usage among coal power plants, as it no longer has to avoid possible deterioration in conversion equipment and variations in the lower heating value of the mixed fuel. As such, in an effort to reduce transport and retrofitting costs, as well as transport emissions, biomass is only transported and used in one of the two active coal power plants. The model which overlooks quality related issues also has significantly lower costs (Million US$ 40,150.17) and emissions (2209.40 kt CO_2). The graphs in Figure 8a,b illustrate the components of costs and emissions for both optimized models.

Without considering biomass quality, costs are lowered in all of its components–purchase, transport, pretreatment, combustion, holding, and capital costs. As explained earlier, this is because of the significantly decreased fuel that flows through the system. Transportation costs are also decreased because the bulk density of the biomass and coal are not accounted for, which entail difficulties in transporting material (e.g., requiring additional trips). Similarly, emissions are considerably lower in this scenario because of the same reasons.

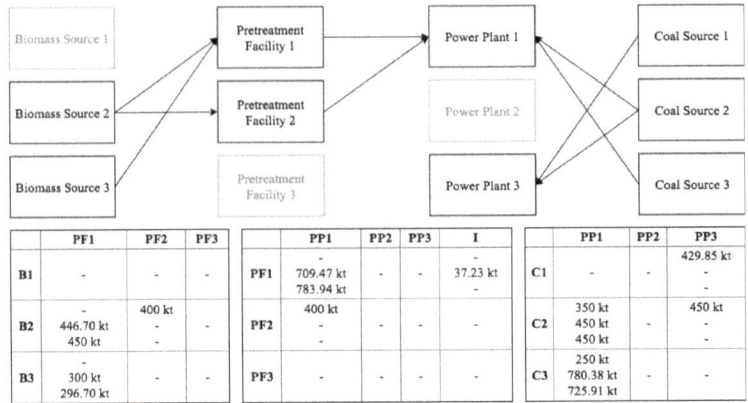

Figure 7. Optimal Biomass Co-firing Network without Quality Considerations.

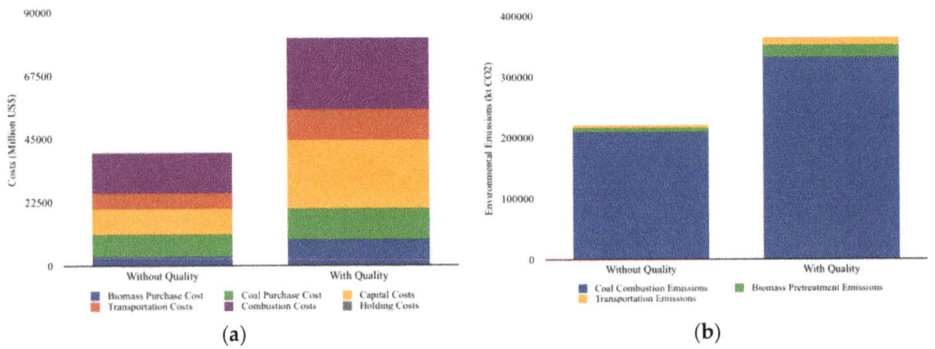

Figure 8. Breakdown of (a) costs and (b) environmental with and without Quality Considerations.

Although this scenario achieved better performance for the financial and environmental objectives, it is an inaccurate and unreliable model of a biomass co-firing network, and will not be useful as a planning or management tool.

6.2. Biomass Properties

Biomass properties show to be a significant consideration in the modelling of biomass supply chains because it influenced network decisions across all activities in the supply chain. Changes in the properties of the biomass may cause an impact in the way the biomass co-firing network is constructed, consequently affecting the network's financial and environmental sustainability.

The biomass properties were improved and worsened across all periods in two different scenarios and are compared with the base scenario. In the improved properties scenario, moisture content and ash content are decreased by 20%, while bulk density is increased by 20%. On the other hand, biomass properties are worsened by increasing moisture content and ash content by 20%, and reducing bulk density by 20%. The cost and environmental emissions performance for the two scenarios and the baseline scenario are shown in Figure 9.

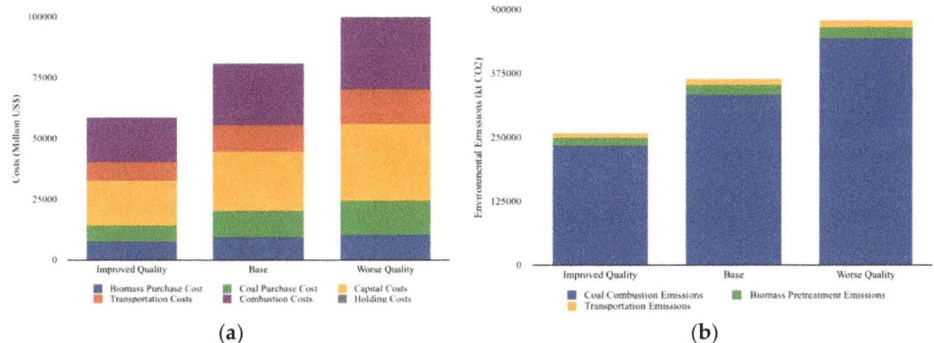

Figure 9. Breakdown of (**a**) costs and (**b**) environmental when Biomass Properties are Varied.

It can be observed that costs and emissions increase as the properties of biomass worsen. The improved properties scenario yielded the least cost and environmental emissions, while the worsened properties scenario resulted in higher costs and emissions.

The breakdown of costs is analysed in Figure 9a to understand why costs increased as properties worsened. With worse biomass quality, efficiency loss experienced by the conversion equipment in coal power plants also worsen, requiring the system to purchase and use more fuel. This increases all cost components, such as fuel purchase costs, transport, pre-treatment, and combustion costs. Transportation costs also increase because of the worse bulk density of the biomass, requiring multiple deliveries. On the other hand, as properties improved, the system would have to rely on less fuel overall because the power plants experience less loss in efficiency. Thus, less biomass and coal are used to reach the required amount of electricity. However, it is noticeable that when properties are worsened, biomass purchase increases only slightly, while the increase in coal purchase is more pronounced. This is because the model attempts to control the damage biomass causes on the conversion equipment and the additional costs needed to handle biomass by diluting biomass properties with more desirable coal properties.

Likewise, Figure 9b also shows an overall increase in environmental emissions as biomass properties worsen. For similar reasons, all components of environmental emissions increase due to the need to process more fuel. Combustion emissions due to coal increases because the model chooses between the corrosion in boiler equipment, which will require additional fuel and cause harmful emissions, and the emissions from burning coal.

Another set of scenarios are analysed. Particularly, a scenario wherein biomass properties are worsened only in the second period and where properties are improved only in the second period. The properties are enhanced and worsened by 50% from their original value.

When biomass properties are relatively stable across periods, storage of the biomass is avoided because it damages the biomass. However, when biomass properties experience increased moisture content, for example during wet season, and increased ash content, purchases are done during earlier periods with sufficient quantity and appropriate quality and stored for future use. When moisture content and ash content were higher and bulk density was lower in the second period, purchase during this period decreased significantly. Instead, the biomass to be used on the second period were purchased during the first and stored. The additional amount purchased in period 1 allotted an extra amount to account for deterioration and loss due to transport and storage. As a result, purchase, transport, and pre-treatment costs and emissions increase because of the additional biomass purchased in period 1, holding costs also increase to store biomass. Changes to combustion costs and emissions, as well as the efficiency loss experienced by the equipment are negligible because the original properties in period 2 are only slightly different from period 1. The optimal network for this scenario is shown in

Figure 10. In the same way, when the properties in period 2 are made significantly better relative to period 3, the model chooses to purchase biomass for period 3 during the second period (Figure 11).

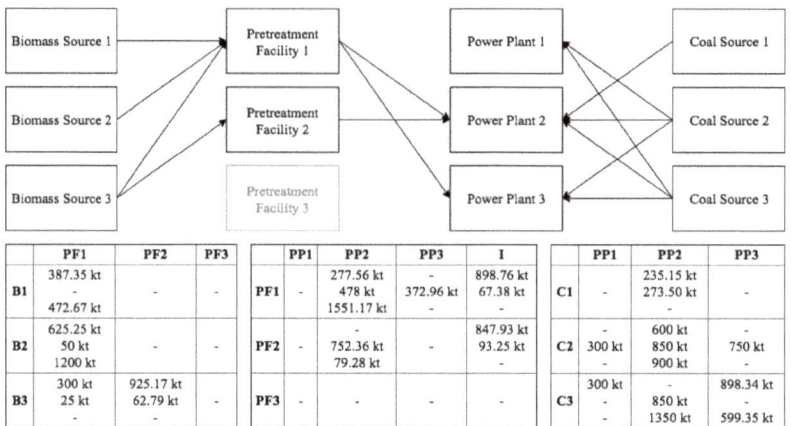

Figure 10. Optimal Network for Worse Biomass Properties in Period 2.

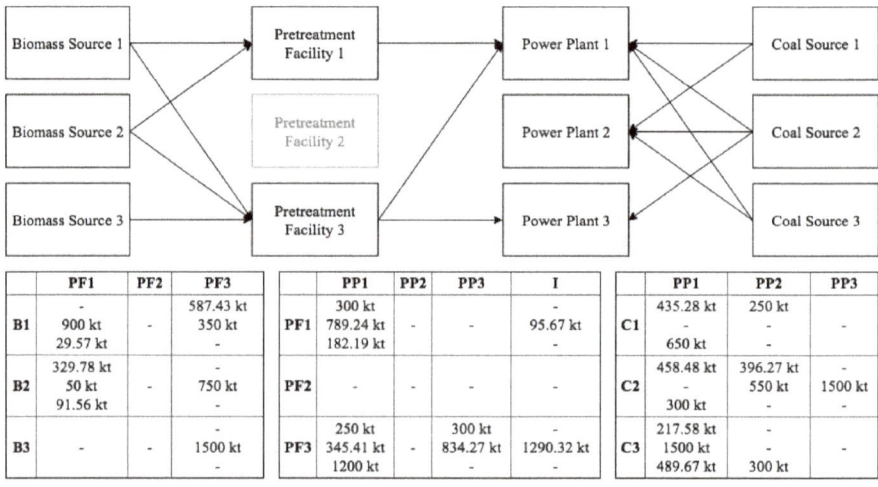

Figure 11. Optimal Network for Improved Biomass Properties in Period 2.

7. Conclusions

A MINLP model for optimizing a biomass co-firing supply chain network has been developed, which integrates feedstock properties considerations while simultaneously minimizing economic costs and environmental emissions through goal programming. This model incorporates the impact of feedstock, transportation, and pre-treatment requirements. Changes in biomass properties as it moves through the network are accounted for, together with the impact of feedstock properties on conversion yield and equipment degradation. The inclusion of these in the model showed to be an important enhancement to traditional models because decisions on how much and when to source biomass and coal, and the use of pretreatment facilities, storage, and combustion in coal power plants were considerably affected by the said considerations.

Minimizing either the financial or environmental objective individually emphasized the conflicting nature of the two objectives. Simultaneously optimizing both objectives created a network which balanced performance on both objectives.

It was also shown that without considerations for feedstock properties, costs and emissions were artificially decreased, leading to the purchase of insufficient fuel and combustion of inappropriate fuel which may result in damage or loss in efficiency of the equipment. Hence, the model proposed in this study is a better fit to design and manage a biomass co-firing network.

Extensions on this research may consider biomass quality and availability as uncertain parameters in a robust multi-objective optimization model. Precise data regarding the quality of the feedstock is not readily available, and any error in estimation requires operational adjustments to be made. As feedstock quality has been proven to be an important inclusion in biomass co-firing networks, these networks should be made robust to such uncertainties. Additionally, this study assumes that the network's nodes are all functional and benign; however, in reality, the presence of faulty and uncooperative components must be considered [31]. Lastly, the parameters used in the validation of the proposed model may be considered too optimistic and difficult to match in real market. Thus, the application of the model to real-world problems may be explored, along with efficient solution strategies for the resulting large-scale problems.

Author Contributions: Conceptualization, C.L.S.; Data curation, J.L.G.S.J.; Formal analysis, J.L.G.S.J.; Supervision, C.L.S., K.B.A. and R.R.T.; Validation, J.L.G.S.J.; Writing—original draft, J.L.G.S.J.; Writing—review & editing, J.L.G.S.J., C.L.S., K.B.A. and R.R.T.

Funding: This research received no external funding.

Conflicts of Interest: The authors declare no conflict of interest.

Appendix A

Table A1. Distance between Biomass Sources and Pretreatment Facilities in km.

	PF 1	PF 2	PF 3
B 1	15	20	25
B 2	25	20	20
B 3	25	18	17

Table A2. Distance between Pretreatment Facilities and Power Plants in km.

	PP 1	PP 2	PP 3
PF 1	10	17	9
PF 2	15	20	15
PF 3	18	18	18

Table A3. Distance between Coal Sources and Power Plants in km.

	PP 1	PP 2	PP 3
C 1	15	15	15
C 2	10	15	12
C 3	15	20	18

Table A4. Biomass and Coal Purchase Costs and Retrofitting Costs.

	B 1	B 2	B 3
Biomass Price (US$/kg)	2.5	1.75	1.5
	C 1	C 2	C 3
Coal Price (US$/kg)	3	1.75	2
	PP 1	PP 2	PP 3
Retrofitting Cost (US$)	5000	7000	4300

Table A5. Power Plant Costs.

	Fixed Operating Cost (US$)			Fixed Biomass Option (US$)		
	Period 1	Period 2	Period 3	Period 1	Period 2	Period 3
PP 1	350	300	350	20	24	35
PP 2	300	450	375	60	45	30
PP 3	385	375	385	35	30	45
	Biomass Combustion Cost (US$/kg)			Coal Combustion Cost (US$/kg)		
	Period 1	Period 2	Period 3	Period 1	Period 2	Period 3
PP 1	2	3	1.5	2	2.25	2.5
PP 2	5	4	1	3	2.75	2.8
PP 3	4	3.5	5	2.6	2.5	3
	Fixed Expansion Costs (US$)			Unit Expansion Costs (US$/kg)		
	Period 1	Period 2	Period 3	Period 1	Period 2	Period 3
PP 1	300	295	325	1	1	1
PP 2	500	375	350	1.25	1.25	1.5
PP 3	475	388	420	1.75	1.5	1

Table A6. Pretreatment Facility Costs.

	Fixed Operating Cost (US$)			Biomass Pretreatment Cost (US$/kg)		
	Period 1	Period 2	Period 3	Period 1	Period 2	Period 3
PF 1	50	75	80	3	1	1.5
PF 2	80	69	78	2	2	2
PF 3	68	80	75	1.7	2.15	1.85
	Fixed Expansion Costs (US$)			Unit Expansion Costs (US$/kg)		
	Period 1	Period 2	Period 3	Period 1	Period 2	Period 3
PF 1	200	350	300	1	1.25	1.5
PF 2	500	400	450	1.75	1.5	1.1
PF 3	200	375	425	1.25	1	1.35
	Fixed Holding Costs (US$)			Unit Holding Costs (US$/kg)		
	Period 1	Period 2	Period 3	Period 1	Period 2	Period 3
PF 1	50	35	75	0.5	0.15	0.3
PF 2	100	65	45	0.15	0.2	0.25
PF 3	50	50	50	0.3	0.25	0.2

Table A7. Emissions Parameters.

Biomass Pretreatment Emissions	0.03 kg CO_2/kg
Transportation Emissions	0.12 kg CO_2/km
Biomass Combustion Emissions	0.08 kg CO_2/kg
Coal Combustion Emissions	0.50 kg CO_2/kg

References

1. International Energy Outlook. 2017; (DOE/EIA-0484(2017)). Available online: https://www.eia.gov/outlooks/ieo/pdf/0484(2017).pdf (accessed on 23 September 2018).
2. Iakovou, E.; Karagiannidis, A.; Vlachos, D.; Toka, A.; Malamakis, A. Waste biomass-to-energy supply chain management: A critical synthesis. *Waste Manag.* **2010**, *30*, 1860–1870. [CrossRef] [PubMed]
3. Ramos, A.; Monteiro, E.; Silva, V.; Rouboa, A. Co-gasification and recent developments on waste-to-energy conversion: A review. *Renew. Sustain. Energy Rev.* **2018**, *81*, 380–398. [CrossRef]
4. Dundar, B.; McGarvey, R.G.; Aguilar, F.X. Identifying Optimal Multi-state collaborations for reducing CO_2 emissions by co-firing biomass in coal-burning power plants. *Comp. Ind. Eng.* **2016**, *101*, 403–415. [CrossRef]
5. Madanayake, B.N.; Gan, S.; Eastwick, C.; Ng, H.K. Biomass as an energy source in coal co-firing and its feasibility enhancement via pre-treatment techniques. *Fuel Process. Technol.* **2017**, *159*, 287–305. [CrossRef]
6. Mauro, C.; Rentizelas, A.A.; Chinese, D. International vs. domestic bioenergy supply chains for co-firing plants: The role of pre-treatment technologies. *Renew. Energy* **2018**, *119*, 712–730. [CrossRef]
7. Zandi Atashbar, N.; Labadie, N.; Prins, C. Modeling and optimization of biomass supply chains: A review and a critical look. *IFAC-PapersOnLine* **2016**, *49*, 604–615. [CrossRef]
8. Shafie, S.; Mahlia, T.; Masjuki, H. Life cycle assessment of rice straw co-firing with coal power generation in Malaysia. *Energy* **2013**, *57*, 284–294. [CrossRef]
9. Oanh, N.T.K.; Tipayarom, A.; Bich, T.L.; Tipayarom, D.; Simpson, C.D.; Hardie, D.; Liu, L.J.S. Characterization of gaseous and semi-volatile organic compounds emitted from field burning of rice straw. *Atmos. Environ.* **2015**, *119*, 182–191. [CrossRef]
10. Ba, B.H.; Prins, C.; Prodhon, C. Models for optimization and performance evaluation of biomass supply chains: An Operations Research perspective. *Renew. Energy* **2016**, *87*, 977–989. [CrossRef]
11. Shang, Y.L. Optimal control strategies for virus spreading in inhomogeneous epidemic dynamics. *Canad. Math. Bull.* **2013**, *56*, 621–629. [CrossRef]
12. Pérez-Fortes, M.; Laínez-Aguirre, J.M.; Bojarski, A.D.; Puigjaner, L. Optimization of pre-treatment selection for the use of woody waste in co-combustion plants. *Chem. Eng. Res. Des.* **2014**, *92*, 1539–1562. [CrossRef]
13. Mohd Idris, M.N.; Hashim, H.; Razak, N.H. Spatial optimisation of oil palm biomass co-firing for emissions reduction in coal-fired power plant. *J. Clean. Prod.* **2018**, *172*, 3428–3447. [CrossRef]
14. Griffin, W.; Michalek, J.; Matthews, H.; Hassan, M. Availability of Biomass Residues for Co-Firing in Peninsular Malaysia: Implications for Cost and GHG Emissions in the Electricity Sector. *Energies* **2014**, *7*, 804–823. [CrossRef]
15. Savic, D. Single-objective vs. Multiobjective Optimisation for Integrated Decision Support. In Proceedings of the 9th International Congress on Environmental Modelling and Software, Lugano, Switzerland, 24–27 June 2002; Volume 119, pp. 7–12.
16. Rollan, C.D.; Li, R.; San Juan, J.L.; Dizon, L.; Ong, K.B. A planning tool for tree species selection and planting schedule in forestation projects considering environmental and socio-economic benefits. *J. Environ. Manag.* **2018**, *206*, 319–329. [CrossRef] [PubMed]
17. Ramos, M.; Boix, M.; Montastruc, L.; Domenech, S. Multiobjective Optimization Using Goal Programming for Industrial Water Network Design. *Ind. Eng. Chem. Am. Chem. Soc.* **2014**, *53*, 17722–17735. [CrossRef]
18. Malladi, K.T.; Sowlati, T. Biomass logistics: A review of important features, optimization modeling and the new trends. *Renew. Sustain. Energy Rev.* **2018**, *94*, 587–599. [CrossRef]
19. Castillo-Villar, K.K.; Eksioglu, S.; Taherkhorsandi, M. Integrating biomass quality variability in stochastic supply chain modeling and optimization for large-scale biofuel production. *J. Clean. Prod.* **2017**, *149*, 904–918. [CrossRef]
20. Boundy, R.G.; Diegel, S.W.; Wright, L.L.; Davis, S.C. *Biomass Energy Data Book*; Oak Ridge National Laboratory: Oak Ridge, TN, USA, 2011.
21. Veijonen, K.; Järvinen, T.; Alakangas, E.; VTT Processes. *Biomass Co-firing—An Efficient Way to Reduce Greenhouse Gas Emissions*; European Bioenergy Networks: Jyväskylä, Finland, 2003.
22. Liu, Z.; Johnson, T.G.; Altman, I. The moderating role of biomass availability in biopower co-firing—A sensitivity analysis. *J. Clean. Prod.* **2016**, *135*, 523–532. [CrossRef]
23. Shang, Y.L. Resilient Multiscale Coordination Control against Adversarial Nodes. *Energies* **2018**, *11*, 1844. [CrossRef]

24. Shang, Y.L. Unveiling robustness and heterogeneity through percolation triggered by random-link breakdown. *Phys. Rev. E* **2014**, *90*. [CrossRef]
25. Hernández, J.J.; Lapuerta, M.; Monedero, E.; Pazo, A. Biomass quality control in power plants: Technical and economical implications. *Renew. Energy* **2018**, *115*, 908–916. [CrossRef]
26. Munby, D. The Assessment of Priorities in Public Expenditure. *Political Q.* **1968**, *39*, 375–383. [CrossRef]
27. Liu, Z.; Xu, A.; Zhao, T. Energy from Combustion of Rice Straw: Status and Challenges to China. *Energy Power Eng.* **2011**, *3*, 325–331. [CrossRef]
28. Kargbo, F.R.; Xing, J.; Zhang, Y. Property analysis and pretreatment of rice straw for energy use in grain drying: A review. *Agric. Biol. J. N. Am.* **2010**, *1*, 195–200. [CrossRef]
29. Bains, M.; Robinson, L. *Material Comparators for End-of-Waste Decisions—Materials for Fuels: Coal*; Environmental Agency: Bristol, UK, 2016.
30. Dai, W.; Chi, Y.; Lu, Z.; Wang, M.; Zhao, Y. Research on Reliability Assessment of Mechanical Equipment Based on the Performance-Feature Model. *Appl. Sci.* **2018**, *8*, 1619. [CrossRef]
31. Shang, Y.L. Localized recovery of complex networks against failure. *Sci. Rep.* **2016**, *6*. [CrossRef] [PubMed]

© 2019 by the authors. Licensee MDPI, Basel, Switzerland. This article is an open access article distributed under the terms and conditions of the Creative Commons Attribution (CC BY) license (http://creativecommons.org/licenses/by/4.0/).

Article

Optimization of Cooling Utility System with Continuous Self-Learning Performance Models

Ron-Hendrik Peesel [1,*], Florian Schlosser [1], Henning Meschede [1], Heiko Dunkelberg [1] and Timothy G. Walmsley [2]

1. Department for Sustainable Products and Processes (upp), University of Kassel, 34125 Kassel, Germany; schlosser@upp-kassel.de (F.S.); meschede@upp-kassel.de (H.M.); dunkelberg@upp-kassel.de (H.D.)
2. Sustainable Process Integration Laboratory—SPIL, NETME Centre, Faculty of Mechanical Engineering, Brno University of Technology—VUT Brno, Technická 2896/2, 616 69 Brno, Czech Republic; walmsley@fme.vutbr.cz
* Correspondence: peesel@upp-kassel.de

Received: 17 April 2019; Accepted: 16 May 2019; Published: 20 May 2019

Abstract: Prerequisite for an efficient cooling energy system is the knowledge and optimal combination of different operating conditions of individual compression and free cooling chillers. The performance of cooling systems depends on their part-load performance and their condensing temperature, which are often not continuously measured. Recorded energy data remain unused, and manufacturers' data differ from the real performance. For this purpose, manufacturer and real data are combined and continuously adapted to form part-load chiller models. This study applied a predictive optimization algorithm to calculate the optimal operating conditions of multiple chillers. A sprinkler tank offers the opportunity to store cold-water for later utilization. This potential is used to show the load shifting potential of the cooling system by using a variable electricity price as an input variable to the optimization. The set points from the optimization have been continuously adjusted throughout a dynamic simulation. A case study of a plastic processing company evaluates different scenarios against the status quo. Applying an optimal chiller sequencing and charging strategy of a sprinkler tank leads to electrical energy savings of up to 43%. Purchasing electricity on the EPEX SPOT market leads to additional costs savings of up to 17%. The total energy savings highly depend on the weather conditions and the prediction horizon.

Keywords: cooling system; mathematical optimization; machine learning; flexible control technology

1. Introduction

Generation of cold utility for industrial processes consumes about 9% of the net electricity in Germany, with similar trends in other countries, and an overall chiller energy efficiency ratio (EER) of less than two [1]. Cold utility systems encompass cooling towers (evaporative coolers), air coolers, and refrigeration. Air coolers incur near-zero running costs (i.e., "free cooling chiller"), cooling towers require water make-up and chemicals, while refrigeration is normally driven by electricity with substantial running expenses that are approximately one order of magnitude higher than a cooling tower system. Maximizing the use of the lowest-cost utility saves running costs and high-quality energy, reducing environmental burdens. The ongoing digitization of many processes in the industry enables the potential to automate and optimize the operation of cold utility units. Coupled with an efficient and flexible utility system, a site can achieve substantially highier levels of system efficiency. Critical to this vision is linking sectors—e.g., process heating and cooling and electricity grid—combined with technologies, processes, and control systems to collect and analyze the data and implement complex energy optimization methods. Prerequisite for an efficient cooling system is the knowledge and the optimal combination of different operating and performance conditions of individual chillers.

The performance of cooling systems depends on their part-load performance and their condensing temperature, which are often overlooked and not continuously measured. Manufacturer performance data often substantially differ from in-plant performance [2].

The optimized operation of refrigeration units in combination with a complex control algorithm can improve the energy efficiency of cooling systems when implemented. Clausen et al. [3] show that the coupling of optimization and simulation of parcel sorting plants lead to efficiency increases. Dynamic simulations support both the validation of efficiency concepts prior to implementation and model-based optimization. Augenstein [4] discusses that a linear optimization of the chiller deployment strategy reduces power consumption by 11%. At the same time, there is also optimization potential for specifically adapted expert control strategies in the sequencing of refrigeration systems. Rule-based controls show a good performance for specific systems under normal conditions [5]. Changes in boundary conditions lead to nonlinear behavior of equipment, and equipment degradation can shift the optimum operating set-points and control of a plant over its life cycle [6].

Several approaches are reported in the literature to optimize cooling system operations. Hovgaard et al. [7] use the example of a supermarket refrigeration supply to show that electrical energy savings of 9%–32% are possible with the help of model predictive control (MPC) with a thermal storage system, predicting the changing load and taking into account varying energy prices. Braun [8] reduced the energy consumption with a component-based and system-based optimization algorithm and correlated the power consumption of the chillers, condenser pump, and cooling tower fans with a quadratic function. Olson et al. [9] describe a mathematical approach to the distribution of loads over several chillers to minimize energy costs. Many different methods for optimal chiller sequencing control have been investigated. Some examples are the branch-and-bound method [10] dynamic programming [11], gradient method [12], genetic algorithms [13], simulated annealing [14], and the particle swarm optimization [15]. These methods generally perform well in a simulation environment.

One of the challenges with practical implementation is system dynamics leading to exponentially increased computational demand. To reduce computing time for dynamic optimization, Powell et al. split up the problem solving into a dynamic one for the total load at each time step and a static one for the optimal chiller loading. For a 24 h time horizon, the energy savings are around 9.4% and the energy costs savings are 17.4%. [16] A literature review reveals that the energy-saving potential varies between 5% [17] and 40% [18] depending on the case study.

Another important factor is the availability of cold storage for a cooling system. Kapoor et al. [19] show that implementing a 15,000 m^3 cold-water storage tank could result in a cost saving of 13%. Cold energy storage presents an opportunity to reduce costs by shifting loads. Cole et al. [20] use an MPC with thermal energy storage and variable electricity prices for peak and off-peak hours to reduce the electrical energy consumption and the operating costs for the cooling system of one building. In the best case, the annual costs are reduced by around 42% while the total energy usage increases by more than 6%. Kircher and Zhang [21] investigate the cooling strategy of an office building with ice and chilled water storages using an MPC and demand response incentives. The optimization horizon is limited to one day, and the non-linear part-load performance of the chillers is neglected. Ma et al. [22] show that an MPC implementation for a building cooling system considering a storage tank, cooling towers, condensing, and evaporation temperature increases the plant coefficient of performance (COP) by around 19%. A direct charging and discharging of the storage by the cooling tower is not analyzed. With a similar idea to the current study, Soler et al. [23] present a performance optimization of a district cooling network with eight chillers and a thermal storage tank for a one day period, assuming the use of chillers according to their efficiency rating and best performance in full load. For minimizing shut-downs and start-ups, no-increase/no-decrease constraints are added.

Many production sites use cold-water storage tanks to smooth the cooling load profile but not for load management and load shifting. A sprinkler tank contains a high volume of water for use in emergencies such as extinguishing fires. There is an opportunity to utilize the sprinkler tank for cold-water storage with load shifting while maintaining a minimum water level for emergency operations. Using a free cooling chiller (i.e., cooling by air or cooling water from a cooling tower) offers the opportunity to provide cold water with a higher energy efficiency compared to a compression chilling machine. However, if ambient temperatures equal or exceed the cold water set point temperature, the free cooling chiller is no longer effective. Extending the operation hours of the free cooling chiller and reducing compression chiller's could save a significant amount of electrical energy. This needs a complex pre-charge and discharge control for the cold-water storages. To this end, a mixed integer linear optimization (MILP) algorithm is advantageous.

In practical applications, the performance is highly dependent on the accuracy of the performance models and the effectiveness of the convergence of the optimization algorithm. As demonstrated in the literature review, a current gap is a lack of properly accounting for in-plant chiller performance and considerations for how performance changes through its lifetime. This weakness can be compensated by implementing continuous self-learning performance models and linking the optimization algorithm and dynamic simulations including part-load (off-design).

The aim of this study is to develop and apply a predictive optimization algorithm for multiple compression chillers, a free cooling chiller, and a sprinkler tank as a cold-water storage tank considering part-load system operation under a dynamic simulation to improve overall energy cost efficiency. The algorithm extends the work of Peesel et al. [2], who introduced a predictive simulation-based optimization for a food processing factory, calculating electricity savings of up to 23% per year. The novelty of this study is the evaluation of the interdependency between part-load ratio, variable electricity prices, and a sprinkler tank using data from an industrial energy monitoring system. Furthermore, the influence of the prediction horizon and the ambient temperature on the optimization result is evaluated. The modeling of the utilities, the load profile of cold water, and the framework conditions are based on the energy data of a plastics processing company. For the performance curves of the utility, a continuous self-learning model is used. Additionally, the initial states of the optimization are reset by the simulation results of the previous timestep. The combination of the linear optimization and the dynamic simulation gains the advantages of both methods. This results in more realistic and accurate results for the energy and costs-saving potential of the predictive simulation-based optimization. The developed optimization algorithm is applied to case studies of plastic manufacturers in Germany and Spain to demonstrate its potential.

The remainder of the article is structured as follows. Section 2 presents the general approach of the optimization, the data pre-processing, modeling of the performance curve, and the simulation model. The case study of a plastic processing company is introduced in Section 3. This is followed by the explanation of the results, including the influences of the prediction time horizon, ambient temperature, and variable electricity price, in Section 4. The conclusion is presented in Section 5 together with an outlook for further research.

2. Method and Simulation Model

This section describes the overall method and simulation modeling approach of the systems using a model-based predictive optimization considering the individual operating performance of the refrigeration supply system by continuously self-learning performance curve models. Figure 1 shows the general optimization method with the energy data from an industrial plant and the self-learning performance curve models. The scheme of the simulation-based optimization is also visualised. The method extends the work of Peesel et al. [24] and comprises four steps. The first step includes collection and pre-processing of the data. The input conditions include the ambient temperature, the residual load, and the electricity price for the production site. At this point, a forecasting model for the weather or the cooling load can be integrated. Weather forecasts are common, whereas load forecasting has been attempted, for example, by statistical methods [25], artificial neural networks [26], and support-vector machines [27]. The energy monitoring system collects the operating data of the cold utility system for all production states. The second step is the modeling of the continuous self-learning performance models for the utilities. This is followed by the predictive optimization for the sequencing of the different chillers. The last step is the simulation of the complete utility system. The different steps are explained in more details in the subsequent subsections.

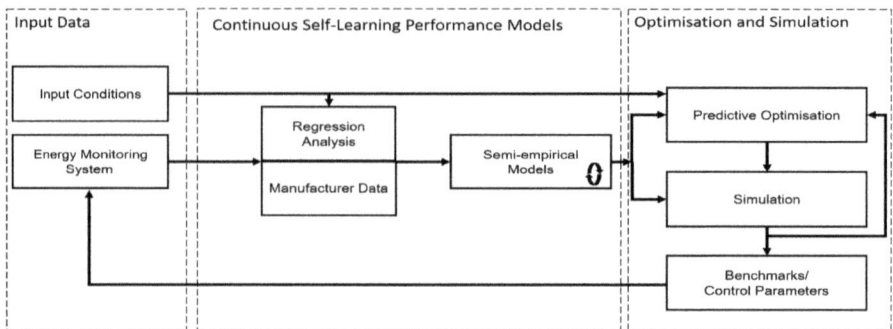

Figure 1. Simulation-based optimization scheme.

2.1. Data Pre-Processing

Energy and performance data of an energy monitoring system often contain some obvious outliers and incorrect values. This requires pre-processing the data to remove these unusable values and extract the reliable data, such as for cold water temperature (T_{cold}), condensing temperature (T_{cond}), part-load ratio (PLR), energy efficiency ratio (EER), electrical power (P_{el}), and thermal energy flow ($\dot{Q}_{cool}$). The definition of the EER and the PLR are defined by Equations (1) and (2).

$$\text{EER} = \frac{\dot{Q}_{cool}}{P_{el,\,CC}} \tag{1}$$

$$\text{PLR} = \frac{\dot{Q}_{cool}}{\dot{Q}_{cool,\,max}(T_{cond})} \tag{2}$$

The data pre-processing is split into two steps—data preparation based on system behavior and the application of statistical methods. The binary operating state is the variable that helps filter data for times, such as when plants are shut-down. These are up to 50% of all values, depending on the time span of the data such as weekly, monthly, or annual data records. Data during plant shut-downs measure electrical power values <1 kW for the chillers in standby, while cooling load data are often erroneous due to thermal inertia in the fluid. By including these data in the characteristic curve generation, an extremely high EER value results that misrepresents the actual situation. A physical method to remove outliers of the EER is the comparison of the ERR calculated from the measurement data with half of the ideal Carnot efficiency. Other important logical criteria are both the chillers' total electrical power consumption and their transferred cooling capacity. Measured values for these variables must not assume negative values and be lower than the nominal electrical power and cooling capacity at the lowest condensing temperature stated by the manufacturer.

Statistical methods are used to identify incorrect values that pass through the analysis of system behavior. The filtering of values is performed for values smaller than a percentile (1% or 5%) or greater than a given percentile from 95% or 99%. This method is also applied to the rate of change of the recorded measurements to filter out extreme changes, especially for values of the electrical power. A special method is a classification from the minimum to the maximum condensing temperature into one Kelvin class. For this transformation, a matrix approach is used. The direction vector with the largest standard deviation is determined for each classified data matrix for PLR and EER. Outliers are determined in linear or non-linear data point distributions. The vector of the largest scatter indicates a transformed coordinate system for the size pairs of EER and PLR. In this coordinate system, further outliers are identified by the interquartile distance (IQR). In statistical studies, 1.5 times the IQR is often used to identify outliers. The analysis of energy monitoring data of six different compression chillers in three sites by the authors shows that this must be selected much smaller to filter out data. In this work, 0.5 times the IQR is used. The distribution of the data points in the respective classes is decisive for filtering out the values. The various methods are run through one after the other, starting with the method that takes the least time to calculate and filters out the most values. The order is important to minimize the total computation time. After the data pre-processing, the scattering and the number of large outliers for different measured variables were considerably reduced, shifting the median mainly due to the removal of the measured data in the switched off state. The change in the measured values after data pre-processing is shown in the following Figure 2.

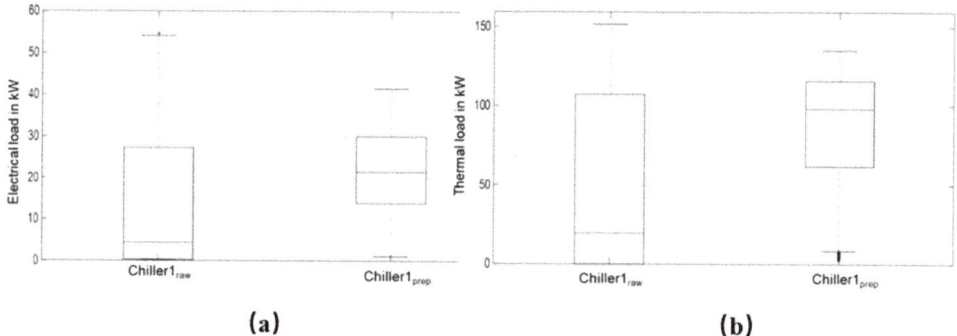

Figure 2. (a) Data overview before and after data pre-processing for the electrical load of chiller 1; (b) data overview before and after data pre-processing for the thermal load of chiller 1.

2.2. Continuous Self-Learning Performance Modeling

After the data pre-processing, the modeling of the load performance curve models is performed. For the free cooling chiller, a second order linear regression model is used. Since the data pre-processing shows that measurement data are available in sufficient quantity and for all operating states, only interpolations are necessary. The following Equation (3) with the regression coefficients B_x defines the model:

$$EER_{FC} = B_0 + B_1 \cdot PLR_{FC} + B_2 \cdot T_{amb} + B_3 \cdot PLR_{FC}^2 + B_4 \cdot PLR_{FC} \cdot T_{amb} + B_5 \cdot T_{amb}^2 \quad (3)$$

The EER_{FC} of the free cooling chiller is dependent on the PLR and the ambient temperature (T_{amb}) for a fixed cold-water supply temperature. For the optimization, the semi-empirical regression chiller model—the Gordon–Ng universal model [28]—based on the evaporator outlet water temperature is used. Equation (4) presents the Gordon–Ng model,

$$\frac{1}{EER} = -1 + \frac{T_{cond}}{T_{cold}} + \frac{-A_0 + A_1 T_{cond} - A_2\left(\frac{T_{cond}}{T_{cold}}\right)}{\dot{Q}_{cool}} \quad (4)$$

where A_0, A_1, A_2 are regression coefficients and $\dot{Q}_{cool}$ is the cooling load of the chiller. In this study, the cold-water temperature (T_{cold}) is kept constant. T_{cond} is the ingoing condensing temperature of the compression chiller.

Former studies show that even two structurally identical compression chillers can perform differently in the same location [29]. This results in a need for empirical models for each utility operation. In operating areas with a significant quantity of energy data, empirical models describe the system performance. In other areas, extrapolating semi-empirical models is more accurate. The linking of physical and empirical modeling ensures an acceptable accuracy in both fields of interpolating and extrapolating. For the modeling of the utility, manufacturer and performance data are combined and continuously adapted to form part-load performance curve models. The existing and new data points are weighted differently in the continuous self-learning model. In considering changes in utility performance, recent data are weighted higher. If the characteristic curves are regularly adjusted over a longer period, the deviation between the previous and new characteristic curve decreases. The disadvantage of this method is that an irregularity in the measurement system or in the utilities leads to significant deviations.

For the validation of the Gordon–Ng universal model, the dataset of 1652 data points is split into a training set (82%) and a validation set (18%). The overall deviation of the EER is 8.3% and the root-mean-square error (RMSE) is 0.355. This is because of a low quantity of data for low and high PLR values. Lee et al. [30] show slightly lower values for the validation of the Gordon–Ng universal model.

2.3. Mathematical Model

The optimization is formulated as a mixed-integer linear programming (MILP) problem. Since the control parameters of the chillers are discrete, the solutions must be integers. The goal of the optimization is to satisfy the cold-water demand by multiple chilling systems for the next N_T timesteps at the lowest energy costs. The sprinkler tank offers the opportunity to store cold water. This potential is used to show the load shifting potential of the cooling system by using a variable electricity price as an input parameter of the optimization. The following equations describe the minimization problem.

$$\min \sum_{t=1}^{N_T} \sum_{n=1}^{N_{cc}} \sum_{st=1}^{N_{St}} (costs_{t,n,st} \cdot X_{t,n,st}) + SC_{t,n} \cdot Z_{t,n} \quad (5)$$

$$\text{subject to } X_{t,n,st} \in \{0,1\}, Z_{t,n} \in \{0,1\} \quad (6)$$

The different part-load operating states (St) are between 0% and 100%, while the smallest step size is 1%. The active chilling machines set the lower bound of the operating states and the step size. The cost term ($costs_{t,n,st}$) includes the costs for each part-load operating state (St) of the chillers. The costs are calculated by multiplying the electrical power consumption of the chiller by the electricity costs for the current time step. The electrical power consumption is based on the regression models—Equation (3) or (4). The number of chillers is defined by N_{cc}. The number of timesteps (N_T) describes the prediction horizon of the optimization in hours. For example, an N_T of 3 by a step size of 3600 s means that the optimal load distribution is found for the sum of loads of the current hour and the following two hours. In this paper, the prediction horizon is varied between 6 and 48 timesteps. Additionally, the starting costs (SC) consider the extra energy demand when starting a machine. SC also helps to prevent an unnecessary starting and stopping of the machines if the gain in efficiency is very low. The active states of the chillers are described by $X_{t,n,st}$, while $Z_{t,n}$ is the binary on/off-variable.

$$T_{res} \cdot \sum_{t=2}^{N_T} \sum_{n=1}^{N_{cc}} \sum_{st=1}^{N_{St}} (CC_{t,n,st} \cdot X_{t,n,st}) \leq Q_{pred} + Q_{plus} \quad (7)$$

$$T_{res} \cdot \sum_{t=2}^{N_T} \sum_{n=1}^{N_{cc}} \sum_{st=1}^{N_{St}} (CC_{t,n,st} \cdot X_{t,n,st}) \geq Q_{pred} - Q_{minus} \quad (8)$$

$$Q_{plus} = m_{Treturn} \cdot c_{p,water} \cdot (T_{return} - T_{cold}) \quad (9)$$

$$Q_{minus} = m_{Tcold} \cdot c_{p,water} \cdot (T_{cold} - T_{return}) \quad (10)$$

Equations (7) and (8) describe the constraints for the lower and upper bounds of the optimization. The total cooling supply by the chillers must be lower or equal to the total cooling demand of all timesteps (Q_{pred}) and the cooling capacity of the sprinkler tank (Q_{plus}). CC is the cooling capacity of the chillers. The cooling capacity of the sprinkler tank Q_{plus} is defined by the product of the mass of water (m) at the cold water return temperature T_{return}, the specific heat capacity of water $c_{p,water}$, and the temperature difference of the cold water T_{cold} and returning water T_{return}. The current potential cooling energy of sprinkler tank Q_{minus} is calculated analogously. By varying the mass of water at the different temperatures, the cooling capacity of the thermal storage can be changed. Additionally, the total supply cannot be lower than the total cooling demand minus the current cooling power of the sprinkler tank. For solving the mixed-integer problem, the Gurobi Optimization 8.1 software is used [31].

2.4. Simulation Model

The results of the optimization are the control parameters for each chilling machine in the current timestep. These parameters represent the operating schedule for the simulation. A dynamic simulation continuously validates the set points in the optimization based on the response and feedback effects from the system. These include thermal inertia of the cooling system, start-up characteristics, as well as the physically modeled heat losses of the storage tank. There is also a difference in the start-up characteristics for a compression chilling machine in the optimization and the simulation. In the optimization, the full cooling capacity of the chiller is immediately available due to the linear programming, whereas, in the simulation, the chiller needs a few minutes to ramp to full capacity. This is modeled by a first-order delay element. Comprehensive measurements have shown that the ramp time can vary from one to five minutes or even longer for larger machines. During starting time, the cooling capacity and the energy efficiency of the compression chillers is reduced according to the part-load of the load performance curve. For part-load values below 20% of the full capacity, the EER is below two. This compensates overestimated EER by the linear programming and integrates lower efficiency in starting and shutting down times in the simulation. Therefore, the simulation compensates the limitation of the linear optimization and verifies the operating schedule. Additionally, the complex heat loss mechanisms of a storage tank are modeled more accurately in a simulation than in the linear model. The applied storage model uses the Multi-Node model of the Matlab/Simulink®CARNOT Blockset. [32] For the calculation of the Multi-Node storage model, the storage volume is divided along its height into n nodes. In each layer, an ideal mixing is assumed, i.e., the temperature is constant in the layer element in both the radial and vertical directions. For each layer and each simulation time step, the balance equation, Equation (11), of the energy transport is solved according to the first law of thermodynamics.

$$\frac{dU}{dt} = \dot{H} + \dot{Q}_{vertical} + \dot{Q}_{loss} \tag{11}$$

The change in the internal energy, U, describes the change in the temperature, T, of the individual layers as a function of entering and leaving enthalpy streams $\dot{H}$ caused by the piston flow in the tank and charging mass flows, direct vertical heat conduction effects $\dot{Q}_{vertical}$ through the layers, and the heat losses $\dot{Q}_{loss}$ through the container wall. To reduce computational time, the nodes in the model are limited to two—one for the cold-water supply temperature and one for the cold water return temperature.

This method eliminates unrealistic starting points. The initial states of the optimization are reset by the simulation results of the previous timestep. The combination of the linear optimization and the dynamic simulation enables the advantages of both methods. This results in more realistic and accurate results for the energy and costs saving potential of the predictive simulation-based optimization.

3. Plastic Processing Case Study

The evaluation of the optimization and the simulation is done by a case study on two plastic processing companies that produce injection-molded parts for the food sector. The purpose of the case study is to establish the level of possible energy cost savings for the considered site. The case study can provide motivation to the site to implement the new optimization approach as well as others to follow a similar procedure. This requires high hygiene standards for the production hall. One is located in Germany and the other in Spain. The variation of the location quantifies the influence of the ambient temperature on the results of the optimization. The injection-molding machines are cooled continuously by two different cooling circuits. The mold-cooling circuit is set at a temperature level of about 14 °C and the machine cooling at about 30 °C. The relatively high process cooling temperature of 14 °C offers the opportunity to integrate free cooling chillers as well as multiple compression chillers. The refrigeration of the molding circuit is carried out by two air-cooled compression chiller machines and a free cooling chiller, which is also used for winter relief. Cooling towers provide the cooling to machine cooling of the plastic processing machines. Moreover, the company has a 900 m^3 sprinkler tank that can be used as a cold-water storage tank at a temperature level of 14 °C. The plastic processing companies often have large warehouses. Fire safety regulations require quick access to a high volume of extinguishing water. Because of the multiple operational states, the predictive simulations-based optimization is well-suited for the mold cooling circuit in the plastic processing industry.

In this study, the prediction horizon is varied between 6 and 48 h and the step size is set to 1 h. The N_{CC} is 3 and the N_{ST} varies from 4 for the compression chillers to 10 for the free cooling chiller. The cooling demand and the ambient temperature are based on the measured data for the year 2017. For this case study, it is assumed that the forecast for the weather and the cooling demand is perfectly known in each time step. The cooling demand varies between 0 and 190 kW. The simulation time (T_{res}) is equal to the step size. The maximum cooling energy stored in the cooling tank is fixed by the maximum temperature difference of 3 K and the 900 m^3 of water, which results in 3140 kWh. The costs for the optimization are calculated by the product of the current energy consumption of the chillers and the current spot market price for electricity. The prices vary between 75 and 200 €/MWh. The fluctuating price for electricity is also compared to a fixed price scenario. The fixed electricity price is 160 €/MWh. Table 1 gives an overview of the installed cooling utilities and Figure 3 shows the scheme of the cooling system of the plastic processing company.

Table 1. Utility system of the plastic processing company.

Utility	Quantity	Temperature Level	Cooling Capacity	Number of Stages	Lower Bound	Step Size
Air-cooled scroll chiller	2	14 °C	200 kW	4	25%	25%
Free cooling chiller	1	14 °C	200 kW	10	10%	10%
Cooling towers	2	30 °C	250 kW	4	25%	25%
Sprinkler tank	1	14 °C	3140 kWh	-	-	-

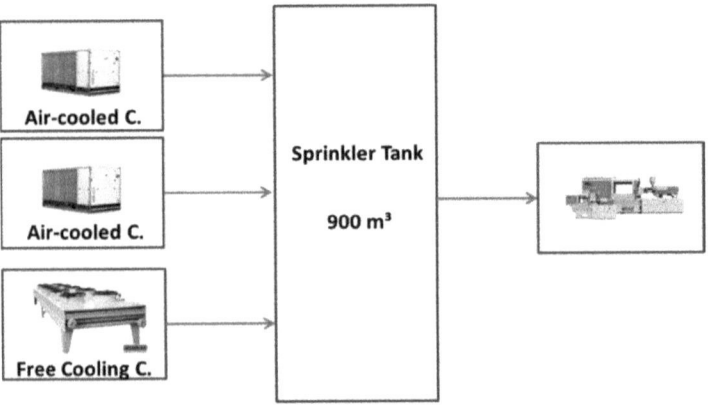

Figure 3. Scheme of cooling system plastic processing company.

The following Figure 4 visualizes the partial load performance curve of the air-cooled chiller, according to the Gordon–Ng model, after data processing. The lower the ambient temperature, the higher the EER because of the lower condensing temperature of the chiller and less energy consumed by the fan of the condenser.

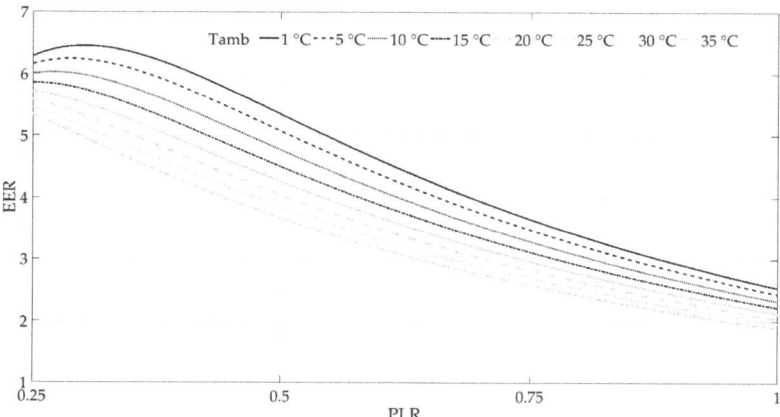

Figure 4. Part-load performance curve for air-cooled compression chiller.

4. Results

The results of the predictive simulation-based optimization are compared in five different scenarios. The first scenario is the reference case, which is based on the energy monitoring data of the plastic processing company. In the reference case, the company is located in Kassel, Germany, and has a fixed price for electricity. In the second scenario, the company is also located in Kassel, and the predictive simulation-based optimization is applied to it. To quantify the influence of the prediction horizon, it is varied between 6 and 48 h. To quantify the influence of the ambient temperature on the predictive simulation-based optimization, scenario 3 is identical to scenario 2, except the company's location is changed from Kassel to Madrid, Spain, and the prediction horizon is set to 48 h. Further, scenario 4 shows the influence of an increased ambient temperature of Kassel by 5 K. Scenario 5 is similar to the second one, except the company purchases its electricity on the EPEX SPOT market and has a variable price. The focus of this scenario is to analyze the influence that a variable electricity price has on the operating schedule of the different chillers. In scenario 6, the company is in Madrid and purchases its electricity from the EPEX SPOT market. Table 2 shows an overview of the different scenarios and the varied parameters of the simulations. The cooling demand for the molding cooling circuit is in all scenarios identically, since hygiene standard forbids any window opening, and therefore the production hall is air-conditioned to a required temperature of 21 °C.

Table 2. Overview of simulated scenarios and parameters.

Scenario	Location	Prediction Horizon (h)	Electricity Price	Optimization
1. Reference case	Kassel	None	Fixed 16 ct/kWh	No
2. Prediction horizon	Kassel	6–48	Fixed 16 ct/kWh	Yes
3. Ambient temperature	Madrid	48	Fixed 16 ct/kWh	Yes
4. Ambient temperature + 5	Kassel + 5 K	48	Fixed 16 ct/kWh	Yes
5. Flexible electricity price	Kassel	48	EPEX SPOT market	Yes
6. Ambient temperature and flexible electricity price	Madrid	48	EPEX SPOT market	Yes

4.1. Prediction Horizon

Figure 5 shows the different operating hours of the three chillers for a prediction horizon n between 6 and 48 h. The bottom bar represents the operating hours of the free cooling chiller. The ascending bars show operating hours of the first compression chiller for each stage and the top bar refers to the second compression chiller. Stage 4 of chiller one and stages 2–4 of chiller two are not used in any of the four cases.

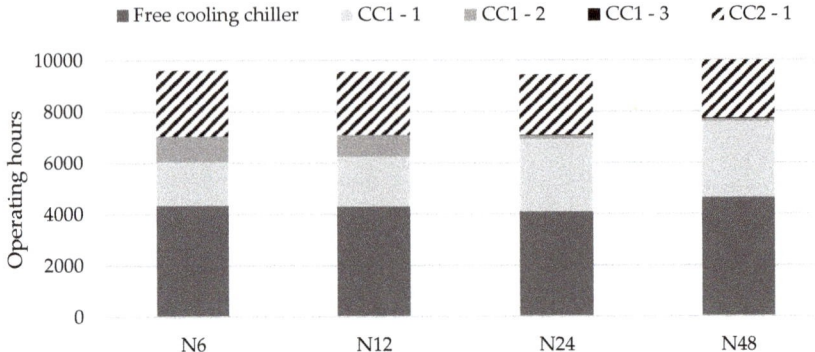

Figure 5. Operating hours of chillers in different stages for varying prediction horizons.

It illustrates that the increase in hours of the prediction horizon leads to a rise of the operating hours of the free cooling chiller and increases the operating hours of the two compressions chillers in stage one. The influence on the operating schedule is discernible.

Due to the better efficiency of the free cooling chiller compared to the compression chillers, the electrical energy consumption decreases proportionally to its operating hours. In comparison, the reference case with standard control is also visualized. Figure 6 illustrates the total energy consumption for the four cases, with differing prediction horizons in scenario 2, the reference scenario, as well as the average EER for the year. The results show that the optimization reduces electrical energy consumption by around 38% (N6–N24). A prediction horizon of $n = 48$ decreases the electrical energy demand by another 5% because of the better charging and discharging strategy of the sprinkler tank.

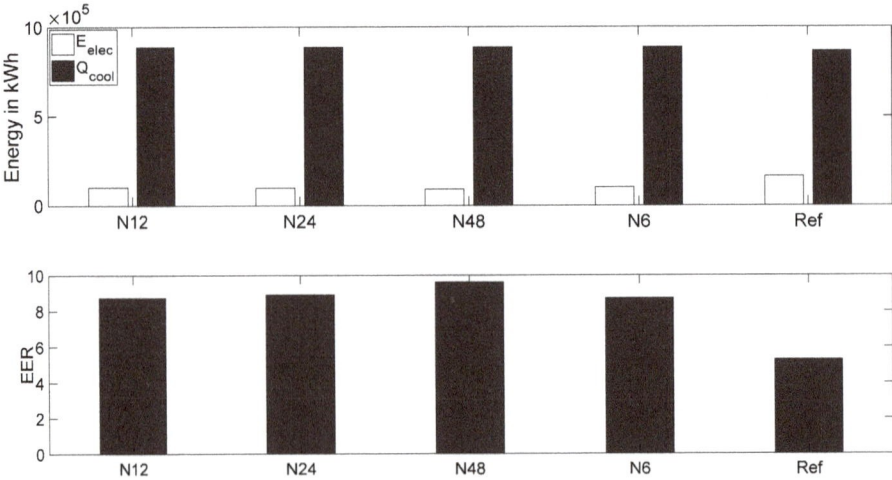

Figure 6. Energy savings and energy efficiency ratio (EER) comparison for varying prediction horizons.

The difference in energy efficiency and the different operating hours of the free cooling chiller and the stages of the compression chillers result from the charging and discharging strategies based on the different prediction horizons. Figure 7 shows the charge level of the storage tank for the low supply temperature at 14 °C (black line) depending on the time of the year in hours for the four different prediction horizons. The charge level of the high temperature water of 17 °C is the mirror image of it. A prediction horizon of 6 or 12 h results in a minor usage of the energy storage potential of the tank. In the case with a prediction horizon of 24 h, the storage is charged and discharged over 50% of its full capacity in the first and fourth quarter of the year. The full capacity of the storage is used in the case with a prediction horizon of 48 h.

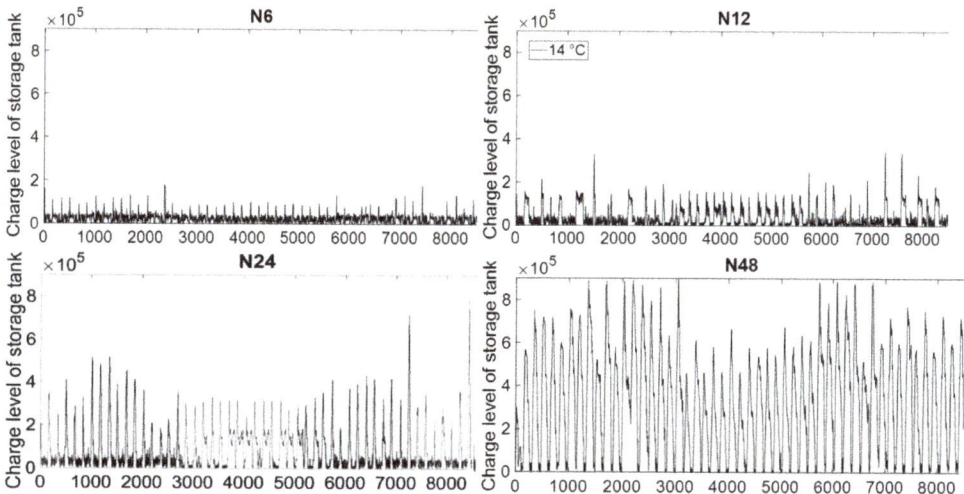

Figure 7. Charge level of the storage tank for varying prediction horizons.

The complete charging and discharging of the tank increase the running hours of the free cooling chillers (Figure 5). With a long prediction horizon, the tank is pre-charged when ambient temperatures fall below 11 °C and discharged in times of high ambient temperatures, minimizing the load on the compression chillers and even placing them in standby mode. This strategy leads to the extra 5% saving potential by extending the prediction horizon to 48 hours.

4.2. Influence of Ambient Temperature

The efficiency of the free cooling chiller and the air-cooled compression chillers depend on the ambient temperature. Lower ambient temperatures allow lower condensing temperatures, resulting in better efficiency. If the ambient temperature is higher than 11 °C, the free cooling chiller does not provide the required cold-water supply temperature. Both effects have an influence on the energy-saving potential of the predictive simulation-based optimization. Hence, in scenario 3, the plastic processing company is moved to Madrid, Spain, to have a different ambient temperature profile. In scenario 4, the German temperature profile is lifted by 5 K. Figure 8 shows the total energy consumption and EER for one year in comparison to the reference case and scenario 2. All optimizations are based on a prediction horizon of 48 h, since the previous analyses in Section 4.1 conclude that this is the best horizon to maximize energy savings.

The change in the ambient temperature reduces the energy-saving potential for Madrid to 23% and, for scenario 4, to 19% compared to the reference case. Because of reduced running hours of the free cooling chiller in scenarios 3 and 4, the energy efficiency is significantly lower compared to scenario 2. However, the analysis of the different ambient temperatures also shows that the predictive simulation-based optimization together with the sprinkler tank reduces the electrical energy consumption in warmer regions. Although the energy-saving potential is reduced for warmer ambient temperatures, there is still a significant reduction compared to the reference case.

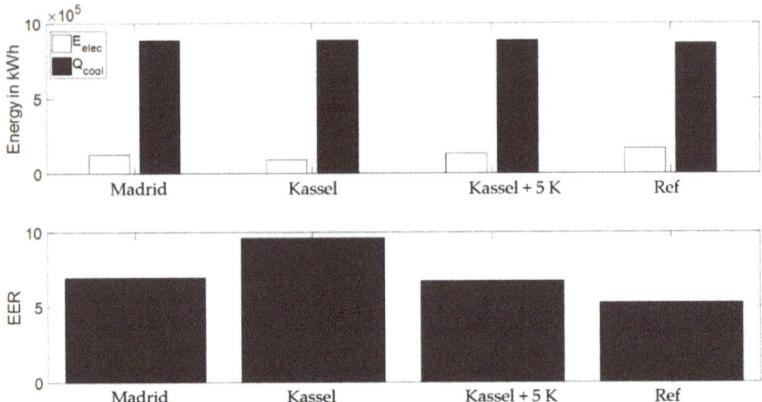

Figure 8. Energy savings and EER comparison for different ambient temperatures.

4.3. Variable Electricity Prices

In the previous scenarios, the electricity price for the plastic processing company is fixed to 16 euro cent/kWh. In scenarios 5 and 6, the electricity procurement is via the day-ahead exchange market in Germany. All taxes, costs, and the German renewable energy levy are included in the model. To benefit from purchasing electricity with a variable price, it is required to shift loads. The sprinkler tank enables the cold utility system to store cold-water in times of little or no demand. This allows the optimization algorithm to start chillers in times of low electricity prices and stop the machines in times of high prices. This increases the optimization options. Figure 9 shows the results of the simulation for the variable cost option in comparison to the fixed price and the reference case. On the left y axis, the thermal and electrical energy consumptions are visualized, and the costs are shown on the right y axis.

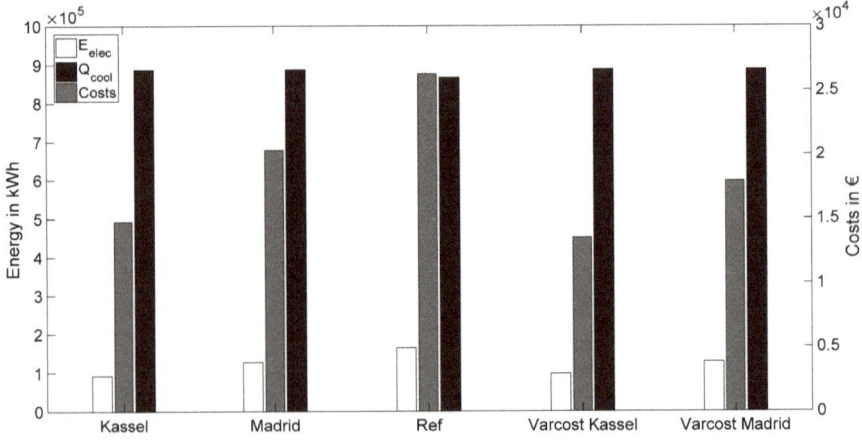

Figure 9. Comparison of variable and fixed price optimization.

Under scenarios 5 and 6, the electrical energy consumption for Kassel increases by 7% and, for Madrid, by 0.4%, but the total energy costs for the cooling systems are reduced. For the plastic processing company in Kassel, the electricity procurement via the spot market reduces the costs by around 9.3%. For the site in Madrid, the costs reduction is about 12%. The increased electrical energy consumption is due to the reduction of the operating hours of the free cooler and the simultaneous increase in the use of the compression chiller. This leads to a decreased overall efficiency of the system. However, the lower efficiency is compensated by the lower prices for the electrical energy on the spot market in comparison to a fixed electricity price. Figure 10 shows the comparison of the charge level of the sprinkler tanks for fixed and variable price scenarios.

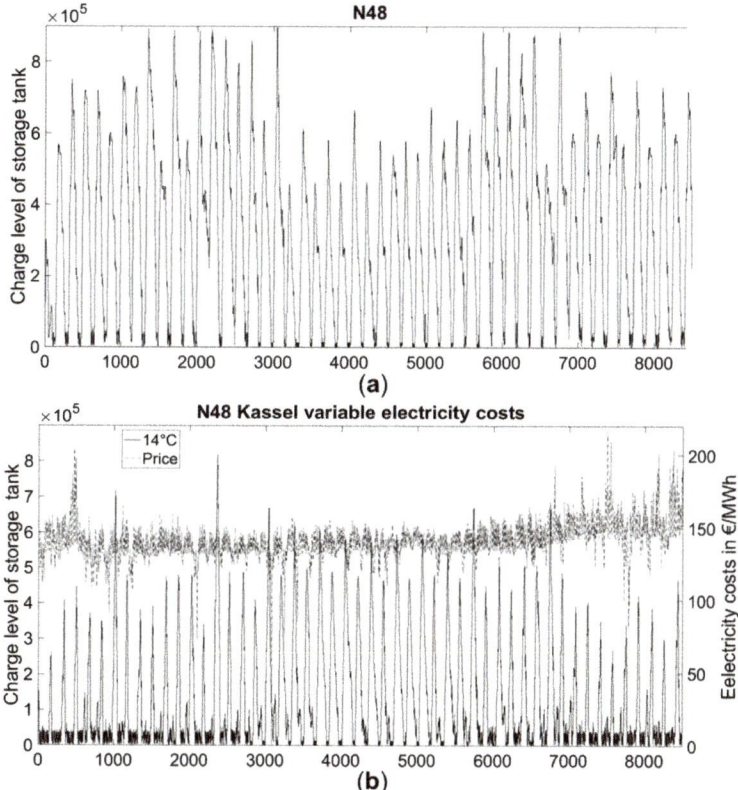

Figure 10. (a) Charge level of storage tank for a fixed electricity price; (b) charge level of storage tank for variable price scenarios.

In Figure 10b, the electricity price is shown on the right y axis. In comparison to the fixed price scenario, the storage tank is never fully charged in the variable price scenario. Although both prediction horizons are equal, the results of the optimization differ for minimizing the costs. The different charging and discharging strategies are the reason for different electrical energy consumption and the different total energy costs.

4.4. Results Summary

Table 3 summarizes the results for the comparison of the different prediction horizons, the influence of the ambient temperature, and the effect of purchasing electricity for a variable price. The savings in percentage for energy and costs use the basis of the reference case.

Table 3. Overview of optimization results.

Scenario	Electrical Energy in kWh	Energy Savings in %	Energy Costs in €	Cost Savings in %
1. Reference case	164,062	0	26,250	0
2.1 Prediction horizon N6	101,760	38.0	16,282	38.0
2.2 Prediction horizon N12	101,540	38.1	16,245	38.1
2.3 Prediction horizon N24	99,485	39.4	15,918	39.4
2.4 Prediction horizon N48	92,248	43.7	14,760	43.8
3. Ambient temperature	127,270	22.4	20,363	22.4
4. Ambient temperature + 5	131,740	19.7	21,078	19.7
5. Flexible electricity price	98,697	39.8	13,727	47.7
6. Ambient temperature and flexible electricity price	127,775	22.1	17,910	31.8

5. Conclusions and Outlook

The study developed a predictive simulation-based optimization of a cooling system with continuous self-learning performance curve models that saves electrical energy by optimal charging and discharging of the sprinkler tank. The optimization reduced electrical energy costs via electricity procurement on the EPEX SPOT market. Results for the case study concluded that the control strategy of the optimization together with the installation of cold-water storage tanks with a significant volume and a free cooling chiller could save over 43% of the electrical energy in comparison to the reference case. To utilize the full potential of the storage tank, a prediction horizon of 48 h was necessary. In comparison, a prediction horizon of 6 h saved up to 38% of electricity use. The difference in the prediction horizon had minimal impact on the computation time. A longer prediction horizon increased the energy savings, but the available thermal storage capacity limits it. In further analyses, the influence of the modeling error and weather forecast changes on the optimal operational strategies as well as the potential for exergy and entropy considerations may be investigated.

The study further investigated the effects of ambient temperature due to different plant locations as well as a variable electricity spot price. The ambient temperature profile had a significant influence on the energy-saving potential. The higher the average ambient temperature, the lower the energy saving—however, the optimization strategy still saved around 20% in warmer regions. Electricity procurement via the EPEX SPOT market led to a slight increase in electrical energy consumption but saved 9.3% to 12% of the energy costs. For the analyzed plastic processing company, the differences in the electricity price were low but still influenced the cooling utility system. If the company should economically support the electricity grid, the price differences need to be higher or include extra incentives. For example, a dynamic tax on electricity based on the instantaneous share of renewable electricity generation presents a chance to increase the differences in the variable prices and make demand-side management more attractive for medium-sized companies.

Author Contributions: Responsible for the conceptualization, R.-H.P. and F.S.; methodology, R.-H.P. and H.M.; validation, R.-H.P.; investigation, R.-H.P. and H.D.; data curation, H.D.; writing—original draft preparation, R.-H.P.; writing—review and editing, F.S., H.M. and T.G.W.; visualization, R.-H.P.; supervision, T.G.W.; funding acquisition, R.-H.P.

Funding: This research received no external funding.

Acknowledgments: The authors gratefully acknowledge the financial support of the Rud. Otto-Meyer-Umwelt-Stiftung and the support of the EU project "Sustainable Process Integration Laboratory – SPIL", project No. CZ.02.1.01/0.0/0.0/15_003/0000456 funded by EU "CZ Operational Programme Research and Development, Education", Priority 1: Strengthening capacity for quality research.

Conflicts of Interest: The authors declare no conflict of interest.

Nomenclature

A_x	Regression coefficients for Gordon–Ng model
B_x	Regression coefficients for free cooling chiller
CC	Cooling capacity
$c_{p,water}$	Specific heat capacity of water
E	Energy
EER	Energy efficiency ratio
FC	Free cooling chiller
$\dot{H}$	Enthalpy flow
MPC	Model predictive control
m_{St}	Mass of water in storage tank
m_{Tcold}	Mass of water in storage tank at cold water temperature of chiller
$M_{Treturn}$	Mass of water in storage tank at cold water return temperature of chiller
N_{CC}	Number of chilling machines
N_T	Timesteps
N_{ST}	Maximum states of chilling machine
PLR	Part-load ratio
$P_{el,CC}$	Electric power of chiller
$\dot{Q}_{cool}$	Cooling load
$\dot{Q}_{vertical}$	Vertical heat conduction
$\dot{Q}_{losses}$	Heat input by storage wall
Q_{minus}	Current cooling energy of sprinkler tank
Q_{pred}	Total cooling demand for all timesteps
Q_{plus}	Current cooling capacity of sprinkler tank
$\dot{Q}_{Rad}$	Heat input by radiation
SC	Starting costs
St	Operating state of chilling machine
T_{amb}	Ambient temperature
T_{cond}	Condensing temperature of chiller
T_{cold}	Cold water temperature of chiller
T_{Res}	Simulation time
T_{St}	Storage temperature
U	Internal energy
X_{tnst}	State of chilling machine
Z_{tn}	On/off Variable

References

1. Heinrich, C.; Witti, S.; Albring, P.; Richter, L.; Safarik, M.; Böhm, U.; Hantsch, A. Nachhaltige Kälteversorgung in Deutschland an den Beispielen Gebäudeklimatisierung und Industrie. 2014. Available online: https://www.umweltbundesamt.de/sites/default/files/medien/378/publikationen/climate_change_25_2014_nachhaltige_kaelteversorgung_in_deutschland_1.pdf (accessed on 2 May 2018).
2. Peesel, R.-H.; Schlosser, F.; Schaumburg, C.; Meschede, H. Prädiktive simulationsgestützte Optimierung von Kältemaschinen im Verbund. In *Simulation in Produktion und Logistik 2017*; Wenzel, S., Peter, T., Eds.; Kassel University Press: Kassel, Germany, 2017; pp. 69–78. ISBN 978-3-7376-0192-4.
3. Clausen, U.; Diekmann, D.; Baudach, J.; Pöting, M. Mathematical Optimisation and Simulation of Parcel Transshipment Terminals—Better Solutions by Linking These Two Complementary Methods. *Simul. Prod. Logist.* **2015**, *2015*, 279–288.
4. Augenstein, E. Betriebsoptimierung von Kältezentralen—Einsparpotentiale durch situationsabhängige Einsatzregeln. KKA Kälte Klima Aktuell Sonderausgabe Großkältetechnik. 2009. Available online: http://www.perpendo.de/files/kka-2009.pdf (accessed on 17 May 2019).
5. Ahn, B.C.; Mitchell, J.W. Optimal control development for chilled water plants using a quadratic representation. *Energy Build.* **2001**, *33*, 371–378. [CrossRef]

6. Mu, B.; Li, Y.; House, J.M.; Salsbury, T.I. Real-time optimization of a chilled water plant with parallel chillers based on extremum seeking control. *Appl. Energy* **2017**, *208*, 766–781. [CrossRef]
7. Hovgaard, T.G.; Edlund, K.; Jorgensen, J.B. The potential of Economic MPC for power management. In Proceedings of the 49th IEEE Conference on Decision and Control (CDC), Atlanta, GA, USA, 15–17 December 2010; pp. 7533–7538.
8. Braun, J.E. Methodologies for the Design and Control of Central Cooling Plants. Ph.D. Thesis, University of Wisconsin, Madison, WI, USA, 1988.
9. Olson, R.T.; Liebman, J.S. Optimization of a chilled water plant using sequential quadratic programming. *Eng. Optim.* **1990**, *15*, 171–191. [CrossRef]
10. Chang, Y.-C.; Lin, F.-A.; Lin, C.H. Optimal chiller sequencing by branch and bound method for saving energy. *Energy Convers. Manag.* **2005**, *46*, 2158–2172. [CrossRef]
11. Chang, Y.-C. An Outstanding Method for Saving Energy—Optimal Chiller Operation. *IEEE Trans. Energy Convers.* **2006**, *21*, 527–532. [CrossRef]
12. Chang, Y.-C.; Chan, T.-S.; Lee, W.-S. Economic dispatch of chiller plant by gradient method for saving energy. *Appl. Energy* **2010**, *87*, 1096–1101. [CrossRef]
13. Chang, Y.-C. Genetic algorithm based optimal chiller loading for energy conservation. *Appl. Therm. Eng.* **2005**, *25*, 2800–2815. [CrossRef]
14. Chang, Y.-C.; Chen, W.-H.; Lee, C.-Y.; Huang, C.-N. Simulated annealing based optimal chiller loading for saving energy. *Energy Convers. Manag.* **2006**, *47*, 2044–2058. [CrossRef]
15. Lee, W.-S.; Lin, L.-C. Optimal chiller loading by particle swarm algorithm for reducing energy consumption. *Appl. Therm. Eng.* **2009**, *29*, 1730–1734. [CrossRef]
16. Powell, K.M.; Cole, W.J.; Ekarika, U.F.; Edgar, T.F. Optimal chiller loading in a district cooling system with thermal energy storage. *Energy* **2013**, *50*, 445–453. [CrossRef]
17. Huang, S.; Zuo, W.; Sohn, M.D. A New Method for The Optimal Chiller Sequencing Control. In Proceedings of the 14th Conference of IBPSA, Hyderabad, India, 7–9 December 2015; pp. 316–323.
18. Thangavelu, S.R.; Myat, A.; Khambadkone, A. Energy optimization methodology of multi-chiller plant in commercial buildings. *Energy* **2017**, *123*, 64–76. [CrossRef]
19. Kapoor, K.; Powell, K.; Cole, W.; Kim, J.; Edgar, T. Improved Large-Scale Process Cooling Operation through Energy Optimization. *Processes* **2013**, *1*, 312–329. [CrossRef]
20. Cole, W.J.; Powell, K.M.; Edgar, T.F. Optimization and advanced control of thermal energy storage systems. *Rev. Chem. Eng.* **2012**, *28*, 81–99. [CrossRef]
21. Kircher, K.J.; Zhang, K.M. Model predictive control of thermal storage for demand response. In Proceedings of the American Control Conference (ACC), Chicago, IL, USA, 1–3 July 2015.
22. Ma, Y.; Borrelli, F.; Hencey, B.; Coffey, B.; Bengea, S.; Haves, P. Model Predictive Control for the Operation of Building Cooling Systems. *IEEE Trans. Contr. Syst. Technol.* **2012**, *20*, 796–803. [CrossRef]
23. Soler, M.S.; Sabaté, C.C.; Santiago, V.B.; Jabbari, F. Optimizing performance of a bank of chillers with thermal energy storage. *Appl. Energy* **2016**, *172*, 275–285. [CrossRef]
24. Peesel, R.-H.; Schlosser, F.; Schaumburg, C.; Meschede, H.; Dunkelberg, H.; Walmsley, T.G. Predictive Simulation-based Optimisation of Cooling System Including a Sprinkler Tank. *Chem. Eng. Trans.* **2018**, *2018*, 349–354.
25. Vaghefi, A.; Jafari, M.A.; Bisse, E.; Lu, Y.; Brouwer, J. Modeling and forecasting of cooling and electricity load demand. *Appl. Energy* **2014**, *136*, 186–196. [CrossRef]
26. Kwok, S.S.K.; Yuen, R.K.K.; Lee, E.W.M. An intelligent approach to assessing the effect of building occupancy on building cooling load prediction. *Build. Environ.* **2011**, *46*, 1681–1690. [CrossRef]
27. Li, Q.; Meng, Q.; Cai, J.; Yoshino, H.; Mochida, A. Predicting hourly cooling load in the building: A comparison of support vector machine and different artificial neural networks. *Energy Convers. Manag.* **2009**, *50*, 90–96. [CrossRef]
28. Gordon, J.M.; Ng, K.C. *Cool Thermodynamics. The Engineering and Physics of Predictive, Diagnostic and Optimization Methods for Cooling Systems*; Cambridge International Science: Cambridge, UK, 2001; ISBN 1898326908.
29. Brenner, A.; Kausch, C.; Kirschbaum, S.; Lepple, H.; Zens, M. Effizienzsteigerung in Einem komplexen Kühlwassersystem. Einsatz von Simulationswerkzeugen. 2014. Available online: http://www.perpendo.de/files/kka-2014.pdf (accessed on 21 October 2016).

30. Lee, T.-S.; Liao, K.-Y.; Lu, W.-C. Evaluation of the suitability of empirically-based models for predicting energy performance of centrifugal water chillers with variable chilled water flow. *Appl. Energy* **2012**, *93*, 583–595. [CrossRef]
31. Gurobi Optimization, L.L.C. Gurobi Optimizer Reference Manual. 2018. Available online: http://www.gurobi.com/documentation/8.1/ (accessed on 17 May 2019).
32. Wemhöner, C.; Hafner, B.; Schwarzer, K. Simulation of Solar Thermal Systemswith Carnot Blockset in the Environment Matlab Simulink. 2000. Available online: http://ptp.irb.hr/upload/mape/kuca/11_Carsten_Wemhoener_SIMULATION_OF_SOLAR_THERMAL_SYSTEMS_WITH.pdf (accessed on 14 June 2016).

© 2019 by the authors. Licensee MDPI, Basel, Switzerland. This article is an open access article distributed under the terms and conditions of the Creative Commons Attribution (CC BY) license (http://creativecommons.org/licenses/by/4.0/).

Article

Computational Model to Evaluate the Effect of Passive Techniques in Tube-In-Tube Helical Heat Exchanger

Miyer Valdes [1], Juan G. Ardila [1], Dario Colorado [2,*] and Beatris A. Escobedo-Trujillo [3]

[1] Instituto Tecnológico Metropolitano, Departamento de Electrómecanica, Calle 54ª No. 30-01, Medellin P.A. 050013, Colombia; miyervaldes@itm.edu.co (M.V.); juanardila@itm.edu.co (J.G.A.)
[2] Centro de Investigación en Recursos Energéticos y Sustentables, Universidad Veracruzana, Av. Universidad km 7.5, Col. Santa Isabel, Coatzacoalcos C.P. 96535, Mexico
[3] Facultad de Ingeniería, Universidad Veracruzana, Av. Universidad km 7.5, Col. Santa Isabel, Coatzacoalcos C.P. 96535, Mexico; bescobedo@uv.mx
* Correspondence: dcolorado@uv.mx; Tel.: +52-(921)-211-5700 (ext. 59230)

Received: 28 December 2018; Accepted: 14 May 2019; Published: 18 May 2019

Abstract: The purpose of this research is to evaluate the effect of twist in the internal tube in a tube-in-tube helical heat exchanger keeping constant one type of ridges. To meet this goal, a Computational Fluid Dynamic (CFD) model was carried out. The effects of the fluid flow rate on the heat transfer were studied in the internal and annular flow. A commercial CFD package was used to predict the flow and thermal development in a tube-in-tube helical heat exchanger. The simulations were carried out in counter-flow mode operation with hot fluid in the internal tube side and cold fluids in the annular flow. The internal tube was modified with a double passive technique to provide high turbulence in the outer region. The numerical results agree with the reported data, the use of only one passive technique in the internal tube increases the heat transfer up to 28.8% compared to smooth tube.

Keywords: computational fluid dynamics; heat transfer; temperature contour

1. Introduction

According to Li et al. [1], two types of techniques have been proposed to improve the heat transfer of heat exchangers: the active and the passive. The active techniques require external power, such as vibration or magnetic fields, whereas the passive techniques require deformations on the tube surface, without external power, as well as on any surface where there is heat transfer. The passive technique was widely recommended by several authors because it considers bent tubes and its ability to compact the heat exchanger. When proposing a combination of passive improvements to improve the heat transfer in a piece of equipment, the problem of knowing the hydrodynamics and temperature profile in the fluid is presented. Passive improvements increase the heat transfer and consequently compact equipment is built.

The contribution of this research is to provide a Computational Fluid Dynamic (CFD) study of a tube-in-tube helical heat exchanger evaluated with two passive techniques implemented. Special interest is put in the effects of the fluid flow rate on the heat transfer while the geometry was changing in the analysis.

The background of this research is described with the following works. Pan et al. [2] described the heat transfer improvements assuming passive techniques, it produced by secondary flows and studied by [3] since 1927. The simple improvement techniques consist of the insertion of coils [4], twist tape [1], staggered tapes in straight tubes, the corrugation of the tube and the curved tube [5]. Another technique is the double improvement, which is a combination of two simple improvements [6], such as a curved tube with springs, corrugated or baffles inserts; or a straight tube with tape inserts and corrugated [7]. Previous research emphasizes the CFD study in the heat exchangers with the aim of validating the numerical results and optimization [8] and control strategies of tubular heat exchangers using neural-networks-based method [9]. A numerical study using CFD approach in a vertical cylindrical tube-in-tank thermal energy storage with helical coils design is presented by [10], in which the effect of the input temperature and flow rate values to the heat transfer device were analyzed. Sharifi et al. [11] presented the influence of coiled wire inserts on the Nusselt number, friction coefficient and overall efficiency in double-pipe heat exchangers using CFD software for their analysis, in which significant improvements are described with regard to the heat exchanger containing coiled wire inserts with pitch length equal to 69 mm, the Nusselt number was enhanced by 1.77 times as compared with the plain heat exchanger. Numerical investigation and fluid flow characteristics of spiral finned-tube heat exchangers considering the effects of location of perforations and RNG $k-\varepsilon$ model was carried out by [12].

Kumar et al. [13] presented the heat transfer characteristics and hydrodynamics behavior of tube-in-tube helical heat exchanger, this research is the basis of our work. The motivation of this research is to present the CFD results to quantify the increase in heat transfer with passive improvements in the helical heat exchanger with the objective of knowing the hydraulic and thermal behavior of the fluid subjected to the proposed operating conditions and the proposed geometries.

This work has been organized as follows: first, the simulation code was developed, and it was carefully verified with numerical results presented by [13]. Second, the simulation of the heat exchanger with four ridges in the internal tube, and the simulation of three heat exchangers increasing the number of twists in the internal tube from one to five. Finally, the Nusselt number was calculated for each case with the aim of assessing the effect of each passive technique on the heat transfer.

2. Methodology

The heat exchanger model consists of two helical concentric tubes, tube-in-tube, according to dimensions and suggestions by Kumar et al. [13]. The hot fluid flows in the internal tube and the cold fluid flows in the annular section in counter-flow using water as working fluid in both flows. The simulation was carried out with ANSYS CFX 17 in a computer of 16 core, 32 GB RAM at the Modeling Laboratory of Materiales Avanzados y Energía (MATyER) research group of Instituto Tecnológico Metropolitano at Medellín-Colombia.

2.1. Modeling of Heat Exchanger

A concentric tube-in-tube heat exchanger was modeled as shown in Figure 1A, considering dimensions and boundary conditions given in Table 1. As can be seen in Figure 1B, four ridges were aggregated in the internal tube conserving it hydraulic diameter. The internal tube was modified increasing the twist, Figure 1C–E illustrate the heat exchanger with one, three, and five turns, respectively. Table 2 showed the information of geometry presented in this work. As can be seen in the Figure 1 and Table 1, this work, models the transfer of heat and hydrodynamic of fluid in a single turn of a heat exchanger, the velocity illustrated in Table 1 is assumed as the average velocity of a modeled velocity profile in the input section of both flows.

Table 1. Geometry and boundary conditions.

Geometry and Boundary Conditions	Internal Tube	External Tube
Outer diameter (m)	0.0254	0.0508
Helical diameter (m)	0.762	0.762
Pitch (m)	0.1	0.1
Velocity (m/s)	0.073	0.32–0.44
Dean number (dimensionless)	1000	4410–6030
Prandtl number (dimensionless)	7	7

Table 2. Geometry of the five models presented in this work.

Geometry	Passive Modification
A	tube-in-tube
B	four ridges without twist
C	four ridges with 1 twist
D	four ridges with 3 twist
E	four ridges with 5 twist

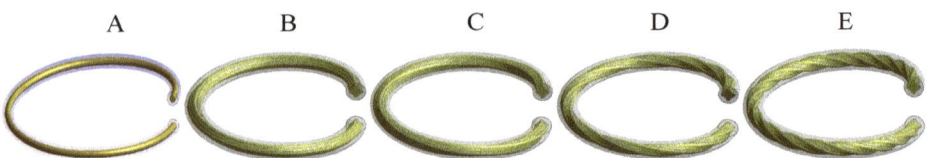

Figure 1. Simulated control volumes: external and internal flows with the second passive technique evolution. A: tube-in-tube. B: four ridges without twist. C: four ridges with 1 twist. D: four ridges with 3 twist. E: four ridges with 5 twist.

The mesh was generated for five heat exchanger geometries described above with element size of 1 mm. Figure 2 showed the mesh for geometry A and B, respectively. Several meshes were created with the aim of comparing the Nusselt number and selecting adequate mesh for simulations. The error between two mesh sizes ϵ_{Nu} was calculated according to Equation (1):

$$\epsilon_{Nu} = \frac{|Nu_U - Nu_L|}{Nu_U} \times 100 \tag{1}$$

where Nu_U and Nu_L mean the Nusselt number for upper mesh and lower mesh, respectively.

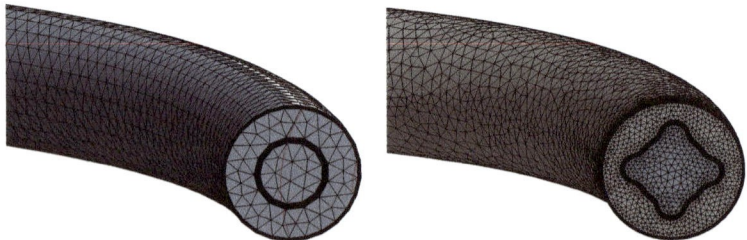

Figure 2. Mesh for geometry A (tube-in-tube) and B (four ridges without twist).

According to the numerical results, the number of elements of each mesh was selected range from between 39.7×10^6 elements for geometry A, up to and 57.5×10^6 elements for geometry E, with the objective of showing independent results of the mesh and errors lower than 1%. An independence mesh analysis was carried out, each of the geometries was evaluated with different numbers of elements and calculating their Nusselt number, the numerical results are shown in Figure 3, and they agree with [14]. As can be seen in Figure 3 the number of elements is greater than 39.5×10^6 elements for geometry B, 40.7×10^6 elements for geometry C and 52.9×10^6 elements for geometry D.

According to the numerical results, the number of elements of each mesh was selected between 40×10^6 and 57×10^6 with the objective of showing independent results of the mesh and errors lower than 1%. An independence mesh analysis was carried out, each of the geometries was evaluated with different numbers of elements and calculating their Nusselt number, the numerical results are shown in Figure 3, and they agree with [14]. As can be seen in Figure 3 the number of elements in the mesh does not affect the results, and they are independent when the number of elements is greater than 15×106 for geometries from B to D, for geometry A is necessary to use a mesh with a smaller element size. Figure 4 shows the evolves of Nusselt number trough the heat exchangers showing that after 180° the change in Nusselt number is minimum.

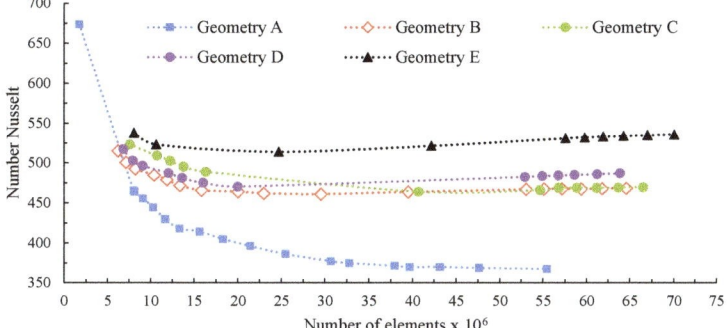

Figure 3. Mesh independence study.

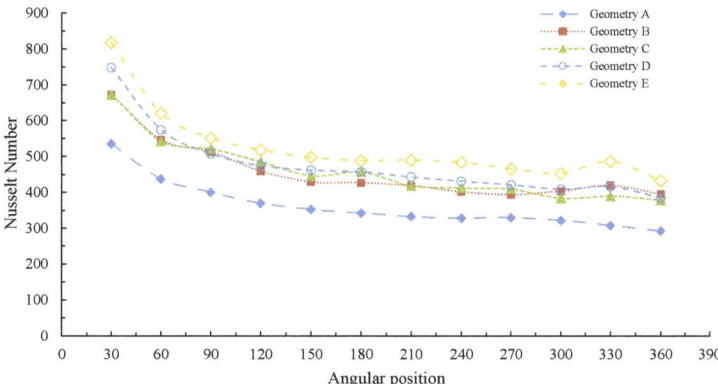

Figure 4. Nusselt number for several positions in the first turn of heat exchanger.

2.2. Numerical Computation

Five Dean numbers were simulated: 4410, 4880, 5340, 5810 and 6030 for an external tube keeping constant the design and increasing the Reynolds number. The inner tube had a constant Dean number of 1000 for all simulations. The inlet and outlet temperatures in the internal tube were fixed at 300 K and 380 K, respectively. The numerical computation was set up at convergence criterion less than 10^{-6} and the total number of iterations were varied from 500 to 600. The Dean number for the external tube was calculated according Equation (2):

$$De = \frac{\rho v D_h}{\mu}\sqrt{\delta} \qquad (2)$$

where v is velocity average, ρ is the density, μ is the dynamic viscosity and D_h is the hydraulic diameter, it was computed as follows:

$$D_h = \frac{4A}{P_h} \qquad (3)$$

δ is the curvature diameter and it was calculated as follows:

$$\delta = \frac{D_c}{D_i} \tag{4}$$

then, A is the cross-sectional area and P_h is the hydraulic diameter, D_c is the helical diameter and D_i is the outer diameter of internal tube.

The equations suggested by [7] for the physical-thermal properties of water such as density (ρ), specific heat (Cp), thermal conductivity (κ) and dynamic viscosity (μ) are computed and used in this investigation.

The differential equations governing the turbulent flow that describe the chaotic movement of the fluid inside heat exchanger can be written in the tensor form according to Equations (5)–(12). The Continuity balance expresses the net mass flow at the scale of the infinitesimal control volume as follows in Equation (5):

$$\frac{\partial \rho u_i}{\partial x_i} = 0 \tag{5}$$

The momentum balance equation (see Equation (6)) states that the temporary increase of the linear momentum plus its net flow at the output must be equal to the sum of the forces acting on the infinitesimal control volume.

$$\frac{\partial(\rho u_i u_j)}{\partial x_j} + \frac{\partial P}{\partial x_i} - \frac{\partial}{\partial x_j}\left[(\mu + \mu_t)\left(\frac{\partial u_i}{\partial x_j} + \frac{\partial u_j}{\partial x_i}\right)\right] - F_i = 0 \tag{6}$$

The first term of Equation (6) represents the change of momentum by convection, the second term is the pressure forces that act on the infinitesimal control volume, the third term indicates the tangential stress caused by the velocity gradients according to the first coefficient of effective viscosity ($\mu + \mu_t$) and the last term (F_i) represents the centrifugal forces that act on the volume of infinitesimal control.

Energy balance equation states that the amount of energy change of the infinitesimal element is equal to the amount of heat added to the element plus the amount of work done on the element. The energy balance equation is shown in Equation (7).

$$\frac{\partial(\rho E u_j)}{\partial x_j} + \frac{\partial(P u_j)}{\partial x_j} - \Phi - \frac{\partial}{\partial x_j}k\left(\frac{\partial T}{\partial x_j}\right) = 0 \tag{7}$$

The first term of Equation (7) expresses the net flow of energy, the second term is the pressure, the third term is the viscous forces and the fourth term represents the net heat flow that enters through the faces of the control volume due to temperature gradients. Where Φ is the viscous heating term in energy equation, its represented by Equation (8).

$$\Phi = \mu \frac{\partial u_i}{\partial x_j}\left(\frac{\partial u_i}{\partial x_j} + \frac{\partial u_j}{\partial x_i}\right) \tag{8}$$

For the $k - \varepsilon$ turbulent model, k represents the kinetic energy associated with turbulence. The turbulent kinetic energy is shown in Equation (9):

$$\frac{\partial(\rho u_j k)}{\partial x_j} - \frac{\partial}{\partial x_j}\left[\left(\mu + \frac{\mu_t}{\sigma_k}\right)\frac{\partial k}{\partial x_j}\right] - P_k + \rho\varepsilon = 0 \tag{9}$$

where σ_k represents the turbulent Prandtl number.

In Equation (9), the first term represents the transport of k by convection, the second term expresses viscous diffusion and turbulent diffusion by pressure-velocity fluctuations, the third term represents the production of energy in large turbulent scales and the last term is the dissipation. The term P_k is the parameter to calculate the generation of turbulent kinetic energy due to the mean velocity gradient and is represented by Equation (10).

$$P_k = \mu_t \left(\frac{\partial u_i}{\partial x_j} + \frac{\partial u_j}{\partial x_i} \right) \frac{\partial u_i}{\partial x_j} \tag{10}$$

While ε represents the dissipation of kinetic energy, it is represented by the turbulent dissipation energy as follows:

$$\frac{\partial (\rho u_j \varepsilon)}{\partial x_j} - \frac{\partial}{\partial x_j} \left[\left(\mu + \frac{\mu_t}{\sigma_\varepsilon} \right) \frac{\partial \varepsilon}{\partial x_j} \right] - \frac{\varepsilon}{k}(C_{\varepsilon 1} P_k - C_{\varepsilon 2} \rho \varepsilon) = 0 \tag{11}$$

The $k - \varepsilon$ model assumes that the turbulence viscosity is coupled to the governing equations via the relation in Equation (12).

$$\mu_t = C_\mu \rho \frac{k^2}{\varepsilon} \tag{12}$$

where ρ is the density and C_μ is a constant that must be determined empirically.

The empirical constants for the turbulence model are assigned the following values: $C_\mu = 0.09$, $C_{\varepsilon 1} = 1.47$, $C_{\varepsilon 2} = 1.92$, $\sigma_k = 1.0$ and $\sigma_\varepsilon = 1.3$. The values of these constants are the same as those of Launder and Spalding [15], the suggestion of [13] for $C_{\varepsilon 1}$ was used.

The Nusselt number (Nu) and friction factor (f) for each simulation was calculated according:

$$Nu = \frac{\alpha D_h}{\kappa} \tag{13}$$

where D_h is the hydraulic diameter, κ is the thermal conductivity and α the convective heat transfer coefficient is determined as:

$$\alpha = \frac{\dot{m} C p (T_{outlet} - T_{inlet})}{A_{wall}(\overline{T_{wall}} - \overline{T_f})} \tag{14}$$

where $\dot{m}$ is mass flow of external fluid, T_{outlet} and T_{inlet} are the outlet and inlet temperature respectively of external fluid, $\overline{T_{wall}}$ is internal tube wall average temperature, $\overline{T_f}$ is external fluid average temperature, and A_{wall} is internal tube wall area.

Darcy-Weisbach factor friction it was calculated as follows:

$$f = \frac{2 \triangle p D_h}{\rho v^2 l} \tag{15}$$

where $\triangle p$ is pressure drop and l is heat exchanger length.

3. Results and Discussion

The computer model has been carefully verified using the numerical results of [13]. Figure 5 shows the Nusselt number reported by [13] and our numerical results for tube-in-tube heat exchanger without passive technique improvements (Geometry A). The models present a similar trend. The discrepancies between the results can be assumed to the incorporation of deflectors in the model by [13]. As can be seen, the model agrees with the numerical (geometry A), numerical results by [13] and experimental data reported by [13]. An average increment of Nusselt number $\triangle_{Nu}$ is calculated to compare the numerical results presented. The following equation was used:

$$\triangle_{Nu} = \frac{\Sigma_1^5 [100 - \frac{Nu_i \times 100}{Nu_A}]}{5} \tag{16}$$

where Nu_i is the Nusselt number of geometry B, C, D, or E and Nu_A is the Nusselt number of geometry A.

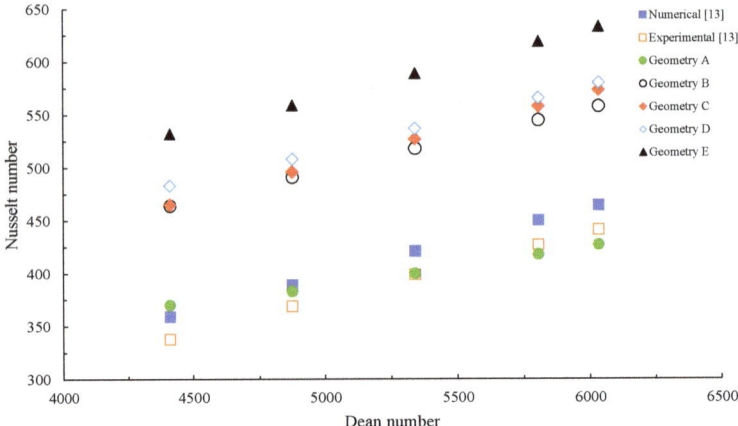

Figure 5. Nusselt number vs. Dean number for outer tube.

The geometry B includes the passive technique without turns into the internal tube, an $\triangle_{Nu}$ approximately 28.8% in the Nusselt number was estimated in comparison with geometry A. Its average increment remains practically constant when rising the Dean number from 4500 to 6000. It is interesting to note that the increases of the Nusselt number $\triangle_{Nu}$ from one to three turns of twist in the internal tube were minor to approximately 3%, nevertheless, the biggest increase was observed when the number of turns changes from three to five in approximately 5.7%. The velocity and temperature contours in the internal and annular flow were analyzed in this section considering a Dean number equal to 4411 in the external fluid. In Figure 6, the relationship between the number of Nusselt of each geometry proposed with twist and the geometry proposed by [13] is calculated. As can be seen, the heat transfer in the heat exchanger is increased with the number of twists in the tube, a slight increase is appreciated as the number of Dean is increased.

Figure 7 illustrates the velocity contours for the internal fluid of five geometries of the heat exchanger. The geometries were arranged in rows, the rows (A), (B), (C), (D) and (E) show the velocity contours considering: a smooth tube, a geometry modified with passive technique without twist, a geometry modified with passive technique with one, three, and five turns, respectively. The columns report different cross-section (30°, 60°, 90°, 180°, 270° and 360°). Arcs located to the left of every field indicate the exterior of the heat exchanger.

As can be seen in the numerical results of model (A), the velocity contours practically remain the same. This behavior is similar to results reported by [13]. In four geometries (B) to (E), the velocity contour changed from the position of 90° appreciating different velocity profiles; however for angles from 180° to 360° small variations are observed in the velocity fields, this is evidence of total developed flow. For the models (A) to (D), the maximum velocity was observed towards the outside of heat exchangers, this effect is attributable to the centrifugal force and secondary flows caused by the curvature of the helical coil, the effect of these forces over the velocity is the same that reported by [3]. As can be seen, for the numerical results of models (D) and (E) it is not possible to observe a pattern in the velocity contour possibly attributed to the effect of twist along the tube.

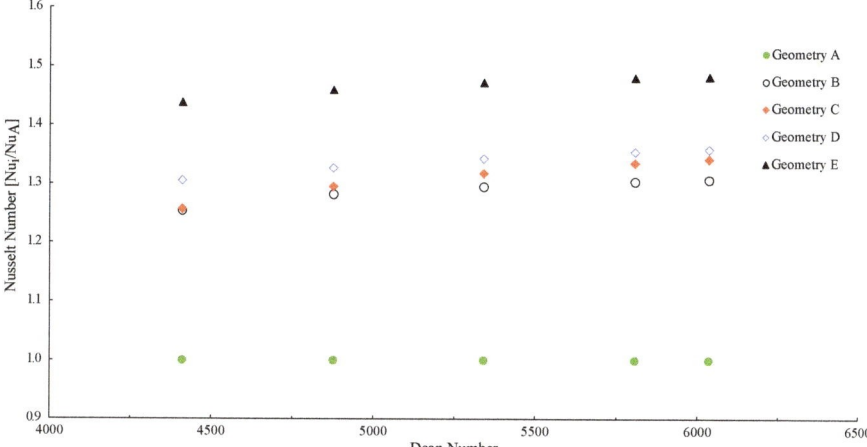

Figure 6. Comparison of Nusselt number for each one of geometries.

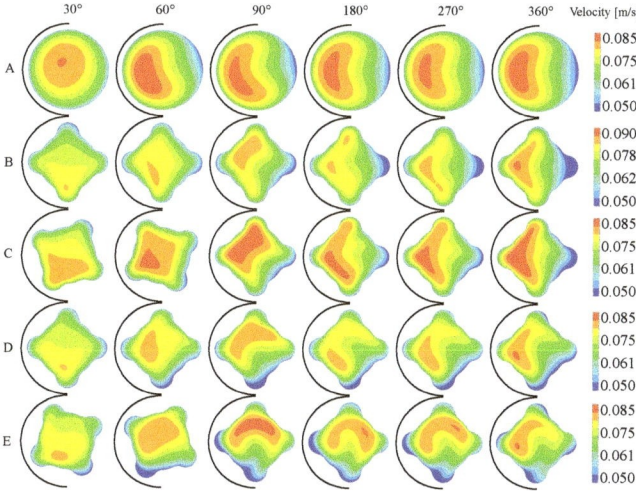

Figure 7. Velocity contours for the inner tube of five heat exchangers with Dean number 1000 for the inner tube and with Dean number 4411 for the outer tube. A: tube-in-tube. B: four ridges without twist. C: four ridges with 1 twist. D: four ridges with 3 twist. E: four ridges with 5 twist.

Figure 8 shows the temperature contours for the five models described previously. In geometries (A) to (D) the maximum temperature was observed towards the exterior of the heat exchanger, this is due to the center of the vortex formed by the secondary flows. In addition, the movement of fluid in the form of vortex produces an increase in the heat transfer mechanism, possibly caused for the increase of velocity contours in these sections (see Figure 7). For the models (D) and (E), the temperature contours were very similar from 30° to 360°, this behavior may be the reason the Nusselt number presents an insignificant increase when the twist increases from three to five (Figure 7). Figure 9 shows the velocity contour for annular flow. For models from (B) to (E), a pattern was not appreciated in velocity fields, small variations were observed in the velocity fields from 270° to 360°, but this is not enough to affirm that there is a developed flow. The secondary flows observable in models from (B) to (E) were more turbulent than model (A), the passive technique benefits the turbulence and consequently the heat transfer. Figure 10 illustrates the temperature contour for annular flow. For models from (B) to (E), the temperature increases uniformly, there are no significant temperature differences between the models in the cross-section selected. The temperature contours correlate with the velocity contours, as previously showed in Figure 9.

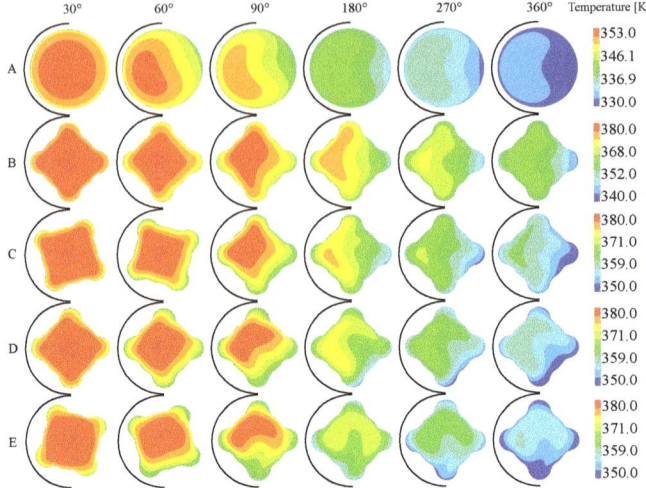

Figure 8. Temperature contours for the inner tube with Dean number 1000 for inner tube and with Dean number 4411 for outer tube. A: tube-in-tube. B: four ridges without twist. C: four ridges with 1 twist. D: four ridges with 3 twist. E: four ridges with 5 twist.

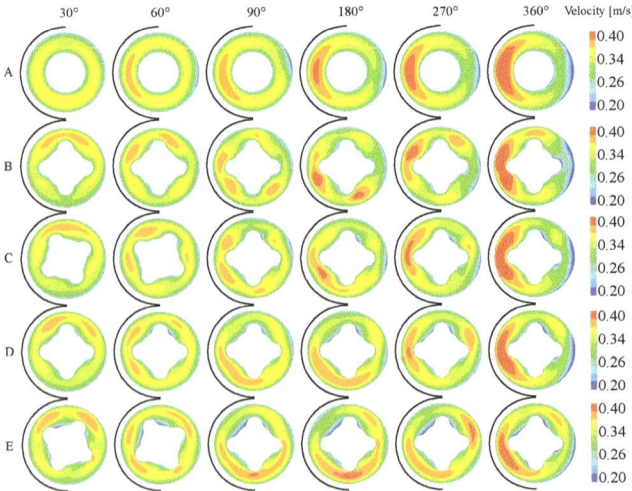

Figure 9. Velocity contours for the outer tube with Dean number 1000 for inner tube and with Dean number 4411 for outer tube. A: tube-in-tube. B: four ridges without twist. C: four ridges with 1 twist. D: four ridges with 3 twist. E: four ridges with 5 twist.

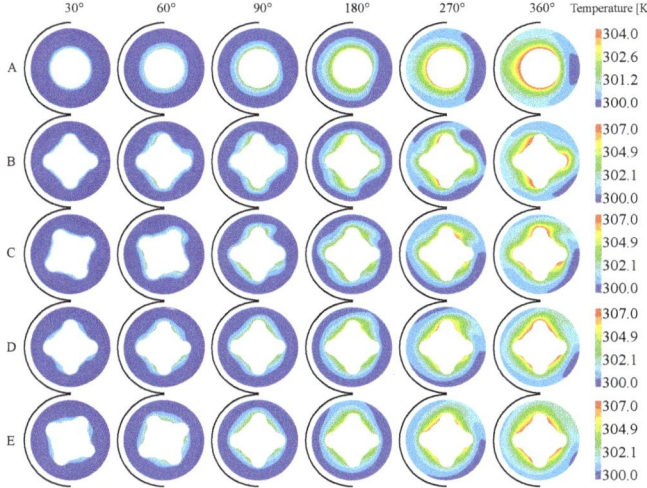

Figure 10. Temperature contours for the outer tube with Dean number 1000 for inner tube and with Dean number 4411 for outer tube. A: tube-in-tube. B: four ridges without twist. C: four ridges with 1 twist. D: four ridges with 3 twist. E: four ridges with 5 twist.

As can be seen, Figures 5–10 provides evidence of the increase of Nusselt number when the Reynolds number was increases. The increases of heat transfer from geometry A to E can be attributed to different reasons: the secondary flow originated by helical coil and twist, centrifugal force, increase in turbulence due to twist and the velocity distribution shown in Figures 7 and 9.

Figure 11 shows the average Darcy friction factor as a function of the Dean number, where a decrease in the friction factor is observed as the Dean number increases. When modifying geometry B to geometry C, the friction factor increased by 0.4%. The reduction of the friction factor indicates a reduction in the pressure drop because these are directly proportional. This reduction can be due to the increase of the cross-sectional area that in the modified geometries is greater than the circular smooth tube. As the twist of the tube increases (geometries D and E) the friction factor continues to increase by 6 and 17% for the geometry D and E respectively, showing that the increase in twist makes the flow of the fluid through the heat exchanger difficult.

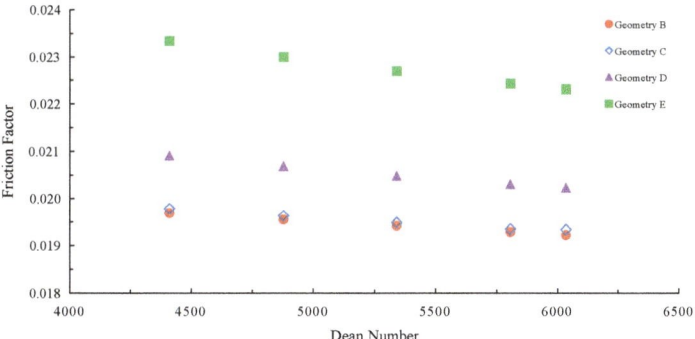

Figure 11. Friction factor vs. Dean number for each one of geometries.

4. Conclusions

The numerical model using CFD of a helical tube-in-tube heat exchanger with and without passive techniques was successfully carried out. The main contributions of the present research are the following:

1. The numerical model of [13] was reproduced considering the assumptions and experimental information described in that research. This evidence gives confidence about the evaluation when the passive techniques were added.
2. With reference to the first passive technique applied the addition of four ridges in the inner tube. Shows an increment up to 28.8% in the Nusselt number were calculated for all cases under study. Then, when the Dean number increased from 4500 to 6000 the Nusselt number increases linearly. This increment can be caused by the velocity contours generated by the addiction of ridges and its influence on heat transfer. In the annular section of heat exchanger, the ridges decrease the centrifugal forces generated by the action of a helical coil.
3. When the twist of the internal tube was added from one to three turns, an increase up to 3% in the Nusselt number was calculated. The biggest increase, up to 9% was calculated when five turns were simulated.
4. The numerical results of this research will be considered to design other passive techniques without twist in tube. The previous suggestion is supported by the increase in the heat transfer, this work assumes that there is no compromise in the mechanical integrity of the tubes caused by twist.

In this work the numerical results, analysis, and discussion were presented; if more information is needed for future research lines, the corresponding author can provide it upon request.

Author Contributions: The authors contribute equally to this manuscript. Investigation, D.C.; Methodology, M.V.; Software, J.G.A.; Writing—review and editing, B.A.E-T.

Funding: This research received no external funding.

Conflicts of Interest: The authors declare no conflict of interest.

References

1. Li, P.; Liu, Z.; Liu, W.; Chen, G. Numerical study on heat transfer enhancement characteristics of tube inserted with centrally hollow narrow twisted tapes. *Int. J. Heat Mass Transf.* **2015**, *88*, 481–491. [CrossRef]
2. Pan, C.; Zhou, Y.; Wang, J. CFD study of heat transfer for oscillating flow in helically coiled tube heat-exchanger. *Comput. Chem. Eng.* **2014**, *69*, 59–65. [CrossRef]
3. Dean, W.R. XVI. Note on the motion of fluid in a curved pipe. *Lond. Edinb. Dublin Philos. Mag. J. Sci.* **1927**, *4*, 208–223. [CrossRef]
4. Kurnia, J.C.; Sasmito, A.P.; Akhtar, S.; Shamim, T.; Mujumdar, A.S. Numerical Investigation of Heat Transfer Performance of Various Coiled Square Tubes for Heat Exchanger Application. *Energy Procedia* **2015**, *75*, 3168–3173. [CrossRef]
5. García, A.; Solano, J.P.; Vicente, P.G.; Viedma, A. The influence of artificial roughness shape on heat transfer enhancement: Corrugated tubes, dimpled tubes and wire coils. *Appl. Therm. Eng.* **2012**, *35*, 196–201. [CrossRef]
6. Rainieri, S.; Bozzoli, F.; Pagliarini, G. Experimental investigation on the convective heat transfer in straight and coiled corrugated tubes for highly viscous fluids: Preliminary results. *Int. J. Heat Mass Transf.* **2012**, *55*, 498–504. [CrossRef]
7. Zachár, A. Analysis of coiled-tube heat exchangers to improve heat transfer rate with spirally corrugated wall. *Int. J. Heat Mass Transf.* **2010**, *53*, 3928–3939. [CrossRef]
8. Salpingidou, C.; Misirlis, D.; Vlahostergios, Z.; Flourous, M.; Donnerhack, S.; Yakinthos, K. Numerical modeling of heat exchangers in gas turbine using CFD computations and thermodynamic cycle analysis tools. *Chem. Eng. Trans.* **2016**, *52*, 517–522.
9. Bakosova, M.; Oravec, J.; Vasickaninova, A.; Meszaros, A. Neural-network-based and robust model-based predictive control of a tubular heat exchanger. *Chem. Eng. Trans.* **2017**, *61*, 301–306.

10. Yang, X.; Xiong, T.; Dong, J.L.; Li, W.X.; Wang, Y. Investigation of the dynamic melting process in a thermal energy storage unit using a helical coil heat exchanger. *Energies* **2017**, *10*, 1129. [CrossRef]
11. Sharifi, K.; Sabeti, M.; Rafiei, M.; Mohammadi, A.H.; Shirazi, L. Computational fluid dynamics (CFD) technique to study the effects of helical wire inserts on heat transfer and pressure drop in a double pipe heat exchanger. *Appl. Therm. Eng.* **2018**, *128*, 898–910. [CrossRef]
12. Ju-Lee, H.; Ryu, J.; Hyuk-Lee, S. Influence of Perforated Fin on Flow Characteristics and Thermal Performance in Spiral Finned-Tube Heat Exchanger. *Energies* **2019**, *12*, 556. [CrossRef]
13. Kumar, V.; Saini, S.; Sharma, M.; Nigam, K.D.P. Pressure drop and heat transfer study in tube-in-tube helical heat exchanger. *Chem. Eng. Sci.* **2006**, *61*, 4403–4416. [CrossRef]
14. Valdés-Ortiz, M.; Ardila-Marin, J.G.; Martínez-Pérez, A.F.; Betancur Gómez, J.D. Via ANSYS Numerical Analysis of Heat Exchangers with Passive Improvement: Case Study Meshing Density and Turbulence Model. *Rev. CINTEX* **2017**, *22*, 59–68.
15. Launder, B.E.; Spalding, D.B. The numerical computation of turbulent flows. *Comput. Methods Appl. Mech. Eng.* **1974**, *3*, 269–289. [CrossRef]

© 2019 by the authors. Licensee MDPI, Basel, Switzerland. This article is an open access article distributed under the terms and conditions of the Creative Commons Attribution (CC BY) license (http://creativecommons.org/licenses/by/4.0/).

Article

Electric Field Effect on the Thermal Decomposition and Co-combustion of Straw with Solid Fuel Pellets

Inesa Barmina [1], Antons Kolmickovs [1,*], Raimonds Valdmanis [1], Maija Zake [1], Sergejs Vostrikovs [1], Harijs Kalis [2] and Uldis Strautins [2]

1. Institute of Physics, University of Latvia, 32 Miera str., 1 LV-2169 Salaspils, Latvia; barmina@sal.lv (I.B.); raimonds.valdmanis@lu.lv (R.V.); mzfi@sal.lv (M.Z.); sergejs.vostrikovs@lu.lv (S.V.)
2. Institute of Mathematics and Computer Science, University of Latvia, 29 Raina blvd, LV-1459 Riga, Latvia; harijs.kalis@lu.lv (H.K.); uldis.strautins@lu.lv (U.S.)
* Correspondence: antons.kolmickovs@gmail.com; Tel.: +371-29-910-674

Received: 11 March 2019; Accepted: 12 April 2019; Published: 22 April 2019

Abstract: The aim of this study was to provide more effective use of straw for energy production by co-firing wheat straw pellets with solid fuels (wood, peat pellets) under additional electric control of the combustion characteristics at thermo-chemical conversion of fuel mixtures. Effects of the DC electric field on the main combustion characteristics were studied experimentally using a fixed-bed experimental setup with a heat output up to 4 kW. An axisymmetric electric field was applied to the flame base between the positively charged electrode and the grounded wall of the combustion chamber. The experimental study includes local measurements of the composition of the gasification gas, flame temperature, heat output, combustion efficiency and of the composition of the flue gas considering the variation of the bias voltage of the electrode. A mathematical model of the field-induced thermo-chemical conversion of combustible volatiles has been built using MATLAB. The results confirm that the electric field-induced processes of heat and mass transfer allow to control and improve the main combustion characteristics thus enhancing the fuel burnout and increasing the heat output from the device up to 14% and the produced heat per mass of burned solid fuel up to 7%.

Keywords: co-firing; wheat straw; softwood; bog peat; pellets; thermal decomposition; combustion; DC electric field

1. Introduction

The present experimental study is in line with the EU-defined energy and environmental objectives, which emphasizes the need to increase by 27% the use of renewable energy sources (biomass, hydropower, geothermal, solar, wind and marine) for energy production, to enhance energy efficiency by 27% and to reduce greenhouse gas emissions into the environment by 40% by 2030 [1]. Among all renewable energy sources, plant biomass is recommended as a source of CO_2- neutral energy, so confirming that the combustion of biomass produces an amount of CO_2 comparable to that this absorbed at photosynthesis, thus reducing greenhouse gas emissions and the impact of energy producers on global warming [2]. However, different types of biomass, for example, wood waste, agriculture residues (straw) and partially bio-decomposed plant biomass (peat), which can be used for energy production, have a dissimilar structure, bulk density, elemental and chemical composition. Therefore, to provide a reliable energy production from plant biomass, the plant biomass should be shaped as pellets or briquettes with controllable density, structure, elemental and chemical composition. Moreover, to convert the biomass into useful forms of energy, thermo-chemical and biochemical conversion of biomass is required [3].

Nowadays, the analysis of the consumption of renewable fuel for energy production for district heating has already revealed the seasonal wood pellets shortage [4]. Therefore, the wider use of alternative biomass fuels, such as agriculture residues (rape straw, wheat straw, rice husk, etc.), must be considered. The feasibility of straw for energy production is limited due to its relatively low heating value and energy density, high nitrogen, moisture and ash content in the biomass [5], as well as due to the ash agglomeration and formation of health harmful polycyclic aromatic hydrocarbons naphthalene and phenanthrene emissions during the combustion of straw [4]. To minimize the negative effects of straw on the heat production and composition of the products, co-combustion/co-firing of straw with different types of solid fuels (coal, wood and peat) is preferable. The results of a systematic study of biomass co-firing give evidence that a promising way to ensure the most efficient use of straw for energy production with no harmful effect on the environment is co-combustion of straw with coal [6]. Therefore, co-firing of wheat straw with granulated wood and peat biomass is studied and analyzed with the aim of obtaining improved main characteristics of the straw thermo-chemical conversion [7,8]. By analogy with the effect of straw co-firing with coal [9], the thermal interaction between the components when straw is co-fired with wood or peat pellets results in enhanced thermal decomposition of the biomass pellet mix, in a faster and more intensive release of the combustible volatiles, their faster ignition and faster formation of the flame reaction zone, which enhances the fuel burnout. Co-firing of straw also increases the heat output from the device, the produced heat per mass of burned pellet mix and the volume fraction of CO_2, decreasing along the air excess in the flue gases. Previous experimental studies show [7–9] that the influence of the straw co-firing on the main combustion characteristics strongly depends on the straw share in the pellet mix, indicating the most effective improvement of the main combustion characteristics when the straw share is about 20–30%. Furthermore, the thermo-chemical conversion of straw depends not only on the straw share and mixture composition, but also depends on the type of air supply in the unit responsible for mixing of the combustible volatiles with air. To obtain better mixing of air with the combustible volatiles, improved combustion conditions and the reduced emission of CO and NOx pollutants, the use of swirling airflow is preferable [10,11], which even at a low swirling number ($S < 0.6$) enhances the reverse flow formation and recirculation of the products with the enhanced mixing of the air and combustible volatiles and enhanced burnout of volatiles.

The results of preliminary studies suggest that additional improvement of the combustion characteristics, heat production and composition of emissions at co-firing of straw with solid fuels can be achieved using the DC electric field effects on the flame [12,13]. The electric field effects on flames can be related to the electric field-induced ion wind formation [14] promoting the processes of field-induced heat/mass transfer in a field direction with direct influence on the flame shape, the local variations in the rate of reactions, the amount of produced heat and the composition of products [15,16]. The applicability of the electric field effects on flames for the control of the heat production and composition of emissions is confirmed by the industrial experiments [17].

With account of applicability of electric field for the control of the main characteristics of the combustion systems, the main goal of the present research is to assess the key factors which influence the development of combustion dynamics during co-firing of wheat straw with softwood or peat pellets, if the electric field is applied to the flame base. In addition, the paper tends to validate the results of the experimental study by means of numerical simulation and mathematical modelling of the field effects on the development of combustion dynamics at co-firing solid fuel mixtures.

2. Experimental Device and Methods

To study the effects of the wheat straw co-firing with biomass pellets of different origin (wood or peat), a batch-size pilot setup with a heat output up to 4 kW has been developed. The experimental setup (Figure 1) consists of a cylindrical 88 Ø × 250 mm biomass gasifier (1), a propane burner (2), a primary air supply (3), a secondary swirling air supply (4) and two cylindrical 88 Ø × 600 mm water-cooled combustion chamber sections (5) and an axially inserted electrode (6).

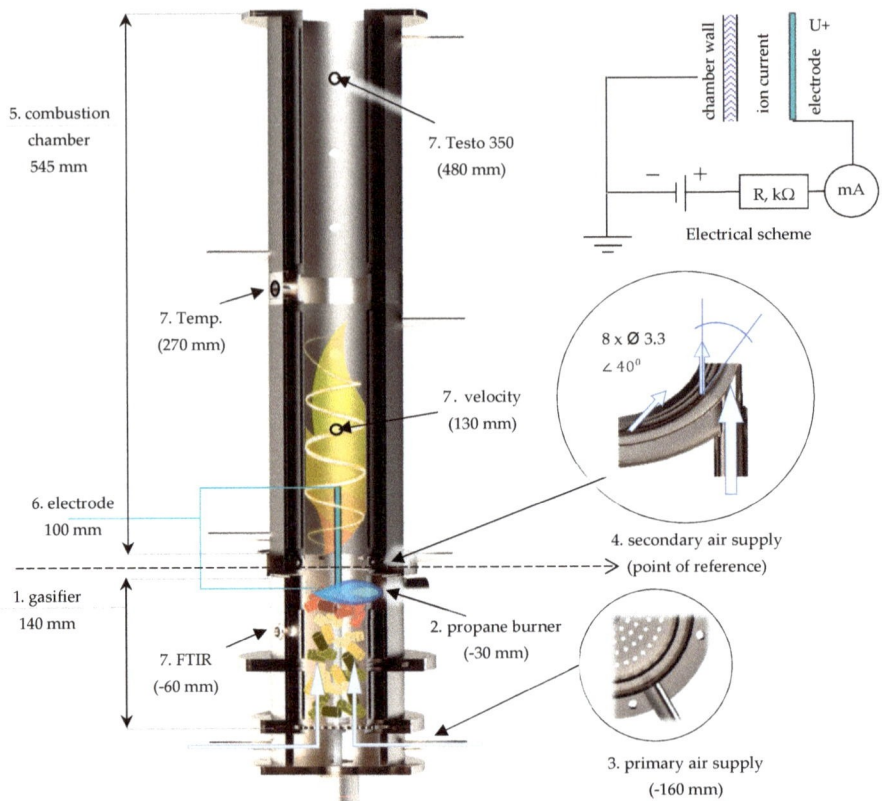

Figure 1. The batch-size pilot setup for experimental studies: 1—gasifier filled with a mixture of biomass pellets; 2—propane flame inlet nozzle; 3—primary air supply at the bottom of the gasifier; 4—secondary swirling air inlet at the combustor bottom; 5—water-cooled sections of the combustion chamber; 6—axially inserted central electrode; 7—openings the diagnostic tools. The technological scheme of main equipment and measurement instruments is available at Figure S2.

The gasifier was filled up to the propane supply nozzle with a discrete portion of a mixture of wheat straw and wood or peat pellets (430–550 g). The wheat straw share in the pellet mix varied from 10 to 100%. The thermal decomposition of biomass pellets was initiated and sustained up to 450 s by an external heat source–propane flame flow with the average heat power 0.86 kW and was switched off after an ignition of volatiles and the formation of self-sustaining thermochemical conversion of biomass pellets, which occurs at 400–500 s. The thermal decomposition of the biomass mixture in the gasifier developed in the fuel-rich conditions (air excess ratio <0.5) and resulted in formation of the axial flow of combustible volatiles (CO, H_2, C_2H_2, C_2H_4, CH_4, etc.) at the flame base. The thermochemical conversion of the biomass mixture lasted about 2600 s if straw is co-fired with wood pellets and about 3600 s if straw is co-fired with peat. The schematic diagram which describes the experiment is available at Figure S1.

The primary air (3), passing through the biomass layer at the average flow rate 30 L·min^{-1}, initiated an axial flow of the volatiles and sustained the char surface oxidation. The secondary swirling airflow, supplied to the bottom of the combustor at the average airflow rate 40 l·min^{-1}, supported the combustion of the volatiles. The air supply flowrate was estimated using Testo 6441 flowmeters (Testo SE & Co. KGaA, Lenzkirch, Germany) with an accuracy ± 3%.

The diagnostic tools (thermocouples, gas-sampling probes, Pitot tube) were introduced through the orifices (7) into the flame reaction zone to make local measurements of the flame temperature, composition of the flue gas and of the axial and tangential components of the flow velocity.

The DC electric field effect on the thermo-chemical conversion of biomass mixtures, the development of combustion dynamics and the composition of emissions were studied using the axially inserted positively biased electrode. The electrode was a 3 mm diameter nichrome wire rod with 100 mm of non-insulated length. The electrode was introduced through the biomass layer upward to the flame reaction zone. The bias voltage of the electrode was switched in after the switch off of the additional heat supply by propane flame and was varied from 0.6 kV up to 1.8 kV, the ion current in the space between the positive electrode and the grounded walls of the combustion chamber was limited to 7 mA at 248 kΩ resistance to avoid the formation of discharge.

The experimental study of the biomass mixture thermal decomposition involves complex time-dependent measurements of the biomass height (L) in the gasifier (dL/dt, mm·s^{-1}) using the moving rod with a pointer. From the measurements of dL/dt and volume density of the biomass mixture, the mixture mass loss rate (dm/dt, g·s^{-1}) was estimated with the accuracy ± 2%. The composition of the volatiles produced at the thermal decomposition of the biomass mixtures was measured at the gasifier outlet by the FTIR spectroscopy method. Gas samples of 50 ml were extracted at a 50 s time interval and analyzed using a Varian Cary 640 spectrometer (Agilent Technologies, Inc., Santa Clara, CA, USA) in the MIR spectral range for CO_2 (668 cm^{-1}), CO (average of 2115 and 2169 cm^{-1}), C_2H_2 (729 cm^{-1}), C_2H_4 (949 cm^{-1}), and CH_4 (3017 cm^{-1}). Additionally, at the gasifier outlet, CO and H_2 were measured by a Testo 350 gas analyzer.

The local measurements of the axial (u) and tangential (w) flow velocity components were made using a Pitot tube and a Testo 435 flowmeter, providing continuous online data monitoring with an accuracy of ± 1%. The air and gas flow swirl number (S, dimensionless), was calculated from the data of the radial measurements of the axial and tangential flow velocity profiles at about 130 mm above the secondary air supply as follows [10]:

$$S \approx \frac{2}{3} \cdot \frac{w_{av}}{u_{av}} \tag{1}$$

where: w_{av}, u_{av}—the average values of the tangential and axial velocity components, (m·s^{-1}).

A Pico logger (Pico Technology, Cambridgeshire, UK) recorded the local temperature data from the Pt/Pt-Rh thermocouples with an accuracy of ±5% providing data online registration. The local measurements of the flue gas composition, i.e., the mass fraction of unburned volatiles (CO, H_2), the volume fraction of the main combustion product (CO_2), the mass fraction of NO_x pollutant, as well as the combustion efficiency (η_{comb}) and the air excess ratio (α) were made by the Testo 350 gas analyzer. In accordance with the Testo 350 specifications, the O_2, CO_2 volume fraction was measured with an accuracy of ±1% and the mass fraction of CO, H_2 and NO_x with an accuracy of about ±5%. The extended description of the measurement data acquisition is at the supplementary materials.

The calorimetric measurements of the cooling water flow involve joint measurements of the cooling water mass flow, which was measured with the accuracy ± 2.5% and of the temperature, which were made with the accuracy ±1% by AD 560 thermo-sensors (Analog Devices, Inc., Norwood, MA, USA), along with online data registration by a Data Translation DT9805 data acquisition module (Data Translation GmbH, Bietigheim-Bissingen, Germany) using Quick DAQ software.

The combustion of the volatiles developed at the average air excess ratio $\alpha \approx 1.5$ in the flame reaction zone at the average value of the inlet airflow swirl number $S \approx 0.7$ (without the electric field applied) which is responsible for the mixing of the axial flow of volatiles with the secondary swirling airflow.

The commercially available wheat straw, softwood (wood working waste—is the mixture of different biological species: pine, spruce, etc.) biomass pellets used for the co-firing studies originated from Latvia, and the bog peat pellets originated from Skrebeļu bog (Latvia).

The elemental composition (C, H, N content) of softwood, wheat straw, bog peat biomass types were measured using the Vario Macro elemental analyzer (Elementar Analysensysteme GmbH, Langenselbold, Germany) and analyzed according to the LVS EN 15104:2011 standard. The ash content was measured as a residue after the treatment at 830 ± 10 K in an ELF 11/6B furnace (Carbolite Gero Limited, Parsons Ln, Hope Valley, UK) in accordance with the LVS EN 14775:2010 standard [18]. The higher heating value (HHV) of the pellets were calculated by the regression equation proposed by Friedl et al. [19]:

$$HHV = (3.55 \cdot C^2 - 232 \cdot C - 2230 \cdot H + 51.2 \cdot C \cdot H + 131 \cdot N + 20600) \cdot 10^{-3}, \; MJ \cdot kg^{-1} \qquad (2)$$

where the elemental composition (C—carbon content, H—hydrogen content, N—nitrogen content and W—moisture) of softwood, wheat straw and peat biomass (Table 1).

Lower heating value (LHV) of the pellets were calculated using the regression equation [20]:

$$LHV = HHV \cdot \left(1 - \frac{W}{100}\right) - 2.447 \cdot \left(\frac{W}{100} - 18.02 \cdot \frac{H}{200} \cdot \left(1 - \frac{W}{100}\right)\right), \; MJ \cdot kg^{-1} \qquad (3)$$

The thermal decomposition of the biomass pellets and their mixtures was studied in the 300–900 K temperature range by differential thermogravimetric and thermal analysis (TGA/DTA) in an oxidative atmosphere (50 ml·min^{-1} air flowrate) using a Star System TGA/DTA 851e (Metter Toledo, Columbus, OH, USA) at a heating rate of 10 K·min^{-1} [18]. The main elemental characteristics of the biomass pellets are summarized in Table 1.

Table 1. The elemental composition and heating values of straw, wood and peat pellets.

Biomass	C *, %	H *, %	O *, %	N *, %	Moisture, %	Ash *, %	LHV, MJ·kg^{-1}
Wheat straw	46.62	5.09	42.72	1.31	9.11	4.26	15.52
Softwood	49.79	5.15	44.24	0.18	6.32	0.64	17.06
Bog peat	53.83	5.11	36.93	1.11	11.44	3.02	17.53

* relative to dry mass.

The main elemental characteristics of the biomass pellets (Table 1) show the highest carbon content as well as the highest LHV for peat, hence, it has also the highest moisture content among the studied biomass samples. Furthermore, peat has the lowest content of oxygen due to the partial bio-decomposition processes occurring in the bog land environment, therefore, it may require more air oxygen to sustain the combustion process.

Although the wheat straw total nitrogen content is very high compared to softwood, bog peat also has a high content of nitrogen in its elemental composition, which may evidence of the peat botanical composition characterized mostly by sphagnum moss, sedge and other acid-water plants [21].

3. Results and Discussion

In order to determine the electric field effects on the thermo-chemical conversion of wheat straw mixtures by co-firing wheat straw with wood or with peat pellets, firstly, the effect of wheat straw share in the mixture (wt.%) on the thermal decomposition of the mixture and heat release was studied.

The TGA and DTA analysis results show that the variations of the straw mass share in the mixtures correlate with the complex variations of the volatiles and char formation and their combustion, depending on the chemical composition of the components (Figure 2a–d). As follows from Figure 2a,b, the most intensive mass loss (1.4 mg·min^{-1}), due to the formation of volatiles, with the most pronounced exothermic heat effect on the thermal decomposition at 598 K was achieved for the softwood samples, which have the highest content of holocellulose (hemicellulose and cellulose together) 75–80% [22]. The less intensive formation of volatiles was observed for the peat samples (0.4 mg·min^{-1} at 578 K), which have the lowest content of polysaccharides (10–30% [23]), however, the volatile carbon content can achieve ≈58% [5]. The middle level of the volatile formation (1.1 mg·min^{-1}) was observed for the wheat straw samples with the highest content of holocellulose (58–60%) [24] and with the lowest thermal decomposition temperature (566 K).

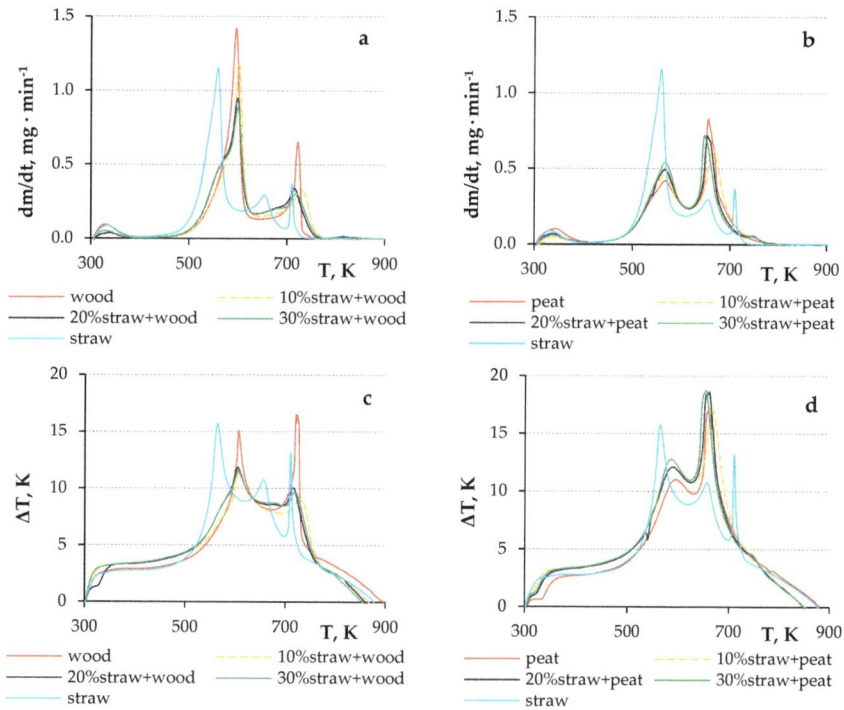

Figure 2. Results of TGA and DTA analysis at thermal decomposition and combustion of straw mixtures with wood (**a,c**) and with peat (**b,d**).

The second peak of the mass loss rate at 665 K on the TGA curve of the wheat straw sample is comparable to the one at the second thermal decomposition stage of the bog peat under analysis (Figure 2b), however the third sharp peak at ≈ 710 K could match the second stage of wood thermal decomposition (Figure 2a).

The DTA analysis, when heating the mixtures in an oxidative atmosphere, also suggests that the combustion of volatiles initiates an intensive heat release at temperatures about 590–620 K. The second pronounced thermal conversion stage at T > 670 K corresponds to the heat release at a less intensive conversion of the aromatic compounds of lignin for wood and straw [25,26] and at the conversion of partially bio-degraded lignin and humic acids for peat [21] (Figure 2c,d).

For the mixtures of straw and wood with 10–30% mass fraction of straw, the heat release corresponds to the wood DTA curve, but has no sharp peaks. The DTA analysis of straw-peat mixtures indicates an increase in heat release at the volatiles combustion stage, due to the higher content of polysaccharides in wheat straw, and at the char combustion stage, which may be due to the differences in components of which the char is formed (Figure 2c,d).

The combustion of biomass discrete portions in the experimental setup allows to eventually analyze different stages of the biomass thermal decomposition and combustion of volatiles. The complete thermo-chemical conversion of discrete portions of wheat straw, softwood and their mixtures lasts about 2900 s, whereas the thermo-chemical conversion of peat and its mixtures with straw takes up to 3600 s (Figure 3a,b). During this period the intensive biomass heating and devolatization processes stimulate the primary flaming combustion of the volatiles with transition to the self-sustained volatiles combustion (at 800–900 s for straw/wood and 900–1000 s for straw/peat), when the exothermic reactions of the thermo-chemical conversion of combustible volatiles maintain a balance with the endothermic processes of biomass heating, dewatering and thermal decomposition.

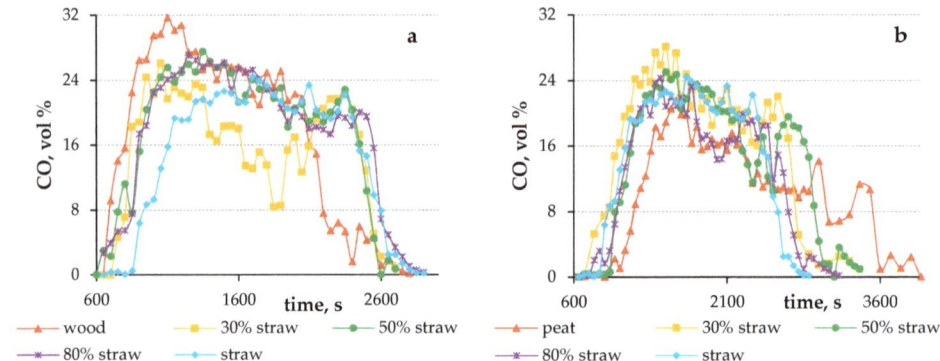

Figure 3. Effect of the wheat straw share wt.% in a mixture with wood (**a**) and with peat (**b**) pellets on the CO formation kinetics in the gasifier (FTIR).

The FTIR spectrum analysis of the CO formation at co-gasification of straw with wood or with peat pellets shows the formation of specific stages of their thermal decomposition. The primary stage of the biomass thermochemical conversion with transition from the endothermic biomass thermal decomposition to the exothermic ignition and flaming combustion of the volatiles lasts from 600 s up to 900 s for wood pellets, from 800 s to 1200 s for straw pellets, and from 800 s to 1400 s for peat pellets, when the flame temperature increases from the minimum value (about 600 K) to the maximum value—1000–1200 K (min-max). The next stage of the self-sustained combustion of volatiles at the thermo-chemical conversion of wood pellets lasts up to ≈1700 s; when co-firing straw with wood it takes up to ≈1800 s and up to ≈2300 s for peat and the flame temperature decreases from the maximum value to the end value of self-sustained combustion—about 800–900 K (max-end). The transition to the char combustion stage results in an enhanced formation of CO, whereas CH_4 is no longer detected. For wood pellets, the transition to the char conversion stage occurs within the 1700 s to 2200 s time interval; for straw at t ≈ 1800–2300 s and for straw/wood mixtures at t ≈ 1800–2300 s. It should be noted that transition to char conversion stage results in a rapid decrease of the flame temperature to 550–650 K, which suggests the development of the endothermic processes, which is confirmed by a correlating increase of the mass fraction of CO and H_2 in the products.

The kinetics of thermochemical conversion of different biomass types are caused by the specific elemental and chemical composition of pellets, but the most characteristic features of the combustion process of different biomass types can be specified. The intensive thermal decomposition of softwood may be explained by the higher heating value of wood pellets, which provides the faster balance between the exothermic and endothermic processes during the thermochemical conversion of pellets. The wheat straw thermochemical conversion is accompanied by the char/ash layer formation on the surface of the raw biomass, which gradually restricts the air access to the reaction zone. The bog peat thermal decomposition is delayed due to the relatively high density of its pellets, high moisture content and low volatiles content, which generally prevents the formation of the primary flaming combustion stage. Nevertheless, the thermochemical conversion of bog peat pellets is a stable and long-lasting process, which is developing with the reduced flame length due to the low content of volatile matter and high content of fixed carbon.

The thermochemical conversion of biomass pellets at co-firing of straw with wood or peat closely links to variations of the mass loss of the mixtures, depending on the mixture composition. During the primary stage of the flame formation ($t < 1000$ s) increasing the mass fraction of straw in the mixture with wood pellets delays the formation and ignition of volatiles (Figure 3a) decreasing to the minimum value the mass loss rate of the mixture (Figure 4a). During the self-sustaining thermochemical conversion of the mixture (max-end) the inverse trend is observed. Increasing the mass fraction of straw in the mixture up to 30% causes the enhanced development of the exothermic reactions at thermo-chemical conversion of the mixture with an increase up to the peak value the mass loss rate of the mixture (Figure 4a). Besides, to the minimum value decreases the mass fraction of combustible volatiles entering the combustor (Figure 4c), which suggests the enhanced burnout of volatiles. Increasing the share of straw in the mixture above 30% promotes a decrease of the weight loss rate of the mixture, whereas increases the volume fraction of combustible volatiles in the flowing gas.

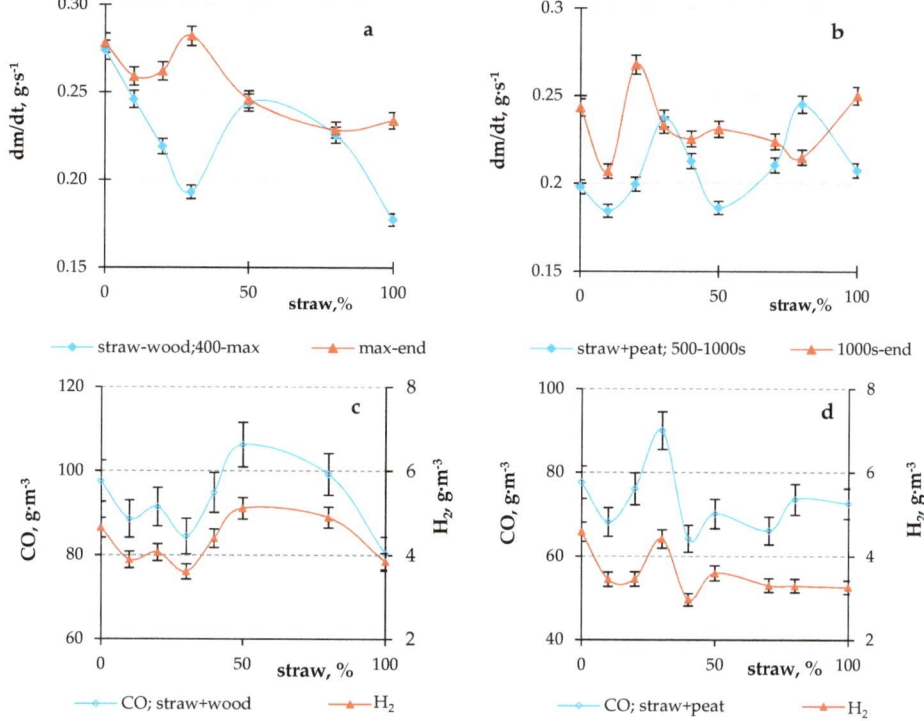

Figure 4. Effect of the wheat straw mass share wt.% in a mixture with wood (**a,c**) and with peat (**b,d**) pellets on the biomass mass loss rates (**a,b**) at different stages of thermo-chemical conversion and on the flowing gas composition at the outlet of the gasifier (**c,d**).

During co-firing of straw with peat increasing the mass fraction of straw in the mixture up to 30% causes the enhanced formation of volatiles (Figure 3b) with a correlating increase up to the peak values the mass loss rate of the mixture and the average values of the volume fraction of combustible volatiles in the gas entering the combustor (Figure 4b,d). Increasing the mass fraction of straw in the mixture above 30% promotes a decrease of the weight loss rate of the mixture with a correlating decrease of the volume fraction of combustible volatiles in the flowing gas.

The research results suggest that the thermal decomposition of the mixtures during co-firing of straw with wood or peat is influenced not only by the variations of the elemental composition and heating values of the components (Table 1), but also by the thermal interaction between the components, which determine their thermo-chemical conversion and composition of the products (Figure 5a,b).

As follows from Figure 5, the enhanced thermal decomposition of the mixture at co-firing of straw with wood or pellets correlates with an increase of the CO_2 volume fraction in the products and flue gas temperature up to peak values, if the mass share wt.% of straw in the mixture is about 30% and starts to decrease if the mass fraction of straw in the mixture exceeds and the further rise of the mixture thermal decomposition is limited by a linear decrease of its heating value. Moreover, with the 30% straw mass share wt.% in the mixture, its co-firing with wood reduces the NO_x mass fraction in the flue gas from 330 ppm to 200 ppm, whereas the straw co-firing with peat reduces the NO_x mass fraction from 290 ppm to 210 ppm. This confirms that the co-firing of straw with wood or with peat assures the cleaner heat production.

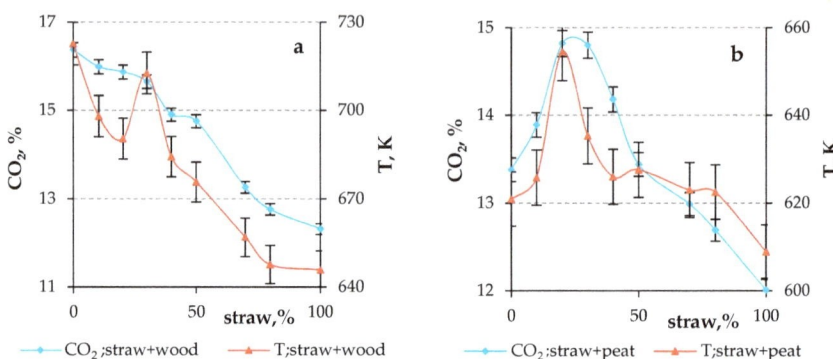

Figure 5. Effect of the straw mass share wt.% on the CO_2 volume fraction in the products and on the flue gas temperature when co-firing straw with wood (**a**) or with peat (**b**).

3.1. Electric Field Effect on the Thermal Decomposition and Combustion Characteristics When Co-Firing Straw with Wood or PEAT Pellets

From the results presented above it follows that co-firing of straw with wood or peat pellets allows the enhanced thermochemical conversion of the mixtures if the mass fraction of straw in the mixture is limited to 20–30%. To assess the potential for the additional improvement of the co-firing of the straw mixture with wood or peat, the electric field effects on the development of thermochemical conversion of the mixtures were studied and analyzed at the fixed mass fraction of straw in the mixture (30%).

When co-combusting straw and peat, the electric field induced variations of the main flame characteristics are determined by the formation of charged flame species, and by the field-induced transport of positive and negative gaseous species from the pyrolysis and reaction zones in the field direction. Elastic collisions between the ions and gaseous compounds result in a momentum transfer from charged particles (predominately positive ions) to neutral species, so enhancing their transport in the field direction and producing the so-called "ionic wind" phenomenon [14] with a direct impact on the swirling flame shape and structure [27,28]. The ion wind formation is strongly influenced by the electric body force $F = q \cdot E$ determining the motion of charged flame species and advancing the heat and mass transfer of neutrals and chemical species. Therefore, to provide effective electric control of the main flame characteristics, it is necessary to apply the electric field to the flame area with a maximum ion density (q). It is generally assumed [29] that the formation of positive ions in hydrocarbon flames can be related to the development of chemical ionization reactions between the flame components:

$$CH + O \rightarrow CHO^+ + \bar{e} \quad (4)$$

$$CH^* + C_2H_2 \rightarrow C_3H_3^+ + \bar{e} \quad (5)$$

$$C_2H + O_2^* \rightarrow CO + CHO^+ + \bar{e} \quad (6)$$

In general, the formation of ions at thermal decomposition of biomass pellets is a result of the consequent release of different hydrocarbons (C_xH_y). The measurements and analysis of the produced gases released at the thermal decomposition of biomass pellets (straw, peat and their mixtures) confirm the intensive formation of the combustible volatiles (H_2, CO) and hydrocarbon traces (C_2H_2, C_2H_4, CH_4), which are responsible for the formation of the flame reaction zone and primary flame ions [12,13]. The measurements of the radial and axial distributions of the flame ions prove that the most intensive formation of the flame ions occurs at the primary stage of flaming combustion (t < 1200 s). The peak value of the ion density (5–6)·10^{17} m^{-3} was observed at the bottom of the combustor (48–60 mm from the biomass surface), close to the flame axis (r/R = 0), where the axial flow of hydrocarbon traces rapidly mixes with the air accelerating the ion formation via the mechanism (4–6). Hence, to obtain the most intensive field-induced variations of the flame characteristics, the electric field must be applied to this part of the flame.

The 30% straw mass share wt.% added to the wood or peat mixture was chosen to investigate the potential of electric control of the combustion dynamics at co-firing straw with wood or peat, with the aim to provide a wide use of straw as a fuel for cleaner heat production, as observed when providing the electric control of the swirling propane flame flow [15,16,28] and the electric field-induced body force enhances the evident variations of the flame shape and length determined by the field-induced variations of flow dynamics.

The complex measurements of the $_i$ electric field effect on the formation of flow dynamics at co-firing of straw with wood or with peat pellets were provided for the positively bias voltage of the axially inserted electrode, which is inserted into the flame reaction zone at 130 mm from the secondary air supply nozzle (Figure 1), where the average flame temperature approaches to 1350K. For the given configuration of the electrode the electric body force acts on the positive flame ions enhancing the processes of radial and reverse axial heat/mass transfer. As a result, to minimum value decreases the axial flow velocity, increasing the swirl intensity of secondary air and enhancing mixing of the reactants along the outside of the flame reaction zone (Figure 6a,b). The field-induced decrease of the axial flow of combustible volatiles increases the residence time of the reactants in the flame reaction zone completing the burnout of the volatiles.

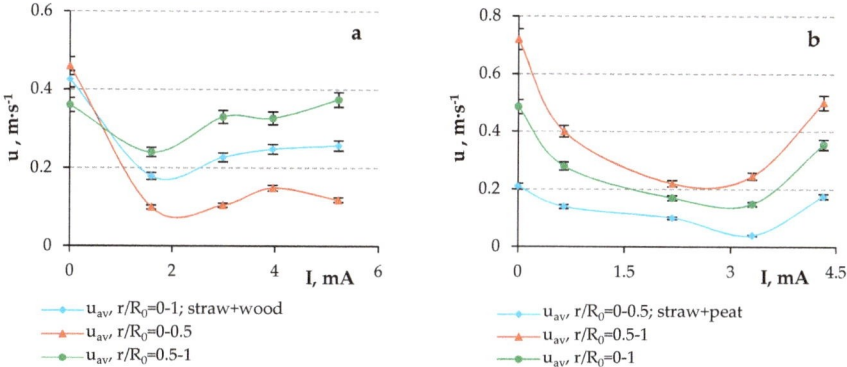

Figure 6. Electric field effect on the average values of the axial flow velocity component of the vortex flow forming in the combustion chamber at co-firing of straw with wood (**a**) or with peat (**b**).

Besides, the field enhanced reverse axial heat mass transfer up to the biomass layer enhances heating and thermal decomposition of the biomass pellets increasing the biomass mass loss rates and the volume density of the volatiles at the gasifier outlet depending on the field-induced ion current (Figure 7a,b).

With the 30% mass share of straw in the mixture and with increasing bias voltage of the positive electrode and the ion current in the space between the electrodes, the mixture mass loss rate tends to decrease during the primary stage of volatiles formation (400 s–max), when the electric field enhances the development of the endothermic processes of the thermal decomposition and it increases during the self-sustained burnout of the volatiles (max–end).

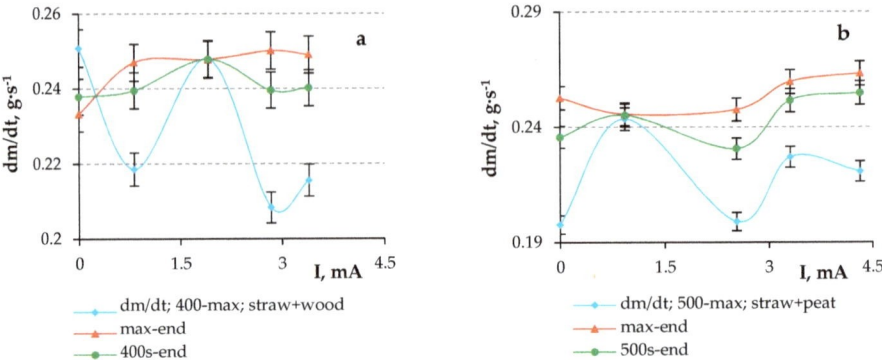

Figure 7. Electric field effect on the biomass pellet mixture mass loss rate at thermal decomposition when co-firing straw with wood (**a**) or with peat (**b**).

The field-enhanced thermal decomposition of the pellet mixtures correlates with the increased volume fraction of the volatiles at the gasifier outlet. The local decrease of the volume fraction of combustible volatiles at 1–3 mA ion current suggests that the field-enhanced reverse axial mass transfer initiates the enhanced mixing of the axial flow of volatiles with the secondary air-flow, thus enhancing the burnout of the volatiles (Figure 8a,b).

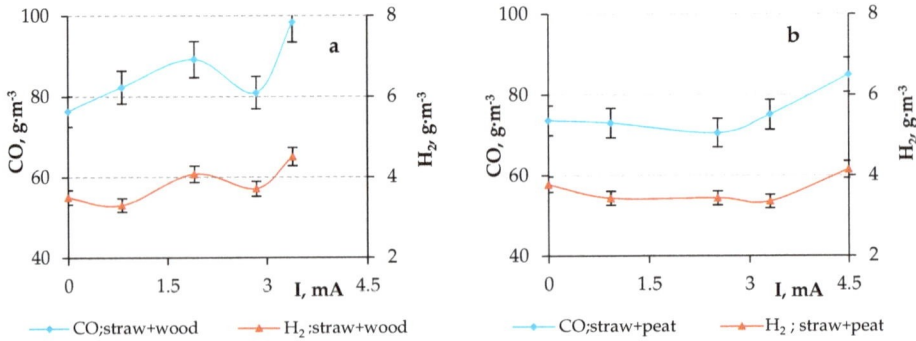

Figure 8. Electric field effect on the formation of volatiles at the gasifier outlet when co-firing straw with wood (**a**) or with peat (**b**).

As the field-enhanced decrease of the axial flow velocity determines the increase of the residence time of the reactants in the primary zone of flame formation completing combustion of volatiles, the heat power ($P_{selfsus}$) from the device at the self-sustained combustion stage increases by ~6–8% with the correlating increase of the total collected heat produced per mass of burned mixture (Q_{sum}) by ~10–12% (Figure 9a–d).

Although the heat power of the device at the stage of self-sustained combustion keeps growing, the total produced heat per mass of burned solid fuel decreases after reaching its peak value at I = 1.9 mA at straw/wood co-firing and at I = 2.5 mA when co-firing straw and peat, which can be theoretically explained by the field-induced increase of the axial flow velocity and decrease of the air swirl intensity thus reducing the reaction residence time and by limiting mixing of the reactants (Figure 6).

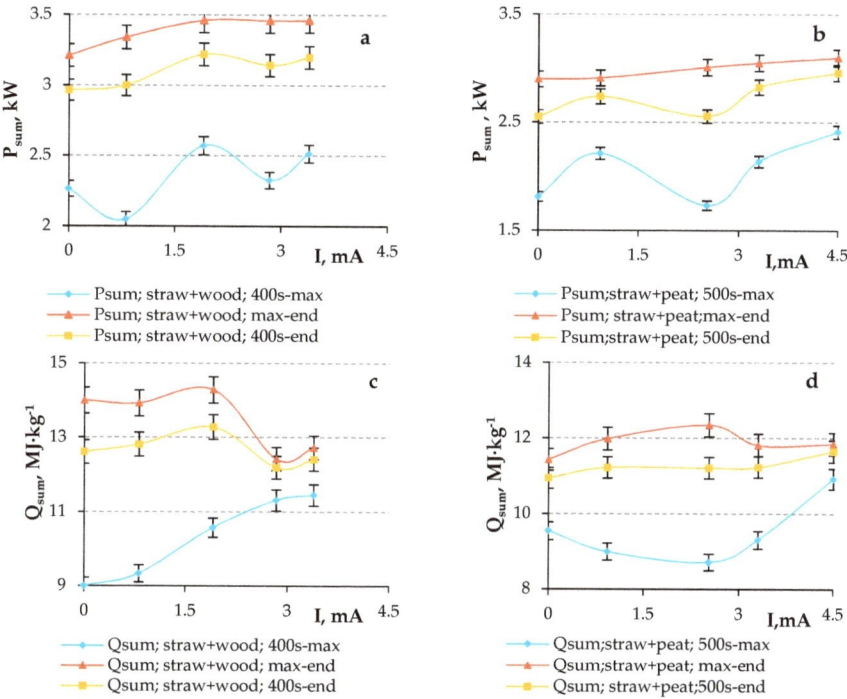

Figure 9. Electric field effect on the heat output from the device (**a**,**b**) and on the produced heat per mass of burned mixture (**c**,**d**) when co-firing straw (30%) with wood (**a**,**c**) or with peat (**b**,**d**) pellets.

Finally, it should be noted however, that the field-enhanced thermo-chemical conversion of the mixtures affects the flue gas composition, increasing the carbon-neutral CO_2 emission and combustion efficiency, whereas the air excess ratio in the products decreases to the minimum value, providing so the cleaner and more efficient heat production (Figure 10a,b).

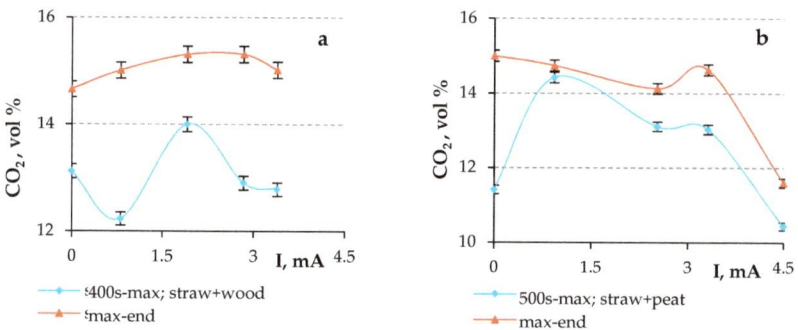

Figure 10. Electric field effect on the CO_2 volume fraction in flue gases when co-firing 30% straw with wood (**a**) or with peat (**b**) pellets.

3.2. Mathematical Modelling of Electric field Effects on Combustion Dynamics When Co-Firing Straw with Wood or with Peat Pellets

In order to analyze in more detail, the processes developing downstream the combustor at the co-combustion of straw and peat pellets, mathematical modelling and numerical simulation of the processes developing when the electric body force is applied to the flame base were performed using two dominant second-order chemical reactions of the volatile (H_2, CO) combustion:

$$H_2 + OH \rightarrow H_2O + H \tag{7}$$

$$CO + OH \rightarrow CO_2 + H \tag{8}$$

The maximum values of the temperature, axial flow velocity and mass fractions of the CO_2 and H_2O species were obtained from a numerical analysis of the systems of nine and eleven parabolic type partial differential equations (PDS) describing the 1D compressible reacting swirling flow [8] at co-firing straw with peat. Numerical modelling was made in accordance with the experimental data assuming the CO and H_2 mass fractions as boundary conditions.

When studying the influence of the electric field on the thermo-chemical conversion of straw pellet mixtures with peat, a simplified model has been proposed [30] which considers the interplay of a 2D compressible, laminar, axisymmetric flow and electrodynamic effects due to the impact of Lorentz forces on the electrons produced during chemical reactions. The axial velocity of the uniform flow in the central part of the combustor inlet $u_o = 0.1$ m·s^{-1}. A simple exothermic chemical reaction A → B was modelled by Arrhenius kinetics using a single step reaction between the fuel and the oxidant. A more plausible model is the A → B ↔ C mechanism, where 'B' represents the intermediate products and 'C' is the final product [31]. This mechanism has been used for the investigation of the straw co-combustion with wood or with peat (1-to-9 ratios), considering the reactions between the chemical substances (CO, H_2, O_2, OH) and the formation of products (CO_2, O, H_2O). In the numerical experiment the applied electric field induces an electric current with the uniform density between the walls of the combustor and the axially inserted electrode at the combustor inlet with no presence of ions. The mathematical model is described by four Euler and three reaction-diffusion and azimuthally induced magnetic field (component B_ϕ) dimensionless equations in the cylindrical coordinates (r, $x = z \cdot r_o^{-1}$) at the time:

$$\begin{cases} \frac{\partial \rho}{\partial t} + M(\rho) + \rho\left(\frac{1}{r} \cdot \frac{\partial (r \cdot u_r)}{\partial r} + \frac{\partial u_{ax}}{\partial x}\right) = 0; \\ \frac{\partial u_r}{\partial t} + M(u_r) - S\frac{u_{tg}^2}{r^3} = -\frac{1}{\rho} \cdot \frac{\partial p}{\partial r} + Re^{-1}\left(\Delta u - \frac{u_r}{r^2}\right) + \frac{1}{\rho} \cdot P_e F_r; \\ \frac{\partial u_{ax}}{\partial t} + M(u_{ax}) = -\frac{1}{\rho} \cdot \frac{\partial p}{\partial x} + Re^{-1} \cdot \Delta u_{ax} + \frac{1}{\rho} \cdot P_e F_r; \\ \frac{\partial u_{tg}}{\partial t} + M(u_{tg}) = Re^{-1} \cdot \Delta_* u_{tg}; \\ \frac{\partial T}{\partial t} + M(T) = P_1 \cdot \frac{1}{\rho} \cdot \Delta T + q_1 A_1 C_1 \cdot exp\left(-\frac{\delta_1}{T}\right) + q_2\left(A_2 C_2 \cdot exp\left(-\frac{\delta_2}{T}\right) + A_3 C_3 \cdot exp\left(-\frac{\delta_3}{T}\right)\right); \\ \frac{\partial C_1}{\partial t} + M(C_1) = P_2 \cdot \Delta C_1 - A_1 C_1 \cdot exp\left(-\frac{\delta_1}{T}\right); \\ \frac{\partial C_2}{\partial t} + M(C_2) = P_2 \cdot \Delta C_2 + A_1 C_1 \cdot exp\left(-\frac{\delta_1}{T}\right) - A_2 C_2 \cdot exp\left(-\frac{\delta_2}{T}\right) + A_3 C_3 \cdot exp\left(-\frac{\delta_3}{T}\right); \end{cases}$$

$$\frac{\partial B_\phi}{\partial t} + M(B_\phi) = \Delta_* \cdot B_\phi; \tag{16}$$

where:

$$\Delta q = \frac{\partial^2 q}{\partial x^2} + \frac{1}{r}\frac{\partial}{\partial r}\left(r\frac{\partial q}{\partial r}\right); \tag{17}$$

$$\Delta q = \frac{\partial^2 q}{\partial x^2} + r\frac{\partial}{\partial r}\left(\frac{1}{r}\frac{\partial q}{\partial r}\right); \tag{18}$$

$$M(q) = u_{ax}\frac{\partial q}{\partial x} + u_r\frac{\partial q}{\partial r}; \tag{19}$$

$$C_3 = 1 - C_1 - C_2 ; \tag{20}$$

where C_1, C_2, C_3 are the mass fractions of volatiles, the intermediate product and the final product. $u_{ax} = u_z \cdot u_0^{-1}$, $u_r = u_{rad} \cdot u_0^{-1}$, $u_{tg} = r \cdot u_\varphi$ are the normalized axial, radial and tangential velocities circulation. q denotes any of the quantities ρ, u_r, u_{ax}, u_{tg}, T, C_1, C_2, B_φ.:

$$P_2 = D/(U_0 \cdot r_0) = 0.01 ; \tag{21}$$

$$P_1 = \lambda/(c_p \cdot \rho_0 \cdot U_0 \cdot r_0) = 0.05 ; \tag{22}$$

$$q_1 = Q_1/(c_p \cdot T_0) = 5 ; \tag{23}$$

$$q_2 = Q_2/(c_p \cdot T_0) = 1 ; \tag{24}$$

where $Q_1 = 1.5 \cdot 10^6$ and $Q_2 = 0.3 \cdot 10^6$ are the heat losses of the reaction (J·kg^{-1}), but

$$\delta_k = E_k/(R \cdot T_0) ; \tag{25}$$

where ($\delta_1 = \delta_3 = 10$, $\delta_2 = 13$) are the scaled activation energies, $R = 8.314$ (J·mol^{-1}·K^{-1}) is the universal gas constant, and $E_1 = E_3 = 2.5 \times 10^5$ (J·mol^{-1}), $E_2 = 3.2 \times 10^5$ (J·mol^{-1}) are the activation energies. $\lambda = 0.25$ (W·m^{-1}·K^{-1}) is the thermal conductivity; $D = 2.5 \times 10^{-4}$ (m^2·s^{-1}) is the molecular diffusivity of species; $A_k = A'_k \cdot r_0 \cdot U_0^{-1}$, ($A_1 = A_3 = 5 \cdot 10^4$, $A_2 = 5 \cdot 10^5$) are the scaled pre-exponential factors ($A'_k = s^{-1}$); $c_p = 1000$ (J·kg^{-1}·K^{-1}) is the specific heat capacity; ρ is the density normalized to the inlet density $\rho_0 = 1$ (kg·m^{-3}), $Re = u_0 \cdot r_0 \cdot \eta^{-1} = 1000$ is the Reynolds number, $\eta = 5 \times 10^{-6}$ (kg·s·m^{-1}) is the viscosity.

The equations were made dimensionless by scaling all the lengths to $r_0 = 0.05$ m, the meridian velocity to $u_0 = 0.1$ m·s^{-1}, the tangential velocity to $u_{tg,0} = 3 \cdot u_0$, the temperature to $T_0 = 300$ K, the pressure to $\rho_0 \cdot u_0^2$ (N·m^{-2}), the current density to:

$$j_r = -\frac{1}{\mu} \frac{\partial B_\phi}{\partial z} ; \tag{26}$$

$$j_z = \frac{1}{\mu r} \frac{\partial (r \cdot B_\phi)}{\partial r} ; \tag{27}$$

$$j_0 = \frac{I}{(2\pi r_0^2)} = \left[\frac{A}{m^2}\right] ; \tag{28}$$

the azimuthal induction B_φ of the magnetic field to:

$$B_0 = \frac{\mu \cdot I}{(2\pi r_0)} = \left[\frac{N}{A \cdot m}\right] ; \tag{29}$$

the electromagnetic forces F_r, F_z to:

$$F_0 = j_0 \cdot B_0 = \left[\frac{N}{m^3}\right] ; \tag{30}$$

where:

$$\mu = 4 \cdot \pi \cdot 10^{-7} = \left[\frac{N}{A^2}\right] ; \tag{31}$$

is the magnetic permeability, $I = 0(0.001)0.01$ [A] is the electric current. The dimensionless radial and axial components of electromagnetic forces:

$$F_r = -\frac{B_\varphi}{r} \frac{\partial (r \cdot B_\varphi)}{\partial r} ; \tag{32}$$

$$F_z = -B_\varphi \frac{\partial B_\varphi}{\partial z} \ ; \tag{33}$$

were quantified by the parameter:

$$P_e = \frac{B_0 \cdot j_0 \cdot r_0}{\rho_0 \cdot u_0^2} \ ; \tag{34}$$

For the dimensionless pressure p a model for perfect gas is used: $p = \rho \cdot T$.
The boundary conditions are the following [31]:

(1) $r = 0$ (along the axis):

$$u_r = u_{ax} = 0; \ \frac{\partial T}{\partial r} = \frac{\partial C_k}{\partial r} = \frac{\partial u_{ax}}{\partial r} = 0; \ B_\phi = 0 \ ; \tag{35}$$

(2) $r = 1$ (at the wall):

$$u_r = u_{ax} = 0; \ \frac{\partial u_{ax}}{\partial r} = \frac{\partial C_k}{\partial r} = 0; \ \frac{\partial T}{\partial r} + B_1(T-1) = 0 \ ; \tag{36}$$

$B_\varphi = B_0 \cdot (1 - x/x_0)$ (consistent with the uniform distribution of j_r = const),

(3) $x = x_0 = 2$ (at the combustor outlet):

$$u_r = 0; \ B_\varphi = 0 \ ; \ \frac{\partial T}{\partial x} = \frac{\partial C}{\partial x} = \frac{\partial u_{ax}}{\partial x} = \frac{\partial u_{tg}}{\partial x} = 0 \ ; \tag{37}$$

(4) $x = 0$ (at the combustor inlet):

$$u_r = 0; \ T = 1; C_2 = 0 \ for \ r \in [0,1] \ ; \tag{38}$$

$$u_{ax} = 1; \ C_1 = 1, \ u_{tg} = 0 \ for \ r \in [0, r_1]; \tag{39}$$

$$u_{tg} = 4 \cdot r \frac{(r-r_1)(1-r)}{(1-r_1)^2}; \ u_{ax} = 0, \ C_1 = 0 \ for \ r > r_1 \tag{40}$$

It should be noted, that a uniform jet flow develops at $r < r_1$ and rotation at $r > r_1$ [30,32]:

$$B_\varphi = \frac{B_0}{r} \ for \ r \in [r^*, 1] \ (j_z = 0) \ ; \tag{41}$$

$$B_\varphi = \frac{B_0}{r}\left(1 - \sqrt{1 - \frac{r^2}{r^{*2}}}\right) \ for \ r \in [0, r^*] \ ; \tag{42}$$

Here:

$$B_i = \frac{h \cdot r_0}{\lambda} = 0.1 \ ; \tag{43}$$

is the Biot number, $r_1 = 0.75$, $r^* = 0.2$, $h = 0.1$ [J·(s·m^2·K)$^{-1}$].

For numerical solutions the stream function Ψ with the following expressions was introduced:

$$r \cdot \rho \cdot u_{ax} = \frac{\partial \Psi}{\partial r} \ ; \ r \cdot \rho \cdot u_r = \frac{\partial \Psi}{\partial x} \ ; \tag{44}$$

where $\Psi_{r=1} = q = 0.5 \cdot r_1^2 = 0.28125$ is the dimensionless fluid volume.

The intensity of vorticity with reverse orientation at the top left corner of the computational domain is expressed as: $I_v = (q - \Psi_{max}) \cdot d^{-1}$, where $d = 1/40$ is the time step of the uniform grid.

To solve a discrete 2D problem with 40 x 80 uniform grid points and the 0.0008 s time step, the ADI method of Douglas and Rachford [33] was used. For the stationary solution with the maximum error 10^{-7}, approximately 20,000-time steps were used (the final time t_f was approximately 10 s). The distribution of the axial, radial and tangential components of velocity, vorticity intensity and temperature were calculated with MATLAB (Table 2).

Table 2. The flame flow parameters: axial velocity ($u_{ax,\,max}$), radial velocity ($u_{r,\,max}$), temperature (T_{max}), mass fraction of intermediate products ($C_{2,\,max}$), radial velocity ($u_{r,\,min}$), mass fraction of final products ($C_{3,\,min}$), vorticity intensity (I_v) and radially-averaged flow temperature (T_{av}), depending on the electromagnetic parameter P_e for the flow swirl number $S = 3$.

P_e	$C_{3,\,min}$	I_v/u_0	$u_{ax,\,max}/u_0$	$u_{r,\,max}/u_0$	$u_{r,\,min}/u_0$	T_{max}/T_0	T_{av}/T_0	$C_{2,\,max}$
0	0.8022	−0.244	4.58	2.61	0	3.650	3.376	0.4056
0.1	0.8025	−0.252	4.59	2.62	−0.11	3.667	3.379	0.4062
0.2	0.8029	−0.256	4.60	2.63	−0.25	3.693	3.384	0.4066
0.5	0.8037	−0.272	4.81	2.66	−0.78	3.782	3.400	0.4066
1.0	0.8013	−0.296	5.31	2.70	−1.73	3.910	3.400	0.4049
2.0	0.7951	−0.328	6.67	2.74	−3.43	4.192	3.333	0.4025
2.5	0.7805	−0.340	7.48	2.76	−4.14	4.288	3.296	0.4048
3.0	0.7781	−0.352	8.92	2.77	−4.85	4.375	3.267	0.4061
4.0	0.7749	−0.380	15.2	2.78	−6.02	4.521	3.222	0.4063

$C_{2\,end,\,max} = 1 - C_{3,\,min}$.

From the results presented in Table 2 it follows that when co-firing straw with peat pellets, the maximum value of the mass fraction of the intermediate product ($C_{2,\,max}$), the minimum value of the mass fraction of the final product ($C_{3,\,min}$) and the average temperature (T_{av}) increase at $P_e < 0.5$ and decrease at $P_e > 0.5$. A similar situation was observed for the maximum value of C_2, i.e., the maximum of C_2 at the gas outlet, $C_{2\,end,\,max} = 1 - C_{3,\,min}$. The temperature maximum values (T_{max}) in the center of the flow and the axial velocity ($u_{ax,\,max}$) increase with the growing P_e. If the electric field is applied to the flame base, the flame vorticity enhances due to the increasing absolute values of the negative $u_{r,\,min}$ and vorticity, which provides the enhanced mixing and combustion of the reactants by increasing the maximum flame temperature, as it follows from the results of the experimental study.

The development of the axial distribution of the radial velocity ($0 < r < 1; 0 < x < 2$) is illustrated in Figure 11. The radial velocity distribution on the z-axis at $r = 0.5$ demonstrates a strong decrease to negative values at the flame base $z/R_0 = [0; 0.75]$, whereas at $r = 0.75$ this distribution shows the most pronounced value decrease at $z/R_0 = [0.5; 1.25]$. Thus, the action of the electric body force at $P_e > 0.5$ leads to a radial reduction of the flame reaction zone at $z/R_0 = [0.5; 1.5]$ with a correlating decrease in radially-averaged flow temperature (Table 2).

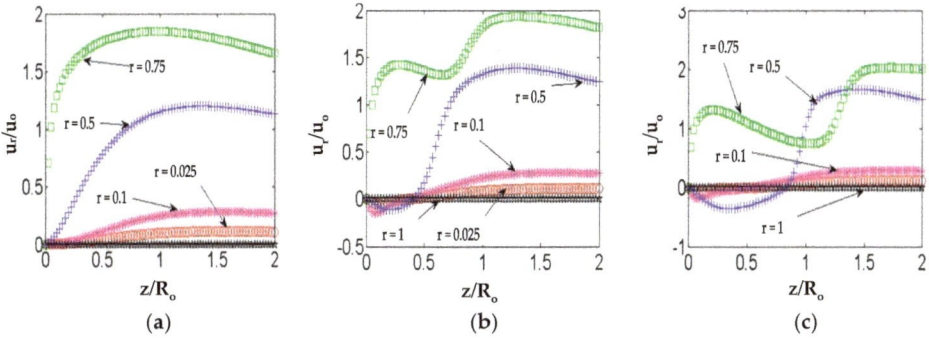

Figure 11. Development of the radial velocity z-axis distribution at $P_e = 0$ (a), $P_e = 0.5$ (b) and $P_e = 1$ (c).

In Figure 12, the represented temperature distribution in the combustion chamber section shows an increase of the maximum temperature T_{max} at the center of the flame base $z/R_0 = [0; 1]$, which may occur due to the field induced radial motion of the reactants towards the center and their more complete burnout. However, the action of the Lorenz force at $P_e > P_0$ leads to a decrease of the radially-average flow temperature T_{av}, which can be accompanied by a decrease of the visible flame radius.

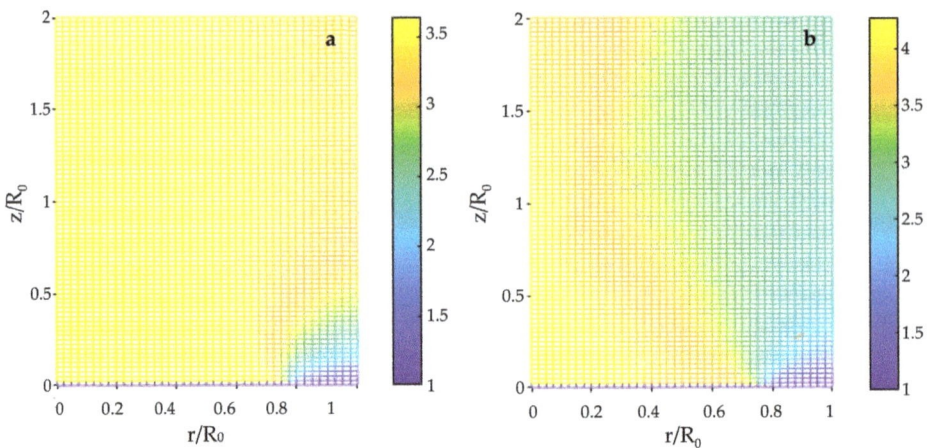

Figure 12. Electric field effect on the temperature development at $P_e = 0$ (a) and $P_e = 2.5$ (b).

Although the radial velocity maximum value $u_{r,\,max}$ at $z/R_0 > 1.5$ keeps growing according to Table 2, the absolute values of the minimum radial velocity $u_{r,\,min}$ increase much faster, which may lead to a contraction of the flame diameter and enlarge of the flame length, considering the dynamic growth of the axial velocity u_{ax} in the center determined by the P_e parameter.

4. Conclusions

The present study was aimed at a more effective use of straw as a fuel for energy production and combined complex experimental study and mathematical modelling of the processes developing when co-firing wheat straw with a solid fuel pellets (wood or peat). In order to assess the electric field applicability for additional control of the main flame characteristics, the electric field effects on the processes developing downstream the combustor were studied and analyzed.

The co-firing of straw with wood or with peat pellets results in the enhanced thermal decomposition of the mixture, which is determined by the mass share wt.% of straw in the mixture, and approaches its peak value if the straw mass share in the mixture is about 20–30%.

The field-induced ion current in the space between the electrodes is responsible for the field-enhanced reverse axial heat/mass transfer of the flame species, which provides the enhanced heating and thermal decomposition of biomass pellets. Increasing the current up to 2–3 mA decreases the average axial velocity at the flame base (70 mm from the secondary air supply nozzle) decreases from 0.36 m·s^{-1} to 0,24 m·s^{-1} for a 30% straw mixture with wood and from 0.49 m·s^{-1} to 0.15% for a 30% straw mixture with peat. In addition, it enhances the swirl intensity close to the flame axis by increasing the axial flow swirl number from 0.03 to 0.5.

The field-enhanced reverse axial heat transfer causes an increase of the average value of the weight loss rate from 0.23 g·s^{-1} to 0.25 g·s^{-1} during the self-sustained burnout of a straw/ wood mixture and from 0.24 to 0.26 g·s^{-1} for a straw/peat mixture with a correlating increase of the volume fraction of the combustible volatiles entering the combustor. For the straw/wood mixture, the volume fraction of CO at the flame base increases from 75 g·m^{-3} to 98 g·m^{-3}, for the straw/peat mixture–from 73 g·m^{-3} to 85 g·m^{-3}.

The enhanced release of the volatiles correlates with the increased heat output from the device. For the straw/wood mixture, the heat output from the device increases from 3.2 kW to 3.5 kW, for the straw/peat mixture–from 2.9 kW up to 3.1 kW. For the straw/wood mixture, the produced heat per mass of burned pellets increased from 13.9 MJ·kg^{-1} to 14.3 MJ·kg^{-1}, for the straw/peat mixture–from 11.4 MJ·kg^{-1} to 12.3 MJ·kg^{-1}.

The field-enhanced combustion of volatiles at thermo-chemical conversion of the mixtures is confirmed by the increase of the CO_2 volume fraction in the products, with dominant increase of CO_2 at the primary stage of volatiles burnout, when CO_2 increases from 13% to 14% in the straw/wood mixture, whereas in the straw/peat mixture from 11.4% up to 14.4%.

The results of the mathematical modelling show that the action of the electric body force at $P_e > 0.5$ leads to a contraction of the flame diameter and enlarges the flame length because of the decrease of the average value of the flame temperature due to the axial expansion of the flame reaction zone and to the correlating increase of the maximum flame temperature at the center. Moreover, if the electric field is applied to the flame, the flame vorticity enhances, which strengthens the mixing and combustion of the volatiles.

The results of the numerical study explain the effect of the electric field applied to the combustion flame flow by the influence of the Lorenz force on the electrons produced in the flame, but the experimental results show the integral effect of the field-induced motion of the positively and negatively charged species in the flame. The mathematical model needs further development, nevertheless, the main combustion parameters obtained by the numerical simulation can be compared to the experimental ones at a higher bias voltage, when the induced current is exceeds 3 mA, and when the field enhanced axial flow of the combustible volatiles is observed, which altogether reduces the combustion efficiency and the total heat collected during the complete combustion of the pelletized biomass mixture samples.

Supplementary Materials: The following are available online at http://www.mdpi.com/1996-1073/12/8/1522/s1, Figure S1: The diagram of a schematic description of the experiment, Figure S2: The technological scheme of the equipment and measurement instruments, discussion about the measurement data acquisition and accuracy, the experiment repeatability and uncertainty.

Author Contributions: Conceptualization, M.Z. and H.K.; Data curation, I.B., A.K., R.V., S.V., H.K. and U.S.; Formal analysis, I.B., A.K., R.V., M.Z, S.V., H.K. and U.S.; Funding acquisition, M.Z.; Investigation, I.B., A.K. and M.Z.; Methodology, I.B.; Project administration, M.Z.; Resources, R.V. and M.Z.; Software, H.K. and U.S.; Supervision, I.B. and M.Z.; Validation, I.B. and M.Z.; Visualization, I.B., A.K., H.K. and U.S.; Writing—original draft, M.Z. and H.K.; Writing—review & editing, I.B. and A.K.

Funding: This research was funded by European Regional Development Funding grant number 1.1.1.1/16/A/004.

Acknowledgments: The authors gratefully acknowledge the European Regional Development Funding, project No.1.1.1.1/16/A/004.

Conflicts of Interest: The authors declare no conflict of interest.

Abbreviations

$A\nu$	absorption on specific wavenumber
C_k	k-th species concentration, g·m$^-$
c_p	heat capacity
D, Ø	diameter, mm
dm/dt	mass loss rate, g·s^{-1}
DTA	differential thermal analysis
TGA	thermogravimetric analysis
FTIR	Fourier Transformation Infrared Spectroscopy
I	ionic current, mA
I_v	vorticity
j_0, j_x, j_z	current density
L	height, mm
LHV	low heat value, MJ/kg

m	mass, g
P	heat power, kW
P_e	electromagnetic force
Q	heat capacity, MJ·kg^{-1}
r	combustion chamber radius, mm
r/R$_0$	dimensionless radius
S	swirl number
t	time, s
T, ΔT	temperature or temperature difference, K
u, u_{ax}	axial velocity, m·s^{-1}
u_0	normalized velocity, m·s^{-1}
u_{rad}, u_r	radial velocity, m·s^{-1}
u_{tg}, w	tangential velocity, m·s^{-1}
α	air excess
η_{comb}	combustion efficiency, %
ν	wave number, cm^{-1}

Subscripts:

av	average
end	the process ending
max	maximum
min	minimum
sum	total or summarized
vol	volumetric

References

1. European Commission. 2030 Climate & Energy Framework. Available online: https://ec.europa.eu/clima/policies/strategies/2030_en (accessed on 11 March 2019).
2. Demirbas, A. Potential applications of renewable energy sources, biomass combustion problems in boiler power systems and combustion related environmental issues. *Prog. Energy Combust. Sci.* **2005**, *31*, 171–192. [CrossRef]
3. McKendry, P. Energy production from biomass (part 2): Conversion technologies. *Bioresour. Technol.* **2002**, *83*, 47–54. [CrossRef]
4. Olsson, M. Wheat straw and peat for fuel pellets—Organic compounds from combustion. *Biomass Bioenergy* **2006**, *30*, 555–564. [CrossRef]
5. Vassilev, S.V.; Baxter, D.; Andersen, L.K.; Vassileva, C.G. An overview of the chemical composition of biomass. *Fuel* **2010**, *89*, 913–933. [CrossRef]
6. Yin, C.; Kær, S.K.; Rosendahl, L.; Hvid, S.L. Co-firing straw with coal in a swirl-stabilized dual-feed burner: Modelling and experimental validation. *Bioresour. Technol.* **2010**, *101*, 4169–4178. [CrossRef] [PubMed]
7. Barmina, I.; Valdmanis, R.; Zake, M. The effects of biomass co-gasification and co-firing on the development of combustion dynamics. *Energy* **2017**, *146*, 4–12. [CrossRef]
8. Barmina, I.; Kolmickovs, A.; Valdmanis, R.; Zake, M.; Kalis, H.; Marinaki, M. Experimental study and mathematical modelling of straw co-firing with peat. *Chem. Eng. Trans.* **2018**, *65*, 91–96. [CrossRef]
9. Barmina, I.; Kolmickovs, A.; Valdmanis, R.; Zake, M.; Kalis, H.; Strautins, U. Kinetic study of thermal decomposition and co-combustion of straw pellets with coal. *Chem. Eng. Trans.* **2018**, *70*, 247–252. [CrossRef]
10. Gupta, A.K.; Lilley, D.G.; Syred, N. *Swirl Flows*; Abacus Press: Tunbridge Wells, UK, 1984; p. 475. ISBN 0856261750.
11. Abricka, M.; Barmina, I.; Valdmanis, R.; Zake, M. Experimental and numerical study of swirling flows and flame dynamics. *Latv J Phys Tech Sci* **2014**, *51*, 25–40. [CrossRef]
12. Barmina, I.; Kolmičkovs, A.; Valdmanis, R.; Zake, M. Co-firing of straw with electrodynamic process control for clean and effective energy production. In Proceedings of the 25th European Biomass Conference and Exhibition, 25 EUBCE, Stockholm, Sweden, 12–15 June 2017; pp. 579–592.

13. Barmina, I.; Kolmickovs, A.; Valdmanis, R.; Zake, M.; Kalis, H.; Strautins, U. Electric field effect on the thermal decomposition and co-combustion of straw pellets with peat. *Chem. Eng. Trans.* **2018**, *70*, 1267–1272. [CrossRef]
14. Lawton, J.; Weinberg, F.J. *Electrical Aspects of Combustion*; Oxford University Press: Oxford, UK, 1970; p. 355. ISBN 0198553412.
15. Zake, M.; Turlajs, D.; Purmals, M. Electric field control of NOx formation in the flame channel flows. *Glob. NEST J.* **2000**, *2*, 99–108.
16. Zake, M.; Barmina, I.; Turlajs, D. Electric field control of polluting emissions from a propane flame. *Glob. NEST J.* **2001**, *3*, 95–108.
17. Colannino, J. *Electrodynamic Combustion Control TM Technology, A Clear Sign White Paper*; Clear Sign Combustion Cooperation: Seattle, WA, USA, 2012; pp. 1–11. Available online: clearsign.com/wp-content/uploads/2012/07/ClearSign-Whitepaper-2012-06-18.pdf (accessed on 11 March 2019).
18. Arshanitsa, A.; Akishin, Y.; Zile, E.; Dizhbite, T.; Solodovnik, V.; Telysheva, G. Microwave treatment combined with conventional heating of plant biomass pellets in a rotated reactor as a high rate process for solid biofuel manufacture. *Renew. Energy* **2016**, *91*, 386–396. [CrossRef]
19. Friedl, A.; Padouvas, H.; Rotter, H.; Varmuza, K. Prediction of heating values of biomass fuel from elemental composition. *Anal. Chim. Acta* **2005**, *544*, 191–198. [CrossRef]
20. Obernberger, I.; Thek, G. *The Pellet Handbook. The Production and Thermal Utilisation of Biomass Pellets*; Routledge: London, UK; Washington, DC, USA, 2010; p. 548.
21. Purmalis, O.; Porsnovs, D.; Klavins, M. Differential thermal analysis of peat and peat humic acids. *Mater. Sci. Appl. Chem.* **2011**, *24*, 89–94.
22. Betts, W.B.; Dart, R.K.; Ball, A.S.; Pedlar, S.L. *Biosynthesis and Structure of Lignocellulose*; Springer: London, UK, 1991; pp. 139–155. [CrossRef]
23. Krumins, J. The Influence of Peat Composition on Metallic Element Accumulation in Fens. Ph.D. Thesis, University of Latvia, Riga, Latvia, 2006. Available online: lu.lv/fileadmin/user_upload/lu_portal/zinas/2016/Krumins_Janis_PD.pdf (accessed on 11 March 2019).
24. Reddy, N.R.; Palmer, J.K.; Pierson, M.D.; Bothast, R.J. Wheat straw hemicelluloses: Composition and fermentation by human colon Bacteroides. *J. Agric. Food Chem.* **1983**, *31*, 1308–1313. [CrossRef]
25. Yang, Q.; Wu, S. Thermogravimetric characteristics of wheat straw lignin. *Cellul. Chem. Technol.* **2009**, *43*, 133–139.
26. Shen, D.K.; Gu, S.; Luo, K.H.; Bridgwater, A.V.; Fang, M.X. Kinetic study on thermal decomposition of woods in oxidative environment. *Fuel* **2009**, *88*, 1024–1030. [CrossRef]
27. Zake, M.; Purmals, M.; Lubane, M. Enhanced Electric Field Effect on a Flame. *J. Enhanc. Heat Transf.* **1998**, *5*, 139–163. [CrossRef]
28. Zaķe, M.; Barmina, I.; Turlajs, D.; Lubane, M.; Krumiņa, A. Swirling flame. Part 2. Electric field effect on the soot formation and greenhouse emissions. *Magnetohydrodynamics* **2004**, *40*, 183–202.
29. Blades, A.T. Ion formation in hydrocarbon flames. *Can. J. Chem.* **1976**, *54*, 2919–2924. [CrossRef]
30. Kalis, H.; Barmina, I.; Zake, M.; Koliskins, A. Mathematical Modelling and Experimental Study of Electrodynamic Control of Swirling combustion. In Proceedings of the 15th International Scientific Conference "Engineering for Rural Development", Jelgava, Latvia, 25–27 May 2016; Latvia University of Agriculture—Faculty of Engineering: Jelgava, Latvia, 2016; pp. 134–141.
31. Mikolaitis, D.W. High temperature extinction of premixed flames. In *Lecture Notes in Physics Vol. 299, Proceedings of Mathematical Modelling in Combustion Science, Juneau, Alaska, 17–21 August 1987*; Buckmaster, J.D., Takeno, T., Eds.; Springer: Berlin, Germany; pp. 67–77. ISBN 978-3-662-13671-3.
32. Boyarevitch, V.V.; Freiberg, Y.Z.; Shilova, E.I.; Scherbinin, E.V. *Electro-Vortex Flows*; Zinatne-Press: Riga, Latvia, 1985; pp. 174–175. (In Russian)
33. Douglas, J.; Rachford, R. On the numerical solution of heat conduction problems in two and three space variables. *Trans. Am. Math. Soc.* **1956**, *83*, 421–439. [CrossRef]

© 2019 by the authors. Licensee MDPI, Basel, Switzerland. This article is an open access article distributed under the terms and conditions of the Creative Commons Attribution (CC BY) license (http://creativecommons.org/licenses/by/4.0/).

Article

Evaluation of the Energy Supply Options of a Manufacturing Plant by the Application of the P-Graph Framework

András Éles [1],*, László Halász [1], István Heckl [1] and Heriberto Cabezas [2]

[1] Department of Computer Science and Systems Technology, University of Pannonia, 8200 Veszprém, Egyetem utca 10., Hungary; halasz@dcs.uni-pannon.hu (L.H.); heckl@dcs.uni-pannon.hu (I.H.)
[2] Institute for Process Systems Engineering and Sustainability, Pázmány Péter Catholic University, 1088 Budapest, Szentkirályi utca 28., Hungary; cabezas.heriberto@ppke.hu
* Correspondence: eles@dcs.uni-pannon.hu

Received: 4 February 2019; Accepted: 13 April 2019; Published: 18 April 2019

Abstract: Industrial applications nowadays are facing the complexity of the problem of finding an optimal energy supply composition. Heating and electricity needs vary throughout a year and need to be addressed. There is usually power available from the market, but a company has other investment options to consider, such as solar power, or utilization of local biomass. Fixed and proportional investment and operational costs must be compared to long-term cost-efficiency. The P-Graph framework is an effective tool in the design and synthesis of process networks, and is capable of showing optimal decisions. In the present work, a new P-Graph model was implemented to address the synthesis of the energy supply options of a manufacturing plant in Hungary. Compared to the original approach, a multi-periodic scheme was applied for heating and electricity demands. Also, the pelletizer and biogas plant investments are modeled in the P-Graph with a new technique that better reflects equipment capacities and flexible input ratios. The best solutions in this case study in terms of total costs are listed. It can be concluded that a long-term investment horizon is needed for the incorporation of sustainable energy sources into the system to be cost-efficient.

Keywords: P-Graph framework; process network synthesis; multi-periodic model; optimization; energy efficiency; sustainability; biomass

1. Introduction

1.1. Overview of the Present Work

Energy supply is one of the most important problems for modern industrial facilities. Usually, when a plant is built, it is connected to the grid allowing it to purchase electricity for its operation. The same holds for heating, water, or other requirements where public services are available. However, environmental regulations play an increasing role, and small to medium scale power plants are becoming popular. Such power sources have the ability to supply the needs of individual residential homes or firms. These can have significant investment costs, but operation in the long-term may make them cost-efficient solutions. There is no absolute winner technology in terms of costs.

The power supply decision can be a complex problem. Several different energy sources have to be taken into account. Renewable energy sources like biomass can have a limited availability and are usually neither economical nor environmentally friendly if they need to be transported over long distances. Different technologies and energy supply methods may coexist, but each having different investment and operational costs, for which capacity is also limited. Energy demands can also vary, not only from one year to the other, but even from month to month, according to the seasons. This is especially true for heating requirements.

The P-Graph framework is a modeling tool with which one can define Process Network Synthesis (PNS) problems. In a PNS, a system of complex possible flow of materials is given, and a cost-optimal selection of possible operating units must be found. The P-Graph framework consists of the mathematical model of P-Graphs, the corresponding theorems and algorithms, and also software in which we can solve PNS problems.

This work presents a case study made for a manufacturing plant, for which decision makers needed to consider alternative energy sources like solar power and local biomass availabilities, instead of purchasing all the electricity and heating need required. Naturally, the method of modeling we present here can be adopted for any plant if the available technology options, energy sources and demands are specified.

The optimization problem for finding the minimal operating cost of the firm during the course of the investments' considered horizon is modeled as a PNS problem, and then solved utilizing the P-Graph framework. The model uses the multi-periodic modeling technique to address fluctuating demands in two different seasons. The pelletizer and biogas plant equipment units are modeled with a new technique allowing mass-based capacities and flexible inputs simultaneously. Several different investment horizons are investigated, and the best solutions for each scenario are presented. In the end, we can conclude that all energy options can be an economical replacement of direct energy purchase, but a long horizon must be assumed to be so.

1.2. Importance of Sustainability

Sustainability is at its core about finding practically possible ways to maintain conditions on Earth suitable for civilized human life. This is considered to be quite a challenge for several reasons:

- The human population is large and it is still increasing [1].
- Simultaneously to increasing population, the consumption of resources shows an increasing trend. A fourfold increase in private consumption expenditures from 1960 to 2000 could be observed [2].
- The human population is using approximately 38% of the world terrestrial net primary consumption [3], leaving a much smaller portion than before to support the planetary ecosystem. Net primary production is the solar energy captured by the ecosystem and made into biologically accessible energy.

Note that manufacturing consumes an enormous amount of energy, for example, 2.2 EJ in 2010 [4]. Energy generation at present still heavily relies on fossil fuels [5], and this has a wide range of environmental impacts. The most widely known is the emission of carbon dioxide which contributes significantly to climate change. Therefore, a possibly effective way in decreasing the human footprint is to target manufacturing. This can be done by the provision of alternative energy supply options for operating plants, especially when shifting to more efficient energy use and to the use of renewable and low environmental impact forms of energy generation. The importance of this is established by the fact that increasing population and consumption will likely result in the increase in manufacturing needs, hence energy consumption. It is a common idea that instead of purely relying on a highly centralized network for electricity or heating, each firm resolves its own energy demands locally. Of course, this implies additional investment costs compared to the ordinary scheme. Particular examples for locally feasible energy supplies are the usage of biomass or solar cells. Nevertheless, decreasing demands by alternative production technologies or more efficient energy usage can also be promising options.

Synthesis of supply chains of renewable energy sources, for example solar, wind, hydropower, and biomass utilization, possibly simultaneously to other sources, until the potential demands of energy or water, is a challenging task in general, and has drawn much attention [6,7]. The complexity of the systems to be designed optimally yields for adequate modeling and optimization tools, regardless of the scope being a single plant or a whole region. Novel, general methods are published [8], using mathematical programming solutions which are a conventional way of modeling. But due to their limitation in solving larger or too special problems, specific case studies may require other model developing techniques.

2. Theory

2.1. Process Network Synthesis and the P-Graph Framework

Process Network Synthesis is the act of designing and making decisions about a complex system of processes consisting of several steps and dependencies. The process under examination is generally the production or achievement of a dedicated material, state, or a set of such. Steps of the process to be modeled must be identified, but the emphasis is on the whole system itself. Usually there is a range of options leading from the available sources, like raw or freely available materials, or conditions, to the desired final products. Most importantly, selection of the actually utilized options must be made, so that other details, like actual material flows can be selected.

The P-Graph is a graph-theoretic model first introduced by Friedler et al. [9]. It is capable of modeling PNS problems unambiguously and gives options to effectively find optimal solutions. The model consists of a directed bipartite graph of a material node set and an operating unit node set. The material nodes resemble the states in the model. A state is usually representing the presence of some material or other physical or virtual property. The operating unit nodes represent transitions of a set of states into another. This is generally some kind of production step, transportation of goods, purchase, but may simply mean some logical consequence of the existence of a state based on others. Arcs from material nodes to operating unit nodes represent consumption. Arcs between operating units to materials represent production. That means the inputs and outputs of an operating unit are the materials for which there exist arcs going towards, and starting from the operating unit. In both cases, arcs are directed towards material flow. Materials can be both inputs and outputs at the same time. In this way, the P-Graph represents the structure of the process, see Figure 1. There are three types of materials in a P-Graph:

- Raw materials are the ones available from external sources, and cannot be produced.
- Final products are those we want to obtain.
- Intermediate material nodes are other materials involved in the system, typically in between the production chain from raw materials to final products. Intermediates can be produced by some operating units and can be consumed by others.

The goal of the PNS problem is to find a structure and operation for the system which is optimal in some manner. A structure of the system involves appropriate selection of available technologies and other activities, and the corresponding material flows. Optimality may refer to different objectives, typically it is cost minimization. A structure of the process network is to be found for which all products are obtained. This is called a solution structure, and it is, in general, a subset of the original P-Graph. Along with the rigorous mathematical definition of P-Graphs, five axioms are described that must hold for a P-Graph in order to be considered as a solution structure [9]:

- All of the final products of the PNS problem are represented in the resulting P-Graph.
- Any material has no input if and only if it is a raw material.
- Every operating unit in the P-graph is also defined in the PNS problem.
- Every operating unit is part of a path leading to a final product.
- Every material in the P-Graph must be the input to or an output from at least one operating unit in the graph.

The P-Graph framework also includes algorithms that can be used to solve PNS problems:

- The Maximal Structure Generation (MSG) is a polynomial-time algorithm which finds the so-called maximal structure of a PNS problem [10]. This structure is the union of all solution structures, and it is itself a solution structure as well, as a consequence of the axioms. The point in finding the maximal structure is that unnecessary parts of the PNS problem can be excluded a priori from optimization.

- The Solution Structure Generation (SSG) is an algorithm for systematically generating all solution structures of the PNS problem [11]. This is useful, because once the structure is fixed, other decisions like exact material flows are easier to determine. For large problems, the number of solution structures may explode.
- The Advanced Branch and Bound (ABB) method is an algorithm which finds the optimal solution of a PNS problem modeled as a P-Graph, driven by underlying MILP model constructed and decisions based on the combinatorial nature of the problem [12]. Note that MSG and SSG only operate on the graph itself, without other problem data like material flow ratios, costs, or capacities, see Figure 2.

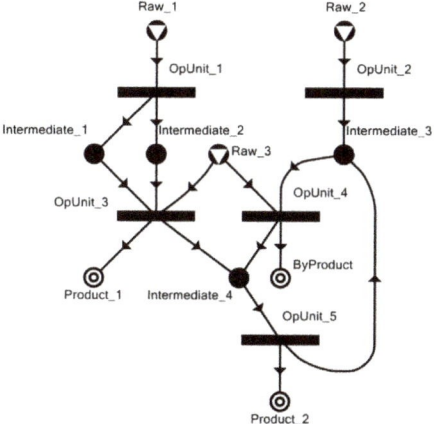

Figure 1. Example P-Graph with five operating units, three possible raw materials, two desired final products, and a byproduct. Note that cycles are possible in P-Graphs.

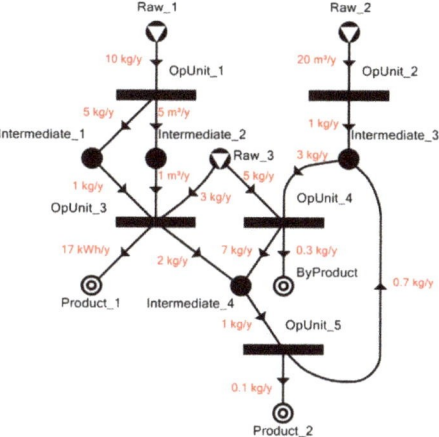

Figure 2. The same example P-Graph with additional data supplied: the consumption and production ratios for each operating unit.

These data also correspond to a PNS problem and the ABB is capable of presenting detailed solutions only when these are also available. PNS Studio is a software toll with which PNS problems can be designed as P-Graphs. These can also be solved using the SSG or ABB methods to find cost-optimal solutions for a process network. The software is freely Available online [13], and was also used in the case study of this work.

2.2. Extensions of the P-Graph Framework

The P-Graph framework originally targeted chemical engineering problems, but the tool is useful for other areas as well. The framework itself was also extended by additional functionalities, either in implementation or application.

Time-Constrained PNS (TCPNS) is a problem where timing constraints on the flows can be added [14]. This gives the possibility to make scheduling decisions by P-Graphs. Note that this was already possible with the framework with certain circumstances. Certain vehicle routing problems where deliveries are fixed in time were also solved by P-Graphs before TCPNS was introduced [15]. Purely scheduling problems of process networks, where there is a fixed set of tasks and orders must be found with various storage policies, can also be addressed [16]. These methods most importantly show that operating units in P-Graphs do not necessarily resemble actual equipment, but possibly logical relationship, like conversion, or precedence relationships.

Separation Network Synthesis (SNS) is a production environment that can separate chemicals into their pure components in multiple possible ways that can be optimized. This is not a utility, but a production optimization problem, which can be transformed to a PNS problem and solved by the ABB algorithm [17].

In a PNS problem, the input and output ratios for operating units are fixed. This is not the case in some real world circumstances, when equipment units can have several different inputs with arbitrary ratios. Pelletizers and furnaces are an example for this. The model complicates if ratios are arbitrary, but also subject to constraints, like minimum or maximum ratios. The P-Graph framework was extended to address such case of input scenarios [18], as the underlying MILP model of the system was extended with linear constraints to obtain an appropriate model. Note that the P-Graph framework may be itself capable of modeling such operating units with several material nodes and operating unit nodes.

The P-Graph methodology can be extended to meet multi-periodic demands and supplies. Multi-periodic means different rates of production of final products or consumption of raw materials and products at various time periods, and optimization of the whole duration in a single model [19]. Periods are an important issue, because assuming average load during a period may lead to underestimating capacity needs during the year. This is especially true if operation should run at an extremely high rate in short periods, while at low rate in general, resulting in an average that is way under the capacity required to be operational at all times. However, the equipment units may have minimal required flow to be working, which complicates the operating unit model of the P-Graph, so the multi-periodic modeling scheme can possible be extended to model such technologies [20]. Although multi-periodic modeling is a logical extension of the P-Graph framework, which means it does not require additional tools to be implemented, the PNS Studio software already has support for making multi-periodic models easier [21]. Note that this usually requires manual addition of a vast amount of data, as each period is usually modeled by its own P-Graph, which is replicated. The PNS problem can get very difficult to understand if the number of periods is high.

2.3. Applications

The P-Graph framework can be applied to a wide range of process network optimization problems, including those for which production, utility, or transport systems are in question. The methodology is a useful alternative to implementation of Mixed-Integer Linear Programming models or other solution methods. For example, pinch analysis [22] is a widely used tool for system design, one possible application of which is energy sector planning with carbon emission constraints [23], but the same problem can also be solved by the utilization of P-Graphs [24]. Note that the ABB algorithm itself relies on linear programming. However, the P-Graph framework has some advantages, for example providing the series of the best solutions, in order, instead of finding just a single optimal solution [25].

A thorough collection of P-Graph applications can be found in [26], showing that the framework is useful in various case studies, including transportation, supply chain system design and plant management. There is a particular review that focuses on P-Graph applications on sustainability projects, like optimization from processing facilities to supply chains [27]. Another review is made by [28], which also notices possible future developments of the framework. Particularly, the theory allows more complex operating unit models than the current implementations can handle.

Polygeneration plant modeling can also be done by the P-Graph framework [29]. The recent study involves satisfying multiple uncertain demands like heating, electricity, cooling, and treated water. Not only the optimal, but the near-optimal solutions can be found. However, risks and possible reactions in case of equipment failure must be taken into account, which can also be handled by the same framework, as an alternative to linear programming models [30]. Uncertainty in supply and demand in general and risks for inoperability generally affects the usability of supply chains. The P-Graph framework was also used in such a scenario to analyze risks in the supply network of bioenergy parks [31]. A more simple case, in the modeling point of view, is when risks can be derived as parameters of the available technological options. This can be the case when probability of losses is attributed to each activity and these can be penalized in the objective [32]. The authors in the example seek to minimize fatalities in the whole bioenergy supply chain.

The framework was used to address a biomass supply chain enhancement [33], where P-Graphs and conventional mathematical programming tools were used simultaneously to obtain a decomposition of the main problem. The authors remark that the results must be regularly revised to reflect real circumstances. Several biomass types were considered in connection with the palm oil industry. Another set of biomass resources were inspected in a work of wood processing residues [34]. One drawback is observed in this case study, particularly, that minor changes to problem data may result in significant changes in the solution structures. This makes the option of finding near-optimal solutions valuable.

Addressing heat and electricity needs simultaneously is a common scenario for plant design. Available biomass is often considered in parallel or as a replacement with purchasing natural gas and electricity directly [35]. The case study, after optimization with the P-Graph framework, pointed out a potential 17% decrease in operating costs if biomass is integrated into the energy supply, while other objectives like ecological footprint can also be taken into account [36].

Spatial distribution of the biomass types and demands points can be simultaneously taken into account to address the optimization of a full renewable energy supply chain [37]. In their work, authors first determine clusters to minimize transportation needs by mathematical programming, and then apply a PNS model to optimize material flow in and between these clusters.

The methodology is also useful in determining bottlenecks in the complete supply chains for biomass utilization [38]. The proposed method also relied on the suboptimal solutions the P-Graph framework was able to find. The method was used to improve sustainability indices in three different, novel scenarios [39].

Synthesis of carbon management networks (CMN) is an important issue in the sustainability point of view, and lead to complex optimization problems. The P-Graph framework was used for the synthesis of biochar-based CMN [40]. The model uses a set of sources of biochar and a set of sinks, that are soils that can contain it, and operating units resemble transportation between sources and sinks. Note that in this case, the number of operating units is the product of the number of sources and sinks, and other limits, for example the contamination of the soils can also be taken into account. Another example for carbon management network optimization used P-Graphs in conjunction with Monte Carlo method [41]. The Monte Carlo simulation was used to test near-optimal solutions reported by the PNS solver, and estimate their robustness.

The multi-periodic modeling technique itself was first applied to a real world case study where corn production was investigated, for which the supplies and demands both change throughout the year [42]. In another model formulation, the method was used to optimize annual electricity production for various demands and sources. Afterwards, a polygeneration case study was performed that also addresses steam, chilled and treated water demands [43]. It can be seen that the multi-periodic scheme can be independently applied to the raw materials, or the final demands of a supply network, based on which of them are fluctuating.

3. Materials and Methods

In this section the P-Graph model to be solved in the case study is described in detail. Our previous work included the case study utilizing a single period model [44], and input data we use is also presented here. The present work has two major improvements. First, electricity and heating demands were modeled as a multi-periodic P-Graph model. Second, the modeling of some equipment units—namely, the pelletizer and the biogas plant—involves a technique that allows mass-based equipment capacities and flexible input of different kinds of biomass.

In the optimization problem, the energy supply needs of the manufacturing plant must be satisfied. This means that some investments are considered against the business as usual solution of purchasing all the electricity and heating power from the public service providers.

The optimization problem is modeled as a PNS problem. This requires that raw material, intermediate, and final product nodes, operating units and connecting arcs are defined. Moreover, problem data like raw material costs, available amounts, operating unit investment and operating costs and capacities, and energy conversion rules must also be explicitly defined. After determining the demands, the underlying single stage model is described. Afterwards, the way this model is extended into a multi-periodic one, is presented.

For the sake of simplicity of the model, all energy quantities are expressed in kWh, regardless of being heat or electricity at a particular point in the process network. Monetary quantities are expressed in HUF. The value of this currency had been fluctuating, one EUR was between 300–330 HUF in the recent years where data into the case study were collected from. In the scenario for the manufacturing plant, expenses were in HUF uniformly. It shall be noted that interpreting results with an exchange rate to another currency only involves a linear factor for all monetary values, including the minimized total costs, but the obtained solution structures and their order would remain the same.

3.1. Electricity and Heating Requirements

The plant has two needs that are subject to the scenario: electricity and heating. These are the demands in the process system, and are modeled as product nodes in the resulting P-Graph. The business as usual solution is that electricity is bought from the grid in the amount needed. Heating is provided by the plant's own furnace, in which natural gas is fed, bought from the public service provider.

Detailed consumption data about the plant's natural gas consumption was provided in Table 1. The plant requires indoors heating. The heating requirements are therefore modeled as the total heating value of gas or other materials and energy consumed, with possible conversion ratios depending on source. A single yearly heat consumption value is assumed to describe the requirement. We can see there are significant fluctuations from year to year, with a clear tendency of decreasing, which is probably due to the company optimizing its operation constantly. Also, the heating requirements are a magnitude larger at winter than at other times of the year. This is the key fact that motivated the multi-periodic modeling scheme to be applied in this case study.

Table 1. Past data for natural gas usage of the manufacturing plant. These data are used to predict future demands.

Gas Used (m³)	2009	2010	2011	2012	2013	2014
January	133,999	128,744	157,085	123,770	75,782	48,635
February	123,836	95,406	137,103	124,305	51,407	49,067
March	120,326	77,536	123,795	83,362	43,560	16,847
April	37,378	58,464	83,305	61,092	15,452	4337
May	35,057	63,719	51,009	37,482	2785	4247
June	37,065	52,094	30,924	17,340	1919	2688
July	30,396	44,485	31,560	12,891	1554	2416
August	34,232	44,628	30,105	20,179	1534	2117
September	28,607	81,730	30,024	19,829	3072	2136
October	82,299	105,612	74,841	25,235	4208	10,982
November	105,599	104,195	125,638	50,535	24,273	43,769
December	116,459	156,139	129,481	73,819	57,240	62,139
Yearly total	885,253	1,012,752	1,004,870	649,839	282,786	249,380

Electricity consumption data was also provided in a similar manner (see Table 2).

Table 2. Past data for electricity usage of the manufacturing plant. These data are used to predict future demands.

Electricity Used (kWh)	2009	2010	2011	2012	2013	2014
January	905,533	796,117	993,044	788,703	453,838	255,517
February	1,128,039	715,508	926,508	769,565	382,042	270,539
March	1,328,232	809,142	1,074,706	736,811	359,696	217,190
April	1,076,030	787,400	963,416	624,634	310,077	176,142
May	1,142,927	918,350	890,317	862,085	228,225	200,673
June	1,176,784	1,021,286	843,147	327,853	251,323	191,459
July	1,215,169	1,170,359	871,462	502,244	254,907	270,710
August	1,281,732	1,089,277	928,240	327,853	241,197	414,119
September	1,183,526	983,531	872,692	454,764	201,446	425,678
October	1,002,034	989,398	868,299	923,389	211,333	439,628
November	926,870	969,743	880,829	346,867	248,156	439,837
December	872,000	901,317	856,199	399,713	289,379	502,415
Yearly total	13,238,876	11,151,428	10,968,859	7,064,481	3,431,619	3,803,907

Electricity consumption also shows a decreasing tendency, but the consumption rates do not seem to clearly depend on the month. They more likely depend on production load of the plant, or other causes for which are unavailable in this case study. It can also be seen that the decreasing tendency is not necessarily rigorous: in the last year with known data, 2014, electricity consumption actually increased. For these reasons, the following decisions were made about the modeling method of the demands:

- Future demands that are used in the model are estimated based on a recursive formula of the data shown.
- Two periods are introduced to be handled differently: winter (from December to February, inclusive), and the other part of the year (from March to November, inclusive). From now on, we call these winter and mid-year periods.

The formula we used in the case study for demands were the following. This is applied to both the natural gas and the electricity consumption:

$$s_{2015} = \frac{0.8 \cdot s_{2012} + 0.9 \cdot s_{2013} + 1.0 \cdot s_{2014}}{2.7} \cdot 1.15 \qquad (1)$$

The first three years are omitted, since the significantly larger consumption in those is due to past changes in the plant operation. For the last three years, we apply a weighted average, with weights of 0.8, 0.9 and 1.0 for 2012, 2013 and 2014, respectively (note that 0.8 + 0.9 + 1.0 = 2.7). It is multiplied by a factor of 1.15 for more security of the energy supply, and to accommodate a slight potential increase.

This single value, obtained for both the electricity and heat demands, are used assuming them to be the constant yearly demands in all consecutive years. Nevertheless, it shall be noted that this is a rough estimation. Actual data could severely affect the results provided by our model. However, the P-Graph model we present can easily be resolved with different data.

To be able to model a multi-periodic scenario, the demands must be estimated for each period individually. Note that even though electricity consumption is somewhat independent of the periods, we have to calculate periodical demands since heat and electricity production can both be done from the same energy sources. The electricity and heating demands we suppose in the case study are summarized in Table 3. Note that it is obtained as a direct application of Equation (1) on data from Tables 1 and 2.

Table 3. Demands of the manufacturing plant assumed in the case study. Note that heating demands are calculated from the natural gas consumed with a 34 MJ/m^3 rate.

Demand	Period	Used Value
Natural gas	yearly (total)	436,045 m^3 ≈ 4,118,206 kWh
	mid-year	248,460 m^3 ≈ 2,346,569 kWh
	winter	187,585 m^3 ≈ 1,771,637 kWh
Electricity	yearly (total)	5,342,793 kWh
	mid-year	3,806,227 kWh
	winter	1,536,566 kWh

In the P-Graph model, the four periodical demands represent the final product nodes. They must be satisfied with a combination of the available technologies, which are able to produce heat, electricity, or both.

3.2. Energy Sources and Their Availability

In the previous section, the final products of the process network were defined. Now, the raw materials are introduced. In the present case study, there are different kinds of energy sources considered. In general, there are more options, but due to the properties and environment of the plant, the following resources were included:

- Public service providers, from which unlimited amount of electricity and natural gas can be purchased.
- Solar power, which is also unlimited, as long as there is solar power plant capacity.
- Several different kinds of biomass of limited availability from the same region.

For purchased electricity and natural gas, and the different kinds of biomass, we assume a unit cost for each resource. Purchased electricity and natural gas are unlimited sources. This means that practically any amount can be purchased as long as the company is willing to spend the money needed. Solar power is freely available. Of course, it does not mean that solar power is a free energy source. Rather, regardless of the amount of solar power plant capacity we want to invest into, there is always a supply of solar radiation. The different kinds of available biomass are purchased locally. Note that these have a limited availability, as these are mostly byproducts from agriculture. Biomass has a drawback of having low energy content. Transportation of biomass over longer distances is not economical, and does not have a sense in the sustainability point of view either. We assume that the biomass types each have an own fixed unit cost, and an upper limit for purchase.

The energy sources are represented in raw material nodes in the P-Graph. Table 4 contains data for these raw materials. We must note that these are estimations we assume in this case study. We also assume that there is no concurrent demand for these materials from other parties. Each of the enlisted raw materials corresponds to a raw material node in the resulting P-Graph.

Note that there are costs associated with the processing of these resources. These are embedded into the operating costs of the operating units instead. For example, maintenance costs of the solar power plants are costs of the power plant, not the solar energy itself. The reason behind this scheme of modeling is that raw material nodes can have a straightforward implementation in the P-Graph based on the unit prices and available amounts.

Table 4. Raw material availabilities in the case study.

Energy Source	Unit Price	Available for Use (per y)
Saw dust	24 HUF/kg	150,000 kg
Wood chips	22 HUF/kg	600,000 kg
Sunflower stem	5 HUF/kg	500,000 kg
Vine stem	7 HUF/kg	600,000 kg
Corn cob	6 HUF/kg	1,200,000 kg
Energy grass	8 HUF/kg	1,600,000 kg
Wood	20 HUF/kg	2,000,000 kg
Solar energy	free	unlimited
Natural gas	114 HUF/m^3	unlimited
Electricity from the grid	38 HUF/kWh	unlimited

3.3. Operating Units

In a P-Graph, all transformations of materials to others are done by operating units. This means that an operating unit may resemble an actual technological step, or market operations like purchase or selling, or only be a modeling tool. In this case study, operating units are introduced to fit into the following roles:

- Direct supply by purchase of electricity or natural gas.
- Solar power plants.
- Biomass processing chain.

Note that these roles are in a one-to-one correspondence with energy source types. Those operating units that correspond to actual equipment units that require investment and operating costs are shown in Table 5, with the following meanings:

- Fixed investment costs are to be paid once, at investment, if the technology is used.
- Proportional investment costs are to be paid once, at investment. The amount is proportional to the yearly capacity of the operating unit, which cannot be changed later.
- Fixed operating costs are paid yearly, for using the operating unit, regardless of rate of utilization.
- Proportional operating costs are paid yearly, for using the operating unit, and are proportional to the actual utilization of the operating unit.

This means we assume that each possible investment has linear costs in terms of capacity, with a starting fixed investment cost. This assumption is required in order for the network to be modeled with the P-Graph framework. More complex cost functions can be estimated, or alternatively, different operating units can be introduced in the model if necessary. Investment costs appear as costs evenly distributed in the investment horizon. That means if a longer horizon is considered, then investment costs are decreasing in the modeling point of view, as the current model seeks to minimize yearly operating costs.

Note that the P-Graph methodology would allow the modeling of any operating units which can operate in parallel to each other from the modeling point of view, as long as they can be properly represented by linear fixed and proportional investment and operating costs and capacities. In this case study, some technologies like heat pumps were initially excluded by the management due to the plant's properties, electricity needs, initial investment costs, and lack of subsidized project possibilities.

Table 5. Possible investments. These are all new equipment units in the case study, also called operating units.

Equipment Costs	Investment Costs		Operating Costs	
	Fixed	Proportional [1,2,3]	Fixed	Proportional
Solar power plant	50,000,000 HUF	353.98 HUF/kWh	none	22.12 HUF/kWh
Pelletizer	5,000,000 HUF	10 HUF/kg	none	4 HUF/kg
Biogas plant	20,000,000 HUF	240 HUF/kg	none	10 HUF/kg
Biogas furnace	10,000,000 HUF	20 HUF/kWh	6,000,000 HUF	4 HUF/kWh
Biogas CHP plant	20,000,000 HUF	36 HUF/kWh	6,000,000 HUF	6 HUF/kWh

[1] The solar power plant has proportional costs given in terms of produced electricity in kWh/y. These values are equal to 400,000 HUF and 25,000 HUF per 1130 kWh/y. [2] The pelletizer and the biogas plant have proportional costs given in terms of input amount, in kg/y. [3] The biogas furnace and CHP plant have proportional costs given in terms of the heating value of the biogas consumed.

Other operating units in the process network are either already present or only represent some activities that are not physical production steps. This means that other operating units have neither investment nor operating costs, nor maximal yearly capacities. We assume that all operating requirements can be expressed in the aforementioned costs, including energy requirements. Note that, for safety reasons, the plant would anyways be connected to the grid. This makes any particular solution easily adjustable if demands unexpectedly increase. It also means that the possibility of self-sustaining energy supply is neglected in this case study.

3.4. Direct Purchase

The electricity consumption is the simplest operation in the process network. An operating unit representing electricity purchase from the grid is introduced. Its single input is the electricity from the grid, and its single output is the electricity demand as depicted in Figure 3.

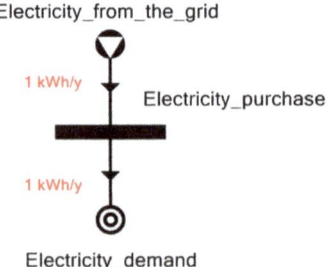

Figure 3. Electricity purchase in the P-Graph model.

The natural gas purchase is done by a single operating unit. This unit has a single input and output: natural gas from the provider, and purchased amount. In contrast to electricity, natural gas purchase is not directly supplying heat. For this reason, the manufacturing plant has a furnace which consumes the natural gas purchased, and produces the heat demand. Transfer ratios represent the heating value of natural gas, which is 34 MJ per m^3, converted to kWh as shown in Figure 4.

Solar power is considered as an alternative energy source to be exploited solely by solar cells. This means that a solar power plant is introduced which transforms solar power into electricity. By estimation, we assume that 8760 kWh solar energy can be converted into 1130 kWh of electricity, or any amounts with this same ratio. Note that although the efficiency of solar cells is important in reality, only the output of 1130 kWh is of interest in this model. The reason for this is that solar energy is free and unlimited anyways, and solar power plant investment and operating costs are given in terms of throughput, see later.

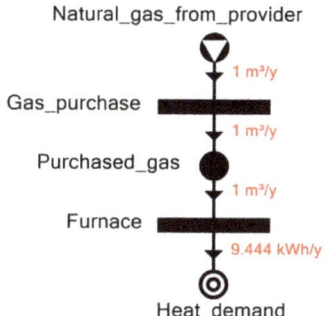

Figure 4. Natural gas purchase and built-in furnace of the plant used in the P-Graph.

The reason for using the number 1130 kWh is that the proportional investment and operating costs were originally available as 400,000 HUF and 25,000 HUF per 1130 kWh/y throughput (see Table 5).

The electricity produced by the solar power plant can be directly fed into the electricity demand. Alternatively, an electric heater can be utilized, which contributes to the heat demand, see Figure 5.

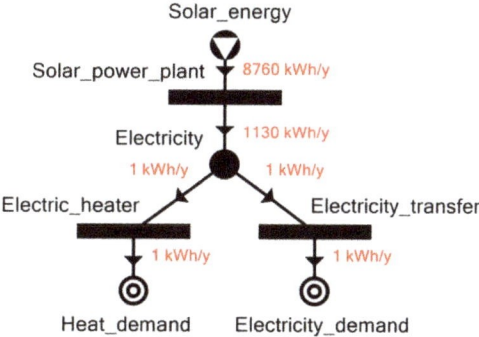

Figure 5. Model for the solar energy utilization.

3.5. Biomass Utilization

The biomass supply is a bit more complicated. First, some kinds of the biomass are needed to be pelletized first. Table 6 summarizes the data for these materials, which are sawdust, wood chips, sunflower stems and vine stems. Note that different pellets have slightly different energy contents. This fact is a key observation for correctly modeling the pelletizer.

Table 6. Biomass types that must be pelletized before usage.

Biomass Type (to be Pelletized)	Heating Value
Saw dust	4.50 kWh/kg
Wood chips	4.25 kWh/kg
Sunflower stem	3.75 kWh/kg
Vine stem	4.10 kWh/kg

Pellets and the rest of the biomass, see Table 7, are considered as feeds into a biogas plant. Note that the heating value of each input type determines the amount of heating power the resulting biogas has.

Table 7. Biomass types that can be directly fed into the biogas plant.

Biomass type (to be Directly Fed)	Heating Value
Corn cob	4.00 kWh/kg
Energy grass	4.80 kWh/kg
Wood	4.16 kWh/kg
Pellets	depends on raw material

The biogas can be fed into either a biogas furnace, or a biogas-based "Combined Heat and Power" (CHP) plant. The former generates heating only, while the latter also produces electricity. We model these operating units in the following way, see Figure 6:

- The input of each operating unit is the total energy content of the biogas available. This is a single material node.
- The biogas furnace consumes 1 unit of biogas, and produces 0.7 kWh heating power, or different amounts with the same conversion rate.
- The biogas CHP plant consumes 1 unit of biogas, and produces 0.4 kWh heating power and simultaneously 0.35 kWh electricity, or different amounts with the same conversion rate.
- Productions are directly fed into the demands of the process network.

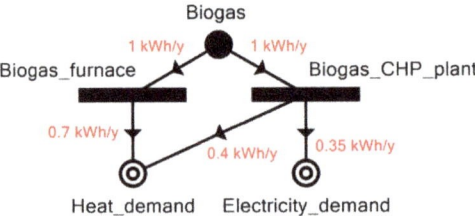

Figure 6. Biogas furnace and CHP plant modeling. Note that the material node "Biogas" is expressed as the heating values of the input materials instead of mass or volume.

Attention is required to correctly model the part of the network from the biomass inputs to the biogas heating power. The key observation is that different inputs yield different energy contents. One way this can be modeled, used in [44], is the following procedure:

- Define a common input material node for both the pelletizer and the biogas plant.
- For all actual inputs, introduce a logical operating unit in the model which transforms the input to its heating power, which is a linear transformation. These are fed into the inputs of the equipment.
- From now on, the operating unit of the pelletizer and the biogas plant consumes the material introduced for the heating power of its inputs, and not their mass.

In this way, the heating content of each individual biomass type is respected and properly transferred until the heating power of the resulting biogas. The P-Graph resulting from this approach is depicted in Figure 7. Note that the appropriate point for modeling energy losses is the conversion factors in this design, so it is not happening at the pelletizer and biogas plant operating units in the model.

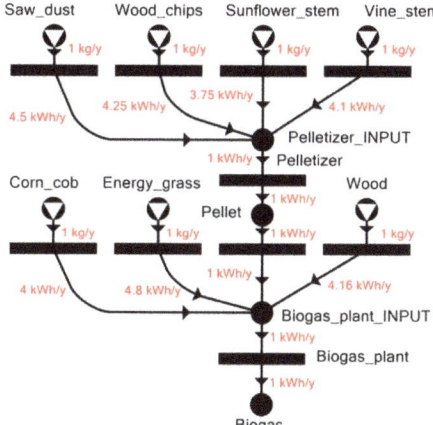

Figure 7. Process network for the production of biogas from the different biomass types. Note that the unnamed operating units are logical operating units representing heating value conversion for each raw material.

3.6. Modeling of Capacity Restrictions and Flexible Input Ratios

The problem with this approach is that the pelletizer and the biogas plant have capacity restrictions based on the mass of their inputs, not their heating power. It is natural, as a pelletizer does not "know" anything about the heating power of its product, only the mass. We must note that the original approach is therefore a modeling simplification which does not introduce a large error on the network, for two reasons. One is that operating costs for the pelletizer and the biogas plant are negligible compared to other terms like material costs and power plant investments. The second reason is that the conversion ratios for all kinds of biomass are similar in magnitude. To correctly model these operating units, a different approach is used in the present case study, summarized as follows:

- An individual logical operating unit is introduced for both the pelletizer and the biogas plant, which represent the investment, have no inputs, and produce a "production capacity" material. This is a logical material that can be distributed among the possible inputs arbitrarily.
- For each input, its pelletizing and biogas production operation is modeled by an individual operating unit each. These consume the production capacity from the corresponding investment.
- The four pelletizing processes produce different materials, for the different pellet types, in a ratio of 1:1 compared to input.
- The seven biogas production processes produce the single biogas heating value material node, in a ratio corresponding to the heating value of each input as in Tables 6 and 7.

In this way, the heating values of the different biomass types are respected in the model, and the pelletizer and biogas plant is constrained by their actual total utilization rates, expressed in total mass of inputs. This approach can be summarized in Figure 8.

Note that simplifications can be made in this design. Observe that pellets have no other usage in the process network than being fed into the biogas plant. This way, the logical operating units of the pelletizer and corresponding ones of the biogas plant can be merged. This means that the pelletizer and the biogas plant together have seven logical operating units that produce the biogas. Each logical operating unit is corresponding to a biomass input type. The final form of the biomass supply part of the P-Graph model is depicted in Figure 9. Energy conversion ratios do not only reflect the heating values of the raw materials, but also the possible losses until the final conversion to heating power.

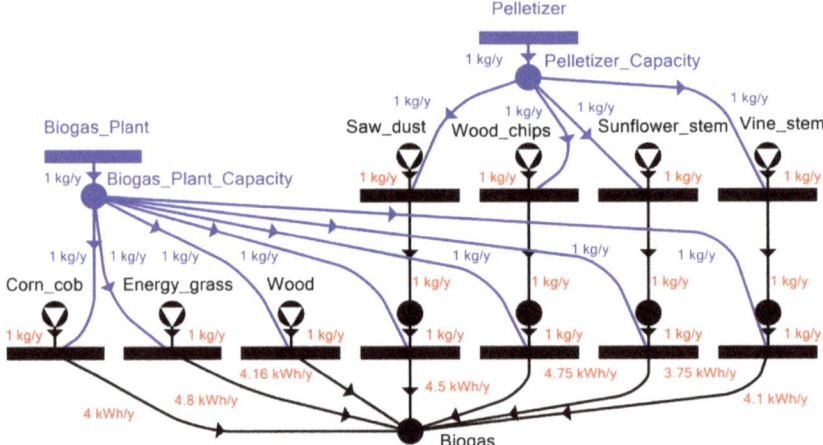

Figure 8. Modeling of the biomass supply part of the process network. With this approach, biogas plant and pelletizer capacities are expressed in terms of total input masses.

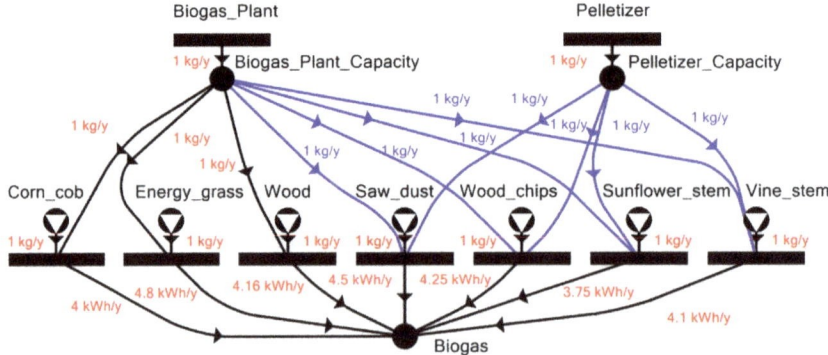

Figure 9. Final model for the potential biomass supply in the manufacturing plant. Note that the simplification resulted in seven logical operating units performing heating value conversions for each available biomass type.

3.7. Multi-Periodic Model

So far, the constructed P-Graph of a process network only considers a single period of operation. In a single period model, all data are given on a yearly basis—that means yearly total demands, and yearly total production rates. It is generally assumed in this case that timing of the production can be arranged to fit the demands arising throughout the period. This can be a valid assumption in some cases. One case is when there are little fluctuations in the supply and the demand, so a constant rate of production and demand satisfaction can be assumed in the whole period. Another case is when there is storage available for the production, and that is used up on demand. The single period P-Graph model is depicted in Figure 10.

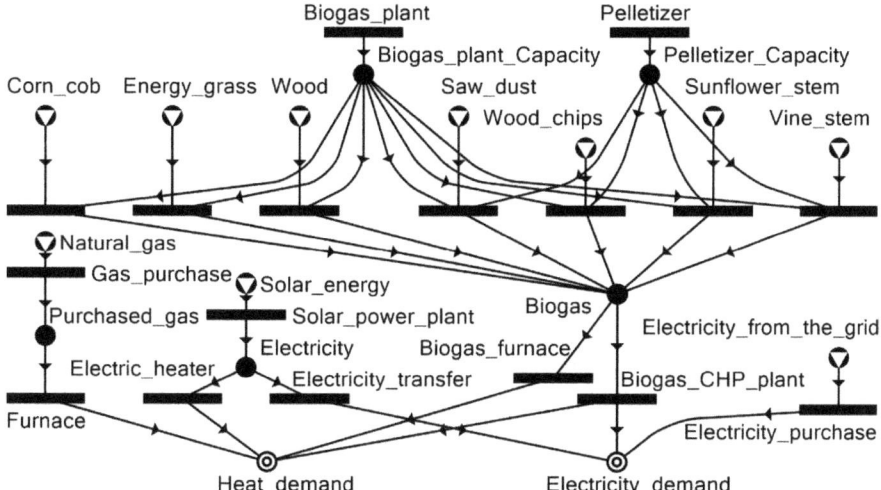

Figure 10. Single period P-Graph model for the energy options of the manufacturing plant.

Into this model, we introduced two different periods instead of the single period of one year. The nine months long mid-year and the three months long winter period are used. Each period has its own demand that must be satisfied. For example, heating demands are considerably higher during winter than at mid-year (see Table 3).

We assume that demands are fixed to the periods, but supplies are not. The required heat must be produced or purchased in the period where it is demanded, and no energy transfer is possible between periods. On the other hand, the same yearly supply of each limited raw material (biomass) is available throughout the year, and can be used up in the periods in any ratios and combinations. This implies that storage from one period to the next is assumed to be available, and its requirements are neglected. Note that the multi-periodic modeling technique would allow binding the supplies to the periods, and also properly modeling significant storage costs and capacities.

Solar energy works a bit differently than raw biomass, because throughput is generally much lower at winter than at mid-year. This is a significant factor that is addressed in the model. Concluding the multi-periodic scheme, demands are calculated for each of the two periods, while supplies are freely distributed between periods—the only exception being the solar radiation.

The data we use about solar power plants is that a production of each yearly 1 kWh of electricity requires 353.98 HUF proportional investment and 22.12 HUF proportional operating costs. These are distributed across a 3 months long winter and a 9 months long mid-year period. If an investment is made, we assume that the solar power plant is utilized throughout the year. For this reason, there is no point in dividing the costs of the plant between the periods. What actually makes sense, is dividing the throughput between the periods. We assume that, on average, solar power plants are $\lambda = 2$ times more efficient in mid-year, than in winter. Provided that a total yearly production of E_{yearly} electricity is given, we can calculate the amounts E_{winter} and $E_{mid-year}$ from the following formulas:

$$E_{winter} = E_{yearly} \cdot \frac{3 \cdot 1}{3 \cdot 1 + 9 \cdot \lambda} = E_{yearly} \cdot \frac{1}{7} \qquad (2)$$

$$E_{mid-year} = E_{yearly} \cdot \frac{9 \cdot \lambda}{3 \cdot 1 + 9 \cdot \lambda} = E_{yearly} \cdot \frac{6}{7} \qquad (3)$$

This means that, in the multi-periodic model we assume that a production of $\frac{1}{7}$ kWh electricity at winter and $\frac{6}{7}$ kWh production at mid-year requires an investment cost of 353.98 HUF and a yearly operating cost of 22.12 HUF. These are both proportional costs.

The way we transform the single period P-Graph model into a multi-periodic one with the aforementioned two periods, is the following:

- The P-Graph is duplicated, each having the same set of raw materials, intermediates, final products, and operating units. One clone is for each period.
- The raw materials are merged between periods, as they are available for consumption at any time.
- The operating units for the biogas plant and the pelletizer, which hold the cost data of these two equipment units, are merged for each, while keeping both of their capacity outputs. The new operating units produce the logical "production capacity" materials in a ratio of $\frac{1}{4}$ for winter, and $\frac{3}{4}$ for mid-year, because of the respective lengths of the periods. These capacities can be freely used in each period, by each of the available logical operating units representing pelletizing and biogas production from a particular input.
- For the biogas furnace, biogas CHP plant and the solar power plant, we replicate the scheme applied to the biogas plant and the pelletizer. A single operating unit is introduced holding the cost data, which produces two logical "production capacity" materials—one used in winter, one used in mid-year. The ratio of capacity generation is $\frac{1}{4}$ for winter and $\frac{3}{4}$ for mid-year, with the exception of the solar power plant, which produces $\frac{1}{7}$ for winter and $\frac{6}{7}$ for mid-year. The production capacity materials are consumed by the only logical operating unit in the period, representing actual production.
- For the rest of the operating units, like the electricity transfer, electric heaters, natural gas furnace, the duplicates remain, as there are no costs or other constraints associated with these operations that would require to be treated globally for the two periods.

The resulting final multi-periodic P-Graph model is shown in Figure 11. We can observe that the same scheme is applied to the five operating units with costs. This scheme had been already done halfway for the pelletizer and the biogas plant, not because of the multi-periodic model, but for the sake of modeling the capacities correctly.

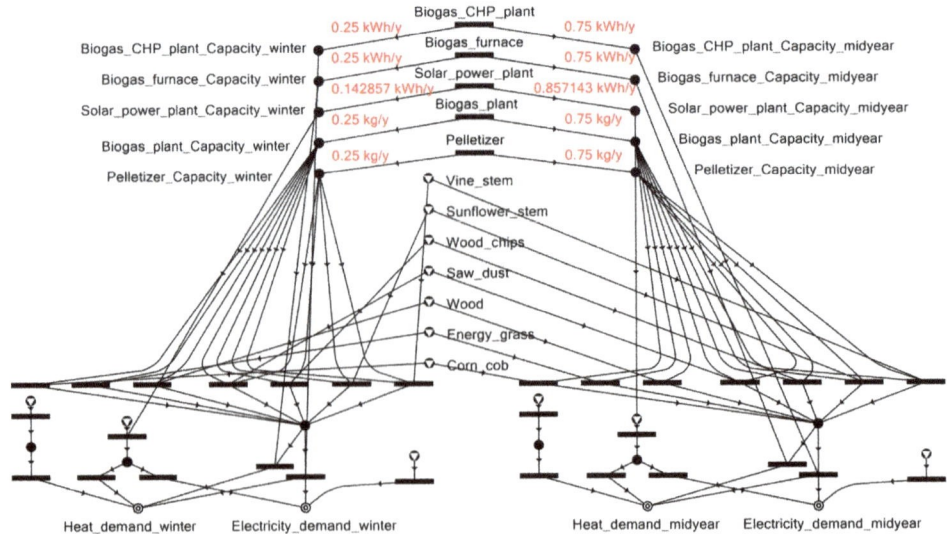

Figure 11. Final multi-periodic P-Graph model for the energy options of the manufacturing plant.

4. Results and Discussion

The single period and multi-periodic models were solved by the PNS Studio program, version 5.2.3.1, on a Lenovo Y50-70 laptop computer equipped with an i7-4710 HQ processor and 8 GB RAM. Note that the problem sizes are relatively small, which means that solution should be fast on any modern computer. First, both models were solved with an assumed investment horizon of 20 years. Then, lower horizons were set. We could observe that the MSG algorithm reports the P-Graphs themselves as maximal structures. The reason behind this is that these P-Graph models do not contain any redundancy. All parts are candidates for an optimal solution. The ABB algorithm was launched, and the first few best solution structures were manually investigated. The algorithm succeeded in 1 s for the single period case, and 2 s for the multi-periodic case. The P-Graph model files for PNS Studio and solutions discussed here are available in the supplementary materials.

4.1. Single Period Model

The optimal solution is 220.709 M HUF/y operating cost for the single period case. The 10 best solution structures (including the first, optimal) are summarized in Table 8. Note that the usage of the Pelletizer and the Biogas plant are a direct consequence of the existence of their respective raw materials in the model. We can see that the first few solution structures differ only slightly in terms of the yearly cost. The difference between the #1 and #10 solutions is 2.72%. Still, a unique set of energy sources is used each time. This means that all of these solutions can be sensible choices, as little change to the data, or other factors considered may justify the selection of an alternative of the optimal solution.

Table 8. The 10 best solution structures for the single period P-Graph model, minimizing yearly operation costs. The objective values as well as the used technologies and biomass types are depicted *.

Obj.	M HUF/y	Cc	Eg	Wd	Sd	Wc	Ss	Vs	Eh	Et	Bf	Bc	Pg	Pe
#1	220.709	X	X									X		X
#2	224.057		X				X	X				X		X
#3	224.325		X	X				X				X		X
#4	224.357		X					X				X	X	X
#5	224.496		X		X			X				X		X
#6	224.526		X			X	X					X		X
#7	225.895		X	X			X					X		X
#8	226.049		X				X					X	X	X
#9	226.380	X	X							X		X		X
#10	226.723		X		X	X	X					X		X

* Columns: Cc—corn cob, Eg—energy grass, Wd—wood, Sd—sawdust, Wc—wood chips, Ss—sunflower stem, Vs—vine stem, Eh—electric heater from the solar power plant, Et—electricity transfer from the solar power plant, Bf—biogas furnace, Bc—Biogas CHP plant, Pg—purchase of natural gas, Pe—purchase of electricity.

The biogas CHP plant is used in all cases, making it a very useful candidate for sustainable energy supply. Contrary, the biogas furnace is not present in the first few solutions, even though it is cheaper. This means that the greatest advantage of the biogas CHP plant is that it can generate electricity. Note that also all seven types of biomass appear in different combinations. Decisions on which is the better and worse are determined by their available amounts, costs and heating values, and whether they must be pelletized or not.

We shall also note that even though solar power plants appear in solution structure #9, it is only used for electricity production. Purchase of natural gas or electricity from the grid is still present in many solutions. Actually, #9 is the only one where purchasing electricity is completely eliminated. There are structures where both additional heating and electricity is required from the providers.

It may be an important question whether the resulting solution structures actually represent good practical choices. One security concern is that if the manufacturing plant is only relying on its own energy supplies, then unforeseen shortages could lead to vast losses. This is not a high risk in our model. In the worst case, electricity and natural gas can still be purchased, even when the scenario is set to completely omit these sources. These options are valid because no investment or operational costs are required for purchase, and the manufacturing plant already has a furnace. If we assume this infrastructure is maintained, losses can be kept minimal.

4.2. Multi-Periodic Model

The optimal solution for the multi-periodic model is 228.942 M HUF/y, see Figure 12. The used options for the 10 best solution structures again are listed in Table 9. The optimal solution for the problem utilizes the biogas plant and a biogas CHP plant for producing part of the requirements at mid-year and at winter. Note that energy grass and corn cobs are utilized at full capacity of the purchased equipment units. Energy grass is the dominant biomass supply, but direct purchase of energy is also needed. It is also worth noting that each period can have an individual supply. In this example, natural gas purchase is not needed at all mid-year, but is utilized at winter. Resource utilization of this solution is detailed in Table 10, and plant capacities in Table 11.

Table 9. The 10 best solution structures for the multi-periodic model. The objective values as well as the usage of technologies and biomass types in each period are depicted *.

Obj.	M HUF/y	Cc	Eg	Wd	Sd	Wc	Ss	Vs	Eh	Et	Bf	Bc	Pg	Pe
#1	228.942	X	X									X	W	X
#2	228.986	M	X									X	W	X
#3	229.205	W	X									X	W	X
#4	229.358		X					X				X	W	X
#5	229.362	W	X					M				X	W	X
#6	229.363	W	X				M	W				X	W	X
#7	229.366	W	X				M					X	W	X
#8	229.378		X				W	M				X	W	X
#9	229.385		X	X								X	W	X
#10	229.391		X				X					X	W	X

* Columns: Cc—corn cob, Eg—energy grass, Wd—wood, Sd—sawdust, Wc—wood chips, Ss—sunflower stem, Vs—vine stem, Eh—electric heater from the solar power plant, Et—electricity transfer from the solar power plant, Bf—biogas furnace, Bc—Biogas CHP plant, Pg—purchase of natural gas, Pe—purchase of electricity. Cell values: W—winter period, M—mid-year period, X—both periods.

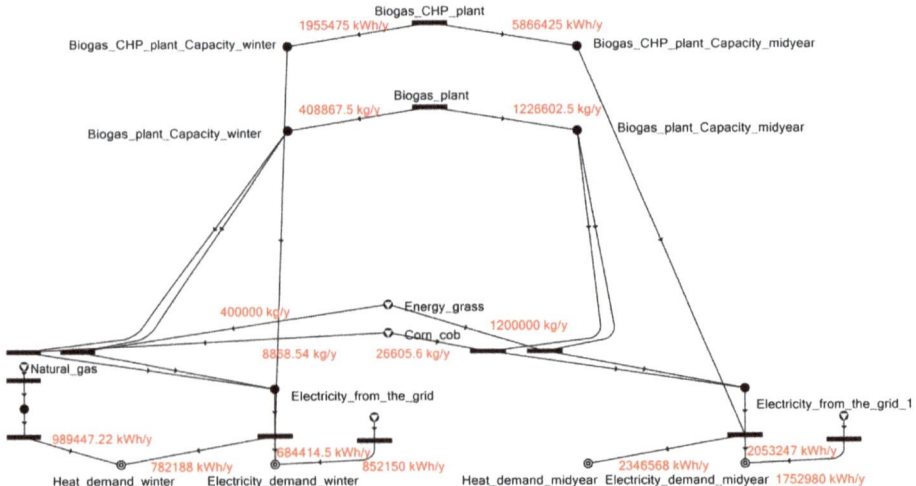

Figure 12. The optimal solution for the multi-periodic scenario, with a 20 y investment horizon, relies on corn cobs and energy grass, as well as direct purchase.

Table 10. Energy supply overview for the optimal solution in the multi-periodic model.

Period	Resource	Usage (per y)	Cost (per y)	Contribution Heat	Contribution Electricity
Mid-year	Corn cob	26,606 kg	159,636 HUF	1.81%	0.98%
	Energy grass	1,200,000 kg	9,600,000 HUF	98.19%	52.97%
	Natural gas	0 m^3	0 HUF	0.00%	N/A
	Electricity	1,752,980 kWh	66,613,200 HUF	N/A	46.06%
Winter	Corn cob	8869 kg	53,214 HUF	0.80%	0.81%
	Energy grass	400,000 kg	3,200,000 HUF	43.35%	43.73%
	Natural gas	104,765 m^3	11,943,200 HUF	55.85%	N/A
	Electricity	852,150 kWh	32,381,700 HUF	N/A	55.46%

Table 11. Utilization of the plant producing biogas, and the CHP power plant producing energy, according to the optimal solution of the multi-periodic model with 20 years investment horizon assumed.

Resource	Period	Capacity	%	Cost (per y)
Biogas plant	yearly (total)	1,635,470 kg	100%	36,980,400 HUF
	mid-year	1,226,603 kg	75%	
	winter	408,868 kg	25%	
Biogas CHP plant	yearly (total)	7,821,900 kWh	100%	68,010,800 HUF
	mid-year	5,866,430 kWh	75%	
	winter	1,955,470 kWh	25%	

Several interesting facts can be deduced from the solutions. First, the objective for the multi-periodic model is slightly worse than the objective for the single period case, by 3.73% at the best structures. This is natural, because the multi-periodic is a more accurate model with additional requirements. This means that not only the total demands must be satisfied, but the demands for each period individually, which is a more restrictive condition.

We can also see that solution structures show little variation, either in structure, or in objective values. Solution #1 and #10 have a difference of just 0.2%. The reason behind this might be that multi-periodic models have much more solution structures in general than their single period counterparts. We suspect that the technologies and options observed in the single period case would turn up if more solution structures of the multi-periodic model were investigated.

The general structure of the first few solutions is that the biogas CHP plant is used as the only investment. It is sufficient to serve the heating requirements in the mid-year period, but natural gas purchase is always required in the winter period. In both periods, the electricity requirements are still not satisfied, so electricity must be purchased. The solar power plant does not appear in the first few solutions. This might be due to the fact that it is especially weak in the winter period. Saw dust and wood chips do not appear in the first 10 solutions, while energy grass is used in all cases, in both periods. There are very small differences between some structures. For example, #1, #2 and #3 only differs on the fact that whether corn cobs are used in winter, mid-year, or both.

Overall, we can observe how the multi-periodic models behave compared to their single period counterparts. The case is more restricted, but there is also more flexibility. The energy supply scenario can be different for the winter period and the mid-year period, even though exactly the same investments are available in both.

4.3. Shorter Investment Horizons

The costs of units in the model have two components, operational cost and investment cost. The investment horizon is used to annualize the investment cost. The horizon in this model means the length in years for which the annually most profitable solution is to be found. Consequently, the investment costs are divided evenly in the investment horizon. Note that this is different from the payback time of a single operating unit, as the investment horizon is a single parameter for the optimization of the whole system. The purchased equipment units are assumed to be in their useful life throughout the horizon, but management often considers only those solution scenarios which are profitable in a much shorter term. As yearly investment costs are vast and are depending on the horizon length, this modeling parameter should be investigated.

The former investigation was done assuming a 20 y horizon, now additional results for shorter ones, 10 y and 5 y are presented. The first two solution structures for each model and in each considered horizon length are shown in Table 12.

What can be seen is that even for 10 y, the business as usual solution is the optimal. This means that all heating and electricity demands are purchased as natural gas and electricity from the grid, rather than being produced, and no additional investments are made. This exact solution is actually the optimum for both the single and multi-periodic cases, for both 10 y and 5 y of investment horizons. It may happen that the two different models, the single and multi-periodic ones, have a common solution, as in this case.

Table 12. Results for different investment horizons.

Investment Horizon	Periods	Obj.	M HUF/y
20 y	Single	#1	220.709
20 y	Single	#2	224.057
20 y	Multi	#1	228.942 [1]
20 y	Multi	#2	228.986
10 y	Single	#1	252.735 [2]
10 y	Single	#2	268.288
10 y	Multi	#1	252.735 [2]
10 y	Multi	#2	264.647 [1]
5 y	Single	#1	252.735 [2]
5 y	Single	#2	342.985 [3]
5 y	Multi	#1	252.735 [2]
5 y	Multi	#2	324.184 [3]

[1] Exactly the same solution, but different objective due to the different horizon length. [2] Exactly the same, business as usual solution, same objective as if there is no investment, so the paybacks in this case are irrelevant. [3] Technically similar solutions.

Note that the best solution for 20 y horizon in the multi-periodic model is the same as the second best for the 10 y horizon case. The second best solutions for 10 y horizon in both models are technically the same. Corn cobs and energy grass are used in a biogas CHP plant. These solutions have a considerable difference from the optimal business as usual solutions, 6.15% at the single period model, and 4.71% at the multi-periodic model. These are larger differences than can be observed in the first 10 solutions of the 20 y horizon cases.

For the 5 y horizon, the situation is much worse; there is a whopping difference of 35.7% for the single period case, and 28.27% for the multi-periodic case. Note that the second best solutions for both models are technically similar. This time, the so far ignored biogas furnace is the option, supported by energy grass consumption. The reason behind this is that probably the biogas furnace is the cheapest working investment, if we want to invest into something and leave the business as usual solution at all costs.

One interesting thing is that the seemingly more restrictive multi-periodic model has a better second best solution than its single period counterpart in the 10 y and 5 y investment horizon cases. The reason for this is that in the multi-periodic solution, gas purchase helps, but only in winter. This is not a possible option in the single period case.

These results have shown that the biomass and solar power availability are not so promising in this particular case study, because these investments would require a long time to become economical. Companies usually plan for much shorter time horizons. However, the situation is constantly changing, better technologies may appear, or more endorsement for sustainable energy supplies. The P-Graph model can be easily adjusted if a scenario with different data and other technologies are introduced, provided that similar assumptions can be made as in this work.

5. Conclusions

A P-Graph model was developed, and the energy supply options of a manufacturing plant in Hungary were investigated. The general PNS was defined, and the P-Graph framework was introduced. Step by step, the P-Graph model of the problem was constructed. Estimates on future demands were made to serve as a basis for our optimization model. We defined the raw materials, the demands, the intermediates, and the operating units. Parameters were gathered to materials and the operating units but the structure of the model has the main focus. Modeling techniques were presented for the P-Graph framework to handle situations like mass-based capacity limitations, multiple potential inputs with arbitrary ratios or activities like purchasing. A multi-periodic P-Graph is implemented which differentiates winter and mid-year consumption. It is capable of modeling operating unit capacities when demands are fluctuating, and also takes into account solar energy supplies.

The PNS Studio software was used to solve the single and multi-periodic models with the ABB algorithm. The multi-periodic scenarios establish that energy supply methods can vary between winter and other parts of the years. It can be observed from the results that significant improvement can be obtained compared to the business as usual solution, where all electricity and heat is purchased from the market. However, this requires that the investments of local energy supply options, biomass and solar energy utilization have a long investment horizon. This means that although there are considerable options for sustainable energy supplies, as they are beneficial in the long-term, the economical environment significantly impacts their efficiency.

The P-Graph model was shown to be capable of determining the best solutions for an energy supply optimization problem. Note that other aspects can be included in the future. Upgrades like insulation, better heating system, energy saving light bulbs, and other investments can be incorporated to yield a more precise model for the demands, as well as other power plant types and resources. Storage and exact availability of raw materials can still be modeled in a multi-periodic manner. Other demands, like the water system of the plant can be governed by a single unified PNS problem. Nevertheless, the current model is capable of handling similar supply and investment options with different data, with minor modifications.

Supplementary Materials: The following are Available online at http://www.mdpi.com/1996-1073/12/8/1484/s1. The P-Graph model files for the case study presented, in PGSX format, with a few solution structures. These can be viewed, edited or resolved by the P-Graph Studio software. Images of the solution structures from the mentioned cases are also provided.

Author Contributions: Introduction, I.H. and H.C.; Materials and Methods, A.É. and I.H. Results, A.É. and H.C.; Discussions, A.É.; Conclusions, I.H. and L.H. Review and editing, H.C. and I.H. Solver programming, L.H.

Funding: We acknowledge the financial support of Széchenyi 2020 under the EFOP-3.6.1-16-2016-00015.

Acknowledgments: We gratefully thank Adrián Szlama for his work on the P-Graph model and data collection and evaluation.

Conflicts of Interest: The authors declare no conflict of interest.

References

1. US Census Bureau. US and World Population Clock. Available online: www.census.gov/popclock (accessed on 16 March 2018).
2. Worldwatch Institute. The State of Consumption Today. Available online: www.worldwatch.org/node/810 (accessed on 16 March 2018).
3. Running, S.W. A Measurable Planetary Boundary for the Biosphere. *Science* **2012**, *337*, 1458–1459. [CrossRef] [PubMed]

4. US Energy Information Administration. First Use of Energy for All Purposes (Fuel and Nonfuel). Available online: www.eia.gov/consumption/manufacturing/data/2010/pdf/Table1_1.pdf (accessed on 16 March 2018).
5. International Energy Agency. Electricity Statistics. Available online: https://www.iea.org/statistics/electricity/ (accessed on 20 February 2019).
6. Saavedra, M.R.; Fontes, C.H.O.; Freires, F.G.M. Sustainable and renewable energy supply chain: A system dynamics overview. *Renew. Sustain. Energy Rev.* **2018**, *82*, 247–259. [CrossRef]
7. Nemet, A.; Klemeš, J.J.; Duić, N.; Yan, J. Improving sustainability development in energy planning and optimisation. *Appl. Energy* **2016**, *184*, 1241–1245. [CrossRef]
8. Kalaitzidou, M.A.; Georgiadis, M.C.; Kopanos, G.M. A General Representation for the Modeling of Energy Supply Chains. *Comput. Aided Chem. Eng.* **2016**, *38*, 781–786.
9. Friedler, F.; Tarjan, K.; Huang, Y.W.; Fan, L.T. Graph-Theoretic Approach to Process Synthesis: Axioms and Theorems. *Chem. Eng. Sci.* **1992**, *47*, 1973–1988. [CrossRef]
10. Friedler, F.; Tarjan, K.; Huang, Y.W.; Fan, L.T. Graph-Theoretic Approach to Process Synthesis: Polynomial Algorithm for the Maximal Structure Generation. *Comput. Chem. Eng.* **1993**, *17*, 929–942. [CrossRef]
11. Friedler, F.; Varga, B.J.; Fan, L.T. Decision-Mapping: A tool for consistent and complete decisions in process synthesis. *Chem. Eng. Sci.* **1995**, *50*, 1755–1768. [CrossRef]
12. Friedler, F.; Varga, B.J.; Fehér, E.; Fan, L.T. Combinatorially Accelerated Branch-and-Bound Method for Solving the MIP Model of Process Network Synthesis. In *State of the Art in Global Optimization*; Floudas, C.A., Pardalos, P.M., Eds.; Kluwer Academic Publishers: Dordrecht, The Netherlands, 1996; pp. 609–626.
13. P-Graph Studio. Available online: http://pgraph.org (accessed on 31 January 2019).
14. Kalauz, K.; Süle, Z.; Bertok, B.; Friedler, F.; Fan, L.T. Extending Process-Network Synthesis Algorithms with Time Bounds for Supply Network Design. *Chem. Eng. Trans.* **2012**, *29*, 259–264.
15. Barany, M.; Bertok, B.; Kovacs, Z.; Friedler, F.; Fan, L.T. Solving vehicle assignment problems by process-network synthesis to minimize cost and environmental impact of transportation. *Clean Technol. Environ.* **2011**, *13*, 637–642. [CrossRef]
16. Frits, M.; Bertok, B. Process Scheduling by Synthesizing Time Constrained Process-Networks. *Comput. Aided Chem. Eng.* **2014**, *33*, 1345–1350.
17. Heckl, I.; Friedler, F.; Fan, L.T. Solution of separation-network synthesis problems by the P-Graph methodology. *Comput. Chem. Eng.* **2010**, *34*, 700–706. [CrossRef]
18. Szlama, A.; Heckl, I.; Cabezas, H. Optimal Design of Renewable Energy Systems with Flexible Inputs and Outputs Using the P-Graph Framework. *AIChE J.* **2016**, *62*, 1143–1153. [CrossRef]
19. Heckl, I.; Halasz, L.; Szlama, A.; Cabezas, H.; Friedler, F. Modeling multi-period operations using the P-graph methodology. *Comput. Aided Chem. Eng.* **2014**, *33*, 979–984.
20. Tan, R.R.; Aviso, K.B. An extended P-Graph approach to process network synthesis for multi-period operations. *Comput. Chem. Eng.* **2016**, *85*, 40–42. [CrossRef]
21. Bertok, B.; Bartos, A. Algorithmic process synthesis and optimisation for multiple time periods including waste treatment: Latest developments in p-graph studio software. *Chem. Eng. Trans.* **2018**, *70*, 97–102.
22. Ebrahim, M.; Kawari, A. Pinch technology: An efficient tool for chemical-plant energy and capital-cost saving. *Appl. Energy* **2000**, *65*, 45–49. [CrossRef]
23. Tan, R.R.; Foo, D.C.Y. Pinch analysis approach to carbon-constrained energy sector planning. *Energy* **2007**, *32*, 1422–1429. [CrossRef]
24. Tan, R.R.; Aviso, K.B.; Foo, D.C.Y. P-Graph Approach to Carbon-Constrained Energy Planning Problems. *Comput. Aided Chem. Eng.* **2016**, *38*, 2385–2390.
25. Varga, V.; Heckl, I.; Friedler, F.; Fan, L.T. PNS Solutions: A P-Graph Based Programming Framework for Process Network Synthesis. *Chem. Eng. Trans.* **2010**, *21*, 1387–1392.
26. Lam, H.L. Extended P-graph applications in supply chain and Process Network Synthesis. *Curr. Opin. Chem. Eng.* **2013**, *2*, 475–486. [CrossRef]
27. Cabezas, H.; Argoti, A.; Friedler, F.; Mizsey, P.; Pimentel, J. Design and engineering of sustainable process systems and supply chains by the P-graph framework. *Environ. Prog.* **2018**, *37*, 624–636. [CrossRef]
28. Klemeš, J.J.; Varbanov, P. Spreading the message: P-Graph enhancements: Implementations and applications. *Chem. Eng. Trans.* **2015**, *45*, 1333–1338.

29. Okusa, J.S.; Dulatre, J.C.R.; Madria, V.R.F.; Aviso, K.B.; Tan, R.R. P-graph Approach to Optimization of Polygeneration Systems Under Uncertainty. In Proceedings of the DLSU Research Congress, Manila, Philippines, 7–9 March 2016.
30. Tan, R.R.; Cayamanda, C.D.; Aviso, K.B. P-Graph approach to optimal operational adjustment in polygeneration plants under conditions of process inoperability. *Appl. Energy* **2014**, *135*, 402–406. [CrossRef]
31. Benjamin, M.F.D. Multi-disruption criticality analysis in bioenergy-based eco-industrial parks via the P-graph approach. *J. Clean. Prod.* **2018**, *186*, 325–334. [CrossRef]
32. Ng, R.T.L.; Tan, R.R.; Hassim, M.H. P-graph methodology for bi-objective optimisation of bioenergy supply chains: Economic and safety perspectives. *Chem. Eng. Trans.* **2015**, *45*, 1357–1362.
33. How, B.S.; Hong, B.H.; Lam, H.L.; Friedler, F. Synthesis of multiple biomass corridor via decomposition approach: A P-Graph application. *J. Clean. Prod.* **2016**, *130*, 45–57. [CrossRef]
34. Atkins, M.J.; Walmsley, T.G.; Ong, B.H.Y.; Walmsley, M.R.W.; Neale, J.R. Application of P-Graph techniques for efficient use of wood processing residues in biorefineries. *Chem. Eng. Tran.* **2016**, *52*, 499–504.
35. Cabezas, H.; Heckl, I.; Bertok, B.; Friedler, F. Use the P-graph framework to design supply chains for sustainability. *Chem. Eng. Prog.* **2015**, *111*, 41–47.
36. Vance, L.; Heckl, I.; Bertok, B.; Cabezas, H.; Friedler, F. Designing sustainable energy supply chains by the P-graph method for minimal cost, environmental burden, energy resources input. *J. Clean. Prod.* **2015**, *94*, 144–154. [CrossRef]
37. Lam, H.L.; Varbanov, P.; Klemeš, J.J. Optimisation of regional energy supply chains utilising renewables: P-Graph approach. *Comput. Chem. Eng.* **2010**, *34*, 782–792. [CrossRef]
38. Lam, H.L.; Chong, K.H.; Tan, T.K.; Ponniah, G.D.; Tin, Y.T.; How, B.S. Debottlenecking of the Integrated Biomass Network with Sustainability Index. *Chem. Eng. Trans.* **2017**, *61*, 1615–1620.
39. How, B.S.; Yeoh, T.T.; Tan, T.K.; Chong, H.K.; Ganga, D.; Lam, H.L. Debottlenecking of sustainability performance for integrated biomass supply chain: P-graph approach. *J. Clean. Prod.* **2018**, *193*, 720–733. [CrossRef]
40. Aviso, K.B.; Belmonte, B.A.; Benjamin, M.F.D.; Arogo, J.I.A.; Coronel, A.L.O.; Janairo, C.M.J.; Foo, D.C.Y.; Tan, R.R. Synthesis of optimal and near-optimal biochar-based Carbon Management Networks with P-graph. *J. Clean. Prod.* **2019**, *214*, 893–901. [CrossRef]
41. Tan, R.R.; Aviso, K.B.; Foo, D.C.Y. P-graph and Monte Carlo simulation approach to planning carbon management networks. *Comput. Chem. Eng.* **2017**, *106*, 872–882. [CrossRef]
42. Heckl, I.; Halasz, L.; Szlama, A.; Cabezas, H.; Friedler, F. Process synthesis involving multi-period operations by the P-graph framework. *Comput. Chem. Eng.* **2015**, *83*, 157–164.
43. Aviso, K.B.; Lee, J.Y.; Dulatre, J.C.; Madria, V.R.; Okusa, J.; Tan, R.R. A P-graph model for multi-period optimization of sustainable energy systems. *J. Clean. Prod.* **2017**, *161*, 1338–1351. [CrossRef]
44. Éles, A.; Halász, L.; Heckl, I.; Cabezas, H. Energy Consumption Optimization of a Manufacturing Plant by the Application of the P-Graph Framework. *Chem. Eng. Trans.* **2018**, *70*, 1783–1788.

© 2019 by the authors. Licensee MDPI, Basel, Switzerland. This article is an open access article distributed under the terms and conditions of the Creative Commons Attribution (CC BY) license (http://creativecommons.org/licenses/by/4.0/).

Article

LCA-Based Comparison of Two Organic Fraction Municipal Solid Waste Collection Systems in Historical Centres in Spain

Jara Laso [1,*], Isabel García-Herrero [1], María Margallo [1], Alba Bala [2], Pere Fullana-i-Palmer [2], Angel Irabien [1] and Rubén Aldaco [1]

1. Department of Chemical and Biomolecular Engineering, University of Cantabria, Avda. de los Castros s/n, 39005 Santander, Spain; isabel.garciaherrero@unican.es (I.G.-H.); maria.margallo@unican.es (M.M.); irabien@unican.es (A.I.); ruben.aldaco@unican.es (R.A.)
2. UNESCO Chair in Life Cycle and Climate Change ESCI-UPF, Universitat Pompeu Fabra, Pg. Pujades 1, 08003 Barcelona, Spain; alba.bala@esci.upf.edu (A.B.); pere.fullana@esci.upf.edu (P.F.-i.P.)
* Correspondence: jara.laso@unican.es

Received: 20 March 2019; Accepted: 10 April 2019; Published: 11 April 2019

Abstract: Municipal solid waste (MSW) collection is an important issue in the development and management of smart cities, having a significant influence on environmental sustainability. Door-to-door and pneumatic collection are two systems that represent a way of arranging waste collection in city´s historic areas in Spain where conventional street-side container collection is not feasible. Since door-to-door collection generates significant direct greenhouse gas emissions from trucks, pneumatic collection emerges as an alternative to the trucking system. While this technology apparently reduces local direct air emissions, it suffers from a large energy demand derived from vacuum production for waste suction. The introduction of new normative frameworks regarding the selective collection of the biodegradable fraction makes necessary a comprehensive analysis to assess the influence of this fraction collection and its subsequent recycling by anaerobic digestion. As a novelty, this work compares both conventional door-to-door and pneumatic collection systems from a life cycle approach focusing on the biodegradable waste. Results indicate that, in spite of the fact electricity production and consumption have a significant influence on the results, the energy savings from the recycling of the organic fraction are higher than the energy requirements. Therefore, the pneumatic collection could be an environmentally-friendly option for MSW management under a circular economy approach in Spanish city´s historic areas, since wastes could be a material or energy source opportunity.

Keywords: anaerobic digestion; biowaste; life cycle assessment; smart city; waste collection

1. Introduction

Cities are expanding and increasing in number. In fact, c.a. 70% of the population will move to urban areas by 2050, leading to vast cities [1]. Such cities will require smart sustainable infrastructures to manage citizens' needs and to offer fundamental and more advanced services [2]. This urbanization process in cities of developing countries has led to an increase in the quantity and complexity in terms of composition of municipal solid waste (MSW) [3]. The worldwide MSW generation in 2016 was approximately 2.01 billion tonnes [4] and it is estimated that, by 2050, the production will rise to 3.40 billion tonnes with approximately 56% of organic content [5]. At a European level, the annual generation of MSW in EU-28 reached 483 kg per person in 2016, with the daily waste production per capita in the European countries ranging from 0.71 to 2.06 kg [6]. Therefore, it is expected that, in the coming years, both the increase of global population and the growth in developing countries will create a boost in MSW production. In this context cities will have to face new challenges, linking technology to the improvement of the quality of citizen's life, moving towards the smart city concept [7]. It is essential to be able to assess the environmental impact of cities by assessing their different components, such as the waste management infrastructure [8].

In response to the challenge posed by the generation and consequently management of MSW, the European Union (EU) has established a legal framework targeting strategies to increase resource efficiency. The most relevant goals for MSW are the recycling rates proposed by the Packaging [9] and Packaging Waste Directive [10] and the Directive on Industrial Emissions (DIE) [11]. The Directive on Landfills [12], which aims to prevent or to reduce as far as possible the negative effects on the environment caused by waste landfilling, has set the amount of biodegradable municipal waste landfilled to be reduced to 50% in 2009 and to 35% in 2016 (compared to 1995 levels). In addition, the Waste Framework Directive [13] establishes a target of recycling and preparing for reuse of 60% by 2025 and 65 % by 2030. To reach these values, the EU fosters selective collection systems and recycling of the MSW fractions, ensuring the sustainability of smart cities moving towards a circular economy approach (Figure 1) [14]. This legal framework contributes to the EU Sustainable Development Goals (SDGs) indicator set. For instance, SDG 7: affordable and clean energy; SDG 11: sustainable cities and communities; SDG 12: responsible production and consumption and SDG 13: climate action [15].

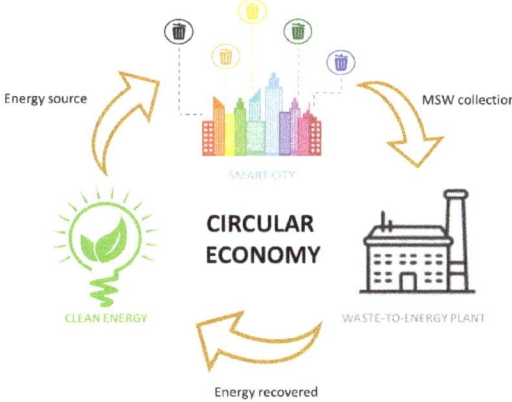

Figure 1. Framework of the study.

At a national level, in most Spanish cities, glass, paper/cardboard and light packaging are currently collected separately, whereas the remaining fraction comprises the organic fraction because biodegradable waste collection system has not been still fully implemented [16]. However, to meet the goals of the European legislation [13], organic waste should be collected separately in a specific container for two main reasons: it represents around 40% of the MSW generated in Europe [17] and a suitable collection allows obtaining a higher quality compost [18]. In fact, one of the main components of the biodegradable fraction is food waste, which has been analysed in terms of prevention and recovery in many studies. Laso et al. [19] combined Life Cycle Assessment (LCA) and Data Envelopment Analysis (DEA) to assess the efficiency of Spanish agri-food system. Garcia et al. [20] quantified the food losses at the distinct stages of the food supply chain in terms of mass, nutrients and economy. On the other hand, Thyberg and Tonjes [21] examined the impacts of food system modernisation on food waste generation. Finally, Gustavsson et al. reported two studies focused on the extent and effects as well as causes and prevention of food losses and food waste, one for medium/high income countries [22] and another for low income countries [23]. In Spain, both European and Spanish legislation have boosted the introduction of selective collection systems for biowaste [24]. In fact, some Spanish regions, such as Catalonia, Madrid and Navarre, have already introduced in their MSW collection systems the "fifth container" or "brown container", exclusive for compostable MSW (see Figure 2). These containers collect kitchen waste, garden rubbish, tree cuts and waste from food market and biodegradable bags [16]. In this way, the amount of MSW disposed in landfills is reduced by means of increasing the percentage of recycling through composting. Nevertheless, there are several disadvantages related to its storage and collection; for instance, the need of specific containers, additional bins and collection points, and the requirement of additional trucks and new routes [17]. All these factors depend on the collection method: street-side or underground containers, door-to-door or pneumatic collection.

Figure 2. Municipal solid waste (MSW) containers in (**a**) Navarre and (**b**) Madrid (Spain). Source: [25,26].

Waste collection accounts for 50–75% of the total MSW cost in developed countries, being one of hots spots from an environmental perspective, due to its energy consumption and related CO_2 emissions [27]. The collection method varies from region to region. For instance, in Spain the conventional management of MSW relies on the collection of the waste from containers placed in the street by means of a fleet of heavy trucks. In particular, in 2015, 92.6% of MSW was collected from street-side containers, whereas underground containers represented 5.2%, door-to-door collection covered 1.5% and pneumatic collection only 0.7% [28]. On the other hand, in Nordic countries such as Sweden, the collection from containers has been replaced by new technological and automated systems such as vacuum waste collection and underground container systems. The use of these systems has increased, particularly in cities and in newly built areas, moving towards more environmentally-friendly cities, but also in city´s historic areas where the conventional street-side collection systems is not feasible due to the difficult access for garbage trucks [29]. Therefore, society seems to be aware about the importance of the selective collection of different waste fractions, but there is more disagreement about the most environmentally suitable selective collection system [30].

In this sense, in the last years, Life Cycle Assessment (LCA) methodology [31] has been applied to evaluate and to compare several MSW collection systems developing ad hoc methodology [32]. Some authors have assessed the most common collection systems, door-to-door and street-side containers under environmental perspectives. Mora et al. [33] used LCA to assess the environmental impact of a waste management system based on kerbside collection. On the other hand, Gilardino et al. [34] combined operational research techniques and LCA to create an effective collection-route system for garbage compactor trucks to attain a reduction in environmental impacts in the city of Lima. Other authors, such as Pérez et al. [35] described a methodology to evaluate the environmental impact of the urban containerization systems by using LCA, whereas Pires et al. [36] also evaluated, apart from the environmental, the economic aspect of a kerbside system and an exclusive bring system. Also, the social perspective of the most common collection systems was assessed under a social perspective [37]. The introduction of new technologies and automated systems boosted the comparison of conventional collection with novel systems such as vacuum collection. In this case, Teerioja et al. [27] compared a hypothetical stationary pneumatic waste collection system with a traditional vehicle-operated door-to-door collection system in an existing, densely populated urban area, from an economic point of view, while Punkkinen et al. [38] and Aranda-Usón et al. [39] applied LCA to compare the environmental sustainability of both collection systems. On the other hand, Iriarte et al. [30] used LCA to quantify and to compare the potential environmental impacts of three selective collection systems: mobile pneumatic, multi-container and door-to-door. The main conclusions previously obtained state that a pneumatic system generates more air emissions due to the consumption of electricity and installation materials. Only when the loads are close to 100%, the vacuum system had the best environmental performance compared to the conventional systems. In addition, under an economic perspective, the pneumatic collection is estimated to be six times more expensive than traditional systems [27].

The mentioned studies consider the management of MSW taking into account glass, paper/cardboard, light packaging and bulky fraction. However, the introduction of the "brown container" makes necessary assessing the energy efficiency of the conventional and new collection systems, in order to determine the influence of the biodegradable fraction and its subsequently treatment by means of anaerobic digestion. The organic fraction of MSW is a substrate of interest due to its availability and characteristics [40]. Therefore, this study compares the energy efficiency of door-to-door system vs pneumatic waste collection, considering two alternatives: (i) the bulky fraction includes the organic fraction and (ii) the organic fraction is collected separately for further anaerobic digestion. The collection using street-side containers was excluded from the assessment since the study is placed in historic areas of Spanish cities.

2. Materials and Methods

The LCA is a tool to assess the potential environmental impacts and resources consumption throughout a product and service life-cycle [41]. In this regard, LCA has become one of the most relevant methodologies to help organisations perform their activities in the most environmentally friendly way along the whole value chain. In this work, LCA is conducted following the recommendations of the ISO 14040 [31] and 14044 [42] international standards in which LCA methodology is divided into four phases: (i) goal and scope, (ii) life cycle inventory (LCI), (iii) life cycle impact assessment (LCIA) and (iv) interpretation of the results.

2.1. Goal and Scope Definition

In this phase an accurate specification of the product or products to be investigated is done, as well as a clear description of the intended application of the study and its scope, in terms of system boundaries and functional unit (FU) [43]. Moreover, allocation procedures, cut-off rules and assumptions are also defined in this phase [44].

The purpose of this study is to assess the primary energy demand (PED) and the environmental efficiency of both conventional door-to-door and the alternative pneumatic waste collection systems considering only the management of the organic fraction. The results are expected to provide an interesting discussion on the suitability of using different collection systems depending on the waste fraction managed.

2.1.1. Function, Functional Unit and System Boundaries

The function of the compared systems is the management of MSW and, in particular, of the organic fraction generated in historic areas of Spanish cities. These areas are characterised by narrow and tortuous streets where, in most cases, the transit of people and vehicles is not feasible, causing traffic jams and the disturbance of the daily routine of citizens. To handle this problem, many cities have pedestrianised and widened the streets making difficult the waste collection by means of street-side containers and garbage trucks. In addition, in the last years, these areas have turned into the shopping and leisure centres of the city, combining banks, company headquarters and public institutions with shops, hotels and restaurants which generate high amounts of MSW that have to be managed properly. To quantify this function, it is necessary to define a FU, to which all the inputs and outputs will be referred. In this case, the FU is described as the collection of one t of MSW with the composition showed in Table 1, in which biowaste is the major fraction (42%), followed by paper/cardboard (15%).

Table 1. Average composition of the Spanish MSW (year 2016) [45].

Municipal Solid Waste Fraction	Percentage
Organic Matter	42%
Paper/Cardboard	15%
Plastic	9%
Glass	8%
Other	8%
Moisture and food debris	7%
Textile	5%
Metals	3%
Wood	2%
Bricks	1%

The system boundaries comprise the stages of the supply chain from cradle to gate, that is to say, the waste collection system, the transport of the collected waste to the sorting plant, the waste classification in the sorting plant and its treatment by means of anaerobic digestion. The waste collection system includes the manufacturing of components (i.e., bins and/or pipes) and the use stage (i.e., operation and maintenance) (see Figure 3).

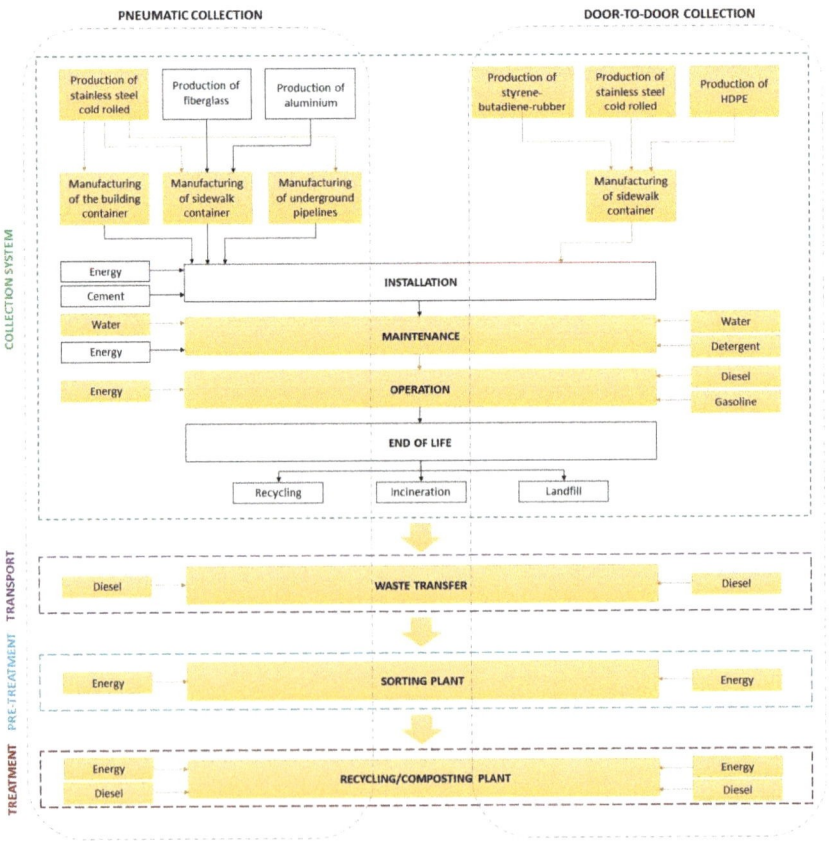

Figure 3. Diagram of selective waste collection boundaries.

2.1.2. Description of Selective Collection Systems

As previously mentioned, door-to-door and pneumatic systems are the waste collection alternatives analysed in this study, that, together with multi-container collection are the waste collection systems most implemented in urban areas. The combination of several collection systems depends on budget limitations, public participation, urbanisation age and municipal and regional planning, among others. In particular, the pneumatic system has been implemented for more than a decade and its future development will depend, just like the other systems, on economic, social and environmental aspects [30].

- *Door-to-door:* in this system, citizens leave each waste fraction outside their house, separated according to a pre-established collection schedule based on daily waste generation. A garbage truck collects waste from door-to-door bins on a specific collection day. Compacting waste trucks collect the waste from the containers (one truck for each collected fraction) in order to deliver it to final treatment. A full service scheme includes the cleaning with hot and cold water and detergent of containers using a mobile container washing vehicle.
- *Pneumatic collection:* this system uses a network of urban pipes, underground storage containers and waste inlets and chutes. Several indoor and outdoor collection points are available for the waste fractions. Waste bags are dropped inside underground containers through a chute and, according to a collecting schedule, waste is transported by a vacuum system to a collection plant. This plant, located in the centre of the collection network, is the heart of the system composed of turbo fans, cyclones, waste compressors, cleaning filters and general equipment, such as conveyor belts, cranes, compressed air and automated control systems. This collection method uses electricity to collect and compact the different waste streams [39].

2.1.3. Assumptions

The following points explain the reason for excluding some processes of the system:

- The production of aluminium and the production of fiberglass have been excluded because, at the time of model construction, no data was available. In addition, the environmental impact associated with the manufacture of these materials is not large enough to cause a deviation of the results due to the small quantities used compared to the stainless steel, the main material of which containers are made. Therefore, it is assumed that all containers are completely made of stainless steel, as well as the pipes.
- The installation of the waste pneumatic system, the manufacture of the waste collection plant, as well as the construction of the sorting plant have not been included in the model. This is due to the lack of available data and irrelevance of the environmental impacts within the pneumatic collection system. It can be reasonably assumed that, when expressed per FU, their contribution to the waste manage life cycle will be minimal. This assumption is based on the environmental impacts caused by other processes involved and the considered lifespan of these infrastructures.
- The end of life is not included because it was outside the system boundaries.
- The manufacture of the waste valves, filters and cyclones are not inventoried because it has been considered that the associated environmental impact is insignificant compared to the waste transported by this system and valves/cyclone lifespan.
- The vehicle manufacture was also excluded from the analysis.

2.1.4. Description of the Scenarios under Study

Separate collection of the organic fraction has been implemented in the northern European countries for several years, and it is now relatively well established there. However, it is not yet widespread in Spain or others countries of the southern Europe, although there is some experience at local or regional level. Therefore, the collection of the organic fraction currently coexists with the collection of the bulky fraction. In this sense, this work assesses four scenarios (see Figure 4), which are based on the two waste collection systems described above and the waste fraction managed. Scenarios 1a and 2a represent the collection of the bulky fraction by means of the door-to-door and pneumatic waste collection systems, respectively, while in scenarios 1b and 2b the organic fraction is collected separately in the fifth container. In these scenarios, sensitivity analyses were carried out varying the effectiveness of the selective collection. Initially the study assumes that 100% of the organic fraction is disposed of in the fifth container. However, this ideal situation does not always occur. Therefore, in the sensitivity analysis this percentage was reduced, increasing the amount of biodegradable fraction in the bulky fraction. This bulky fraction is conducted to a sorting plant in order to recover the largest amount of organic matter. In this study, only 9.8% of the organic fraction was considered to be recovered from the bulky fraction, and the collection truck is considered to cover an average distance of 35 km to the sorting plant [46]. Finally, since the aim of the selective collection of the organic fraction is to convert it into high quality compost, the final destination of the organic fraction is a composting plant, which is located next to the sorting plant.

Scenario 1a. DOOR-TO-DOOR COLLECTION OF THE BULKY FRACTION

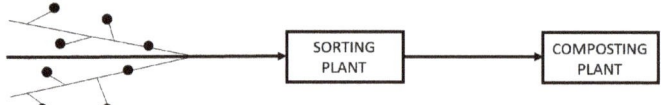

Scenario 1b. DOOR-TO-DOOR COLLECTION OF THE ORGANIC FRACTION

Scenario 2a. PNEUMATIC COLLECTION OF THE BULKY FRACTION

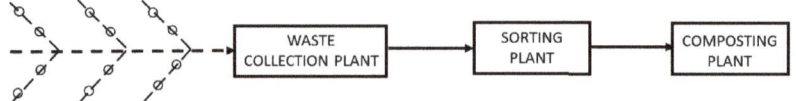

Scenario 2b. PNEUMATIC COLLECTION OF THE ORGANIC FRACTION

- ● Containers and their associated infrastructure located in collection points
- ○ Pneumatic suction points
- ⟶ Truck urban and inter-city transport
- --▸ Pneumatic urban transport

Figure 4. Diagram of selective waste collection scenarios under study.

2.1.5. Allocations

The scenarios under study are multi-outputs processes in which the management of MSW is the main function of the system and the production of electricity and compost are additional functions. To handle this problem ISO [31] establishes a specific allocation procedure in which system expansion is the first option. In this case, in the anaerobic digestion stage, methane is assumed to be combusted with a 25% efficiency of the low heating value of the biogas to generate electricity. The delivering residue of the anaerobic digestion, i.e., digestate, is transferred to a composting plant for the production of biocompost. While the compost is assumed to replace mineral fertilizer, with a substitution ratio of 20 kg N equivalent per ton of compost, the energy generated from biogas is considered to replace the Spanish electric mix [47].

2.2. Life Cycle Inventory (LCI)

LCI consists on the collection of the relevant input and output data for the assessed systems [48], and is one of the most effort-consuming phases of an LCA, both in terms of work and dedicated time [44]. In this section, data sources and main assumptions for the scenarios under study are detailed. In this case, the LCI was collected by means of an input-output analysis. Regarding the processes shared by the two scenarios, such as the sorting plant and the composting plant, data were obtained from Cobo et al. [46] and Righi et al. [49], respectively. Since the amount of MSW entering pre-treatment and treatment stages depends on the scenario analysed, the data of the sorting plant and composting plant collected in Table 2 are expressed per tonne of waste treated.

Table 2. Life cycle inventory for door-to-door collection and pneumatic collection systems.

		Units	Quantity
	Door-to-door collection		
	Container	$m^3 \cdot FU^{-1}$	0.11
	HDPE	$kg \cdot m^3$ container	4.61
	Stainless steel cold rolled	$kg \cdot m^3$ container	1.33
Inputs	Styrene-butadiene-rubber	$kg \cdot m^3$ container	0.23
	Detergent	$kg \cdot m^3$ container	2.22
	Water	$m^3 \cdot m^3$ container	84.7
	Diesel	$L \cdot FU^{-1}$	3.98
	Gasoline	$L \cdot FU^{-1}$	0.07
	Pneumatic collection		
	Electricity	$MJ \cdot FU^{-1}$	438
Inputs	Stainless steel cold rolled	$kg \cdot FU^{-1}$	0.05
	Water	$m^3 \cdot FU^{-1}$	1.00×10^{-4}
	Sorting plant		
Inputs	Electricity	$MJ \cdot t\ waste^{-1}$	4.86
	Composting plant		
	Electricity	$MJ \cdot t\ waste^{-1}$	446
Inputs	Lubricant	$kg \cdot FU^{-1}\ t\ waste^{-1}$	0.10
	Water	$kg \cdot t\ waste^{-1}$	186
	Diesel	$kg \cdot t\ waste^{-1}$	0.64
	Biogas	$kg \cdot t\ waste^{-1}$	333
	Compost	$kg \cdot t\ waste^{-1}$	210
	NH_3	$kg \cdot t\ waste^{-1}$	0.04
Outputs	CH_4	$kg \cdot t\ waste^{-1}$	2.34×10^3
	CO	$kg \cdot t\ waste^{-1}$	0.40
	HCl	$kg \cdot t\ waste^{-1}$	6.00×10^{-3}
	NOx	$kg \cdot t\ waste^{-1}$	0.30
	NMVOC	$kg \cdot t\ waste^{-1}$	0.30

2.2.1. Door-to-Door Collection

Primary data were obtained from the Spanish Ministry of Environment and Rural and Marine Affairs (MERMA) [50] and from the UNESCO Chair in Life Cycle and Climate Change (by means of a personal communication from Ecoembes in the framework of the Life-FENIX project), while secondary data were sourced from the thinkstep database [51]. The fuel consumption rates were estimated at 3.98 and 0.07 L of diesel and gasoline per collected t of waste [50]. Energy demand and emission factors for fuel production were taken from the thinkstep database [51]. Based on Ecoembes information, containers requirements were estimated at 0.107 m^3 per FU. All fractions were assumed to be deposited in a high density polyethylene (HDPE) container, which was modelled assuming the following composition: 74.75% HDPE, 21.57% steel and 3.68% styrene-butadiene-rubber. A lifetime of 7.5 years was assumed for all the containers. They were considered to be washed six times/y, using 0.35 kg of detergent per m^3 of container. The LCI of the system is shown in Table 2.

2.2.2. Pneumatic Collection

The main sources of information were Ecoembes and the two main companies that manage the pneumatic system in Portugal and Spain: Envac Iberia [52] and Ros Roca [53]. The authors personally visited their plants aiming to have a better perception of the system performance and to get real and representative data allowing the construction of the model as close as possible to reality. Secondary data, such as the production of stainless steel cold rolled for the manufacture of bins and pipes, were taken from the Thinkstep database [51]. Data collection is explained according to three main infrastructure stages of the system: waste collection plant, central collection points and underground pipes. The LCI data is listed in Table 2.

- Waste collection plant

All the waste gathered in the collection points is carried to the waste collection plant through the network of underground pipes. Then, waste is compacted and deposited in a container until it is full. The energy consumed by the system is linked to the suction and compaction operations at the waste collection plant. The suction stage is estimated to work 3,500 h/y with a power of 220 kW. The average compaction demand is established at 15 kW for 1,100 h/y. From these data, the consumptions per metric ton of waste collected for both processes has been calculated, obtaining the total consumption of this waste collection system. The estimation is explained in detail in Section S1 of the Supplementary Material.

- Central collection points

The central collection points are composed of a number of bins where citizens can deposit the wastes. Below these boxes, there are the waste valve and an air valve which are responsible for connecting the bin of the central collection point to the overall network of underground pipes. These components have been excluded from the system because of their irrelevance comparing to the whole environmental impact of the system. According to Envac Iberia [52] and Ros Roca [53], the number of bins per collection point depends on the amount of waste fractions managed. In the model under study, it has been considered only one fraction at time (the organic fraction). A lifetime of 9 years is assumed for the bins [54]. Two different containers composed the system: a sidewalk container, entirely made of stainless steel, and an indoor bin, made of stainless steel and a small fraction of fiberglass and aluminium. However, both containers have been assumed to be made entirely of stainless steel because the contribution of fiberglass and aluminium is negligible compared to the stainless steel. According to Ecoembes [28], there are 70 collection points per kilometre and, in this case, one bin per collection point having a weight of 40 kg each [55]. In Section S2 of the Supplementary Material the procedure for the calculation of the weight of bins is described.

- Underground pipes

The underground network of pipes is made of stainless steel pipes with an average length of 1 km per station line and an inner diameter of 50 cm and 12.5 mm of thickness. A 30 years lifetime is considered for them. Inside of these pipes, a stream of air transports waste bags at an average speed of 25 m/s. [55]. In Section S3 of the Supplementary Material the weight of pipe per tonne of waste collected is estimated.

3. Results and Discussion

3.1. Comparison of the Two Waste Collection Systems

The four scenarios were assessed following the LCA methodology. Figure 5 shows the primary energy demand (PED) results per FU, also cited as cumulative energy demand (CED), an indicator commonly used to assess waste management systems [56]. The results are divided into the four stages of the life cycle (collection, transport, pre-treatment and treatment). As it can be observed, the results change significantly when the organic fraction is collected separately (scenarios 1b and 2b) compared to scenarios 1a and 2a where the bulky fraction includes the biodegradable fraction. In this case, the PED of scenarios 1b and 2b is similar: −3,128 MJ and −2,701 MJ, respectively. The negative values are associated to the energy savings of the anaerobic digestion process. This is because the environmental benefits of electricity and compost production displace the environmental impact of production of electricity from the Spanish grid mix of 2016 and the production of fertilizers, and overcomes the energy inputs of the collection system. It is important to correctly calculate the credits obtained through material recycling and energy recovery [32]. Those results indicate that, when the biowaste is collected separately and, considering the composting of these residues, the use of a pneumatic system could be a suitable option, as well as the door-to-door collection was. However, when the bulky fraction includes the biodegradable fraction, pneumatic collection (scenario 2a) exhibits the largest PED, estimated at 405 MJ per FU, whereas the door-to-door collection presents negative PED values (−245 MJ). The reason is that the energy demand for the vacuum system is higher than the energy recovered from the composting process. Authors such as Iriarte et al. [30] and Punkkinen et al. [38] stated that the system with the greatest environmental impact is the pneumatic collection compared to door-to-door and multi-container collection. Nevertheless, these studies only include the collection stage, without the subsequent waste valorisation treatment and environmental benefits. On the other hand, considering the CED indicator, Iriarte et al. [30] stated that the door-to-door system had the greatest energy demand, in particular 38% higher than the pneumatic system. In our study, this difference between door-to-door and pneumatic collection is lower, around 14%, due to the energy requirement for the anaerobic process. Finally, the contribution of the MSW transportation to the pre-treatment installation and the contribution of the pre-treatment in the sorting plant was negligible, around 1% (in case of scenarios 1b and 2b the pre-treatment stage is not necessary, therefore, its contribution to the total PED is zero). These results reinforce the idea that the use of the pneumatic system could be an appropriate waste collection system in the development and the implementation of smart city technologies in historic areas of cities [57]. As mentioned previously, official data sets that street-side containers is the most common MSW collection system used. In addition, some authors such as Iriarte et al. [30] and Aranda-Usón et al. [39] have stated that this collection system presents the lowest environmental impact. However, since the main objective of this study is assessing the management of the organic fraction generated in historic areas of cities where the conventional street-side collection systems is not feasible due to traffic restrictions, this type of collection was excluded from the study. In this sense, there is a debate about which is the best collection system for these areas. Currently, there are not studies that address this problem from an environmental point of view and, in particular, under an energy and climate change perspective. In this sense, historic areas of cities are suitable zones to apply smart city technologies where monitoring and information are essential for the correct management of these areas. The management of MSW is an example for this purpose.

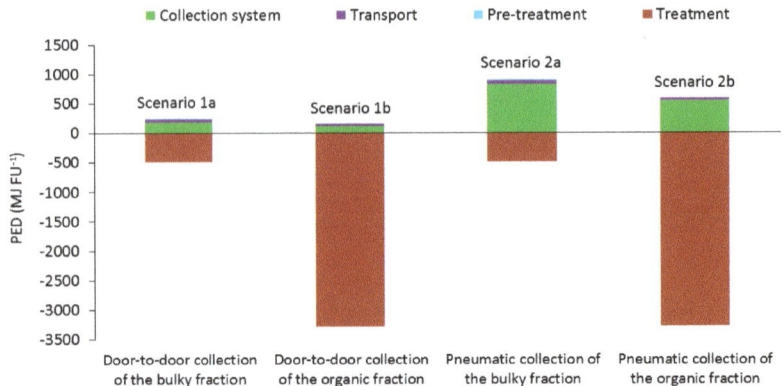

Figure 5. Primary energy demand (PED) per functional unit of each scenario.

Figure 6 focuses on the collection system stage and displays the PED consumption of each process of the waste collection system (manufacturing of components, maintenance and operation). As can be observed in Figure 6, when the waste collection system is considered isolated, the pneumatic collection consumes 5.0 times more primary energy than the door-to-door collection. This is in agreement with Iriarte et al. [30] and Punkkinen et al. [38] and highlights the importance of considering the subsequent waste valorisation treatment and environmental benefits. Regarding the contribution of the different processes, in the conventional door-to-door collection (scenarios 1a and 1b), fuel production for garbage trucks in the operation step was responsible for 64% of the PED. The rest of PED is attributed to the production and maintenance of containers (36%). On the other hand, in the pneumatic system, the production of electricity used in the process accounts for almost 100% of the total PED, whereas the manufacturing of bins and pipelines and its maintenance is negligible compared to the consumption of energy.

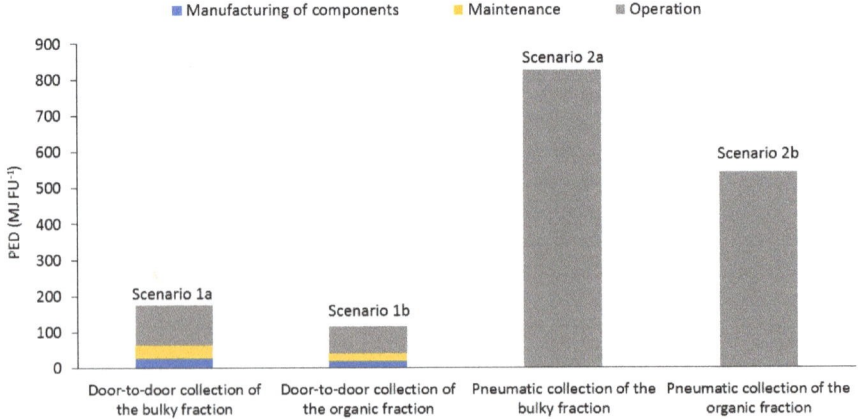

Figure 6. Primary energy demand (PED) of each process of the MSW collection system.

According to the results of the study, the electricity requirements for vacuum production is the item with the most impact. Therefore, reducing energy consumption for waste transport through the underground system and using a more environmentally friendly energy source, such as renewable energies [58], are the best improvement measures to ensure the sustainability of the pneumatic process, since waste management and energy supply systems are becoming more inter-connected [59]. This study contributes to decision-making in waste management strategies and enables the introduction of the environmental variable in the design model of smart cities. In addition, this study comprises the basis for the future optimisation of pneumatic collection as hybrid systems feed by renewable energies, such as photovoltaic installations.

3.2. Sensitivity Analysis

Our results in the baseline case study depend on many assumptions concerning the installation of the pneumatic system. Particularly, the effectiveness of the biodegradable collection, the number of waste collection points in the pneumatic system and the population density were examined.

3.2.1. Effectiveness of the Biodegradable Collection

The energy efficiency of the pneumatic collection system depends on the attitude of the citizens towards the introduction of the selective collection of the organic fraction. In this sense, a sensitivity analysis was performed varying the effectiveness of the biodegradable collection, considering the best scenarios (scenarios 1b and 2b), in which 100% of the organic waste generated is collected separately, together with less-efficient scenarios in which part of the organic fraction is collected separately and the rest remains in the bulky fraction. Figure 7 shows the results of the sensitivity analysis.

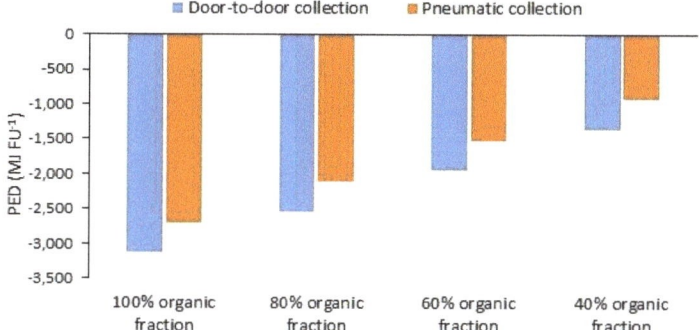

Figure 7. Results of the sensitivity analysis varying the effectiveness of the selective collection of the organic waste (100%, 80%, 60% and 40%).

As previously mentioned, the negative values are associated with the energy savings from the composting process. When the efficiency varies from 100% to 40%, the energy savings are reduced to 57% for door-to-door collection and to 66% for the pneumatic system. Therefore, these results highlight the importance of people awareness and information campaigns about the benefits of the biodegradable waste being recycled and aim at emphasizing the role that consumers play in biowaste separation at source [17].

3.2.2. Population Density

Taking as reference 20,000 citizens per km^2, to see how the environmental impact is influenced by higher densities and, therefore, larger waste volumes, the number of citizens was increased up to 80,000 citizens per km^2. Given the fixed capacity of the collection system, the PED will be established by the increase in the emptying frequency due to the increase in the waste volumes. For the sake of comparison, the management of higher waste volumes in the door-to-door collection requires either a fourfold increase in waste containers, a four times higher emptying frequency, or an intermediate solution. In this case, a higher emptying frequency is considered [27].

The increase in the population density in a specific zone is mainly due to tourism. In Spain, cities such as Fuengirola (Málaga), Benidorm (Alicante), Ibiza (Balearic Islands) and Salou (Tarragona) experiment an increase of between 65–75% in their population density due to tourism [60]. The tourism can sustain high levels of employment and incomes in the economies of many regions, but the sector is a source of environmental impacts, resource consumption and public health problems [61]. In particular, one of the most important impacts of tourism is the generation of MSW. According to Mateu-Sbert et al. [61], on average, an increase of 1% in the tourist population in Menorca causes an overall MSW raise of 0.282%. Following this statement and considering that a population density of 20,000 citizens/km^2 generate around 1,936 metric tons of MSW per year [27], the increase of the waste volume due to the raise of the population density was estimated. On the other hand, the composition of generated MSW is extremely variable as a consequence of seasonal and lifestyle impacts, in particular in historic areas of cities, which are a tourist attraction [62]. In this sense, it is considered that the increase in the population density affects the waste composition. A population density of 20,000 citizens/km^2 in winter, could corresponds to 40,000 citizens/km^2 in autumn period, reaching 60,000 and 80,000 citizens/km^2 in spring and summer, respectively. Figure 8 displays the different waste generation and composition for each population density considered.

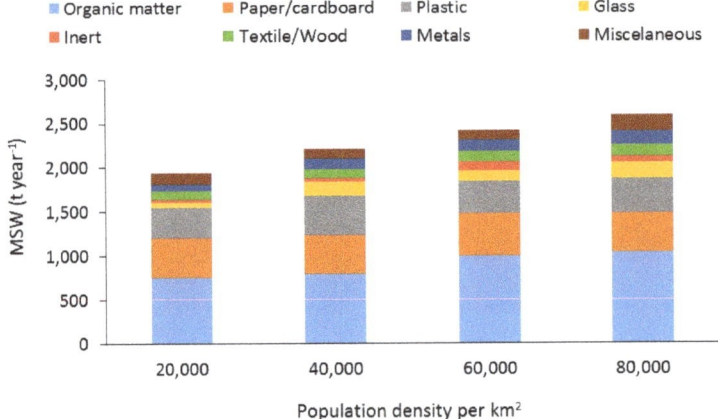

Figure 8. Variation in the municipal solid waste (MSW) generation and composition due to the increase in the population density.

The amount of MSW increases from 1,936 metric tons with 20,000 citizens per km^2 to 2,586 metric tons when the population density reaches 80,000 citizens per km^2. Regarding the waste composition, there is no a great difference in the organic content, varying from 36% in winter to 41% in spring. These results were introduced into the environmental model previously described to analyse the model sensitivity to changes in the population density. Figure 9 displays the results of PED obtained for door-to-door collection and pneumatic collection of the biodegradable fraction. It can be observed that the higher increase in the population density, the greater negative value of the PED, which means a higher recovery of energy from the treatment of the MSW. An increase of 75% of the population density means 27% of PED savings for both collection systems. These results indicate that, even though the increase of MSW generation implies a higher emptying frequency for the door-to-door collection, and a higher use of energy for the pneumatic collection, since more daily pneumatic transportation is needed, the savings from the anaerobic digestion treatment are higher.

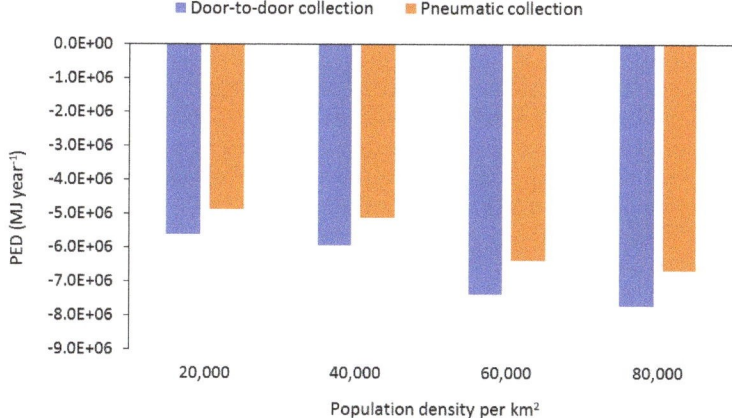

Figure 9. Results of the sensitivity analysis varying the population density of the area considered.

4. Conclusions

In this study, pneumatic collection is presented as an alternative to conventional trucking systems, reducing noise and odour effects, presenting potential space savings, increasing pedestrian areas, and apparently decreasing resource consumption. The energy assessment of door-to-door and pneumatic collection, focusing on the biodegradable fraction collection and its subsequent recycling by anaerobic digestion indicated that, when the organic fraction is collected separately, out of the bulky fraction, the pneumatic collection could be a suitable alternative because the energy requirements are balanced with the savings from the anaerobic digestion process. These results highlight the idea that energy recovery from MSW is a feasible alternative to face the challenge to move towards smart cities. Moreover, the combination of this technology with renewable energy supply, such as solar and wind (when high insolation and land availability are present), and new information and communications technologies, such as sensors to retrieve the fill level of containers, could be a good option to handle the cities transition along with the implementation of encouraging renewable technologies.

In the performed sensitivity analysis, one of the variables that potentially have a significant impact on the results is the effectiveness of the biodegradable fraction collection since when the efficiency varies from 100% to 40%, the energy savings are reduced to 57% for door-to-door collection and to 66% for the pneumatic system. This result highlights the importance of people awareness and information campaigns about the benefits of the biodegradable waste selective collection. On the other hand, other variable with potential influence on the results is the population density. An increase of 75% of the population density means 27% of PED savings for both collection systems due to the avoided impacts from the anaerobic digestion treatment.

This study aims to develop a general suitable model for Spain, based on average values, facilitating the decision-making process for the MSW collection and management. It is important to highlight the need of carrying out specific case studies with the aim of validating the model and to support the results of this work.

5. Proclamations

Due to the rapid growth of the world population, improvements in urban environments, structures and services are necessary to meet citizen's needs. New technologies open the door to innovative systems, to move conventional cities towards becoming smart cities. The generation and management of MSW is an important issue in the development and management of smart cities, where waste collection has a significant influence on the environmental sustainability. European Directives enhance selective collection systems and recycling of different MSW fractions, ensuring the capability of smart cities to move towards a circular economy. In addition, the current trend of decreasing the amount of waste disposed in landfills fosters the selective collection of the biodegradable fraction and its subsequent treatment by means of anaerobic digestion. The selective collection of MSW, in particular the organic waste, in historic areas of cities is an issue which has opened the door to a discussion in Spain about which is the best collection system for these zones, since the conventional street-side collection systems is not feasible due to traffic and access restrictions. In those areas where the lack of historical remains that can compromise the installation of the system is guaranteed, and from an energy point of view, this study encourages the use of pneumatic collection. However, in order to make a more comprehensive study, the energy assessment of the collection systems should be complemented including social and economic aspects, whose deployment can shift the balance towards one or another management options. In addition, an environmental sustainability evaluation of the collection systems could confirm the benefits of pneumatic collection mainly due to the environmental benefit deriving from the anaerobic digestion of the organic fraction, as stated some authors such as Ranieri et al. [63] and Digiesi et al. [64].

Supplementary Materials: The following are available online at http://www.mdpi.com/1996-1073/12/7/1407/s1, Section S1: Waste collection plant, Section S2: Central collection points, Section S3: Underground pipes.

Author Contributions: Conceptualization, R.A.; Formal analysis, A.B., P.F.-i.-P., A.I. and R.A.; Investigation, J.L. and I.G.-H.; Methodology, A.B., P.F.-i.-P., A.I. and R.A.; Supervision, M.M. and R.A.; Writing–original draft, J.L. and I.G.-H.; Writing–review & editing, J.L., I.G.-H. and A.B.

Funding: This work has been made under the financial support of the Project Ceres-Procom: Food production and consumption strategies for climate change mitigation (CTM2016-76176-C2-1-R) (AEI/FEDER, UE) financed by the Ministry of Economy and Competitiveness of the Government of Spain.

Acknowledgments: The authors want to acknowledge the earlier contributions within the LIFE+ project: FENIX-Giving Packaging a New Life (LIFE08ENV/E/000135) by Bala, A. and Fullana-i-Palmer, P.; and to the UNESCO Chair in Life Cycle and Climate Change. In this respect, the authors are responsible for the choice and presentation of the information contained in this paper as well as for the opinions expressed therein, which are not necessarily those of UNESCO and do not commit this Organization.

Conflicts of Interest: The authors declare no conflicts of interest.

References

1. Anagnostopoulos, T.; Zaslavsky, A.; Kolomvatsos, K.; Medvedev, A.; Amirian, P.; Morley, J.; Hadjieftymiades, S. Challenges and opportunities of waste management in IoT-enabled smart cities: A survey. *IEEE Trans. Sustain. Comput.* **2017**, *2*, 275–289. [CrossRef]
2. Talari, S.; Shafie-Khah, M.; Siano, P.; Loia, V.; Tommasetti, A.; Catalao, J.P.S. A review of smart cities based on the internet of things concept. *Energies* **2017**, *10*, 421. [CrossRef]
3. Taheri, M.; Gholamalifard, M.; Ghazizade, M.J.; Rahimoghli, S. Environmental impact assessment of municipal solid waste disposal site in Tabriz, Iran using rapid impact assessment matrix. *Impact Assess. Proj. Apprais.* **2014**, *32*, 162–169. [CrossRef]
4. Kaza, S.; Yao, L.; Bhada-Tata, P.; Van Woerden, F.V. *What a Waste 2.0. A Global Snapshot of Solid Waste Management to 2050. Urban Development*; The World Bank Group: Washington, DC, USA, 2018.
5. Capuzano, R.; González-Martínez, S. Characteristics of the organic fraction of municipal solid waste and methane production: A review. *Waste Manag.* **2016**, *54*, 3–12. [CrossRef] [PubMed]
6. EUROSTAT. Statistical Office of the European Union. 2018. Available online: www.ec.europa.eu/eurostat/web/waste/municipal-waste-generation-and-treatment-by-treatment- (accessed on 19 October 2018).
7. Díaz-Díaz, R.; Muñoz, L.; Pérez-González, D. The business model evaluation tool for smart cities: Application to SmartSantander use case. *Energies* **2017**, *10*, 262. [CrossRef]
8. Albertí, J.; Balaguera, A.; Brodhag, C.; Fullana-i-Palmer, P. Towards life cycle sustainability assessment of cities. A review of background knowledge. *Sci. Total Environ.* **2017**, *609*, 1049–1063. [CrossRef]
9. EC. European Parliament and Council Directive 1994/62/EC on Packaging and Packaging Waste. 1994. Available online: https://eur-lex.europa.eu/legal-content/EN/TXT/PDF/?uri=CELEX:31994L0062&from=EN (accessed on 19 October 2018).

10. EC. European Parliament and Council Directive 2004/62/EC Amending Directive 94/62/EC on Packaging and Packaging Waste. 2004. Available online: https://eur-lex.europa.eu/resource.html?uri=cellar:f8128bcf-ee21-4b9c-b506-e0eaf56868e6.0004.02/DOC_1&format=PDF (accessed on 19 October 2018).
11. EC. European Parliament and Council Directive 2010/75/EU on Industrial Emissions (Integrated Pollution Prevention and Control). 2010. Available online: https://eur-lex.europa.eu/legal-content/EN/TXT/PDF/?uri=CELEX:32010L0075&from=EN (accessed on 19 October 2018).
12. EC. European Council Directive 1999/31/EC on the Landfill of Waste. 1999. Available online: https://eur-lex.europa.eu/legal-content/EN/TXT/PDF/?uri=CELEX:31999L0031&from=EN (accessed on 19 October 2018).
13. EC. European Parliament and Council Directive 2008/98/EC on Waste and Repealing Certain Directives. 2008. Available online: https://eur-lex.europa.eu/legal-content/EN/TXT/PDF/?uri=CELEX:32008L0098&from=EN (accessed on 19 October 2018).
14. Arushanyan, Y.; Björklund, A.; Eriksson, O.; Finnveden, G.; Söderman, M.L.; Sundqvist, J.; Stenmarck, A. Environmental assessment of possible future waste management scenarios. *Energies* **2017**, *10*, 247. [CrossRef]
15. UNDP. Sustainable Development Goals. 2015. Available online: https://www.un.org/sustainabledevelopment/sustainable-development-goals/ (accessed on 19 October 2018).
16. MITECO. Introducción a los Modelos de Gestión de Residuos (Introduction to the Residues Management Models). 2018. Available online: /www.miteco.gob.es/es/calidad-y-evaluacion-ambiental/temas/prevencion-y-gestion-residuos/flujos-domesticos/gestion/modelo_gestion/Default.aspx (accessed on 19 October 2018). (In Spanish)
17. Bernad-Beltrán, D.; Simó, A.; Bovea, M.D. Attitude towards the incorporation of the selective collection of biowaste in a municipal solid waste management system. A case study. *Waste Manag.* **2014**, *34*, 2434–2444. [CrossRef]
18. Di Mateo, U.; Nastasi, B.; Albo, A.; Astiaso-Garcia, D. Energy contribution of OFMSW (organic Fraction of Municipal Solid Waste) to energy-environmental sustainability in urban areas at small scale. *Energies* **2017**, *10*, 229. [CrossRef]
19. Laso, J.; Hoehn, D.; Margallo, M.; García-Herrero, I.; Batlle-Bayer, L.; Bala, A.; Fullana-i-Palmer, P.; Vázquez-Rowe, I.; Irabien, A.; Aldaco, R. Assessing energy and environmental efficiency of the Spanish agri-food system using the LCA/DEA methodology. *Energies* **2018**, *11*, 3395. [CrossRef]
20. Garcia-Herrero, I.; Hoeh, D.; Margallo, M.; Laso, J.; Bala, A.; Batlle-Bayer, I.; Fullana, P.; Vázquez-Rowe, I.; Gonzalez, M.J.; Durá, M.J.; et al. On the estimation of potential food waste reduction to support sustainable production and consumption policies. *Food Policy* **2018**, *80*, 24–38. [CrossRef]
21. Thyberg, K.L.; Tonjes, D.J. Drivers of food waste and their implications for sustainable policy development. *Resour. Conserv. Recycl.* **2016**, *106*, 110–123. [CrossRef]
22. Gustavsson, J.; Cederberg, C.; Sonesson, U.; van Otterdijk, R.; Meybeck, A. *Global Food Losses and Food Waste: Extent, Causes and Prevention*; Food and Agriculture Organization of the United Nations (FAO): Rome, Italy, 2011.
23. Gustavsson, J.; Cederberg, C.; Sonesson, U.; Emanuelsson, A. *The Methodology of the FAO Study: "Global Food Losses and Food Waste–Extent, Causes and Prevention"–FAO*; The Swedish Institute for Food and Biotechnology (SIK): Göteborg, Sweden, 2013.
24. Ley 22/2011. de 28 de Julio, de Residuos y Suelos Contaminados. BOE-A-2011-13046. 2011. Available online: https://www.boe.es/buscar/act.php?id=BOE-A-2011-13046 (accessed on 19 October 2018).
25. Residuos Profesional. Galdakao Implanta la Recogida Neumática de Materia Orgánica (Galdakao Set up the Pneumatic Collection of Organic Residues). 2014. Available online: www.residuosprofesional.com/galdakao-implanta-la-recogida-neumatica-de-materia-organica/ (accessed on 22 December 2018). (In Spanish)
26. Madridfree. Madrid Estrena el Contenedor Marrón de Materia Orgánica (Madrid Launches the Brown Container of Organic Residues). 2017. Available online: www.madridfree.org/madrid-contenedor-marron/ (accessed on 19 October 2018). (In Spanish)
27. Teerioja, N.; Moliis, K.; Kuvaja, E.; Ollikainen, M.; Punkkinen, H.; Merta, E. Pneumatic vs. door-to-door waste collection systems in existing urban areas: A comparison of economic performance. *Waste Manag.* **2012**, *32*, 1782–1791. [CrossRef]
28. Ecoembes. 2018. Available online: www.ecoembes.com (accessed on 15 October 2018).
29. Avfall Sverige. *Swedish Waste Management*; Swedish Waste Management and Recycling Association: Malmö, Sweden, 2016. Available online: https://www.avfallsverige.se/in-english/ (accessed on 15 October 2018).

30. Iriarte, A.; Gabarrell, X.; Rieradevall, J. LCA of selective waste collection systems in dense urban areas. *Waste Manag.* **2009**, *29*, 903–914. [CrossRef]
31. ISO 14040. *Environmental Management–Life Cycle Assessment–Principles and Framework*; International Organization for Standardization: Geneva, Switzerland, 2006.
32. Bala, A.; Raugei, M.; Fullana-i-Palmer, P. Introducing a new method for calculating the environmental credits of end-of-life material recovery in attributional LCA. *Int. J. Life Cycle Assess.* **2015**, *20*, 645–654.
33. Mora, C.; Manzini, R.; Gamberi, M.; Cascini, A. Environmental and economic assessment for the optimal configuration of a sustainable solid waste collection system: A "kerbside" case study. *Prod. Plan. Control* **2014**, *25*, 737–761. [CrossRef]
34. Gilardino, A.; Rojas, J.; Mattos, H.; Larrea-Gallegos, G.; Vázquez-Rowe, I. Combining operational research and Life Cycle Assessment to optimize municipal solid waste collection in a district in Lima (Peru). *J. Clean. Prod.* **2017**, *156*, 589–603. [CrossRef]
35. Pérez, J.; Lumbreras, J.; de la Paz, D.; Rodríguez, E. Methodology to evaluate the environmental impact to urban solid waste containerization system: A case study. *J. Clean. Prod.* **2017**, *150*, 197–213. [CrossRef]
36. Pires, A.; Sargedas, J.; Miguel, M.; Pina, J.; Martinho, G. A case study of packaging waste collection systems in Portugal–Part II: Environmental and economic analysis. *Waste Manag.* **2017**, *61*, 108–116. [CrossRef]
37. Yildiz-Geyhan, E.; Altun-Ciftçioglu, G.A.; Neset Kadirgan, M.A. Social life cycle assessment of different packaging waste collection system. *Resour. Conserv. Recycl.* **2017**, *124*, 1–12. [CrossRef]
38. Punkkinen, H.; Merta, E.; Teerioja, N.; Moliis, K.; Kuvaja, E. Environmental sustainability comparison of a hypothetical pneumatic waste collection system. *Waste Manag.* **2012**, *32*, 1775–1781. [CrossRef]
39. Aranda-Usón, A.; Ferreira, G.; Zambrana-Vázquez, D.; Zabalza-Bribián, I.; Llera-Sastresa, E. Environmental-benefit analysis of two urban waste collection systems. *Sci. Total Environ.* **2013**, *463–464*, 72–77. [CrossRef]
40. Keucken, A.; Habagil, M.; Batstone, D.; Jeppsson, U.; Arnell, M. Anaerobic co-digestion of sludge and organic food waste-performance, inhibition and impact on the microbial community. *Energies* **2018**, *11*, 2325. [CrossRef]
41. Margallo, M.; Onandía, R.; Aldaco, R.; Irabien, A. When life cycle thinking is necessary for decision making: Emerging cleaner technologies in the chlor-alkali industry. *Chem. Eng. Trans.* **2016**, *52*, 475–480.
42. ISO 14044. *Environmental Management–Life Cycle Assessment–Requirements and Guidelines*; International Organization for Standardization: Geneva, Switzerland, 2006.
43. Guinée, J.B.; Udo de Haes, H.A.; Huppes, G. Quantitative life cycle assessment of products. 1: Goal definition and inventory. *J. Clean. Prod.* **1993**, *1*, 3–13. [CrossRef]
44. Rebitzer, G.; Ekvall, T.; Frischknecht, R.; Hunkeler, D.; Norris, G.; Rydberg, T.; Schmidt, W.P.; Suh, S.; Weidema, B.P.; Pennington, D.W. Life cycle assessment–Part 1: Framework, goal and scope definition, inventory analysis, and applications. *Environ. Int.* **2004**, *30*, 701–720. [CrossRef]
45. Ministry of Agriculture, Food and Environment of Spain. *PEMAR: Plan Estatal Marco de Gestión de Residuos 2016–2022 (Waste Management Plan 2016–2022)*; Ministerio de Agricultura, Alimentación y Medio Ambiente de España. Ministry of Agriculture, Food and Environment of Spain: Madrid, Spain, 2015. (In Spanish)
46. Cobo, S.; Dominguez-Ramos, A.; Irabien, A. Trade-offs between nutrient circularty and environmental impacts in the management of organic waste. *Environ. Sci. Technol.* **2018**, *52*, 10923–10933. [CrossRef]
47. Hoehn, D.; Margallo, M.; Laso, J.; García-Herrero, I.; Bala, A.; Fullana-i-Palmer, P.; Irabien, A.; Aldaco, R. Energy Embedded in Food Loss Management and in the Production of Uneaten Food: Seeking a Sustainable Pathway. *Energies* **2019**, *12*, 767. [CrossRef]
48. Garcia-Herrero, I.; Margallo, M.; Onandía, R.; Aldaco, R.; Irabien, A. Environmental challenges of the chlor-alkali production: Seeking answers from a life cycle approach. *Sci. Total Environ.* **2017**, *580*, 147–157. [CrossRef]
49. Righi, S.; Oliviero, L.; Pedrini, M.; Buscaroli, A.; Della Casa, C. Life Cycle Assessment of management systems for sewage sludge and food waste: Centralized and decentralized approaches. *J. Clean. Prod.* **2013**, *44*, 8–17. [CrossRef]

50. MERMA, Spanish Ministry of Environment and Rural and Marine Affairs. Agencia d'Ecologia Urbana de Barcelona. Diagnóstico de la Gestión de Residuos de Competencia Municipal. Caso 2: Modelo 4 Contenedores. Aplicación del Programa SIMUR (Municipal Waste Management Diagnosis. Case 2: 4 Containers Model. SIMUR Programme Application). 2011. Available online: www.mapama.gob.es/ca/calidad-y-evaluacion-ambiental/temas/prevencion-y-gestion-residuos/simur_marm_caso_5__4cont_tcm8-230480.pdf (accessed on 14 March 2018). (In Spanish).
51. Thinkstep. *Gabi 6 Software and Database on Life Cycle Assessment*; Thinkstep: Leinfelden-Echterdingen, Germany, 2017.
52. Envac. Situación de la Recogida Automática o Neumática de Residuos en España y en la Región sur de Europa, Sistemas Sostenibles de Recogida Neumática de Residuos. (State of the Art of the Automatic or Pneumatic Collection of Residues in Spain and South Europe Region, Sustainable Systems of Pneumatic Collection of Residues. 2010. Available online: https://www.envac.es/ (accessed on 14 March 2018). (In Spanish)
53. Ros Roca. Recogida Neumática de Residuos, Descripción General del Sistema, Fichas Técnicas Centrales de Recogida. (Pneumatic Collection of Residues, General Description of the System, Collection Data Sheets). 2011. Available online: www.rosroca.es/es/ (accessed on 19 October 2018). (In Spanish)
54. Hernandez, C. *Recogida Neumática de Residuos Sólidos Urbanos, España. Conama10, Congreso Nacional del Medio Ambiente. (National Congress of Environment. Pneumatic Collection of Municipal Solid Waste, Spain)*; Congress Communication: Madrid, Spain, 2010. Available online: http://www.conama10.conama.org/conama10/download/files/CT%202010/40864.pdf (accessed on 18 October 2018). (In Spanish)
55. Medina-Díaz, R. *Ecodiseño y Sostenibilidad en el Sistema de Recogida de los RSU. (Eco-Designing and Sustainability in the Collection of MSW). Proyecto de fin de Carrera*; Universidad Pontificia Comillas–Escuela Técnica Superior de Ingeniería (ICAI): Madrid, Spain, 2009. Available online: https://www.iit.comillas.edu/pfc/resumenes/4a30edc029ec1.pdf (accessed on 18 October 2018). (In Spanish)
56. Puig, R.; Fullana-i-Palmer, P.; Baquero, G.; Riba, J.; Bala, A. A Cumulative Energy Demand indicator (CED), Life Cycle based, for Industrial Waste Management decision making. *Waste Manag.* **2013**, *33*, 2789–2797. [CrossRef]
57. Popa, C.L.; Carutasu, G.; Cotet, C.E.; Carutasu, N.L.; Dobrescu, T. Smart city platform development for an automated waste collection system. *Sustainability* **2017**, *9*, 2064. [CrossRef]
58. Zhang, T.; Wang, M.; Yang, H. A review of the energy performance and life-cycle assessment of buildings-integrated photovoltaic (BIPV) systems. *Energies* **2018**, *11*, 3157. [CrossRef]
59. Eriksson, O. Energy and waste management. *Energies* **2017**, *10*, 1072. [CrossRef]
60. INE. Instituto Nacional de Estadística (Spanish Statical Office). 2018. Available online: https://www.ine.es/ (accessed on 11 March 2019).
61. Mateu-Sbert, J.; Ricci-Cabello, I.; Villalonga-Olives, E.; Cabeza-Irigoyen, E. The impact of tourism on municipal solid waste generation: The case of Menorca Island (Spain). *Waste Manag.* **2013**, *33*, 2589–2593. [CrossRef]
62. Gidarakos, E.; Havas, G.; Ntzamilis, P. Municipal solid waste composition determination supporting the integrated solid waste management system in the island of Crete. *Waste Manag.* **2006**, *26*, 668–679. [CrossRef]
63. Ranieri, L.; Mossa, G.; Pellegrino, R.; Digiesi, S. Energy recovery from the organic fraction of municipal solid waste: A real options-based facility assessment. *Sustainability* **2018**, *10*, 368–383. [CrossRef]
64. Digiesi, S.; Facchini, F.; Mossa, G.; Mummolo, G.; Verriello, R. A Cyber-based DSS for a Low Carbon Integrated Waste Management System in a Smart City. *IFAC-PapersOnLine* **2015**, *48*, 2356–2361. [CrossRef]

© 2019 by the authors. Licensee MDPI, Basel, Switzerland. This article is an open access article distributed under the terms and conditions of the Creative Commons Attribution (CC BY) license (http://creativecommons.org/licenses/by/4.0/).

Article

Evaluation of Large-Scale Production of Chitosan Microbeads Modified with Nanoparticles Based on Exergy Analysis

Samir Meramo-Hurtado, Adriana Herrera-Barros and Ángel González-Delgado *

Nanomaterials and Computer Aided Process Engineering Research Group (NIPAC), University of Cartagena, Cartagena 130015, Colombia; smeramoh@unicartagena.edu.co (S.M.-H.); aherrerab2@unicartagena.edu.co (A.H.-B.)
* Correspondence: agonzalezd1@unicartagena.edu.co

Received: 23 January 2019; Accepted: 22 March 2019; Published: 28 March 2019

Abstract: Novel technologies for bio-adsorbent production are being evaluated on the lab-scale in order to find the most adequate processing alternative under technical parameters. However, the poor energy efficiency of promising technologies can be a drawback for large-scale production of these bio-adsorbents. In this work, exergy analysis was used as a computer-aided tool to evaluate from the energy point of view, the behavior of three bio-adsorbent production topologies at large scale for obtaining chitosan microbeads modified with magnetic and photocatalytic nanoparticles. The routes were modeled using an industrial process simulation software, based on experimental results and information reported in literature. Mass, energy and exergy balances were performed for each alternative, physical and chemical exergies of streams and chemical species were calculated according to the thermodynamic properties of biomass components and operating conditions of stages. Exergy efficiencies, total process irreversibilities, energy consumption, and exergy destruction were calculated for all routes. Route 2 presents the highest process irreversibilities and route 3 has the highest exergy of utilities. Exergy efficiencies were similar for all simulated cases, which did not allow to choose the best alternative under energy viewpoint. Exergy sinks for each topology were detected. As values of exergy efficiency were under 3%, it was shown that there are process improvement opportunities in product drying stages and washing water recovery for the three routes.

Keywords: bio-adsorbents; exergy analysis; chitosan microbeads; nanoparticles

1. Introduction

Nowadays, there is an increasing interest in developing sustainable and green-chemistry-based ways to synthesize novel materials for applications in emerging industries. Industrial wastewater treatment is a major global problem, mainly due to the restricted quantities of water that can be re-used along with the high cost of purification. In recent years, appropriate technologies for the treatment of wastewater and other effluents have raised great interest due to more strict laws and regulations [1]. The literature reports a variety of processes to water treatment and there are a lot of technologies used for this purpose. Among these methods, adsorption stands out as an effective, efficient and low-cost technology [2]. The adsorbents may be of mineral, organic or biological origin and are selected according to their applications. Polymeric adsorbents have been widely used to remove organic pollutants from industrial wastewater, but in recent years, the use of adsorbents produced from biomasses, organic residues and biopolymers have attracted interest due to its high availability and environmental issues related to disposal of residual biomasses and wastes [3]. In this sense, chitosan-based adsorbents have been evaluated showing high removal yield for several substances such as heavy metals, polycyclic aromatic hydrocarbons, among others [4].

Green chemistry is related to a new idea which has developed in the industry and research contexts as a natural evolution of pollution-prevention strategies [5]. The use of these types of processes brings the concept of developing chemical plants that can reduce waste generation and the resource demands. Recently, the green chemistry concept is also related to those processes that use smaller amounts of energy maintaining process yield/efficiency while providing reasonably market-valuable products [6]. On the other hand, nanotechnology is an innovative science/field associated with the manipulation of components and materials at the atomic/molecular scale and explains the behavior of these substances when used on the nanoscale. Recently, metal oxide nanoparticles have attracted interested by their potential application in different fields [7].

Modification of bio-adsorbents produced from natural polymers as chitosan microbeads with chemical species as thiourea or magnetite nanoparticles gives to these materials a higher adsorption capacity and selectivity to several heavy metals and liquid hydrocarbons, and higher efficiency in recovery processes via magnetic field application, which was studied at the lab-scale [8], however, the behavior at large scale of these emerging technologies is unknown, so the computer-aided simulation of these technologies at large-scale is relevant for further industrial applications [9].

Biomass-based processing technologies (and many chemical processes) commonly require huge amounts of energy and water, therefore, there are some published studies addressing the application of exergy analysis to measure the process performance from an energy viewpoint. Ojeda et al. [10] applied the exergy analysis to assess process alternatives for producing ethanol from lignocellulosic biomass. Peralta-Ruiz et al. [11] used exergy analysis as a tool for screening process alternatives for microalgae oil extraction. Aghbashlo et al. [12] applied an exergy analysis of a lignocellulosic-based biorefinery along with a sugarcane mill for simultaneous lactic acid and electricity production. Meramo-Hurtado et al. [13] developed an exergy analysis of bioethanol production from rice residues, the results for this study showing that pretreatment stage was the unit with the lowest exergy efficiency. They pointed out that this subprocess can be potentially improved to obtain a better energy performance. Restrepo-Serna et al. [14] developed a study related to the energy efficiency of various biorefinery schemes using sugarcane bagasse as raw material. They performed the assessment based on process simulation and exergy analysis, so it is shown in these studies that exergy analysis can be considered a useful evaluation instrument for the diagnosis of novel technologies involving sustainability goals.

In this work, three production processes for chitosan microbeads modified with nanoparticles were evaluated using exergy analysis as an instrument for screening design alternatives and as a decision-making tool for evaluation of novel technologies from energy viewpoint. Indicators as exergy efficiency, total irreversibilities, exergy utilities, and exergy destruction were determined for each processing alternative evaluated.

2. Materials and Methods

In this study, chitosan obtained from shrimp exoskeletons is used as feedstock to synthesize bio-adsorbents which are further modified with nanoparticles. Chitosan is a polymer present in many organic structures in Nature. It is commonly obtained from crustacean wastes through chitin deacetylation techniques [15]. The design and modeling of the processing routes for bio-adsorbents production is developed using a computer-aided process engineering software. This tool requires process information as mass and energy balances, operational conditions, reaction yields, among others. It is important to point out that the information required for the simulations were obtained from literature and experimental results previously published by the authors [16]. Another issue that must be considered is the input of an adequate thermodynamic model which allows estimating most accurate physical-chemical properties of the substances. The simulations give the extended energy and mass balances considering the operational conditions and processing routes of each design. The above parameters are important due to their uses to perform the exergy analysis for each alternative. Finally, exergy indicators are evaluated to compare each chitosan microbeads processing technology with the aim to identify improvement opportunities and screen the most suitable design from an exergy basis.

2.1. Processes Simulation

Process simulation mainly implies selecting chemical components used in the process, choosing an appropriate thermodynamic model, setting processing capacity, using suitable operating units and setting up input conditions such as mass flow rates, temperature, pressure, among others [17]. For the simulation of large-scale production processes of chitosan microbeads was used the industrial process software Aspen Plus, the chemical species required for the simulations were taken from the software database. For those compounds which are not available in the software database, molecules were created using the Aspen Property Estimator based on the properties reported for these species in the literature [18]. It was necessary to create components such as titanium isopropoxide and chitosan during the development of this work. The thermodynamic model Non-Random Two Liquids (NRTL) was selected for process simulations taking into account its good performance in the prediction of thermodynamic properties for polar/non-polar mixtures.

2.1.1. Process 1: Production of Chitosan Microbeads Modified with TiO_2 Nanoparticles via Green Chemistry

The production process of chitosan microbeads modified with TiO_2 nanoparticles ($CMTiO_2$) is developed through three specific subprocesses. The first subprocess is the lemongrass oil extraction where organic compounds such as myrcene, undecyne, neral, among others are obtained based on a green chemistry synthesis. These substances are used as surfactants to guarantee the nanosize of the particles. Lemongrass is first pretreated for cellulosic material removal, then it is sent to drying and crushing to reduce the particle size. Thus, the lemongrass oil is obtained in a liquid-solid extract. The second subprocess is the synthesis of TiO_2 nanoparticles. This stage uses titanium tetraisopropoxide (TTIP) as a precursor for TiO_2 nanoparticles through hydrolysis reactions. Propanol is formed as a by-product of the reaction. Hydrolysis reaction stoichiometry is described as follows:

$$Ti(OC_3H_7)_4 + 2H_2O \rightarrow TiO_2 + 4C_3H_7OH$$

As mentioned earlier, this stage involves the formation of TiO_2 nanoparticles via hydrolysis of TTIP. This substance is first diluted to a concentration of 100 mM under controlled conditions and sent to the hydrolysis reactor system. On the other hand, the oil extract (obtained from the first subprocess) is added and mixed with the main stream in the reactor [9]. Figure 1 shows the process diagram for the large-scale production of chitosan microbeads modified with TiO_2 nanoparticles. After TiO_2 synthesis the main stream is sent to a separation train composed of three washing units where the pH is stabilized. In the second unit, ethanol is used for washing, while for the first- and third-units water is used. In this separation stage, moisture content of the product is reduced. Finally, the product is sent to a drying unit to obtain the bio-adsorbents with water content equal to zero. The third subprocess begins after TiO_2 nanoparticles formation. Chitosan microbeads are prepared from chitosan (main raw material). In the first tank, a chitosan dissolution is prepared at a concentration of 2 w/v %. The solution is mixed in a second vessel with acetic acid (2 w/v %) obtaining a chitosan gel solution. This gel mixture is sent to another mixing stage where TiO_2 nanoparticles are added. The mass ratio between chitosan and TiO_2 nanoparticles is 1:2. In this stage, chitosan microbeads modified with TiO_2 nanoparticles are formed. For product formation is necessary to set alkaline conditions, hence, NaOH (at a concentration of 2.5 M) is used to reach the pH required [16]. Finally, the microbeads are physically mixed through an ultrasound-assisted agitation system. It is important to point out that the above procedure is developed at a temperature of 28 °C. The resulting microbeads have high moisture content so it is necessary to perform a separation stage where a train of washing and drying units are used to purify the bio-adsorbents. For all process units the operational pressure is 1 atm.

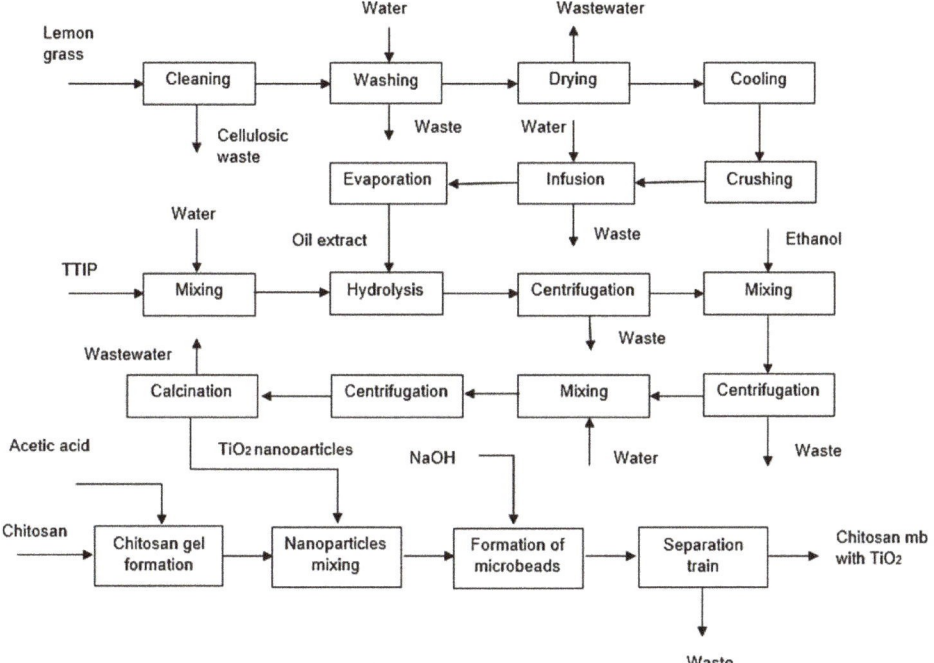

Figure 1. Process diagram of large-scale production of chitosan microbeads modified with TiO_2 nanoparticles.

2.1.2. Process 2: Production of Chitosan Microbeads Modified with TiO_2 and Magnetite Nanoparticles

Production of chitosan microbeads modified with TiO_2 and magnetite nanoparticles ($CMTiO_2$-Mag) is divided into two main subprocesses. Subprocess (1) is related to the synthesis of magnetite nanoparticles while subprocess (2) to the microbeads formation. For magnetite nanoparticles synthesis, the first stage implies the preparation of iron oxides solutions ($FeCl_3 \cdot H_2O$ and $FeCl_2 \cdot 4H_2O$) at a temperature of 301.15 K [19]. Subsequently, these solutions are mixed and sent to a heat exchanger unit to reach a temperature of 353.15 K. Thus, the resulting stream is fed into the iron oxides reactor where magnetite is produced along with NaCl and H_2O. In order to form the nanoparticles, NaOH is added at a concentration of 2% v/v. The outlet stream of the reactor is cooled to 301.15 K, and sent to a separation stage. In this stage, the nanoparticles are separated using magnetic fields. The above allows removing non-ferrous material. To reach higher purity, the magnetite nanoparticles are fed into a separation train composed of three consecutive washing units (an analog system used for $CMTiO_2$ process). Finally, the main stream is sent to a furnace unit where the microbeads are completely dried. For the final product, it is desired to obtain a mass ratio for chitosan, TiO_2 and magnetite of 2:1:1, respectively. On the other hand, second subprocess $CMTiO_2$-Mag alternative is similar to subprocess (3) of $CMTiO_2$ design described in Section 2.1.1 but the modification with magnetite is incorporated after TiO_2 nanoparticles modification. Thus, it is important to mention that these modifications are related to physical mixing processes. Figure 2 shows the process diagram for the large-scale production of chitosan microbeads modified with TiO_2 and magnetite nanoparticles.

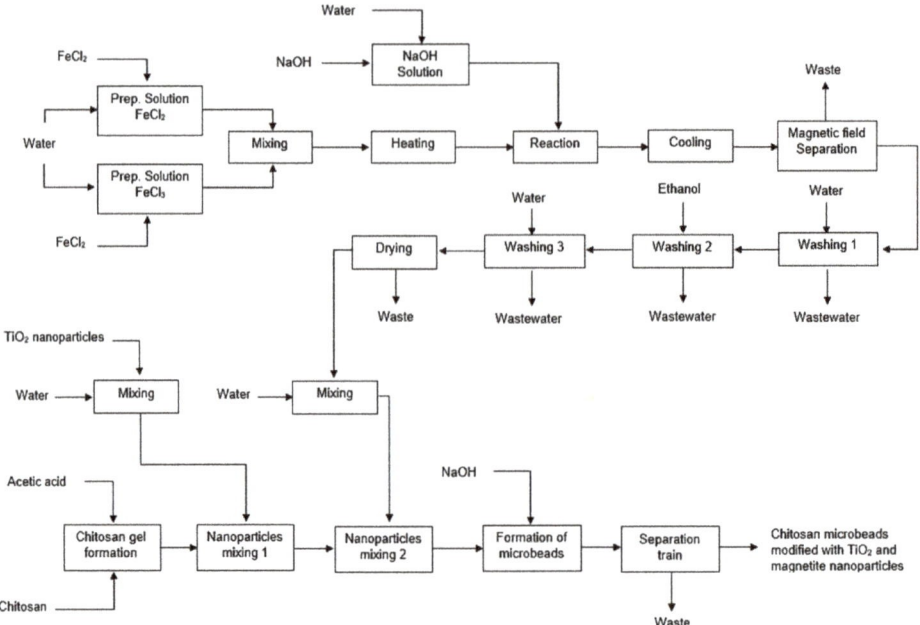

Figure 2. Process diagram of large-scale production of chitosan microbeads modified with TiO$_2$ and magnetite nanoparticles.

2.1.3. Process 3: Production of Chitosan Microbeads Modified with Thiourea

The third process evaluated consists in the large-scale production of chitosan microbeads modified with thiourea. In this process, thiourea is mixed with the main stream after chitosan gel formation. The process follows to the microbeads formation using diluted NaOH. Finally, the microbeads are purified through washing and drying stages. The mass ratio for chitosan and thiourea is 1:1. Figure 3 shows the process diagram for the large-scale production of chitosan microbeads modified with thiourea

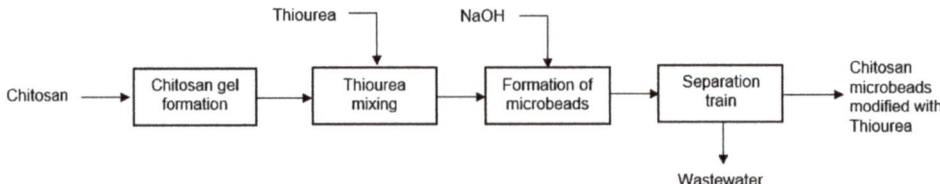

Figure 3. Process diagram of large-scale production of chitosan microbeads modified with thiourea.

2.2. Exergy Analysis

Thermodynamics-based tools such as energy analysis, exergy analysis, emergy analysis, among others have been applied for evaluation of industrial systems and thermal energy storage processes [20]. The first and second law of thermodynamics serve as a theoretical basis for energy analysis, along with the calculation of energy efficiencies for the studied processes. However, an energy balance not necessary provides information associated with the energy losses or quantifies the quality of the mass and energy streams of the evaluated routes. Exergy analysis is presented as an alternative tool which allows surpass the limitations of the laws of thermodynamics for a desired process [21]. Exergy analysis also shows the source of energy degradation in a process and can help to optimize an operation, a technology or a processing unit [22]. In addition, exergy analysis allows evaluating several process technologies to improve the design of a process. The above means that the exergy analysis is an appropriate tool to assess novel technologies for any chemical process. The exergy is defined as the maximum theoretical useful work that could be obtained from a system that interacts only with the environment. Considering that the exergy of a system depends on the selected state of reference, it is usual to choose standard environmental conditions as the reference state. Based on steady-state conditions, four balance equations have to be addressed to determine work/heat interactions. The first equation is the mass/matter conservation principle given by Equation (1), the second equation refers to the first law of thermodynamics given by Equation (2). The third equation refers to the second law of thermodynamics given by Equation (3):

$$\sum_i (m_l)_{in} = \sum_i (m_l)_{out} \qquad (1)$$

$$\sum_i (m_l h_i)_{in} = \sum_i (m_l h_i)_{out} + Q - W = 0 \qquad (2)$$

$$\sum_i (m_l S_i)_{in} = \sum_i (m_l S_i)_{out} + \sum_i \frac{Q_l}{T_i} + Q - W = 0 \qquad (3)$$

A fourth equation is addressed to formulate the exergy balance for around the system during a finite time interval given by Equation (4):

$$\text{Exergy input} - \text{Exergy output} - \text{Exergy consumption} = \text{Exergy accumulation} \qquad (4)$$

Exergy consumed is the product between the entropy generated and the environmental temperature, as follows in Equation (5) [9]:

$$\text{Exergy consumed} = \text{Environmental temperature} \times \text{Entropy} \qquad (5)$$

The Equation (4) can be expressed in terms of the mass exergy entering or leaving the system ($Ex_{mass,in}$ and $Ex_{mass,out}$), exergy flow by heat (Ex_{Heat}), exergy flow by work (Ex_{work}), and exergy loss (Ex_{loss}), in mathematical terms as follows:

$$Ex_{mass,in} - Ex_{mass,out} + Ex_{Heat} - Ex_{work} = Ex_{loss} \qquad (6)$$

Calculation of exergy flow by mass transfer is shown in Equation (6), it can be defined as a sum of physical exergy (Ex_{phy}), chemical exergy (Ex_{che}), potential exergy (Ex_{pot}), and kinetic exergy (Ex_{kin}), according to Equation (7):

$$Ex_{mass} = Ex_{phy} + Ex_{che} + Ex_{pot} + Ex_{kin} \qquad (7)$$

The chemical exergy is calculated using the chemical exergy per mole of each component. The estimation of the chemical exergy per mole is performed based on the free energy of formation (ΔG_i) and the specific chemical exergy of each atom presented in the molecule ($Ex_{chem,j}$):

$$Ex_{chem,i} = \Delta G_i + \sum_j n_{ele} Ex_{chem,j} \qquad (8)$$

The exergy of a heat stream Q is determined by the Carnot factor as follows:

$$Ex_{heat} = Q + \left(1 - \frac{T_o}{T}\right) \qquad (9)$$

where T_o is the environment (or reference) temperature and T is the temperature at which Q is available. Finally, the last term in the exergy balance (Ex_{work}) is associated for losses by work. Equation (10) gives the relation between these parameters:

$$Ex_{work} = W \qquad (10)$$

There is exergy destruction in operational processes due to its irreversibilities, the thermodynamic efficiency is associated with the amount of exergy loss. For measuring these losses, process exergy efficiency (Ψ) is formulated. This term is defined as an indicator to determine the degree of used exergy, given by Equation (11):

$$\Psi = 1 - \left(\frac{\text{Exergy loss}}{\text{Exergy input}}\right) \qquad (11)$$

Sorin et al. [23] mentioned that it is possible to perform the exergy balance considering all incoming and out-going streams in a defined system. The total exergy input of any real process/system is ever higher than its output. The above is related to the fact that always an amount of exergy is irreversibly lost within the system.

3. Results

The chemical composition of lemongrass oil used in process 1 is summarized in Table 1. This raw material reports high content of moisture, and organic compounds as myrcene, undecyne, nerol, geranial, among other [24]. These substances are the key compounds in the lemongrass extract and are used to guarantee the nanosize of the TiO_2 nanoparticles. The oil content in the lemongrass is about 1.10% of the total mass while the moisture and solid contents are approximately 71.23% and 27.67%, respectively. The above implies that the oil extraction method has to be highly efficient to reach acceptable/required yields. Most of the solid contents in lemongrass is cellulosic biomass, therefore it was assumed (for simulation purpose) that all solid content is cellulose.

Table 1. Chemical composition of lemongrass dry basis.

Compound	Average Concentration
β-Myrcene	5.5%
3-Undecyne	3.2%
Neral	30.5%
Geranial	42.5%
Nerol	3.5%
Others	14.8%

3.1. Processes Simulation Results

3.1.1. Simulation of Large-Scale Production of Chitosan Microbeads Modified with TiO_2 Nanoparticles via Green Chemistry

Process simulation flowsheet of large-scale production of chitosan microbeads modified with TiO_2 nanoparticles is shown in Figure 4. And the pressure, temperature, composition and flows of main process streams are shown in Table 2. As is described in Section 2.2, the system is composed of three sections: (1) Lemongrass oil extraction; (2) TiO_2 nanoparticles synthesis, and (3) $CMTiO_2$ production. For subprocess (1), stream 1 represents the inlet flow of the main feedstock (lemongrass). This stream is sent to washing and drying stages for cellulosic material removal and moisture reduction, respectively. After feedstock cleaning, the material is sent to a crusher unit for size-reduction. The above is performed with the aim of increasing the surface contact between the lemongrass and the extraction media (water). The liquid-solid extraction (infusion stage) is developed by a heating-sedimentation stage where the oil extract is separated in the upper site of the unit while the residual solids remain at the bottom. The oil extract is obtained with high moisture content (99% w/w) so the stream passes through an evaporation unit to reduce the water content close to 3% w/w. Finally, the oil extract is cooled until 28 °C, and it is stored. Subprocess (2) starts with the preparation of TTIP solution and the hydrolysis reaction where the TiO_2 nanoparticles are synthesized (see stream 39). In the case of subprocess (3), this stage starts with the production of the chitosan gel considering inlet flow according to the composition of stream 43. Stream 39 is added to the mixing tank where the $CMTiO_2$ are formed maintaining the required proportion of 1:1 for Chitosan and TiO_2, respectively. Finally, the product obtained with a flow of 232.03 kg/h according to stream 56.

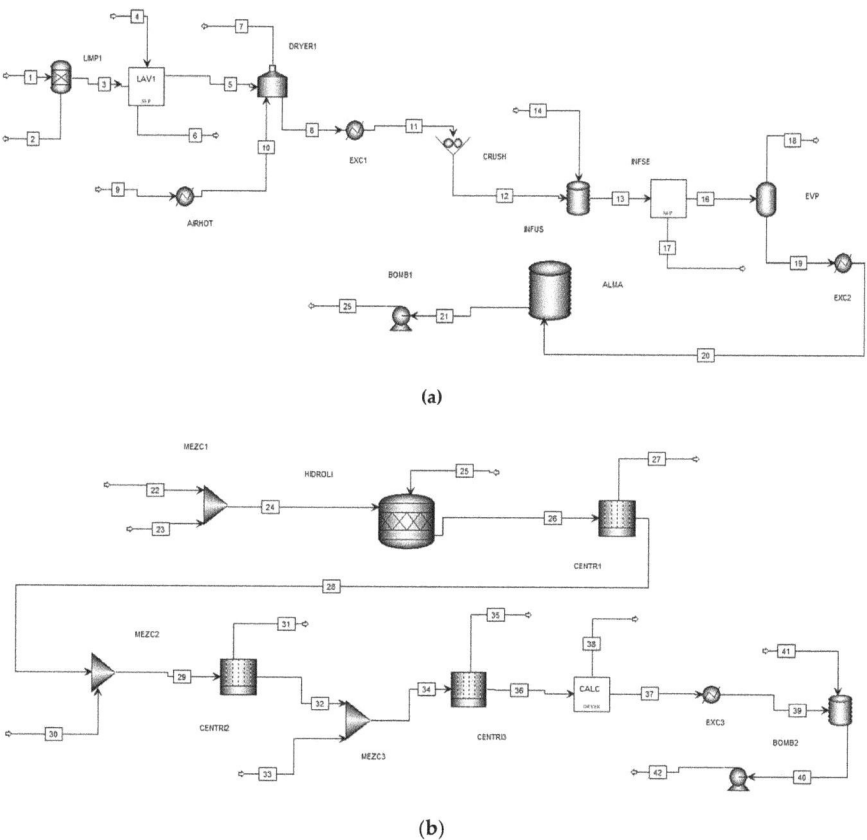

Figure 4. *Cont.*

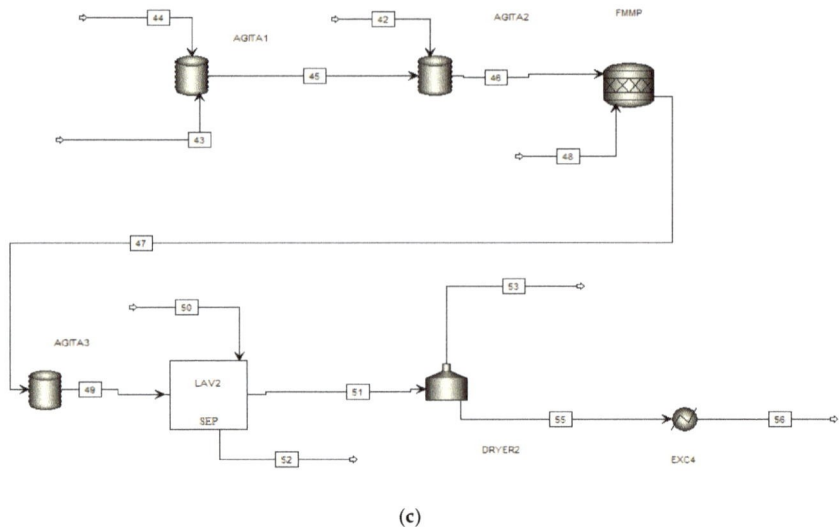

(c)

Figure 4. Process flowsheet of large-scale production of chitosan microbeads modified with TiO$_2$ nanoparticles via a green chemistry method: (**a**) Lemongrass oil extraction (**b**) TiO$_2$ nanoparticles synthesis; (**c**) chitosan microbeads production.

Table 2. Main process streams for Chitosan microbeads modified with TiO$_2$ nanoparticles.

Streams	1	19	22	39	43	48	56
Temperature (K)	301.15	373.15	301.15	823.15	301.15	301.15	301.15
Pressure (kPa)	101.32	101.32	101.32	101.32	101.32	101.32	101.32
Component mass flow (kg/h)							
β-Myrcene	2.04	0.47	0.00	0.00	0.00	0.00	0.00
3-Undecyne	1.19	0.55	0.00	0.00	0.00	0.00	0.00
Gerenial	27.10	22.46	0.00	0.00	0.00	0.00	0.00
Nerol	1.28	1.086	0.00	0.00	0.00	0.00	0.00
Cellulose	1882.97	0.00	0.00	0.00	0.00	0.00	0.00
Water	1466.9	608.58	0.00	0.00	3782.31	17,367.7	0.00
TTIP	0.00	0.00	592.90	0.00	0.00	0.00	0.00
TiO$_2$	0.00	0.00	0.00	154.84	0.00	0.00	154.84
Chitosan	0.00	0.00	0.00	0.00	77.20	0.00	77.20
NaOH	0.00	0.00	0.00	0.00	0.00	1929.75	0.00
Total	3381.47	633.15	592.90	154.84	3859.50	19,297.50	232.03

3.1.2. Simulation of Large-Scale Production of Chitosan Microbeads Modified with TiO$_2$ and Magnetite Nanoparticles

The simulation for the second processing route of bio-adsorbents production was performed based on general mass/energy balances, and operational conditions obtained from literature and lab experiments. Figure 5 shows the simulated process flowsheet of large-scale production of chitosan microbeads modified with TiO$_2$ nanoparticles and magnetite. Table 3 summarizes the main streams and operational conditions of this process. As it was described, this designed system is constituted by two main subprocesses: (1) Magnetite nanoparticles synthesis, and (2) chitosan microbeads production (with nanoparticles modification). For magnetite nanoparticles synthesis, stream 1 and 2 represent the inlet flow of main feedstocks: $FeCl_2 \cdot 4H_2O$ and $FeCL_3 \cdot 6H_2O$, respectively. The reaction between these iron chlorides and NaOH generates magnetite (desired product) along with hematite, water and sodium chloride. For simulation purpose, it was assumed that all produced hematite is magnetite. After reaction stage, the stream is sent to separations processes where all impurities are removed from the product. The magnetic properties of these nanoparticles allow separating them using a magnetic field or a magnet [19]. Finally, the dried and purified magnetite nanoparticles are sent to the second subprocess, the chitosan microbeads modified with the nanoparticles (TiO$_2$ and magnetite) are sent to purification stage where the desired product is obtained dried and with high purity with a mass flow rate of 307.86 kg/h.

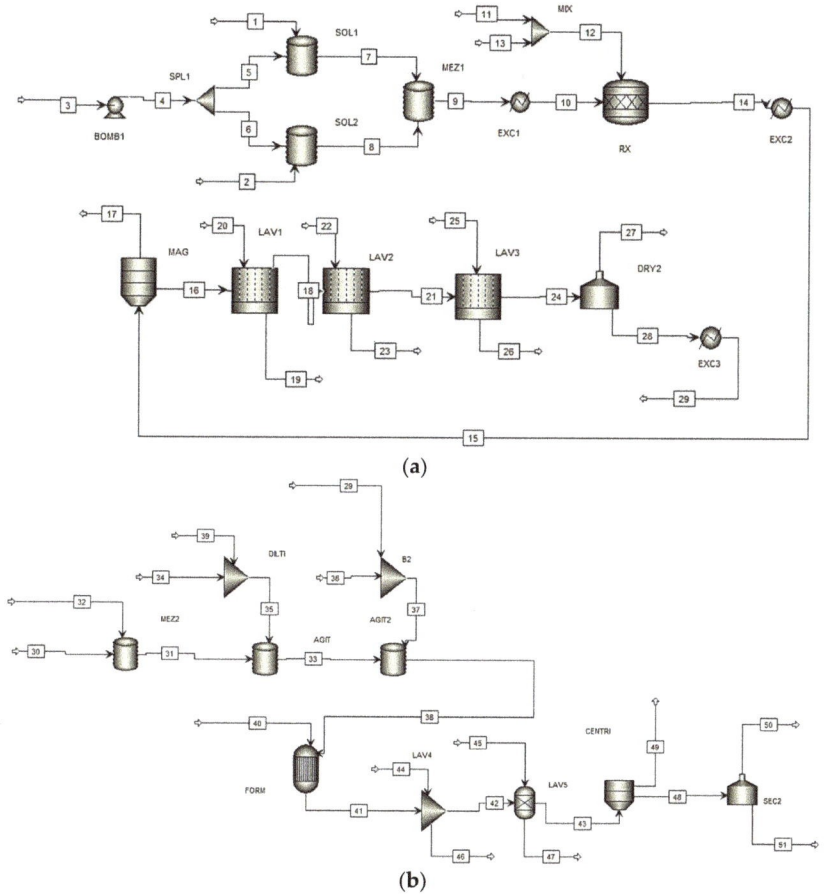

Figure 5. Process flowsheet of Large-scale production of chitosan microbeads modified with TiO$_2$ and magnetite nanoparticles: (**a**) Magnetite nanoparticles synthesis; (**b**) chitosan microbeads with modifications subprocess.

Table 3. Main process streams for chitosan microbeads modified with TiO$_2$ and magnetite nanoparticles.

Streams	1	2	11	29	30	34	51
Temperature (K)	301.15	301.15	301.15	301.15	301.15	301.15	301.15
Pressure (kPa)	101.32	101.32	101.32	101.32	101.32	101.32	101.32
Component mass flow (kg/h)							
Water	0.00	0.00	0.00	0.00	10,907.70	0.00	0.00
FeCl$_3$·6H$_2$O	200.00	0.00	0.00	4.00	0.00	0.00	2.87
FeCl$_2$·4H$_2$O	0.00	79.17	0.00	7.09	0.00	0.00	5.08
NaOH	0.00	0.00	134.23	0.00	0.00	0.00	0.00
Magnetite	0.00	0.00	0.00	83.94	0.00	0.00	60.24
NaCl	0.00	0.00	0.00	0.01	0.00	0.00	0.01
Chitosan	0.00	0.00	0.00	0.00	222.607	0.00	159.74
TiO$_2$	0.00	0.00	0.00	0.00	0.00	111.35	79.90
Acetic acid	0.00	0.00	0.00	0.00	0.00	0.00	0.00
Total	200.00	79.17	134.23	95.05	11,130.30	111.35	307.86

3.1.3. Simulation of Large-Scale Production of Chitosan Microbeads Modified with Thiourea

The simulation of the third processing route was developed according to the subprocess stage for chitosan microbeads production described for CMTiO$_2$ and CMTiO$_2$-Mag processes. In this case, the modification of the microbeads does not require the synthesis of any nanoparticles because the thiourea is introduced to the process as an available raw material. After this stage, the stream is sent to microbeads production and purification stages. Table 4 reports the main mass flows and operational conditions of this process. Figure 6 shows the simulated process flowsheet of large-scale production of chitosan microbeads modified thiourea. Finally, in this process alternative are produced 155.23 kg/h of chitosan microbeads modified with thiourea according to stream 13.

Table 4. Main process streams for chitosan microbeads modified with thiourea.

Streams	1	2	3	6	7	11	13
Temperature (K)	301.15	301.15	301.15	301.15	301.15	301.15	301.15
Pressure (kPa)	101.32	101.32	101.32	101.32	101.32	101.32	101.32
Component mass flow (kg/h)							
Water	0.00	2,857.54	0.00	34,520.80	37,452.40	53,653.80	0.00
Chitosan	77.61	0.00	0.00	0.00	77.61	0.00	77.61
Thiourea	0.00	0.00	77.61	0.00	77.61	0.00	77.61
NaOH	0.00	0.00	0.00	1700.40	1535.8	0.00	0.01
Sodium acetate	0.00	0.00	0.00	0.00	337.60	0.00	0.00
Acetic acid	0.00	274.13	0.00	0.00	0.00	0.00	0.00
Total	77.61	3104.67	77.61	36,221.20	39481.10	53,653.80	155.23

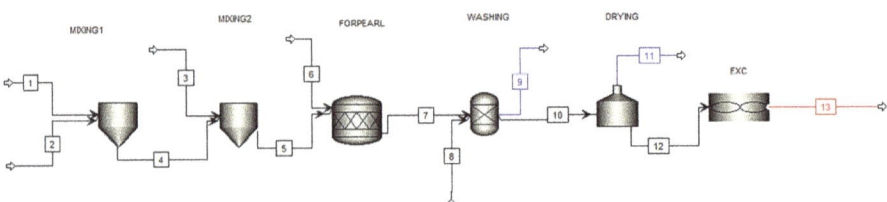

Figure 6. Process flowsheet of large-scale production of chitosan microbeads modified with thiourea.

3.2. Exergy Analysis of the Routes Simulated

These large-scale production processes can be generally divided into two main stages: nanoparticles preparation and chitosan microbeads formation. Considering the results of simulations and composition of each stream, chemical and physical exergies were estimated. Table 5 shows the estimated and reported chemical exergies of main components for bio-adsorbent processing routes.

Table 5. Chemical exergy of main components for bio-adsorbent production processes.

Component	Chemical Exergy (MJ/kmol)	Component	Chemical Exergy (MJ/kmol)
Myrcene	9670.10	NaOH	74.90
Undecyne	11,397.10	Magnetite	116.30
Gerenial	9,708.60	Thiourea	112.10
Nerol	10,348.10	TiO_2	21.73
TTIP	7,141.20	Water	0.90
Chitosan	12,462.78	Acetic acid	907.20

The higher chemical exergies correspond to those compounds with the longest carbon chains in its molecular structures (undecyne, myrcene and chitosan). Chemical exergy for common components as water, NaOH, or acetic acid was found in literature [25]. Table 6 reports the results obtained for exergy performance indicators for each processing route.

Table 6. Results for exergy performance of each processing route.

Process	Irreversibilities (MJ/h)	Exergy Efficiency (%)	Exergy of Residues (MJ/h)	Exergy Utilities (MJ/h)
$CMTiO_2$	181,665.63	0.04	144,445.08	91,033.74
$CMTiO_2$-Mag	216,795.71	2.83	182,698.34	88,030.46
CMThi	116,991.60	2.50	33,654.48	105,964.41

$CMTiO_2$ route presents the lowest exergy efficiency performance with a 0.04%. For $CMTiO_2$-Mag and CMThi processes were obtained a corresponding exergy efficiency of 2.83% and 2.50%, respectively. From a general viewpoint, the performance of the exergy efficiency was significantly low for all bio-adsorbent processes. This result implies that these designs might require technological improvements to reach better exergy and energy performances. In this sense, $CMTiO_2$ requires special attention due to its exergy efficiency shows that almost 100% of the inlet exergy is lost through the operation.

Figure 7 shows the exergy destruction of each bio-adsorbent processing route. The process with higher irreversibilities is $CMTiO_2$-Mag with an exergy flow of 182,698.34 MJ/h, followed by $CMTiO_2$ route with destroyed exergy of 144,445.08 MJ/h. The performance of this parameter is congruent respect to exergy efficiencies obtained for all processes. For $CMTiO_2$ route, it was found that the stage with the highest irreversibilities was the separation train. This stage is composed of three consecutive centrifuges representing approximately 53.00% of total irreversibilities for this process. The use of other separation technologies may contribute to reduce exergy losses. For the case of $CMTiO_2$-Mag route, it was obtained that microbeads-drying unit was the stage with the highest irreversibilities with a contribution of 41.05%, followed by washing unit in the separation stage (see "Lav4" in Figure 5b) with a 24.20%. In the case of CMThi route, drying unit was the stage with the highest exergy destruction representing 92.48% of total irreversibilities. The above results (for all cases) imply that these designs require better/improved separation technologies/stages to avoid several irreversibilities. This also could contribute to obtaining higher exergy efficiencies for each process.

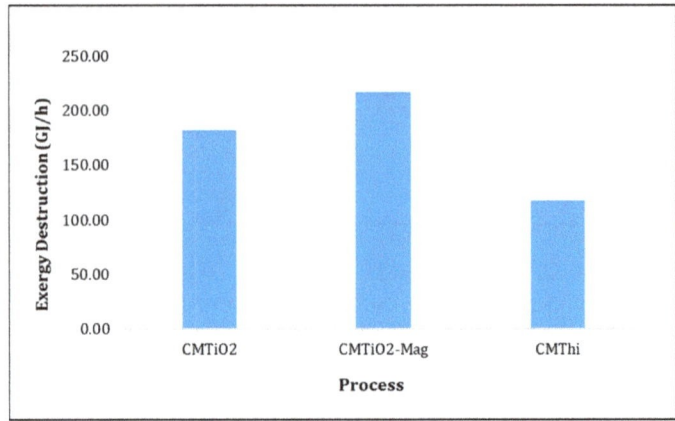

Figure 7. Comparison of exergy destruction for each processing route.

On the other hand, the exergy of residues was higher for CMTiO$_2$-Mag route, which was expected due to this process presents higher mass inventory respect to the other processing routes. Figure 8 shows the exergy of residues for each assessed process. CMThi process destroys less exergy due to residues with a flow of 33,654.48 MJ/h, while for CMTiO$_2$ and CMTiO$_2$-Mag were obtained exergy flows of 144,405.08 MJ/h and 182,698.34 MJ/h, respectively.

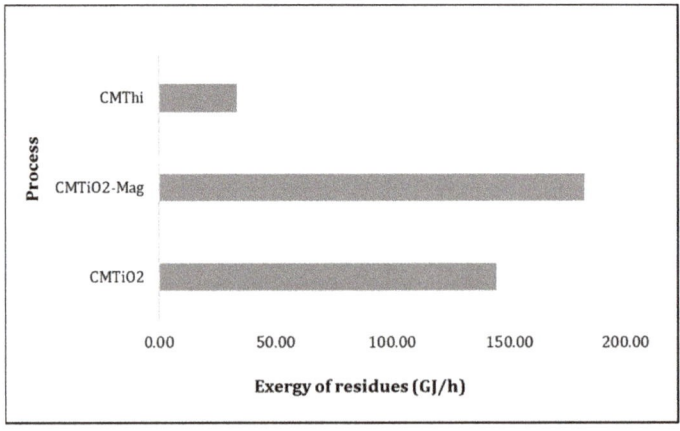

Figure 8. Comparison of exergy of residues for each processing route.

Finally, the exergy of utilities was estimated for each processing route. Figure 9 shows the comparison of this parameter for each bio-adsorbent processing route. The results for exergy of utilities present a similar performance for the three alternatives, obtaining an exergy flow of 91,033.74 MJ/h for CMTiO$_2$, 88,030.46 MJ/h for CMTiO$_2$-Mag, and 105,964.41 MJ/h for CMThio. For the case of CMTiO$_2$-Mag route, drying unit was the most significant stage representing an around 99% of the total exergy by utilities for this process. The above result indicates that this unit probably has important energy requirements that implies a high demand of industrial utilities. In this sense, the application of process optimization techniques could contribute to decrease the energy requirements, or obtain a better energy distribution. For CMTiO$_2$ process was found that hydrolysis reactor is a critical stage due to the most of exergy utility is consumed in this unit with a contribution of 47.22% of the total. It is explained by studying the thermodynamics of this reaction (hydrolysis of TTIP) because it is highly exothermic, thus, a cooling system is needed to maintain the reaction temperature constant.

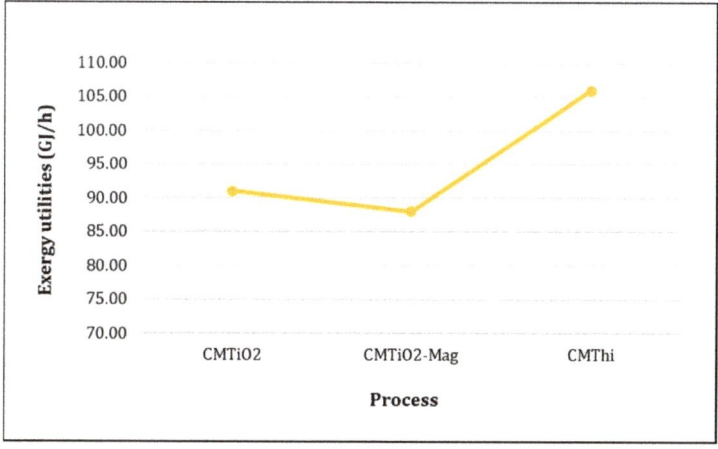

Figure 9. Comparison of exergy utilities for each processing route.

As described by CMTiO$_2$-Mag route, for CMThio alternative, the drying stage was also the most critical unit for exergy of utilities parameter with an exergy flow of 105,942.41 MJ/h which represents almost a 100% of the total. This result confirms the described behavior for exergy destruction performance which was previously explained for this bio-absorbent route.

4. Conclusions

Exergy analysis was used for the comparative evaluation of emerging alternatives of chitosan-based bio-absorbent production. The exergy analysis results were not conclusive for the selection of a promising alternative given the similar values obtained for the exergy efficiencies of the three routes evaluated (0.04%, 2.83% and 2.50%), being the production of chitosan microbeads modified with TiO$_2$ and magnetite nanoparticles, the route with the highest overall exergy efficiency (2.83%). However, the analysis presented in this work allowed to diagnose the emerging production processes from the exergetic point of view as sustainability criteria related to resources conservation. All the routes evaluated showed a similar poor exergetic performance. This behavior of the topologies assessed is related to the huge amount of water that are lost (and not recovered) in the bio-adsorbent purification stages along with the low energetic efficiency of the separation units. The use of a neutralization unit instead of washing stages might contribute to save large quantities of water, and improve the exergy efficiency of the processes. Route 2 also showed the highest irreversibilities of the three cases evaluated. The highest exergy destruction for production of chitosan microbeads modified with TiO$_2$ and magnetite nanoparticles was found in the microbeads-drying unit, but also, this alternative presents the lowest exergy flow associated to utilities. On the other hand, CMTiO$_2$ route presented the lowest exergy efficiency (0.04%), and an exergy loss of 181,665.63 MJ/h. For the case of CMThio route, it was determined that this process has high energy demands in the drying stage, resulting in a high exergy flow by utilities. For all routes evaluated is recommendable to apply process optimization techniques as mass and energy integration to decrease the requirement of utilities for separation/purification stages. For future work, it is recommended to develop studies related to economic and environmental impacts of the routes evaluated in order to obtain more sustainability parameters for selection of the most promising alternative under parameters evaluated.

Author Contributions: S.M.-H., A.H.-B. and A.G.-D. conceived and designed the paper, and wrote the Introduction and Materials and Methods. S.M.-H. and A.G.-D. wrote the Results and prepared figures and tables. Discussions and Conclusions were the collective work of all authors. The writing-review & editing was performed by A.G.-D. and A.H.-B., A.G.-D. supervised the development of this paper.

Funding: This research was funded by the Colombian Administrative Department of Science, Technology and Innovation COLCIENCIAS, code 110748593351 CT069/17. The APC was funded by COLCIENCIAS.

Acknowledgments: The authors thank to the Colombian Administrative Department of Science, Technology and Innovation COLCIENCIAS and the Doctoral Engineering program of the University of Cartagena, for its support with project "Removal of polycyclic aromatic hydrocarbons (PAHs), present in coastal waters Cartagena bay by using shrimp exoskeleton as a source of nanoparticle-modified bioadsorbents", code 1107748593351 CT069/17.

Conflicts of Interest: The authors declare no conflict of interest. The funders had no role in the design of the study; in the collection, analyses, or interpretation of data; in the writing of the manuscript, or in the decision to publish the results.

References

1. Bhojwani, S.; Topolski, K.; Mukherjee, R.; Sengupta, D.; El-Halwagi, M.M. Technology review and data analysis for cost assessment of water treatment systems. *Sci. Total Environ.* **2019**, *651*, 2749–2761. [CrossRef]
2. De Gisi, G.; Lofrano, M.; Grassi, M.; Notarnicola, M. Characteristics and adsorption capacities of low-cost sorbents for wastewater treatment: A review. *Sustain. Mater. Technol.* **2016**, *9*, 10–40. [CrossRef]
3. Sun, X.; Liu, Q.; Yang, L.; Liu, H. Chemically modified magnetic chitosan microspheres for Cr (VI) removal from acidic aqueous solution. *Particuology* **2016**, *26*, 79–86. [CrossRef]
4. Zhou, L.; Liu, J.; Liu, Z. Adsorption of platinum (IV) and palladium (II) from aqueous solution by thiourea-modified chitosan microspheres. *J. Hazard. Mater.* **2009**, *172*, 39–446. [CrossRef] [PubMed]
5. Righi, S.; Baioli, F.; Dal Pozzo, A.; Tugnoli, A. Integrating Life Cycle Inventory and Process Design Techniques for the Early Estimate of Energy and Material Consumption Data. *Energies* **2018**, *11*, 970. [CrossRef]
6. Hendershot, D. Green chemistry and process safety. *J. Chem. Health Saf.* **2015**, *22*, 39. [CrossRef]
7. Wang, P.; Lombi, E.; Zhao, F.; Kopittke, P. Nanotechnology: A New Opportunity in Plant Sciences. *Trends Plant Sci.* **2016**, *21*, 699–712. [CrossRef]
8. Karimi, M.H.; Mahdavinia, G.R.; Massoumi, B.; Baghban, A.; Saraei, M. Ionically crosslinked magnetic chitosan/κ-carrageenan bioadsorbents for removal of anionic eriochrome black-T. *Int. J. Boil. Macromol.* **2018**, *113*, 361–375. [CrossRef]
9. Meramo-Hurtado, S.; Bonfante, H.; De Avila-Montiel, G.; Herrera-Barros, A.; González-Delgado, A. Environmental assessment of a large-scale production of TiO_2 nanoparticles via green chemistry. *Chem. Eng. Trans.* **2018**, *70*, 1063–1068.
10. Ojeda, K.; Sanchez, E.; Kafarov, V. Sustainable ethanol production from lignocellulosic biomass—Application of exergy analysis. *Energy* **2011**, *36*, 2119–2128. [CrossRef]
11. Peralta-Ruiz, Y.; Gonzalez-Delgado, A.; Kafarov, V. Evaluation of alternatives for microalgae oil extraction based on exergy analysis. *Appl. Energy* **2013**, *101*, 226–236. [CrossRef]
12. Aghbashlo, M.; Mandegari, M.; Tabatabaei, M.; Farzad, S.; Mojarab Soufiyan, M.; Görgens, J.F. Exergy analysis of a lignocellulosic-based biorefinery annexed to a sugarcane mill for simultaneous lactic acid and electricity production. *Energy* **2018**, *149*, 623–638. [CrossRef]
13. Meramo-Hurtado, S.; Ojeda-Delgado, K.; Sánchez-Tuírán, E. Exergy analysis of bioethanol production from rice residues. *Comtemp. Eng. Sci.* **2018**, *11*, 467–474. [CrossRef]
14. Restrepo-Serna, D.L.; Martinez-Ruano, J.A.; Cardona-Alzate, C.A. Energy Efficiency of Biorefinery Schemes Using Sugarcane Bagasse as Raw Material. *Energies* **2018**, *11*, 3474. [CrossRef]
15. Goméz-Ríos, D.; Barrera-Zapata, R.; Ríos-Estepa, R. Comparison of process technologies for chitosan production from shrimp shell waste: A techno-economic approach using Aspen Plus®. *Food Bioprod. Process.* **2017**, *103*, 49–57. [CrossRef]
16. Bonfante-Álvarez, H.; De Avila-Montiel, G.; Cogollo-Herrera, K.; Herrera-Barros, A.; González-Delgado, A. Evaluation of Five Chitosan Production Routes with Astaxanthin Recovery from Shrimp Exoskeletons. *Chem. Eng. Trans.* **2018**, *70*, 1969–1974. [CrossRef]
17. Meramo-Hurtado, S.; Ojeda-Delgado, K.; Sánchez-Tuírán, E. Environmental assessment of a biorefinery: A case study of a purification stage in biomass gasification. *Comtemp. Eng. Sci.* **2018**, *11*, 113–120. [CrossRef]

18. Wang, L.; Yang, Z.; Sharma, S.; Mian, A.; Lin, T.; Tsatsaronis, G.; Maréchal, F.; Yang, Y. A Review of Evaluation, Optimization and Synthesis of Energy Systems: Methodology and Application to Thermal Power Plants. *Energies* **2018**, *12*, 73. [CrossRef]
19. Wei, Y.; Han, B.; Hu, X.; Lin, Y.; Wang, X.; Den, X. Synthesis of Fe_3O_4 nanoparticles and their magnetic properties. *Eng. Procedia* **2012**, *27*, 632–637. [CrossRef]
20. Querol, E.; González-Regueral, B.; Pérez-Benedito, J.L. *Practical Approach to Exergy and Thermoeconomic Analyses of Industrial Processes*, 1st ed.; Springer: Berlin, Germany, 2013.
21. Wang, J.J.; Yang, K.; Xu, Z.L.; Fu, C. Energy and exergy analyses of an integrated CCHP system with biomass air gasification. *Appl. Energy* **2015**, *142*, 317–327. [CrossRef]
22. Abosoglu, A.; Kanoglu, M. Exergetic and thermoeconomic analyses of diesel engine powered cogeneration: Part 1–Formulations. *Appl. Therm. Eng.* **2009**, *29*, 234–241. [CrossRef]
23. Sorin, M.; Lambert, J.; Paris, J. Exergy flows analysis in chemical reactors. *Chem. Eng. Res. Des.* **1998**, *76*, 389–395. [CrossRef]
24. Tajidin, N.E. Chemical composition and citral content in lemongrass (*Cymbopogon citratus*) essential oil at three maturity stages. *Afr. J. Biotechnol.* **2012**, *11*, 2685–2693. [CrossRef]
25. Kaushik, S.C.; Singh, O.K. Estimation of chemical exergy of solid, liquid and gaseous fuels used in thermal power plants. *J. Therm. Anal. Calorim.* **2014**, *115*, 903–908. [CrossRef]

© 2019 by the authors. Licensee MDPI, Basel, Switzerland. This article is an open access article distributed under the terms and conditions of the Creative Commons Attribution (CC BY) license (http://creativecommons.org/licenses/by/4.0/).

Article

Utilization of an Air-PCM Heat Exchanger in Passive Cooling of Buildings: A Simulation Study on the Energy Saving Potential in Different European Climates

Pavel Charvát [1], Lubomír Klimeš [2,*] and Martin Zálešák [1]

1. Department of Thermodynamics and Environmental Engineering, Brno University of Technology, Technická 2896/2, 61669 Brno, Czech Republic; charvat@fme.vutbr.cz (P.C.); 162098@vutbr.cz (M.Z.)
2. Sustainable Process Integration Laboratory—SPIL, NETME Centre, Brno University of Technology, Technická 2896/2, 61669 Brno, Czech Republic
* Correspondence: klimes@fme.vutbr.cz; Tel.: +420-54114-3241

Received: 4 February 2019; Accepted: 19 March 2019; Published: 22 March 2019

Abstract: The energy saving potential (ESP) of passive cooling of buildings with the use of an air-PCM heat exchanger (cold storage unit) was investigated through numerical simulations. One of the goals of the study was to identify the phase change temperature of a PCM that would provide the highest energy saving potential under the specific climate and operating conditions. The considered air-PCM heat exchanger contained 100 aluminum panels filled with a PCM. The PCM had a thermal storage capacity of 200 kJ/kg in the phase change temperature range of 4 °C. The air-PCM heat exchanger was used to cool down the outdoor air supplied to a building during the day, and the heat accumulated in the PCM was rejected to the outdoors at night. The simulations were conducted for 16 locations in Europe with the investigated time period from 1 May–30 September. The outdoor temperature set point of 20 °C was used for the utilization of stored cold. In the case of the location with the highest ESP, the scenarios with the temperature set point and without the set point (which provides maximum theoretical ESP) were compared under various air flow rates. The average utilization rate of the heat of fusion did not exceed 50% in any of the investigated scenarios.

Keywords: energy conservation; latent heat thermal energy storage; phase change materials; passive cooling

---

1. Introduction

Many countries, especially in Europe, have adopted a number of requirements on the energy performance of buildings in the last several decades in order to make the building stock more energy efficient [1]. The early energy saving policies focused mainly on energy consumption for space heating. The requirements on the thermal resistance of building envelopes have been repeatedly tightened. As a result, the transmission heat loss of newly-built and renovated buildings has decreased significantly and so has the amount of energy needed for space heating. As the transmission of heat loss decreased, the ventilation heat loss and its impact on the energy consumption of buildings became more important. Consequently, various approaches to the reduction of ventilation heat loss have been introduced. Mechanical ventilation with heat recovery has become more common in many types of buildings, and various energy saving strategies for building ventilation were developed. "Build tight—ventilate right" [2] is becoming a widely-adopted approach in the building sector. The air tightness testing is now a rather standard procedure during building construction, particularly in the case of passive buildings.

Well thermally-insulated buildings with a high air tightness level brought about problems with thermal comfort during the warm part of year. This is one of the factors contributing to the increasing demand for space cooling in Europe [3]. The actual energy consumption for space cooling in buildings is more difficult to determine than the energy consumption for space heating. The energy consumption for space heating can often be obtained or estimated based on the amount of consumed heat or fuels. On the other hand, the energy used for space cooling is mostly in the form of electricity and needs to be obtained from consumption of electricity [4]. Nonetheless, the consumption of energy for space cooling in buildings is becoming a concern even in climates where space cooling was not an issue several decades ago [5]. This development brought about an interest in passive cooling of buildings. Many studies have been conducted in this area since the beginning of the 21th Century. Artmann et al. [6] assessed the climatic potential for passive cooling of buildings by night-time ventilation. The analysis was carried out for 259 locations in Europe. The authors concluded that there was a "very significant" potential for passive cooling by night-time ventilation in Northern Europe. In Central and Eastern Europe and some parts of Southern Europe, the authors assessed the potential of this passive cooling technique as "significant". However, the authors pointed out that "floating building temperature is an inherent necessity of this passive-cooling concept". Passive cooling by night-time ventilation is thus mostly suitable for the buildings that are not occupied at night and where low air temperatures and relatively high air velocities during night-time ventilation would not cause discomfort.

Another issue with passive cooling of buildings by night-time ventilation is thermal energy storage. For this passive cooling technique to work effectively, cold needs to be stored at night in order to be available for the next day. Cold is usually stored in the building structures, and thus, night-time ventilation works best in buildings with high thermal mass. As many new buildings consist of light-weight building structures, additional thermal mass needs to be provided for passive cooling to work effectively [7].

Some of the problems of passive cooling of buildings (e.g., lightweight construction or large swings in indoor air temperature) could be mitigated by the use of cold storage units. The cold can be stored in the cold storage unit at night and then utilized during the day. This approach has been used in mechanical cooling systems for decades in the form of ice storage. The use of cold storage in passive cooling makes it possible to avoid supplying large volumes of cool outdoor air into the building at night and thus compromising thermal comfort. Cold can be stored in the cold storage unit at night and utilized the following day.

A number of studies have been conducted in the area of energy saving with latent heat thermal energy storage (LHTES) under various climatic conditions. Arkar and Medved [8] reported investigations of PCM-based LHTES integrated into the ventilation system for free cooling of a building. The LHTES unit was a cylindrically-packed bed comprising spherical containers filled with the encapsulated RT20 paraffin-based PCM (Phase Change Material). A mathematical model of the LHTES unit was created as a 2D continuous-solid-phase packed-bed model. The model was validated with the experimental results obtained for the packed bed with spheres of two diameters; 50 mm and 37.6 mm. The developed model was then used to obtain the multi-parametric temperature-response function of the LHTES unit in the form of a Fourier series. The temperature-response function allowed for calculation of the outlet air temperature of LHTES in the case that the inlet air temperature was approximated with a smooth periodic function. The thermal-response function was then used in the TRNSYS simulation tool for the investigation of free cooling in the case of a low-energy house. In the studied case, 6.4 kg of PCM per m^2 of floor area was found as an optimum value. The same authors [9] investigated the correlation between the local climate and the free-cooling potential of LHTES. The same packed bed LHTES, as in the previous paper [8], was considered. The climatic data of six cities in Europe were used in the study, and the simulated time period was from 1 June–31 August. The free-cooling potential was assessed in terms of the cooling degree hours (CDH). As no building characteristics or cooling set point were considered, the adopted approach provided the maximum theoretical cooling potential for the specific climatic conditions and the ratio between the air flow rate and the mass of the PCM. The authors concluded that the wide melting range of a PCM was not a disadvantage in free cooling of buildings. The optimum melting temperature of PCM was 2 °C above the average outdoor air temperature at the location in the investigated time period, and the optimum ratio between the mass of PCM and the air flow rate was 1–1.5 $\frac{\text{kg}}{\text{m}^3/\text{h}}$.

Chen et al. [10] investigated the energy saving potential of a ventilation system with a latent heat thermal energy storage (LHTES) unit under different climates in China. The authors used an in-house-developed model of the LHTES unit (air-PCM heat exchanger). The LHTES unit consisted of 20 parallel PCM slabs, which were 1.5 m long, 0.5 m wide, and 0.01 m thick. The air gaps between the slabs were 0.01 m high. The thermophysical properties of $CaCl_2 \cdot 6 H_2O$ were considered in the study with the exception of the melting temperature. The optimal melting temperature for each location was obtained by simulations. An isothermal phase change of the PCM was considered. The simulations were done for eight cities in China for the time period from 1 June–31 August. The average outdoor air temperature in the studied locations varied from 21.3 °C (city of Harbin) to 28.0 °C (city of Guangzhou). The study showed that the optimal PCM melting temperature varied from 21 °C in the case of Harbin to 29 °C in the case of Guangzhou. The average utilization rate of the heat of fusion of the PCM ranged from 0.09 in Shanghai to 0.24 in Beijing. Beijing was also the location with the highest electricity savings (87 kWh).

Costanzo et al. [11] used the EnergyPlus simulation tool to investigate the contribution of the PCM mats, installed behind the plasterboards, on the energy need for cooling and also thermal comfort in an air-conditioned office building with a large glass area. The PCM considered in the study was a fatty acid-based organic material. Three commercially-available variants of the material with the peak melting temperatures of 23 °C, 25 °C, and 27 °C were considered. The non-isothermal phase change without hysteresis with the melting range of about 2 °C was assumed. The summer climatic conditions of three cities in Europe (Rome, Vienna, and London) were used in the study. The authors concluded that the peak values of the operative temperature were reduced by about 0.2 °C when the PCM was used in comparison to a basic scenario without the PCM. The average reduction of the peak cooling load was around 10%, and the cooling energy need was reduced by 6%–12%.

An inverse method to estimate the average air flow rate through the air-PCM heat exchanger in free-cooling operation was proposed in [12]. The air-PCM heat exchanger (HEX) consisted of PCM slabs separated by air gaps. The air flow rate was estimated with the use of experimentally-obtained temperatures at the exit of the heat exchanger. MATLAB and COMSOL Multiphysics were used in the simulations. Considering the plane symmetry, only a half of the PCM slab thickness and a half of the air gap were modeled. Different apparent heat capacity curves were considered in melting and congealing of the PCM (i.e., phase change hysteresis). The authors reported fairly good agreement between the measured air flow rates and the air flow rates obtained by the proposed inverse method. The investigation was the first step in the optimization of the air flow rates with regard to the melting and congealing processes.

The goal of the present simulation study was to assess the energy saving potential of passive cooling of buildings, with the use of an air-PCM heat exchanger, when considering a cooling set point that would allow addressing both the ventilation cooling load and the internal cooling loads. In the study conducted by Chen et al. [10], the temperature set point for the utilization of stored cold was 26 °C as the ventilation cooling load was addressed. Moreover, an isothermal phase change of the PCM was assumed, which is rather rare in the case of most real-life PCMs. A lower temperature set point makes it possible to utilize the stored cold for the removal of indoor cooling loads, and it leads to higher utilization rates of the available cold storage capacity of the air-PCM HEX. In the simulation study presented by Medved and Arkar [9], no temperature set point was considered for the utilization of stored cold. Such an approach provides the maximum theoretical utilization rate of cold storage capacity, but it overestimates the real-life potential of passive cooling (particularly in cold climates). These issues were the motivation for the present simulation study, and the authors aimed at taking them into account. The air-PCM HEX considered in the present study was different from the one (a packed bed) considered in [9]. The present study revealed significant difference between the "optimal" ratio between the mass of the PCM and the air flow rate in the case of the air-PCM HEX consisting of parallel PCM slabs (panels) and the cylindrically-packed bed of spherical containers filled with the encapsulated PCM [9].

2. Simulated Scenario

The need for space cooling in buildings depends on many factors; most of them building-specific (thermal resistance of the building envelope, area of glazing, ventilation rates, internal heat gains, cooling set point, etc.). Certain assumptions about the operation of cold storage had to be made in order to assess the energy saving potential of passive cooling with an air-PCM heat exchanger (air-PCM HEX) without considering building-specific parameters. The actual amount of cold that can be stored in and released from an air-PCM heat exchanger depends on the temperature of air passing through the air-PCM HEX. In the case of passive cooling, the outdoor air flows through the air-PCM HEX during both rejection of heat (at night) and utilization of cold (during the day). For that reason, the outdoor air temperature was the most important factor in the study. The lower the outdoor air temperature, the larger the amount of cold that can be stored in the air-PCM HEX. On the other hand, when the outdoor air temperature is below a certain value during the day, the outdoor air does not need to be cooled down in the air-PCM HEX. For these reasons, it was assumed that the cold stored in the air-PCM HEX would be utilized only when the outdoor air temperature is above 20 °C. Though the set point of 20 °C seems arbitrary, there are justifiable reasons for choosing this value in the case of passive cooling. Many buildings may need space cooling when the outdoor air temperature is above 20 °C because of the internal heat gains, solar heat gains, and other factors. On the other hand, when the outdoor air temperature is below 20 °C, unconditioned outdoor air can be supplied to a building to provide passive cooling. With the upper limit of a good thermal comfort level at about 26 °C, the supply air temperature below 20 °C provides sufficient cooling capacity in many cases. Once again, cooling loads are very building-specific, and a thorough analysis of a particular case would be needed before actual application of passive cooling with cold storage.

The energy saving potential of the air-PCM HEX in passive cooling was determined from the mass flow rate of air and the difference between the outdoor air temperature and the outlet air temperature of the air-PCM HEX. The investigations were performed for 16 locations in Europe (described in Section 4) and for the time period from 1 May–30 September. Six nominal phase change temperatures of PCM were considered (16 °C, 18 °C, 20 °C, 22 °C, 24 °C, and 26 °C). The phase change temperature of the PCM providing the highest energy saving potential for a particular location could thus be identified.

An air-PCM heat exchanger, described in detail in the next section, was considered in the study. The simulated scenario was as follows. The rejection (discharge) of heat from the air-PCM HEX took place at night between midnight and 6:00. The air flow rate through the heat exchanger was 800 m^3/h during the discharge phase. The air passing through the air-PCM HEX during the discharge (rejection) of heat was assumed to return to the outdoor environment. The cold stored in the air-PCM HEX was utilized during the day. Several conditions were set for the utilization of the stored cold. The outdoor air would be supplied to a building through the air-PCM HEX between 8:00 and 20:00, but only if the outdoor air temperature was above 20 °C and the outlet air temperature of the air-PCM HEX was lower than the outdoor air temperature. When the air was supplied to a building through the air-PCM HEX, the air flow rate was 400 m^3/h (Figure 1). The simulation model of the air-PCM HEX allows for the calculation of the heat loss/gain to/from the surrounding environment, but the adiabatic walls of the exchanger were considered in the study. Considering a relatively small temperature difference between the inside of the air-PCM HEX and the ambient environment in the passive cooling operation, almost adiabatic walls can be achieved in practice with an appropriate level of (inexpensive) thermal insulation.

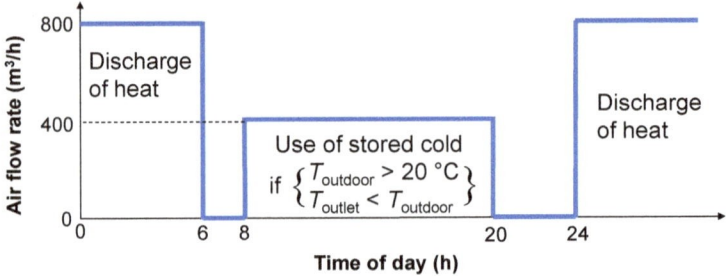

Figure 1. Operation of cold storage.

3. Model of the Air-PCM Heat Exchanger

The model of the air-PCM HEX, previously developed by the authors [13], was used in the study. The exchanger (heat storage unit) consisted of aluminum panels filled with a PCM. The unit contained a total of 100 CSM panels (compact storage modules) in five rows of 20 panels with each CSM panel containing 0.5 kg of the PCM (the total weight of the PCM was 50 kg). A schematic of the air-PCM heat exchanger can be seen in Figure 2. The configuration with 100 panels arranged in five rows of twenty panels was chosen because it was previously investigated both experimentally and numerically [13].

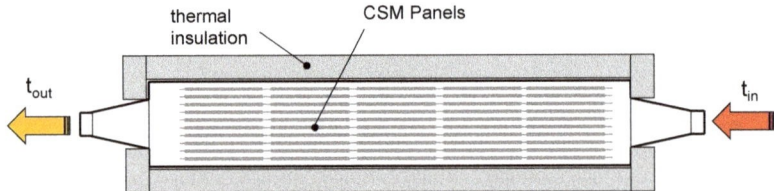

Figure 2. Schematic of the air-PCM heat exchanger. CSM, compact storage module.

The application of such a or similar designs of the latent heat thermal energy storage (LHTES) unit for cold storage has been investigated by many researchers; see, e.g., [14]. The main advantage of the design is its flexibility. The overall thermal energy storage capacity of the air-PCM HEX increases with the number of compact storage modules (panels). The doubling of the number of CSM panels in the air-PCM HEX theoretically doubles its overall thermal energy storage capacity. However, not all arrangements of the CSM panels provide the same performance of the air-PCM HEX. The heat charging and discharging characteristics can be adjusted by the arrangement of the panels in columns and rows.

A PCM exhibiting non-isothermal phase change, but without a phase change hysteresis was considered in the study. The thermophysical properties of the PCM are shown in Table 1. Though no specific PCM was considered in the study, the thermophysical properties were similar to those of paraffin-based PCMs [15]. The model of the heat storage unit was devised as quasi-two-dimensional. It consisted of a set of sub-models of the CSM panels, which were coupled with an energy-balance sub-model for the air flowing between the panels. The sub-model of the CSM panel was devised for the 1D solution of heat conduction in the CSM panel. The conductive heat transfer in the PCM was considered only in the direction of the thickness of the CSM panel (perpendicular to the air flow), while the conductive heat transfer in the other two dimensions was not taken into consideration (as the dominant temperature gradient was in the direction of the thickness of the CSM panel). The effective heat capacity method was used for the phase change modeling. The governing heat transfer equation solved by the sub-model of the CSM panel reads [16]:

$$\varrho c_{\text{eff}} \frac{\partial T}{\partial \tau} = \frac{\partial}{\partial x}\left(k \frac{\partial T}{\partial x}\right) \tag{1}$$

where ϱ is the density, c_{eff} is the effective heat capacity, T is the temperature, τ denotes time, k is the thermal conductivity, and x is the spatial coordinate. The bell-shaped effective heat capacity defined as a function of the temperature was adopted (e.g., [17]):

$$c_{\text{eff}}(T) = c_0 + c_m \exp\left\{-\frac{(T - T_{\text{mpc}})^2}{\sigma}\right\} \qquad (2)$$

where c_0 is the heat capacity outside the temperature range of the phase change, $c_m = 110\,\text{kJ/kg}$ is the increment of the heat capacity corresponding to the latent heat of the phase change, T_{mpc} is the mean phase change temperature, and $\sigma = 1.05\,(°C)^2$ is a parameter characterizing the range of the phase change. The curve of the effective heat capacity according to Equation (2) for the PCM with these parameters, with $T_{\text{mpc}} = 22\,°C$ and other properties shown in Table 1, can be seen in Figure 3.

Table 1. Thermophysical properties of the PCM considered in the CSM panels

Parameter	Value
Phase change temperature range	$\langle T_{\text{mpc}} - 2, T_{\text{mpc}} + 2 \rangle\,°C$
Mean phase change temperature T_{mpc}	$\{16, 18, 20, 22, 24, 26\}\,°C$
Amount of latent heat	$200\,\text{kJ/kg}$
Specific heat	$2\,\text{kJ/kg}\,°C$
Thermal conductivity	$0.2\,\text{W/m}\,°C$
Density	$730\,\text{kg/m}^3$

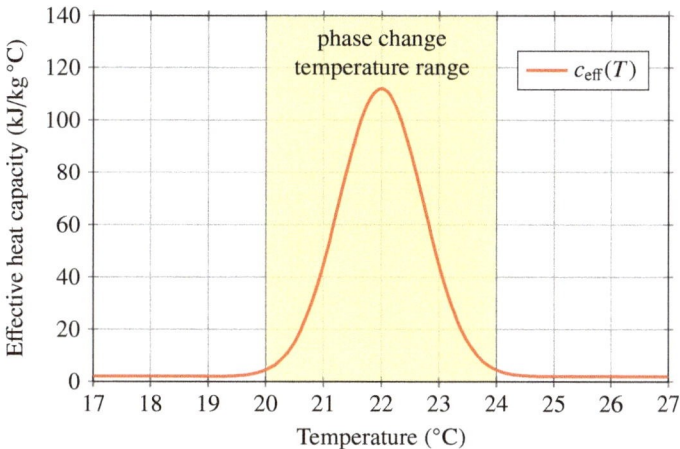

Figure 3. The curve of the effective heat capacity for the considered PCM.

Equation (2) provides a flexible approximation of the distribution of the effective heat capacity of PCMs. Even though most PCMs have an asymmetrical distribution of the effective heat capacity with regard to the mean phase change temperature [18], the approximation by Equation (2) is far more realistic than the assumption of an isothermal phase change considered in some phase change models [19].

The 1D sub-models of heat conduction in the CSM panels interact with each other, as well as with the air flowing in the channel by means of the sub-model of the air flow in the channel. The idea is that each 1D sub-model of the CSM panel is coupled with one node of the air flow sub-model by means of the CSM-air boundary condition. The sub-model of the air flow is devised for the solution of the movement of air in the channel including heat transfer interactions with the sub-models of the CSM panels. A description of this approach was already published by the authors of the paper, and is not repeated here. Readers interested in a detailed description are referred to [20].

4. Considered Locations

The climate in Europe varies significantly, not only with the latitude, but also with other factors such as altitude or the proximity of the sea or ocean. Considering the assumptions and simplifications adopted in the simulated passive cooling scenario, the main factor in the study was the ambient (outdoor) air temperature, which influences the potential of LHTES in passive cooling of buildings. Since the total energy saving potential increases with the number of buildings adopting a particular energy saving measure, only the locations of densely-populated areas were considered in the study. To keep the selection of the locations relatively simple, four cities with their latitude above 55° N were chosen to represent the cold European climate; eight cities with their latitude between 45° N and 55° N represented the mild to cool regions, where the majority of the European population lives; and four cities with their latitude below 45 ° N represented warm European climates.

Figure 4 shows the boxplots of outdoor air temperature of four cities located above the 55th parallel. A box plot is a graphical representation of numerical data by means of their quartiles. The red horizontal line indicates the median, and the blue box represents values in the second and third quartile (between the 25th and 75th percentiles). The whiskers reach the most extreme values (maximum and minimum values), which are not considered outliers, while the outliers are plotted individually as red plus signs. The outliers had values that were 1.5-times the interquartile range below the 25th percentile or above 75th percentile.

The locations above the 55th parallel provide very good potential for passive cooling of buildings by the supply of unconditioned outdoor air into a building. As can be seen, the outdoor temperature rarely exceeds 20 °C. For that reason, the potential for the use of cold storage in passive cooling is rather limited, as will be further demonstrated by the results of the study.

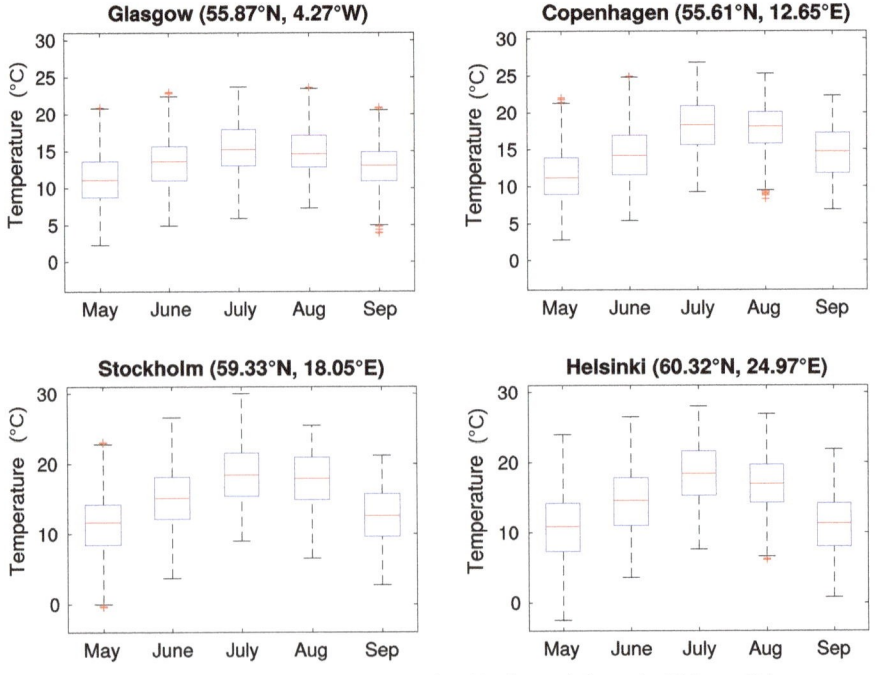

Figure 4. Outdoor air temperatures in the cities located above the 55th parallel.

The box plots of outdoor air temperatures for eight cities located between the 45th and 55th parallels are shown in Figures 5 and 6.

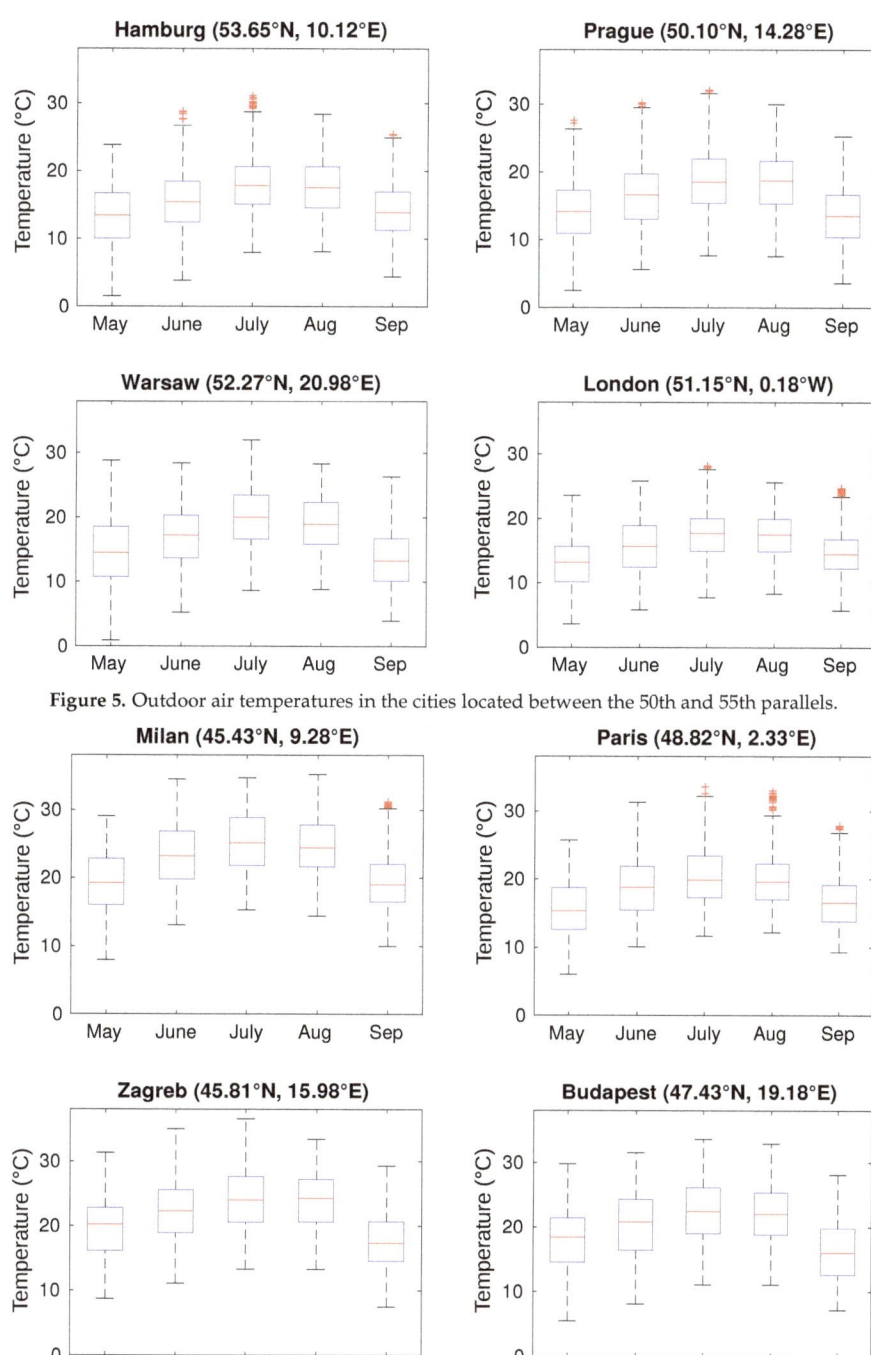

Figure 5. Outdoor air temperatures in the cities located between the 50th and 55th parallels.

Figure 6. Outdoor air temperatures in the cities located between the 45th and 50th parallels.

The outdoor air temperatures for the four cities located below the 45th parallel are shown in Figure 7. Unlike in case of the four cities above the 55th parallel, there is a high potential for utilization of cold storage in the south of Europe. The problem is the rejection of heat at night as the air temperatures remain rather high all day long.

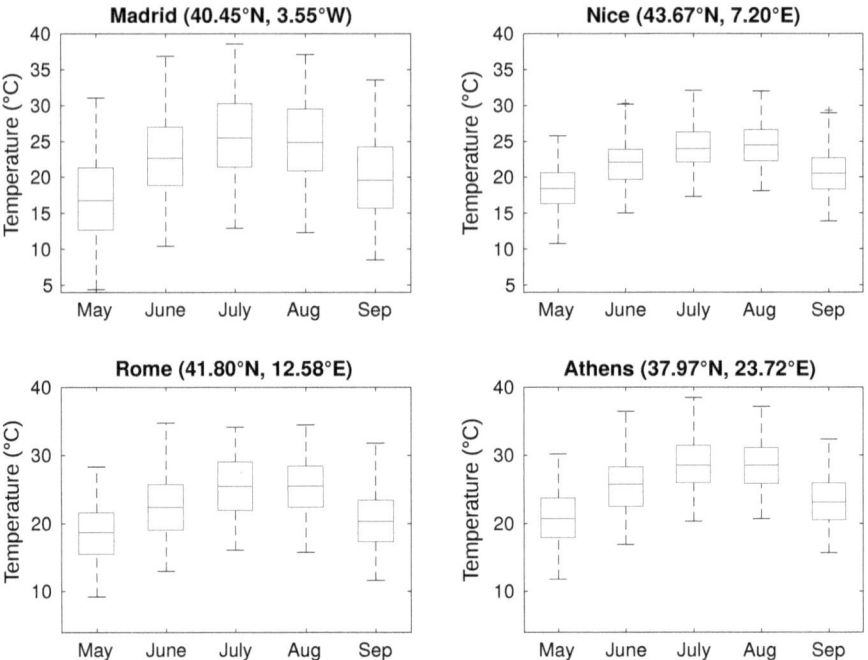

Figure 7. Outdoor air temperatures in the cities located below the 45th parallel.

5. Results and Discussion

The considered air-PCM HEX could store 10,000 kJ (2.78 kWh) of heat or cold in the phase change temperature range of the PCM (Figure 3). The specific heat of the PCM was 2 kJ/kg °C and, thus, 100 kJ/°C could be stored in the form of sensible heat outside of the phase change temperature range. Cold storage in the air-PCM HEX worked in daily cycles, and this means that the maximum amount of cold theoretically available for passive cooling on one day was about 3 kWh. This total heat storage capacity could not be exploited every day as passive cooling was not needed on some days or the heat accumulated in the PCM could not be fully rejected at night.

The amount of cold available for passive cooling of a building was evaluated from the mass flow rate of air $\dot{m}_{air}$ and the temperature difference between the outlet air temperature of the air-PCM HEX T_{outlet} and the outdoor air temperature $T_{outdoor}$ according to:

$$\dot{Q} = \dot{m}_{air} c_{p,air} (T_{outdoor} - T_{outlet}) \qquad (3)$$

where $\dot{Q}$ is the time-dependent heat transfer rate in which cold is transferred from the HEX to the building. The amount of cold (energy saving potential (ESP)) was obtained as:

$$ESP = \int_{t \in \mathcal{T}} \dot{Q} \, dt \qquad (4)$$

where $\mathcal{T}$ is the set of time instances during the day satisfying the conditions $T_{outdoor} > 20\,°C$ and $T_{outlet} < T_{outdoor}$ for the utilization of cold from the air-PCM HEX, as explained in Section 2.

The utilization rate of the heat of fusion utilization rate of the heat of fusion (URHF) was obtained as the ratio between the ESP in a 24-h cold storage cycle and the heat of fusion of the PCM in the air-PCM HEX:

$$\text{URHF} = \frac{\text{ESP}_{24}}{m_{\text{PCM}} H_{\text{m}}} \quad (5)$$

where m_{PCM} is the weight of the PCM in the air-PCM HX and H_{m} is the enthalpy of fusion of the PCM. Table 2 shows the results for the cities located above 55° N. The largest energy saving potential was in case of Helsinki (82.4 kWh) for the PCM with the nominal phase change temperature of 18 °C. This phase change temperature provided also the largest energy saving potential at other two locations, Copenhagen and Stockholm. Despite a lower latitude than Helsinki and Stockholm and a similar latitude to Copenhagen, the ESP in the case of Glasgow was rather small (proximity of the ocean influencing summer temperatures, as could be seen in Figure 4).

Table 2. Energy saving potential in kWh (cities above 55° N).

		Nominal Phase Change Temperature of PCM (°C)					
		16	18	20	22	24	26
Glasgow	May	2.0	1.3	0.7	0.2	0.1	0.1
	June	6.1	5.3	3.0	1.1	0.3	0.2
	July	12.0	11.0	7.0	2.3	0.8	0.6
	August	15.5	13.4	8.4	3.3	1.2	0.9
	September	1.2	0.8	0.4	0.1	0.1	0.1
	Total	36.8	31.8	19.4	7.0	2.4	1.9
Copenhagen	May	3.3	2.3	1.2	0.4	0.2	0.2
	June	13.1	14.5	10.6	4.9	1.5	0.7
	July	17.4	24.4	25.2	17.1	7.5	2.9
	August	14.8	21.4	20.2	11.5	3.9	1.3
	September	4.3	5.4	3.3	1.0	0.3	0.3
	Total	52.9	68.0	60.5	35.0	13.5	5.3
Stockholm	May	7.1	5.2	3.0	1.2	0.4	0.3
	June	13.8	16.8	17.1	10.5	4.7	1.8
	July	20.6	28.7	30.7	24.1	14.0	6.6
	August	18.9	26.3	25.5	14.1	5.2	1.6
	September	4.0	2.7	1.4	0.4	0.2	0.2
	Total	64.4	79.7	77.6	50.4	24.5	10.6
Helsinki	May	7.0	5.3	3.1	1.3	0.5	0.3
	June	14.1	14.2	10.8	5.8	2.8	1.3
	July	25.6	33.7	33.5	23.3	11.4	4.3
	August	21.4	26.1	19.9	11.1	4.3	1.8
	September	4.4	3.0	1.7	0.7	0.3	0.3
	Total	72.5	82.4	69.0	42.2	19.3	7.9

The energy saving potentials for the cities between 50° N and 55° N are shown in Table 3. Warsaw was the location with the highest energy saving potential of LHTES in passive cooling; 126.5 kWh for $T_{\text{mpc}} = 20\,°\text{C}$. The situation for Prague, another city located inland, was rather similar. The energy saving potential in the case of London and Hamburg was influenced by the proximity of the sea, which reduced daily temperature swings.

The energy saving potential of LHTES did not improve very much in the case of the cities located between 45° N and 50° N as demonstrated in Table 4. All four cities in this latitude range were located inland. The highest energy saving potential was reached in the case of Milan; 148.5 kWh for $T_{\text{mpc}} = 24\,°\text{C}$. The phase change temperature providing the largest energy potential shifted to higher values.

Table 3. Energy saving potential in kWh (cities between 50° N and 55° N).

		Nominal Phase Change Temperature of PCM (°C)					
		16	18	20	22	24	26
London	May	8.7	6.8	3.9	1.5	0.6	0.4
	June	21.2	21.3	16.4	8.0	2.6	1.3
	July	19.9	22.6	19.6	12.0	6.4	3.2
	August	22.9	26.3	19.6	10.4	4.2	1.8
	September	8.2	10.5	8.1	4.1	1.6	0.6
	Total	80.9	87.5	67.6	36.1	15.4	7.3
Hamburg	May	10.0	8.2	4.7	1.7	0.7	0.5
	June	15.8	17.1	15.4	10.3	5.2	2.2
	July	20.0	25.9	25.8	18.9	11.6	6.3
	August	19.9	26.8	26.9	19.0	10.2	4.5
	September	8.7	9.2	7.3	3.8	1.7	0.7
	Total	74.4	87.1	80.1	53.6	29.3	14.3
Prague	May	17.4	18.5	14.3	7.7	3.4	1.6
	June	17.8	21.3	21.2	17.1	10.6	5.7
	July	23.2	29.2	31.2	24.7	16.3	8.9
	August	24.5	32.6	34.1	25.3	14.9	7.3
	September	13.0	14.6	12.3	6.5	2.7	1.1
	Total	96.0	116.3	113.1	81.4	47.9	24.5
Warsaw	May	20.1	20.9	18.1	12.6	7.5	3.9
	June	22.9	24.4	22.2	17.3	9.2	4.1
	July	23.0	30.1	35.9	33.9	24.6	14.5
	August	23.0	32.4	38.3	29.1	14.9	5.5
	September	13.3	13.9	12.1	6.5	2.8	1.2
	Total	102.3	121.8	126.5	99.4	59.1	29.2

The best energy saving potential among all investigated locations was obtained for Madrid; 188.7 kWh for $T_{mpc} = 24\,°C$ (Table 5). Other locations below the 45th parallel did not provide much better energy saving potential than the studied locations with the latitude between 45° N and 55° N, but for different reasons. Rejection of heat accumulated in the PCM was a problem in the case of Nice, Rome, and Athens, as can be seen from the rather narrow interquartile temperature ranges (Figure 7).

As mentioned earlier, the overall thermal energy storage capacity of the air-PCM HEX in the phase change temperature range of the PCM was 2.78 kWh. For the time period from 1 May–September 30, this means the theoretical ESP of about 425 kWh. This potential can only be exploited if the PCM completely melts and solidifies in every cold storage cycle (this means every single day). This cannot be achieved in the considered passive cooling operation, where the daily outdoor temperature swings do not always overlap with the phase change temperature range of the PCM. As a result, the heat of fusion cannot be fully taken advantage of in each cold storage cycle. The largest amount of stored cold utilized during the entire investigated time period was 188.7 kWh in the case of Madrid. This means a utilization rate of the heat of fusion (URHF) of about 44%. This value of URHF is higher than the URHF reported for different climates in China [10]. However, the authors considered an isothermal phase change of PCM, which is not an advantage in passive cooling according to [9]. Furthermore, the authors considered a higher temperature set point for the utilization of stored cold, as they focused on the reduction of ventilation cooling load.

Table 4. Energy saving potential in kWh (cities between 45° N and 50° N).

		Nominal Phase Change Temperature of PCM (°C)					
		16	18	20	22	24	26
Paris	May	11.4	14.5	14.2	8.6	3.5	1.2
	June	16.0	21.4	20.8	16.0	10.1	5.5
	July	10.1	19.1	25.4	24.7	18.7	11.6
	August	11.7	19.2	24.3	21.6	15.1	9.4
	September	13.3	12.0	10.4	8.0	3.9	1.7
	Total	62.5	86.3	95.1	78.7	51.2	29.4
Milan	May	21.8	32.6	35.7	30.7	20.2	10.1
	June	7.5	15.3	25.8	34.3	35.8	30.2
	July	2.8	5.8	13.1	26.4	37.2	42.1
	August	4.2	9.2	17.1	26.7	35.7	37.6
	September	14.4	24.1	30.5	25.7	19.6	12.3
	Total	50.7	87.0	122.3	143.9	148.5	132.3
Zagreb	May	16.3	21.2	23.5	21.5	17.3	11.5
	June	9.8	15.0	19.5	22.8	21.4	18.9
	July	4.6	9.0	15.8	23.9	30.3	30.3
	August	5.3	10.8	16.8	21.7	25.2	26.9
	September	12.5	16.3	14.9	12.0	7.8	3.9
	Total	48.4	72.2	90.5	102.0	101.9	91.6
Budapest	May	24.8	26.7	26.9	23.9	14.6	7.2
	June	23.3	31.7	35.7	30.5	23.0	16.4
	July	11.8	20.3	29.4	35.7	36.6	30.6
	August	11.9	21.3	29.7	34.5	34.0	24.7
	September	21.7	21.7	19.0	13.6	6.6	3.1
	Total	93.5	121.6	140.7	138.2	114.7	81.9

The largest amount of cold utilized in one cold storage cycle (this means during one day) in all locations during the entire investigated time period was 2.67 kWh, meaning URHF of about 95% (Milan, $T_{mpc} = 16\,°C$, 13 May). In the vast majority of the cold storage cycles, only a small fraction of the available heat storage capacity was utilized. If the considered air-PCM HEX was used for cold storage in an all-air air-conditioning system, it would be theoretically possible to utilize the entire amount of heat of fusion in the cold storage cycle. Another interesting result of the study is the relatively small difference in the energy saving potential between locations at markedly different latitudes: Helsinki (82.4 kWh) and Athens (107 kWh). Helsinki (latitude 60.32° N) was the northernmost location considered in the study, while Athens (latitude 37.97° N) was the southernmost location.

The amount of cold that can be stored in the air-PCM HEX (and later utilized) depends on the mass flow rate of air. Small mass flow rates of air mean that the available cold storage capacity cannot be fully utilized. On the other hand, large airflow rates lead to the increases of the necessary fan power and thus decrease the energy efficiency of cold storage. The fan power was not considered in the study, as it depends on the actual integration of the air-PCM HEX in the ventilation or air-conditioning systems.

Table 5. Energy saving potential in kWh (cities below 45° N).

		Nominal Phase Change Temperature of PCM (°C)					
		16	18	20	22	24	26
Madrid	May	27.6	31.2	31.7	26.0	17.2	9.1
	June	15.9	25.6	34.6	42.3	44.6	38.1
	July	5.6	10.4	19.6	33.1	45.8	54.4
	August	6.9	13.0	23.0	36.2	48.6	53.7
	September	22.8	32.9	41.3	39.4	32.5	22.3
	Total	78.9	113.2	150.1	176.9	188.7	177.7
Nice	May	12.5	20.0	20.5	12.1	4.4	1.6
	June	3.1	10.1	21.2	24.0	16.1	10.1
	July	1.4	2.4	7.5	18.7	29.7	23.8
	August	1.4	2.0	6.5	17.5	27.2	24.7
	September	5.4	13.9	23.8	21.9	12.9	5.3
	Total	23.8	48.4	79.4	94.1	90.3	65.4
Rome	May	24.2	32.3	33.0	26.2	14.7	6.6
	June	7.9	17.1	27.6	33.5	30.4	25.2
	July	2.5	6.2	14.4	27.9	40.6	44.3
	August	2.8	6.4	13.1	24.8	36.8	41.5
	September	12.4	22.3	32.3	30.7	21.3	12.5
	Total	49.9	84.2	120.4	143.1	143.9	130.2
Athens	May	12.0	22.4	30.1	30.3	22.8	11.3
	June	1.9	4.3	10.7	21.8	30.3	29.2
	July	1.8	1.8	2.0	3.6	10.3	23.3
	August	1.8	1.8	1.9	3.8	10.4	24.6
	September	2.2	6.0	15.5	29.1	28.0	18.6
	Total	19.7	36.3	60.2	88.5	101.8	107.0

Table 6 shows the ESP and the URHF for Madrid under various air flow rates in the considered scenario with the temperature set point of 20 °C for the utilization of cold. Both the ESP and the URHF increase with the increasing air flow rates. Table 7 shows the results for the same scenario, but without the temperature set point. The energy saving potential was obtained from Equation (4), but without the condition $T_{outdoor} > 20\,°C$. The obtained values of the ESP and URHF were the theoretical maximums that can be reached under the considered conditions. As can be seen, neither the ESP nor the URHF increased very much. The main reason is the frequent occurrence of outdoor air temperatures above 20 °C during the daytime (8:00–20:00) in the studied time period.

Table 6. Energy saving potential (ESP) in kWh for Madrid at various air flow rates for the temperature set point $T_{outdoor} > 20\,°C$. The percentage in brackets is the utilization rate of the heat of fusion (URHF).

Air Flow Rate (m³/h)		Nominal Phase Change Temperature of PCM (°C)					
Daytime	Night	16	18	20	22	24	26
100	200	71.4 (16.8%)	94.7 (22.3%)	118.0 (27.7%)	133.8 (31.5%)	139.2 (32.7%)	130.1 (30.6%)
200	400	76.5 (18.0%)	106.7 (25.1%)	138.3 (32.5%)	161.0 (37.9%)	170.0 (40.0%)	159.9 (37.6%)
400	800	78.9 (18.5%)	113.2 (26.6%)	150.1 (35.3%)	176.9 (41.6%)	188.7 (44.4%)	177.7 (41.8%)
800	1600	79.8 (18.9%)	116.3 (27.4%)	155.9 (36.7%)	185.2 (43.5%)	197.7 (46.5%)	186.4 (43.8%)

Table 7. Energy saving potential (ESP) in kWh for Madrid at various air flow rates without the temperature set point. The percentage in brackets is URHF.

Air Flow Rate (m³/h)		Nominal Phase Change Temperature of PCM (°C)					
Daytime	Night	16	18	20	22	24	26
100	200	81.9 (19.3%)	105.5 (24.8%)	127.1 (29.9%)	140.1 (32.9%)	142.8 (33.6%)	131.8 (31.0%)
200	400	90.4 (21.3%)	120.6 (28.4%)	150.2 (35.3%)	168.7 (39.7%)	174.2 (41.0%)	161.3 (37.9%)
400	800	94.6 (22.3%)	128.7 (30.3%)	163.1 (38.3%)	185.0 (43.5%)	192.9 (45.4%)	178.7 (42.0%)
800	1600	96.4 (22.7%)	132.5 (31.2%)	169.1 (39.8%)	193.0 (45.4%)	201.9 (47.5%)	187.1 (44.0%)

Figure 8 shows the comparison of the ESP and the URHF, in the case of Madrid, for the daytime air from rates from 50 m³/h–1000 m³/h (the air flow rates at night were twice higher, from 1000 m³/h–2000 m³/h) for the scenarios with and without the temperature set point. The results in Figure 8 are shown for the nominal phase change temperature $T_{mpc} = 24\,°C$. The chart shows a rather significant influence of the air flow rate below about 400 m³/h (0.125 kg of PCM per 1 m³/h of the air flow rate). This PCM mass to air flow rate ratio is significantly lower than the 1–1.5 $\frac{kg}{m^3/h}$ reported in [9]. However, it needs to be emphasized that a very different type of air-PCM HEX (cylindrical packed bed) was used in that study. Furthermore, the PCM considered in the study had a smaller melting enthalpy and a wider temperature phase change range than the PCM considered in the present study. This only demonstrates the difficulty of providing quantitative recommendations that would be generally applicable in a latent heat thermal energy storage. Fortunately, with the current state of knowledge and many available models and simulation tools, it is possible to investigate the performance of a particular design of LHTES under a given set of operating conditions.

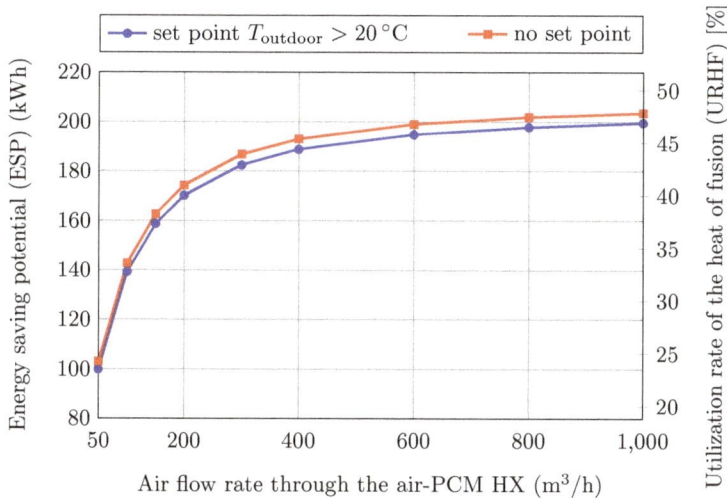

Figure 8. Energy saving potential (ESP) and the utilization rate of the heat of fusion (URHF) for Madrid and the nominal phase change temperature $T_{mpc} = 24\,°C$.

The comparison of the URHF for Madrid in the case of the scenarios with and without the temperature set point, in the form of histograms, is shown in Figure 9. Each bar of the histogram represents a URHF range of 5%. The histograms are for the daytime air flow rate of 400 m³/h and the nominal phase change temperature of the PCM $T_{mpc} = 24\,°C$. As can be seen, the URHF did not exceed 95% in either of the scenarios, but on about 10 days, the URHF was over 90%. As already mentioned earlier, the outdoor temperature set point had only a small impact on both the ESP and the URHF in the case of Madrid.

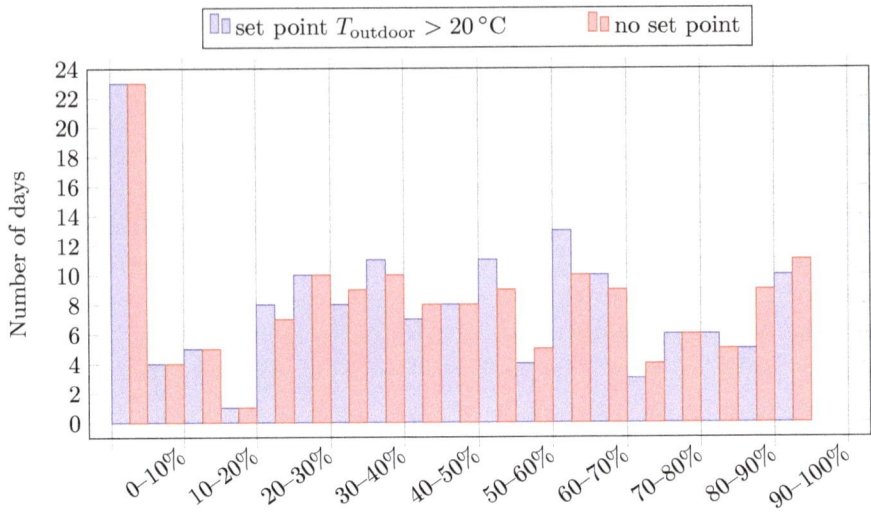

Figure 9. Histogram of the utilization rate of the heat of fusion (URHF) for Madrid and the nominal phase change temperature $T_{mpc} = 24\,°C$; each bar represents a range of 5%.

6. Economic Considerations

Despite many studies demonstrating the energy-saving potential of the PCMs in heat and cold storage, the real-life application of PCM-based LHTES is still lagging behind its potential. One of the reasons is the current economic viability of many solutions employing the PCMs. The economic considerations presented in this section concern only the system investigated in the present paper. A thorough analysis needs to be carried out in each particular case.

Economically, passive cooling is most cost effective if it makes it possible to avoid installation of a mechanical cooling system. In such a case, the benefit of both lower capital costs and lower operating costs can be exploited. If mechanical cooling needs to be installed anyway, the economic benefits of passive cooling are related to lower operating costs and possibly the downsizing of the mechanical cooling system, which can save some capital costs. The more complex the passive cooling system is (e.g., when it includes latent heat thermal energy storage), the less advantage in terms of capital costs it offers in comparison to relatively inexpensive mass-produced air-conditioners.

In the present study, the energy saving potential of passive cooling of buildings with the use of an air-PCM HEX (cold storage unit) was investigated. The economic question was whether the potential energy savings could pay for the cost of the considered air-PCM HEX. The most expensive part of the heat exchanger are the CSM panels. Considering the cost of 10 Euro per panel the total cost of the panels in the air-PCM HEX would be 1000 Euro. The shell of the heat exchanger, thermal insulation, dampers, and other fittings for connecting the heat exchanger to a ventilation system might cost about 200 Euro. The largest amount of stored cold utilized in the entire investigated time period was 188.7 kWh. If a vapor compression cooling system with the coefficient of performance of three (COP = 3) were used for space cooling, the reduction of cooling demand due to LHTES would translate to a savings of about 63 kWh of electricity. With the average price of electricity in the EU being about

0.2 Euro/kWh, the simple payback time would be well beyond the expected lifespan of the air-PCM heat exchanger.

7. Conclusions

As expected, the energy saving potential (ESP) of the air-PCM heat exchanger for cold storage in passive cooling of buildings depends on the climatic conditions. However, the differences between some locations were not as significant as could be expected. The energy saving potential in the case of Athens was only about 30% higher than the ESP in the case of Helsinki. The influence of the air flow rates on the ESP was rather significant. In the considered configuration of the air-PCM HEX and the investigated operating conditions, the ESP decreased rather significantly below about 400 m^3/h.

The average utilization rate of the heat of fusion (URHF) was lower than 50% in all investigated cases. The main reason for that was the daily outdoor temperature swing that does not always overlap (straddle) the phase change temperature range of the PCM. Nonetheless, the URHF exceeded 90% on some days. In the case of Madrid (the location with the highest average ESP), the outdoor temperature set point of 20 °C for the utilization of cold during daytime had only a small influence on the overall ESP, which was very close to the theoretical maximum (without considering any set point). The main reason was the frequent occurrence of outdoor air temperatures above 20 °C during the utilization of cold, which made the set point condition inconsequential most of the time.

From the economic point of view, the considered way of cold storage (an air-PCM heat exchanger) was not economically viable even in the climates with the most favorable conditions.

Author Contributions: Conceptualization, P.C.; methodology, P.C. and L.K.; software, L.K. and M.Z.; formal analysis, P.C., L.K., and M.Z.; investigation, P.C., L.K., and M.Z.; writing, original draft preparation, P.C. and L.K.; writing, review and editing, P.C. and L.K.; visualization, L.K. and M.Z.; supervision, P.C.; funding acquisition, P.C. and L.K.

Funding: This research was funded by the project Sustainable Process Integration Laboratory (SPIL), funded as Project No. CZ.02.1.01/0.0/15_003/0000456, by the European Research Development Fund and by the project Computer Simulations for Effective Low-Emission Energy funded as Project No. CZ.02.1.01/0.0/0.0/16_026/0008392 by the Operational Programme Research, Development and Education, Priority Axis 1: Strengthening capacity for high-quality research.

Conflicts of Interest: The authors declare no conflict of interest.

References

1. Mata, E.; Sasic Kalagasidis, A.; Filip Johnsson, F. Contributions of building retrofitting in five member states to EU targets for energy savings. *Renew. Sustain. Energy Rev.* **2018** *93*, 759–774. [CrossRef]
2. Perera, E.; Parkins, L. *Build Tight–Ventilate Right*; Building Services, Chartered Institution of Building Services Engineers: London, UK, 1992; pp. 37–38.
3. Connolly, D. Heat Roadmap Europe: Quantitative comparison between the electricity, heating, and cooling sectors for different European countries. *Energy* **2017**, *139*, 580–593. [CrossRef]
4. Gouveia, J.P.; Seixas, J.; Mestre, A. Daily electricity consumption profiles from smart meters—Proxies of behavior for space heating and cooling. *Energy* **2017**, *141*, 108–122. [CrossRef]
5. Patronen, J.; Kaura, E.; Torvestad, C. Nordic heating and cooling: Nordic approach to EU's Heating and Cooling Strategy. *TemaNord* **2017**, *532*, 1–110. [CrossRef]
6. Artmann, N.; Manz, H.; Heiselberg, P. Climatic potential for passive cooling of buildings by night-time ventilation in Europe. *Appl. Energy* **2007**, *84*, 187–201. [CrossRef]
7. Del Pero, C.; Aste, N.; Paksoy, H.; Haghighat, F.; Grillo, S.; Leonforte, F. Energy storage key performance indicators for building application. *Sustain. Cities Soc.* **2018**, *40*, 54–65. [CrossRef]
8. Arkar, C.; Medved, S. Free cooling of a building using PCM heat storage integrated into the ventilation system. *Sol. Energy* **2007**, *81*, 1078–1087. [CrossRef]
9. Medved, S.; Arkar, C. Correlation between the local climate and the free-cooling potential of latent heat storage. *Energy Build.* **2008**, *40*, 429–437. [CrossRef]

10. Chen, X.; Zhang, Q.; Zhai, Z. Energy saving potential of a ventilation system with a latent heat thermal energy storage unit under different climatic conditions. *Energy Build.* **2016**, *118*, 339–349. [CrossRef]
11. Costanzo, V.; Evola, G.; Marletta, L.; Nocera, F. The effectiveness of phase change materials in relation to summer thermal comfort in air-conditioned office buildings. *Build. Simul.* **2018**, *11*, 1145–1161. [CrossRef]
12. Ousegui, A.; Marcos, B.; Havet, M. Inverse method to estimate air flow rate during free cooling using PCM-air heat exchanger. *Appl. Therm. Eng.* **2019**, *146*, 432–439. [CrossRef]
13. Charvát, P.; Klimeš, L.; Ostrý, M. Numerical and experimental investigation of a PCM-based thermal storage unit for solar air systems. *Energy Build.* **2014**, *68*, 488–497. [CrossRef]
14. Osterman, E.; Butala, V.; Stritih, U. PCM thermal storage system for 'free' heating and cooling of buildings. *Energy Build.* **2015**, *106*, 125–133. [CrossRef]
15. Pielichowska, K.; Pielichowski, K. Phase change materials for thermal energy storage. *Prog. Mater. Sci.* **2014**, *65*, 67–123. [CrossRef]
16. Agyenim, F.; Hewitt, N.; Eames, P.; Smyth, M. A review of materials, heat transfer and phase change problem formulation for latent heat thermal energy storage systems (LHTESS). *Renew. Sustain. Energy Rev.* **2010**, *14*, 615–628. [CrossRef]
17. Kuznik, F.; Virgone, J.; Roux, J.J. Energetic efficiency of room wall containing PCM wallboard: A full-scale experimental investigation. *Energy Build.* **2008**, *40*, 148–156. [CrossRef]
18. Iten, M.; Liu, S.; Shukla, A. Experimental validation of an air-PCM storage unit comparing the effective heat capacity and enthalpy methods through CFD simulations. *Energy* **2018**, *155*, 495–503. [CrossRef]
19. Muhieddine, M.; Canot, E.; March, R. Various approaches for solving problems in heat conduction with phase change. *Int. J. Finite Vol.* **2009**, *6*, 1–20.
20. Stritih, U.; Charvát, P.; Koželj, R.; Klimeš, L.; Osterman, E.; Ostrý, M.; Butala, V. PCM thermal energy storage in solar heating of ventilation air—Experimental and numerical investigations. *Sustain. Cities Soc.* **2018**, *37*, 104–115. [CrossRef]

© 2019 by the authors. Licensee MDPI, Basel, Switzerland. This article is an open access article distributed under the terms and conditions of the Creative Commons Attribution (CC BY) license (http://creativecommons.org/licenses/by/4.0/).

Article

A Process Integration Method for Total Site Cooling, Heating and Power Optimisation with Trigeneration Systems

Khairulnadzmi Jamaluddin [1,2], Sharifah Rafidah Wan Alwi [1,2,*], Zainuddin Abdul Manan [1,2], Khaidzir Hamzah [2] and Jiří Jaromír Klemeš [3]

[1] Process Systems Engineering Centre (PROSPECT), Research Institute for Sustainable Environment, Universiti Teknologi Malaysia, Johor Bahru 81310 UTM, Malaysia; nadzmi2009@gmail.com (K.J.); dr.zain@utm.my (Z.A.M.)
[2] School of Chemical and Energy Engineering, Faculty of Engineering, Universiti Teknologi Malaysia, Johor Bahru 81310 UTM, Malaysia; khaidzir@utm.my
[3] Sustainable Process Integration Laboratory—SPIL, NETME Centre, Faculty of Mechanical Engineering, Brno University of Technology—VUT BRNO, Technická 2896/2, 616 69 Brno, Czech Republic; klemes@fme.vutbr.cz
* Correspondence: syarifah@utm.my; Tel.: +6019-868-3085

Received: 1 February 2019; Accepted: 11 March 2019; Published: 16 March 2019

Abstract: Research and development on integrated energy systems such as cogeneration and trigeneration to improve the efficiency of thermal energy as well as fuel utilisation have been a key focus of attention by researchers. Total Site Utility Integration is an established methodology for the synergy and integration of utility recovery among multiple processes. However, Total Site Cooling, Heating and Power (TSCHP) integration methods involving trigeneration systems for industrial plants have been much less emphasised. This paper proposes a novel methodology for developing an insight-based numerical Pinch Analysis technique to simultaneously target the minimum cooling, heating and power requirements for a total site energy system. It enables the design of an integrated centralised trigeneration system involving several industrial sites generating the same utilities. The new method is called the Trigeneration System Cascade Analysis (TriGenSCA). The procedure for TriGenSCA involves data extraction, constructions of a Problem Table Algorithm (PTA), Multiple Utility Problem Table Algorithm (MU PTA), Total Site Problem Table Algorithm (TS PTA) and estimation of energy sources by a trigeneration system followed by construction of TriGenSCA, Trigeneration Storage Cascade Table (TriGenSCT) and construction of a Total Site Utility Distribution (TSUD) Table. The TriGenSCA tool is vital for users to determine the optimal size of utilities for generating power, heating and cooling in a trigeneration power plant. Based on the case study, the base fuel source for power, heating and cooling is nuclear energy with a demand load of 72 GWh/d supplied by 10.8 t of Uranium-235. Comparison between conventional PWR producing power, heating and cooling seperately, and trigeneration PWR system with and without integration have been made. The results prove that PWR as a trigeneration system is the most cost-effective, enabling 28% and 17% energy savings as compared to conventional PWR producing power, heating and cooling separately.

Keywords: trigeneration system; process integration; pinch analysis; co-generation; storage system; trigeneration system cascade analysis; total site heat integration

1. Introduction

Rapid industrialisation and rising global population contribute to the rapid depletion of energy resources, environmental pollution and climate change. The International Energy Agency [1]

has predicted increasing CO_2 emissions from 0.15×10^{12} MWh in 2008 to 0.23×10^{12} MWh in 2035 as well as a rising crude oil price from 60 USD/barrel in 2011 to 120–140 USD/barrel from 2020 onwards. These challenges have become the key drivers to improve the energy efficiency of power plants. Zhang et al. [2] summarised strategies to reduce greenhouse gas emissions that include utilisation of a mixture of energy generation technologies in one location, development of highly-efficient energy production and re-use methods as well as implementing the use of incentives, technologies, taxes and quotas. Abdul Manan et al. [3], on the other hand, proposed a methodology that provides clear visualisation insights for CO_2 emission planning as well as good target estimation for problems involving resource planning and conservation towards achieving cleaner production goals. Implementation of integrated energy systems such as cogeneration and trigeneration systems as a centralised power plant can improve its energy efficiency by reuse of waste heat produced for other applications such as distillation process, district heating and cooling. Cogeneration systems, which are also known as Combined Heating and Power (CHP) systems is a technology whereby electricity and heat are produced simultaneously from a single fuel source. Trigeneration systems, on the other hand, are an advanced cogeneration system technology which produces cooling, heating and electricity at the same time from a primary source of energy. Production of cooling by using absorption chillers is an advantage in a trigeneration system. Khamis et al. [4] stated that an improvement in energy efficiency could translate into lower operating cost, reduced emissions and reduced usage of fossil fuels.

Process Integration (PI) is a process to reduce the consumption of resources as well as environmental emissions. Pinch Analysis (PA) is one of the PI methodologies which has been widely applied for designing and obtaining optimal targets for various resource conservation networks. Recent studies show that various resources proposed for PA such as heat, water, mass, carbon, property and gas were progressively developed, see Klemeš et al. [5]. The progressive development of PA in various resource networks proved that the methodology had gained acceptance by the public due to its simple insightful approaches using graphical or numerical techniques. The latest studies related to Power Pinch Analysis (PoPA) approaches have been included in this paper. PoPA which is introduced by Wan Alwi et al. [6] helps designers obtain the amount of excess electricity as well as minimum targets for outsourced electricity. Mohammad Rozali et al. [7] extended the application of PoPa by including losses analysis associated with power conversion, transfer and storage. Ho et al. [8] proposed a new numerical method based on PoPA approaches which were called Electricity System Cascading Analysis (ESCA). The method is developed for designing and optimising non-intermittent power generator such as biomass, biogas, natural gas, nuclear and diesel as well as energy storage systems. Liu et al. [9] combined both methods developed by Mohammad Rozali et al. [7] and Ho et al. [8] to obtain optimal design and sizing of multiple decentralised energy systems and a centralised energy system. Jamaluddin et al. [10] then extended the PoPA method from Mohammad Rozali et al. [7] to determine the minimum targets for outsourced power, heating and cooling, amount of excess power, heating and cooling during the first day as well as for continuous 24 h operations simultaneously; and to determine the maximum storage capacity in a trigeneration system. Jamaluddin et al. [11] then included safety considerations in PoPA for designing safe and resilient hybrid power systems. Recent studies had been done by Hoang et al. [12] to obtain an optimal hybrid renewable energy system which can sustainably meet the electricity demand by using the PoPA method.

Initially, Dhole and Linnhoff [13] introduced Total Site Integration of industrial systems. The Total Site Integration concept developed by Dhole and Linnhoff [13] is based on the ideas of the Site Heat Source and Site Heat Sink. Total Site Heat Integration (TSHI), developed by Klemeš et al. [14], is a tool which focuses on integrating heat at multiple sites. TSHI can be very beneficial in terms of cost effectiveness since the new and existing plant piping systems can be used to indirectly transfer heat through utility systems. The concept of Total Site was extended by Perry et al. [15] to a broader spectrum of processes in addition to the industrial process. Integration of renewable energy sources was included in the analysis to reduce the carbon footprint of a Locally Integrated Energy Sector (LIES). Heat sources and sinks from small scale industrial plants, offices, residential areas and large building complexes such as hotels and hospitals can be analysed by using LIES. Matsuda et al. [16] applied Total Site Integration in a number of chemical industrial sites and heterogeneous Total Site involving a brewery and several commercial energy users. Varbanov and Klemeš [17] improved the concept of Total Site by introducing a set of time slices to meet the variation of energy supply and demand. Varbanov and Klemeš [18] then extended the Total Site concept by including heat storage, waste heat minimisation and carbon footprint reduction as well as the Total Site heat cascade. Next, Liew et al. [19] introduced a new numerical approach to allow designers and engineers to assess the sensitivity of a whole site with respect to operational changes using a Total Site Sensitivity Table (TSST) as well as to assess the impact of sensitivity changes on a cogeneration system, determine the optimum utility generation system size, assess the need for backup piping and estimate the amount of external utilities needed. Total Site Integration can also be extended to cogeneration targeting. Shamsi and Omidkhah [20] developed a thermo-economically-based approach for optimisation of steam levels in steam production as well as reduction of total cost of the utility system in Total Site. Chew et al. [21] extended TSHI by including pressure drop on utility. Klemeš et al. [22] reviewed Total Site Integration methodologies on cogeneration. The representation of cogeneration potential has been firstly documented by Raissi [23]. Site Utility Grand Composite Curve (SUGCC) developed by Klemeš et al. [14] allows thermodynamic targets for cogeneration with targets for site-scope Heat Recovery minimising the cost of utilities. Varbanov et al. [24] introduced improvements to the model of back-pressure steam turbine performance. Boldyrev et al. [25] calculated capital cost assessment for power cogeneration and evaluated the potential steam turbine placement for various steam pressure levels. Liew et al. [26] later improved the TSST for planning the TSHI centralised utility system. Liew et al. [27] further improved the methodology by incorporating absorption and electric chillers. A new TSHI is proposed by Tarighaleslami et al. [28] to optimise both non-isothermal and isothermal utilities. Ren et al. [29] then proposed a simulation to target the cogeneration potential of Total Site utility systems. Recent studies proposed by Pirmohamadi et al. [30] to obtain the optimum design of cogeneration systems in Total Site by using exergy approach.

A numerical tool based on PA approach called Problem Table Algorithm (PTA) was developed by Linnhoff and Flower [31] for intra-process heat integration. This tool has the same application as the Composite Curves and Grand Composite Curve but provides more accurate values for the Pinch Points. Costa and Queiroz [32] extended the concept of PTA by implementing multiple utility targets. Unified Targeting Algorithm (UTA) proposed by Shenoy [33] is used as a powerful tool to determine maximum resource recovery for PI. The methods proposed by Costa and Queiroz [32] and later by Shenoy [33] had a weakness. The UTA developed by Shenoy [33] cannot be used for TSHI problems whereas the method developed by Costa and Quiroz [32] involves complex calculations. Liew et al. [19] developed a new numerical for targeting TSHI which known as the Total Site Problem Table Algorithm (TS-PTA) to tackle the weaknesses in Costa and Quiroz's approach [32] and Shenoy's [33] works. The methodology has been improved by including time slide due to variation on demands and sources [26], absorption and electric chillers for production of chilled water [27] and incorporating long and short terms heat energy supply and demand variation problem [34].

Until now, the published extensions of Pinch Analysis have yet to provide a complete solution for trigeneration systems. The Total Site Heat Integration should be extended for Cooling, Heating and Power (TSCHP), and the benefits of power, heating and cooling targeting related to the actual trigeneration system design need to be emphasised. The objective of this work is to develop an insight-based numerical Pinch Analysis methodology to minimise the heating, cooling and power requirements as well as to determine the capacity of energy storage systems of a trigeneration system for TSCHP. The intermittency from the demands can greatly affect the performance of the system because the energy should be continuously produced and supplied based on demand needs. The development of this systematic methodology is very important for users to determine allocation and targets of power, heating and cooling in a trigeneration system as well as to optimise the trigeneration system.

2. Methodology and Case Study

Trigeneration System Cascade Analysis (TriGenSCA) is a new numerical method being developed in this paper to minimise power, heating and cooling targeting as well as to optimise sizing of the turbine, absorption chiller, cooling tower and steam generator. TSCHP integration method is an extension of TSHI which focuses on intra-processes of integrating heating, cooling and power for multiple sites. Summary of the overall methodology of this paper is shown in Figure 1. Based on the figure shown, the overall methodology can be categorised into eight steps which are data extraction, Problem Table Algorithm (PTA) [19] for an individual process, Multiple Utility Problem Table Algorithm (MU-PTA) [19] for an individual process, Total Site Problem Table Algorithm (TS-PTA) for all processes, estimation of energy source from trigeneration system, Trigeneration System Cascade Analysis (TriGenSCA), Trigeneration Storage Cascade Table (TriGenSCT) and Total Site Utility Distribution (TSUD) to obtain optimal size of trigeneration system.

The trigeneration system is implemented as a centralised energy system to supply power, heating and cooling applications to the demand as shown in Figure 2. Based on the figure shown, Very High-Pressure Steam (VHPS) is produced from steam generator (acting as the same function as a boiler) and passes through a double extraction turbine simultaneously producing power and lower pressure steams such as High-Pressure Steam (HPS) and Low-Pressure Steam (LPS). The HPS which is produced from the double extraction turbine can be supplied to meet the demands directly or stepped down to LPS using a relief valve. Excess HPS and LPS can be cooled down by using CW or condensed by using the condensing turbine to generate more power. Condensing turbines have an advantage whereby they can adjust their electrical output by altering the proportion of steam passing through the turbine. Hot Water (HW), on the other hand, is generated by using the condensation system. HW can then be used either directly to the demand or converted to cooling utilities such as Cooling Water (CW) and Chilled Water (ChW). CW is produced through the cooling tower and ChW by using an absorption chiller.

The cooling tower is generally used to cool process water via evaporation [35]. Operation of cooling tower starts with HW is pumped to enter at the top through nozzles. The HW flowing through the nozzles is dispersed onto a large surface area which is also known as a fill. The fill is used to delay water from reaching the bottom of the tower and allow more time for the air to interact with the HW. The water then slowly makes its way through the fill tanks via gravity and a fan forces air across the water path until it reaches the bottom of the tower. CW is then produced at the bottom of the tower and supplied to the demands. The absorption chiller, on the other hand, consists of four main components which are generator, condenser, evaporator and absorber [36]. The process of producing of ChW by using absorption chiller is summarised below:

1) Generator—the HW produces refrigeration vapour from a strong refrigerate solution by transferring heat from HW to coll the solution. The refrigeration vapour needs to pass through a rectifier for dehydration before it enters the condenser.
2) Condenser—dehydrated and high-pressure refrigerant enters the condenser where it is condensed. The refrigerant goes through an expansion valve after cooling. Expansion valve reduces the pressure and temperature of the refrigerant. The new values of refrigerant must be below than in evaporator.
3) Evaporator—the cold refrigerated space is appearing in the evaporator. The cooled refrigerant enters the evaporator, absorbs heat and then leaves as saturated refrigeration vapour.
4) Absorber—the refrigeration vapour exposed to a spray of the weak refrigerant-absorbent solution. The weak solution changes to a strong solution. The new solution passes through regenerator which is also known as a heat exchanger. The solution arrives at the generator has the same pressure as before. The process is then repeated.

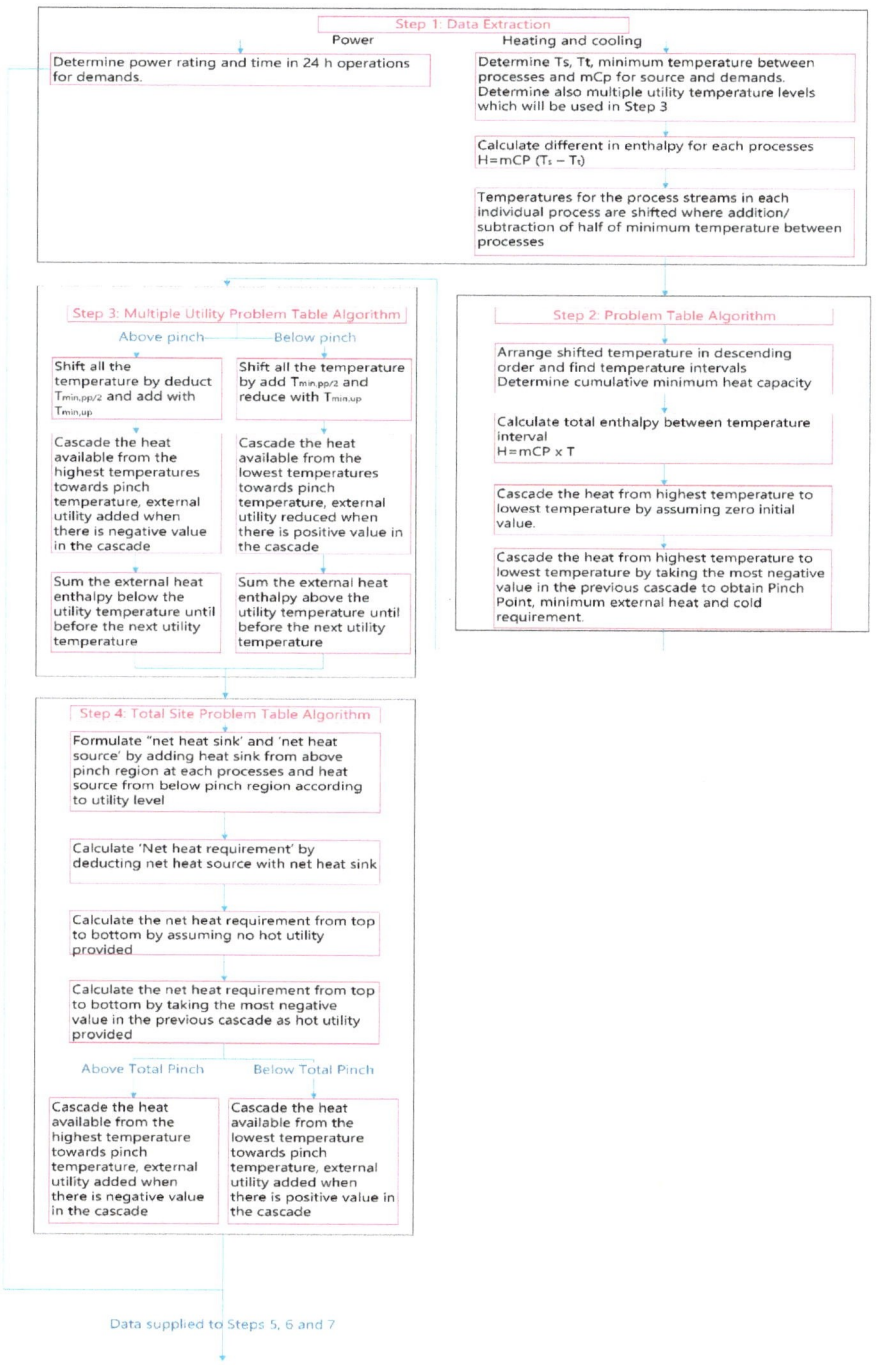

(a)

Figure 1. *Cont.*

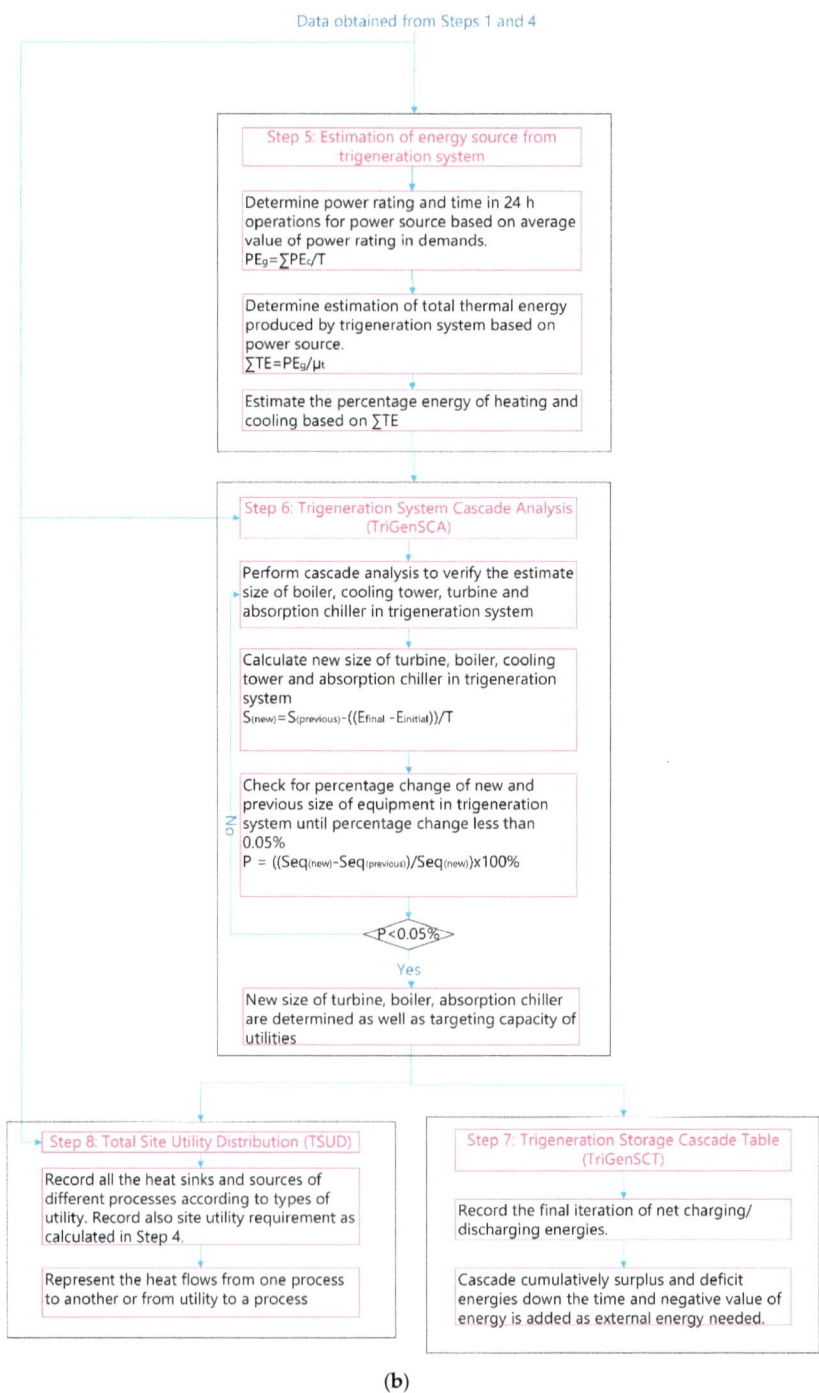

Figure 1. (a,b) Overview of the proposed methodology.

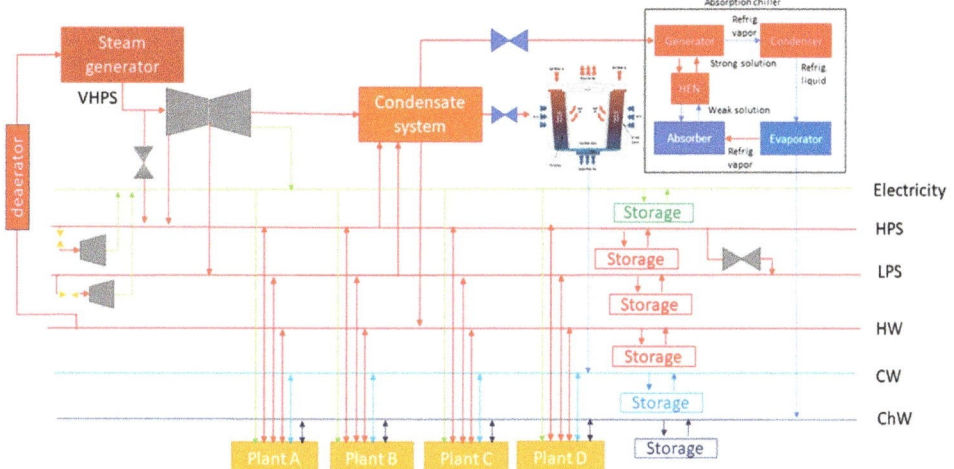

Figure 2. Schematic of a power, heating and cooling system for a Total Site.

Some simplifying assumptions are listed below:

(i) Energy loss due to transmission has not been considered at this stage.
(ii) Energy loss from storage system is considered where the lead-acid battery is used as a power storage system with charging and discharging efficiencies of 90% [37]. Thermo-chemical storage, on the other hand, is used as a heating and cooling thermal storage systems with charging and discharging efficiencies of 58% [38]. Conversions of power from AC to DC and from DC to AC are also considered with an inverter efficiency at 90% [37].
(iii) Energy conversion is taken into consideration where the efficiency of double extraction turbine is assumed to be 25% [39], the efficiency of condensing turbine is 33% [40], the efficiency of the condensation system is 30% [41], and the efficiency of the cooling tower and absorption chiller are 30% [42].
(iv) Trigeneration system and industrial plants are in continuous 24 h operations, and fluctuation due to weather change and demands have not been accounted for.
(v) Energy consumption remains unchanged regardless of changes of the topology in the industrial plants.

2.1. Step 1: Data Extraction

In the first step, the local energy supply and demand data of the trigeneration system and industrial plants are needed. The data extraction is separated into two sides which are power and heating/cooling sides. Figure 3 shows the hourly average electricity demands for four industrial. Figure 4 summarises the total electricity demands for the four industrial plants.

Meanwhile, heating/cooling data extraction requires a supply temperature, T_s, target temperature, 10 °C T_t, the minimum temperature difference between utility and process streams, $\Delta T_{min,up}$ and minimum flowrate heat capacity, mCP. The difference in enthalpy ΔH is obtained using Equation (1):

$$\Delta H = mCP \times (T_s - T_t) \tag{1}$$

where ΔH = the difference in enthalpy in MW; mCP = Minimum flowrate heat capacity in MW/°C; supply temperature in °C; T_t = Target temperature in °C.

Stream data for four industrial plants are obtained from Perry et al. [15] and has been modified. The stream data for four industrial plants are shown in Tables 1–4. Calculation of shifted temperatures for the process streams in each individual process are necessary where temperatures of cold streams, T_c, are shifted to perform shifted cold stream temperatures (T_c') by adding half of the minimum temperature between processes, $\Delta T_{min,pp}$, whereas the temperatures of hot streams, T_h, are shifted to perform shifted hot stream temperatures (T_h') by reducing half of the $\Delta T_{min,pp}$. The value of $\Delta T_{min,pp}$ for Plants A and C is assumed to be 20 °C whereas $\Delta T_{min,pp}$ for Plants B and D are assumed to be 10 °C [19]. Multiple utility temperature levels data available at the plants is described in Table 5.

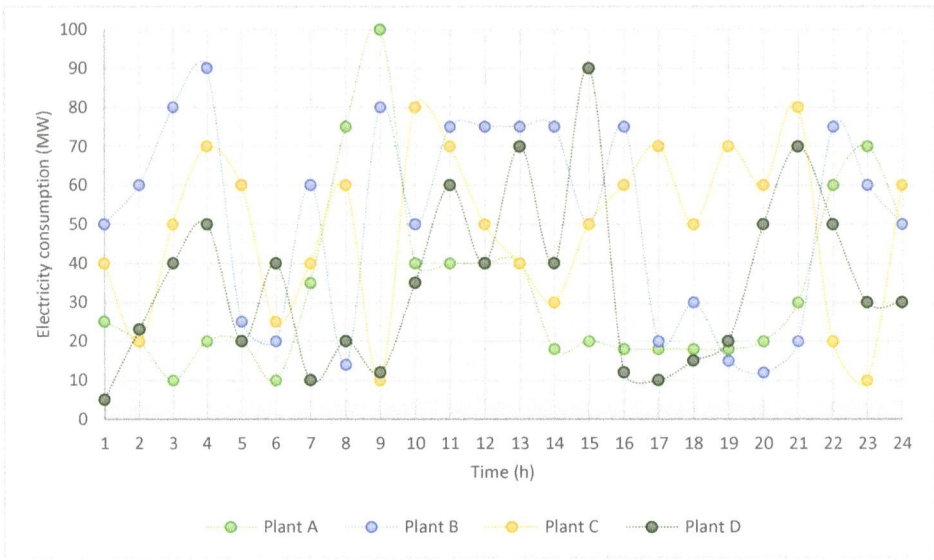

Figure 3. Power variations for Industrial Plants A to D in 24 h operations [8,43].

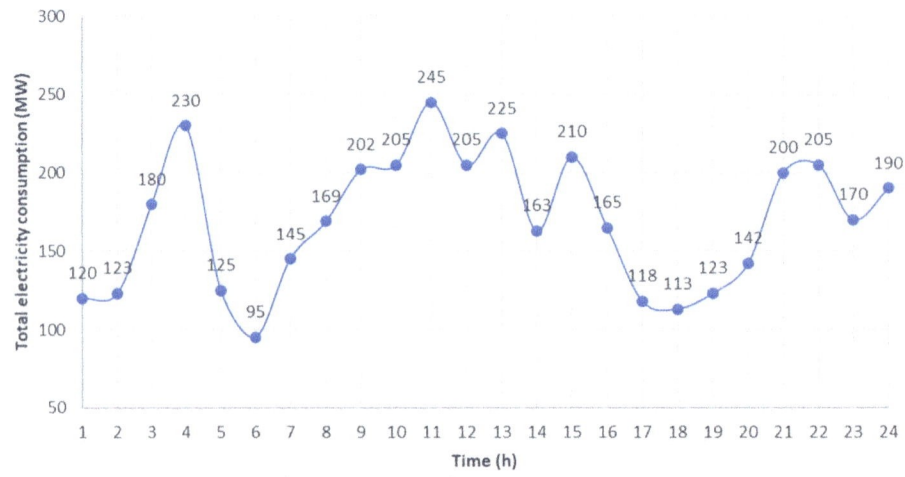

Figure 4. Total electricity consumption for the four industrial plants.

Table 1. Stream data for Industrial Plant A with $\Delta T_{min,pp}$ 20 °C [19].

Stream	T_s (°C)	T_t (°C)	ΔH (MW)	mCP (MW/°C)	T_s' (°C)	T_t' (°C)
A1 Hot	170	80	5	0.06	160	70
A2 Hot	150	55	6.48	0.07	140	45
A3 Cold	25	100	15	0.2	35	110
A4 Cold	70	100	1.05	0.04	80	110
A5 Cold	30	65	5.25	0.15	40	75

Table 2. Stream data for Industrial Plant B with $\Delta T_{min,pp}$ 10 °C [19].

Stream	T_s (°C)	T_t (°C)	ΔH (MW)	mCP (MW/°C)	T_s' (°C)	T_t' (°C)
B1 Hot	200	20	0.0005	0.08	195	15
B2 Cold	10	100	4	0.04	15	105
B3 Cold	100	120	10	0.5	105	125
B4 Hot	150	40	8.443	0.08	145	35
B5 Cold	60	110	1	0.02	65	115
B6 Cold	75	150	7	0.09	80	155

Table 3. Stream data for Industrial Plant C with $\Delta T_{min,pp}$ 20 °C [19].

Stream	T_s (°C)	T_t (°C)	ΔH (MW)	mCP (MW/°C)	T_s' (°C)	T_t' (°C)
C1 Hot	85	40	225	5	75	30
C2 Hot	80	40	400	10	70	30
C3 Hot	41	38	105.3	35.1	31	28
C4 Cold	25	65	23.6	0.59	35	75
C5 Cold	55	65	25.8	2.58	65	75
C6 Cold	33	60	6.48	0.24	43	70
C7 Cold	25	60	77	2.2	35	70
C8 Cold	30	240	29.4	0.14	40	250
C9 Cold	25	28	150	50	35	38
C10 Cold	30	100	59.5	0.85	40	110
C11 Cold	18	50	224	7	28	60
C12 Cold	21	200	8.95	0.05	31	210

Table 4. Stream data for Industrial Plant D with $\Delta T_{min,pp}$ 10 °C [19].

Stream	T_s (°C)	T_t (°C)	ΔH (MW)	mCP (MW/°C)	T_s' (°C)	T_t' (°C)
D1 Cold	15	60	149.85	3.33	20	65
D2 Cold	15	80	515	7.92	20	85

Table 5. Stream data for Industrial Plant D with $\Delta T_{min,pp}$ 10 °C [19].

Multiple Utilities	Temperature (°C)
HPS	240
LPS	150
HW	50
CW	20
ChW	10

2.2. Step 2: Construct Problem Table Algorithm (PTA) for Each Plant

Heating and cooling streams data need to be further analysed by using PTA. PTA is a numerical method proposed by Linnhoff and Flower [31] to obtain Temperature Pinch Point, minimum external heat, QH_{min} and minimum external cold, QC_{min}, required. The PTA has similar functions as Composite Curves (CCs) and Grand Composite Curves (GCCs) in a graphical approach and also provides more precise values at crucial points. For details on the construction of PTA, readers may refer to Linnhoff and Flower [31]. Tables S1–S4 (in Supplementary) Material show completed PTA on Plants A to D. The construction of PTA is shown below:

1) Column 1 presents shifted temperature in descending order which is obtained from Step 1 whereas Column 2 shows temperature intervals.
2) Column 3 shows minimum heat capacity from Step 1. Downside arrow represents a hot stream which is a positive value whereas upside arrow represents cold stream which is a negative value. On the other hand, Column 4 presents the cumulative minimum heat capacity obtained from Column 3.
3) Column 5 shows total enthalpy between temperature interval which shown in Equation (2). Enthalpy represents energy cumulated in the steam:

$$\Delta H = mCP \times \Delta T \tag{2}$$

where ΔH = Total enthalpy between temperature interval; mCP = minimum heat capacity; ΔT = Temperature intervals.

4) Single utility heat cascade shown in Column 7 follows the same equation as in Equation (3). The initial value is taken from the highest negative value in initial heat cascade (from Column 6) but make the value in positive. Values of QH_{min} and QC_{min} are obtained from the first and last row of Column 7.

$$H_i = H_{i-1} + \Delta H \tag{3}$$

where H_i = Current initial heat; H_{i-1} = Previous initial heat; ΔH = Total enthalpy on temperature interval.

5) Single utility heat cascade shown in Column 7 follows the same equation as in Equation (3). The initial value is taken from the highest negative value in initial heat cascade (from Column 6) but make the value in positive. Values of QH_{min} and QC_{min} are obtained from the first and last row of Column 7.

PTA results are summarised in Table 6. Based on Table 6, Temperature Pinch Points of all industrial plants are obtained. The Temperature Pinch Point for Plants A, B, C and D are 35 °C, 105 °C, 75 °C and 20 °C. The Temperature Pinch Point will be used in the next step. Meanwhile, the value of QH_{min} in Plant A is 9.82 MW and minimum external cold required is unnecessary as the value of QC_{min} is zero. Plant B, on the other hand, requires 4.30 MW and 5.74 MW of QH_{min} and QC_{min}. Value of QH_{min} for Plants C and D are 61 MW and 664.85 MW and value of QC_{min} for Plant C is 111.42 MW. Minimum external cooling for Plant D is unnecessary.

Table 6. Summary of PTA for Industrial Plants A to D.

	Plant A	Plant B	Plant C	Plant D
QH_{min} (MW)	9.82	4.30	61	664.85
QC_{min} (MW)	0	5.74	111.42	0
Temperature Pinch Point	35 °C	105 °C	75 °C	20 °C

2.3. Step 3: Construct the Multiple Utility Problem Table Algorithm (MU PTA) for Each Plant

MU-PTA developed by Liew et al. [19] is an extension of PTA where four columns are added to target the amounts of various utility levels selected as potential sinks and sources for use in TSCHP. MU PTA has been used to identify pockets and target the exact amounts of utilities required within a given utility temperature interval. Multiple utility cascades are performed in two separate regions which are above and below regions of the Temperature Pinch Point obtained from Step 2 for each plant. For further details on the construction of MU PTA readers may refer to Liew et al.'s [19] work. Tables S5–S8 show MU PTA for all industrial plants.

2.3.1. Multiple Utility Cascades in the Region above the Pinch of Each Plant

At the above region of the Pinch Point, all shifted temperatures (T') are reduced by $\Delta T_{min,pp}/2$ for returning the temperature back to normal and $\Delta T_{min,up}$ is added, as shown in Column 2 (in Supplementary Material). The resulting temperature is shown as T". The temperature for multiple utility shown in Table 5 is added to Column 2 as well. Implementation of multiple utility temperature in Column 2 will ease the user to determine the utility distribution at a later stage.

Columns 3 until 6 follow the same method as shown in Step 2. Column 7 shows heat is cascaded from the highest temperature to the temperature Pinch Point. The cascading is different as compared with that in Step 2 because the cascading process is done interval-by-interval. The external utility is immediately added as soon as a negative value is encountered. This cascading process is known as 'multiple utility heat cascade'. The amount of external utility added is listed in Column 8 where the amount of external utility is equal to the negative value in Column 7. Once the amount of external utility is added, heat cascade in Column 7 becomes zero. The procedure is then repeated until reach to the Pinch Temperature.

The amounts of each type of utility consumed can be obtained once the multiple utility heat cascades are completed. The heat utility sink or source is shown in Column 9 is obtained by adding the utility consumed below the utility temperature before the next utility temperature.

2.3.2. Multiple Utility Cascades in the Region below the Pinch of Each Plant

The same methodology is used for multiple utility cascading below the Pinch temperature. Below the Pinch region, all shifted temperatures (T') are added $\Delta T_{min,pp}/2$ and $\Delta T_{min,up}$ subtracted from them to obtain the temperatures in the utility temperature scale. Multiple utilities are also added in Column 2. Multiple utilities in Column 7, however, start from the bottom temperature to the Pinch temperature. Positive heat value must be zeroed out by generating utilities. Negative values are encountered during multiple utility cascading, as shown in Column 8, and they represent pockets in the GCC.

The amount of utility can be obtained by addition of the amounts of excess heat from above the utility temperature to the next utility temperature level. Column 9 presents the amounts of utility based on different utility temperature.

A summary of MU PTA for all industrial plants is shown in Table 7. Based on Table 7, Plant A required 5.66 MW of LPS and 4.16 MW of HW as a heat sink to the streams. Plant B required 4.30 MW of LPS as a heat sink to the streams whereas heat source in the form of HW is generated which has a value of 3.23 MW. CW and ChW also in a deficit by 2.32 MW and 0.20 MW to cool down the streams. Plant C required 17 MW of HPS as a heat sink to the streams whereas Plants C and D needed 44 MW and 327.26 MW of LPS as a heat sink. 100.7 MW of HW in Plant C is in surplus whereas Plant D required 337.59 MW of HW. CW in Plant C is in deficit by 117 MW.

Table 7. Summary of MU PTA for Industrial Plants A to D.

Utility	Plant A	Plant B	Plant C	Plant D
HPS	0 MW	0 MW	17 MW	0 MW
LPS	5.66 MW	4.30 MW	44 MW	327.26 MW
HW	4.16 MW	−3.23 MW	−100.7 MW	337.59 MW
CW	0 MW	−2.32 MW	−117 MW	0 MW
ChW	0 MW	−0.19 MW	0 MW	0 MW
Temperature Pinch Point	35 °C	105 °C	75 °C	20 °C

2.4. Step 4: Construct Total Site Problem Table Algorithm (TS PTA)

Development of TS PTA is an extension of PTA where this step represents the site CC in Total Site. TS PTA, proposed by Liew et al. [19], is used to determine the amounts of utilities which can be exchanged among processes. Table 8 shows the completed development of TS PTA in industrial plants. The utilities obtained from Step 3 is arranged from highest to lowest temperature as presented in Column 2. The utilities generated below the Pinch Temperature in Step 2 are added as a net source as shown in Column 3. Meanwhile, utilities consumed above the Pinch temperature in Step 2 are determined as a net sink as shown in Column 4. Net heat requirement in Column 5 is the subtraction of net heat source with the net heat sink. A negative value of the net heat requirement represents a heat deficit whereas a positive value represents a heat surplus. Column 6 shows cascading of heat transferred from higher to lower temperatures which follows the Second Law of Thermodynamics. The heat surplus at a higher temperatures utility can be cascaded to heat deficits at a lower temperature utility. The initial heat cascade is started from zero, and the net heat requirement is cascaded from top to bottom. The most negative value in Column 6 is used to investigate the amount of external heat utility required for the system by making it positive and cascading the net heat requirement again as shown in Column 7. The Total Site Pinch Point can be obtained where the zero value is the Pinch point location in this column.

The utilities can be separated into two parts which are regions above and below the Total Site Pinch Point. The same methods as Step 3 are used and shown in Columns 8 and 9 were at above Total Site Pinch Point, net heat requirement (in Column 5) is cascaded from the top to the Pinch Point by assuming no external heat supplied at a temperature above the HPS. The same amount of external heating utility is added in Column 9 as there is a negative value in the cascade. Below region of Total Site Pinch Point, net heat requirement of multiple utilities is cascaded from the bottom to the Pinch Point. Cooling utilities are added as there is a positive value in the cascade until it reaches zero and represented by negative numbers.

Table 8. TS PTA for all industrial plants.

1	2	3	4	5	6	7	8	9
Utility	Utility Temperature (°C)	Net Heat Source (MW)	Net Heat Sink (MW)	Net Heat Requirement (MW)	Initial Heat Cascade	Final Single Heat Cascade	Multiple Utility Heat Cascade	External Utility Requirement (MW)
					0	636.04	0	
HPS	240	0	17	−17				17
					−17	619.04	0	
LPS	150	0	381.22	−381.22				381.22
					−398.22	237.82	0	
HW	50	103.93	341.75	−237.82				237.82
					−636.04	0	0	Pinch
CW	20	119.32	0	119.32				−119.32
					−516.72	119.32	0	
ChW	10	0.197	0	0.197				−0.197
					−516.52	0.197	0	

2.5. Step 5: Estimation of Energy Source from a Trigeneration System

In this step, the energy source from a trigeneration system is preliminarily estimated to show values of energy required to supply to the demands. Various fuels such as coal, natural gas, diesel and nuclear as well as renewables can be applied in trigeneration systems. In this work, nuclear energy is suggested as a trigeneration system fuel since it is zero CO_2 emissions. Nuclear energy is a good supplier of energy for non-electrical applications such as heating and cooling processes for the same reason as electricity [4]. Figure 5 shows the range of applicability for nuclear reactors. Double extraction turbine requires the maximum steam temperature of 275.6 °C to generate power [39]. A Water Cooled Reactor (WCR) is thus chosen as the best nuclear reactor because it has the highest temperature of 320 °C. Pressurised Water Reactor Nuclear Power Plant (PWR NPP) is one of the types of WCR. Uranium-235 is used as a fuel to generate nuclear energy for PWR NPP.

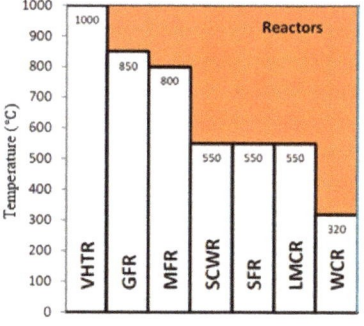

Figure 5. The range of applicability for nuclear reactors [4].

Taking PWR NPP as a trigeneration system, calculation for power rating in the source is shown in Equation (4). The total daily power consumption is 4068 MWh/day (from Figure 4). The total power consumption is the cumulative power rating in a day. Generation of power from trigeneration system is assumed to be 169.5 MW and operate in 24 h since the nuclear power plant is a stable system:

$$PE_g = \frac{\sum PE_c}{T} \tag{4}$$

where PE_g = Average power generation in MW; $\sum PE_c$ = Total power consumption in MWh; T = total time in h.

Total thermal energy produced by the trigeneration system then can be estimated based on power generation, PE_g. Equation (5) shows the estimation of total thermal energy produced by the trigeneration system. Average power generation for the trigeneration system is 169.5 MW and double extracting turbine efficiency is assumed to be 25% [39]. Based on a calculation by using Equation (5), the total thermal energy produced by the steam generator is 678 MW. The total thermal energy produced by the steam generator does not include any additional energy from extra fuel. This means that all production of Very High-Pressure Steam (VHPS) is used directly to the turbine without undergo relief valve to the lower temperature of utilities:

$$\sum TE = \frac{PE_g}{\mu_t} \tag{5}$$

where $\sum TE$ = Total thermal energy produced by trigeneration system in MW; PE_g = Average power generation in MW; μ_t = Double extracting turbine efficiency.

Remaining waste energy can be determined by using Equation (6) and energy losses are assumed to be 10% [44]. Based on Equation (6), the remaining waste energy is 440.7 MW:

$$E_{waste} = TE - PE_g - (TE \times 10\%) \tag{6}$$

where E_{waste} = Remaining waste energy in MW; PE_g = Average power generation in MW; TE = total thermal energy produced by the trigeneration system in MW.

The division of remaining waste thermal energy produced by the trigeneration system is based on the highest temperature of utilities to the lowest temperature of utilities. The remaining waste energy produced by the trigeneration system starts with HPS and follows with LPS and HW. This means that 440.7 MW is divided into; (1) 15 MW of HPS, (2) 380 MW of LPS and (3) remaining 45.7 MW of waste heat which will be converted into HW by using the condensation system. Taking efficiency of condensation system into consideration, 13.71 MW of HW is produced from 45.7 MW of waste heat. Out of 13.71 MW, 10.5 MW can be used directly to meet demands whereas the remaining 3.21 MW of HW can be converted into CW and ChW through the cooling tower and an absorption chiller.

Production of 0.197 MW of ChW from absorption chiller requires 0.657 MW of HW and the remaining 2.553 MW of HW can produce 0.7659 MW of CW by using the cooling tower by taking the values of the absorption chiller and cooling tower efficiencies which are 30% into account. Figure 6 shows a summary of the energy that is formed by the trigeneration system.

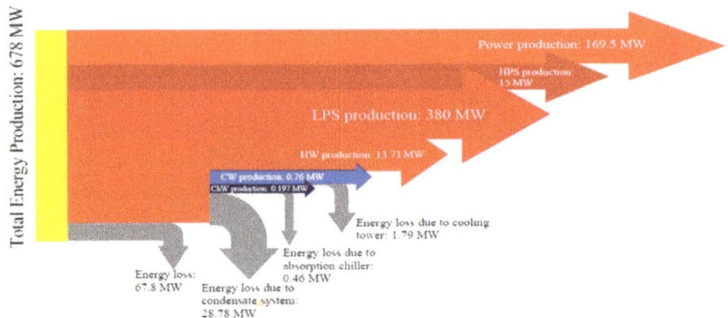

Figure 6. Energy balance for a trigeneration system.

2.6. Step 6: Trigeneration System Cascade Analysis (TriGenSCA)

TriGenSCA is introduced in this paper to further target the minimum power, heating and cooling as well as to optimise the size of utilities in trigeneration system considering storage system is available to store surplus energy at one time and utilise when there is deficit energy requirement. The construction of TriGenSCA consists of three major steps which are cascade analysis, calculation of the new size of the trigeneration system and percentage change between previous and new trigeneration system size.

2.6.1. Cascade Analysis

Cascade analysis is the first step in developing TriGenSCA. The cascade analysis is used to verify the estimated size of utilities in trigeneration system. Tables A1–A8 show cascade analysis for the TSCHP before iteration whereas Tables A9–A16 show cascade analysis for the TSCHP after iteration. The Table for cascade analysis can be constructed as shown below:

1) Column 1 shows time for 24 h operations with 1 h interval.
2) Column 2 shows power, heating and cooling demands, whereas Column 3 shows power, heating and cooling generations from the trigeneration system. Power demand data is obtained from data extraction in Step 1 whereas heating and cooling demands are obtained from TS PTA in Step 4. For power, heating and cooling generations data are obtained from Step 5.
3) Column 4 presents the net energy requirement where power, heating and cooling generations in Column 3 is subtracted the power, heating and cooling demands in Column 2. This also indicates that any available power, heating and cooling source are supplied to the respective power, heating and cooling demand at the same time interval first and according to the utility types. A positive value of net energy requirement means that the energy is in surplus whereas a negative value of net energy requirement means that the energy is in deficit.
4) Column 5 shows the new net energy requirement which presents the transfer of surplus energy at higher utility temperature to deficit energy at lower utility temperature. The value of surplus energy at higher utility temperature will not be the same when transferring to deficit energy at lower utility temperature. This is because the energy is lost due to the efficiency of utility. HPS and LPS can be condensed by using the condensing turbine to produce power. Energy losses due to energy conversion are taken into consideration where the efficiency of a condensing turbine is 33% [40], and absorption chiller and cooling tower are 30% [42]. For example,

in Tables A11 and A12, the deficit value of 49.79 MW for CW can be reduced by converting the HW to CW through the cooling tower. Since the efficiency of the cooling tower is 30%, the energy of 14.94 MW is lost. Positive values in this column show energy in surplus whereas negative values show energy in deficit.

5) Column 6 shows the analysed system's energy excesses and deficits through charging and discharging efficiencies of storage systems by referring to the new net energy requirement. Excess energy is charged and stored into the storage systems and deficit energy indicates insufficient energy which requires an additional source of energy from the storage systems (discharged from storage systems). Charging and discharging efficiencies of power and thermal storage systems are taken into consideration to indicate energy losses due to charging and discharging energy using energy storage systems. Inverter efficiency for power is also included to show the conversion of AC to DC and vice versa. Assumptions have been made where charging and discharging efficiencies of power by using the lead-acid battery is 90% [37] whereas charging and discharging efficiencies of heat and cool energy by using thermo-chemical storage system is 58% [38]. Inverter efficiency for power, on the other hand, is 90% [37]. Charging energy for power in Column 6 (positive value) is obtained by multiplying the positive new net surplus power in Column 5 by the storage charging and inverter efficiencies. Discharging energy for power, on the other hand, is obtained by dividing the new deficit power in Column 5 by the storage discharging and inverter efficiencies. Charging energy for heating is calculated by multiplying the new net surplus energy by the storage charging efficiency whereas charging cooling energy is obtained by dividing the new surplus cooling by the charging efficiency. Discharging heat energy is obtained by dividing the new net deficit heat energy in Column 5 by the discharging efficiency, and for cooling energy it is obtained by multiplying the new deficit energy by the discharging efficiency. The calculation of heating and cooling is the contrary due to the opposite storage concept of a thermal storage system in heating and cooling applications. As stated by [45], excess heat energy is stored by extraction from the energy producer to the storage whereas cool energy is stored by extracting heat from the storage to the energy producer.

6) Next, the surplus energy at the time interval can be stored by using power or thermal storage systems to allow energy to be used in the following time interval. Initial energy for a start-up is assumed to be zero. Surplus energy is accumulated from highest to lowest time intervals. Cumulative energy is shown in Column 7 which follows Equation (7). Negative values in Column 7 represent deficit energy whereas positive values show surplus energy. The cumulative energy can determine the highest deficits of energy by searching for the highest negative value in this column:

$$E_{i+1} = E_i + E_{nr} \quad (7)$$

where E_{i+1} = Cumulative energy for the next time interval; E_i = Cumulative energy on time interval; E_{nr} = New net energy requirement.

7) Column 8 shows new cumulative energy which also follows Equation (6). The initial cumulative energy in this column is taken from the highest negative value in Column 7 and making the value positive to represents external energy required in the storage tank to supply the demand. The last row of this column, on the other hand, shows excess energy available in the storage tank for the next day.

2.6.2. Calculate the Size of Utility in a Trigeneration System

From the analysis of the data presented in Table A1, it was determined that outsourced energy required to start-up the system for power, HPS, LPS, HW and CW are 227.5 MWh, 82.76 MWh, 50.43 MWh, 9,407.6 MWh and 1,650.3 MWh. ChW has zero initial energy content which means that no external energy is required. The outsourced heating and cooling energy needed to start up the system can be bought from other plants whereas outsourced power can be bought from the grid. Excess energy available at t = 24 h can be transferred to the next day to reduce the initial energy required at t = 0 h. The final energy content at t = 24 h for power is 43.66 MWh. The cascade analysis between trigeneration system and industrial plants shows imbalance energy between utilities. These energy surpluses can be reduced if the energy gap between initial energy at t = 0 h and excess energy available at t = 24 h could be minimised. Two conclusions can be drawn from this analysis. Firstly, the final content of energy is more than the initial amount of energy shows the capacity of utility in trigeneration system is oversized. If the final content of energy is less than the initial amount of energy, the capacity of utility in trigeneration system is undersized.

Equation (8) is derived to calculate the new size of utility (turbine, steam generator, condensation system, absorption chiller and cooling water) in a trigeneration system:

$$S_{eq(new)} = S_{eq} - \frac{(E_{final} - E_{initial})}{T} \tag{8}$$

where $S_{eq(new)}$ = New estimate size of utility in trigeneration system in MW; S_{eq} = Previous estimate size of utility in trigeneration system in MW; E_{final} = Final energy content in MWh; $E_{initial}$ = Initial energy content in MWh; T = total time duration in h.

By using this formula, the new estimated size of utilities is determined. Power, HPS, LPS, HW and CW generation produced in the trigeneration system has been increased from 169.5 MW to 177.16 MW, from 15 MW to 18.45 MW, from 380 MW to 382.1 MW, from 10.5 MW to 402.48 MW and from 0.77 MW to 69.53 MW. The size of the absorption chiller producing ChW remains unchanged.

2.6.3. Percentage Change between the Previous and New Size of a Trigeneration System

The percentage change is derived by using Equation (9) to determine the optimal size of the trigeneration system which reduces the energy gap between the initial energy required to start up the system and available excess energy that can be supplied to the next day. An iteration method is involved in this step. The target of 0.05% is set as a tolerance to make sure the accuracy of the results [8]:

$$P = \frac{\left|S_{eq(new)} - S_{eq}\right|}{S_{eq(new)}} \times 100\% \tag{9}$$

where P = Percentage change between the previous and new size of trigeneration system; $S_{eq(new)}$ = New estimate size of utility in trigeneration system; S_{eq} = Previous estimate size of utility in the trigeneration system.

From the calculation, the percentage changes for the first iteration are 4.32% for power, 18.69% for HPS, 0.55% for LPS, 97.39% for HW and 98.9% for CW. Since the iteration is larger than 0.05%, the calculation is repeated using the new size of utilities in the trigeneration system. The iteration is stopped when the percentage change of each utility is less or equal than 0.05%. According to the case study, the calculation stops at the 12th iteration since all percentage changes of utility are less than 0.05%.

Tables A9–A16 show TriGenSCA after the final iteration. Based on the table data, the outsourced energy needed for power is 112.68 MWh which means that external power is needed to supply the demand. The outsourced energy for HPS, LPS, HW, CW and ChW are zero. This means that no external energy required to supply in the storage tank. On the other hand, final energy content for the power of 109.75 MWh shows values of available excess energy which can be transferred to the next day operations. This means that 2.93 MWh of power is in deficit. On the other hand, HPS, LPS and HW are in deficits of 0.12 MWh, 3.13 MWh and 1.35 MWh. The deficit power can be obtained by converting the excesses of HPS and LPS through condensing turbines. Excess HW, on the other hand, will be delivered to the steam generator through a deaerator. Figure 7 shows the TriGenSCA results before and after iterations in a graphical approach to offer more visualisation insights.

Production of CW and ChW are based on converting HW by using a cooling tower and an absorption chiller. Equation (10) shows the HW needed to be converted into CW or ChW by using a cooling tower and an absorption chiller. The efficiency of the cooling tower and absorption chiller are assumed to be 30% [42]. Energy production for CW and ChW are 119.32 MW and 0.197 MW. Based on Equation (10), the energy of HW required for producing CW and ChW are 397.73 MW and 0.657 MW:

$$E_{HW(CW/ChW)} = \frac{E_{(CW/ChW)}}{\mu_{CW/ChW}} \quad (10)$$

where $E_{HW(CW/ChW)}$ = Additional HW required to produce CW or ChW in MW; $E_{(CW/ChW)}$ = CW or ChW energy production in MW; $\mu_{CW/ChW}$ = Efficiency of absorption chiller or cooling tower.

Total steam energy required to produce HW in the whole system to supply to the demands is shown in Equation (11). The HW can be directly supplied to the demands or converted into CW and ChW by using a cooling tower and an absorption chiller. Production of HW is achieved by using the condensation system, and the efficiency of the condensation system is assumed to be 30% [41]. The energy production for HW application is 302.68MW, whereas HW required to produce CW and ChW are 397.73 MW and 0.657 MW. The total steam energy required to produce HW is 2,336.89 MW:

$$E_{ST \to HW} = \frac{(E_{HW} + E_{HW(CW/ChW)})}{\mu_{condenser}} \quad (11)$$

where $E_{ST \to HW}$ = Total steam energy required to produce HW in MW; E_{HW} = Energy production of HW in MW; $E_{HW(CW/ChW)}$ = Energy of HW required to converting CW or ChW in MW; $\mu_{condenser}$ = Efficiency of the condensation system.

The power generation after the final iteration is 176.79 MW. Based on Equation (5), the total energy produced is 707.16 MW. The remaining waste energy is 459.65 MW by using Equation (6). The division of remaining waste energy is from the highest temperature of utility to the lowest temperature of the utility. This means that any remaining waste heat energy is divided into: (1) 17.01 MW of HPS, (2) 381.63 MW of LPS and (3) 61.014 MW of waste heat is converted into HW by using the condensation system. The production of HW from waste heat is 18.304 MW. Based on Equations (10) and (11), the total steam required to produce HW for supplying it directly to the demands as well as converting it to CW and ChW through the cooling tower and absorption chiller is 2,336.89 MW. This means that excess steam energy of 2,275.88 MW (2,336.89 MW − 61.014 MW = 2,275.88 MW) is required from the steam generator.

2.7. Step 7: Trigeneration Storage Cascade Table (TriGenSCT)

TriGenSCT is introduced in this step to determine the amount of energy that can be transferred by the trigeneration system, the amount of energy available for storage and the maximum capacity of the power and thermal storage systems. The table of TriGenSCT at the final iteration is shown in Tables A17 and A18 and can be constructed as follows:

1) Columns 1 to 6 follow the same method as cascade analysis in TriGenSCA (as shown in Tables A9–A16).
2) Column 7 shows the storage capacity of the power and thermal storage systems whereas Column 8 presents the outsourced energy required in the system. The surplus and deficit power and thermal energy from Column 6 are cascaded cumulatively down the time interval starting at t = 0 h. The energy is cascading down the time to show the cumulative energy has been stored in the storage systems. However, when the energy is in deficit, the energy is discharged from storage systems until no energy is left in the storage systems. The net cascaded energy surpluses are recorded in the storage capacity in Column 7. Once there is no energy available in the storage systems, external energy needs to be supplied to meet the energy demands as shown in Column 8 to represent the total amount of external energy supplies needed. The largest cumulative energy surplus in Column 7 represents the maximum capacity of the power and thermal storage systems.

Based on Tables A17 and A18, the maximum storage systems for power, HPS, LPS and HW are 180.59 MWh, 0.12 MWh, 3.13 MWh and 1.41 MWh. The values of CW and ChW are zero to show no storage is needed as all of the CW and ChW energies have been supplied to the demands. The total amount of external energy supplied needed is obtained from the last row of Column 8. Based on the case study, only power needs external energy which is 112.7 MWh. The external power can be bought from the grid.

2.8. Step 8: Total Site Utility Distribution (TSUD)

TSUD was proposed by Liew et al. [19] to visualise the utility flow in the sites. The SCC does not show the utility distribution when there are several processes involved on the integrated site. Table 9 shows TSUD based on case study performed in this research. Values of energy source and energy sink of heating and cooling are taken from results obtained in TS PTA (in Step 4). Positive values of external heat requirement in TS PTA shows the energy sink in Column 4 whereas negative values of external heat requirement in TS PTA shows the energy source in Column 3. Average power for all industrial plants, on the other hand, is obtained from Step 1. Power, heating and cooling sources from trigeneration system are obtained from the final iteration of Step 6. Arrows within the table present that heat sources can be transferred to heat sinks for the same type of utility. For example, 381.43 MW of LPS from the trigeneration system is distributed to deficits of energy in all industrial plants. If there are extra heat sources on higher utility levels, heat can be supplied to the lower utility levels. Heat energy loss efficiency is taken into consideration in this step where absorption chiller and cooling tower are assumed to have an efficiency of 30% [42]. Additional energy is transferred to meet demand load. For example, Plants B and C required a total of 119.32 MW of CW and Plant B needed 0.195 MW of ChW to cool down the streams. The trigeneration system then needs to supply 397.73 MW of HW (excess heat supply of 278.41 MW due to energy loss) to Plants B and C in a form of CW through the cooling tower. On the other hand, 0.66 MW of HW is supplied to Plant B to form ChW through the absorption chiller (excess of heat supplied of 0.462 MW due to energy loss).

3. Discussion

TriGenSCA is used to determine the minimised power, heating and cooling targets and optimise the sizing of utilities in the trigeneration system. The final iteration of TriGenSCA shows PWR required energy of 3,000 MW (72 GWh/d) to overcome a deficit of demand load. The VHPS from the steam generator needs to be transferred to the lower temperature of utilities. Reference [46] has stated that 0.45 t of Uranium-235 can create 3,000 MWh/day of thermal energy. This means that PWR as a trigeneration system with integration requires 10.8 t of Uranium-235 as a fuel. Figures 8–10 show that the final network of three different systems in the same demand load. Analysis has been made by comparing between conventional PWR producing power, heating and cooling in a separate system, PWR as a trigeneration system without integration and PWR as a trigeneration system with integration.

The highest value of the power demand is 245 MW. The power that needs to be generated in a conventional PWR and PWR as a trigeneration system without integration is the same as the highest value of the power demand because any HPS and LPS surpluses cannot be cascaded to produce additional power in a conventional PWR and PWR as a trigeneration system without integration.

As a result, PWR as a trigeneration system without integration and conventional PWR producing power, heating and cooling in a separate system dump an excess power of 1,656 MWh/day. On the other hand, PWR as a trigeneration system with integration can only produce excess HPS, LPS and HW of 0.12 MWh, 3.13 MWh and 1.35 MWh. The power deficit of 1.07 MWh can be obtained by converting 3.13 MWh of HPS and 0.12 MWh of LPS using a condensing turbine with an efficiency of 33% [40]. The remaining power deficit of 1.86 MWh can be bought from the local grid. The excess 1.35 MWh of HW, on the other hand, is sent back to the steam generator through the deaerator. Production of power, heating and cooling in a separate PWR system needed more Uranium-235 as a fuel for energy source as compared with PWR as trigeneration with and without integration.

Table 9. Total Site Utility Distribution (TSUD).

Utility	Temperature (°C)	Energy Source (MW)				Energy Sink (MW)				
		Trigen Plant	Plant A	Plant B	Plant C	Plant D	Trigen Plant	Plant A	Plant B	Plant C	Plant D
Power	-	176.79			33.96		51.5	33.96	51.5	48.96	35.08
HPS	240	17.022								17	
LPS	150	381.43						5.663	4.2953	44	327.26
HW	50	636.24		3.23	100.7			4.159	0.929	99.77	337.59
CW	20								2.319	117	390
ChW	10					0.657				0.197	

The values in bold represents energy that has been transferred from energy source to energy demand. Green arrow: power; Red arrows: heat energy; Blue arrows: cool energy; Purple arrows: chilled energy.

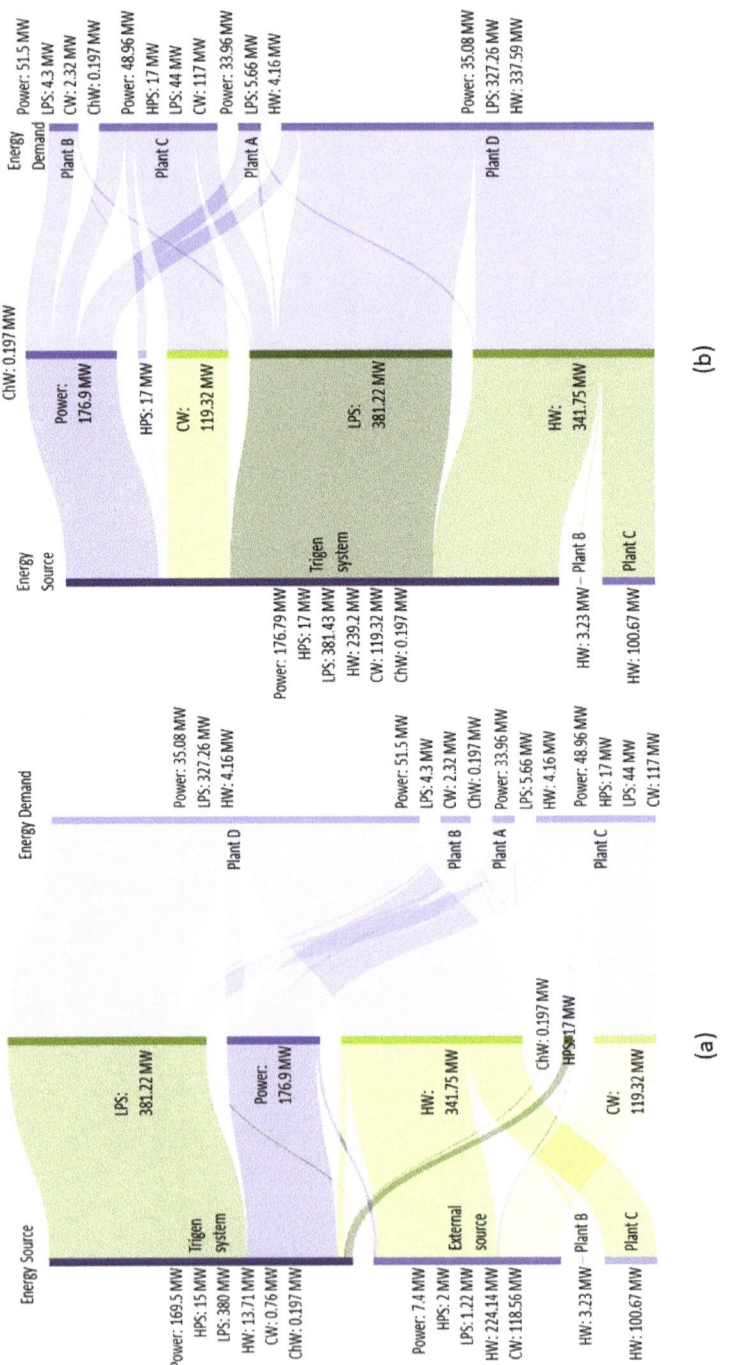

Figure 7. Graphical TriGenSCA (**a**) before, and (**b**) after iterations.

The amount of Uranium-235 required as a fuel in the conventional PWR system is 12.9 t whereas PWR as a trigeneration system without integration requires 10.95 t of Uranium-234 as a fuel. The total energy required for a conventional PWR system to produce the same energy as the PWR with trigeneration system is 3,598 MW or 86,352 MWh/day. PWR as a trigeneration system without integration requires a total energy of 3,040 MW that translates into approximately 73,000 MWh/day. Energy losses on the whole system are shown in Equation (12). Based on Equation (12), the energy losses per day for a conventional PWR producing power, heating and cooling is the highest value which is 64,000 MWh, followed by PWR as a trigeneration system without integration (53,000 MWh) and PWR as a trigeneration system with integration (50,000 MWh):

$$E_{loss} = (TE - E_{useful}) \times 24 \text{ hours} + E_{excess} \tag{12}$$

where E_{loss} = Energy losses MWh/day; TE = Total energy required in MW; E_{useful} = Useful energy produced by trigeneration system in MW; E_{excess} = Excess energy MWh/day.

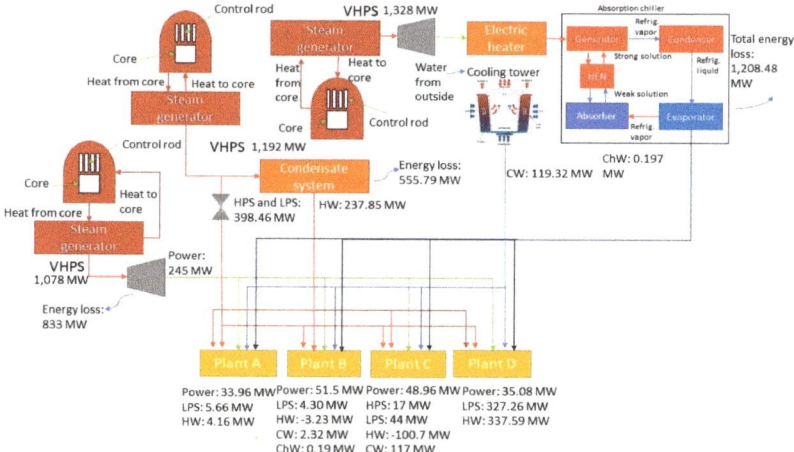

Figure 8. Final network for a conventional PWR that separately produces power, heating and cooling.

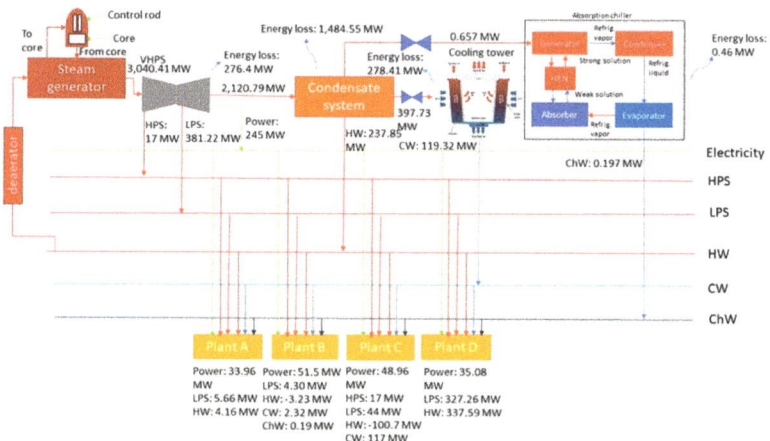

Figure 9. Final network for a PWR as a trigeneration system without integration.

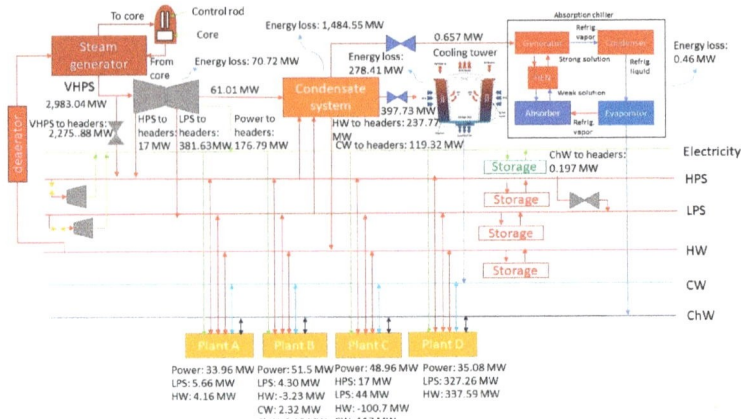

Figure 10. Final network for a PWR as a trigeneration system.

In terms of cost, the equivalent annual cost is calculated as the annual cost of owning, operating and maintaining an asset over its entire life. The equivalent annual cost can evaluate the cost of each system as well as optimise the systems for the lowest life-cycle cost. Equation (13) shows the equivalent annual cost. Operational and maintenance costs for fuel and non-fuel of PWR are 0.49 USD/kWh and 1.37 USD/kWh [47]. The initial investment for PWR is assumed to be 770 USD/kW, and life-cycle of PWR is 30 y [48]. The rate of return is assumed to be 10%. For PWR as a trigeneration system with integration, the initial investment for power and HW storages are taken into consideration where storage for HW is assumed to be 70 USD/kWh by using thermo-chemical storage and power is assumed to be 100 USD/kWh by using lead-acid battery [49]. Based on Table A18, maximum storage systems needed for power, HPS, LPS and HW are 180.59 MWh, 0.12 MWh, 3.13 MWh and 1.41 MWh. Cost for buying power from the local grid due to insufficient power in the trigeneration system with integration is also considered in an operational cost where the price of power is assumed to be 0.1 USD/kWh [50]. Based on the calculation using Equation (13), conventional PWR has the highest value of equivalent annual cost which is MUSD 364/year followed by PWR as a trigeneration system without integration (MUSD 307/year) and PWR as a trigeneration system with integration (MUSD 262/year):

$$EAC = IC_{initial} \times \frac{i(1+i)^n}{(1+i)^{n-1}} + OM \times 365 \text{ days} \tag{13}$$

where EAC = Equivalent annual cost in USD/year; i = the rate of return; n = life-cycle of PWR in year; $IC_{initial}$ = Initial investment cost in USD; OM = Operation and maintenance costs in USD.

Table 10 summarises the three different systems at the same demand load. Based on the table, PWR as a trigeneration system with integration is the best choice in terms of cost and also energy as compared with PWR as a trigeneration system without integration and conventional PWR producing separate power, heating and cooling. PWR as a trigeneration system with integration can create a savings of 28% for the equivalent annual cost and 17% of energy production as compared with conventional PWR producing power, heating and cooling. For energy loss, trigeneration PWR with integration can save up to 22% whereas trigeneration PWR without integration can save up to 17%. PWR as a trigeneration system without integration can only create a saving of 16% for equivalent annual cost and 16% for energy production when a comparison is made with a conventional PWR system. Moreover, trigeneration PWR with integration only required external power of 1.86 MWh as compared with trigeneration PWR without integration and conventional PWR which produce an excess power of 1,656 MWh. Excess power in trigeneration PWR without integration and conventional must be dumped as it cannot be used to serve a load [51]. This will create a waste if the power is in excess. As compared with trigeneration PWR with integration, deficit power can be bought from the power grid. Trigeneration PWR with integration is the best choice in term of cost and energy saving as compared with trigeneration PWR without integration and conventional PWR.

Table 10. Comparison between conventional PWR, PWR as a trigeneration system with and without integration.

	Trigeneration PWR without Integration	Trigeneration PWR with Integration	Conventional PWR
Energy loss (MWh/day)	53,000	50,000	64,000
Energy production (MWh/day)	73,000	72,000	86,352
Amount of Uranium-235 (t)	10.95	10.83	12.9
Power Energy in excess (+)/deficit (−) (MWh)	1,656	−1.86	1,656
Equivalent annual cost (MUSD/year)	307	262	364

4. Conclusions

This paper described a novel insight-based Pinch Analysis numerical technique for targeting the minimum cooling, heating and power for a centralised trigeneration system integrated with several industrial sites generating the same utilities. A summary of the contributions of this work is listed below:

1) A new methodology has been developed to minimise power, heating and cooling targets as well as to optimise sizing of utilities in trigeneration system which called as Trigeneration System Cascade Analysis (TriGenSCA). The TriGenSCA is capable of determining the power, heating and cooling capacity of utilities in a trigeneration system. Moreover, TriGenSCA can also determine the maximum energy storage systems of power, heating and cooling, simultaneously. Development of this tool also enables designers to reduce the utility size when it is oversized and increase the utility size when it is undersized. This gives a benefit on distributing heat sources to the heat sinks under optimum conditions. With the predicted size of utilities that can be obtained, the users could perform simple costing on the system in the preliminary design phase.

2) TriGenSCT has been successfully developed to determine the amount of energy that can be transferred by a trigeneration system, the amount of energy available for storage and the maximum capacity of storage systems for power and thermal energy can be obtained. The TriGenSCT provides useful tools for energy managers, electrical and power engineers to design an optimal trigeneration system.

3) TSUD has been modified by considering power, heating and cooling utilities to visualise the utility flow in the sites. TSUD shows that higher temperature of utilities can be transferred to lower the temperature of utilities as well as converting HPS/LPS to power and HW to CW and ChW. This tool also able to give a visual on the transition of energy from the heat source to the heat sink.

The current work only considers non-intermittent centralised trigeneration system. It is envisioned that the TrigenSCA framework can be extended to cater for at least two other vital applications. First, it can be further developed for designing intermittent systems involving wind and solar energy. Secondly, it can be extended to consider transmission energy losses in Total Site system to achieve an actual value of utility in a trigeneration system.

Supplementary Materials: The following are available online at http://www.mdpi.com/1996-1073/12/6/1030/s1, Table S1: Problem Table Algorithm for Industrial Plant A, Table S2: Problem Table Algorithm for Industrial Plant B, Table S3: Problem Table Algorithm for Industrial Plant C, Table S4: Problem Table Algorithm for Industrial Plant D, Table S5: Multiple Utility Problem Table Algorithm for Industrial Plant A, Table S6: Multiple Utility Problem Table Algorithm for Industrial Plant B, Table S7: Multiple Utility Problem Table Algorithm for Industrial Plant C, Table S8: Multiple Utility Problem Table Algorithm for Industrial Plant D.

Author Contributions: Conceptualization, K.J., S.R.W.A., Z.A.M., J.J.K. and K.H.; methodology, K.J.; S.R.W.A., Z.A.M. and J.J.K.; formal analysis, K.J. and S.R.W.A.; investigation, K.J.; writing—original draft preparation, K.J.; writing—review and editing, S.R.W.A., Z.A.M., K.H. and J.J.K.; visualization, S.R.W.A., Z.A.M., K.H. and J.J.K.; supervision, K.H., S.R.W.A., Z.A.M. and J.J.K.; funding acquisition, S.R.W.A., Z.A.M., K.H. and J.J.K.

Funding: This research was funded by the Malaysia Ministry of Education (MOE), Fundamental Research Grant (FRGS) titled 'Integrated Industrial Site Planning for Minimising Resource Utilisation and Pollution' under vote number R.J130000.7809.4F918, Universiti Teknologi Malaysia Research University Fund under Vote Numbers Q.J130000.3509.05G96, Q.J130000.2509.19H34, Q.J130000.2508.17H16, and Q.J130000.2546.18H90; and Skim Latihan Akademik Bumiputera (SLAB) under MOE and by the EC project Sustainable Process Integration Laboratory–SPIL, funded as project No CZ.02.1.01/0.0/0.0/15 003/0000456 by the Czech Republic Operational Programme Research and Development, Education, Priority 1: Strengthening capacity for quality research under the collaboration agreement with the UTM, Johor Bahru, Malaysia.

Conflicts of Interest: The authors declare no conflict of interest.

Appendix A

Table A1. TriGenSCA before iteration from time 1 to 12 h.

1	2						3					
	Demand (MW)						Generation (MW)					
Time (h)	Power	Heating		Cooling			Power	Heating		Cooling		
		HPS	LPS	HW	CW	ChW		HPS	LPS	HW	CW	ChW
1	120	17	381.22	237.82	119.32	0.197	169.5	15	380	10.5	0.76	0.197
2	123	17	381.22	237.82	119.32	0.197	169.5	15	380	10.5	0.76	0.197
3	180	17	381.22	237.82	119.32	0.197	169.5	15	380	10.5	0.76	0.197
4	230	17	381.22	237.82	119.32	0.197	169.5	15	380	10.5	0.76	0.197
5	125	17	381.22	237.82	119.32	0.197	169.5	15	380	10.5	0.76	0.197
6	95	17	381.22	237.82	119.32	0.197	169.5	15	380	10.5	0.76	0.197
7	145	17	381.22	237.82	119.32	0.197	169.5	15	380	10.5	0.76	0.197
8	169	17	381.22	237.82	119.32	0.197	169.5	15	380	10.5	0.76	0.197
9	202	17	381.22	237.82	119.32	0.197	169.5	15	380	10.5	0.76	0.197
10	205	17	381.22	237.82	119.32	0.197	169.5	15	380	10.5	0.76	0.197
11	245	17	381.22	237.82	119.32	0.197	169.5	15	380	10.5	0.76	0.197
12	205	17	381.22	237.82	119.32	0.197	169.5	15	380	10.5	0.76	0.197

Table A2. TriGenSCA before iteration from time 13 to 24 h.

1	2						3					
	Demand (MW)						Generation (MW)					
Time (h)	Power	Heating		Cooling			Power	Heating		Cooling		
		HPS	LPS	HW	CW	ChW		HPS	LPS	HW	CW	ChW
13	225	17	381.22	237.82	119.32	0.197	169.5	15	380	10.5	0.76	0.197
14	163	17	381.22	237.82	119.32	0.197	169.5	15	380	10.5	0.76	0.197
15	210	17	381.22	237.82	119.32	0.197	169.5	15	380	10.5	0.76	0.197
16	165	17	381.22	237.82	119.32	0.197	169.5	15	380	10.5	0.76	0.197
17	118	17	381.22	237.82	119.32	0.197	169.5	15	380	10.5	0.76	0.197
18	113	17	381.22	237.82	119.32	0.197	169.5	15	380	10.5	0.76	0.197
19	123	17	381.22	237.82	119.32	0.197	169.5	15	380	10.5	0.76	0.197
20	142	17	381.22	237.82	119.32	0.197	169.5	15	380	10.5	0.76	0.197
21	200	17	381.22	237.82	119.32	0.197	169.5	15	380	10.5	0.76	0.197
22	205	17	381.22	237.82	119.32	0.197	169.5	15	380	10.5	0.76	0.197
23	170	17	381.22	237.82	119.32	0.197	169.5	15	380	10.5	0.76	0.197
24	190	17	381.22	237.82	119.32	0.197	169.5	15	380	10.5	0.76	0.197

Table A3. TriGenSCA before iteration from time 1 to 12 h.

1	4					5					
	Net Energy Requirement (MWh)					New Net Energy Requirement (MWh)					
Time (h)	Power	Heating			Cooling		Power	Heating			Cooling	
		HPS	LPS	HW	CW	ChW		HPS	LPS	HW	CW	ChW
1	49.5	−2	−1.22	−227.35	−118.55	0	49.5	−2	−1.22	−227.35	−118.55	0
2	46.5	−2	−1.22	−227.35	−118.55	0	46.5	−2	−1.22	−227.35	−118.55	0
3	−10.5	−2	−1.22	−227.35	−118.55	0	−10.5	−2	−1.22	−227.35	−118.55	0
4	−60.5	−2	−1.22	−227.35	−118.55	0	−60.5	−2	−1.22	−227.35	−118.55	0
5	44.5	−2	−1.22	−227.35	−118.55	0	44.5	−2	−1.22	−227.35	−118.55	0
6	74.5	−2	−1.22	−227.35	−118.55	0	74.5	−2	−1.22	−227.35	−118.55	0
7	24.5	−2	−1.22	−227.35	−118.55	0	24.5	−2	−1.22	−227.35	−118.55	0
8	0.5	−2	−1.22	−227.35	−118.55	0	0.5	−2	−1.22	−227.35	−118.55	0
9	−32.5	−2	−1.22	−227.35	−118.55	0	−32.5	−2	−1.22	−227.35	−118.55	0
10	−35.5	−2	−1.22	−227.35	−118.55	0	−35.5	−2	−1.22	−227.35	−118.55	0
11	−75.5	−2	−1.22	−227.35	−118.55	0	−75.5	−2	−1.22	−227.35	−118.55	0
12	−35.5	−2	−1.22	−227.35	−118.55	0	−35.5	−2	−1.22	−227.35	−118.55	0

Table A4. TriGenSCA before iteration from time 13 to 24 h.

1	4					5					
	Net Energy Requirement (MWh)					New Net Energy Requirement (MWh)					
Time (h)	Power	Heating			Cooling		Power	Heating			Cooling	
		HPS	LPS	HW	CW	ChW		HPS	LPS	HW	CW	ChW
13	−55.5	−2	−1.22	−227.35	−118.55	0	−55.5	−2	−1.22	−227.35	−118.55	0
14	6.5	−2	−1.22	−227.35	−118.55	0	6.5	−2	−1.22	−227.35	−118.55	0
15	−40.5	−2	−1.22	−227.35	−118.55	0	−40.5	−2	−1.22	−227.35	−118.55	0
16	4.5	−2	−1.22	−227.35	−118.55	0	4.5	−2	−1.22	−227.35	−118.55	0
17	51.5	−2	−1.22	−227.35	−118.55	0	51.5	−2	−1.22	−227.35	−118.55	0
18	56.5	−2	−1.22	−227.35	−118.55	0	56.5	−2	−1.22	−227.35	−118.55	0
19	46.5	−2	−1.22	−227.35	−118.55	0	46.5	−2	−1.22	−227.35	−118.55	0
20	27.5	−2	−1.22	−227.35	−118.55	0	27.5	−2	−1.22	−227.35	−118.55	0
21	−30.5	−2	−1.22	−227.35	−118.55	0	−30.5	−2	−1.22	−227.35	−118.55	0
22	−35.5	−2	−1.22	−227.35	−118.55	0	−35.5	−2	−1.22	−227.35	−118.55	0
23	−0.5	−2	−1.22	−227.35	−118.55	0	−0.5	−2	−1.22	−227.35	−118.55	0
24	−20.5	−2	−1.22	−227.35	−118.55	0	−20.5	−2	−1.22	−227.35	−118.55	0

Table A5. TriGenSCA before iteration from time 1 to 12 h.

1	6					7					
	Charging (+) and Discharging (−) Energies (MWh)					Cumulative Energy (MWh)					
Time (h)	Power	Heating			Cooling		Power	Heating			Cooling	
		HPS	LPS	HW	CW	ChW		HPS	LPS	HW	CW	ChW
							0	0	0	0	0	0
1	40.10	−3.45	−2.10	−391.98	−68.76	0						
							40.10	−3.45	−2.10	−391.98	−68.76	0
2	37.67	−3.45	−2.10	−391.98	−68.76	0						
							77.76	−6.89	−4.20	−783.96	−137.52	0
3	−12.96	−3.45	−2.10	−391.98	−68.76	0						
							64.79	−10.35	−6.30	−1175.94	−206.28	0
4	−74.69	−3.45	−2.10	−391.98	−68.76	0						
							−9.89	−13.79	−8.41	−1567.93	−275.04	0
5	36.05	−3.45	−2.10	−391.98	−68.76	0						
							26.15	−17.24	−10.51	−1959.91	−343.81	0
6	60.35	−3.45	−2.10	−391.98	−68.76	0						
							86.50	−20.69	−12.61	−2351.89	−412.57	0
7	19.85	−3.45	−2.10	−391.98	−68.76	0						
							106.34	−24.14	−14.71	−2743.87	−481.33	0
8	0.41	−3.45	−2.10	−391.98	−68.76	0						
							106.75	−27.59	−16.81	−3135.85	−550.09	0

Table A5. Cont.

1	6					7					
	Charging (+) and Discharging (−) Energies (MWh)					Cumulative Energy (MWh)					
Time (h)	Power	Heating		Cooling		Power	Heating			Cooling	
		HPS	LPS	HW	CW	ChW		HPS	LPS	HW	CW	ChW
9	−40.12	−3.45	−2.10	−391.98	−68.76	0						
10	−43.83	−3.45	−2.10	−391.98	−68.76	0	66.62	−31.04	−18.91	−3527.83	−618.85	0
11	−93.21	−3.45	−2.10	−391.98	−68.76	0	22.79	−34.48	−21.01	−3919.81	−687.61	0
12	−43.83	−3.45	−2.10	−391.98	−68.76	0	−70.42	−37.93	−23.11	−4311.8	−756.37	0

Table A6. TriGenSCA before iteration from time 13 to 24 h.

1	6					7					
	Charging (+) and Discharging (−) Energies (MWh)					Cumulative Energy (MWh)					
Time (h)	Power	Heating		Cooling		Power	Heating			Cooling	
		HPS	LPS	HW	CW	ChW		HPS	LPS	HW	CW	ChW
							−114.24	−41.38	−25.22	−4,703.78	−825.13	0
13	−68.52	−3.45	−2.10	−391.98	−68.76	0	−182.76	−44.83	−27.32	−5,095.76	−893.89	0
14	5.27	−3.45	−2.10	−391.98	−68.76	0	−177.50	−48.28	−29.42	−5,487.74	−962.65	0
15	−50	−3.45	−2.10	−391.98	−68.76	0	−227.50	−51.72	−31.52	−5,879.72	−1031.41	0
16	3.65	−3.45	−2.10	−391.98	−68.76	0	−223.85	−55.17	−33.62	−6,271.7	−1,100.17	0
17	41.72	−3.45	−2.10	−391.98	−68.76	0	−182.14	−58.62	−35.72	−6,663.68	−1,168.94	0
18	45.77	−3.45	−2.10	−391.98	−68.76	0	−136.37	−62.07	−37.82	−7,055.66	−1,237.7	0
19	37.66	−3.45	−2.10	−391.98	−68.76	0	−98.71	−65.52	−39.92	−7,447.65	−1,306.46	0
20	22.28	−3.45	−2.10	−391.98	−68.76	0	−76.43	−68.97	−42.02	−7,839.63	−1,375.22	0
21	−37.65	−3.45	−2.10	−391.98	−68.76	0	−114.09	−72.41	−44.13	−8,231.61	−1,443.98	0
22	−43.83	−3.45	−2.10	−391.98	−68.76	0	−157.91	−75.86	−46.23	−8,623.59	−1,512.74	0
23	−0.62	−3.45	−2.10	−391.98	−68.76	0	−158.53	−79.31	−48.33	−9,015.57	−1,581.5	0
24	−25.31	−3.45	−2.10	−391.98	−68.76	0	−183.84	−82.76	−50.43	−9,407.55	−1,650.26	0

Table A7. TriGenSCA before iteration from 1 to 12 h.

1	8					
	New Cumulative Energy (MWh)					
Time (h)	Power	Heating			Cooling	
		HPS	LPS	HW	CW	ChW
1	227.49	82.76	50.43	9,407.55	1,650.26	0
2	267.59	79.31	48.33	9,015.57	1,581.50	0
3	305.26	75.86	46.23	8,623.59	1,512.74	0
4	292.29	72.41	44.13	8,231.61	1,443.98	0
	217.60	68.97	42.02	7,839.63	1,375.22	0

Table A7. Cont.

Time (h)	1	8				
		New Cumulative Energy (MWh)				
	Power	Heating			Cooling	
		HPS	LPS	HW	CW	ChW
5	253.65	65.52	39.92	7,447.65	1,306.46	0
6	313.99	62.07	37.82	7,055.67	1,237.69	0
7	333.84	58.62	35.72	6,663.68	1,168.94	0
8	334.24	55.17	33.62	6,271.70	1,100.18	0
9	294.19	51.72	31.52	5,879.72	1,031.41	0
10	250.29	48.28	29.42	5,487.74	962.65	0
11	157.08	44.83	27.32	5,095.76	893.89	0
12						

Table A8. TriGenSCA before iteration from 13 to 24 h.

Time (h)	1	8				
		New Cumulative Energy (MWh)				
	Power	Heating			Cooling	
		HPS	LPS	HW	CW	ChW
13	113.25	41.38	25.21	4703.78	825.13	0
14	44.74	37.93	23.11	4,311.79	756.37	0
15	50	34.48	21.01	3,919.81	687.61	0
16	0	31.03	18.91	3,527.83	618.85	0
17	3.645	27.59	16.81	3,135.85	550.09	0
18	45.36	24.14	14.71	2,743.87	481.33	0
19	91.13	20.69	12.61	2,351.89	412.57	0
20	128.79	17.24	10.51	1,959.91	343.81	0
21	151.07	13.79	8.41	1,567.93	275.04	0
22	113.41	10.35	6.30	1,175.94	206.28	0
23	69.58	6.89	4.20	783.96	137.52	0
24	68.97	3.45	2.10	391.98	68.76	0
	43.66	0	0	0	0	0

Appendix B

Table A9. TriGenSCA after iteration from 1 to 12 h.

1	2						3					
	Demand (MW)						Generation (MW)					
Time (h)	Power	Heating			Cooling		Power	Heating			Cooling	
		HPS	LPS	HW	CW	ChW		HPS	LPS	HW	CW	ChW
1	120	17	381.22	237.82	119.32	0.197	176.79	17.02	381.63	302.68	69.53	0.197
2	123	17	381.22	237.82	119.32	0.197	176.79	17.02	381.63	302.68	69.53	0.197
3	180	17	381.22	237.82	119.32	0.197	176.79	17.02	381.63	302.68	69.53	0.197
4	230	17	381.22	237.82	119.32	0.197	176.79	17.02	381.63	302.68	69.53	0.197
5	125	17	381.22	237.82	119.32	0.197	176.79	17.02	381.63	302.68	69.53	0.197
6	95	17	381.22	237.82	119.32	0.197	176.79	17.02	381.63	302.68	69.53	0.197
7	145	17	381.22	237.82	119.32	0.197	176.79	17.02	381.63	302.68	69.53	0.197
8	169	17	381.22	237.82	119.32	0.197	176.79	17.02	381.63	302.68	69.53	0.197
9	202	17	381.22	237.82	119.32	0.197	176.79	17.02	381.63	302.68	69.53	0.197
10	205	17	381.22	237.82	119.32	0.197	176.79	17.02	381.63	302.68	69.53	0.197
11	245	17	381.22	237.82	119.32	0.197	176.79	17.02	381.63	302.68	69.53	0.197
12	205	17	381.22	237.82	119.32	0.197	176.79	17.02	381.63	302.68	69.53	0.197

Table A10. TriGenSCA after iteration from 13 to 24 h.

1	2						3					
	Demand (MW)						Generation (MW)					
Time (h)	Power	Heating			Cooling		Power	Heating			Cooling	
		HPS	LPS	HW	CW	ChW		HPS	LPS	HW	CW	ChW
13	225	17	381.22	237.82	119.32	0.197	176.79	17.02	381.63	302.68	69.53	0.197
14	163	17	381.22	237.82	119.32	0.197	176.79	17.02	381.63	302.68	69.53	0.197
15	210	17	381.22	237.82	119.32	0.197	176.79	17.02	381.63	302.68	69.53	0.197
16	165	17	381.22	237.82	119.32	0.197	176.79	17.02	381.63	302.68	69.53	0.197
17	118	17	381.22	237.82	119.32	0.197	176.79	17.02	381.63	302.68	69.53	0.197
18	113	17	381.22	237.82	119.32	0.197	176.79	17.02	381.63	302.68	69.53	0.197
19	123	17	381.22	237.82	119.32	0.197	176.79	17.02	381.63	302.68	69.53	0.197
20	142	17	381.22	237.82	119.32	0.197	176.79	17.02	381.63	302.68	69.53	0.197
21	200	17	381.22	237.82	119.32	0.197	176.79	17.02	381.63	302.68	69.53	0.197
22	205	17	381.22	237.82	119.32	0.197	176.79	17.02	381.63	302.68	69.53	0.197
23	170	17	381.22	237.82	119.32	0.197	176.79	17.02	381.63	302.68	69.53	0.197
24	190	17	381.22	237.82	119.32	0.197	176.79	17.02	381.63	302.68	69.53	0.197

Table A11. TriGenSCA after iteration from 1 to 12 h.

1	4						5					
	Net Energy Requirement (MWh)						New Net Energy Requirement (MWh)					
Time (h)	Power	Heating			Cooling		Power	Heating			Cooling	
		HPS	LPS	HW	CW	ChW		HPS	LPS	HW	CW	ChW
1	56.79	0.02	0.42	64.83	−49.79	0	56.79	0.02	0.42	0.10	0	0
2	53.79	0.02	0.42	64.83	−49.79	0	53.79	0.02	0.42	0.10	0	0
3	−3.21	0.02	0.42	64.83	−49.79	0	−3.06	0	0	0.10	0	0
4	−53.21	0.02	0.42	64.83	−49.79	0	−53.06	0	0	0.10	0	0
5	51.79	0.02	0.42	64.83	−49.79	0	51.79	0.02	0.42	0.10	0	0
6	81.79	0.02	0.42	64.83	−49.79	0	81.79	0.02	0.42	0.10	0	0
7	31.79	0.02	0.42	64.83	−49.79	0	31.79	0.02	0.42	0.10	0	0
8	7.79	0.02	0.42	64.83	−49.79	0	7.79	0.02	0.42	0.10	0	0
9	−25.21	0.02	0.42	64.83	−49.79	0	−25.06	0	0	0.10	0	0
10	−28.21	0.02	0.42	64.83	−49.79	0	−28.06	0	0	0.10	0	0
11	−68.21	0.02	0.42	64.83	−49.79	0	−68.06	0	0	0.10	0	0
12	−28.21	0.02	0.42	64.83	−49.79	0	−28.06	0	0	0.10	0	0

Table A12. TriGenSCA after iteration from 13 to 24 h.

1	4					5					
	Net Energy Requirement (MWh)					New Net Energy Requirement (MWh)					
Time (h)	Power	Heating			Cooling		Power	Heating			Cooling	
		HPS	LPS	HW	CW	ChW		HPS	LPS	HW	CW	ChW
13	−48.21	0.02	0.42	64.83	−49.79	0	−48.06	0	0	0.10	0	0
14	13.79	0.02	0.42	64.83	−49.79	0	13.79	0.02	0.42	0.10	0	0
15	−33.21	0.02	0.42	64.83	−49.79	0	−33.06	0	0	0.10	0	0
16	11.79	0.02	0.42	64.83	−49.79	0	11.79	0.02	0.42	0.10	0	0
17	58.79	0.02	0.42	64.83	−49.79	0	58.79	0.02	0.42	0.10	0	0
18	63.79	0.02	0.42	64.83	−49.79	0	63.79	0.02	0.42	0.10	0	0
19	53.79	0.02	0.42	64.83	−49.79	0	53.79	0.02	0.42	0.10	0	0
20	34.79	0.02	0.42	64.83	−49.79	0	34.79	0.02	0.42	0.10	0	0
21	−23.21	0.02	0.42	64.83	−49.79	0	−23.06	0	0	0.10	0	0
22	−28.21	0.02	0.42	64.83	−49.79	0	−28.06	0	0	0.10	0	0
23	6.79	0.02	0.42	64.83	−49.79	0	6.79	0.02	0.42	0.10	0	0
24	−13.21	0.02	0.42	64.83	−49.79	0	−13.06	0	0	0.10	0	0

Table A13. TriGenSCA after iteration from 1 to 12 h.

1	6						7					
	Charging (+) and Discharging (−) Energy (MWh)						Cumulative Energy (MWh)					
Time (h)	Power	Heating			Cooling		Power	Heating			Cooling	
		HPS	LPS	HW	CW	ChW		HPS	LPS	HW	CW	ChW
							0	0	0	0	0	0
1	46	0.01	0.24	0.06	0	0						
							46	0.01	0.24	0.06	0	0
2	43.57	0.01	0.24	0.06	0	0						
							89.58	0.02	0.48	0.11	0	0
3	−3.78	0	0	0.06	0	0						
							85.79	0.02	0.48	0.17	0	0
4	−65.51	0	0	0.06	0	0						
							20.29	0.02	0.48	0.22	0	0
5	41.95	0.01	0.24	0.06	0	0						
							62.24	0.03	0.72	0.28	0	0
6	66.25	0.01	0.24	0.06	0	0						
							128.49	0.04	0.96	0.34	0	0
7	25.75	0.01	0.24	0.06	0	0						
							154.25	0.05	1.20	0.39	0	0
8	6.31	0.01	0.24	0.06	0	0						
							160.56	0.06	1.45	0.45	0	0
9	−30.94	0	0	0.06	0	0						
							129.62	0.06	1.45	0.51	0	0
10	−34.65	0	0	0.06	0	0						
							94.97	0.06	1.45	0.56	0	0
11	−84.03	0	0	0.06	0	0						
							10.94	0.06	1.45	0.62	0	0
12	−34.65	0	0	0.06	0	0						

Table A14. TriGenSCA after iteration from 13 to 24 h.

1	6						7					
	Charging (+) and Discharging (−) Energy (MWh)						Cumulative Energy (MWh)					
Time (h)	Power	Heating			Cooling		Power	Heating			Cooling	
		HPS	LPS	HW	CW	ChW		HPS	LPS	HW	CW	ChW
							−23.7	0.06	1.45	0.67	0	0
13	−59.34	0	0	0.06	0	0						
							−83.04	0.06	1.45	0.73	0	0

Table A14. Cont.

1	6						7					
	Charging (+) and Discharging (−) Energy (MWh)						Cumulative Energy (MWh)					
Time (h)	Power	Heating			Cooling		Power	Heating			Cooling	
		HPS	LPS	HW	CW	ChW		HPS	LPS	HW	CW	ChW
14	11.17	0.01	0.24	0.06	0	0	−71.87	0.07	1.69	0.79	0	0
15	−40.82	0	0	0.06	0	0	−112.68	0.07	1.69	0.84	0	0
16	9.55	0.01	0.24	0.06	0	0	−103.13	0.08	1.93	0.9	0	0
17	47.62	0.01	0.24	0.06	0	0	−55.51	0.08	2.17	0.95	0	0
18	51.67	0.01	0.24	0.06	0	0	−3.83	0.09	2.41	1.01	0	0
19	43.57	0.01	0.24	0.06	0	0	39.74	0.1	2.65	1.07	0	0
20	28.18	0.01	0.24	0.06	0	0	67.93	0.11	2.89	1.12	0	0
21	−28.47	0	0	0.06	0	0	39.45	0.11	2.89	1.18	0	0
22	−34.65	0	0	0.06	0	0	4.81	0.11	2.89	1.23	0	0
23	8.39	0.01	0.24	0.06	0	0	13.19	0.12	3.13	1.29	0	0
24	−16.13	0	0	0.06	0	0	−2.93	0.12	3.13	1.35	0	0

Table A15. TriGenSCA after iteration from 1 to 12 h.

1	8					
	New Cumulative Energy (MWh)					
Time (h)	Power	Heating			Cooling	
		HPS	LPS	HW	CW	ChW
1	112.68	0	0	0	0	0
2	158.69	0.01	0.24	0.06	0	0
3	202.26	0.02	0.48	0.11	0	0
4	198.48	0.02	0.48	0.17	0	0
5	132.97	0.02	0.48	0.22	0	0
6	174.92	0.03	0.72	0.28	0	0
7	241.18	0.04	0.96	0.34	0	0
8	266.93	0.05	1.2	0.39	0	0
9	273.24	0.06	1.45	0.45	0	0
10	242.3	0.06	1.45	0.51	0	0
11	207.66	0.06	1.45	0.56	0	0
12	123.63	0.06	1.45	0.62	0	0

Table A16. TriGenSCA after iteration from 13 to 24 h.

Time (h)	Power	Heating			Cooling	
		HPS	LPS	HW	CW	ChW
		New Cumulative Energy (MWh)				
13	88.98	0.06	1.45	0.67	0	0
14	29.65	0.06	1.45	0.73	0	0
15	40.82	0.07	1.69	0.79	0	0
16	0	0.07	1.69	0.84	0	0
17	9.55	0.08	1.93	0.9	0	0
18	57.18	0.08	2.17	0.95	0	0
19	108.85	0.09	2.41	1.01	0	0
20	152.43	0.1	2.65	1.07	0	0
21	180.61	0.11	2.89	1.12	0	0
22	152.14	0.11	2.89	1.18	0	0
23	117.49	0.11	2.89	1.23	0	0
24	125.88	0.12	3.13	1.29	0	0
	109.75	0.12	3.13	1.35	0	0

Table A17. TriGenSCT after iteration from 1 to 12 h.

Time (h)	Storage Capacity (MWh)						Outsourced Energy Needed (MWh)					
	Power	Heating			Cooling		Power	Heating			Cooling	
		HPS	LPS	HW	CW	ChW		HPS	LPS	HW	CW	ChW
1	46	0.01	0.24	0.06								
2	89.57	0.02	0.48	0.12								
3	85.78	0.02	0.48	0.18								
4	20.26	0.02	0.48	0.23								
5	62.21	0.03	0.72	0.29								
6	128.46	0.04	0.95	0.35								
7	154.21	0.05	1.19	0.41								
8	160.52	0.06	1.43	0.47								
9	129.57	0.06	1.43	0.53								
10	94.92	0.06	1.43	0.59								
11	10.89	0.06	1.43	0.64								
12		0.06	1.43	0.7			−23.77					

Table A18. TriGenSCT after iteration from time 13 to 24 h.

1	7						8					
	Storage Capacity (MWh)						Outsourced Energy Needed (MWh)					
Time (h)	Power	Heating			Cooling		Power	Heating			Cooling	
		HPS	LPS	HW	CW	ChW		HPS	LPS	HW	CW	ChW
13		0.06	1.43	0.76			−59.34					
14	11.17	0.07	1.67	0.82								
15		0.07	1.67	0.88			−29.66					
16	9.55	0.07	1.91	0.94								
17	57.17	0.08	2.15	0.1								
18	108.84	0.09	2.39	1.05								
19	152.41	0.1	2.62	1.11								
20	180.59[a]	0.11	2.86	1.17								
21	152.11	0.11	2.86	1.23								
22	117.46	0.11	2.86	1.29								
23	125.84	0.12	3.1	1.35								
24	109.7[b]	0.12[c,d]	3.1[c,d]	1.41[c,e]								
	Total external energy needed						−112.7					

[a] maximum power energy storage; [b] power supply to the next day operation; [c] maximum thermal energy storage; [d] convert to power; [e] to steam generator.

References

1. International Energy Agency. *World Energy Outlook 2017*; International Energy Agency: Paris, France, 2018.
2. Zhang, H.I.; Baeyens, J.; Degreve, J.; Caceres, G. Concentrated solar power plants: Review and design methodology. *Renew. Sustain. Energy Rev.* **2013**, *22*, 466–481. [CrossRef]
3. Abdul Manan, Z.; Mohd Nawi, W.N.R.; Wan Alwi, S.R.; Klemeš, J.J. Advances in Process Integration research for CO_2 emission reduction—A review. *J. Clean. Prod.* **2017**, *167*, 1–13. [CrossRef]
4. Khamis, I.; Koshy, T.; Kavvadias, K.C. Opportunity for cogeneration in Nuclear Power Plants. In Proceedings of the 2013 World Congress on Advances in Nano Biomechanics, Robotics, and Energy Research, Seoul, Korea, 25–28 August 2013; pp. 455–462.
5. Klemeš, J.J.; Varbanov, P.S.; Walmsley, T.G.; Jia, X. New directions in the implementation of Pinch Methodology (PM). *Renew. Sustain. Energy Rev.* **2018**, *98*, 439–468. [CrossRef]
6. Wan Alwi, S.R.; Mohammad Rozali, N.E.; Abdul Manan, Z.; Klemeš, J.J. A process integration targeting method for hybrid power systems. *Energy* **2012**, *44*, 6–10. [CrossRef]
7. Mohammad Rozali, N.E.; Wan Alwi, S.R.; Manan, Z.A.; Klemeš, J.J.; Hassan, M.Y. Process Integration techniques for optimal design of hybrid power systems. *Appl. Therm. Eng.* **2013**, *61*, 26–35. [CrossRef]
8. Ho, W.S.; Hashim, H.; Muis, Z.A.; Shamsuddin, N.L.M. Design of distributed energy system through Electric System Cascade Analysis (ESCA). *Appl. Energy* **2012**, *99*, 309–315. [CrossRef]
9. Liu, W.H.; Wan Alwi, S.R.; Hashim, H.; Muis, Z.A.; Klemeš, J.J.; Mohammad Rozali, N.E.; Lim, J.S.; Ho, W.S. Optimal design and sizing of integrated centralized and decentralized energy systems. *Energy Procedia* **2017**, *105*, 3733–3740. [CrossRef]
10. Jamaluddin, K.; Wan Alwi, S.R.; Manan, Z.A.; Klemeš, J.J. Pinch Analysis Methodology for trigeneration with energy storage system design. *Chem. Eng. Trans.* **2018**, *70*, 1885–1890.
11. Jamaluddin, K.; Wan Alwi, S.R.; Abdul Manan, Z.; Hamzah, K.; Klemeš, J.J. Hybrid power systems design considering safety and resilience. *Process Saf. Environ. Prot.* **2018**, *120*, 256–267. [CrossRef]
12. Hoang, T.-V.; Ifaei, P.; Nam, K.; Rashidi, J.; Hwangbo, S.; Oh, J.-M.; Yoo, C.K. Optimal management of a hybrid renewable energy system coupled with a membrane bioreactor using enviro-economic and power pinch analyses for sustainable climate change adaption. *Sustainability* **2019**, *11*, 66. [CrossRef]
13. Dhole, V.R.; Linnhoff, B. Total Site targets for fuel, co-generation, emissions, and cooling. *Comput. Chem. Eng.* **1993**, *17* (Suppl. 1), S101–S109. [CrossRef]
14. Klemeš, J.J.; Dhole, V.R.; Raissi, K.; Perry, S.J.; Puigjaner, L. Targeting and design methodology for reduction of fuel, power and CO_2 on total site. *Appl. Therm. Eng.* **1997**, *7*, 993–1003. [CrossRef]

15. Perry, S.; Klemeš, J.J.; Bulatov, I. Integrating waste and renewable energy to reduce the carbon footprint of locally integrated energy sectors. *Energy* **2008**, *33*, 1489–1497. [CrossRef]
16. Matsuda, K.; Hirochi, Y.; Tatsumi, H.; Shire, T. Applying heat integration total site based pinch technology to a large industrial area in Japan to further improve performance of highly efficient process plants. *Energy* **2009**, *34*, 1687–1692. [CrossRef]
17. Varbanov, P.S.; Klemeš, J.J. Total site integrating renewables with extended heat transfer and recovery. *Heat Transf. Eng.* **2010**, *31*, 733–741. [CrossRef]
18. Varbanov, P.S.; Klemeš, J.J. Integration and management of renewables into total slice with variable supply and demand. *Comput. Chem. Eng.* **2011**, *35*, 1815–1826. [CrossRef]
19. Liew, P.Y.; Wan Alwi, S.R.; Varbanov, P.S.; Manan, Z.A.; Klemeš, J.J. A numerical technique for Total Site sensitivity analysis. *Appl. Therm. Eng.* **2012**, *40*, 397–408. [CrossRef]
20. Shamsi, S.; Omidkhah, M.R. Optimization of steam pressure levels in a Total Site using a thermoeconomic method. *Energies* **2012**, *5*, 702–717. [CrossRef]
21. Chew, K.H.; Klemeš, J.J.; Wan Alwi, S.R.; Abdul Manan, Z.; Reverberi, A.P. Total Site Heat Integration considering pressure drops. *Energies* **2015**, *8*, 1114–1137. [CrossRef]
22. Klemeš, J.J.; Varbanov, P.S.; Wan Alwi, S.R.; Abdul Manan, Z. *Process Saving Energy Integration and Intensification*, 2nd ed.; Water and Resources, De Gruyter: Berlin, Germany, 2018.
23. Raissi, K. Total Site Integration. PhD Thesis, UMIST, Manchester, UK, 1994.
24. Varbanov, P.S.; Doyle, S.; Smith, R. Modelling and optimisation of utility systems. *Chem. Eng. Res. Des.* **2004**, *82*, 561–578. [CrossRef]
25. Boldyryev, S.; Varbanov, P.S.; Nemet, A.; Klemeš, J.J.; Kapustenko, P. Capital cost assessment for Total Site power cogeneration. *Comput. Aided Chem. Eng.* **2013**, *32*, 361–366.
26. Liew, P.Y.; Wan Alwi, S.R.; Varbanov, P.S.; Manan, Z.A.; Klemeš, J.J. Centralised utility system planning for a Total Site Heat Integration network. *Comput. Chem. Eng.* **2013**, *57*, 104–111. [CrossRef]
27. Liew, P.Y.; Walmsley, T.G.; Wan Alwi, S.R.; Manan, Z.A.; Klemeš, J.J.; Varbanov, P.S. Integrating district cooling systems in Locally Integrated Energy Sectors through Total Site Heat Integration. *Appl. Energy* **2016**, *184*, 1350–1363. [CrossRef]
28. Tarighaleslami, A.H.; Walmsley, T.G.; Atkins, M.J.; Walmsley, M.R.W.; Neale, J.R. Total Site Heat Integration: Utility selection and optimisation using cost and exergy derivative analysis. *Energy* **2017**, *141*, 949–963. [CrossRef]
29. Ren, X.Y.; Jia, X.X.; Varbanov, P.S.; Klemeš, J.J.; Liu, Z.Y. Targeting the cogeneration potential for Total Site utility systems. *J. Clean. Prod.* **2018**, *170*, 625–635. [CrossRef]
30. Primohamadi, A.; Ghazi, M.; Nikkian, M. Optimal design of cogeneration systems in total site using exergy approach. *Energy* **2019**, *166*, 1291–1302. [CrossRef]
31. Linnhoff, B.; Flower, J.R. Synthesis of heat exchanger networks: I. Systematic generation of energy optimal networks. *AIChE J.* **1978**, *24*, 633–642. [CrossRef]
32. Costa, A.L.H.; Queiroz, E.M. An extension of the problem table algorithm for multiple utilities targeting. *Energy Convers. Manag.* **2009**, *50*, 1124–1128. [CrossRef]
33. Shenoy, U.V. Unified targeting algorithm for diverse process integration problems of resource conservation networks. *Chem. Eng. Resour. Des.* **2011**, *89*, 2686–2705. [CrossRef]
34. Liew, P.Y.; Wan Alwi, S.R.; Ho, W.S.; Manan, Z.A.; Varbanov, P.S.; Klemeš, J.J. Multi-period energy targeting for Total Site and Locally Integrated Energy Sectors with cascade Pinch Analysis. *Energy* **2018**, *155*, 370–380. [CrossRef]
35. Pitcher, J. Part 1: How Heat Loads Affect Evaporative Cooling Tower Efficiency. Available online: https://www.flowcontrolnetwork.com/part-i-how-heat-loads-affect-evaporative-cooling-tower-efficiency/ (accessed on 30 December 2018).
36. ARANER. How Do Absorption Chillers Work? Available online: https://www.araner.com/blog/how-do-absorption-chillers-work/ (accessed on 30 December 2018).
37. Dondi, P.; Bayoumi, D.; Haederli, C.; Julian, D.; Suter, M. Network integration of distributed power generation. *J. Power Sources* **2002**, *106*, 1–9. [CrossRef]
38. Pomianowski, M.; Heiselberg, P.; Zhang, Y. Review of thermal energy storage technologies based on PCM application in buildings. *Energy Build.* **2013**, *67*, 56–69. [CrossRef]

39. Types of Steam Turbine. Available online: https://www.sciencedirect.com/topics/engineering/types-of-steam-turbine (accessed on 8 January 2019).
40. Zhu, K.; Chen, X.; Dai, B.; Wang, Y.; Li, X.; Li, L. Experimental study on the thermal performance improvement of a new designed condenser with liquid separator. *Energy Procedia* **2016**, *104*, 269–274. [CrossRef]
41. Regulagadda, P.; Dincer, I.; Naterer, G.F. Exergy analysis of a thermal power plant with measured boiler and turbine losses. *Appl. Therm. Eng.* **2010**, *30*, 970–976. [CrossRef]
42. Srikirin, P.; Aphornratana, S.; Chungpalbulpatana, S. A review of absorption refrigeration technologies. *Renew. Sustain. Energy Rev.* **2001**, *5*, 343–372. [CrossRef]
43. Hobby, J.D.; Tucci, G.H. Analysis of the residential, commercial and industrial electricity consumption. In Proceedings of the 2011 IEEE PES Innovative Smart Grid Technologies, Perth, WA, Australia, 13–16 November 2011.
44. Wu, D.W.; Wang, R.Z. Combined cooling, heating and power: A review. *Prog. Energy Combust. Sci.* **2006**, *32*, 459–495. [CrossRef]
45. Iglesias, F.; Palensky, P.; Cantos, S.; Kupzog, F. Demand side management for stand-alone hybrid power systems based on load identification. *Energies* **2012**, *5*, 4517–4532. [CrossRef]
46. European Nuclear Society. Fuel Comparison. Available online: https://www.euronuclear.org/info/encyclopedia/f/fuelcomparison.htm (accessed on 30 December 2018).
47. Adams, R. Nuclear Energy Is Cheap and Disruptive; Controlling the Initial Cost of Nuclear Power Plants Is a Solvable Problem. Available online: https://atomicinsights.com/nuclear-energy-is-cheap-and-disruptive-controlling-the-initial-cost-of-nuclear-power-plants-is-a-solvable-problem/ (accessed on 11 January 2019).
48. Woite, G. Capital Investment Costs of Nuclear Power Plants. Available online: https://www.iaea.org/sites/default/files/publications/magazines/bulletin/bull20-1/20104781123.pdf (accessed on 11 January 2019).
49. Zhang, H.; Baeyens, J.; Caceres, G.; Degreve, J.; Lv, Y. Thermal energy storage: Recent developments and practical aspects. *Prog. Energy Combust. Sci.* **2016**, *53*, 1–40. [CrossRef]
50. Zhang, H.F.; Ge, X.S.; Ye, H. Modelling of a space heating and cooling system with seasonal energy storage. *Energy* **2007**, *32*, 51–58. [CrossRef]
51. HOMER. Excess Electricity. Available online: https://www.homerenergy.com/products/pro/docs/3.10/excess_electricity.html (accessed on 11 January 2019).

© 2019 by the authors. Licensee MDPI, Basel, Switzerland. This article is an open access article distributed under the terms and conditions of the Creative Commons Attribution (CC BY) license (http://creativecommons.org/licenses/by/4.0/).

Article

Problems Related to Gasification of Biomass—Properties of Solid Pollutants in Raw Gas

Jan Najser [1], Petr Buryan [2], Sergej Skoblia [2], Jaroslav Frantik [1], Jan Kielar [1] and Vaclav Peer [1,*]

[1] ENET Centre, VSB-Technical University of Ostrava, 17. listopadu 15/2172, 70833 Ostrava-Poruba, Czech Republic; jan.najser@vsb.cz (J.N.); jaroslav.frantik@vsb.cz (J.F.); jan.kielar@vsb.cz (J.K.)
[2] Department of Gaseous and Solid Fuels and Air Protection, UCT Prague, Technicka 5, 16628 Prague 6-Dejvice, Czech Republic; petr.buryan@vscht.cz (P.B.); sergej.skoblia@vscht.cz (S.S.)
* Correspondence: vaclav.peer@vsb.cz; Tel.: +42-0597-327-156

Received: 31 December 2018; Accepted: 8 March 2019; Published: 13 March 2019

Abstract: Nowadays, thermochemical biomass conversion appears to be a very promising way to process heat and steam generation, for use in a cogeneration unit engine, or for example in gas turbines producing electrical energy. The biggest problem regarding using the syngas in internal combustion engines, are pollutants, which have quite an inauspicious influence on their proper working. This article deals with the establishment of the distribution size of solid particles captured by the fiber filters in the syngas with a suitable cleaning design. Gas was produced in the fixed-bed "Imbert" type generator. Filter cake, which contained pollutants, was captured on a filter and then analyzed. Based on single total solid particles (TSP) components, we conclude that this includes its partial elimination.

Keywords: gasification; biomass; total solid particle

1. Gasification of Biomass

In recent years, biomass, as a renewable source of energy, has attracted rising attention around the world [1].

The thermochemical transformation of biomass (pyrolysis, gasification) seems to be one of the most promising routes towards the utilization of renewable sources of energy to obtain energy. These renewable forms of energy present many ecological benefits. With respect to thermochemical conversion technologies, fixed-bed [2] or fluidised bed [3] gasification attracts the greatest attention because it offers higher process efficiency than combustion and pyrolysis.

The process of conversion of solid carbonaceous fuel to flammable gas by partial oxidation is known as gasification. The resultant gas, which is sometimes called process gas, is much more versatile in its use than the original solid biomass [4]. The great advantages are that the synthesis gas can be stored and transported, and its final use can be independent of the production process [5]. During final processing the synthesis gas can be combusted to produce process heat and steam or used in gas turbines for the production of electricity.

In terms of a gasification medium, the technology can be divided into gasification by air, oxygen, water vapors, CO_2, or other various combinations [6].

2. Problems of Gasification Technologies and Their Solutions

Currently, during the development of new technologies for transforming the energy content of biomass as a renewable energy source to its more sophisticated form, attention is focused on the development of new technological equipment for more efficient gasification, or thermochemical conversion. In this technological segment, the majority of research work is aimed at the cleaning of raw generator gas, because tar and solid pollutants (SP) present in the raw gas are the main undesirable components in the gas, limiting its use in combustion motors or cogeneration unit turbines [7].

The increase in production efficiency sets ever growing demands for gas cleaning [8]. The benefits of gasification also include the possible reduction of emissions, emissions of compounds of Sulphur, chlorine, and nitrogen, but also persistent organic pollutants (POP), e.g., polychlorinated benzodioxines and benzodifuranes (PCDD, PCDF). Reduction of emissions is achieved by removing these compounds and their precursors directly from the produced gas prior to combustion. The volume of gas is smaller compared to the volume of produced flue gases; the concentration of pollutants is higher, and more efficient removal in smaller technological equipment is possible. Pollutants are present in a reduced form; their aggressiveness towards equipment is significantly lower.

2.1. Undesirable Substances

The choice of a suitable procedure for cleaning depends on the technology selected for its properties and content of undesirable components [9]. Undesirable components are formed from inorganic components of fuel or incomplete conversion of material. They cause abrasion, corrosion, formation of sediments, and degradation reactions, e.g., in catalyzers, or they represent ecological loads [8].

The volume of contaminating substances in the gas generated during gasification is directly dependent on the contamination of solid fuel. The volume of undesirable substances in biomass and the conversion process can be significant, e.g., N, P, K, Si, Ca, Mg, S, Na, Cl [6]. The main portion of heavy and alkali metals (potassium 80%) tend to stay in the solid phase during fuel conversion [10].

The following pollutants are formed during the gasification process:

- Solid pollutants (SP) are defined as substances in the resultant gas formed by unreacted biomass fraction and ash matter (inorganic substances). A certain proportion is also formed by soot, and its removal is problematic due to the presence of tar [11]. When the gas is cooled, there is a risk of more solid particles forming. Wet scrubbing, high-temperatures barrier filters [12], cyclones, and electrostatic separators are used for removal purposes [13].
- Tars can be defined as organic substances which are produced on the basis of thermal or partial oxidation of any organic material [14]. Generally it is presumed that tars are largely composed of polycyclic aromatic hydrocarbons (PAH). Tar substances can be classified as primary, secondary, or tertiary, according to the operating conditions under which they originated. Primary tars, which contain oxygen in substantial volumes, are formed by the decomposition of biomass building blocks. Secondary and tertiary tars are formed by the destruction of primary tar substances and the recombination of fragments [15]. During these processes, oxygen and particles of hydrogen are removed [16].
- Compounds containing nitrogen
- Alkaline compounds
- Sulphur
- Chlorine

2.2. Gas Utilization

Gas can be combusted in a burner, combustion engine, or gas turbine. If gas is combusted in a burner, only heat is generated. The dust is removed from gas in cyclones. During combustion in an engine or turbine, the benefit is electricity and waste heat. If the gas is a fuel for a combustion turbine, it must be de-dusted. Tars do not need to be removed because the temperature in the turbine chamber is way above the tar dew point. Tar removal is required when the gas is used in a combustion engine. The demands for gas quality are increasing from gas motors to fuel cells. Tables 1 and 2 show the general requirements for gas quality for a gas turbine and for compression ignition and spark ignition engines. Table 3 shows the limits of pollutants for Molten Carbonate Fuel Cells (MCFC) operation.

Table 1. Requirements for gas quality for gas turbines [17].

Parameter	Value
Minimum heating value (MJ·m_n^{-3})	4–6
Minimum hydrogen content (% vol.)	10–20
Maximum supply temperature (°C)	60–450
Maximum concentration of alkali (ppb)	20–1000
Tar at inlet temperature	In gaseous phase or none
Sulphur (ppm)	<1
HCl (ppm)	<0.5
Maximum content of solid particles (ppm)	
Average >20 mm	<0.1
10–20 mm	<1
4–10 mm	<10

Table 2. Requirements for gas quality for compression ignition and spark ignition engines [17].

Parameter	Value
Maximum hydrogen content (% vol.)	7–10
Maximum relative humidity (%)	80
Maximum supply temperature (°C)	40
Maximum ammonia content (mg/10 kWh)	55
Maximum tar content (mg·m_n^{-3})	<100
Maximum halogens content (mg/10 kWh)	<100
Maximum content of sulfur recalculated to H_2S (ppm)	2000
Maximum content of solid particles (mg·m_n^{-3})	5–50

Table 3. Pollutant limits for Molten Carbonate Fuel Cells (MCFC) type fuel cells [18].

Parameter	Value
Maximum sulfur content (ppm)	<0.1
Maximum ammonia content (% vol.)	<0.1
Maximum hydrogen cyanide content (ppm)	<0.1
Maximum chlorides content (ppm)	<0.1
Maximum fluoride content (ppm)	<0.1
Maximum tar content (ppm)	<2000
Maximum particle size (μm)	<0.01
Maximum lead content (ppm)	<1

2.3. SP Removal

To remove mechanical impurities and tar droplets originating from condensation, separators are used, with gravitational settlement, centrifugal forces, filtration, scrubbing, and electrostatic trapping in a high-tension electric field. According to the principle of their action, they are divided into:

- Inertial (momentum),
- Gravitational (horizontal, vertical and spherical),
- Centrifugal (cyclone),
- Filtration, combined.

The mutual comparison of the respective types of equipment with the size of the particles of trapped aerosols is shown in Figure 1. When choosing the type of separator, it is necessary to take into account that certain types of particles agglomerate in larger aggregates, which facilitate the cleaning of gases.

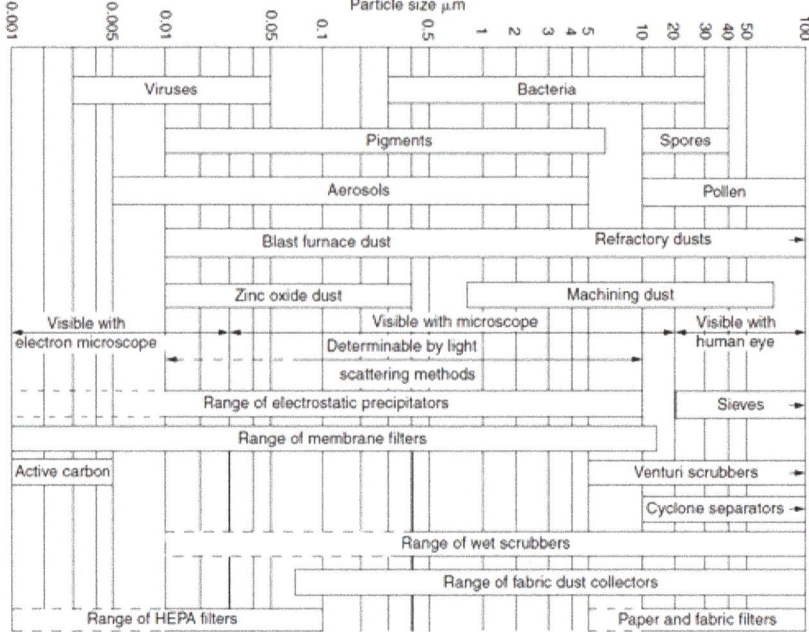

Figure 1. Comparison of types of separators for trapping aerosol particles [19].

Today, the removal of tar from generator gas, due to the physiochemical properties of organic substances which form it, is often performed in technological practice by applying modified physiochemical processes developed in the past for gasification and carbonization of coal. Organic substances, which are similar in nature to the compounds forming tars, are used as the washing liquid in these processes. The advantage of applying organic substances in comparison to water washing is the fact that at temperatures of around 50 °C, a much lower tar content in the gas can be achieved downstream of the washer without condensation of water vapor from the gas and problematic water solutions or emulsions.

However, in these cases, pursuant to the rules of thermodynamic equilibrium, the increasing content of tar in the washing liquid increases the residual concentration of tar in the gas. This means that in the case of concentration of tar in the absorption medium, the concentration of residual tar components in the gas starts to increase. During gasification of biomass, the saturated absorption solution in small equipment typically does not regenerate but is combusted or gasified. This leads to increased operating costs. Several practical applications under various trade abbreviations are known. The best known and longest used application is cleaning of gas from an 8 MWt fluid gasification generator with a circulation bed of cogeneration units in Güssing (Austria), where the content of tar in the gas is reduced from 1500 mg·m^{-3} to 10–40 mg·m^{-3} (at a temperature under 50 °C). Such cleaned gas is combusted in the cogeneration unit engine with an output of 2 MWe, and the used bio-diesel is combusted in the combustion section of the generator [20].

Reduction of the tar content in gas also enables optimization of the operating parameters of generators with a fluid bed with catalytically active material operating at temperatures over 900 °C.

The generation of gas with low tar content can also be achieved during operation of a so-called dual-stage generator, utilizing the principle of partial oxidation of pyrolysis products represented by a high content of tar components released in the pyrolysis section of the generator [21]. In addition, a concurrent generator such as the "Imbert" type, produces gas with low tar content, provided that technological conditions mainly respecting the biomass properties are complied with. This generator in its various modifications, is the most common generator in small cogeneration units in the Czech Republic.

An often-underestimated factor during the exploitation of similar units is the ingress of fine SP into the cylinder space of the combustion engine from where their residues after combustion are mainly pulled into the oil, thereby compromising its lubricating properties, alkalinity, and shortening its replacement interval. Quantity of SP which exceeds limits set by engine manufacturers also increases wear of the intake ducting and cylinders and, in the case of the use of a turbocharger, it also causes its accelerated wear.

The synchronous impact of the above-mentioned factors is one of the reasons for insufficient economic efficiency of gasification units operated today in the Czech Republic. An improvement in the situation can be achieved mainly by using more efficient SP removal, preferably using barrier filters operated above the dew point of tars and water vapors in the gas. In cases where the removal of tar and SP is performed simultaneously, there is a high risk of formation of a sticky filtration cake, causing significant problems in its subsequent removal from the filter surface. This process has a progressive character and the degree of filter surface degradation gradually increases during operation. A worse condition can only be the penetration of condensed tar into the actual filtration material. The described problem occurs quite often during starting-up of the gasification technology. Raw hot gas enters the cold space with the filtration elements, which contains amongst others, much more tar at engine start-up than during the stabilized operation. Upon contact with the cold filter the gas immediately cools, tar and water vapor condense, penetrate the filter structure and particles bond to the filter. After a certain operation period the filter heats up to the necessary temperature, but due to the polycondensation and polymerization reactions of reactive tar components the filter is permanently clogged, increasing its pressure loss and preventing its further use, often leading to the failure of the cleaning system, which then requires replacement of the filtration elements. A similar "result" can also be achieved quite easily by using cold pressure media for regeneration of the filter during reverse impulse purging.

The corresponding design of filtration materials and filter designs require knowledge of the amount, composition, and distribution of SP particles, which must be removed from the generated gas in certain cases, or the limit of the content of organic substances adsorbed by the trapped solid particles must be determined. Well known solutions of SP and tars removal from big facilities including hot filters for removing SP, catalytic decomposition of tars, or other sophisticated devices are not suitable for small energy units because of their high investment costs.

A certain problem with the operation of generators gasifying wood is the formation of incrusts in the generator's fire grate, which limits the discharge of solid residues from the gasifying generator, as well as the actual operation of the gasification process.

3. Experimental Section

3.1. Gasification Technology

The subject of the study was the determination of distribution of the size of solid particles trapped by the sleeve filter from gas produced by a gasification generator with a solid bed, type "Imbert", necessary for the objective design of the filtration material of the developed filtration sleeve equipment for removal of SP, their composition, quantity of adsorbed organic substances, and compositions of incrust from the generator's fire grate.

The operated generator was fitted with gasification air preheating by waste heat from the reactor in the installed heat exchanger. This enabled achievement of a gasification air temperature of 200 to 240 °C and gasification temperature of 780–830 °C. The average consumption of woody biomass was 82 kg/h, the amount of air flow was not measured. The whole experiment lasted for 16 h.

The offtake of gas from this generator was performed from the top section from the tube leading produced gas from the generator, to the gas cleaning system. A closure valve is located in the top section of the generator for its automatic filling using a high-lift container. Gas rising from the generator through the outlet channels (piping), with a temperature of approximately 250 °C, was routed to a collector connected to the gas pipe, which lead to the hot multi-cyclone and then to the technological line where the other equipment was located, including the monitored sleeve filter equipped with 15 textile sleeves (Figure 2) with a diameter of 20 cm and a length of 1.2 m. The total amount of trapped SP during the whole experiment was 10,282.1 g, which corresponds that 0.78 wt.% of fuel was converted into SP.

Figure 2. Sleeve filter for trapping solid particles (SP).

In compliance with the objectives of this task, one textile sleeve, which was tested for the cleaning of the generator gas for about one month, was taken for detailed analysis from the tested dry filter (Figure 3). The solid pollutants were carefully removed mechanically from its outer surface.

Figure 3. Fragment of sample of SP trapped by the filter.

3.2. Fuel

The raw material processed in the generator were coniferous cuttings—maximum length 10 to 15 cm, maximum width 5 cm, and a maximum thickness of 2 cm. Their typical physiochemical properties are shown in Table 4. The proportion of fine particles and particles with bark in the gasified material was negligible. This fuel was used long-term in this unit. It was available, affordable, and had good quality and properties for gasification. This type of material had a medium agglomeration-slagging propensity [22].

Table 4. Properties of used fuel.

Parameter	Value
Moisture W^r (wt.%)	16.73
Volatile combustibles V^d (wt.%)	82.99
Fixed carbon F^d (wt.%)	16.71
Ash A^d (wt.%)	0.30
C (wt.%)	49.86
H (wt.%)	6.14
O (wt.%)	43.38
N (wt.%)	0.31
S (wt.%)	0.01
HHV, wet (MJ·kg^{-1})	17.59
HHV, dry (MJ·kg^{-1})	20.52

^{r} values of original wood sample; ^{d} dry sample without moisture.

3.3. Analytical Methods

Parameters of the fuel from Table 4 (moisture, volatile, and fixed carbon and ash) were measured by thermogravimetric analyzer Netzsch STA 449 F1 Jupiter. The heating rate was 10 °C/min in a nitrogen atmosphere up to 900 °C. For the amount of ash, there was a change in nitrogen by oxygen and heating continued up to 1200 °C. Ultimative analysis of fuel (amount of C, H, O, N and S) was measured by analyzer LECO CHN 628 with added module 628 S. Combustion heat was measured by the calorimeter LECO AC600.

Thermogravimetric analysis (TG) was measured by the analyzer Netzsch STA 449 F1 Jupiter. The heating rate was 10 °C/min in nitrogen atmosphere. The differential thermal analysis DTA was measured simultaneously, which measures the temperature difference between the standard sample (aluminium oxide) and the sample during heating.

Composition of the filtration cake inorganic fraction and incrust was measured with apparatus ARL 9400 XP+ equipped with a Rh lamp with a head Be-oxides window.

The distribution of the trapped particle size was performed by a Fritsch Particle Sizer Analysette 22 apparatus, where the light source is an He-Ne-laser with a wavelength of 633 nm and maximum measuring range of 0.1–1250 μm. To ensure the homogeneity of the measured sample and reliable statistics in the illuminated volume, the suspension was permanently mixed in an external mixing device during measurement using ultrasound, which provided for the separation of the agglomerate.

The measuring of organic substances composition trapped by the filtration cake was necessary to solve these substances in the solvent. Mechanically separated cake from the surface of the outer filter layer was inserted into the Soxhlet extractor and washed for four hours by acetone. The obtained solution was injected into the gas chromatograph Hewlett Packard HP 6890 with a mass detector Hewlett Packard MSD 5973 (GC-MS).

4. Results

4.1. Analysis of the Composition of the Filtration Cake

The solid pollutants were carefully removed mechanically from its outer surface. The sample of SP was tested by thermogravimetric analysis—see Figure 4—until 100 °C 1.5% by wt. was released into the stream of carrier gas, until 200 °C about 2.5% by wt., until 300 °C about 7.5% by wt., and until 400 °C about 8% by wt. The simultaneously performed DTA analysis found that there is only one significant endothermal effect, detected at approximately 725 °C. This effect is probably the melting of the inorganic fraction of the filtration cake.

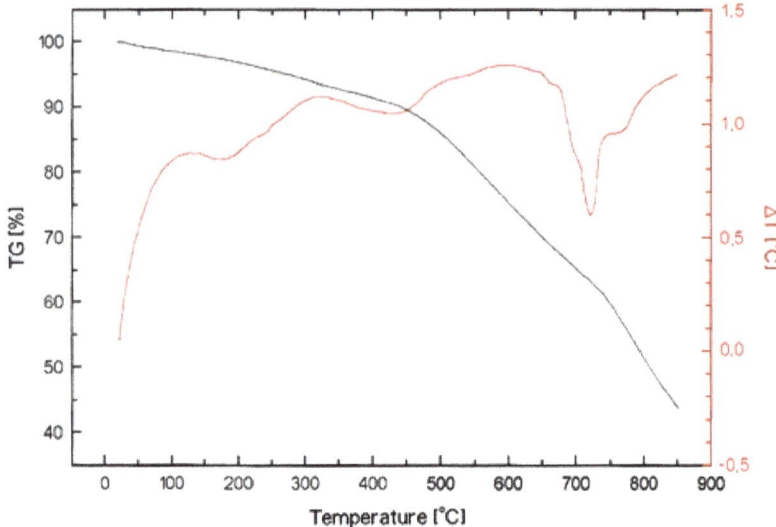

Figure 4. Thermogravimetric analysis (TG)/DTA analysis of filtration cake.

The quantity of ash in the SP sample determined by its combustion at 850 °C was 37.28% by wt. The quantity of carbon dioxide in the sample determined using HCl corresponding to the detected carbonates was 2.7% by wt.

4.2. Composition of Inorganic Fraction and SP Size of the Filtration Cake

The results of the composition of the filtration cake inorganic fraction were performed using fluorescent X-ray analysis and are summarized in Table 5.

Table 5. Composition of the filtration cake inorganic fraction.

Substance	Content (wt.%)	Substance	Content (wt.%)
Al_2O_3	16.73	NiO	0.07
BaO	0.83	P_2O_5	5.54
CaO	16.71	SO_3	15.70
CdO	0.30	SiO_2	12.93
Cl	0.49	SrO	0.14
Cr_2O_3	6.14	TiO_2	0.08
CuO	9.85	Fe_2O_3	2.68
K_2O	0.31	PbO	0.06
MgO_3	0.01	ZnO	1.47
MnO	7.59	Na_2O	2.37

Measurement of the distribution of trapped particle size was performed for samples taken in the middle of four sleeves using the laser technique Analysette. The results are shown in Table 6. Particle sizes were in the range of 10 μm to 3 mm.

Table 6. Cumulative particle size distribution of SP trapped by filtration cake.

%	Sleeve 1 (g)	Sleeve 2 (g)	Sleeve 3 (g)	Sleeve 4 (g)
1	0.935	0.665	0.692	0.676
2	1.366	0.943	0.976	0.963
5	2.628	1.624	1.670	1.654
10	4.846	2.606	2.795	2.659
15	7.043	3.574	3.921	3.664
20	9.234	4.572	5.000	4.703
25	11.497	5.585	6.704	5.762
30	13.939	6.622	7.110	6.811
35	16.598	7.707	8.159	7.872
40	19.526	8.804	9.252	8.925
45	22.631	9.977	10.389	10.017
50	25.934	11.252	11.628	11.168
55	29.517	12.646	12.972	12.381
60	33.293	14.244	14.481	13.745
65	37.052	16.121	16.228	15.3
70	41.191	18.487	18.325	17.204
75	45.754	21.618	21.000	19.776
80	50.654	25.909	24.461	21.029
85	56.602	31.498	29.001	35.043
90	63.754	37.17	35.792	47.127
95	74.544	46.917	46.012	58.692
98	85.998	55.385	56.700	69.138
99	93.76	60.362	63.296	75.364
100	183.788	102.15	106.200	112.657

4.3. Analysis of Organic Substances Trapped by Filtration Cake

The next part of the research work was related to the identification and quantification of the components trapped in the filtration cake by the sleeves (Table 7).

Table 7. Composition of organic substances captured on the filtration sleeve.

Identified Substance	Content (µg/g)
Benzene	30.4
Toluene	35.9
m + p + o-xylene + ethylbenzene	44.3
Styrene	0
C3-benzene total (sat. + unsat.)	0
Others [1]	78.7
BTX total	189.3
Oxygen containing substances total	180.1
Phenol	36.9
Methylphenols	15.1
Dibenzofuranes *	128.2
Nitrogen containing substances	0.0
Indene+indan	2.5
Naphthalene	63.6
Methylnaphthalenes	20.0
Alkylnaphthalenes (alkyl >= C2)	11.8
Biphenyl	15.5
Acenaphthylene	129.7
Acenaphthene	6.6
Fluorene	42.9
PAH; m/z = 165–166	11.6
Phenanthrene	1470.0
Anthracene	314.7

Table 7. *Cont.*

Identified Substance	Content (µg/g)
Methylfenathrenes+4H-cyclopenta[def]fenanthrene	315.7
Phenylnaphthalenes	118.0
Fluoranthene [2]	1987.3
Pyrene [3]	1857.6
Benzfluorenes	46.3
Methylfluorantene+methylpyrene	164.7
PAH of 4 circles ** (m/z = 226.228)	4201.4
PAH of 5 circles *** (m/z = 252)	5851.0
PAH of 6 circles **** (m/z = 276)	2214.9
other substances (TAR)	538.5
total TAR (excl. BTX)	19564.5

[1] this group includes other substances from the BTX and alkylbenzenes group. [2] together with fluorantene also includes fenantrylene M = 202 eluted from GC column immediately after it. [3] together with pyrene also includes aceantrylene M = 202 eluted from GC column immediately before it. * benzofurane, dibenzofuranes, methylbenzofuranes, naphtobenzofuranes. ** benz[c]fenanthrene, benzo[ghi]fluorantene,3,4-dihydrocyclopenta[cd]pyrene, cyclopenta[cd]pyrene. *** benzo[j]fluoranthene,benzo[k]fluoranthene,benzo[e]pyrene,benzo[a]pyrene,perylene. **** indeno[1,2,3-cd]pyrene,dibenzo[a,h]antracene,benzo[ghi]perylene,dibenzo[def,mno]chrysene and other PAH with M = 278–302.

4.4. Chemical Composition of Incrust

The composition of the incrust (Figure 5) shown in Table 8, where the concentration of the 16 most significant components from the 79 monitor elements is shown.

Figure 5. Incrust from the grate section of generator.

Table 8. Incrust composition.

Oxide	Content (wt.%)	Oxide	Content (wt.%)
MgO	3.11	MnO	2.02
Al_2O_3	5.67	Fe_2O_3	2.71
SiO_2	41.78	CuO	0.01
P_2O_5	4.69	SrO	0.11
SO_3	0.06	BaO	0.16
Cl	0.03	Na_2O	0.52
K_2O	3.73	TiO_2	0.53
CaO	34.82	NiO	0.02

Evaluation of X-ray diffraction analysis assessing only the composition of the crystalline phase of the incrust is shown in Table 9.

Table 9. X-ray diffraction analysis results.

Score	Name	Summary Chemical Formula	Content (%)
58	Silica	SiO_2	23
46	Calcite	$CaCO_3$	13
46	Calcium lime	CaO	17
20	Potassium Aluminium Silicate	$K_2SiAl_2O_6$	8
22	Carbon	C	4
31	Silica (other than above)	SiO_2	1
22	Sodium Calcium Aluminium Carbonate Silicate	$Ca_2Na_2Si_6Al_6O_{12}(CO_3)_{0,5}$	33

5. Discussion

Fuel from woody biomass is a common commercial fuel but it has a higher amount of ash, e.g., woody pellets have only 0.15 wt.% of ash but coniferous cuttings could be contaminated by bark or pieces of soil.

From Table 5, it clearly follows that the dominant oxide of the inorganic fraction of SP are aluminum, calcium, and silicon oxide. The second most abundant component is potassium oxide, followed by manganese oxide, silica, ferric oxide, and sodium oxide. This table also clearly shows that phosphates and sulphates form an inconsiderable part of ash matter from biomass trapped in the filter. Chlorides represent only 0.1% by wt. in the monitored sample. A surprising result was the determined concentrations of heavy metals in the analyzed matter, particularly in terms of the toxicity of fly ash and its disposal and storage related to transport etc. This involves mainly copper, chrome, zinc, and manganese oxides.

The determined distribution of particle sizes of solid pollutants trapped by the sleeve filter from the raw generator gas produced by gasifying wood chips by air, demonstrated that their absolute size ranges from approximately 0.4 to 185 µm, whereas 99% by wt. of monitored particles were smaller than 100 µm. The proportion with a diameter up to 1 µm was approximately 1%, and up to 10 µm was approximately 22%.

The distribution of the individual mass fractions from 10% to 80% did not demonstrate any significant differences. The difference in particle size between 99 and 100% cannot be considered decisive due to the genesis of the sample.

Comparison of TG and GC-MS data shows that the results of thermogravimetric analysis do not correspond fully with the results of the performed identification. The sum of the weight of the individual identified tar components was substantially lower than the weight of substances released from the heated filtration cake in a nitrogen atmosphere. This could be caused by chemical sorption of organic substances on the SP, which could not be removed by the solvent.

The performed analyses of the incrust show that this is made up of mainly inorganic components of ash from the raw material gasified in the generator [23]. Its formation is defined significantly by potassium, calcium, and silicon oxides; and also phosphorus, iron, and magnesium oxides in the amorphous part [24].

Many of the identified oxides are typical components of ash matter forming low melting ash during combustion, or wood gasification. The relatively high silica content shows that this component was dosed into the generator with fuel during its preparation. One of its undesirable effects in the generator's oxidation zone, where relatively high temperatures are present, is the formation of crystalline substances creating incrusts. An example of a K_2O-CaO-SiO_2 phase diagram is shown in Figure 6, where even the following eutectics can be found [25]:

$K_2Si_4O_9$	765 °C
$K_8CaSi_{10}O_{25}$	940 °C
$K_4CaSi_3O_9$	950 °C
$K_4CaSi_6O_{15}$	950 °C
K_2SiO_3	976 °C
$K_2Si_2O_5$	1036 °C

To prevent similar undesirable conditions during the operation of cogeneration units, it is necessary to substantially limit its presence in the woody matter dosed into the generator. This applies to silica itself contaminating the tree bark or silica present in soil. The thermal conditions of decomposition of clay minerals must be considered: Kaolinite 450–700 °C, montmorillonite 600–700 °C, 800–900 °C, and illite 450–700 °C, 850–950 °C.

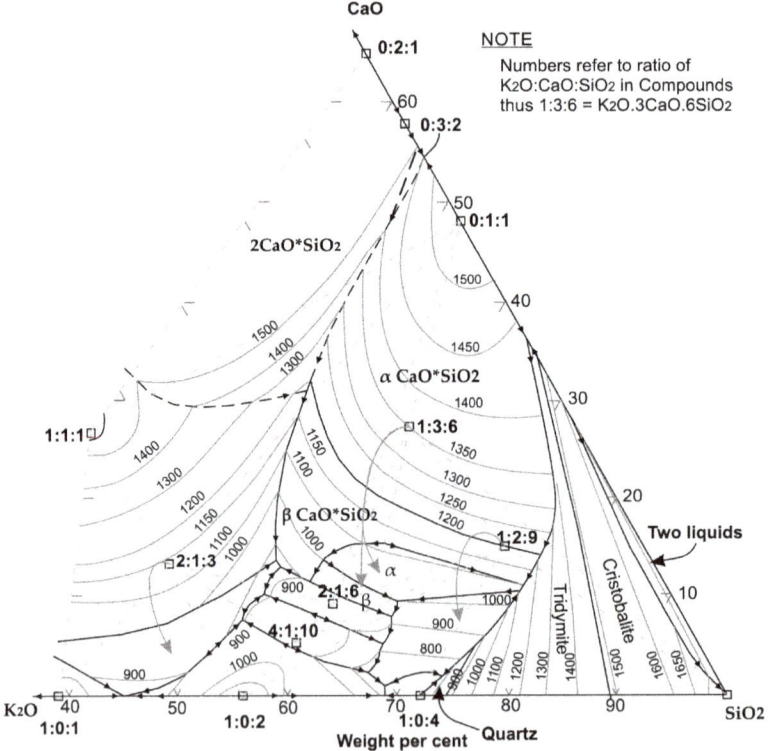

Figure 6. Detail of the K_2O-CaO-SiO_2 ternary system [25].

Author Contributions: Conceptualization, J.N. and P.B.; Data curation, J.N.; Funding acquisition, J.F.; Investigation, P.B. and S.S.; Methodology, S.S.; Supervision, J.N. and J.F.; Validation, J.K.; Visualization, V.P.; Writing—original draft, V.P.

Funding: This research was funded by Ministry of Education, Youth and Sports of Czech republic, grant number CZ.1.05/2.1.00/19.0389: Research Infrastructure Development of the Centre ENET.

Conflicts of Interest: The authors declare no conflict of interest. The funders had no role in the design of the study; in the collection, analyses, or interpretation of data; in the writing of the manuscript, or in the decision to publish the results.

References

1. French, S.J.; Czernik, S. Catalytic pyrolysis of biomass for biofuels production. *Fuel Process Technol.* **2010**, *91*, 25–32. [CrossRef]
2. Sheth, P.N.; Babu, P.V. Experimental studies on producer gas generation from wood waste in a downdraft biomass gasifier. *Bioresource Technol.* **2009**, *100*, 3127–3133. [CrossRef] [PubMed]
3. Basu, P. *Combustion and Gasification in Fluidized Beds*, 1st ed.; CRC Press, Taylor and Francis Group: London, UK, 2006; ISBN 0-8493-3396.
4. Babu, B.V.; Sheth, P.N. Modelling and simulation of reduction zone of downdraft biomass gasifier: Effect of char reactivity factor. *Energy Convers. Manag.* **2006**, *47*, 2602–2611. [CrossRef]
5. Asadullah, M. Biomass gasification gas cleaning for downstream applications: A comparative critical review. *Renew. Sustain. Energy Rev.* **2014**, *40*, 118–132. [CrossRef]
6. Chlond, R. Inhibitors in the gasification process. In Proceedings of the Conference Energy from Biomass X, Brno University of Technolog, Brno, Czech Republic, 25–26 November 2009; ISBN 978-80-214-4027-2.
7. Reed, T.B.; Das, A. *Handbook of Biomass Downdraft Gasifier Engine Systems*; The Biomass Energy Foundation Press: Golden, CO, USA, 1988. Available online: https://www.nrel.gov/docs/legosti/old/3022.pdf (accessed on 15 December 2018).
8. Lisý, M.; Baláš, M.; Špilácek, M.; Skála, Z. Operating specifications of catalytic cleaning of gas from biomass gasification. *Acta Polytech.* **2015**, *55*, 401–406. [CrossRef]
9. Skoblja, S. Modification of gas composition from biomass gasification. Doctoral Thesis, Institute of Chemistry and Technology Prague, Prague, Czech Republic, 2005.
10. Vakalis, S.; Moustakas, K.; Sénéchal, U.; Schneider, R.; Salomo, B.; Kurz, M.; Malamis, D.; Zschunke, T. Assessment of potassium concentration in biochar before and after the after-burner of a biomass gasifier. *Chem. Eng. Trans.* **2017**, *56*, 631–636. [CrossRef]
11. Radina, O. Energy utilization of wood gas. Bachelor's Thesis, University of West Bohemia, Pilsen, Czech Republic, 2013.
12. Balas, M.; Lisy, M.; Moskalik, J. Biomass Gasification: Gas for Cogeneration Unit. Available online: https://www.researchgate.net/publication/281547834_Biomass_gasification_gas_for_cogeneration_units (accessed on 12 November 2018).
13. Higman, C. *Gasification*; Elsevier: Amsterdam, The Netherlands, 2003; 391p, ISBN 07-506-7707-4.
14. Morioka, H.; Shimizu, Y.; Sukenobu, M.; Ito, K.; Tanabe, E.; Shishido, T.; Takehira, K. Partial oxidation of methane to synthesis gas over supported Ni catalysts prepared from Ni-Ca/Al-layered double hydroxide. *Appl. Catal. A Gen.* **2001**, *215*, 11–19. [CrossRef]
15. Li, D.; Wang, L.; Koike, M.; Nakagawa, Y.; Tomishige, K. Steam reforming of tar from pyrolysis of biomass over Ni/Mg/Al catalysts prepared from hydrotalcite-like precursors. *Appl. Catal. B Environ.* **2011**, *102*, 528–538. [CrossRef]
16. Zhang, X.; Yang, S.; Xie, X.; Chen, L.; Sun, L.; Zhao, B.; Si, H. Stoichiometric synthesis of Fe/CaxO catalysts from tailored layered double hydroxide precursors for syngas production and tar removal in biomass gasification. *J. Anal. Appl. Pyrolysis* **2016**, *120*, 371–378. [CrossRef]
17. Pravda, L. Energy gas–product of gasification. In Proceedings of the Energy from Biomass III, Brno, Czech Republic, 2–3 December 2004; ISBN 80-214-2805-8.
18. Iaquaniello, G.; Mangiapane, A. Integration of biomass gasification with MCFC. *Int. J. Hydrogen Energy* **2006**, *21*, 399–404. [CrossRef]
19. Sutherland, K. *Filters and Filtration Handbook*, 5th ed.; Butterworth-Heinemann is an imprint of Elsevier: Burlington, MA, USA, 2008; ISBN 978-1-8561-7464-0.
20. Brandin, J.; Tuner, M.; Odenbrand, I. *Small Scale Gasification: Gas Engine CHP for Biofuels*; Linnaeus University: Växjö, Sweden, 2011; pp. 101–106. ISBN 978-91-86983-07-9.
21. Henriksen, U.; Ahrenfeldt, J.; Jensen, T.K.; Gøbel, B.; Benzen, J.D.; Hindsgaul, C.; Sørensen, L.H. The design, construction and operation of a 75 kW two-stage gasifier. *Energy* **2006**, *31*, 1542–1553. [CrossRef]
22. De Fusco, L.; Jeanmart, H.; Contino, F.; Blondeau, J. Advanced characterization of available not conventional mediterranean biomass solid fuels for ash related issues in thermal processes. *Chem. Eng. Trans.* **2016**, *50*, 229–234. [CrossRef]
23. Šatava, V. *Introduction to Physical Chemistry of Silicates*, 8th ed.; SNTL Prague: Bratislava, Slovakia, 1965.

24. Basu, P. *Biomass Gasification and Pyrolysis: Practical Design and Theory*; Elsevier: Amserdam, The Netherlands, 2010; ISBN 978-0-12-374988-8.
25. Roedder, E. Silicate melt systems. *Phys. Chem. Earth* **1959**, *3*, 224–297. [CrossRef]

© 2019 by the authors. Licensee MDPI, Basel, Switzerland. This article is an open access article distributed under the terms and conditions of the Creative Commons Attribution (CC BY) license (http://creativecommons.org/licenses/by/4.0/).

Article

Design of Robust Total Site Heat Recovery Loops via Monte Carlo Simulation

Florian Schlosser [1,*], Ron-Hendrik Peesel [1], Henning Meschede [1], Matthias Philipp [2], Timothy G. Walmsley [3], Michael R. W. Walmsley [4] and Martin J. Atkins [4]

1. Dep. Umweltgerechte Produkte und Prozesse, Universität Kassel, Kurt-Wolters-Straße 3, 34125 Kassel, Germany; peesel@upp-kassel.de (R.-H.P.); meschede@upp-kassel.de (H.M.)
2. Bayernwerk Natur GmbH, Carl-von-Linde-Straße 38, 85716 Unterschleißheim, Germany; matthias.philipp@bayernwerk.de
3. Sustainable Process Integration Laboratory–SPIL, NETME Centre, Faculty of Mechanical Engineering, Brno University of Technology-VUT Brno, Technická 2896/2, 616 69 Brno, Czech Republic; walmsley@fme.vutbr.cz
4. Energy Research Centre, School of Engineering, University of Waikato, Private Bag 3105, Hamilton 3240, New Zealand; walmsley@waikato.ac.nz (M.R.W.W.); matkins@waikato.ac.nz (M.J.A.)
* Correspondence: schlosser@upp-kassel.de; Tel.: +49-561-804-3442

Received: 31 January 2019; Accepted: 27 February 2019; Published: 10 March 2019

Abstract: For increased total site heat integration, the optimal sizing and robust operation of a heat recovery loop (HRL) are prerequisites for economic efficiency. However, sizing based on one representative time series, not considering the variability of process streams due to their discontinuous operation, often leads to oversizing. The sensitive evaluation of the performance of an HRL by Monte Carlo (MC) simulation requires sufficient historical data and performance models. Stochastic time series are generated by distribution functions of measured data. With these inputs, one can then model and reliably assess the benefits of installing a new HRL. A key element of the HRL is a stratified heat storage tank. Validation tests of a stratified tank (ST) showed sufficient accuracy with acceptable simulation time for the variable layer height (VLH) multi-node (MN) modelling approach. The results of the MC simulation of the HRL system show only minor yield losses in terms of heat recovery rate (HRR) for smaller tanks. In this way, costs due to oversizing equipment can be reduced by better understanding the energy-capital trade-off.

Keywords: total site heat integration; heat recovery loop (HRL); heat storage; Monte Carlo (MC) simulation; data farming

1. Introduction

In Europe, the food and beverage industry is the fifth largest energy consumer [1], consuming the highest share of low-temperature heat demand of the total process heat demand [2]. Using renewable energies and increasing energy efficiency are ways to reduce greenhouse gas (GHG) emissions [3]. First, the potential to reduce process energy demand can be identified using process integration methods [4]. If the possibilities of increasing the energy efficiency at the process level through direct heat recovery are exhausted, further energy saving is possible by means of total site heat integration [5], with a specific focus on low-temperature processes [6] and heat recovery loops (HRL) [7].

In many industries, such as the dairy and beverage industry, the individual processes are operated non-continuously for product changes and cleaning reasons. This poses additional challenges to heat recovery, often requiring the use of energy storage. Two approaches to characterizing batch processes and its flows are the time average model (TAM) and the time slice model (TSM). The TAM indicates heat recovery potential, which can be achieved by indirect heat transfer and storage based on an HRL.

Olsen et al. [8] present a systematic thermal energy storage integration method based on the TAM approach. Their indirect source and sink profile (ISSP) method enables the concrete identification of suitable sink and source profiles and the dimensioning of intermediate circuits and heat storage [9]. Due to the integration of multiple sink and source streams, the utilization of the storage capacity increases the overall availability and performance of sources and sinks. For sensible heat storage, there are closed stratified tanks (ST) and fixed temperature variable mass (FTVM) open storage tanks. FTVM tanks require a larger storage volume than ST tanks, but they are easier to operate and suitable for very large circulating volumes and small temperature differences. ST tanks need only half the volume and can operate at temperatures above 100 °C when placed under pressure. Due to the temperature dependency of fluid density, higher temperature fluid (lower density) forms a hot zone at the top of the tank, while lower temperature fluid forms one at the bottom. These zones are separated by a narrow transition area (between h_{Tu} and h_{Tl})—the thermocline. For varying source and sink streams, the volume of fluid in each temperature zone increases or decreases, moving the thermocline vertically within the storage tank (see Figure 1). A completely loaded or unloaded storage tank, in addition to other mixing effects, results in a loss of stratification, loss of heat recovery, and even temporary shutdown of the HRL [10].

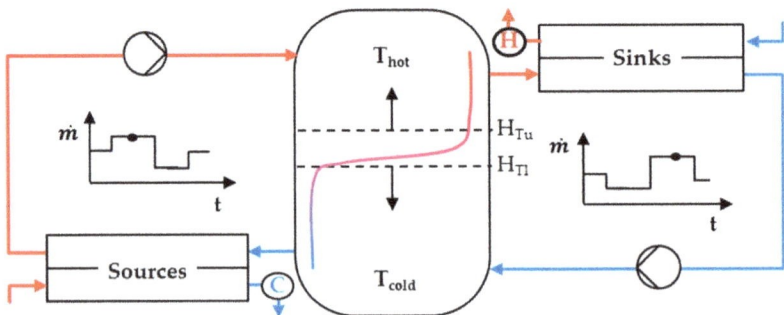

Figure 1. Heat recovery loop (HRL) based on a stratified tank (ST) and thermocline control.

Dynamic simulation is necessary for evaluating and proving the performance of an HRL system in any scenario. This is due to arbitrary production and logistic timing, the changing operation time of process states, thermal inertias, as well as dynamic interaction. Past models have focused on either the heat recovery side of the HRL [11] or the ST unit. These models assumed that there was a perfect interface between the cold and hot area of the ST [12], which is a simplification of practical installations. Walmsley et al. [13] studied an experimental scale model of an industrial ST, including the interaction between flows and thermocline movement and growth, but did not incorporate this element into their HRL models. The growth and movement of the thermocline accelerate the loss of stratification, reducing the effective heat recovery capacity of the tank over time and leading to reduced energy savings. Baeten et al. [14] presented a model that incorporates the buoyancy and mixing in one-dimensional stratified storage tanks in building energy simulations to estimate the uncertainty of parameters as a function of the storage capacity caused by the storage tank model. Campos Celador et al. [15] showed that the storage tank model used in such simulations can have major effects on calculated annual savings and design decisions. Powell and Edgar [16] introduced an adaptive-grid model with a high-resolution grid in the region of thermal stratification. This ensures higher accuracy and simulation performance than other one-dimensional models, and has speed advantages over computational fluid dynamics (CFD) models in repetitive simulations (e.g., for optimization, validation, control, prediction). Schlosser et al. [17] demonstrated a multi-node (MN) approach can represent a good compromise between accuracy and simulation time for the high-fidelity simulation of HRL systems. Numerical diffusion errors were reduced by approximately 35% using a variable layer height

(VLH) model. The stratification of the physical thermocline growth could be re-established while still operating the ST by siphoning the mixed thermocline area from the tank [13]. Furthermore, a control system consisting of a single flow and system control for maintaining stratification was applied. The system control operated depending on the thermocline position, enabling restratification after being completely charged or discharged and a single flow control adjusted the individual streams to the target temperatures T_{hot} or T_{cold}. This previous simulation study [17] included a typical week of production for tank sizing and evaluation, taking no account of stochastic changes (e.g., order quantity) or deterministic influences, such as recipe sequences. As an extension, the present work simulates several weeks of production, based on synthetic time series with probabilistic influence (e.g. changing recipes and order amounts) by Monte Carlo (MC) simulation to generate a sensitivity evaluation of the HRL system. Atkins et al. [10] highlighted the optimal sizing of an ST as a prerequisite for economic efficiency and operability. Moreover, Meschede et al. [18] applied MC methods to highlight the risk caused by using a single reference time series for designing energy systems. The MC method helps to define the mean of the heat recovery rate (HRR) for different tank sizes more precisely, and to better assess the risk of fluctuating heating and cooling load due to stochastic influences [19]. A prerequisite for this, however, is an accurate and high-performance VLH model.

The aim of the paper is to improve the design and operation of industrial HRLs by constructing a comprehensive model of storage and heat exchanger operations, simulated with a MC approach. Therefore, the first step of this study is to validate the performance and accuracy of the VLH model. Secondly, the effect of the variability of streams on the sizing approach in terms of heat integration potential and operability is evaluated by MC simulation, with real process data from the dairy industry. The MC method considers the sensitivity of stochastic influences (order quantity, sequence of recipes) and deterministic relations (production process: idling, operation, cleaning in place (CIP)) on the volatility of sink and source profiles. The results are probability distributions of the HRR for different storage dimensions of the present case study.

The article essentially contains two objectives: (1) the validation of the accuracy and performance of the VLH multi-node (MN) model and (2) a risk assessment of the robustness of the ST design and control by MC simulation, based on the validated model. The validation of the VLH model, developed by Schlosser et al. [17], is carried out on the basis of the validation experiments introduced by Walmsley [20] in comparison with a real laboratory tank and a CFD model (Sections 2.1 and 3.1). The simulation of a VLH model-based HRL, designed according to the ISSP method (Section 2.3), is realized by stochastically generated time series as input following the MC approach (Section 3.3). The generation based on cumulative distribution functions (CDF) was developed in the context of this work (Section 2.2) and applied (Section 3.2) for measurement data of a case study [12]. The results of Section 3 are summarized in Section 4.

2. Materials and Methods

The following sections describe the methods for validating the VLH model and the evaluation approach for the HRL dimensioning. The first section presents the different validation experiments proving the accuracy and performance of the VLH model. Subsequently, the generation of stochastic time series based on random influences and deterministic correlations is described. Finally, the evaluation approach of a robust storage sizing based on dynamic simulation is presented.

2.1. Verification and Validation of the Variable Layer Height Model

Verification and validation tests confirm if a model reproduces a certain (physical) behavior qualitatively and quantitatively correct, or with tolerable deviations. In this context, verification means checking whether the model correctly implements the derived equations. Verification can be understood as a qualitative examination of the transformation and the functionality of the model. Validation quantitatively checks the representation of the original system by the model.

The quantitatively correct representation of the reference system is determined by subjectively defined tolerances, so that a model can be described as sufficiently accurate within a certain error deviation [21].

The model is examined in two dimensions–performance and accuracy. To calculate a large number of probable time series for a production week in an appropriate time frame, sufficient simulation speed is required. In terms of verification, the functionality is given by using universal physical descriptions. The validation is based on data from a CFD model, which is validated by a lab-scale, insulated ST (6.44 L) made of Perspex [20]. In this way, sufficient accuracy of the VLH model can be demonstrated by various experiments. After the numerical diffusion errors are minimized [17], the position of the thermocline, characterizing the heat recovery capacity, is used as a measure of the error deviation. As a validation quality, the following unitless key metrics are developed. The percentage of ideal cases (PIC) represents the share of the ideal ST case A_0 compared to the areas of cold A_1 and hot A_2 losses of stratification, as illustrated in Figure 2. PIC quantifies the level of stratification. The PIC number is defined as:

$$PIC = 1 - \frac{A_1 + A_2}{\left(\frac{A_0}{2}\right)} \qquad (1)$$

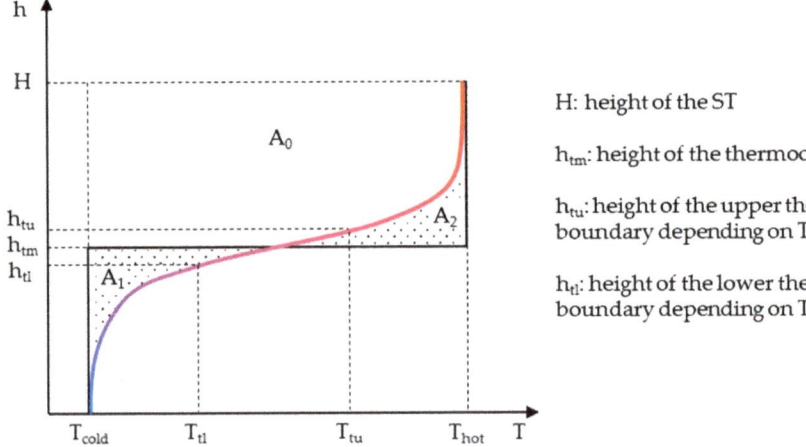

Figure 2. Illustration of variables used to define the validation criteria in a ST.

This results in the following operating ranges [20]:

- The tank is perfectly stratified: $PIC = 1$
- The tank is fully mixed, all hot or cold fluid: $PIC = 0$
- The tank is normal operating: $0 \leq PIC \leq 1$

The position and height of the thermocline h_{tm} gives information about the available heat recovery capacity. The dimensionless height $h_{tm,i}$ of the thermocline is determined by the division of the actual height h_{tm} and the total tank height H. Associated with the thermocline thickness H_t the degradation of the stratification is quantified.

$$H_t = h_{tu}(T_{tu}) - h_{tl}(T_{tl}) \qquad (2)$$

This work is now based on the validation parameters of the PIC, characterizing the degree of thermocline degradation. The deviation of the PIC from the experimental values in different experiments is quantitatively assessed. The height of thermocline $h_{tm,i}$ is qualitatively discussed by means of a targeted loading and unloading profile. The following validation measures are conducted. The settings and boundary conditions of each experiment are presented in Table 1.

(1) The **Static Mode** investigates heat transfer mechanisms during a certain simulation time in a ST, with ideal stratification at the beginning, without inlet or outlet flows, and with (a) and without (b) heat losses. The loss of stratification by the numerical and physical diffusion effects is described by PIC.
(2) Establishment of the thermocline by **Charging** studies the effect of the inlet flow rate of a charging process on the level of thermal stratification. In particular, the thermocline thickness and the location are observed. The tank is completely filled with cold water and warm water is supplied with different flow rates.
(3) **Operating Mode Movement** studies the effect a changing charging and discharging flow rates on the thermal thermocline position based on h_{tm}. Moreover, the degradation of stratification was measured by PIC. The storage tank is alternately loaded and unloaded with 2 L cold, 4 L warm, and again 4 L cold water at a flow rate of 0.4 L/min.

Table 1. Settings and boundary conditions of validation.

Number	Validation Experiment	Physical Modelling	$\dot{V}_i$ L/min	$T_{0,hot}$ °C	$T_{0,cold}$ °C	T_{amb} °C	h_{0tm} -	H_{0tm} m	Simulation Time min
1a	Static Mode	Adiabatic	-	40	20	-	0.5	0	200
1b		Non-adiabatic	-	40	20	15	0.5	0	200
2a	Charging	Constant charging with 40 °C	0.15	20	20	15	1.0	0	26.7
2b			0.4						10.0
2c			1.0						4.0
2d			2.0						2.0
3	Movement	Sequentially charging	0.4	40	20	15	0.5	0	25

2.2. Generation of Stochastic Time Series for Monte Carlo Simulation

Understanding the variability of the available sources and sinks is prerequisite for an optimized volume of the required thermal storage and to achieve a balanced system. As already outlined, several different factors have an influence on the height and temporal course of the heat source and heat sink processes. Within the framework of a simulation study, reliable and comparable energy savings considering the dynamic system's behavior, can be calculated. A sensitivity analysis considers the influence of fluctuating factors on the target variables robustness of the HRL dimensioning and HRR. If probabilities are assigned to the stochastic influencing factors, time series that occur with a certain probability can be generated from the deterministic correlations. The generated time series serve as input data for the dynamic simulation. Thus, the simulation validates the resulting heat recovery potential, depending on the tank size, regarding probability. The methodology for generating time series used in this paper follows the steps shown in Figure 3.

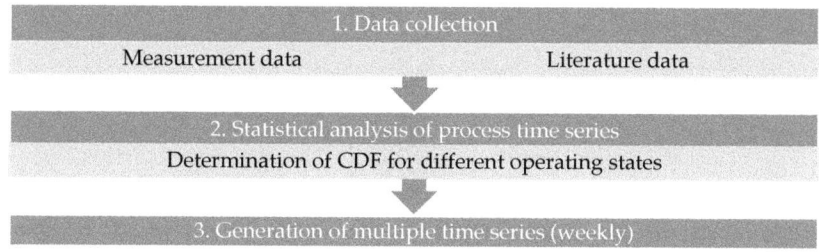

Figure 3. Flow chart of the approach for generating weekly time series of the heat source and sink processes.

The first step is to collect data. These data are usually measured energy flows and production quantities. Data from a case study in New Zealand are used for this purpose [12]. The analysis of all essential influencing parameters of a production week requires a recording of the measured data with a suitable high temporal resolution. Seasonal effects (e.g., the weather, or availability of milk) is neglected for the heat recovery problem. If no measurement data is available (e.g., for greenfield planning), the necessary parameters could be assumed as normally distributed around average values from literature data or expert opinion.

In the second step, the time series are analyzed for possible deterministic correlations. Since the flow rate of the streams varies due to process fluctuations and discontinuous operation of the plants, which are caused by regular cleaning, maintenance work, plant failure, product change, and changing milk supply. Supply temperatures are generally quite constant, due to robust process control, and do not represent a major cause of variability in this case. Thus, the sequence of idling, production, and CIP is mandatory.

Moreover, the probabilistic components of the weekly values are described by CDF based on cumulative frequency plots. Based on the result of the statistical analysis, multiple weekly time series are generated in step three.

Due to the interaction of product changes and cleaning constraints on the energy demand, the streams have different demand profiles. Generally, two operating states can be distinguished, especially in the food industry: "on-production", associated with an energy demand, and "off-production" (CIP or breakdown), without any energy demand. Duration, frequency, and amplitude (i.e., production rate) of these operating conditions can be determined by statistical analysis of process time series. Constant and fluctuating energy demands can be detected. Based on that, the following three characteristic time series can be derived.

Using dairy case study data [11], the method for generating time series is demonstrated. The measurement data of two processing months are statistically evaluated to derive the weekly demand time series. Based on the load profiles in Figure 4, the frequency distributions of the heat flow rate and of the durations t for the operating states "on-production" and "off-production" are represented in the form of histograms (Figure 5a). The determinations of the frequency and cumulative probability distributions are shown in Figure 5, using the time series type (b) for stream Whey A, as an example. The computations are carried out with MATLAB®(9.3.0.713579 (R2017b), The MathWorks Inc., Natick, MA, USA).

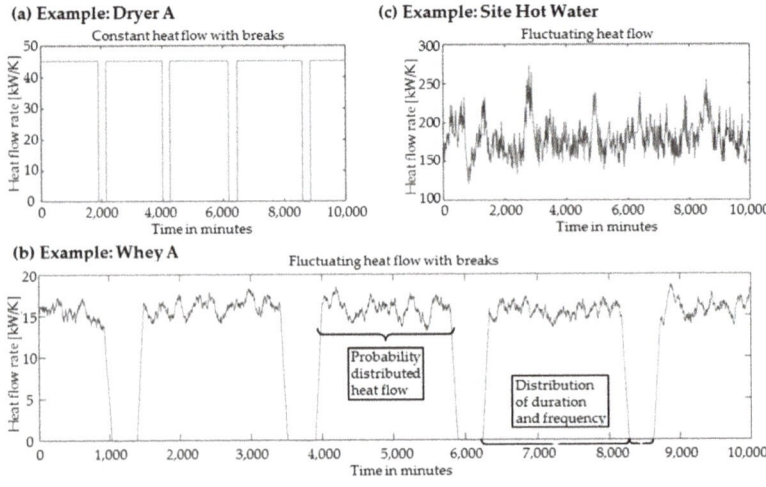

Figure 4. Characteristic time series for a (**a**) constant and (**b**) fluctuating heat flow with time intervals in off-production and (**c**) a continuously fluctuating heat flow.

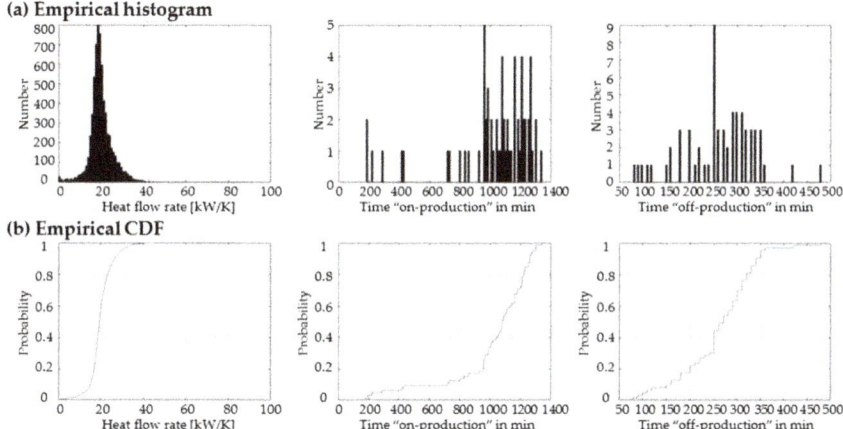

Figure 5. (a) Histograms and (b) empirical cumulative distribution functions (CDF) of the heat flow rate, duration of "on-production" and duration of "off-production" for the example of Whey A.

The generation of the time series begins with the decision of whether production is "on" or "off". The probability for this decision is calculated by the mean value from the mean duration of the "on" and "off" periods. The CDF based on the cumulative frequency plots of the heat flow rate (Figure 5b) over the production period is used to set probable heat flow rate for every time step during the operating state "on-production", by which the demand is scattered with the corresponding variance. The distributed durations of the operating states t_{on} for "on-production" and t_{off} for "off-production" can also be determined from the CDF functions (Figure 5b).

In this context, MC simulations allow the consideration of numerous probabilistic inputs to determine the distribution of probabilistic results. In line with previous studies [22–24] on the robustness of energy systems, probabilistic times series based on probability density functions (PDF) and CDF are used. In this way, representative data can be generated for multiple purposes, e.g., validation and sensitivity analysis.

2.3. Evaluation of the Robust Heat Recovery Loop-Dimensioning via Dynamic Monte Carlo Simulation

The stochastic time series of the heat sources and sink processes are used as input data for the thermal simulation of the HRL from Figure 1. In the sensitive simulation study, the tank size is designed for the average stream data according to the ISSP method. Furthermore, the target temperatures T_{hot} and T_{cold} of the intermediate circuits will be varied. Finally, the HRRs for different scenarios are presented depending on their probability of occurrence. The HRL (Figure 1) simulation model based on MATLAB/Simulink® consists of heat exchangers, a stratified storage tank, and a control system. The heat exchangers are modelled according to the ε-NTU [25]. The VLH model based on the MN approach is chosen as the modelling approach for the ST [17]. The system control is divided into two parts, as shown below.

The position and width of the thermocline describe the hot and cold zone capacity and therefore serve as control criteria in the system control. The system control (Figure 6a) specifies various operating states (i.e., normal operation, loading, unloading) depending on thermocline location in the ST. If the middle of the thermocline reaches the lower or upper end of the storage tank, this is considered to be completely loaded or unloaded. The source and sink circuits are then deactivated to regenerate the stratification. For hysteresis reasons, the sources and sinks remain deactivated until the corresponding zone again occupies 10% of the storage volume. The single mass flow control (Figure 6b) adjusts the streams to the target temperatures T_{hot} or T_{cold} by mass flow control. Moreover, the growth of thermocline is contained to a certain extent $H_{t,max}$ by siphoning while the HRL and ST are still in full operation.

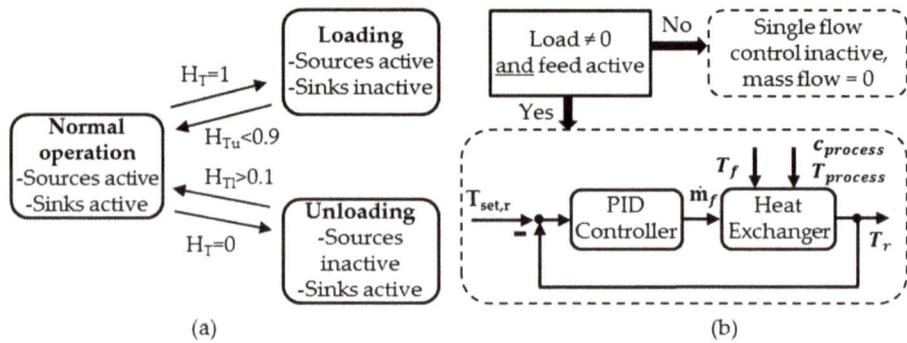

Figure 6. (**a**) System controls depending on the thermocline position and (**b**) single mass flow control adjusting T_{hot} and T_{cold}.

For evaluating the quality of the HRL system, an energetic target value for the maximum heat recovery (MHR) potential is determined according to the TAM method. The design of the HRL is carried out according to the ISSP method. Based on average stream data of a repetitive stream-wise repeat operation period (SROP) of one week, the hot and cold streams are defined. The stream data consist of average load, supply, and target temperatures. Considering a minimum temperature difference ΔT_{min}, the hot and cold composite curves (CC) are formed. The overlapping area indicates the heat recovery potential (Figure 7a).

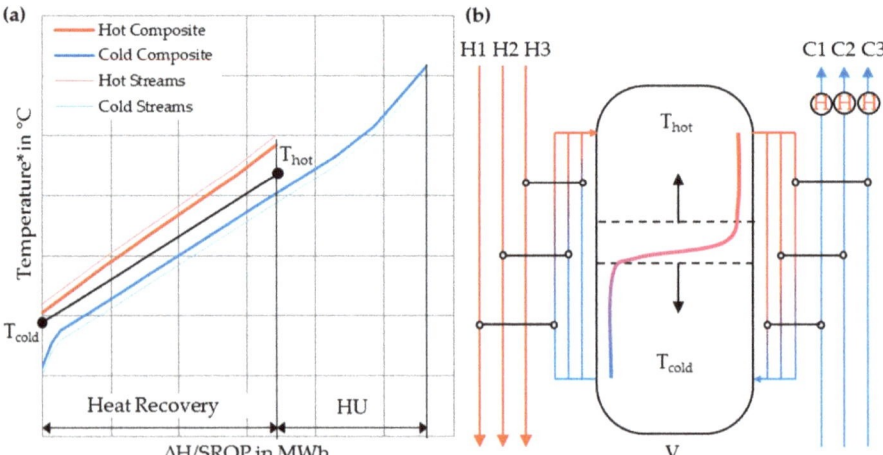

Figure 7. From (**a**) the indirect source and sink profile (ISSP) to (**b**) the HRL dimensioning.

Depending on the form and position of the CCs, the temperatures of the intermediate circuits, as well as T_{hot} and T_{cold}, can be read directly, in addition to the number of intermediate circuits (Figure 7b). With known temperature difference ($T_{hot} - T_{cold}$) between the hot and the cold side of the tank and the maximum product of the difference between heat sources $\dot{Q}_{Sources}$ and heat sinks $\dot{Q}_{Sinks}$ for a certain time t, the water-based storage volume can be calculated.

$$V = \frac{\max((\sum \dot{Q}_{Sources} - \sum \dot{Q}_{Sinks}) \cdot t)}{1.16 \frac{kWh}{m^3 \cdot K} \cdot (T_{hot} - T_{cold})} \tag{3}$$

The resulting tank size is highly dependent on the time of the occurrence of sinks and sources. Therefore, the design on the basis of one measuring profile can lead to a decrease of the HRR, or worse, to a standstill. A reliable estimation of energy savings and assessment of the tank size is only possible with a sensitivity analysis. The simulation of multiple weekly time series with random thermocline positions at the start of production evaluates the reliability and robustness of the chosen tank size. The model is also simulated applying different tank sizes. Finally, the probability distribution of the heat recovery potential for different tank sizes is evaluated based on the MC analysis.

3. Results

In the following sections, the results of the validation, the HRL sizing, and the sensitive simulation study applied to the dairy case study are presented.

3.1. Validation of the Variable Layer Height Model

The experiments described in Section 2.1 are performed for the laboratory tank, the CFD, and the VLH multi-node (MN) model. For each experiment and each model, temperature curves are compared for characteristic points over the dimensionless storage height. The degree of destratification is represented by the PIC value.

(1) (a) **Static Mode:** Numerical diffusion effects are reduced to an acceptable degree, as shown in the previous study [13]. For the lab tests, the heat losses cannot be eliminated completely by the thermal insulation. For all experiments (Figure 11a–c), the thermocline grows with the simulation duration due to numerical diffusion effects. The deviation from the ideal stratification (black line) measured by the PIC values thus also increases, as shown in Figure 11d. Using the VLH model, the influence of numerical diffusion is even lower than when using the CFD model.
(b) Incoming or outgoing mass flows are still deactivated. Heat losses, on the other hand, are taken into account, which leads to the following destratification for different simulation times in Figure 8a–c. The destratification of the three models takes place to a similar extent in Figure 8d. Since no vertical heat conduction effects are taken into account for the MN approach, the stratification can only degrade evenly over the height (Figure 8b).

(2) Establishment of the Thermocline by **Charging:** From a flow rate of over 1 L/min, turbulence causes destratification (Figure 10a,c). Therefore, the Courant condition has to be met, which states that a fluid particle should not move further than a node per time step. The destratification of the three models takes place to a similar extent until 1.0 L/min (Figure 10d). The degradation of the stratification of the CFD model and the lab-scale tank grows above this. On the other hand, the VLH model does not take turbulent flow forms into account, as seen in Figure 10b. For this reason, it can only be operated without restrictions up to a maximum inflow velocity of 0.002 m/s.

(3) **Operating Mode Movement:** The thermocline initially moves upwards from its initial state by being filled with cold water (2 L). All models reach the expected position for the ideal behavior (black dotted lines in Figure 9a–c). After filling with warm water (4 L) the thermocline moves to the expected lower position. The storage tank is now filled to 81% with warm water. Subsequently, the predetermined heights are reached again when the tank is filled with cold water (4 L). The degradation of stratification generated by the flow rate can best be mapped by the VLH MN model (Figure 9d).

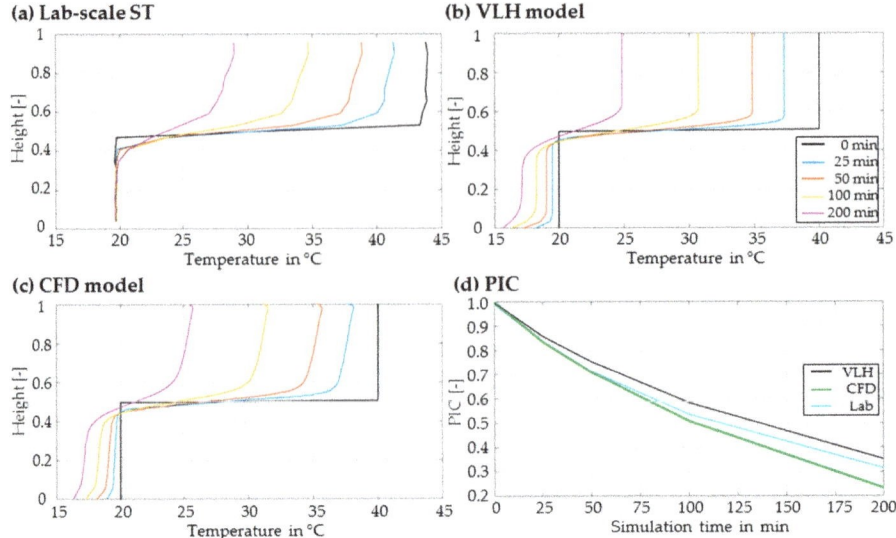

Figure 8. Destratification of the (**a**) lab, (**b**) VLH model and (**c**) CFD model due to heat losses for the **Static Mode**, compared based on the (**d**) PIC value.

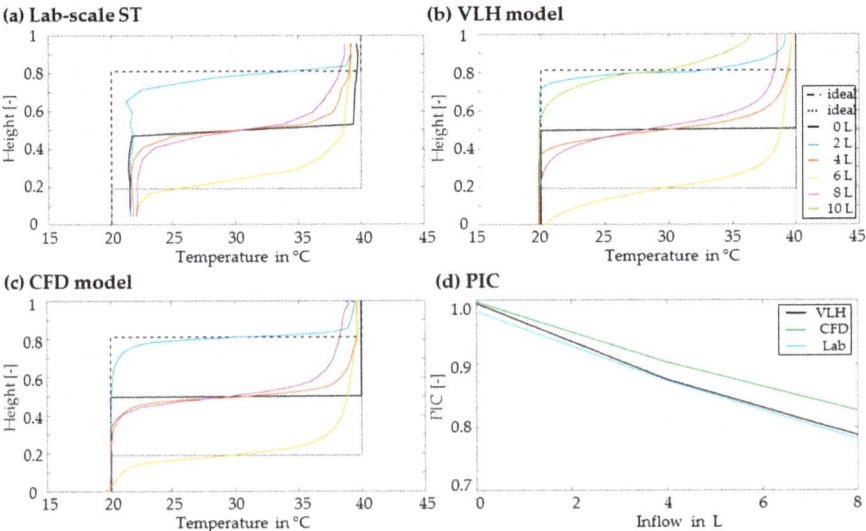

Figure 9. Thermocline **Movement** for a targeted tank charging sequence in comparison to the ideal position for the (**a**) lab, (**b**) VLH model and (**c**) CFD model, compared based on the (**d**) PIC value.

Moreover, the scalability from lab-scale sizes to industrial-scale has to be shown. Therefore, the lab-scale tank and the VLH model have the same aspect ratio (height/width) as common industrial applications. Moreover, the stratification is influenced by the flow characteristics, especially the residence time in the tank. The maximum flow velocity for the laboratory tank was 0.002 m/s, which can also be found in industrial applications. The modelled ST should be in between this range. For this reason, sufficient scalability is given.

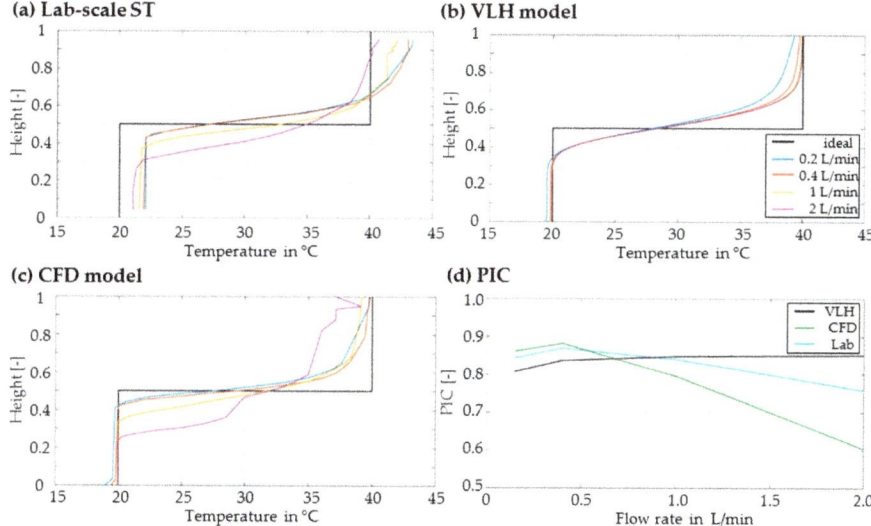

Figure 10. Influence of turbulent flows on stratification of the (**a**) lab, (**b**) VLH model and (**c**) CFD model during **Charging**, compared based on the (**d**) PIC value.

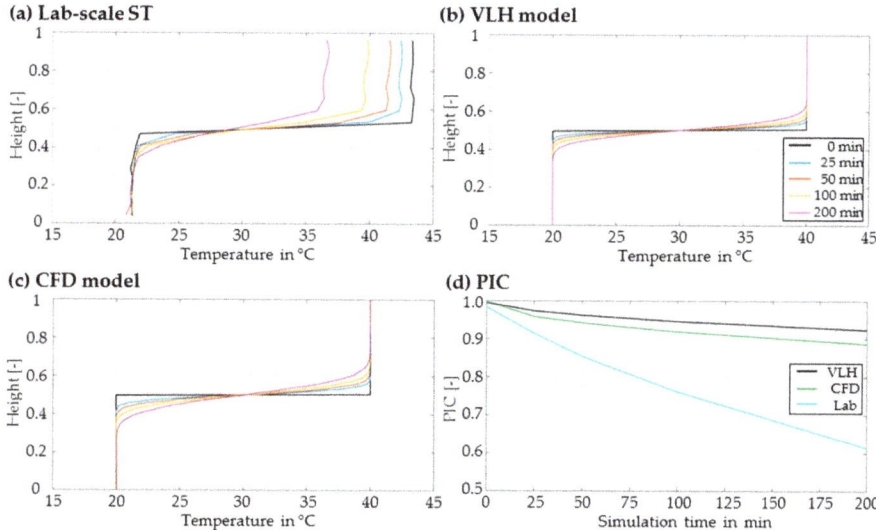

Figure 11. Comparison of the (**a**) lab tank, (**b**) variable layer height (VLH) multi-node (MN) model and (**c**) computational fluid dynamics (CFD)-modelled tank in terms of the (**d**) PIC value for an adiabatic **Static Mode**.

3.2. Stochastic Time Series of Dairy Plant

For the case study, representative process data of the milk processing industry have been assembled into an exemplary complete system. This results in load profiles of the sources and sinks for a mean SROP, as seen in Table 2.

Table 2. Stream table of dairy case study, producing milk powder and cream [11].

Process	Type	T_{supply} °C	T_{target} °C	$\dot{W}_{operating}$ kW/K	$\dot{H}_{operating}$ kW	$\dot{W}_{average}$ kW/K	$\dot{H}_{average}$ kW
Utility	Hot 1	50	30	10	195	8	160
Casein	Hot 2	50	20	49	1477	32	956
Dryer A	Hot 3	55	10	143	6415	139	6266
Dryer B	Hot 4	55	10	75	3368	73	3290
Dryer C	Hot 5	55	10	45	2020	44	1973
Milk Treatment	Cold 1	10	55	90	4057	90	4057
Whey	Cold 2	12	50	20	763	16	601
Site Hot Water	Cold 3	15	65	160	7987	160	7987

Using the approach for generating time series, explained in Section 2.2, various load profiles can be generated for a SROP, as presented in Figure 12 for the example of Whey A.

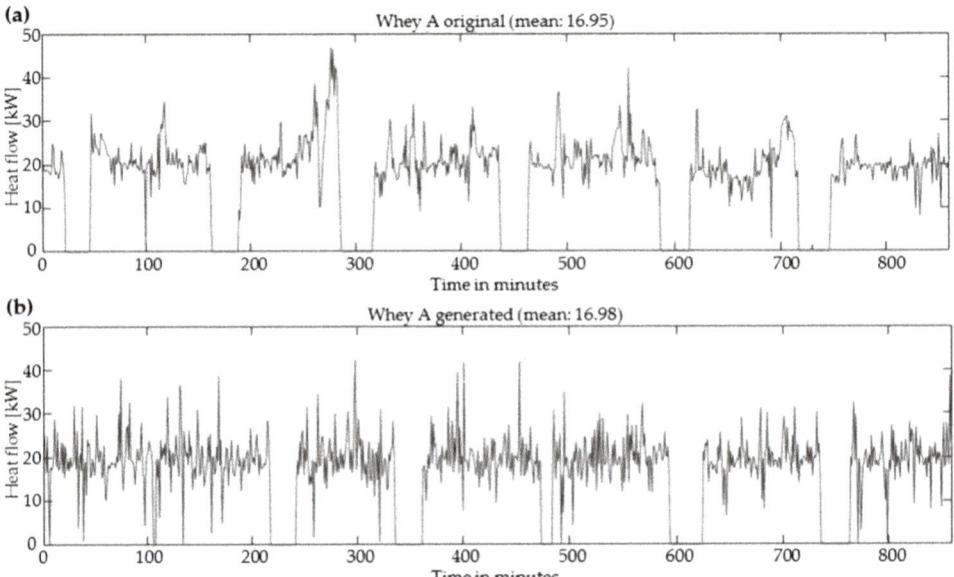

Figure 12. (a) Original time series of Whey A compared to the (b) generated one.

Profiles can be generated in the same way for all streams from Table 2. Figure 13 shows four examples of the accumulation of total sources and sink streams. This results in multiple total sink and source streams, which represent randomly arranged operating states and loads, as seen in Figure 13. These can overlap in any imaginable way.

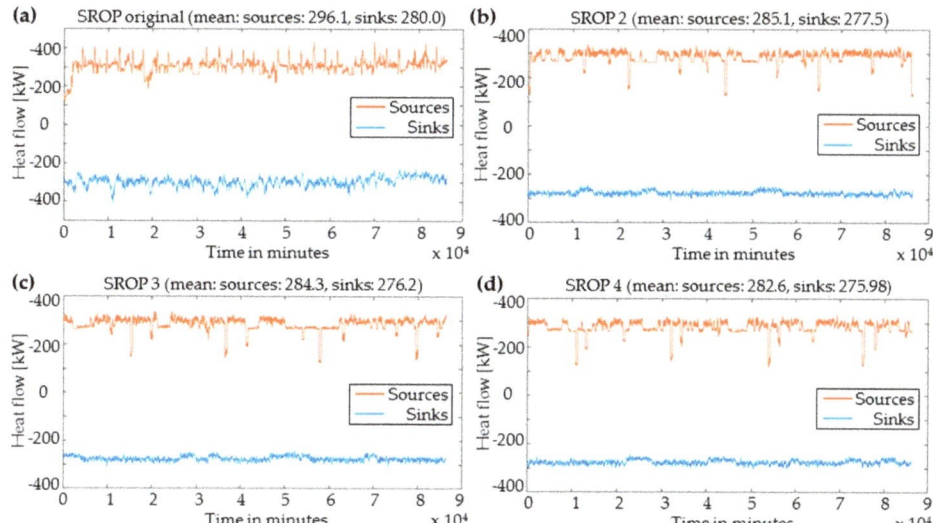

Figure 13. (**a**) Original and (**b**–**d**) exemplary total available source and sink streams of one stream-wise repeat operation periods (SROPs).

3.3. Evaluation of the Robust Heat Recovery Loop-Dimensioning via Monte Carlo Simulation

Figure 14 shows the exemplary dimensioning of the HRL (temperature and tank size) by ISSP for the mean values. A tank volume of 1000 m³ is calculated. Since the concurrency of the sources and sinks is highly variable, the sizes 50, 100, 300, 500, and 2000 m³ are investigated by the sensitivity analysis. For all simulations, a height/diameter aspect ratio of 3/1 is chosen.

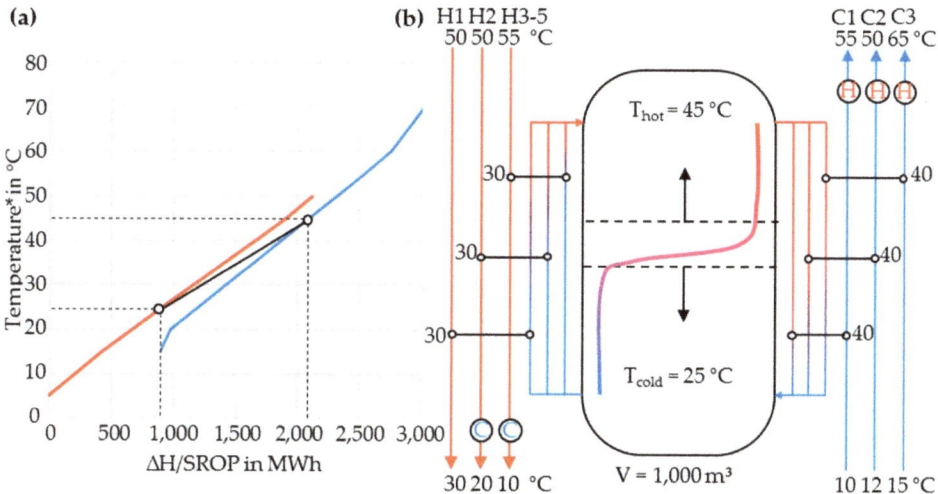

Figure 14. (**a**) Heat recovery potential according to the time average model (TAM) and (**b**) HRL sizing by the ISSP method.

The simulation tool uses the storage temperatures, size, and loop temperatures targeted from the ISSP design. Next, the HRL is simulated for different sink and source profiles dictated by the control system. The simulation of multiple time series for each tank size with random thermocline positions at production start takes place. The distribution of the resulting HRRs for the different storage sizes and simulation runs are shown in Figure 15. The HRR is calculated by the ratio of recovered heat to the target value. For the case study with a minimum temperature difference of 5 K, a heat recovery target of 1250 MWh corresponds to the overlapping area of the hot and cold CC in Figure 14a.

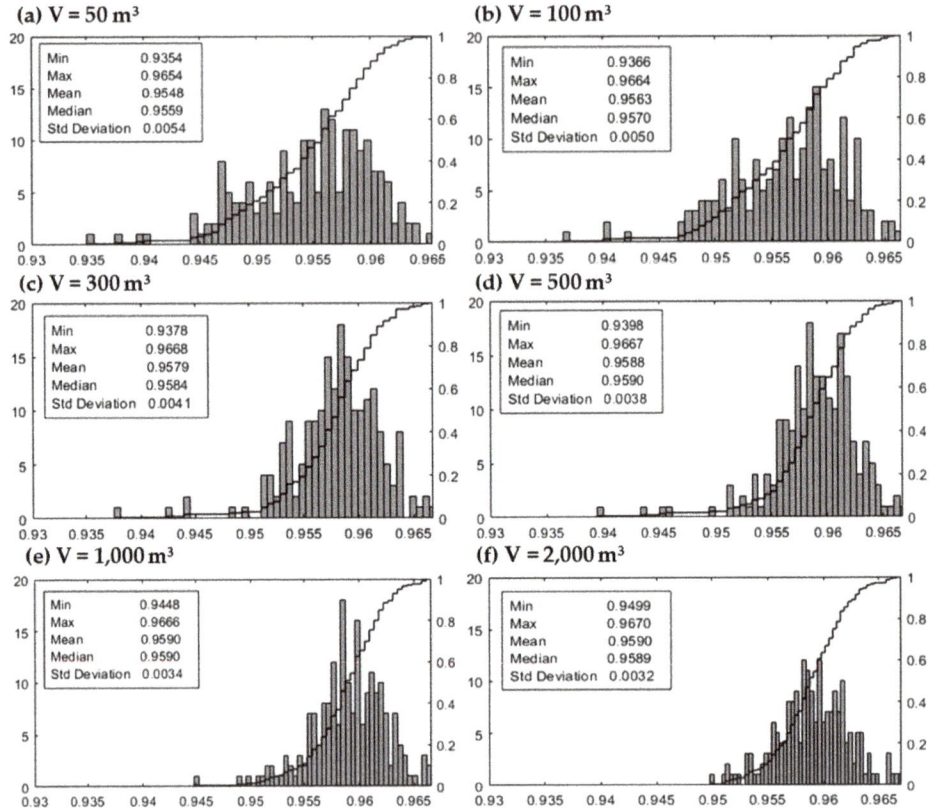

Figure 15. Distribution (bars) and CDF (stairs) of the heart recovery rate (HRR) for (**a**–**f**) different tank sizes from 50 m³ to 2000 m³.

Regarding the number of runs that must be carried out to achieve a distinctive probability distribution, some approaches use a convergence criterion to determine the number of simulations. In this paper, 200 simulations are run to generate probable time series, since this number shows reasonable distributions within an appropriate computing time. Storage tanks with > 1000 m³ (Figure 15f) do not achieve any further heat recovery. A high HRR can be achieved with much smaller tanks. Only the standard deviation increases significantly with decreasing size.

In total, all tank sizes in combination with the system control are suitable to achieve HHR over 90%. With decreasing storage volume at the same aspect ratio, tank height and diameter are reduced. As a result, the incoming and outgoing flow velocities temporarily rise above the limits of the validity range, which is avoided by limiting the mass flow by the control system. For this reason, the yields decrease further for smaller tank sizes.

4. Discussion

Previous studies [17] showed that a VLH MN modelling approach reduces thermocline growth, due to numerical diffusion effects while performing with suitable execution time and accuracy. The conducted validation tests confirm this and show lower thermocline growth than a CFD model. However, it should be noted that vertical heat conduction effects are not considered in the MN approach. Furthermore, the validity range of the MN approach must be restricted regarding the flow velocity, since turbulent flow effects are not taken into account. However, these operating points are to be avoided anyway for a stable HRL operation. In addition, scalability from lab-scale sizes to industrial-scale can be assumed based on the same aspect ratio (height/width).

The recording of thermal energy data for the design and evaluation of HRLs often requires a good deal of effort and is not representative of the whole operating range. One way to compensate for this is to generate one's own time series data, also called data farming. The representativeness of the generated time series increases with the increase in basis data. Figure 5a shows a left-shifted distribution of the loads. The distributions of the durations for "on-production" and "off-production" states, on the other hand, scatter wider. If no measurement data is available, input data can be randomly distributed around expert values.

Recent study [19] had shown the relevance of risk evaluation due to the variability of streams by MC simulation methods when assessing the sensitivity of the results. The results of this study confirm this. Although the design methods provide a storage capacity of 1000 m^3, the results for 50 m^3 only show a HRR reduced by less than 1%. Even the insignificant further variance of the values suggests a stable operation of the HRL in combination with the existing control system over a wide operating range. A prerequisite for the extensive simulation runs is an accurate and high-performance VLH model. Based on this work, an evaluation method for robust system sizing, stable operation control, and reliable energy-saving prognosis is developed, which is applicable also to other thermal systems. Future work will investigate the robustness of heat recovery systems based on heat pumps with STs.

5. Conclusions and Outlook

The validation shows suitable accuracy and performance of the VLH model with reduced numerical diffusion effects, restricted by the validity range for non-turbulent piston flow. Furthermore, a method is developed that generates stochastic time series for a sensitivity analysis with few input values. A suitable form of sensitivity analysis is the MC simulation assessing the risk. The robustness of the HRL system is aided by the control system. The method also avoids oversizing in any of the probable operation states. The simulations show that the tank size has only a small influence on the HRR, since the mean value varies only between 95.48% and 95.90% for tank sizes of 50–2000 m^3. The standard deviations are very small (< 0.54%) but increase with decreasing storage size. This leads to the conclusion that a 50 m^3 tank reduces the HRR by less than 1%, with a minimally larger variance, compared to the 1000 m^3 and 2000 m^3 tanks. Furthermore, it is now possible to better forecast reliable energy saving.

The usage of the VLH model for a model predictive control strategy is planned for future works. In addition, an evaluation benchmark for the verification of a sufficiently large sample of time series for a convergent result should be introduced.

Author Contributions: Responsible for the conceptualization, F.S.; methodology, F.S. and H.M.; validation, F.S., M.R.W.W. and M.J.A.; investigation, F.S.; data curation, F.S., T.G.W. and M.R.W.W.; writing—original draft preparation, F.S.; writing—review and editing, M.P., H.M., R.-H.P. and T.G.W.; visualization, F.S.; supervision, T.G.W.

Funding: This research has been supported by the EU Project "Sustainable Process Integration Laboratory—SPIL", Project No. CZ.02.1.01/0.0/0.0/15_003/0000456 funded by EU "CZ Operational Programme Research, Development and Education", Priority 1: Strengthening capacity for quality research, in collaboration with University of Kassel (DE), Technische Hochschule Ingolstadt - Institute of new Energy Systems (DE), and University of Waikato (NZ).

Conflicts of Interest: The authors declare no conflict of interest.

Nomenclature

A	Area
CC	Composite curve
CFD	Computational fluid dynamics
CDF	Cumulative distribution functions
CIP	Cleaning in place
FTVM	Fixed temperature, variable mass
$\dot{H}$	Heat flow
HRL	Heat recovery loop
HRR	Heat recovery rate
H/h_{tm}	Medium height of the thermocline
H/h_{tl}	Lower height of the thermocline
H/h_{tu}	Upper height of the thermocline
ISSP	Indirect source and sink profile
$\dot{m}$	Mass flow
MC	Monte Carlo
MHR	Maximum heat recovery
MN	Multi-node
NTU	Number of transfer units
PIC	Percentage of ideal case
$\dot{Q}_i$	Thermal load
SROP	Stream-wise repeat operation period
ST	Stratified tank
t	Time
TAM	Time average model
T_{cold}	Temperature of the cold-water reservoir of a stratified tank
T_{hot}	Temperature of the hot-water reservoir of a stratified tank
TSM	Time slice model
V	Volume
$\dot{V}$	Volume flow
VLH	Variable layer height
$\dot{W}$	Heat flow rate

References

1. Eurostat. *Energy Balance Sheets. 2015 Data. Statistical Books*; Publications Office of the European Union: Luxembourg, 2017; ISBN 978-92-79-69844-6.
2. Chan, Y.; Kantamaneni, R. Study on Energy Efficiency and Energy Saving Potential in Industry from possible Policy Mechanisms. 2015. Available online: https://ec.europa.eu/energy/sites/ener/files/documents/151201%20DG%20ENER%20Industrial%20EE%20study%20-%20final%20report_clean_stc.pdf (accessed on 28 November 2017).
3. Philipp, M.; Schumm, G.; Peesel, R.-H.; Walmsley, T.G.; Atkins, M.J.; Schlosser, F.; Hesselbach, J. Optimal energy supply structures for industrial food processing sites in different countries considering energy transitions. *Energy* **2018**, *146*, 112–123. [CrossRef]
4. Klemeš, J.J. *Handbook of Process Integration: Minimisation of Energy and Water Use, Waste and Emissions*; Woodhead Publishing: Cambridge, UK, 2013; ISBN 9780857095930.
5. Klemeš, J.; Dhole, V.R.; Raissi, K.; Perry, S.J.; Puigjaner, L. Targeting and design methodology for reduction of fuel, power and CO_2 on total sites. *Appl. Therm. Eng.* **1997**, *17*, 993–1003. [CrossRef]
6. Schumm, G.; Philipp, M.; Schlosser, F.; Hesselbach, J.; Walmsley, T.G.; Atkins, M.J. Hybrid-heating-systems for optimized integration of low-temperature-heat and renewable energy. *Chem. Eng. Trans.* **2016**, *52*, 1087–1092. [CrossRef]

7. Walmsley, T.G.; Walmsley, M.R.W.; Tarighaleslami, A.H.; Atkins, M.J.; Neale, J.R. Integration options for solar thermal with low temperature industrial heat recovery loops. *Energy* **2015**, *90*, 113–121. [CrossRef]
8. Olsen, D.; Liem, P.; Abdelouadoud, Y.; Wellig, B. Thermal energy storage integration based on pinch analysis—Methodology and application. *Chem. Ing. Tech.* **2017**, *89*, 598–606. [CrossRef]
9. Krummenacher, P. Contribution to the Heat Integration of Batch Processes (with or without Heat Storage). Ph.D. Thesis, École Polytechnique Fédérale de Lausanne, Lausanne, Switzerland, 2002.
10. Atkins, M.J.; Walmsley, M.R.W.; Neale, J.R. The challenge of integrating non-continuous processes—Milk powder plant case study. *J. Clean. Prod.* **2010**, *18*, 927–934. [CrossRef]
11. Walmsley, T.G.; Walmsley, M.R.W.; Atkins, M.J.; Neale, J.R. Integration of industrial solar and gaseous waste heat into heat recovery loops using constant and variable temperature storage. *Energy* **2014**, *75*, 53–67. [CrossRef]
12. Atkins, M.J.; Walmsley, M.R.W.; Neale, J.R. Process integration between individual plants at a large dairy factory by the application of heat recovery loops and transient stream analysis. *J. Clean. Prod.* **2012**, *34*, 21–28. [CrossRef]
13. Walmsley, M.R.W.; Atkins, M.J.; Riley, J. Thermocline management of stratified tanks for heat storage. *Chem. Eng. Trans.* **2009**, *18*, 231–236.
14. Baeten, B.; Confrey, T.; Pecceu, S.; Rogiers, F.; Helsen, L. A validated model for mixing and buoyancy in stratified hot water storage tanks for use in building energy simulations. *Appl. Energy* **2016**, *172*, 217–229. [CrossRef]
15. Campos Celador, A.; Odriozola, M.; Sala, J.M. Implications of the modelling of stratified hot water storage tanks in the simulation of CHP plants. *Energy Convers. Manag.* **2011**, *52*, 3018–3026. [CrossRef]
16. Powell, K.M.; Edgar, T.F. An adaptive-grid model for dynamic simulation of thermocline thermal energy storage systems. *Energy Convers. Manag.* **2013**, *76*, 865–873. [CrossRef]
17. Schlosser, F.; Peesel, R.-H.; Meschede, H.; Philipp, M.; Walmsley, T.G. Evaluation of a stratified tank based heat recovery loop via dynamic simulation. *Chem. Eng. Trans.* **2018**, *70*, 403–408. [CrossRef]
18. Meschede, H.; Dunkelberg, H.; Stöhr, F.; Peesel, R.-H.; Hesselbach, J. Assessment of probabilistic distributed factors influencing renewable energy supply for hotels using Monte-Carlo methods. *Energy* **2017**, *128*, 86–100. [CrossRef]
19. Lal, N.S.; Atkins, M.J.; Walmsley, M.R.W.; Walmsley, T.G. Accounting for stream variability in retrofit problems using Monte Carlo simulation. *Chem. Eng. Trans.* **2018**, *70*, 1015–1020. [CrossRef]
20. Walmsley, M.R.W.; Atkins, M.J.; Linder, J.; Neale, J.R. Thermocline movement dynamics and thermocline growth in stratified tanks for heat storage. *Chem. Eng. Trans.* **2010**, *21*, 991–996. [CrossRef]
21. Rabe, M.; Spieckermann, S.; Wenzel, S. *Verifikation und Validierung für die Simulation in Produktion und Logistik. Vorgehensmodelle und Techniken*; Springer: Berlin, Germany, 2008; ISBN 978-3-540-35282-2.
22. Dufo-López, R.; Pérez-Cebollada, E.; Bernal-Agustín, J.L.; Martínez-Ruiz, I. Optimisation of energy supply at off-grid healthcare facilities using Monte Carlo simulation. *Energy Convers. Manag.* **2016**, *113*, 321–330. [CrossRef]
23. Dunkelberg, H.; Sondermann, M.; Meschede, H.; Hesselbach, J. Assessment of flexibilisation potential by changing energy sources using Monte Carlo simulation. *Energies* **2019**, *12*, 711. [CrossRef]
24. Nijhuis, M.; Gibescu, M.; Cobben, J.F.G. Bottom-up Markov chain Monte Carlo approach for scenario based residential load modelling with publicly available data. *Energy Build.* **2016**, *112*, 121–129. [CrossRef]
25. VDI. *VDI-Wärmeatlas. Mit 320 Tabellen*; Springer Vieweg: Berlin, Germany, 2013; ISBN 3642199801.

 © 2019 by the authors. Licensee MDPI, Basel, Switzerland. This article is an open access article distributed under the terms and conditions of the Creative Commons Attribution (CC BY) license (http://creativecommons.org/licenses/by/4.0/).

Article

Flow Boiling Heat Transfer Characteristics in Horizontal, Three-Dimensional Enhanced Tubes

Zhi-Chuan Sun [1], Xiang Ma [2], Lian-Xiang Ma [2], Wei Li [1,*] and David J. Kukulka [3,*]

1. Department of Energy Engineering, Zhejiang University, 38 Zheda Road, Hangzhou 310027, China; sunzhichuan@zju.edu.cn
2. Department of Mechanical and Electrical Engineering, Qingdao University of Science and Technology, 99 Songling Road, Qingdao 266061, China; maxiang7632@126.com (X.M.); oldhorse@qust.edu.cn (L.-X.M.)
3. Department of Mechanical Engineering Technology, State University of New York College at Buffalo, 1300 Elmwood Avenue, Buffalo, NY 14222, USA
* Correspondence: weili96@zju.edu.cn (W.L.); kukulkdj@buffalostate.edu (D.J.K.)

Received: 10 January 2019; Accepted: 2 March 2019; Published: 10 March 2019

Abstract: An experimental investigation was conducted to explore the flow boiling heat transfer characteristics of refrigerants R134A and R410A inside a smooth tube, as well as inside two newly developed surface-enhanced tubes. The internal surface structures of the two enhanced tubes are comprised of protrusions/dimples and petal-shaped bumps/cavities. The equivalent inner diameter of all tested tubes is 11.5 mm, and the tube length is 2 m. The experimental test conditions included saturation temperatures of 6 °C and 10 °C; mass velocities ranging from 70 to 200 kg/(m^{2}s); and heat fluxes ranging from 10 to 35 kW/m^2, with inlet and outlet vapor quality of 0.2 and 0.8. It was observed that the enhanced tubes exhibit excellent flow boiling heat transfer performance. This can be attributed to the complex surface patterns of dimples and petal arrays that increase the active heat transfer area; in addition, more nucleation sites are produced, and there is also an increased interfacial turbulence. Results showed that the boiling heat transfer coefficient of the enhanced surface tubes was 1.15–1.66 times that of the smooth tubing. Also, effects of the flow pattern and saturated temperature are discussed. Finally, a comparison of several existing flow boiling heat transfer models using the data from the current study is presented.

Keywords: flow boiling; surface-enhanced tube; heat transfer coefficient; flow pattern

1. Introduction

Heat transfer enhancement technologies offer more design options for increasing the thermal efficiency of a heat transfer unit. Nowadays, high-efficiency compact heat exchangers have received significant attention in a wide variety of industrial applications. Evaporators and condensers are the important components of a variety of heating and cooling systems. Hence, high-performance heat exchange pipes with enhanced surface structures need to be designed as the basic element of a heat exchanger. Besides, Thermodynamic characteristics of refrigerants also play a vital role in the flow boiling heat-transfer process. R134A is a widely-used working fluid in refrigerator and automobile air conditioning, and it is recognized as the best substitute for R12. R410A has replaced R22 in many applications and it is a kind of near-azeotropic refrigerant (R32/R125 mixture). The thermophysical properties and environmental protection indexes of R134A and R410A are given in Table 1.

Table 1. Thermophysical and environment properties of refrigerants R134A and R410A.

Refrigerant	R134A (Pure Refrigerant)	R410A (Near-Azeotropic Refrigerant)	
Composition	$C_2H_2F_4$	R32, R125, 50/50 (weight percent)	
ASHRAE safety	A1	A1	
ODP	0	0	
GWP	1430	2100	
Molecular	102	72.6	
P_c (kPa)	4066	4950	
T_c (°C)	101.1	72.5	
Saturation Properties of Refrigerants			
T_{sat}	6 °C	6 °C	10 °C
P_{sar} (kPa)	361.98	965.29	1088.4
P_r (-), P_{sar}/P_c	0.089	0.195	0.220
ρ_l (kg/m^3)	1274.7	1145.4	1128.4
ρ_v (kg/m^3)	17.72	36.35	41.177
μ_l (Pa·s)	2.47×10^{-4}	1.50×10^{-4}	1.43×10^{-4}
μ_v (Pa·s)	1.09×10^{-5}	1.25×10^{-5}	1.27×10^{-5}
k_l (W/m·K)	0.089	0.100	0.097
Pr_l	3.753	1.183	2.315
σ (N/m)	0.01060	0.00813	0.00753
h_{lv} (kJ/kg)	194	219	213

Previously reported studies related to in-tube heat transfer enhancement have been typically passive enhancement techniques, which modify the surface structures, material composition, or fluid type to enhance the two-phase heat transfer performance of a single tube or a tube bundle. In this study, heat transfer enhancement was obtained by using surface-enhanced tubes with surface modifications. Investigations of enhanced tubes with two-dimensional roughness are common in the open literature, such as studies on microfin tubes with small helical internal fins [1–6]. However, researches on three-dimensional enhanced tubes are relatively scarce, to the authors' knowledge.

Kukulka et al. [7] tested the overall thermal characteristics of four types of three-dimensional (3-D) enhanced tubes with staggered dimples and petal arrays. These enhanced surface tubes show superior heat transfer characteristics through the mixed effects of surface structures, which include increased heat transfer areas and interfacial turbulence, secondary flow generation, and boundary layer disruption. After that, Kukulka et al. [8,9] experimentally studied the tube-side condensation and evaporation characteristics of flows in these surface-enhanced tubes (namely EHT series tubes). Guo et al. [10] evaluated the evaporation heat transfer of R22/R32/R410A inside a plain tube, a herringbone micro-fin tube and a dimpled tube enhanced by petal-shaped patterns (1EHT). Their results indicate that the 1EHT tube presents good evaporation heat transfer performance for the entire mass flux range, mainly due to the large number of nucleation sites generated by the special surface structures. Li and Chen [11,12] studied the condensation and evaporation characteristics of R410A inside two EHT tubes (2EHT) and one smooth tube. According to their experimental results, an increase of mass flux results in a rise in the heat transfer coefficient and frictional pressure loss. The 2EHT tubes exhibited superior heat transfer performance under the same operating conditions. In addition, the higher evaporation coefficient was found at a relatively low wall superheat. Shafaee et al. [13] discussed the flow boiling characteristics inside smooth and helically dimpled tubes with R600a as the working fluid. Ayub et al. [14] investigated the flow boiling heat transfer of refrigerant R134A in a dimpled tube. In order to create the in-tube annular flow passage, a round plastic rod was inserted in the test tube. The enhanced tube having the rod exhibited the higher heat transfer coefficient three times as that of an equivalent smooth tube. Kundu et al. [15] measured the boiling heat transfer coefficient and pressure loss of R134A and R407C in a 9.52-mm OD smooth tube. Tests were carried out over the mass flux range of 100–400 kg/m^{2}s, with heat fluxes changing from 3 to 10 kW/m^2.

They found that the flow boiling coefficient raised as the mass flux or heat flux increased. It was also found that the measured coefficient for R134A was higher than that for R407C at the same mass fluxes. Lillo et al. [16] analyzed the flow boiling in a stainless-steel smooth tube with an inside diameter of 6.0 mm using R32 and R410A. They noticed that the evaporating coefficient for R32 was larger than that for R410A under the same test conditions. Greco and Vanoil [17] tested the boiling heat-transfer coefficients of a horizontal smooth tube using different refrigerants (R22, R134A, R507, R404A and R410A). Results indicated an increase in heat transfer coefficient with the increment of saturation temperature and heat flux.

Additionally, the size of channel and flow orientations also play an important role on the flow boiling heat transfer. Li and Wu [18] presented a micro/mini channel criterion for evaporation heat transfer. They reported that saturated-flow boiling characteristics in micro/mini-channels could be different from those in conventional channels. Jige et al. [19] performed an experimental research on flow boiling in small-diameter tubes using refrigerant R32. Their results show that the heat transfer coefficient increases with the decreasing tube diameter. Taking into consideration the effect of tube diameter, Saitoh et al. [20] developed a general correlation for in-tube flow boiling heat transfer by predicting the dry-out quality, which is based on the Chen-type correlation [21]. Recently, Sira et al. [22] studied the flow regimes and evaporation characteristics of R134A in a mini-channel having an internal diameter of 0.53 mm respectively for horizontal and vertical flow orientations. Their results revealed the importance of flow direction. The higher evaporating coefficient can be obtained when the two-phase refrigerant flows towards the vertical downward direction. A summary of previous literature is given in Table 2.

Table 2. Summary of previously published studies on evaporation inside a tube.

Authors	Tube	d_o (mm)	Refrigerant	G (kg/m^{2}s)	q (kW/m^2)	x (-)
Yu te al. [1]	Smooth tube/Micro-fin tube	10.7	R134A	163–408	2.2–56	0.1–0.9
Spindler and Müller-Steinhagen [2]	Micro-fin tube	9.52	R134A/R404A	25–150	1–15	0.1–0.7
Rollmann and Spindler [3]	Micro-fin tube	9.52	R407C/R410A	25–300	1–20	0.1–1.0
Padovan et al. [4]	Micro-fin tube	7.69	R134A/R410A	80–600	14–83.5	0.1–0.99
Celen et al. [5]	Smooth tube/Micro-fin tube	9.52	R134A	190–381	10	0.2–0.77
Wu et al. [6]	Micro-fin tube	5.00	R22/R410A	100–620	5–31	0.1–0.8
Kukulka et al. [8]	1EHT tube/3EHT tube	12.7	R410A	80–180	-	0.2–0.9
Kukulka et al. [9]	1EHT tubes/4EHT tube	9.52	R410A	160–390	-	0.2–0.8
Guo et al. [10]	Smooth tube/Herringbone micro-fin tube/1EHTtube	12.7	R22/R32/R410A	50–150	13.9–36	0.1–0.9
Li and Chen [11,12]	Smooth tube/2EHT tube	12.7	R410A	60–175	-	0.1–0.9
Shafaee et al. [13]	Smooth tube/Helical dimpled tube	9.45	R600a	155–470	15.8	0–0.8
Ayub et al. [14]	Enhanced dimpled tube	19.05	R134A	80–200	2.5–15	0.12–0.72
Kundu et al. [15]	Smooth tube	9.52	R134A/R407C	100–400	3–10	0.1–0.9
Lillo et al. [16]	Smooth tube	6.00	R32	146–507	2.4–41.2	0.02–0.99
Greco and Vanoli [17]	Smooth tube	6.00	R22/R134A/R507C/R404A/R410A	360	11–21	0–1

Table 2. *Cont.*

Authors	Tube	d_o (mm)	Refrigerant	G (kg/m^{2}s)	q (kW/m^2)	x (-)
Jije et al. [19]	Horizontal small-diameter tube	1.00/2.20/3.50	R32	50–600	5–40	0–1
Saitoh et al. [20]	Horizontal circular mini-channel	1.75	R134A	200–1000	1–83	0–1
Sira et al. [22]	Horizontal and vertical mini-channels	1.00	R134A	250–820	1–60	0.1–0.9

Only a few previous studies exist for flow boiling inside the 3-D enhanced heat transfer tubes that are considered in this research. In contrast to micro-fin tubes, these test tubes are two-layer, two-sided, enhanced surface tubes that are designed using shallow and deep cavities/protrusions, as can be seen in Figure 1. These surface-enhanced tubes, made of copper, can produce more nucleation sites, mainly owing to the petal-shaped cavities/protrusions in staggered arrangement. Specifically, the EHT tube has shallow, petal-shaped cavities in a web-like structure, and staggered deep dimples on the external surface, while the same enhanced patterns are located on the internal surface of the Re-EHT tube. Similarly, both the EHT concave and Re-EHT convex exhibit the dimpled protrusions and raised petal-shaped patterns in staggered rows. As also shown in Figure 1, the primary surface structures (dimple/protrusion) of the EHT tube and the Re-EHT tube have a height of 1.71 mm/1.81 mm and a projection diameter of 4.4 mm/4.0 mm. The pitch of dimple is 9.86 mm with a helix angle of 60°, because of the staggered arrangement of the dimples/protrusions. Using the Nanovea ST400 non-contact profilometer, the EHT tube was found to have a 20% increase in inner surface area compared with the smooth tube, and the Re-EHT tube indicates a 34% surface area increase. Details of these test tubes are listed in Table 3.

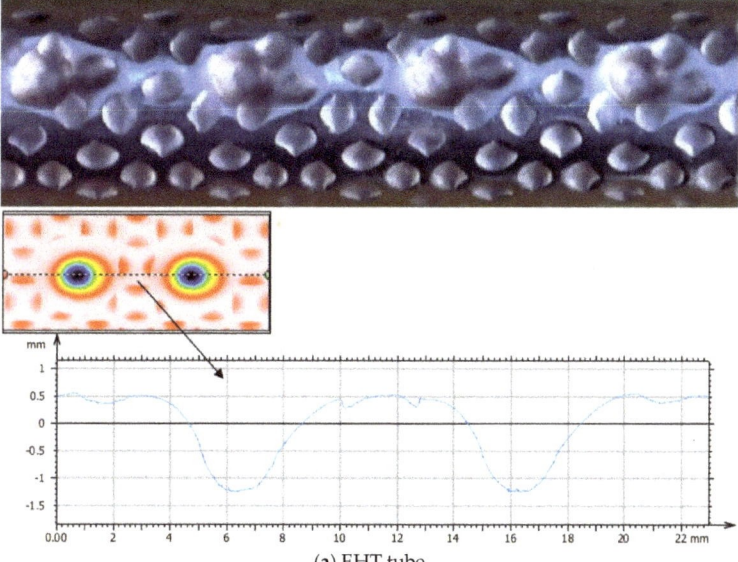

(a) EHT tube

Figure 1. *Cont.*

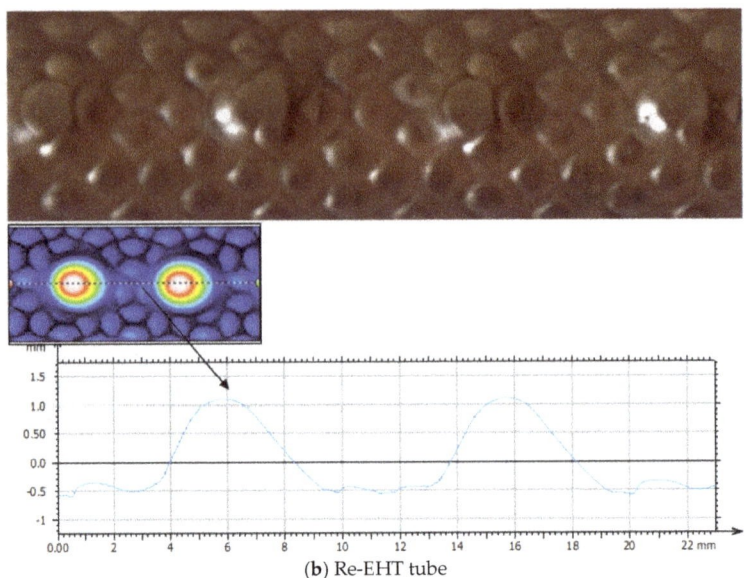

Figure 1. The external surface structure of the (a) EHT tube and (b) Re-EHT tube.

Table 3. Details of the test tubes.

Tube Type	EHT Tube	Re-EHT Tube	Smooth Tube
Inside diameter of inner tube d_i (mm)		11.5	
Outside diameter of inner tube d_o (mm)		12.7	
Average wall thickness (mm)		0.6	
Height of dimple/protrusion (mm)	1.71	1.81	-
Projection diameter (mm)	4.4	4.0	-
Dimpled/protruded pitch (mm)	9.86	9.86	-
Helix angle, deg	60	60	-
Ratio of actual heat transfer area, A_E/A_S	1.20	1.34	1
Inside diameter of outer tube D_i (mm)		17.0	
Tube length L (m)		2.0	
Tube material		Copper	
Thermal conductivity (W/m^2·K)		379	

The main objective of this work is to experimentally study the heat transfer characteristics of R134A and R410A during flow boiling in two horizontal, surface-enhanced tubes and one smooth tube. In addition, the effects of flow pattern, mass velocity, and saturation temperature on the flow boiling is also analyzed and discussed.

2. Experimental Procedure

2.1. Test Apparatus

The schematic diagram of the test apparatus utilized to evaluate the flow boiling heat transfer characteristics of R134A and R410A inside circular tubes is shown in Figure 2a. It was composed of three closed circuits: (1) a refrigerant circuit, the major component of the test system; (2) a recycled water circuit used to exchange heat with the refrigerant and regulate the outlet vapor quality of the test section by controlling the inlet temperature and mass flow rate of water; and (3) a condensation section, which is used to cool the saturated refrigerant leaving the test tube at a given temperature. The refrigerant circuit comprised of a storage tank, a digital gear pump, a Coriolis mass flow meter, a preheating section, a test section, and several flow regulating valves. Sub-cooling refrigerant in the reservoir was sent to the test system by a gear pump regulated by a frequency converter. An oil separator was used to decrease the mass friction of lubricating oil in the liquid refrigerant. After that, a Coriolis mass flow meter (with a test accuracy of 0.2% of reading) was fixed to monitor the refrigerant mass flux. The inlet vapor quality of the test section can be calculated by measuring the water mass flow rate and water temperatures at the entrance and exit to the pre-heater.

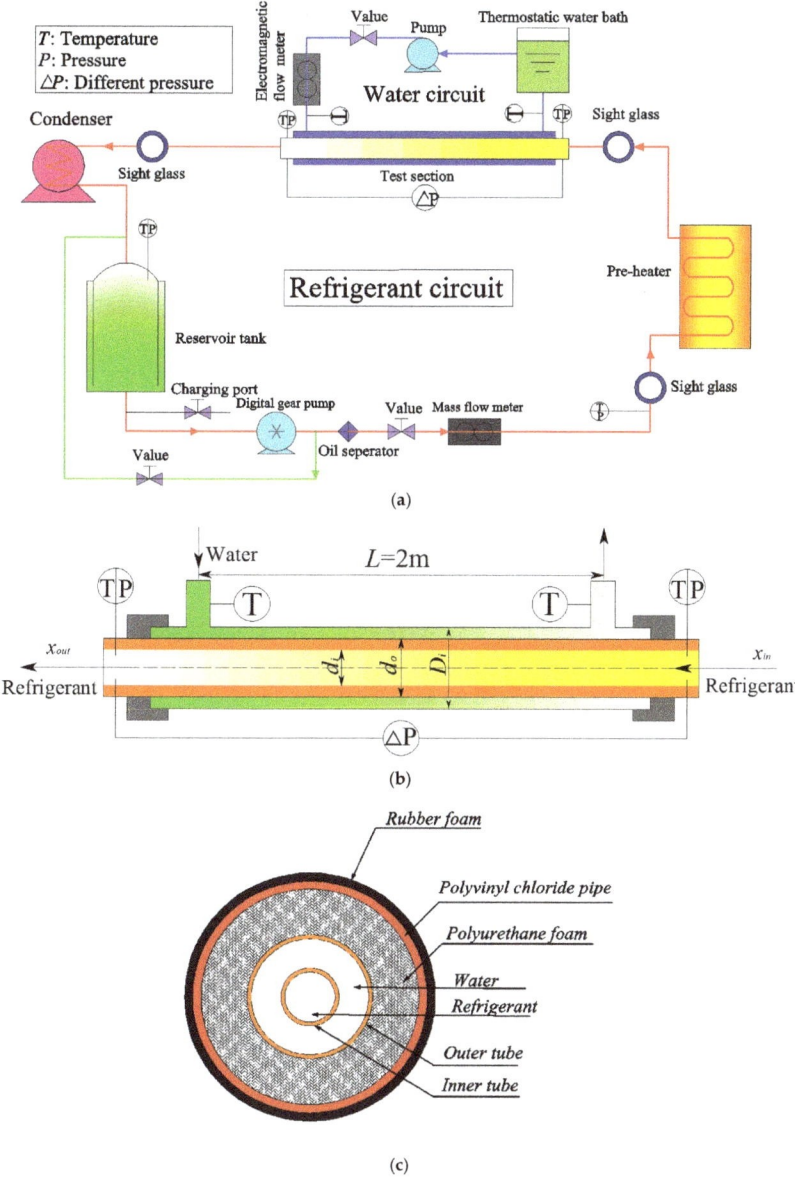

Figure 2. Schematic diagram of the (**a**) experimental system, (**b**) test section, (**c**) cross-section of the test section.

The two-phase refrigerant entered the test section and was evaporated immediately. The water circuit consisted of a thermostatic water bath, a centrifugal water pump, a magnetic flow meter, and several valves. As described in Figure 2b, the test section is a typical horizontal counter-flow-type, double-pipe heat exchanger with a heated length of 2.0 m. Water flowing in the annulus side of the test section provided heat energy for the refrigerant from the preheating section. Water mass flux is measured by a Coriolis mass flow meter with an accuracy of 0.2% of real-time reading. Meanwhile, the inlet and outlet temperatures of refrigerant were measured with Platinum RTD-100 temperature transducers with a testing precision of ± 0.1 °C. In addition, the inlet and outlet absolute pressure of the refrigerant side of the test section were measured by two pressure transducers, and the total pressure loss across the test tube was obtained by a differential pressure gauge. All pressure measure instruments have a test accuracy of 0.075% of the reading. Then, the two-phase refrigerant entered the condensation section, where it was sub-cooled to at least 10 °C lower than the given saturated temperature. Lastly, liquid refrigerant was sent into a reservoir tank with a 50-L capacity.

Figure 2c provides the details of a cross-section of the tested tube. The average wall thickness of the enhanced surface tube is 0.6 mm. All the test tubes have the same outside diameter of 12.7 mm, and the maximum inside diameter is 11.5 mm. The outer tube is a smooth copper tube with an inside diameter of 17 mm. Table 3 lists the main dimensional parameters of the two enhanced tubes and one plain tube. To minimize the heat loss to surroundings, the entire test section was insulated in a large PVC circle pipe with an outer diameter of 110 mm. Polyurethane foam (approximately 90 mm thick) was filled into the gap between the PVC pipe and the outer tube to provide an insulation layer. Furthermore, a 10-mm-thick rubber foam was used to tightly wrap the PVC pipe.

To evaluate the heat insulation of the entire test section, two single-phase tests were performed to investigate heat loss using R134A and R410A. Figure 3 illustrates the results of single-phase heat balance measurements for the test section. It can be seen that the deviations between the water-side heat flow rate ($Q_{w,ts}$) and refrigerant-side heat flow rate ($Q_{ref,ts}$) are lower than 5%. This ensures that the heat loss in the experimental apparatus can be neglected, due to its insignificant influence on the flow boiling heat transfer.

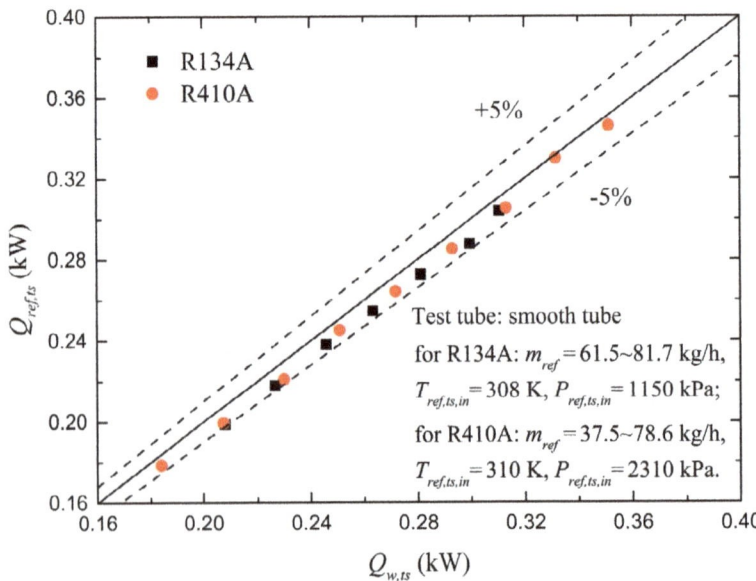

Figure 3. Heat balance measurements for single-phase flow in the smooth tube.

2.2. Experimental Test Conditions

For every test, the refrigerant mass velocities (determined by the actual cross-sectional area of the inner tube, $A_{c,ref}$) were varied, while the saturated pressure, water mass flux, inlet, and outlet vapor qualities of the test section were kept constant. The test range of flow boiling conditions are summarized in Table 4. All temperature and pressure signals were stored by a 16-bit 20-channel data collection card, and then the collected data were relayed in real time to a host computer. In order to ensure the steady state conditions, data points were collected over 200 s with 20-s intervals. During this period, the deviations of temperature, pressure, and vapor quality were below 0.1 °C, 5 kPa, and 0.05, respectively.

Table 4. Experimental conditions.

Parameters	Range	
Refrigerant	R134A	R410A
Saturation temperature T_{sat}, (°C)	6	
Refrigerant mass velocity G, (kg/m²s)	70–200	
Heat flux q, (kW/m²)	10–40	
Inlet vapor quality x_{in}	0.2	
Outlet vapor quality x_{out}	0.8	

3. Data Reduction and Uncertainty Analysis

3.1. Data Reduction

In this paper, heat transfer data was reduced in order to calculate the vapor quality, heat flow rate, and evaporation coefficient. For the test section, the water-side heat transfer rate was calculated by the heat balance equation

$$Q_{w,ts} = c_{p,w,ts} m_{w,ts} (T_{w,ts,in} - T_{w,ts,out}) \quad (1)$$

Here, $c_{p,w,ts}$, $m_{w,ts}$, $T_{w,ts,in}$, and $T_{w,ts,out}$ represent the specific heat of water taken at the mean bulk temperature, the mass flow rate of the recycled water, and the water temperatures at the entrance and exit to the annular channel, respectively. The heat flux q was calculated from Equation (2), by using the inner surface area A_i based on the maximum diameter d_i:

$$q = Q_w / A_i \quad (2)$$

Vapor quality at the test section inlet, x_{in}, can be determined by the heat energy conservation in the preheating section. The total heat transfer rate in the pre-heater, Q_{pre}, was calculated in Equation (3), and was composed of sensible heat (Q_{sens}) and latent heat (Q_{lat}):

$$Q_{pre} = c_{p,w,pre} m_{w,pre} (T_{w,pre,in} - T_{w,pre,out}) = Q_{sens} + Q_{lat} \quad (3)$$

$$Q_{sens} = c_{p,ref,pre} m_{ref} (T_{sat} - T_{ref,pre,in}) \quad (4)$$

$$Q_{lat} = m_{ref} h_{lv} x_{in} \quad (5)$$

$$x_{in} = \frac{Q_{pre} - c_{p,ref,pre} m_{ref} (T_{sat} - T_{ref,pre,in})}{m_{ref} h_{lv}} \quad (6)$$

where $c_{p,w,pre}$ and $m_{w,pre}$ represent the specific heat and mass flow rate, respectively, of hot water flowing across the pre-heater. In addition, $T_{w,pre,in}$, $T_{w,pre,out}$, and T_{sat} are defined as the water temperatures at the preheating section inlet and outlet, and the saturation temperature of refrigerant. Additionally, $c_{p,ref,pre}$, $T_{ref,pre,in}$, m_{ref}, and h_{lv} are the special heat taken at the mean bulk temperature, inlet temperature, mass flow rate, and latent heat of vaporization of the refrigerant flowing through the pre-heater coils, respectively. As a consequence, the outlet vapor quality of the test section (x_{out}) is defined as follows:

$$x_{out} = x_{in} + \frac{Q_{w,ts}}{m_{ref} h_{lv}} \quad (7)$$

The logarithmic temperature difference for a tube-in-tube heat exchanger was calculated using the water and refrigerant temperatures at the inlet and outlet:

$$LMTD = \frac{\left(T_{w,ts,in} - T_{ref,ts,out}\right) - \left(T_{w,ts,out} - T_{ref,ts,in}\right)}{\ln\left[\left(T_{w,ts,in} - T_{ref,ts,out}\right)/\left(T_{w,ts,out} - T_{ref,ts,in}\right)\right]} \quad (8)$$

Here, $T_{ref,ts,in}$ and $T_{ref,ts,out}$ represent the refrigerant temperatures at the test section inlet and outlet, respectively. Assuming no fouling thermal resistance, the tube-side evaporating coefficient (h_i) can be calculated using the following correlation:

$$h_i = \frac{1}{A_i \left[\frac{LMTD}{Q_{w,exp}} - \frac{1}{h_o A_o} - \frac{\ln(d_o/d_i)}{2\pi L \cdot k}\right]} \quad (9)$$

where k is the thermal conductivity of wall material, and h_o is the water-side heat transfer coefficient. It is worth noting that A_o is the external surface area, decided by the nominal outside diameter (d_o).

Gnielinski [23] presents a classical correlation that is widely used to predict the single-phase heat transfer coefficient for smooth tubing or annuli. This correlation is applicable for $3000 < Re_w < 5 \times 10^6$ and $0.5 < Pr_w < 2000$:

$$h_o = \frac{(f/2)(Re_w - 1000)Pr_w}{1 + 12.7(f/2)^{0.5}\left(Pr_w^{2/3} - 1\right)} \left(\frac{\mu_{bulk}}{\mu_{wall}}\right)^{0.14} \frac{k_w}{d_w} \quad (10)$$

The dynamic viscosity ratio, $(\mu_{bulk}/\mu_{wall})^{0.14}$, can be evaluated using the average value of the bulk temperatures of water and inner wall. The results show the property differences in this study to be no more than 1%. In addition, d_w is the water-side hydraulic diameter, which is equal to the inside diameter. The fanning friction factor (f) can be determined from the Petukhov correlation [24] given by Equation (11). For a smooth tube, the range of application of the prediction correlation is $3000 < Re_w < 5 \times 10^6$:

$$f = (1.58 \ln Re_w - 3.28)^{-2} \quad (11)$$

Since the internal and external surfaces of the EHT tubes are rough, due to the special surface structures, and the Gnielinski correlation [23] only applies to the smooth tubing, a water-side heat transfer enhancement factor C, decided by the Wilson plot method [25], was utilized to modify the Gnielinski correlation [23]. In fact, the factor C is the heat transfer coefficient ratio of the enhanced surface tubes to an equivalent plain tube. Then, the overall thermal resistance of the double-pipe heat exchanger for enhanced tubes can be calculated by following equation:

$$\frac{1}{C \cdot h_o} = \frac{1}{U} - \frac{d_o}{d_i h_i} - \frac{d_o \ln(d_o/d_i)}{2k} \quad (12)$$

where U is the overall heat transfer coefficient of the test section.

At a given large mass flow rate of refrigerant, the inner thermal resistance and the wall thermal resistance can be considered as a constant value. Therefore, the water-side/shell-side heat transfer coefficient (h_o) can be determined by varying the temperature and mass flux of the recycled water.

As depicted in Figure 4, the Wilson plot tests were done using the data points of refrigerants R134A and R410A. The water-side heat transfer enhancement factor C can be calculated directly by applying a linear regression method. It is found that the term C is only related to the special surface structures, and does not depend on the working fluid. According to the experimental results, the appropriate value of C is 2.70 for the EHT tube and 2.29 for the Re-EHT tube. This heat transfer enhancement is attributed to the dimples/protrusions on the tube wall. Something else to note is that all thermodynamic properties of R134A and R410A were obtained from REFPROP 9.1 software [26].

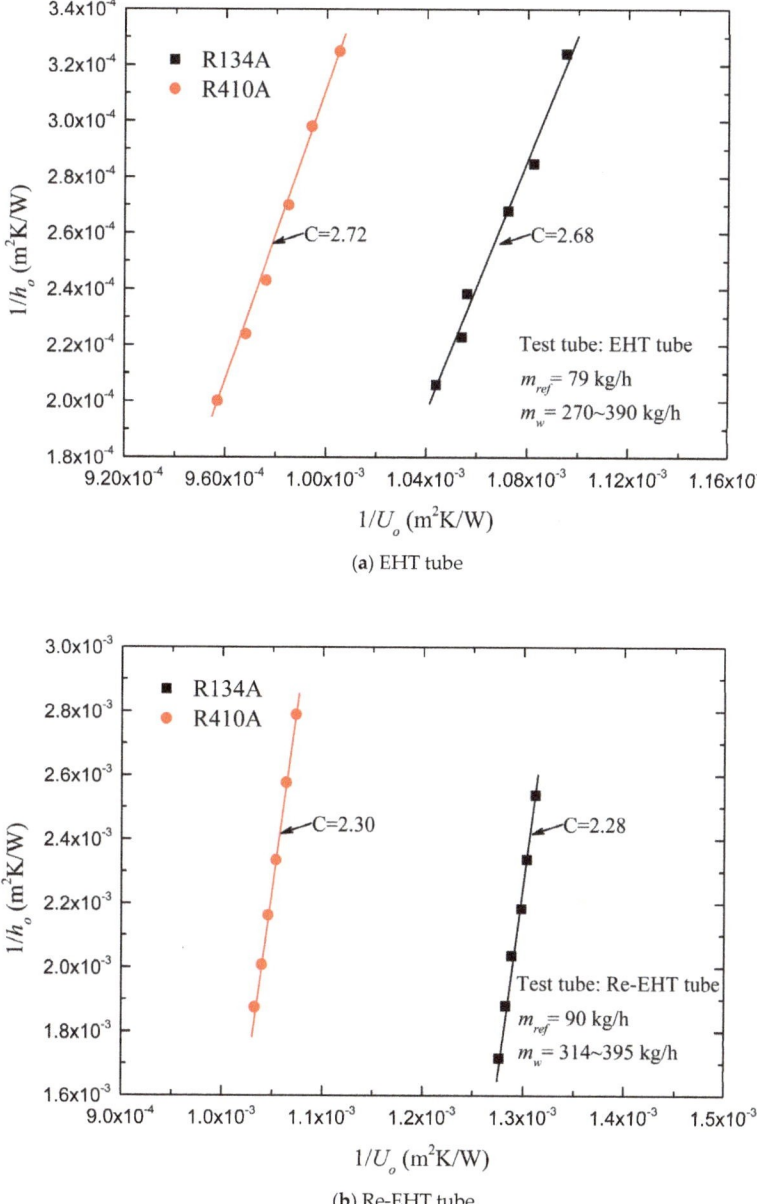

Figure 4. Test results of Wilson plot [23] of the (**a**) EHT tube and (**b**) Re-EHT tube.

3.2. Experimental Uncertainty Analysis

The measurement error strongly depends on the flow boiling conditions and the accuracy of the thermocouples, pressure gauges and flowmeter. Uncertainties in the measured and calculated parameters were estimated by an error-delivering algorithm, as described in Moffat [27]. According to previously published papers, the relative uncertainty (U_R) of the experimental parameter can be calculated using the following equation:

$$U_R = \left[\sum_{i=1}^{n}\left(\frac{\partial R}{\partial X_i}U(X_i)\right)^2\right]^{1/2} \quad (13)$$

The relative error of the heat transfer rate supplied by the hot water flowing in the annulus side, $U(Q_{ts})$, can be calculated from the energy conservation in the test section:

$$U(Q_{ts}) = \sqrt{\frac{\partial^2(m_{w,ts})}{m_{w,ts}^2} + \frac{\partial^2(T_{w,ts,in}) + \partial^2(T_{w,ts,out})}{(T_{w,ts,in} - T_{w,ts,out})^2}} \quad (14)$$

Then, Equation (15) gives the relative uncertainty of the refrigerant mass velocity $U(G_{ref})$:

$$U(G_{ref}) = \sqrt{\frac{\partial^2(m_{ref})}{m_{ref}^2} + 4\frac{\partial^2(d_i)}{d_i^2}} \quad (15)$$

As a result, the measurement uncertainty in boiling heat-transfer coefficient can be expressed in the following form:

$$U(h_{ref}) = \sqrt{U^2(R) + U^2(A_i)} \quad (16)$$

with

$$R = \frac{LMTD}{Q_{ts}} + \frac{1}{h_w A_o} + \frac{\ln(d_o/d_i)}{2\pi L k_{wall}} \quad (17)$$

where R is the overall thermal resistance and $LMTD$ is the logarithmic mean temperature difference. Besides, k_{wall} is the thermal conductivity of the wall material.

On the basis of the previous results in open literature, the Gnielinski correlation [23] usually leads to a deviation of up to 10%. Table 5 gives a summary of the experimental uncertainties of the measured and calculated parameters. The results indicate that the maximal error of the in-tube heat transfer coefficient is estimated to be 8.34%. Thus, the test system is proven to be reliable and stable.

Table 5. Uncertainties of measured and calculated parameters.

Measured Parameters	Uncertainty
Diameter	±0.05 mm
Length	±0.5 mm
Temperature	±0.1 K
Pressure, range: 0–5000 kPa	±0.075% of full scale
Differential pressure, range: 0–50 kPa	±0.075% of full scale
Refrigerant mass flow rate, range: 0–120 kg/h	±0.2% of reading
Water mass flow rate, range: 0–600 kg/h	±0.2% of reading
Calculated Parameters	**Uncertainty**
Mass velocity G, (kg/m^{2}s)	±1.17%
Heat flux q (kW/m^2)	±2.64%
Vapor quality x	±3.96%
Heat transfer coefficient h (W/m^{2}K)	±8.34%

4. Results and Discussion

4.1. Single-Phase Heat Transfer

Figure 5 shows the experimental Nusselt number (Nu) of R134A during flow boiling in the two enhanced tubes and one smooth tube, versus the Reynolds number (Re). Results indicate that the Nusselt number of the Re-EHT tube is about 34% higher than that of the smooth tube, and the EHT tube exhibits an impressive 50% increase in Nusselt number. The enhancement of the heat transfer characteristics can be attributed to the stronger interfacial turbulence effect and boundary layer disruption caused by the protrusions/dimples and staggered petal arrays. The solid line represents the Nu-Re curve of the smooth tube predicted by the Dittus-Boelter correlation [28], and the dashed line represents the predictions provided by the Gnielinski correlation [23]. As a result, both the two widely-used single-phase heat transfer models show well-accepted predictive ability, showing a maximum deviation error of 10%.

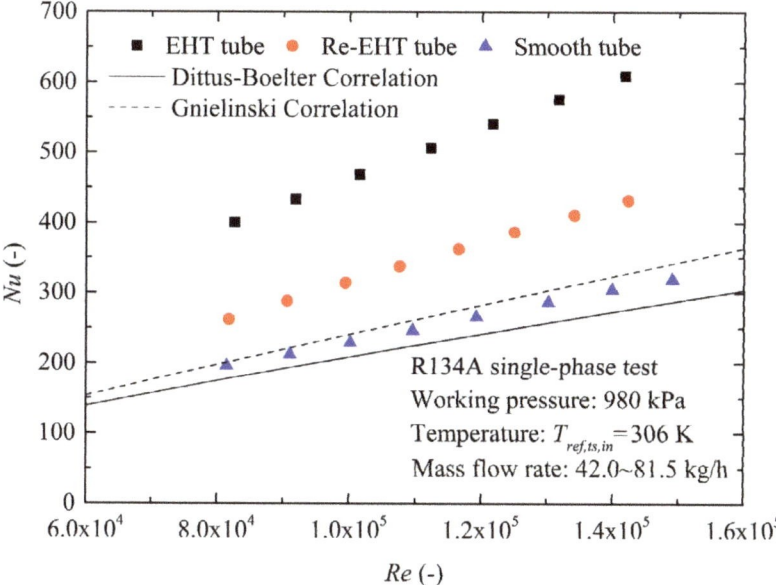

Figure 5. Variation of Nusselt number with Reynolds number for single-phase heat transfer in two enhanced tubes and one smooth tube.

4.2. Flow Pattern Analysis

The Wojtan et al. [29] flow pattern map has been widely used for boiling flow heat transfer in horizontal smooth tubes. Figure 6 shows the predicted flow pattern in this study for R134A and R410A, at G_{ref} = 100 kg/m^{2}s, T_{sat} = 6 °C, and d_i = 11.5 mm. Obviously, it can be inferred that the main flow regimes are slug flow, stratified-wavy flow, intermittent flow, and annular flow. In the smooth tube tested, the flow patterns are the slug flow and stratified-wavy flow, using R134A and R410A as the working fluid at low mass fluxes. When G_{ref} > 150 kg/m^{2}s, the intermittent low and annular flow will occur according to the flow pattern map in the vapor quality range of 0.2–0.8. The local dry-out could appear as the mass velocity and vapor quality increases, when G_{ref} > 200 kg/m^{2}s and x > 0.9. Similar to the results reported in [4,16,29], evaporating coefficients tend to decrease with the increment of the vapor quality. For the boiling heat transfer process, nucleate boiling is dominant in the low-quality region, while the contribution of convective boiling increases as the mass velocity increases. Intermittent and annular flow patterns are usually considered as the optimal heat transfer patterns due to the smaller internal thermal resistance caused by the thin liquid film. The transition vapor quality from intermittent flow to annular flow, x_{IA}, can be determined by the Kattan-Thome model [30] and is given by:

$$x_{IA} = \left\{ \left[0.34^{1/0.875} \left(\frac{\rho_v}{\rho_l} \right)^{-1/1.75} \left(\frac{\mu_l}{\mu_v} \right)^{-1/7} \right] + 1 \right\}^{-1} \tag{18}$$

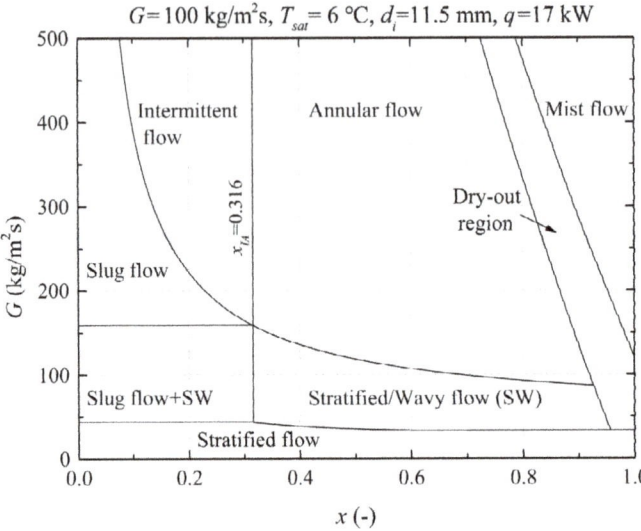

(a) R134A evaporation with G_{ref} = 100 kg/m²s and q = 17 kW at T_{sat} = 6 °C

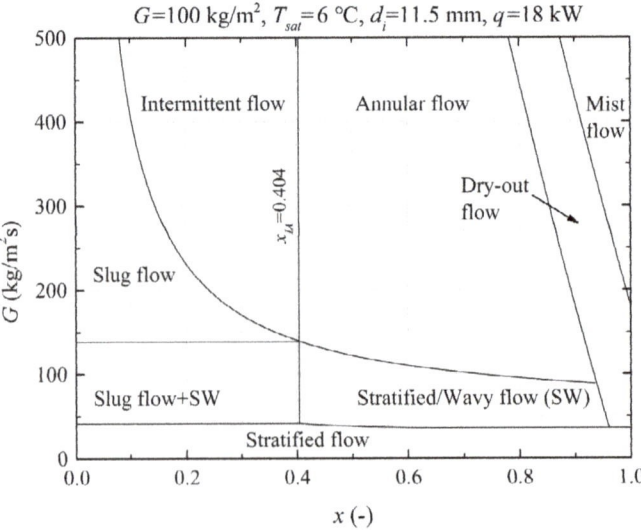

(b) R410A evaporation with G_{ref} = 100 kg/m²s and q = 18 kW at T_{sat} = 6 °C

Figure 6. Flow pattern map of Wojtan et al. [29] for flow boiling in a smooth tube with an inside diameter of 11.5 mm.

As a result, the transition quality x_{IA} is 0.316 for R134A and 0.404 for R410A. For R134A, flow boiling in a horizontal smooth tube, slug flow, and stratified-wavy flow dominate the flow boiling condition over the mass flux range of 70–150 kg/m²s. In this case, intermittent flow and annular flow may appear only in the high-quality region. For the case of R410A, the flow mechanism will be dominated by slug flow and stratified-wavy flow when G_{ref} is less than 140 kg/m²s, while annular flow and intermittent flow occur at higher vapor quality values when G_{ref} > 140 kg/m²s.

For the EHT and Re-EHT tubes, there is no available information of flow pattern map from the previous studies. A flow pattern analysis that predicts the flow boiling inside the enhanced tubes was made. When compared to smooth tubing, the transition from intermittent flow to annular flow is expected to happen at a lower vapor quality, while the transition line between stratified-wavy flow and annular flow trends to appear at lower mass velocities and vapor qualities. The primary dimples/protrusions and secondary petal arrays on the heated surface can pull the liquid film to distribute around the circumference, and force the bubbles to move towards the center of the tube. Previous studies [1,2,31] have reported flow boiling flow patterns of horizontal micro-fin tubes. It was found that the intermittent-to-annular-flow transition quality is lower than that of the smooth tube under the same operating conditions. The earlier transition of flow regimes in the micro-fin tube is mainly due to the liquid droplet entrainment effect of spiral mini-channels on the inner wall. Mashouf et al. [32] carried out a visualization study to observe evaporation flow patterns of R600a in horizontal dimpled and smooth tubes. At the same mass flux, the flow pattern transition in the dimpled tube occurred at a lower vapor quality in comparison with the equivalent smooth tube. It can be concluded that the dimples/protrusions are beneficial to the decrease of transition quality. Moreover, it is inferred that the intermittent/annular flow regimes dominate the entire test range for R134A and R410A flow boiling in the enhanced tubes. In a future study, a visualization observation will be performed to determine the flow patterns of working fluid in the EHT enhanced tubes.

4.3. Flow Boiling Heat Transfer Coefficient

Flow boiling heat transfer characteristics of R134A and R410A inside the three test tubes were evaluated. Figure 7 depicts the measured heat transfer coefficient and heat flux as a function of mass velocity at a saturation temperature of 6 °C. Experimental results indicate that the evaporating coefficients increase with a rise in mass flux. The greater vapor velocities enhance the convective boiling heat transfer with the increasing shear stress on the gas-liquid interface and inner wall, and the reduced liquid film thickness. In addition, the interaction between the dimples/protrusions and liquid film near the tube wall also enhances the heat transfer coefficient. Consequently, the boiling heat transfer coefficient increases with the increasing mass flux.

Compared to the smooth tube, the evaporation coefficients of enhanced tubes are significantly higher. As indicated in Figure 7a, the heat transfer coefficient of the EHT tube is about 1.25–1.32 times that of the tested smooth tube, for mass velocities varying from 70–150 kg/m²s, and 1.58–1.66 times that of the Re-EHT tube. Dimples/protrusions generate periodic vortexes, continue to separate the boundary layer, and enhance the turbulence between the fluid and wall surface. The strong gas-phase shear stress caused by the low gas-phase viscosity may drive liquid droplets into the vapor, thereby generating flow separation and mixing. Moreover, these enhanced surface structures produce more nucleation sites, causing higher boiling heat transfer coefficients than those found in a smooth tube. These reasons result in a higher heat transfer coefficient for the EHT tube. However, the Re-EHT tube shows a superior heat transfer performance under the same mass flux conditions. This higher efficiency is partially attributed to the larger internal surface area. On the other hand, the surface tension also plays a vital role in thinning the film thickness in the dimpled tubes. As a consequence, the heat transfer enhancement of 3-D surface structures of the Re-EHT tube is more efficient than that of the EHT tube.

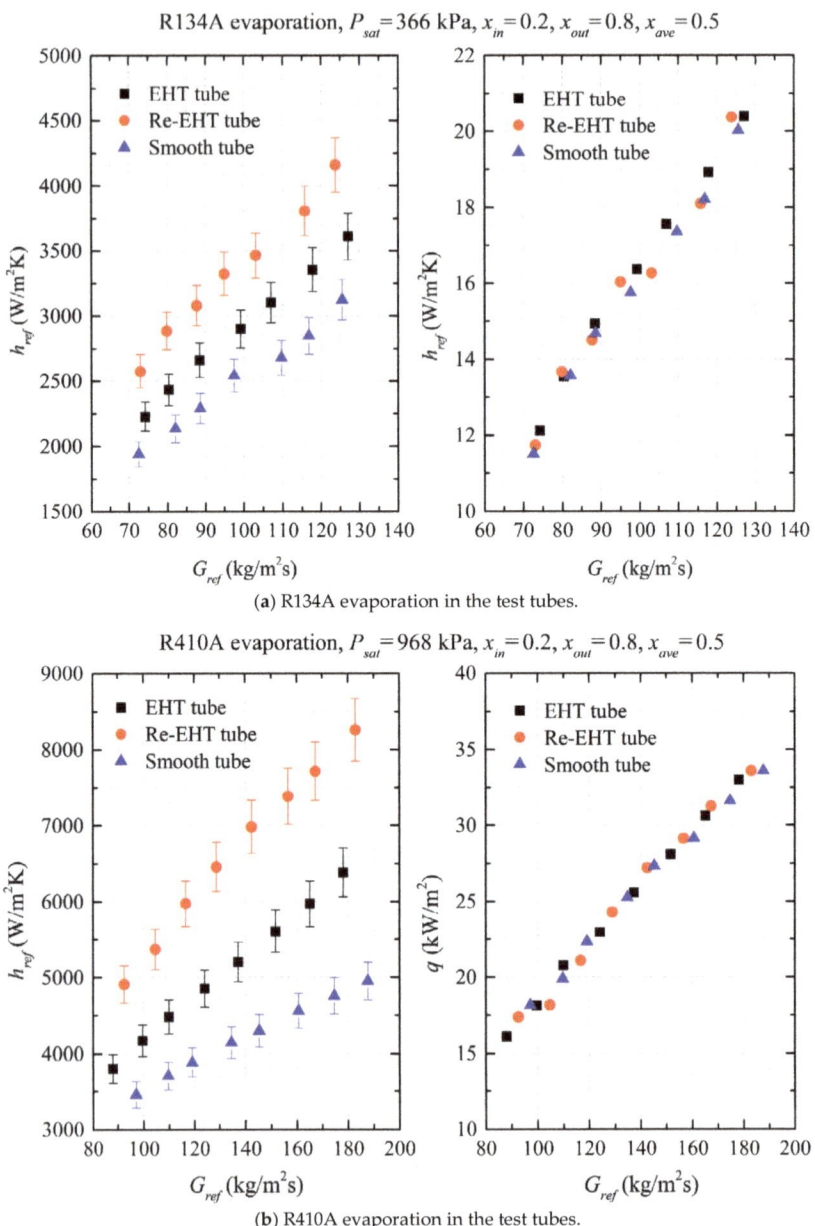

Figure 7. Comparison of flow boiling heat transfer coefficient and heat flux between enhanced and smooth tubes using (**a**) R134A and (**b**) R410A.

A comparison of evaporation heat transfer coefficients of R410A between the tested enhanced and smooth tubes are shown in Figure 7b. The heat transfer enhancement ratio of the EHT tube to the smooth tube is in the range of 1.15–1.28, and 1.45–1.65 times for the Re-EHT tube. It can be seen from Figure 7 that the measured coefficient of R134A is lower than that of R410A at the same mass flux. This phenomenon can be explained by the fact that the liquid-phase thermal conductivity and latent heat of R410A are larger when compared with R134A. Additionally, R410A has the higher imposed heat flux compared to R134A. Steiner and Taborek [33] used the onset of a nucleate boiling (ONB) criterion to develop a correlation for predicting the minimum heat flux to achieve the ONB during in-tube evaporation:

$$q_{ONB} = \frac{2\sigma T_{sat} h_{cb}}{r \rho_v h_{lv}} \qquad (19)$$

In view of the fact that the minimum heat flux at ONB for R410A is lower than that for R134A, owing to the larger latent heat of vaporization, the importance of nucleate boiling heat transfer for R410A case exceeds that for R134A case in the present study. Therefore, the heat transfer coefficient of the Re-EHT tube raises rapidly with the increasing mass flux. This can be explained by the fact that the forced convective boiling component is more and more important, and that high mass flux induces liquid entrainment, thereby weakening the thermal resistance. In order to avoid sub-boiling in the entrance section of test tube, the inlet vapor quality was maintained at 0.2. Besides, the vapor outlet quality was controlled to 0.8 to prevent local dry-out in the exit section.

4.4. Effect of Saturation Temperature on Flow Boiling Heat Transfer Characteristics

Figure 8 compares the effect of saturated temperature on the average flow boiling coefficients for a constant inlet quality of 0.2 and outlet quality of 0.8, over the mass flux range of 80–200 kg/m²s, with heat flux varying between 16 and 35 kW/m². Tests were conducted at two different saturation temperatures (6 °C and 10 °C). Under the same flow boiling conditions, the heat transfer curve of the Re-EHT tube is higher than that of the EHT tube and the smooth tube. On other hand, these results indicate that the boiling heat transfer coefficients measured at a saturated temperature of 6 °C are higher than those at T_{sat} = 10 °C under the boiling conditions. The weakened wall shear stress and gas-liquid interfacial stress may be responsible. In addition, the vapor-phase density increases with an increase in saturation temperature, which leads to a lower vapor velocity. Furthermore, the liquid-phase heat conductivity decreases as the saturated temperature increases. This leads to a rise in internal thermal resistance. For these reasons, a lower saturation temperature is beneficial for the boiling heat-transfer coefficient. Lima et al. [34] also observed the similar experimental results that higher heat transfer coefficients were found at lower saturation temperatures.

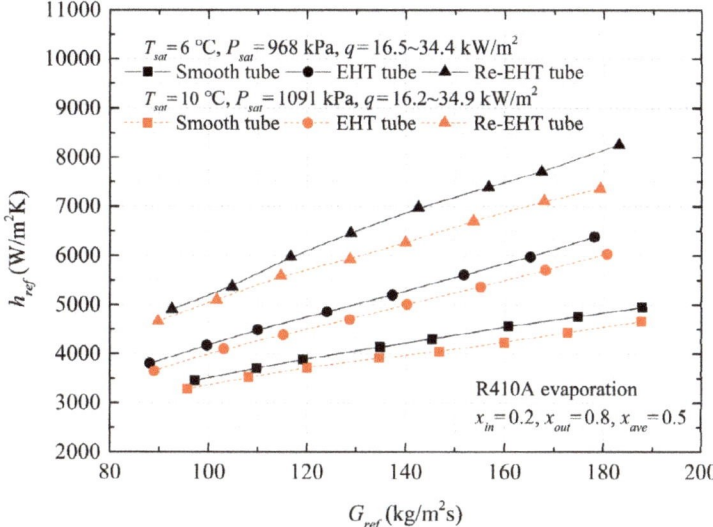

Figure 8. The effect of saturated temperature on the evaporating heat transfer in the test tubes.

4.5. Evaluation of Flow Boiling Heat Transfer Correlations

Figure 9 shows the comparison of experimental data and predicted values calculated by four well-known correlations (Liu and Winterton [35], Gungor and Winterton [36], Kandlikar [37], and Wojtan et al. [38]). Table 6 summarizes the detailed parameters of these correlations [35–38]. Table 7 lists the deviations between the experimental and predicted results by adopting the mean absolute error (MAE) and mean relative error (MRE), which are given by

$$MAE = \frac{1}{N}\sum_{i=1}^{N}\left|\frac{h_{exp} - h_{cal}}{h_{exp}}\right| \times 100\% \tag{20}$$

$$MRE = \frac{1}{N}\sum_{i=1}^{N}\left(\frac{h_{exp} - h_{cal}}{h_{exp}}\right) \times 100\% \tag{21}$$

where N is the total number of experimental data points. In addition, h_{exp} is the experimental heat transfer coefficient and h_{cal} is the calculated value by using the prediction correlations.

The Gungor and Winterton model [36] tends to over-estimate the experimental results, with a mean absolute error of 42.26% and mean relative error of −45.43%; the Kandlikar correlation [37] also exhibits a larger predictive error. Thus, the Gungor and Winterton correlation [36] and the Kandlikar correlation [37] are not applicable for our experiments. In fact, the refrigerant used in this study is different from those used in the study of Kandlikar [37], where working fluids only include the water, R11, R12, R22, R113, R114, R152, ethylene glycol, and nitrogen. Better agreement can be presented by the prediction correlation of Wojtan et al. [38] and Liu and Winterton [35]. Both of the correlations could predict 80% of experimental data points within a ±30% error band. The Wojtan et al. correlation [38] was built based on the mathematical analysis of the liquid film thickness (δ) and dry-out angle (θ_{dry}). The convective boiling heat transfer coefficient (h_{cb}) was developed from the Dittus–Boelter model [28] by replacing Re_l and d_h with Re_δ and δ. The Liu and Winterton correlation [35] was based on the first general model developed by Chen [21] for saturated boiling heat transfer, by considering the force convective term and nucleate boiling term. This correlation is valid for flow boiling heat transfer in channels with the hydraulic diameters in the range from 2.95 to 32 mm.

$$h_{tp} = Eh_{cb} + Sh_{nb} \tag{22}$$

Here, h_{tp} is the two-phase heat transfer coefficient, E is the enhanced factor, S is suppression factor, and h_{nb} is the nucleate boiling component. It is noticed that the prediction correlation still cannot accurately estimate the boiling heat transfer coefficient for the present study, since it ignores the effect of the surface roughness (R_p).

Based on the Liu and Winterton correlation [35], Cooper [39] developed a nucleate boiling heat transfer correlation, by considering the effect of surface roughness on the interfacial turbulence and nucleation sites. The modified term is given by

$$h_{nb} = 55 P_r^{0.12 - 0.2\lg R_p}(-\lg P_r)^{-0.55} M^{-0.5} q^{0.67} \tag{23}$$

For the smooth tube, the surface roughness R_p is considered to be in the range 0.3–0.4 µm, due to the higher flatness. For the enhanced tubes tested, the three-dimensional surface roughness is in the range of 1.5–2.5 µm for the EHT tube, and 6–7 µm for the Re-EHT tube.

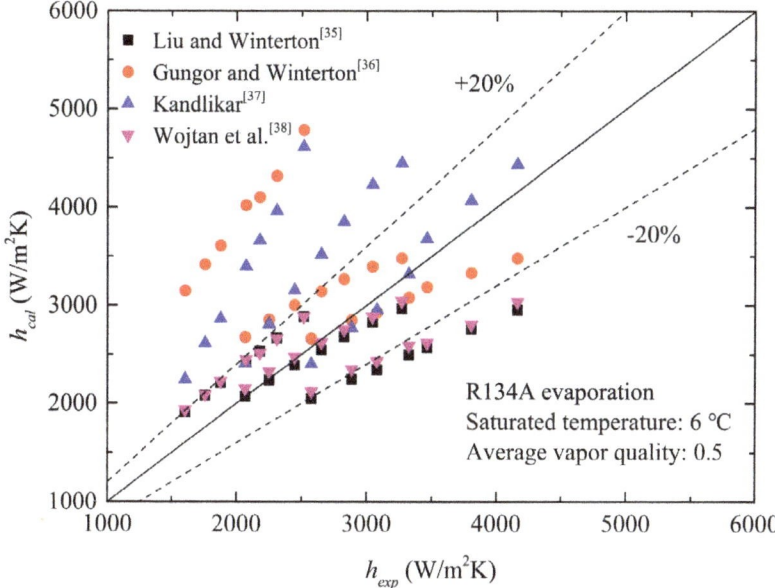

(a) Comparison of experimental and predicted results for R134A evaporation in the test tubes.

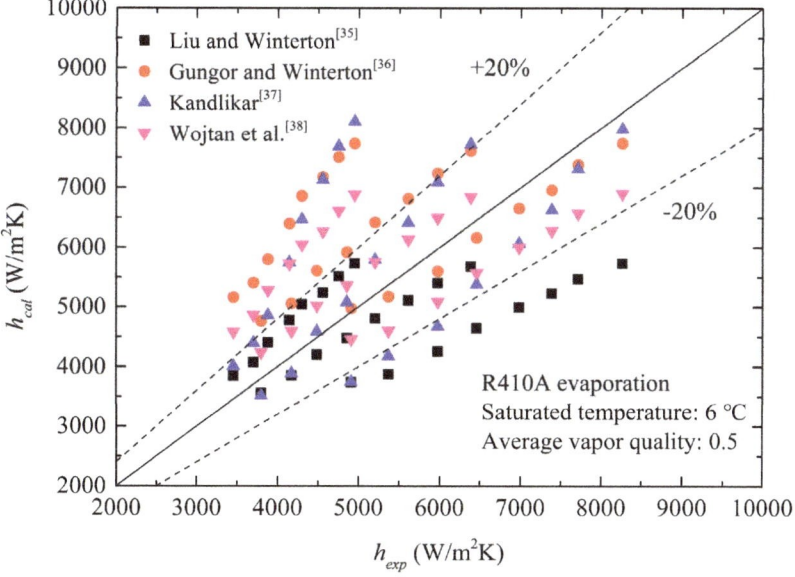

(b) Comparison of experimental and predicted results for R410A evaporation in the test tubes.

Figure 9. Evaluation of four heat transfer correlations for evaporation heat transfer in the tube.

Table 6. Details of four existing correlations for evaporation heat transfer.

Author	Correlation
Liu and Winterton [35]	$h_{tp} = \left[(Eh_{cb})^2 + (Sh_{nb})^2\right]^{0.5}, h_{cb} = 0.023 Re_l^{0.8} Pr_l^{0.4} \frac{k_l}{d_h}$ $E = \left[1 + xPr_l\left(\frac{\rho_g}{\rho_l} - 1\right)\right]^{0.35}, Re_l = \frac{Gd_h}{\mu_l}$ $h_{nb} = 55 P_r^{0.12}(-\log_{10} P_r)^{-0.55} M^{-0.5} q^{0.67}, S = \left(1 + 0.055 E^{0.1} Re_l^{0.16}\right)^{-1}$
Gungor and Winterton [36]	$h_{tp} = Eh_l + Sh_{nb}, h_l = 0.023 Re_l^{0.8} Pr_l^{0.4} \frac{k_l}{d_h}$ $h_{nb} = 55 P_r^{0.12}(-\log P_r)^{-0.55} M^{-0.5} q^{0.67}, E = 1 + 24000 Bo^{1.16} + 1.37 \left(\frac{1}{X_{tt}}\right)^{0.86}$ $S = \frac{1}{1+1.15 \times 10^{-6} E^2 Re_l^{1.17}}, X_{tt} = \left(\frac{1-x}{x}\right)^{0.9} \left(\frac{\rho_v}{\rho_l}\right)^{0.5} \left(\frac{\mu_l}{\mu_v}\right)^{0.1}, Bo = \frac{q}{Gh_{lv}}$
Kandlikar [37]	$h_{tp} = MAX(h_{cb}, h_{nb})$ $h_{cb} = \left[1.136 Co^{-0.9}(25 Fr_l)^{0.3} + 667.2 Bo^{0.7}\right] h_l$ $h_{nb} = \left[0.6683 Co^{-0.2}(25 Fr_l)^{0.3} + 1058 Bo^{0.7}\right] h_l$ $h_l = 0.023 Re_l^{0.8} Pr_l^{0.4} \frac{k_l}{d_i}, Fr_l = \frac{G^2}{g d_i \rho_l^2}, Bo = \frac{q}{Gh_{lv}}, Co = \left(\frac{1-x}{x}\right)^{0.8} \left(\frac{\rho_v}{\rho_l}\right)^{0.5}$
Wojtan et al. [38]	$h_{tp} = \frac{\theta_{dry} h_v + (2\pi - \theta_{dry}) h_{wet}}{2\pi}, h_v = 0.023 Re_v^{0.8} Pr_v^{0.4} \frac{k_v}{d_i}$ $h_{wet} = \left[(h_{cb})^3 + (h_{nb})^3\right]^{1/3}, h_{nb} = 55 P_r^{0.12}(-\log P_r)^{-0.55} M^{-0.5} q^{0.67}$ $h_{cb} = 0.0133 Re_\delta^{0.69} Pr_l^{0.4} \frac{k_l}{d_i}, Re_\delta = \frac{4 G \delta (1-x)}{\mu_l (1-\varepsilon)}, \delta = \frac{\pi d (1-\varepsilon)}{2(2\pi - \theta_{dry})}$ $slug/intermittent/annular: \theta_{dry} = 0$ $slug - stratified\ wavy\ flow: \theta_{dry} = \theta_{strat} \frac{x}{x_{IA}} \left[\frac{(G_{wavy} - G)}{(G_{wavy} - G_{strat})}\right]^{0.61}$ $stratified - wavy\ flow: \theta_{dry} = \theta_{strat} \left[\frac{(G_{wavy} - G)}{(G_{wavy} - G_{strat})}\right]^{0.61}$

Table 7. Prediction accuracy of the heat transfer correlations.

Correlations	MAE (%)	MRE (%)
Liu and Winterton [35]	18.84	1.98
Gungor and Winterton [36]	42.26	−45.43
Kandlikar [37]	30.00	−27.97
Wojtan et al. [38]	23.81	−15.83
Modified correlation	3.64	0.85

A comparison of the predictions of the modified correlation and experimental data are shown in Figure 10. All data points of the enhanced and smooth tubes tested are predicted within a ±10% error band. The modified correlation may not entirely be suitable for all test conditions. For example, the effects of tube diameter and effective heated length need to be examined further to enlarge the application range of the modified correlation.

4.6. Performance Factor

On account of the fact that the actual internal surface areas of the tested tubes were different, a performance evaluation factor (*PF*) was adopted to evaluate the thermal efficiency of the enhanced tubes, which can be written as

$$PF = \frac{h_e}{h_s} \cdot \frac{A_s}{A_e} \tag{24}$$

where h_S and A_S are the heat transfer coefficient and the actual inner surface area of the smooth tube. Similarly, h_e and A_e represent the evaporating coefficient and internal surface area of the enhanced tubes, respectively.

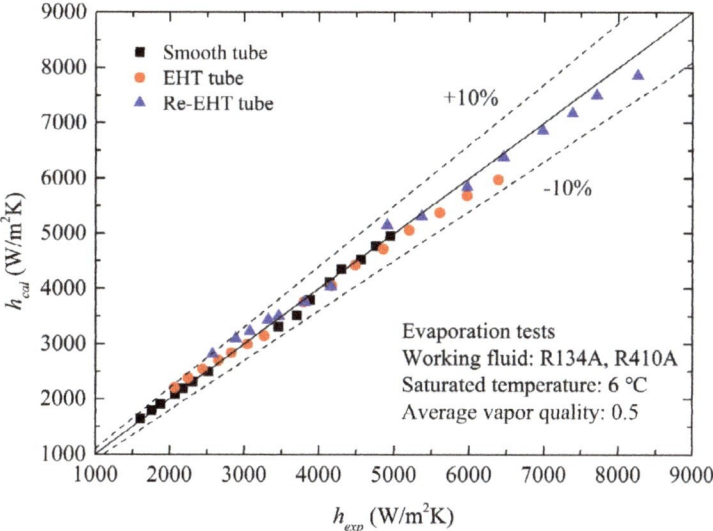

Figure 10. Evaluation of the modified correlation for the boiling heat transfer coefficient of the tested tubes.

Figure 11a details the variation of the performance factor of R134A during flow boiling in the EHT and Re-EHT tubes with mass flux. Over the entire test range, the *PF* values of the enhanced tubes are kept steady. The performance factor values are about 15% for the EHT tube and 30% for the Re-EHT tube. Different from the results of R134A case, the *PF* value of R410A case increases as the mass flux increases at first, and then gradually tends to be flat when G_{ref} > 150 kg/m²s. In summary, the *PF* values of the EHT and Re-EHT tubes are larger than unity, showing a good heat transfer performance.

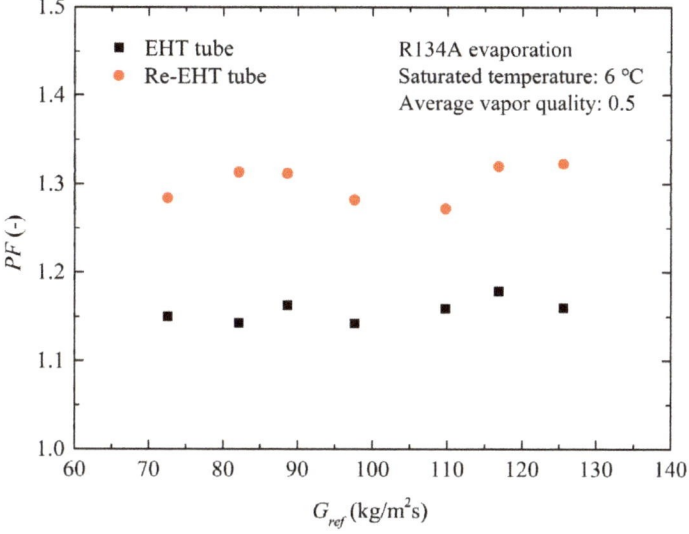

(**a**) Performance factor for the enhanced tubes using R134A.

Figure 11. *Cont.*

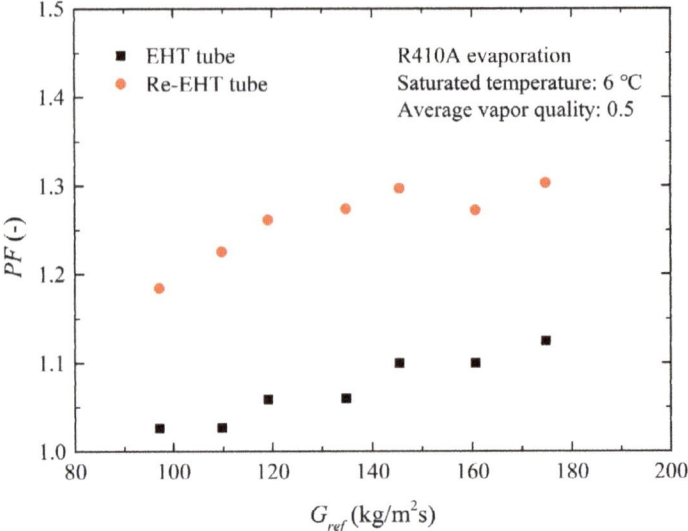

(**b**) Performance factor for the enhanced tubes using R410A.

Figure 11. Performance factor versus mass flux for boiling heat transfer in the enhanced tubes using (**a**) R134A and (**b**) R410A.

5. Conclusions

An investigation on the flow boiling of refrigerants R134A and R410A in two enhanced surface tubes and one smooth tube was carried out. The effects of flow pattern, mass flux, and saturation temperature on the boiling heat transfer were discussed. A comparison between experimental data and predictions calculated by several existing correlations for in-tube evaporation was conducted. The main conclusions can be summarized as follows:

1. Intermittent/annular flows could be the major flow patterns over the entire experimental range for R134A and R410A flow boiling in the enhanced tubes in this study (as shown in Wojtan et al.'s map [29]).
2. The boiling heat transfer coefficients increase with the increasing mass flux. Two enhanced tubes present better heat transfer performance than the smooth tube. This can be attributed to the complex two-layer surface structures of enhanced tubes. The dimples/protrusions enhance the in-tube heat transfer by increasing heat transfer surface area, promoting the interfacial turbulence, providing more nucleation sites and destroying the boundary layer.
3. An increase in heat transfer coefficient is found at a lower saturation temperature, in view of the fact that the gas-liquid interfacial and wall shear stresses are weakened as the saturation temperature increases.
4. Boiling heat transfer coefficients are evaluated using the four widely-used correlations. The correlations of Wojtan et al. [38] and Liu and Winterton [35] show a good predictive ability. Considering the effect of the surface roughness, an optimization model was presented, which can predict all data points within a $\pm 10\%$ error band.
5. The enhanced tubes showed a good performance factor. Hence, the dimples/protrusions and petal arrays are the effective surface structures for enhancing the tube-side evaporation. The Re-EHT tube has the largest potential for boiling heat-transfer enhancement.

Author Contributions: Conceptualization, Z.-C.S. and D.J.K.; data curation and investigation, Z.-C.S., W.L. and X.M.; methodology and formal analysis, W.L. and D.J.K.; project administration, L.M.; resources and visualization, D.J.K.; funding acquisition, W.L. and L.-X.M.; writing—original draft preparation, Z.-C.S. and X.M.; writing—review and editing, D.J.K. and Z.-C.S. All authors read and approved the final manuscript.

Funding: This work was supported by the National Science Foundation of Zhejiang Province (LY19E060004).

Conflicts of Interest: The authors declare no conflict of interest.

Nomenclature

A_c	cross sectional area, m^2
A_i	inner surface area of test tube, m^2
A_o	outer surface area of test tube, m^2
Bo	Boiling number, [-]
c_p	specific heat, J/(kg·K)
C	enhancement factor, [-]
Co	convective number, [-]
d_i	inner diameter of test tube, m
d_o	outer diameter of test tube, m
d_h	hydraulic diameter, m
E	enhancement factor, [-]
f	Fanning friction factor, [-]
Fr	Froude number, [-]
G	mass flux, kg/(m^{2}s)
h	heat transfer coefficient, W/(m^2·K)
h_{lv}	latent heat of vaporization, J/kg
k	thermal conductivity, W/(m·K)
L	tube length, m
$LMTD$	logarithmic mean temperature, K
m	mass flow rate, kg/s
M	molecular weight, g/mol
P_w	wetted perimeter, m
PF	performance factor, [-]
Pr	Prandtl number, [-]
P_r	reduce pressure, [-]
Q	heat transfer amount, W
q	heat flux, W/m^2
r	bubble radius, m
R_p	surface roughness, m
Re	Reynolds number, [-]
S	suppression factor, [-]
T	temperature, K
U	total heat transfer coefficient, W/(m^2·K)
x	vapor quality, [-]
x_{IA}	Transition quality from intermittent flow to annular flow, [-]
X_{tt}	Martinelli parameter, [-]

Greek symbols

σ	surface tension, N/m
μ	dynamic viscosity, Pa·s
θ	angle, [°]
ρ	density, kg/m^3
ε	void fraction, [-]
δ	liquid film thickness, m

Subscripts

ave	average
cal	calculation
cb	convective boiling
dry	dry-out
e	enhanced
exp	experiment
i	inner
in	inlet
l	liquid phase
lat	latent heat
nb	nucleate boiling
o	outer
ONB	onset of nucleate boiling
out	outlet
pre	Preheating section
ref	refrigerant
s	smooth
sat	saturated
sens	sensible heat
tp	two-phase
ts	test section
v	vapor phase
wall	tube wall
w	water

References

1. Yu, M.H.; Lin, T.K.; Tseng, C.C. Heat transfer and flow pattern during two-phase flow boiling of R-134a in horizontal smooth and microfin tubes. *Int. J. Refrig.* **2002**, *25*, 789–798. [CrossRef]
2. Spindler, K.; Müller-Steinhagen, H. Flow boiling heat transfer of R134a and R404A in a microfin tube at low mass fluxes and low heat fluxes. *Heat Mass Transf.* **2009**, *45*, 967–977. [CrossRef]
3. Rollmann, P.; Spindler, K. New models for heat transfer and pressure drop during flow boiling of R407C and R410A in a horizontal microfin tube. *Int. J. Therm. Sci.* **2016**, *103*, 57–66. [CrossRef]
4. Padovan, A.; Col, D.D.; Rossetto, L. Experimental study on flow boiling of R134a and R410A in a horizontal microfin tube at high saturation temperatures. *Appl. Therm. Eng.* **2011**, *31*, 3814–3826. [CrossRef]
5. Celen, A.; Çebi, A.; Ahmet, S.D. Experimental investigation of flow boiling heat transfer characteristics of R134a flowing in smooth and microfin tubes. *Int. Commun. Heat Mass Transf.* **2018**, *93*, 21–33. [CrossRef]
6. Wu, Z.; Wu, Y.; Sundén, B.; Li, W. Convective vaporization in micro-fin tubes of different geometries. *Exp. Therm. Fluid Sci.* **2013**, *44*, 398–408. [CrossRef]
7. Kukulka, D.J.; Smith, R.; Fuller, K.G. Development and evaluation of enhanced heat transfer tubes. *Appl. Therm. Eng.* **2011**, *31*, 2141–2145. [CrossRef]
8. Kukulka, D.J.; Yan, H.; Smith, R.; Li, W. Condensation and evaporation characteristics of flows inside three dimensional Vipertex enhanced heat transfer tubes. *Chem. Eng. Trans.* **2017**, *61*, 1777–1782.
9. Kukulka, D.J.; Smith, R.; Li, W.; Zhang, A.; Yan, H. Condensation and evaporation characteristics of flows inside Vipertex 1EHT and 4EHT small diameter enhanced heat transfer tubes. *Chem. Eng. Trans.* **2018**, *70*, 13–18.
10. Guo, S.P.; Wu, Z.; Li, W.; Kukulka, D.J.; Sundén, B.; Zhou, X.P.; Wei, J.J.; Simon, T. Condensation and evaporation heat transfer characteristics in horizontal smooth, herringbone and enhanced surface EHT tubes. *Int. J. Heat Mass Transf.* **2015**, *85*, 281–291. [CrossRef]
11. Li, W.; Chen, J.X.; Zhu, H.; Kukulka, D.J.; Minkowycz, W.J. Experimental study on condensation and evaporation flow inside horizontal three dimensional enhanced tubes. *Int. Commun. Heat Mass Transf.* **2017**, *80*, 30–40. [CrossRef]

12. Chen, J.; Li, W. Local flow boiling heat transfer characteristics in three-dimensional enhanced tubes. *Int. J. Heat Mass Transf.* **2018**, *121*, 1021–1032. [CrossRef]
13. Shafaee, M.; Mashouf, H.; Sarmadian, A.; Mohseni, S.G. Evaporation heat transfer and pressure drop characteristics of R-600a in horizontal smooth and helically dimpled tubes. *Appl. Therm. Eng.* **2016**, *107*, 28–36. [CrossRef]
14. Ayub, Z.H.; Ayub, A.H.; Ribatski, G.; Moreira, T.A.; Khan, T.S. Two-phase pressure drop and flow boiling heat transfer in an enhanced dimpled tube with a solid round rod insert. *Int. J. Refrig.* **2017**, *75*, 1–13. [CrossRef]
15. Kundu, A.; Kumar, R.; Gupta, A. Heat transfer characteristics and flow pattern during two-phase flow boiling of R134a and R407C in a horizontal smooth tube. *Exp. Therm. Fluid Sci.* **2014**, *57*, 344–352. [CrossRef]
16. Lillo, G.; Mastrullo, R.; Mauro, A.W.; Viscito, L. Flow boiling of R32 in a horizontal stainless steel tube with 6.00 mm ID. Experiments, assessment of correlations and comparison with refrigerant R410A. *Int. J. Refrig.* **2018**. [CrossRef]
17. Greco, A.; Vanoli, G.P. Flow-boiling of R22, R134a, R507, R404A and R410A inside a smooth horizontal tube. *Int. J. Refrig.* **2005**, *28*, 872–880. [CrossRef]
18. Li, W.; Wu, Z. A general criterion for evaporative heat transfer in micro/mini-channels. *Int. J. Heat Mass Transf.* **2010**, *53*, 1778–1787. [CrossRef]
19. Jige, D.; Sagawa, K.; Inoue, N. Effect of tube diameter on boiling heat transfer and flow characteristic of refrigerant R32 in horizontal small-diameter tubes. *Int. J. Refrig.* **2017**, *76*, 206–218. [CrossRef]
20. Saitoh, S.; Daiguji, H.; Hihara, E. Correlation for boiling heat transfer of R-134a in horizontal tubes including effect of tube diameter. *Int. J. Heat Mass Transf.* **2007**, *50*, 5215–5225. [CrossRef]
21. Chen, J.C. Correlation for boiling heat transfer to saturated fluids in convective flow. *Ind. Eng. Chem. Process Des. Dev.* **1966**, *5*, 322–329. [CrossRef]
22. Saisorn, S.; Wongpromma, P.; Wongwises, S. The difference in flow pattern, heat transfer and pressure drop characteristics of mini-channel flow boiling in horizontal and vertical orientations. *Int. J. Multiph. Flow* **2018**, *101*, 97–112. [CrossRef]
23. Gnielinski, V. New equations for heat and mass transfer in turbulent pipe and channel Flow. *Int. Chem. Eng.* **1976**, *16*, 359–368.
24. Petukhov, B.S. Heat transfer and friction in turbulent pipe flow with variable physical properties. *Adv. Heat Transf.* **1970**, *6*, 503–564.
25. José Fernández-Seara Francisco, J.; Uhía Sieres, J.; Campo, A. A general review of the Wilson plot method and its modifications to determine convection coefficients in heat exchange devices. *Appl. Therm. Eng.* **2007**, *27*, 2745–2757. [CrossRef]
26. Lemmon, E.W.; Huber, M.L.; Mclinden, M.O. *NIST Standard Reference Database 23, Reference Fluid Thermodynamic and Transport Properties—REFPROP, Version 9.0, Standard Reference Data Program*; National Institute of Standards and Technology: Gaithersburg, MD, USA, 2010.
27. Moffat, R.J. Describing the uncertainties in experimental results. *Exp. Therm. Fluid Sci.* **1988**, *1*, 3–17. [CrossRef]
28. Dittus, F.W.; Boelter LM, K. Heat transfer in automobile radiators of the tubular type. *Int. Commun. Heat Mass Transf.* **1985**, *12*, 3–22. [CrossRef]
29. Wojtan, L.; Ursenbacher, T.; Thome, J.R. Investigation of flow boiling in horizontal tubes: Part I—A new diabatic two-phase flow pattern map. *Int. J. Heat Mass Transf.* **2005**, *48*, 2955–2969. [CrossRef]
30. Kattan, N.; Thome, J.R.; Favrat, D. Flow boiling in horizontal tubes. Part 1—development of a diabatic two-phase flow pattern map. *J. Heat Transf.* **1998**, *120*, 140–147. [CrossRef]
31. Singh, A.; Ohadi, M.M.; Dessiatoun, S. Flow boiling heat transfer coefficients of R-134a in a microfin tube. *J. Heat Transf.* **1996**, *118*, 497–499. [CrossRef]
32. Mashouf, H.; Shafaee, M.; Sarmadian, A.; Mohseni, S.G. Visual study of flow patterns during evaporation and condensation of R-600a inside horizontal smooth and helically dimpled tubes. *Appl. Therm. Eng.* **2017**, *124*, 1392–1400. [CrossRef]
33. Steiner, D.; Taborek, J. Flow boiling heat transfer in vertical tubes correlated by an asymptotic model. *Heat Transf. Eng.* **1992**, *13*, 43–69. [CrossRef]
34. Lima RJ, D.S.; Jesús Moreno Quibén Thome, J.R. Flow boiling in horizontal smooth tubes: New heat transfer results for R-134a at three saturation temperatures. *Appl. Therm. Eng.* **2009**, *29*, 1289–1298. [CrossRef]

35. Liu, Z.; Winterton RH, S. A general correlation for saturated and subcooled flow boiling in tubes and annuli, based on a nucleate pool boiling equation. *Int. J. Heat Mass Transf.* **1991**, *34*, 2759–2766. [CrossRef]
36. Gungor, K.E.; Winterton, R.H.S. A general correlation for flow boiling in tubes and annuli. *Int. J. Heat Mass Transf.* **1986**, *29*, 351–358. [CrossRef]
37. Kandlikar, S.G. A general correlation for saturated two-phase flow boiling heat transfer inside horizontal and vertical tubes. *J. Heat Transf.* **1990**, *112*, 219–228. [CrossRef]
38. Wojtan, L.; Ursenbacher, T.; Thome, J.R. Investigation of flow boiling in horizontal tubes: Part II—Development of a new heat transfer model for stratified-wavy, dryout and mist flow regimes. *Int. J. Heat Mass Transf.* **2005**, *48*, 2970–2985. [CrossRef]
39. Cooper, M.G. Saturated nucleate pool boiling: A simple correlation. *Inst. Chem. Eng. Symp. Ser.* **1984**, *86*, 785–792.

© 2019 by the authors. Licensee MDPI, Basel, Switzerland. This article is an open access article distributed under the terms and conditions of the Creative Commons Attribution (CC BY) license (http://creativecommons.org/licenses/by/4.0/).

Article

A Numerical and Experimental Investigation of Dimple Effects on Heat Transfer Enhancement with Impinging Jets

Parkpoom Sriromreun [1] and Paranee Sriromreun [2],*

[1] Department of Mechanical Engineering, Faculty of Engineering, Srinakharinwirot University, Ongkharak, Nakhonnayok 26120, Thailand; prakpum@g.swu.ac.th
[2] Department of Chemical Engineering, Faculty of Engineering, Srinakharinwirot University, Ongkharak, Nakhonnayok 26120, Thailand
* Correspondence: paranee@g.swu.ac.th; Tel.: +66-859-104-426

Received: 7 November 2018; Accepted: 25 February 2019; Published: 1 March 2019

Abstract: This research was aimed at studying the numerical and experimental characteristics of the air flow impinging on a dimpled surface. Heat transfer enhancement between a hot surface and the air is supposed to be obtained from a dimple effect. In the experiment, 15 types of test plate were investigated at different distances between the jet and test plate (B), dimple diameter (d) and dimple distance (E_r and E_θ). The testing fluid was air presented in an impinging jet flowing at Re = 1500 to 14,600. A comparison of the heat transfer coefficient was performed between the jet impingement on the dimpled surface and the flat plate. The velocity vector and the temperature contour showed the different air flow characteristics from different test plates. The highest thermal enhancement factor (TEF) was observed under the conditions of B = 2 d, d = 1 cm, E_r = 2 d, E_θ = 1.5 d and Re = 1500. This TEF was obtained from the dimpled surface and was 5.5 times higher than that observed in the flat plate.

Keywords: impinging jet; dimple; Nusselt number; heat transfer; heat exchanger

1. Introduction

Heat exchangers, whether large or small, play an important role not only in industry but also in everyday life. A heat exchanger is any device that can transfer heat from one medium to another. To achieve a high heat-transfer rate over a smaller area, the impinging technique is quite interesting. The fluid will be impinged to attack the wall. This can be used to increase or decrease the temperature of a device, such as by transferring the heat out from electronic hardware. The heat produced from a CPU can be passed along the fins and out into the surrounding environment. The development is interesting in terms of the impinging jet characteristics and heat exchanger surfaces. In this research, our study was focused on the improvement of the heat exchanger surface. This has been developed from a flat plate to a dimpled surface for generating a vortex and high turbulence intensity on the surface. This can increase the heat transfer and reduce the pressure drop. In the conventional technique, ribs or baffles are set on the heat exchanger plate for increasing the heat transfer. However, this can also increase the pressure drop and the friction factor, leading to the production of a low thermal enhancement factor (TEF). The TEF indicates the heat-transfer performance as calculated from the Nusselt number (Nu) and friction factor (f). As such, the TEF represents the ratio of heat transfer and the power ability of the fluid to flow through the heat exchanger device. A higher TEF cannot be obtained at a higher heat-transfer rate with a high pressure drop.

Investigations have been conducted widely into the effect of those parameters of an impinging air jet on the heat transfer in many roughened-surface geometries. Recently, Ries et al. [1] studied the heat transfer of a turbulent jet impinging on a 45° incline by using a direct numerical simulation (DNS) method. The results showed that the highest heat-transfer position was not the fluid impinging point (the stagnation point) but was in fact a little shifted to above the impinging point. Ries et al. [2] investigated the flow and heat characteristics by experiment of an air jet impinging on a flat plate inclined at different angles. The highest Nu was found at the stagnation point for the flat plat inclined at 90°. Meola et al. [3] investigated convective heat transfer coefficients on a flat plate. They were focused on the effects of shear layer dynamics. An infrared scanning radiometer and the heated thin foil technique were used to measure the temperature and determine the Nusselt number. Guerra et al. [4] found that the local velocity and the temperature distributions were presented as well as longitudinal turbulence profiles on the impinged plate. Chaudhari et al. [5] measured temperature and pressure on the impinged plate to find the heat transfer and pressure drop on this plate along in the radial line. The experiments were focused on the following parameters: Re in the range of 1500–4200. The ratio of the axial distance between the heated plate and the jet to the jet diameter was in the range of 0–25 in this study. The results of the maximum heat transfer coefficient as a Nusselt number ($\overline{Nu}$) were 11 times higher with the synthetic jet than with natural convection. Draksler and Končar [6] numerically analyzed heat transfer rates using the turbulence models for predictions of heat transfer and the flow characteristics of an axis-symmetric impinging jet. The results were validated based on the free jet impingement experiment. Nanan et al. [7] studied the flow and heat-transfer characteristics of swirling impinging jets on an impinged plate. The results showed that the swirling impinging jets offered higher heat-transfer rates on impinged surfaces than the conventional impinging jets. Nuntadusit et al. [8] investigated the effect of using multiple swirling impinging jets (MSIJs) and found that the Nu from using MSIJs was higher than from using multiple conventional impinging jets. Qiang et al. [9] studied the low viscosity fluid (non-Newtonian fluid) impinging to the flat plate. The results indicate that non-Newtonian fluid is the optimum choice to obtain high heat transfer rate for laminar flow.

Tong [10] studied the solving of Navier-Stokes equations by using a finite-volume for investigating the hydrodynamics and heat transfer of the impingement process. The results showed that a high maximum Nusselt number also provided the maximum pressure drop on the surface. M. Goodro et al. [11] showed the effects of array jets impinging on the flat plate in Reynolds numbers ranging from 8200 to 30,500 and Mach numbers from 0.1 to 0.6. The heat transfer also significantly increased as the Reynolds number and Mach number increased. Pakhomov and Terekhov [12] presented the results of numerical investigation into the flow structure and heat transfer of impact mist jets with a low concentration of droplets.

Ekkad and Kontrovitz [13] focused on heat transfer distributions over a jet impingement target surface with dimples. They used the transient liquid crystal technique to measure the heat transfer. Results showed that the presence of dimples on the target surface, in-line or staggered with respect to jet location, produced lower heat transfer coefficients than the non-dimpled target surface. Lienhart et al. [14] carried out an experimental and numerical investigation into how the turbulent flow over dimpled surfaces has an effect on the friction drag. The results showed that the friction factors of dimpled surfaces were higher than those of flat plates. Kanokjaruvijit and Martinez-Botas [15] studied heat transfer in an inline array of round jets impinging on a staggered array of hemispherical dimples. They considered various parametric effects, such as Reynolds number, jet-to-plate distance, dimple depth and ratio of jet diameter to dimple-projected diameter. A transient wide-band liquid crystal method was used to measure the heat transfer rate. At a dimple depth of 0.15 and jet-to-plate distance twice that of the dimple diameter, the results showed that the heat transfer under these conditions was a significant 70% higher than with a flat plate. Xing and Weigand [16] studied a jet array impinging on flat and dimpled plates by using the transient liquid crystal method. The best heat transfer performance was obtained with the minimum cross flow and narrow jet-to-plate spacing, whether on a flat or dimpled plate. The dimples on the target plate resulted in higher heat transfer coefficients than with the flat plate. Won et al. [17] combined PIV experiments and modeling to obtain two-and three-dimensional (3-D) microjet flows. From the 3-D velocity fields, the researchers could quantify the flow physics around the impingement between the orifice and the bottom surface. Kwon et al. [18] used a naphthalene sublimation technique for obtaining local and average Nusselt numbers on a dimple. The average Nusselt number increased as the turbulence intensity in the mixing layer over the dimple increased, which affected the dimple depth and the increase in Reynolds number. Turnow et al. [19] studied the use of Large Eddy Simulation (LES) and Laser Doppler Velocimetry (LDV) to find vortex structures and heat transfer enhancement mechanisms of turbulence flow over a staggered array of dimples. LES was used to calculate the flow and temperature fields. The vortex structure of dimple packages was shown from LDV and LES. The heat transfer rate from a dimpled plate could be about 200% higher than that of a flat plate.

De Bonis and Ruocco [20] applied the impingement technique for food drying or dehydration. The heat transfer rate, water activity and moisture depletion were measured in a food substrate using a turbulent air jet impingement for food heating. Parida et al. [21] studied the jet impingement for cooling high heat flux (i.e., cutting-edge electronic technologies). The results showed that developments in cooling methodologies were required to avoid unacceptable temperature rise and to maintain a high performance. The overall improvement in quality of 150–200%, based on the maximum Nu recorded both experimentally and numerically, was affected by impingement and associated swirl. Na-pompet and Boonsupthip [22] were interested in cooling rate impingements of food. They demonstrated the comparison between experimental data from related publications and numerical simulations. The results showed that a proper design of the plate position significantly enhanced the overall energy transfer on the target surface and improved the cooling rate of food model. Alenezi et al. [23] studied the flow structure and heat transfer of jet impingement on a rib-roughened flat plat. The results indicates that at the smooth surface with the same rib position, the maximum $\overline{Nu}$ for each location was obtained when the rib height was close to the corresponding boundary layer thickness.

According to the previous work, the half-sphere dimpled surface of the heat exchanger plate has been studied and presented a moderate heat transfer rate. However, the half-sphere shape is difficult to machine, so the cylindrical dimpled surface was modified for use in this work. The cylindrical shape not only presents a high heat transfer rate but also presents the vortex of the fluid flow, which can occur around the dimple edge in order to promote heat transfer between the air and the test-section plate. In this research, the air is used as an impinging jet for Re = 1500 to14,600. A total of 15 case studies were investigated, based on different dimple sizes and positions on the impingement plate.

This research work aimed to continue the research carried out in previous work [24]. The researchers aimed to develop a heat exchanger plate to improve the heat transfer from the impinging jet. The surface was designed and tested in many different cases. However, the experimental results failed to clearly explain the flow behavior on the test-section plate. This is because a sensor could not be set atop the plate as doing so would obstruct the flow. Instead, numerical studies were used to analyze and describe the experiment. In the simulation part, the friction factor (f) value on the test-section plate was the key variable utilized to present the flow obstacle of that device. In this work, the test-section plate with a higher f value required a higher pump power to drive the air flow through the device compared to the test-section plate with a lower f value. However, the TEF was the most important variable used to compare the heat exchange efficiency of the test-section plate. The TEF represents both the heat exchange efficiency and the pump power, which were analyzed in terms of Nu and f, respectively.

2. Experimental Setup

In the experiment, an air compressor with a tank was used to blow the air impinging on the test-section plate. The air continuously flowed through the orifice. The manometer was connected in parallel to the orifice for measuring the pressure drop. The pressure drop was converted to air flow velocity. The adjustable velocity valve was set in front of the test section. Ten Reynolds numbers of air flow velocity were analyzed in the range of Re = 1500 to 14,600. The air was then injected through the nozzle jet (diameter = 1 cm) impinging on the test plate (diameter plate = 30 cm). The test plate was covered by the flat plate on the top to protect the impinging jet air moving out. The three different distances between the nozzle jet and test plate were 2, 4 and 6 cm.

The heater was installed on the bottom wall of the test plate to maintain a uniform surface of heat flux. The electrical output power was controlled by a Variac transformer to obtain a constant heat flux. In Figure 1, the diameter of the plate (D) is 30 cm. The diameters of the cylindrical dimples are 1 and 2 cm (d) with radial distances between dimples, E_r = 2 d and 3 d, and circumferential dimple distances, E_θ = 1.5 d and 3 d. The depth of the dimples is 0.5 cm. The confinement plates, which are 2, 4 and 6 cm (B), are above the test plate. The details of the geometry of the impingement plate with dimples and tested conditions are shown in Table 1. There were 15 case studies, shown in Table 2.

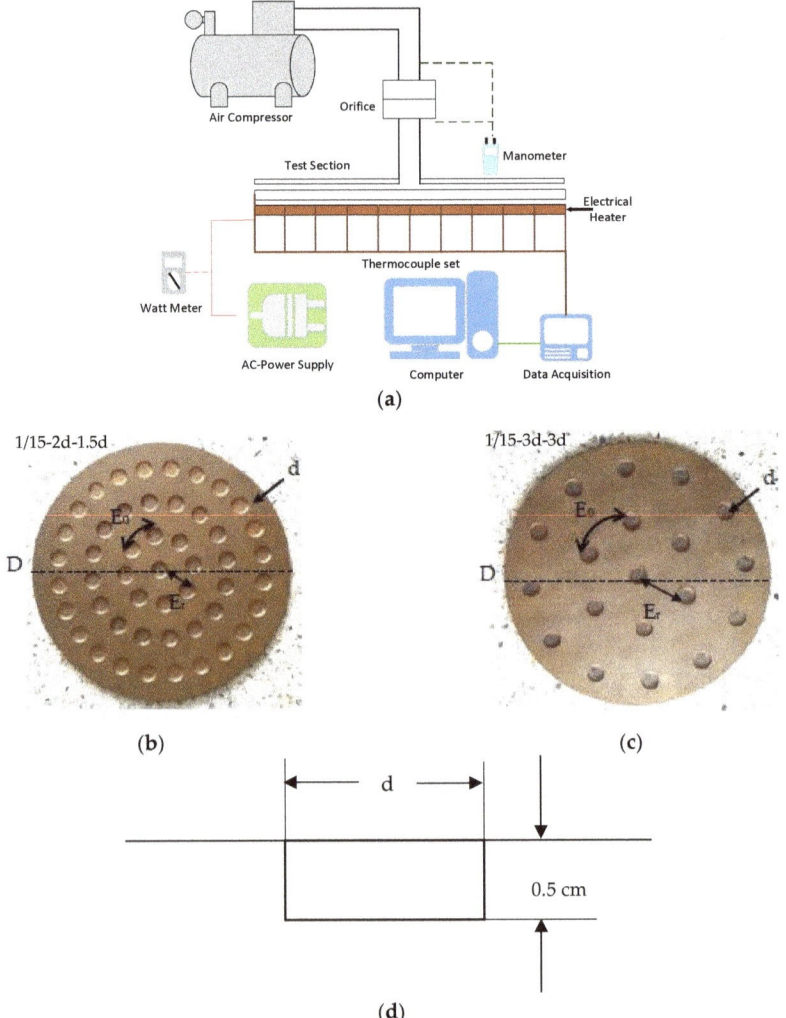

Figure 1. (a) Schematic diagram of experimental apparatus, (b) Details of the impingement dimples on the surface at E_r = 2 d and E_θ = 1.5 d, (c) at E_r = 3 d and E_θ = 3 d, (d) The depth of the dimples.

The 16 thermocouples were set in three different radii under the test plate for recording the average value of temperature at the surface. The experimental apparatus is shown in Figure 2.

Figure 2. Test section.

Table 1. The detail of the geometry of the impingement plate with dimples and tested conditions.

Working Fluid	Air
Reynolds number (Re)	1500 to 14,600
Jet diameter (D_j)	1 cm
Plate diameter (D)	30 cm
Depth of dimple	0.5 cm
Dimple diameter (d)	D_j and 2 D_j
Dimple distance in the radius (E_r)	2 d, 3 d
Dimple distance in the circumference (E_θ)	1.5 d, 3 d
The distance between test plates and jet (B)	2 D_j, 4 D_j and 6 D_j

Table 2. The 15 case studies in detail.

Dimple Diameter/ Plate Diameter (d/D)	Dimple Distance in the Radius (E_r)	Dimple Distance in the Circumference (E_θ)	The Distance between Test Plates and Jet (B)	Case Study Name (d/D-E_r-E_θ-B)
Flat	-	-	2	Flat-2
Flat	-	-	4	Flat-4 [24]
Flat	-	-	6	Flat-6
1/5	3 d	3 d	2	1/15-3 d-3 d-2
1/5	3 d	3 d	4	1/15-3 d-3 d-4
1/5	3 d	3 d	6	1/15-3 d-3 d-6
1/5	2 d	1.5 d	2	1/15-2 d-1.5 d-2
1/5	2 d	1.5 d	4	1/15-2 d-1.5 d-4 [24]
1/5	2 d	1.5 d	6	1/15-2 d-1.5 d-6
1/30	3 d	3 d	2	1/30-3 d-3 d-2
1/30	3 d	3 d	4	1/30-3 d-3 d-4
1/30	3 d	3 d	6	1/30-3 d-3 d-6
1/30	2 d	1.5 d	2	1/30-2 d-1.5 d-2 [24]
1/30	2 d	1.5 d	4	1/30-2 d-1.5 d-4 [24]
1/30	2 d	1.5 d	6	1/30-2 d-1.5 d-6

The thermocouple was put in the lower-test section plate. The two RTDs were used to measure the average temperature of the air outlet. The error of the temperature sensor was not over ±0.2 °C. All temperature sensors were connected to the signal converter. The data was reported to the monitor and recorded in the computer. The next section was heat at the test-section plate. This was generated from the heater, in which the constant heat flux could be controlled by variable voltage transformer. The last section was the insulator, which was installed at the bottom to control the heat direction so that it only flowed into the test-section plate. The uncertainty in velocity measurement was estimated to be less than ±7%, whereas that of wall temperature measurement was about ±0.5%. The maximum uncertainties of non-dimensional parameters were ±5% for Nusselt number, and ±5% for Reynolds number [25].

3. Computational Models and Numerical Method

The computational models used in this research can be written as follows:

Transfer of energy:

$$\rho C_p \frac{\partial T}{\partial t} + \rho C_p \vec{V} \cdot \nabla T = \nabla \cdot (K \nabla T) \tag{1}$$

Flow continuity:

$$\nabla \cdot \vec{V} = 0 \tag{2}$$

Equations (1) and (2) are based on the assumptions of a steady-state air flow and incompressible flow. The turbulent models used the Navier–Stokes equations with k-ε and RNG, and the energy equations for solving this present research problems.

Momentum transfer:

$$\rho \frac{\partial \vec{V}}{\partial t} + \rho \vec{V} \cdot \nabla \vec{V} = -\nabla p + \nabla \cdot (\mu + \mu_t) \left[\nabla \vec{V} + \left(\nabla \vec{V} \right)^T \right] \tag{3}$$

Transfer of turbulent kinetic energy:

$$\rho \frac{\partial k}{\partial t} + \rho \vec{V} \cdot \nabla k = \nabla \cdot \left[\left(\mu + \frac{\mu_t}{\sigma_k} \right) \nabla k \right] + \frac{\mu_t}{2} \left[\nabla \vec{V} + \left(\nabla \vec{V} \right)^T \right]^2 - \rho \varepsilon \tag{4}$$

Transfer of turbulent energy dissipation rate:

$$\rho \frac{\partial \varepsilon}{\partial t} + \rho \vec{V} \cdot \nabla \varepsilon = \nabla \cdot \left[\left(\mu + \frac{\mu_t}{\sigma_\varepsilon} \right) \nabla \varepsilon \right] + \frac{C_{1\varepsilon} \varepsilon \mu_t}{2k} \left[\nabla \vec{V} + \left(\nabla \vec{V} \right)^T \right]^2 - \frac{C_{2\varepsilon} \rho \varepsilon^2}{k} \tag{5}$$

The turbulent viscosity:

$$\mu_t = \frac{\rho C_\mu k^2}{\varepsilon} \tag{6}$$

Equations (1)–(6) [26] were solved using a finite volume and were discretized by the Quadratic Upstream Interpolation for Convective Kinetics (QUICK) and the Semi-Implicit method for Pressure-Linked Equations (SIMPLE) algorithms [27]. All the results converge with the residual values, being less than 10^{-6}.

The friction factor (f), Nusselt number (Nu), and thermal enhancement factor (TEF) are the key variables in this research. The friction factor can be calculated from the pressure drop on the test plate:

$$\Delta p = f \frac{\rho L \overline{v}^2}{2D} \tag{7}$$

The Nusselt number can be written as:

$$Nu_D = \frac{hD}{k} \tag{8}$$

The TEF is defined as:

$$TEF = \left(\frac{Nu_b}{Nu_0} \right) \left(\frac{f_b}{f_0} \right)^{-1/3} \tag{9}$$

Nu_b and f_b stand for the values for the bulk fluid in the dimple plate, while Nu_0 and f_0 stand for the values for the bulk fluid in the flat plate [28].

Reynolds Number in this research is calculated from diameter of jet (D_j):

$$Re = \frac{VD_j}{\upsilon} \qquad (10)$$

4. Simulation of Flow Configuration

The total area on the circular plate was tested in the experiment, while only a quarter of the total area was analyzed in the simulation because of the symmetry, as shown in Figure 3. The air at 300 K flows through the inlet section and impinges on the test plate and then flows out at the outlet. The constant heat flux is set at the test plate. The inlet air flow velocity is set corresponding to Reynolds numbers of 1500, 4400, 7300, 10,200 and 14,600.

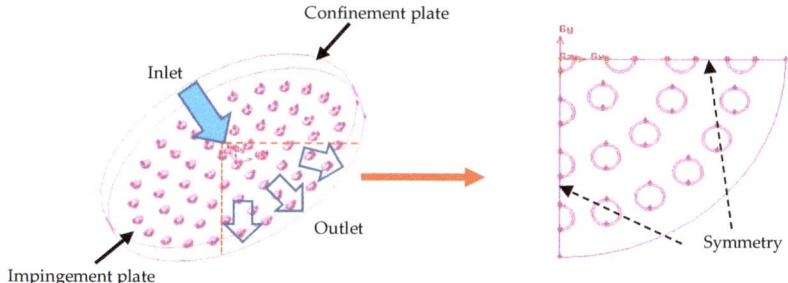

Figure 3. Channel geometry and Computational domain of flow.

Fluent in Ansys was used for the simulations in this work. Grids were manually created by the researcher for defining the gridlines. To resolve the near test plate region, much finer grids were set near the test plate. The mesh configuration was tested to be sufficient to provide grid-independent results.

5. Results and Discussion

5.1. Effect of Dimples Shape

Figure 4 presents the temperatures on the test plate, air inlet and outlet at an Re of 10,200 (experiment). The lowest temperature was shown to be at the center of test plate and the temperature slightly increases from the center to the edge along the radius of the test plate. The air from the inlet flowed onto the test plate and then flowed out. The outlet temperature of the air increased because it had received the heat from test plate. The flat plate showed a significantly higher temperature at the surface than the dimpled plate because of the lower levels of heat transferred to the air.

The average Nusselt number ($\overline{Nu}$) was calculated from the experiment results and converted to the heat transfer rate. A higher $\overline{Nu}$ resulted in a higher heat transfer rate. The results of the flat plate and the dimple plate (at dimple diameters of (d) = 1 and 2 cm) were compared alongside the Reynolds numbers, as shown in Figure 5. The results show that $\overline{Nu}$ increases as the Reynolds number increases. The $\overline{Nu}$ of the dimpled plate was higher than that of the flat plate for all Reynolds numbers in the range. At Re = 14,600, E_r = 2 d, E_θ = 1.5 d and B = 2, the Nusselt numbers of the test plate at d = 1 cm was 100 and d = 2 cm was 60. The $\overline{Nu}$ at d = 1 cm was about 170% higher than at d = 2 cm.

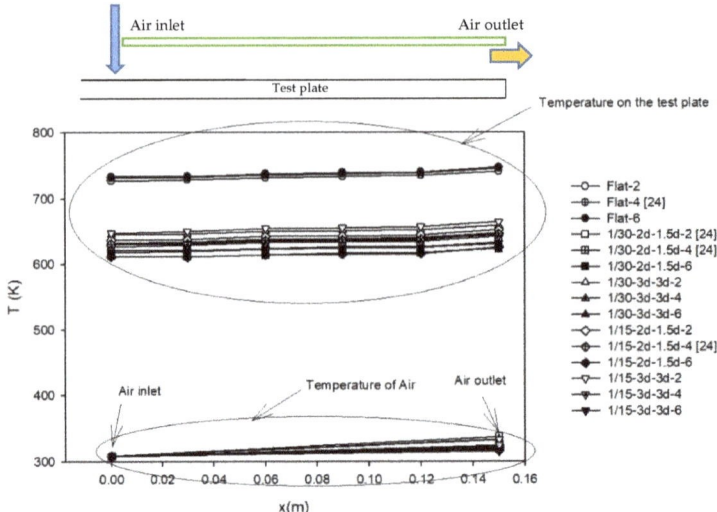

Figure 4. Temperature on the test plate, air inlet and outlet of Re at 10,200 (experiment).

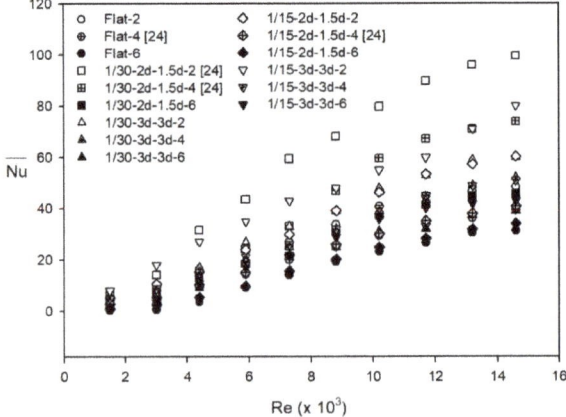

Figure 5. Variation of $\overline{N}_u$ with Re for flat plate and dimpled plate (experiment).

From Figure 6, it can be seen that the local Nusselt number (Nu) distributions on the test plate are the highest around the center of the plate and they tend to decrease with increasing distance from the center of the test plate (E_r). The local Nusselt numbers of dimpled surfaces were also higher than those of the flat plate, because the dimpled shape resulted in the vortex, an increase in turbulent intensity and a larger heat transfer area.

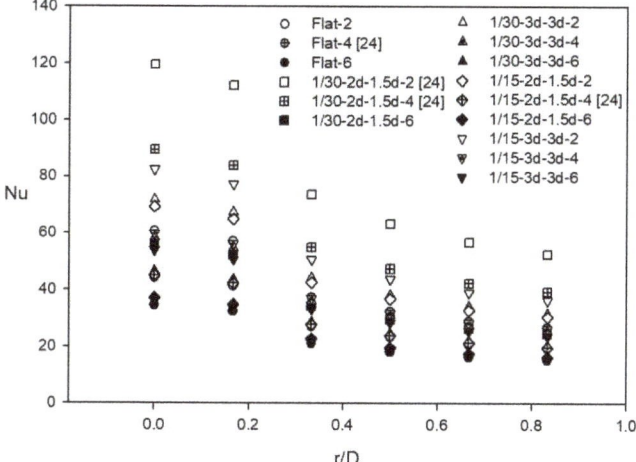

Figure 6. Nusselt number distribution at an Re of 10,200 for the flat plate and dimpled plate (experiment).

5.2. Effect on the Distance Between Test Plate and Jet (B)

There were three distances between the nozzle jet and test plate (B), which were 2 d, 4 d and 6 d. The results show that the shorter the distance between the nozzle jet and test plate, the higher the $\overline{Nu}$ and Nu, as shown in Figures 5 and 6. Nu values on the same test plate (d = 1 cm, E_r = 2 d, E_θ = 1.5 d and Re = 14,600) were 100, 75 and 45 at B = 2 d, 4 d and 6 d, respectively.

In Figure 7, it can be seen that the local Nusselt number (Nu) of the shortest B (B = 2 d) is the highest. This is due to the fact that the shortest distance between the nozzle jet and test plate (B) caused the strongest impingement on the test plate and the highest turbulence intensity.

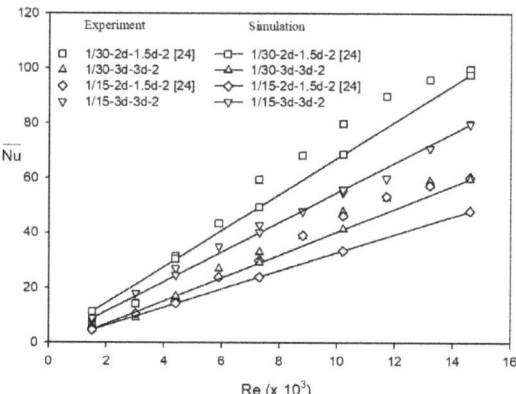

Figure 7. Variation of $\overline{Nu}$ with Re for the flat plate and dimpled plate (experiment and simulation).

5.3. Effect of the Distance between Dimples (E_r, E_θ)

There were two types of distance between dimples (E_r, E_θ), which were E_r = 2 d, E_θ = 1.5 d and E_r = 3 d, E_θ = 3 d. In Figures 5 and 6, at the same dimple diameter (d) and the same distance between nozzle jet and test plate (B), the results show that the $\overline{Nu}$ at E_r = 2 d, E_θ = 1.5 was higher than at

$E_r = 3$ d, $E_\theta = 3$ d. It means that a lower E_r and E_θ indicates a higher $\overline{Nu}$, which is similar to the trend seen for Nu (see Figure 6).

5.4. Simulation of Impinging Jet on Dimple Surface

The flow behavior can be more clearly explained by the simulation data than by the experiment results, which allows us obtain highly efficient research results. Moreover, the data that cannot be measured in the experiment can be obtained from a simulation. In this work, the pressure drop on the test plate was very low and did not really differ in each experiment, which made it difficult to interpret the data. Therefore, the simulation was used to analyze the results. However, the precision of the simulation result was the most important factor. In this work, the comparisons between the experimental and simulation work are shown in Figure 7. It is worth noting that the numerical results are in good agreement with measurements. The discrepancies in the results were less than $\pm 10\%$ for all 15 case studies. The pressure drop was the important piece of data and it was converted to friction factor (f). The friction factor represents how easy or hard it is for the fluid to flow through the equipment. The equipment that uses high power to move the fluid will present with a high f, which shows the lower thermal enhancement factor (TEF). The TEF represents the total heat transfer capacity and can be calculated from Nu and f. A high TEF will be observed in test plates with a high heat transfer rate.

The friction factor ratios (f/f_0) of the dimpled surface and flat plate are shown in Figure 8. The results show that f/f_0 increases as the Re increases. At the same Re, a higher f/f_0 will be obtained at a shorter distance between the nozzle jet and test plate (B), and a higher dimple distance (E_r and E_θ). For example, the highest f/f_0 value of 7 (TEF = 0.9) was observed in the test section plate at d = 2 cm, B = 1 d, E_r= 3 d, E_θ = 3 d and Re = 14,600. On the other hand, the highest TEF value of 5.5 ($f/f_0 = 4$) was observed at d = 1 cm, B = 2 d, E_r= 2 d, E_θ = 1.5 d and Re = 1500 (see Figure 9).

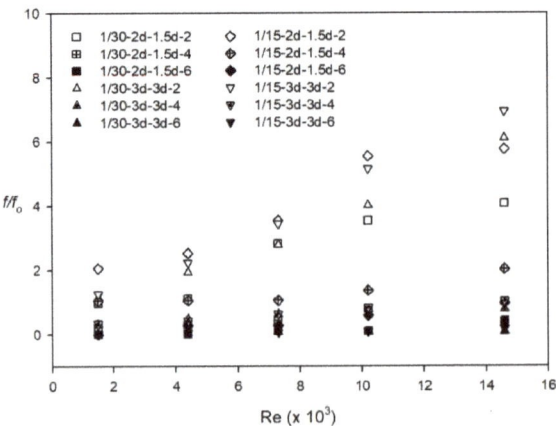

Figure 8. Variation of f/f_0 with Re of the dimpled plate (simulation).

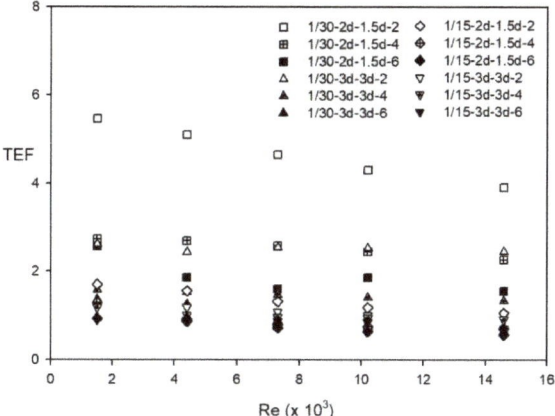

Figure 9. Variation of TEF with Re for the dimpled plate (simulation).

The air flow characteristics and heat transfer results can be analyzed by using a simulation technique. Figure 10 shows the 15 kinds of air flow, impinging on the test-section plate. The highest velocity air flow (long arrow sign) presented itself in the distance between the nozzle jet and test-section plate. The vortex then occurred near the center of test plate. The shorter B indicates turbulent fluid flow. The vortex and turbulent flow occur more often on the dimpled plate than the flat plate, which leads to better heat transfer. On a plate with a bigger dimple diameter (2 cm), the small vortex is observed above the dimple area at the outlet of the test plate. The cold air cannot be exchanged with the hot air because the vortex flow blocks the air flow. A small dimple diameter (1 cm) results in a higher TEF.

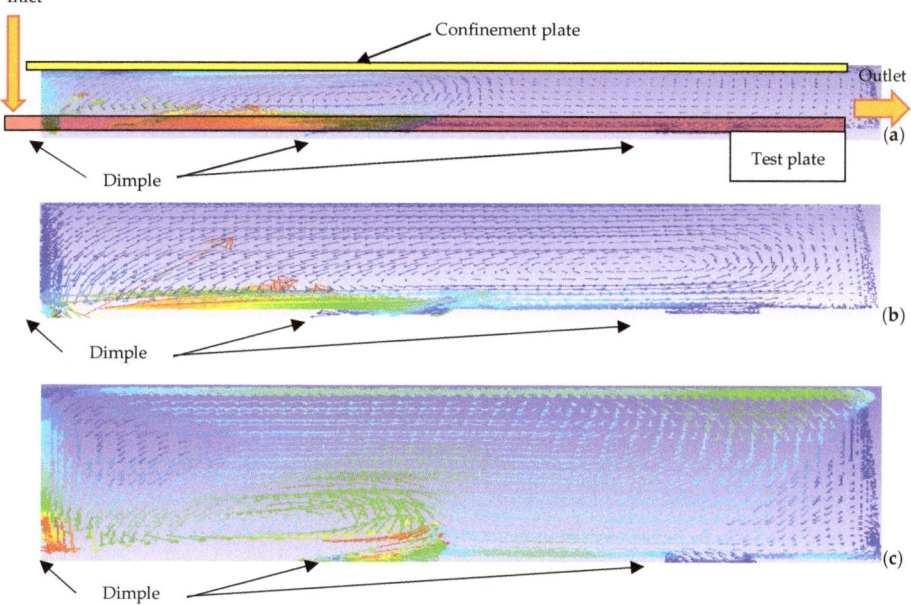

Figure 10. Cont.

Energies **2019**, *12*, 813

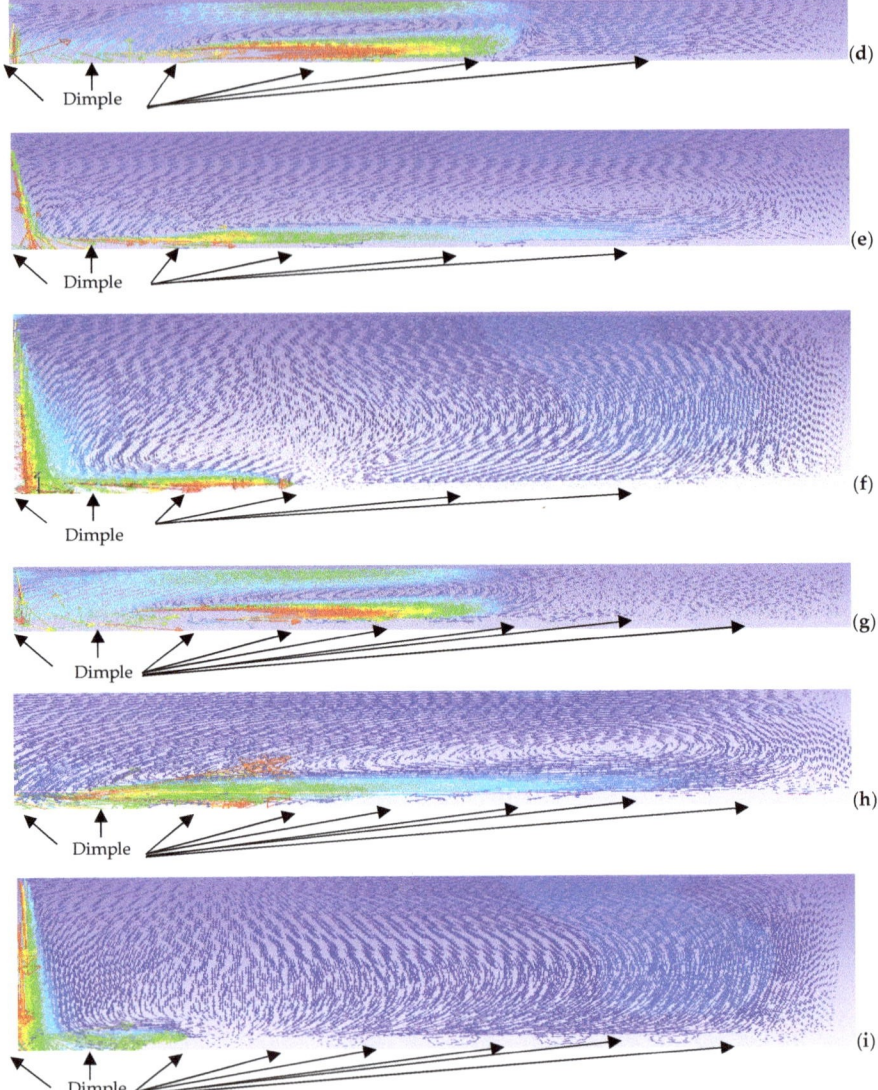

Figure 10. Streamlines at Re = 14,600: (d/D-E$_r$-E$_\theta$-B) (**a**) 1/15-3-3-2, (**b**) 1/15-3-3-4, (**c**) 1/15-3-3-6, (**d**) 1/30-3-3-2, (**e**) 1/30-3-3-4, (**f**) 1/30-3-3-6, (**g**) 1/30-2-1.5-2, (**h**) 1/30-2-1.5-4 [24] and (**i**) 1/30-2-1.5-6.

The air flow characteristics which resulted from the different Re values in five of the cases studied (Re = 1500, 4400, 7300, 10,200 and 14,600), at d = 1 cm, B = 2 d, E$_r$ = 2 d, and E$_\theta$ = 1.5 d, are shown in Figure 11. The results show that the vortex flow increases as Re increases, resulting in a higher TEF.

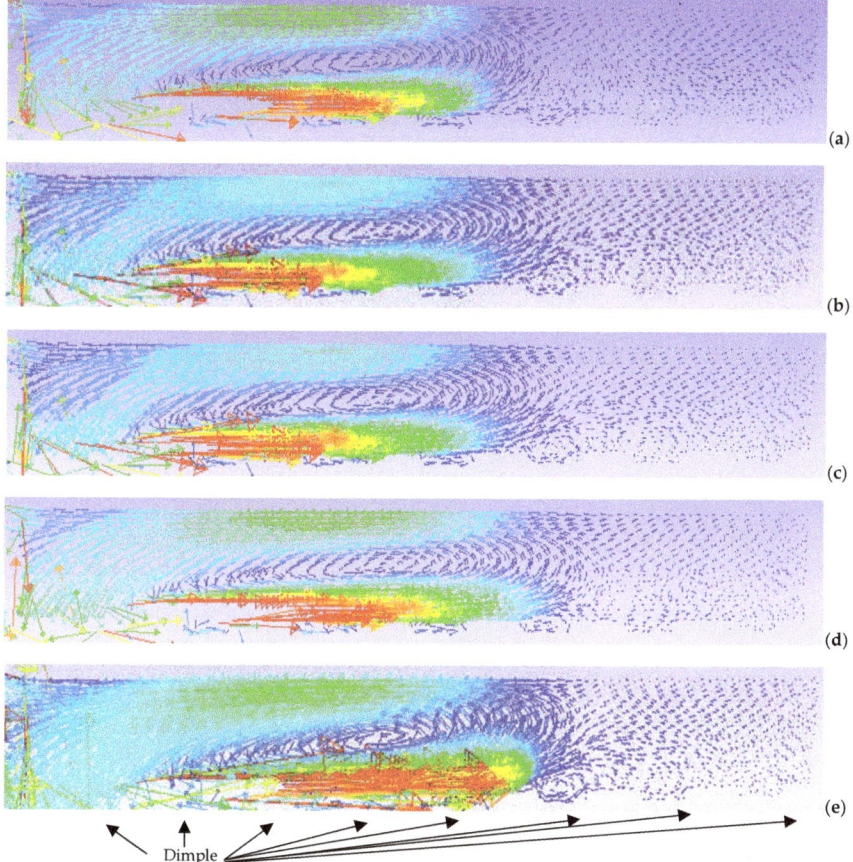

Figure 11. Streamlines with d = D_j and B = 2 D_j at (**a**) Re = 1500, (**b**) Re = 4400, (**c**) Re = 7300, (**d**) Re = 10,200 and (**e**) Re = 14,600 [24].

The temperature contours on the test section plate are shown in Figure 12. A higher TEF occurs at a lower temperature because of the heat transfer of the air from the wall to the air. The lower temperature of the test section results in a higher TEF. At d = 1 cm, B = 2 d, E_r = 2 d, E_θ = 1.5 d, a higher TEF is observed at the lowest temperature contour.

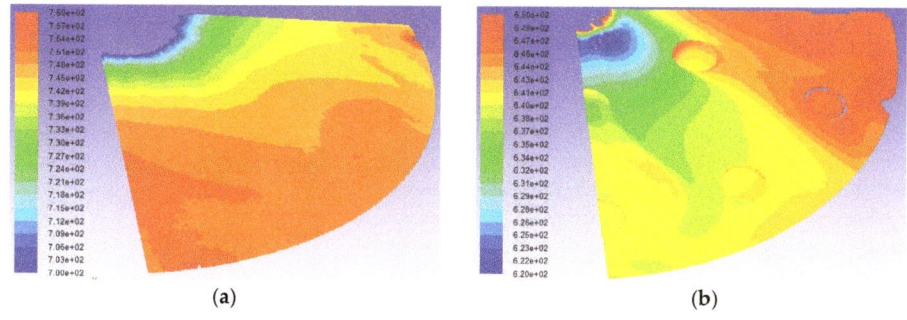

Figure 12. Cont.

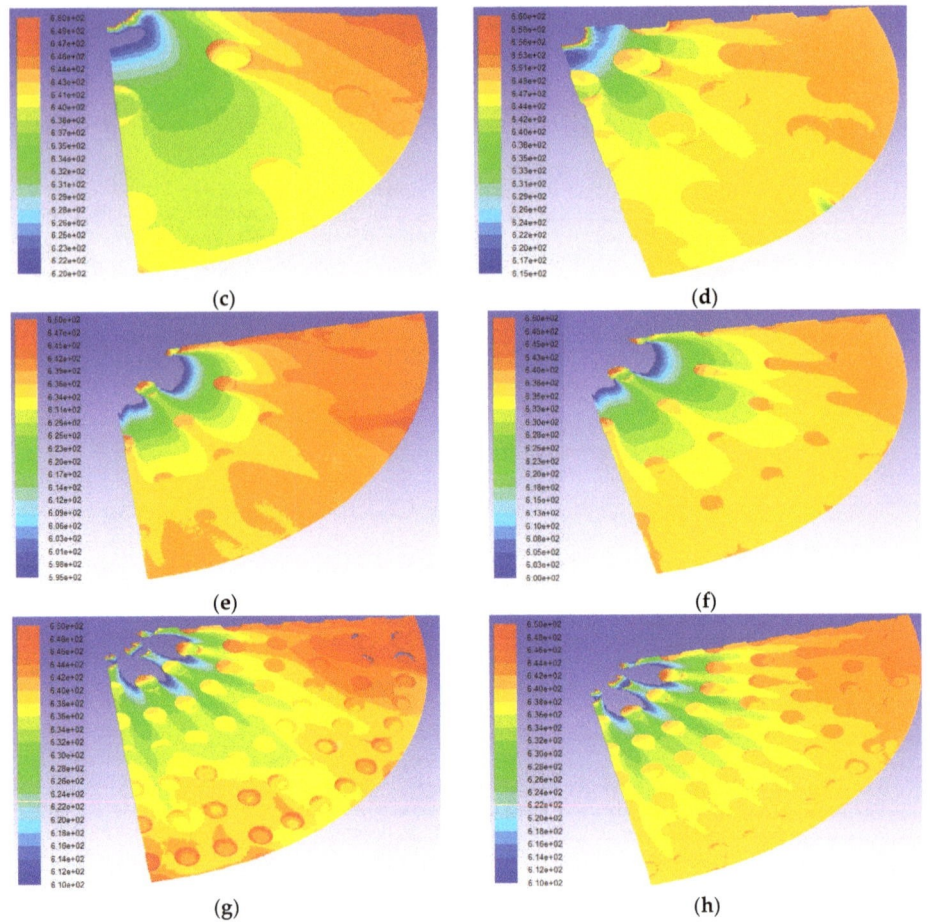

Figure 12. Temperature contours (K) at Re = 14,600: (d/D-E_r-E_θ-B) (**a**) F-2, (**b**) 1/15-3-3-2, (**c**) 1/15-3-3-4, (**d**) 1/15-2-1.5-4 [24], (**e**) 1/30-3-3-2, (**f**) 1/30-3-3-4, (**g**) 1/30-2-1.5-2 [24], and (**h**) 1/30-2-1.5-4 [24].

6. Conclusions

Actual experiments were carried out and the results utilized as part of the analysis in the simulation part. In the simulation, the heat transfer behavior and friction factor value on the test plate were investigated for the thermal enhancement factor (TEF). TEF can be explained by the heat transfer efficiency of an impinging jet on the test plate. The optimum conditions in terms of the fluid pump power and heat transfer value could be obtained.

From the experiment and simulation results, one can conclude that the cylindrical dimpled surface enhances the TEF. The highest TEF (= 5.5) was found for the configuration with a dimple diameter of d = 1 cm, a radial dimple distance of E_r = 2 d, a circumferential dimple distance of E_θ = 1.5 d, a distance between test plate and jet of B = 2, and Reynolds number of Re = 1500. The TEF which was obtained in this work was higher than that obtained in other research, which was about 4.0. This is because the friction factor ratio in this work (f/f_0) was very low, even though the $\overline{Nu}$ was not different from the $\overline{Nu}$ in other work. The most suitable test plate conditions which could be used to obtain a high TEF are d = 1 cm, B = 2 d, E_r = 2 d and E_θ = 1.5 d for Re=1500–14,600.

Author Contributions: For research articles with several authors, a short paragraph specifying their individual contributions must be provided. The following statements should be used "conceptualization, supervision, and methodology, Parkpoom Sriromreun; formal analysis, and writing—review and editing, Paranee Sriromreun".

Funding: This research was funded by the Research Grants of Faculty of Engineering, Srinakharinwirot University and The Strategic Wisdom and Research Institute, Srinakharinwirot University.

Acknowledgments: The funding of this research work is supported by the Research Grants of Faculty of Engineering, Srinakharinwirot University and The Strategic Wisdom and Research Institute, Srinakharinwirot University.

Conflicts of Interest: The authors declare no conflict of interest.

References

1. Ries, F.; Li, Y.; Klingenberg, D.; Nishad, K.; Janicka, J.; Sadiki, A. Near-Wall Thermal Processes in an Inclined Impinging Jet: Analysis of Heat Transport and Entropy Generation Mechanisms. *Energies* **2018**, *11*, 1354. [CrossRef]
2. Ries, F.; Li, Y.; Nishad, K.; Janicka, J.; Sadiki, A. Entropy Generation Analysis and Thermodynamic Optimization of Jet Impingement Cooling Using Large Eddy Simulation. *Entropy* **2019**, *21*, 129. [CrossRef]
3. Meola, C.; de Luca, L.; Carlomagno, G.M. Influence of shear layer dynamics on impingement heat transfer. *Exp. Therm. Fluid Sci.* **1996**, *13*, 29–37. [CrossRef]
4. Guerra, D.R.S.; Su, J.; Silva Freire, A.P. The near wall behavior of an impinging jet. *Int. J. Heat Mass Transf.* **2005**, *48*, 2829–2840. [CrossRef]
5. Chaudhari, M.; Puranik, B.; Agrawal, A. Heat transfer characteristics of synthetic jet impingement cooling. *Int. J. Heat Mass Transf.* **2010**, *53*, 1057–1069. [CrossRef]
6. Draksler, M.; Končar, B. Analysis of heat transfer and flow characteristics in turbulent impinging jet. *Nucl. Eng. Des.* **2011**, *241*, 1248–1254. [CrossRef]
7. Nanan, K.; Wongcharee, K.; Nuntadusit, C.; Eiamsa-ard, S. Forced convective heat transfer by swirling impinging jets issuing from nozzles equipped with twisted tapes. *Int. Commun. Heat Mass Transf.* **2012**, *39*, 844–852. [CrossRef]
8. Nuntadusit, C.; Wae-hayee, M.; Bunyajitradulya, A.; Eiamsa-ard, S. Heat transfer enhancement by multiple swirling impinging jets with twisted-tape swirl generators. *Int. Commun. Heat Mass Transf.* **2012**, *39*, 102–107. [CrossRef]
9. Qiang, Y.; Wei, L.; Luo, X.; Jian, H.; Wang, W.; Li, F. Heat Transfer and Flow Structures of Laminar Confined Slot Impingement Jet with Power-Law Non-Newtonian Fluid. *Entropy* **2018**, *20*, 800. [CrossRef]
10. Tong, A.Y. On the impingement heat transfer of an oblique free surface plane jet. *Int. J. Heat Mass Transf.* **2003**, *46*, 2077–2085. [CrossRef]
11. Goodro, M.; Park, J.; Ligrani, P.; Fox, M.; Moon, H.-K. Effects of hole spacing on spatially-resolved jet array impingement heat transfer. *Int. J. Heat Mass Transf.* **2008**, *51*, 6243–6253. [CrossRef]
12. Pakhomov, M.A.; Terekhov, V.I. Enhancement of an impingement heat transfer between turbulent mist jet and flat surface. *Int. J. Heat Mass Transf.* **2010**, *53*, 3156–3165. [CrossRef]
13. Ekkad, S.V.; Kontrovitz, D. Jet impingement heat transfer on dimpled target surfaces. *Int. J. Heat Fluid Flow* **2002**, *23*, 22–28. [CrossRef]
14. Lienhart, H.; Breuer, M.; Köksoy, C. Drag reduction by dimples? A complementary experimental/numerical investigation. *Int. J. Heat Fluid Flow* **2008**, *29*, 783–791. [CrossRef]
15. Kanokjaruvijit, K.; Martinez-Botas, R.F. Heat transfer correlations of perpendicularly impinging jets on a hemispherical-dimpled surface. *Int. J. Heat Mass Transf.* **2010**, *53*, 3045–3056. [CrossRef]
16. Xing, Y.; Weigand, B. Experimental investigation of impingement heat transfer on a flat and dimpled plate with different crossflow schemes. *Int. J. Heat Mass Transf.* **2010**, *53*, 3874–3886. [CrossRef]
17. Won, Y.; Wang, E.N.; Goodson, K.E.; Kenny, T.W. 3-D visualization of flow in microscale jet impingement systems. *Int. J. Therm. Sci.* **2011**, *50*, 325–331. [CrossRef]
18. Kwon, H.G.; Hwang, S.D.; Cho, H.H. Measurement of local heat/mass transfer coefficients on a dimple using naphthalene sublimation. *Int. J. Heat Mass Transf.* **2011**, *54*, 1071–1080. [CrossRef]
19. Turnow, J.; Kornev, N.; Zhdanov, V.; Hassel, E. Flow structures and heat transfer on dimples in a staggered arrangement. *Int. J. Heat Fluid Flow* **2012**, *35*, 168–175. [CrossRef]

20. De Bonis, M.V.; Ruocco, G. An experimental study of the local evolution of moist substrates under jet impingement drying. *Int. J. Therm. Sci.* **2011**, *50*, 81–87. [CrossRef]
21. Parida, P.R.; Ekkad, S.V.; Ngo, K. Experimental and numerical investigation of confined oblique impingement configurations for high heat flux applications. *Int. J. Therm. Sci.* **2011**, *50*, 1037–1050. [CrossRef]
22. Na-pompet, K.; Boonsupthip, W. Effect of a narrow channel on heat transfer enhancement of a slot-jet impingement system. *J. Food Eng.* **2011**, *103*, 366–376. [CrossRef]
23. Alenezi, A.; Almutairi, A.; Alhajeri, H.; Addali, A.; Gamil, A. Flow Structure and Heat Transfer of Jet Impingement on a Rib-Roughened Flat Plate. *Energies* **2018**, *11*, 1550. [CrossRef]
24. Sriromreun, P.K.; Sriromreun, P.N. Experimental and Numerical Studies of Heat Transfer Characteristics for Impinging Jet on Dimple Surfaces. *Chem. Eng. Trans.* **2018**, *70*, 1273–1278. [CrossRef]
25. ANSI/ASME. Test Uncertainty. Available online: http://gost-snip.su/download/asme_ptc_19_1,2005_test_uncertainty (accessed on 28 February 2019).
26. Patankar, S.V. *Numerical Heat Transfer and Fluid Flow*; McGraw-Hill: New York, NY, USA, 1980.
27. Versteeg, H.K.; Malalasekera, W. *An Introductionto ComputationalFluid Dynamics, The Finite Volume Method*, 2nd ed.; PEARSON Prentice Hall: London, UK, 2007.
28. Incropera, F.P.; Dewitt, D.B. *Introduction to Heat Transfer*, 5th ed.; John Wiley and Sons, Inc.: Hoboken, NJ, USA, 2007.

© 2019 by the authors. Licensee MDPI, Basel, Switzerland. This article is an open access article distributed under the terms and conditions of the Creative Commons Attribution (CC BY) license (http://creativecommons.org/licenses/by/4.0/).

Article

Cost-Optimal Heat Exchanger Network Synthesis Based on a Flexible Cost Functions Framework

Matthias Rathjens * and Georg Fieg

Hamburg University of Technology, Institute of Process and Plant Engineering, Am Schwarzenberg-Campus 4, 21073 Hamburg, Germany; g.fieg@tuhh.de
* Correspondence: matthias.rathjens@tuhh.de

Received: 17 January 2019; Accepted: 21 February 2019; Published: 26 February 2019

Abstract: In this article an approach to incorporate a flexible cost functions framework into the cost-optimal design of heat exchanger networks (HENs) is presented. This framework allows the definition of different cost functions for each connection of heat source and sink independent of process stream or utility stream. Therefore, it is possible to use match-based individual factors to account for different fluid properties and resulting engineering costs. Layout-based factors for piping and pumping costs play an important role here as cost driver. The optimization of the resulting complex mixed integer nonlinear programming (MINLP) problem is solved with a genetic algorithm coupled with deterministic local optimization techniques. In order to show the functionality of the chosen approach one well studied HEN synthesis example from literature for direct heat integration is studied with standard cost functions and also considering additional piping costs. Another example is presented which incorporates indirect heat integration and related pumping and piping costs. The versatile applicability of the chosen approach is shown. The results represent designs with lower total annual costs (TAC) compared to literature.

Keywords: heat exchanger network (HEN); synthesis; optimization; direct heat integration; indirect heat integration; piping; pumping

1. Introduction

The application of heat integration strategies can have a significant impact on reducing the amount of utility used by a process and thus improve its economic performance. Against the background of increasing global competition, environmental specifications, climate change and assumedly increasing energy costs heat integration using heat exchanger networks (HENs) have a significant importance [1].

Heat integration strategies have been developed to reduce both, capital and operating costs since decades by now. Pinch technology [2] and mathematical programming [3] have been the two main approaches and have been improved numerous times by many researchers [4]. Not only single processes but also total site heat integration has been considered. Initial works have been carried out by Dhole and Linnhoff [5]. Due to new challenges, in recent years publications have covered relevant practical issues in a higher degree of detail. As a consequence the problem complexity increased. The main issues influencing practical implementation of total site integration have been formulated by Chew et al. [6]. The consideration of further impact factors like safety related issues has been a major topic in heat integration during the last years [1]. The identification of critical risk equipment and respective streams for total site heat integration was developed by Liu et al. [7]. Nemet et al. [8,9] developed approaches for including risk assessment already during the HEN synthesis. Multiperiod HEN synthesis as well as controllability and disturbance propagation have been studied [10–12]. Due to operational issues and safety concerns direct heat integration is not always practical to realize [13,14]. Therefore, Wang et al. [14] developed a graphical methodology to investigate different connection patterns for total site heat integration. This methodology was developed further by applying mathematical models to determine the optimal solution for multi-plant heat integration [15]. Multi-plant heat integration has been further considered by Chang et al. [13,16]. The consideration of plant layout issues is an important factor during optimization. Liew et al. [17] introduced an improved heat cascade algorithm considering pressure drop and heat loss for utility targeting in total site heat integration. Pouransari and Maréchal [18] took into account individual priority levels for different possible connections and Souza et al. [19] included pressure drops in piping as well as in heat exchangers.

The review given above shows that the various demands on HEN optimization in the literature are manifold and a huge variety of optimization models are used. In this work our aim is to incorporate a flexible consideration of cost functions into the cost-optimal HEN synthesis to account for various fields of application. The most important part is the definition of individual factors for each possible connection of heat source and sink. Therefore it is possible to represent a multitude of practical implementation requirements with the same mixed integer nonlinear programming (MINLP) model. For example, it is possible to consider individual cost functions for different types of heat exchangers required for different operation conditions and properties of the involved process streams. Depending on the properties of the process streams the material and thus the costs of the installed heat exchanger can vary significantly. Peripheral equipment, layout constraints as well as cost of premises can be taken into account. These cost functions can be directly incorporated without changing the model or the solution algorithm itself. Concerning the algorithm performance it is the clear aim to be able to generate valid network structures that are competitive to the best solutions published in literature by now. Therefore, different approaches are combined. This model was primarily developed for direct heat integration [1], but is also shown to be applicable for the cost optimization of indirect heat integration problems.

2. Methodology

The utilized simultaneous cost optimization model is mainly based on a superstructure MINLP formulation and a hybrid genetic algorithm developed by Luo et al. [20] to solve the problem resulting from this formulation. A common HEN optimization problem statement is described by a set of N_h hot process streams, N_c cold process streams, their respective heat capacity flow rates $\dot{W}$ and heat transfer coefficients h. The supply and target temperatures (T' and T_{out}) of the process streams are given as well as the hot utility HU and cold utility streams CU with their respective temperature levels. Furthermore, the cost parameters should be known to perform a cost optimization.

2.1. Fundamentals

The model for HEN optimization in this work is based on counterflow heat exchangers for the process-to-process heat exchangers as well as the utility heat exchangers. The heat load $\dot{Q}$ in kW is calculated as the product of overall heat transfer coefficient U in kW/(m²K), the heat exchanger area A in m² and the logarithmic mean temperature difference (LMTD) ΔT_m in K:

$$\dot{Q} = U \cdot A \cdot \Delta T_m \qquad (1)$$

The overall heat transfer coefficient is calculated via the individual heat transfer coefficients h_h and h_c in kW/(m²K) of the connected hot and cold process streams. The thermal resistance of the wall is neglected:

$$U = \left(\frac{1}{h_h} + \frac{1}{h_c} \right)^{-1} \qquad (2)$$

Due to the consideration of counterflow heat exchangers, the LMTD is used as temperature driving force for heat transfer:

$$\Delta T_m = \frac{(T'_h - T''_c) - (T''_h - T'_c)}{\ln\left(\frac{T'_h - T''_c}{T''_h - T'_c}\right)} \tag{3}$$

The correlations for hot and cold outlet temperatures for the heat exchangers are as follows:

$$T''_h = T'_h - \frac{\dot{Q}}{\dot{W}_h} \tag{4}$$

and:

$$T''_c = T'_c + \frac{\dot{Q}}{\dot{W}_c} \tag{5}$$

In order to represent different HEN solutions, a stage-wise superstructure proposed by Yee et al. [21] is used. As an example, a superstructure for two hot process streams j and three cold process streams k with stages i is shown in Figure 1 including the nomenclature used in the following equations.

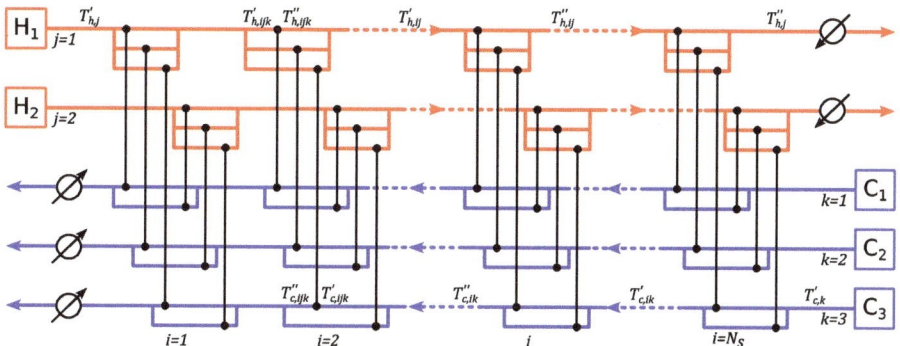

Figure 1. Stage-wise superstructure of a heat exchanger network (HEN) with two hot and three cold process streams.

In each stage of the superstructure each hot stream is possibly connected with each cold stream. At the end of the streams a utility heat exchanger may be placed to ensure the desired target temperatures to get reached. Within the superstructure every heat exchanger has a specific index ijk depending on the positioning:

$$ijk = (i-1)\cdot N_h \cdot N_c + (j-1)\cdot N_c + k \tag{6}$$

Rearranging Equations (1)–(5) according to the outlet temperatures leads to a formulation which allows for the numerical calculation of the HEN temperatures:

$$\begin{bmatrix} T''_{h,ijk} \\ T''_{c,ijk} \end{bmatrix} = \begin{bmatrix} \frac{(1-R_{h,ijk})e^{-NTU_{ijk}(1-R_{h,ijk})}}{1-R_{h,ijk}e^{-NTU_{ijk}(1-R_{h,ijk})}} & \frac{1-e^{-NTU_{ijk}(1-R_{h,ijk})}}{1-R_{h,ijk}e^{-NTU_{ijk}(1-R_{h,ijk})}} \\ \frac{R_{h,ijk}(1-e^{-NTU_{ijk}(1-R_{h,ijk})})}{1-R_{h,ijk}e^{-NTU_{ijk}(1-R_{h,ijk})}} & \frac{1-R_{h,ijk}}{1-R_{h,ijk}e^{-NTU_{ijk}(1-R_{h,ijk})}} \end{bmatrix} \begin{bmatrix} T'_{h,ijk} \\ T'_{c,ijk} \end{bmatrix} \tag{7}$$

with $R_{h,ijk}$ being:

$$R_{h,ijk} = \frac{\dot{W}_{h,ijk}}{\dot{W}_{c,ijk}} \tag{8}$$

and:

$$NTU_{ijk} = \frac{(UA)_{ijk}}{\dot{W}_{h,ijk}} \tag{9}$$

The corresponding non-isothermal mixing temperature for each process stream after each stage is calculated based on an energy balance [22]. The maximum number of stages to represent a HEN during optimization is set to $N_s = max\{N_h, N_c\}$ [22].

2.2. Objective Function

The objective function is structured in such way that a flexible cost functions framework can get incorporated in order to consider different implementation related factors like piping and pumping. A different cost function for every coupling of possible hot and cold process stream matches can be defined. In addition, different cost functions can get added for each utility heat exchanger. If necessary they can be further refined for every stage in the superstructure representation of the HEN. The objective function is based on the proposed structure by Rathjens and Fieg [1] and was further developed for the present work:

$$min\left\{ \sum_{n=1}^{N_h+N_c} \left[C_{CU} \cdot max\left\{ \dot{W}_n \left(T_n'' - T_{out,n}^+\right), 0 \right\} + C_{HU} \cdot max\left\{ \dot{W}_n \left(T_{out,n}^- - T_n''\right), 0 \right\} \right] \right.$$
$$\left. + \sum_{i=1}^{N_s} \sum_{j=1}^{N_h} \sum_{k=1}^{N_c} z_{ijk} X_{ijk} + \sum_{n=1}^{N_h+N_c} \left[z_{CU,n} X_{CU,n} + z_{HU,n} X_{HU,n} \right] \right\} \tag{10}$$

Hot utility and cold utility are used to reach the desired target outlet temperatures $T_{out,n}$ and cause costs based on their individual supply costs C_{HU} and C_{CU}. During optimization hot and cold utility can potentially be applied to each process stream depending on the maximum positive or negative deviation from the specified bounds of target temperatures. The respective utility which remains unused will make no contribution to the total annual costs (TAC). The binary variables z state the existence of a heat exchanger within the superstructure. The match-based costs X are calculated based on different cost function formulations with a wide range of possible correlations that can get implemented. The common cases of using a power function with the coefficients a_0, a_1 and a_2 as well as a general polynomial representation up to 4th degree (coefficients b_0, b_1, b_2, b_3 and b_4) are included. More specialized dependencies are covered using exponential (coefficients c_0 and c_1) or logarithmic expressions (coefficients d_0 and d_1):

$$X_{ijk} = \begin{array}{l} a_{0,ijk} + a_{1,ijk} \xi^{a_{2,ijk}} + \\ b_{0,ijk} + b_{1,ijk} \xi + b_{2,ijk} \xi^2 + b_{3,ijk} \xi^3 + b_{4,ijk} \xi^4 \\ c_{0,ijk} + e^{c_{1,ijk} \xi} + \\ d_{0,ijk} \ln(\xi) + d_{1,ijk} \end{array} \tag{11}$$

The different cost function formulations given in Equation (11) are chosen because they were able to represent the cost correlations for practical implementation that were faced when working together with our industrial partners. In order to be used in the objective function, the dependencies given in Equation (11) have to get expressed through the individual heat exchanger areas A, the temperature levels T and heat capacity flow rates $\dot{W}$ of the involved streams and the heat load $\dot{Q}$:

$$\zeta \epsilon \{A_{ijk}, \dot{W}_{h,ijk}, \dot{W}_{c,ijk}, \min\{\dot{W}_{h,ijk}, \dot{W}_{c,ijk}\}, \max\{\dot{W}_{h,ijk}, \dot{W}_{c,ijk}\}, T'_{h,ijk}, T'_{c,ijk}, T''_{h,ijk}, T''_{c,ijk}, \dot{Q}_{ijk}\} \quad (12)$$

The expression of cost dependencies through distinct variables was chosen due to ensuring the universal applicability of the shown approach without extending the programming work for each different problem. Furthermore these variables can be used with minimal computational overhead. Because ζ can be chosen out of an amount of alternatives given in Equation (12) the combinatorial possibilities of representing different cost dependencies are immense. The costs of utility heat exchangers $X_{HU,n}$ and $X_{CU,n}$ are structured equivalently to X_{ijk}.

2.3. Genetic Algorithm

The genetic algorithm uses the common genetic operations like selection, crossover and mutation. As optimization variables the parameters $(UA)_{ijk}$, $\dot{W}_{h,ijk}$ and $\dot{W}_{c,ijk}$ are used like proposed by Fieg et al. [22]. The selection is based on the fitness value F. The fitness value represents the quality of an individual with respect to the objective function. It is based on the relative relation towards the average costs among all individuals and the minimum TAC of all individuals. The calculation of the fitness value is given in the following equation:

$$F = \frac{C_{TAC}^{-1} + C_{TAC,min}^{-1} - 2C_{TAC,avg}^{-1}}{C_{TAC,min}^{-1} - C_{TAC,avg}^{-1}} \quad (13)$$

The selection is carried out according to roulette wheel selection. Crossover is split in parameter crossover and structure crossover. Mutation is carried out on the variable parameters stated above. For details about the explicit formulations please refer to Fieg et al. [22]. The probability for the application of a crossover operation is 89% in this work and the probability for parameter crossover was chosen as 23% according to Brandt [23]. The probability for the general mutation is 1%, for parameter mutation it is 50% with a respective gene mutation probability of 1% [23].

Because of the optimization of the parameter $(UA)_{ijk}$ using the explicit temperature solution depicted in Equation (7) instead of the heat loads of the heat exchangers, every solution is thermodynamically feasible. Otherwise occurring temperature or heat load constraints can be omitted. The binary variable handling and corresponding constraints considering the continuous variables are adopted from Luo et al. [20]. During optimization the strategy of excessive use of utilities is used [24]. If the specified hot utility temperature is higher than any target temperature of all process streams and cold utility temperature is lower than any cold target temperature of all process streams, each network generated throughout optimization is feasible. As a result no outlet temperature constraints are considered here. Because $\dot{W}_{h,ijk}$ and $\dot{W}_{c,ijk}$ are also used as optimization variables, these values have to get constrained to be within a feasible region to hold the constraints given in Equations (14) and (15):

$$\sum_{k=1}^{N_c} \dot{W}_{h,ijk} = \dot{W}_{h,j} \quad (14)$$

$$\sum_{j=1}^{N_h} \dot{W}_{c,ijk} = \dot{W}_{c,k} \quad (15)$$

During optimization $\dot{W}^*_{h,ijk}$ and $\dot{W}^*_{c,ijk}$ can occur that violate the above mentioned constraints. In order to keep these constraints, the normalization strategy from Fieg et al. is applied [22], which is shown by the following equations:

$$\dot{W}_{h,ijk} = \frac{\dot{W}_{h,j}}{\sum_{k=1}^{N_c} \dot{W}^*_{h,ijk}} \dot{W}^*_{h,ijk} \tag{16}$$

$$\dot{W}_{c,ijk} = \frac{\dot{W}_{c,k}}{\sum_{j=1}^{N_h} \dot{W}^*_{c,ijk}} \dot{W}^*_{c,ijk} \tag{17}$$

Furthermore, the structural control strategy proposed by Luo et al. [20] was adapted to ensure heterogeneity in the population and avoid local optimal solutions.

Structurally forbidden matches are removed automatically from a possible occurrence in the superstructure by setting the respective $(UA)_{ijk}$ to zero. A simple assignment of high costs for the specific match is not efficient against the background of fast convergence. This is particularly true for the second example shown below.

2.4. Local Optimization

Due to the strong nonlinearity of the stated objective function, an algorithm relying only on genetic operations would take too long to find an acceptable solution. Therefore, deterministic methods are used for local optimization. In order to generate promising HEN structures in the initialization step the approach of enhanced vertical heat transfer proposed by Stegner et al. [25] is used. The method proposed by Stegner et al. [25] adopts ideas known from the conventional vertical heat transfer concepts from traditional pinch approach but a new form of graphic depiction was implemented. Loop breaking as one of the heuristic approaches known from pinch technology is considered as well. This approach was successfully combined with stochastic optimization by Brandt et al. [26].

Furthermore, Newton's method is utilized for local parameter optimization [22]. Equation (18) shows the nomenclature in Newton's method for the variables $(UA)^*_{ijk}$, $\dot{W}^*_{h,ijk}$ and $\dot{W}^*_{c,ijk}$ to get the optimized expression:

$$\left\{(UA)_{ijk}, \dot{W}_{h,ijk}, \dot{W}_{c,ijk}\right\} = \left\{(UA)^*_{ijk}, \dot{W}^*_{h,ijk}, \dot{W}^*_{c,ijk}\right\} - \frac{\frac{\partial C_{TAC}}{\partial \left\{(UA)^*_{ijk}, \dot{W}^*_{h,ijk}, \dot{W}^*_{c,ijk}\right\}}}{\frac{\partial^2 C_{TAC}}{\partial \left\{(UA)^*_{ijk}, \dot{W}^*_{h,ijk}, \dot{W}^*_{c,ijk}\right\}^2}} \tag{18}$$

The partial derivatives needed in Equation (18) are calculated numerically using a step size of 10^{-4}.

3. Examples and Results

In order to show the application of the presented approach two examples have been chosen. Example 1 is a mid-sized optimization problem for direct heat integration. Example 2 is an optimization problem for indirect heat integration using a heat recovery loop (HRL).

3.1. Example 1

Example 1 was first stated by Pho and Lapidus [27] and was named 10SP1. It consists of five hot and five cold streams. In the first part of the analysis in our work we focus on the cost functions already used in other publications to achieve comparable results (see Section 3.1.1). In the second part we focus on additional costs caused by the consideration of piping (see Section 3.1.2).

3.1.1. Example 1a

The 10SP1 problem is well studied and has been used as an optimization case several times so far with decreasing costs throughout the years by many researchers (e.g., Nishida et al. [28] or Flower and Linnhoff [29]).

During the last two decades the 10SP1 problem has still attracted a considerable amount of attention. The associated problem data is given in Table 1. Two different formulations for the cost calculation of the annual costs of heat exchangers for the 10SP1 problem are given in the literature. The first cost formulation is given in Equation (19):

$$X_{ijk/HU_n/CU_n} = 140 \cdot A^{0.6}_{ijk/HU_n/CU_n} \tag{19}$$

The second cost formulation is given in Equation (20):

$$X_{ijk/HU_n/CU_n} = 145.63 \cdot A^{0.6}_{ijk/HU_n/CU_n} \tag{20}$$

A summary of the achieved TAC is given in Table 2. Several different approaches were used to tackle the 10SP1 problem. Lewin et al. [24] used a two-level approach consisting of a genetic algorithm and a lower level transformation into a linear parametric optimization problem. Lewin [30] also used an alternative nonlinear programming approach for the lower level optimization. Lin and Miller [31] used a tabu search algorithm to tackle the 10SP1 problem. Pariyani et al. [32] used a randomized algorithm for problems with stream splitting and a modified version of a previous work from Chakraborty and Ghosh [33] for designing networks without stream splitting. Yerramsetty and Murty [34] used a differential evolution algorithm and Peng and Cui [35] utilized a two-level simulated annealing algorithm. The latest work of Aguitoni et al. [36] found the best solution for the cost formulation 1 with a combination of a genetic algorithm and differential evolution.

Table 1. Problem data for example 1a.

Stream	T' (°C)	T_{out} (°C)	h ((kW/(m²K))	$\dot{W}$ (kW/K)	C_U ($/(kWyr))
H1	160	93	1.704	8.79	-
H2	249	138	1.704	10.55	-
H3	227	66	1.704	14.77	-
H4	271	149	1.704	12.56	-
H5	199	66	1.704	17.73	-
C1	60	160	1.704	7.62	-
C2	116	222	1.704	6.08	-
C3	38	221	1.704	8.44	-
C4	82	177	1.704	17.28	-
C5	93	205	1.704	13.90	-
HU	236	236	3.408	-	37.64
CU	38	82	1.704	-	18.12

Table 2. Results comparison for example 1a.

Sources	Reported TAC ($/yr)	
	Cost Formulation 1	Cost Formulation 2
Lewin et al. 1998 [24]	-	43,452 [1] (43,752 [1,3])
Lewin 1998 [30]	-	43,799 [1]
Lin and Miller 2004 [31]	43,329 [2]	-
Pariyani et al. 2006 [32]	-	43,439 [1] (43,611 [2])
Yerramsetty and Murty 2008 [34]	-	43,538 [1]
Peng and Cui 2015 [35]	-	43,411 [1]
Aguitoni et al. 2018 [36]	43,227 [2]	43,596 [2]
This work	42,963 [2]	43,321 [2]

[1] Solution without stream splits. [2] Solution with stream splits. [3] Revised by Pariyani et al. [32].

The optimization procedure with the chosen approach incorporates 375 optimization variables considering five hot streams, five cold streams and five stages for the superstructure. Each heat exchanger has the same area-related cost function.

In our work cost formulation 1 was used for the optimization. The minimum TAC found was 42,963 $/yr. The solution is depicted in Figure 2 and shows a configuration of two stream splits as well as the use of two utility heat exchangers.

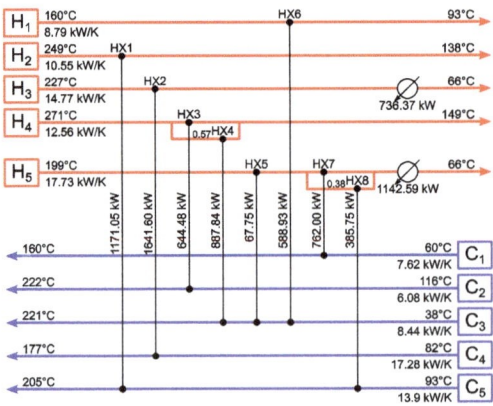

Figure 2. Grid-diagram representation of the optimal HEN configuration of example 1a (TAC: 42,963 $/yr).

The cost formulation for the optimization is favoring maximum energy recovery (MER) networks. Therefore, the obtained network recovers the maximum amount of energy and only cold utility is used. The TAC obtained for example 1a is significantly lower than in the publications cited in relation to the progress in TAC made over the years. The HEN shown in Figure 2 was also evaluated using cost formulation 2 and the resulting costs are 43,321 $/a, which is the lowest TAC found for formulation 2 so far. The results show that the chosen approach is capable of generating competitive results compared to other approaches presented in literature. Therefore, the consideration of even more complex problem formulations should be manageable, which is shown in the following parts.

3.1.2. Example 1b

Example 1b is built on the problem definition from example 1a. In addition, piping costs are considered. The required coordinates for the corresponding streams are taken from Pouransari and Maréchal [18]. They added arbitrary Cartesian coordinates to the 10SP1 problem stated above. For the hot utility coordinates the values from Rathjens and Fieg [1] are used. The additional data is given in Table 3.

Table 3. Additional Cartesian coordinates for the 10SP1 problem.

Stream	x (m)	y (m)	z (m)
H1	4	3	8
H2	6	7	4
H3	9	8	7
H4	2	2	5
H5	2	8	2
C1	7	4	1
C2	9	3	10
C3	1	4	2
C4	8	6	9
C5	4	5	3
HU	5	5	0
CU	6	9	5

The required inner pipe diameter D_i in m for each possible connecting pipe between a heat source and heat sink is calculated with the correlation for optimal diameter given by Peters et al. [37] for turbulent flow and $D_i \geq 0.0254$ m:

$$D_i = 0.363 \cdot \dot{V}^{0.45} \varrho^{0.13} \tag{21}$$

with the volumetric flow rate $\dot{V}$ in m^3/s, the fluid density ϱ of 983 kg/m^3 and a specific heat capacity of 4.18 kJ/(kgK).

Peters et al. [37] give a diagram for the correlation between pipe diameter and the related costs per meter of pipe. Due to manually selecting and reading out values from this diagram there might be a slight deviation from the book values which cannot be further specified. Considering stainless-steel welded pipe of type 304 [37] results in the following cost correlations with respect to the stream heat capacity flow rates and heat exchanger areas:

$$X_{ijk} = 140 \cdot A_{ijk}^{0.6} + L_{jk} \cdot \left(-0.174 \cdot \left(min\{ \dot{W}_{h,ijk}, \dot{W}_{c,ijk} \} \right)^2 + 17.773 \cdot \left(min\{ \dot{W}_{h,ijk}, \dot{W}_{c,ijk} \} \right) + 27.426 \right) \tag{22}$$

and:

$$X_{CU_n} = 140 \cdot A_{CU_n}^{0.6} + L_{CU_n,j} \left(-0.174 \cdot \dot{W}_{CU_n}^2 + 17.773 \cdot \dot{W}_{CU_n} + 27.426 \right) \tag{23}$$

with the distance L being:

$$L_{jk/CU_n,j} = 2 \cdot \left(|x_j - x_{k/CU_n}| + |y_j - y_{k/CU_n}| + |z_j - z_{k/CU_n}| \right) \tag{24}$$

The number of optimization variables is 375 as in example 1a. However, the number of potentially different cost functions is 25 (each match between hot and cold stream). The number of potentially different cost functions for the hot and cold utility usage is 10 for each utility. The real number of different cost functions is lower due to the fact that some distances between multiple hot and cold streams as well as utilities are equal.

The relatively short distances between the heat sources and heat sinks are compensated by the annualization of the piping costs over a one-year period. As a result the portion of piping costs becomes around 35% of TAC.

The solution with the lowest TAC found in this work is shown in Figure 3a. It has a TAC of 68,476 $/yr. A structurally different solution which was found during optimization with a less complex structure but the same utility configuration is shown in Figure 3b.

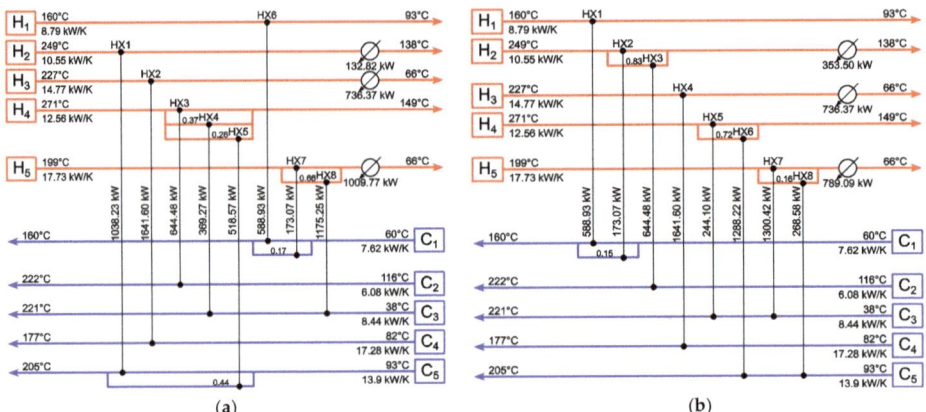

Figure 3. Grid-diagram representations of: (**a**) the optimal HEN configuration of example 1b (TAC: 68,476 $/yr); and (**b**) an alternative HEN configuration of example 1b (TAC: 68,843 $/yr).

As opposed to the solution of example 1a only two stages within the superstructure are taken by heat exchangers. Furthermore, the number of utility heat exchangers increased to three compared to the case not considering piping. The results still represent MER networks. Figures 4 and 5 show the relative orientation of the heat sources and heat sinks towards each other.

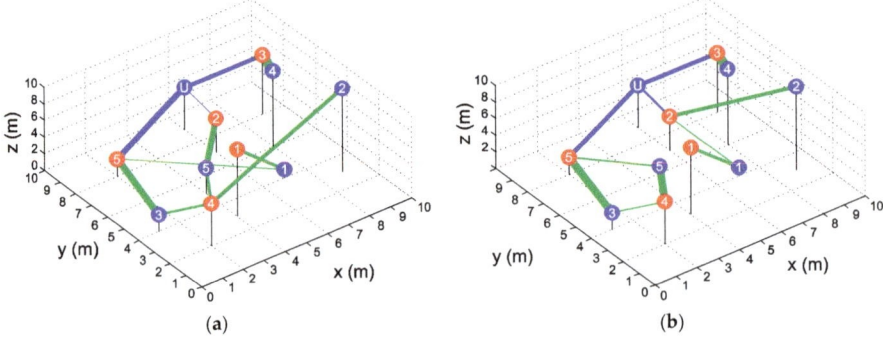

Figure 4. (**a**) 3D layout representation of the optimal HEN configuration of example 1b, the chosen line thickness is proportional to the corresponding heat exchanger heat load; and (**b**) 3D layout representation of the alternative HEN configuration of example 1b, the chosen line thickness is proportional to the corresponding heat exchanger heat load.

The configuration containing three utilities was dominant throughout the optimizations. The consideration of piping costs and thus the changed objective function had a strong influence on the heat load distribution within the HENs. The comparison between the optimal results of example 1b and example 1a shows about 14.2% (4016 $/yr) less piping costs, which is counterbalanced by an increase in installed heat exchanger area of about 16.2% (39.4 m^2) and thus an increase in capital costs for the heat exchangers of 14.7% (1307 $/yr).

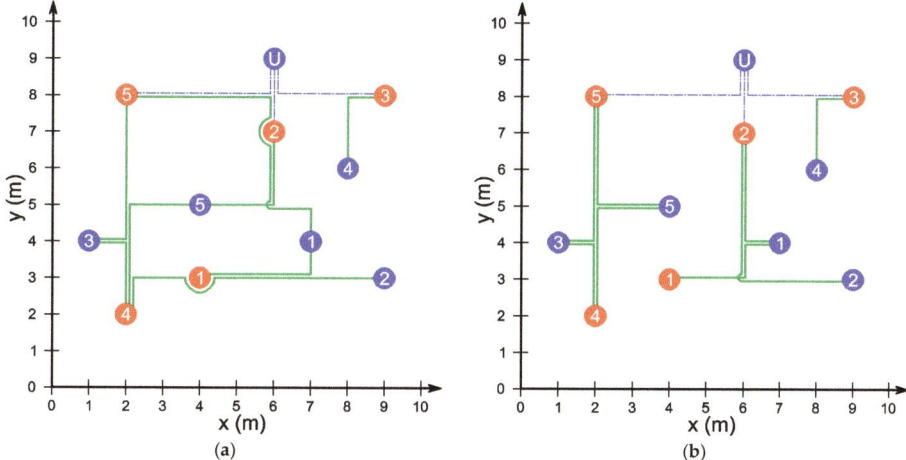

Figure 5. (a) 2D layout representation of the optimal HEN configuration of example 1b; (b) 2D layout representation of the alternative HEN configuration of example 1b.

The favorable design of using cold utility for the streams H3 and H5 is still preferred when considering piping costs during optimization due to the coordinates chosen by Pouransari and Maréchal [18]. Furthermore the optimization is dominated by the close arrangement of the streams H3 and C4 as well as the streams H4, H5 and C3. Despite that, the solutions found considering piping costs already during the optimization are structurally different. Rathjens and Fieg [1] used a different cost functions formulation and were able to show significantly more local clustering of heat exchange similar to Figure 5b.

3.2. Example 2

The second example is a heat integration case with indirect heat integration utilizing a HRL with an intermediate fluid. It was taken from Chang et al. [13]. It is a case study of a heat integration project in the southern part of China. The example comprises two different plants with a distance L of 1000 m. One plant is the heat source plant and the other one the heat sink plant with seven process streams each. The problem data is given in Table 4.

Table 4. Problem data for example 2.

Stream	T' (°C)	T_{out} (°C)	h ((kW/(m²K))	$\dot{W}$ (kW/K)
H1 (plant1)	148.1	114.7	1.642	311.9
H2 (plant1)	145.4	105.6	1.451	303.3
H3 (plant1)	141.9	98.4	1.754	302.6
H4 (plant1)	140.8	75.5	1.411	307.4
H5 (plant1)	135.3	55.3	1.531	335.4
H6 (plant1)	133.9	42.2	1.721	330.2
H7 (plant1)	131.9	41.2	1.713	331.3
C1 (plant2)	78.2	135.7	1.518	335.4
C2 (plant2)	69.3	108.5	1.631	323.3
C3 (plant2)	60.5	95.6	1.108	305.6
C4 (plant2)	59.5	90.3	1.501	321.5
C5 (plant2)	50.2	79.5	1.203	381.5
C6 (plant2)	45.9	71.4	1.102	311.5
C7 (plant2)	42.9	65.4	1.102	301.5

$C_{HU} = 20\ \$/(\text{kWyr})$, $C_{CU} = 8\ \$/(\text{kWyr})$ [38]; $C_{HX} = 11{,}000 + 150 \cdot A_{HX}$ [39].

In addition to the utility costs and the capital costs for the heat exchangers, the costs for pumping and piping are considered. The detailed equations are given in the following part.

For the calculation of the inner pipe diameter D_i for the HRL, Equation (21) is used. The fluid density of 960 kg/m³ and a specific heat capacity flow rate of 4.2 kJ/(kgK) is assumed.

The outer diameter D_{out} in m, the specific pipe weight wt_{pipe} in kg/m and the resulting pipe capital costs $Pcul$ in $/m for schedule 80 steel pipes are calculated according to Stijepovic and Linke [39]:

$$D_{out} = 1.101 \cdot D_{in} + 0.006349 \tag{25}$$

$$wt_{pipe} = 1330 \cdot D_{in}^2 + 75.18 \cdot D_{in} + 0.9268 \tag{26}$$

$$Pcul = 0.82 \cdot wt_{pipe} + 185 D_{out}^{0.48} + 6.8 + 295 \cdot D_{out} \tag{27}$$

The resulting capital costs for piping in $ are:

$$C_{pipe} = 2 \cdot L \cdot P_{cul} \tag{28}$$

For the calculation of the pumping capital costs the fanning friction factor f, the Reynolds number Re and the fluid velocity u in m/s are used to estimate the pressure drop Δp in Pa and the resulting costs [13,39]. The fluid velocity is calculated according to Equation (29) using the fluids mass flow rate $\dot{m}$ in kg/s of the intermediate fluid:

$$u = \frac{4 \cdot \dot{m}}{\varrho \cdot \pi \cdot D_{in}^2} \tag{29}$$

The Reynolds number is calculated assuming a dynamic viscosity μ of 0.0002834 Pa s:

$$Re = \frac{\varrho \cdot u \cdot D_{in}}{\mu} \tag{30}$$

The resulting pressure drop Δp is given by:

$$\Delta p = 4 \cdot f \cdot \frac{L \cdot \varrho \cdot u^2}{2 \cdot D_{in}} \tag{31}$$

with the fanning friction factor of:

$$f = \frac{0.046}{Re^{0.2}} \tag{32}$$

The cumulated capital costs for the two pumps C_{pump} in $ for the HRL are calculated according to Equation (33), utilizing the corresponding parameters for centrifugal pumps given in Jabbari et al. [40]:

$$C_{pump} = 2 \cdot \left(8600 + 7310 \cdot \left(\frac{\dot{m} \cdot \Delta p}{\varrho} \right)^{0.2} \right) \tag{33}$$

The operating costs for pumping $C_{pump,op}$ in $/yr are composed of the pump efficiency η_{pump} of 0.7, the price for electricity C_{el} of 0.1 $/(kWh) [13], an operation duration t of 8000 h/yr and the parameters calculated beforehand [13]:

$$C_{pump,op} = 2 \cdot C_{el} \cdot t \cdot \frac{\dot{m} \cdot \Delta p}{\varrho \cdot \eta_{pump}} \tag{34}$$

The annualization of the capital costs for the heat exchangers, piping and the required pumps is done using an annualization factor AF in $1/yr$. The annualization factor is calculated according to Chang et al. [13] with an operation time n of five years and a fractional interest rate I of 10% per year:

$$AF = \frac{I \cdot (1+I)^n}{(1+I)^n - 1} \tag{35}$$

In order to use the superstructure of the utilized optimization model in this work, two pseudo-streams are added to the problem definition to emulate the HRL. The respective heat transfer coefficients for the two streams are $1 \text{ kW}/(\text{m}^2\text{K})$ each [13,14]. The resulting superstructure has eight stages as well as eight hot streams and eight cold streams. In order to represent the cost correlations described beforehand, they are transferred to a dependency in heat capacity flow rate to get integrated in the cost functions framework proposed in this work. Therefore, the costs for piping and pumping are added up and calculated for different mass flow rates. The resulting costs are plotted against heat capacity flow rates calculated based on the respective mass flow rates. A curve fitting is carried out afterwards. The deviation from calculating the exact costs based on the equations given above is below 0.0012% considering rounded values for the piping and pumping costs. The resulting correlations are shown in Equation (36):

$$X_{ijk} = \begin{cases} 2910.77 + 39.57 \cdot A_{ijk}{}^1, & 2 \leq i \leq 7, 1 \leq j \leq 7, k = 1 \\ (109,107.62 + 300.52 \cdot \dot{W}_{h,j} - 0.1019 \cdot \dot{W}_{h,j}{}^2 & ijk = \{57, 505\} \\ +3.7486 \cdot 10^{-5} \cdot \dot{W}_{h,j}{}^3 - 6.1293 \cdot 10^{-9} \cdot \dot{W}_{h,j}{}^4)/2, & \\ 2910.77 + 39.57 \cdot A_{ijk}{}^1, & 2 \leq i \leq 7, j = 8, 2 \leq k \leq 8 \\ forbidden, & otherwise \end{cases} \tag{36}$$

The utility heat exchangers for the two plants are considered as already installed like it was done by Chang et al. [13]. Therefore, $X_{HU,n}$ and $X_{CU,n}$ are zero for this example. The first and last stages are used to include the costs for pumping and piping. The areas of the heat exchangers to emulate the HRL ($ijk = \{57, 505\}$) are chosen to be near infinite by the algorithm to avoid utility usage, thus create a self-adjusting temperature level and balance the heat transferred from the heat source plant to the heat sink plant. In order to prohibit direct heat integration between the two plants, all direct matches are forbidden. Altogether the applied superstructure yields 1536 optimization variables (compared to 1029 for direct heat integration without the pseudo streams). Due to the large number of forbidden matches (426) in relation to the number of possible matches (512), the number of optimization variables considered is reduced equivalently. The whole problem is described with only three different cost functions (see Equation (36)).

Iteration over the mass flow rate yields the optimal configuration for the HRL for example 2 which is shown in Figure 6. The superstructure representation of the solution in Figure 6 is shown in Figure A1 in Appendix A.

The TAC achieved for the solution shown in Figure 6 is 1.54 M\$/yr. The associated annualized capital costs for the heat exchangers are 0.574 M\$/yr (37.2% of TAC) and the utility consumption causes annual costs of 0.603 M\$/yr (39.1% of TAC). The overall piping costs are 0.326 M\$/yr (21.1% of TAC) and pumping costs are 0.038 M\$/yr (2.5% of TAC). The solution obtained in this work is therefore 4.4% cheaper than the solution reported by Chang et al. [13]. These differences are explained by the non-isothermal mixing model used and the relaxed temperature constraints in this work.

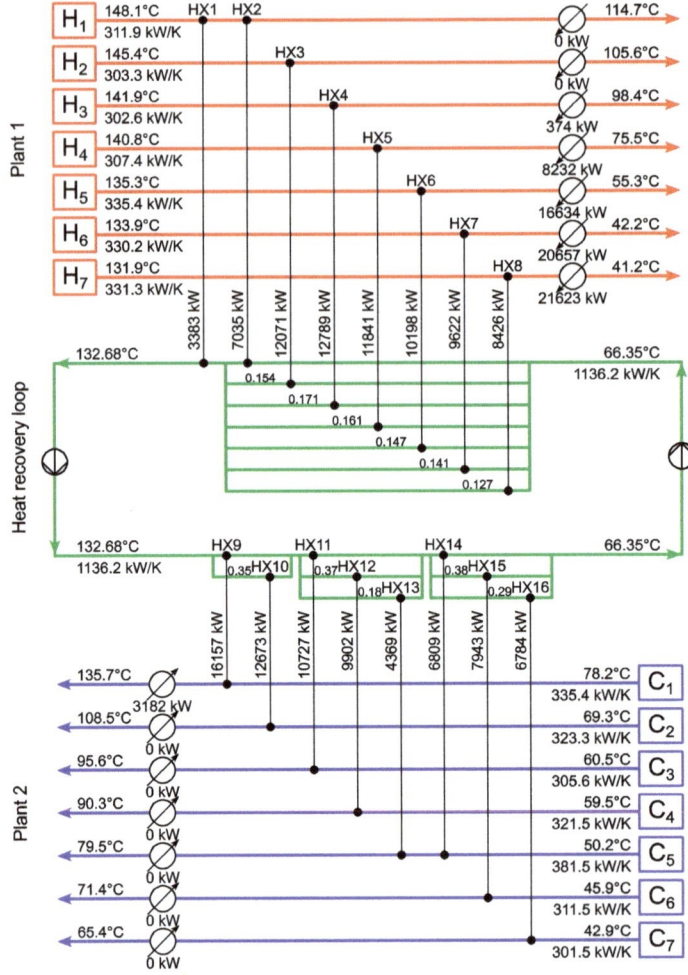

Figure 6. Plant representation of the optimal HEN configuration found of example 2 including a HRL (TAC: 1.54 M$/yr).

4. Conclusions

The presented approach to incorporate a flexible cost functions framework to synthesize cost-optimal heat exchanger networks (HENs) was carried out successfully and showed promising results. The flexible structured objective function allows for the integration of individual, match-dependent cost functions. The introduction of pseudo streams in combination with the flexible cost functions framework allow for the application for various problems. Corresponding optimizations utilizing different parametrizations have been carried out successfully. The universal applicability was shown by the execution of optimizations with different areas of application. The presented approach is applicable for direct as well as indirect heat integration utilizing the same superstructure and the same genetic algorithm for solving the problems. The use for combined direct and indirect heat integration is possible if initially forbidden matches get extended with cost information. This work presents results with lower TAC than other results published in literature beforehand.

For the practical implementation it is advisable to incorporate factors like piping already during optimization. The consideration of these factors can have a huge impact of the overall HEN structure and can lead towards significantly different and more efficient solutions. Local clustering was observed for some solutions like already reported by Rathjens and Fieg [1].

Author Contributions: Conceptualization, M.R.; methodology, M.R.; software, M.R.; validation, M.R. and G.F.; formal analysis, M.R.; investigation, M.R.; resources, M.R. and G.F.; data curation, M.R.; Writing—Original Draft preparation, M.R.; Writing—Review and Editing, M.R. and G.F.; visualization, M.R.; supervision, G.F.; project administration, M.R. and G.F.; funding acquisition, M.R. and G.F.

Funding: Authors would like to acknowledge the financial support from German Federal Ministry for Economic Affairs and Energy through ZIM program (Zentrales Innovationsprogramm Mittelstand, project number: ZF4025905CL7). The publication was funded by the Deutsche Forschungsgemeinschaft (DFG, German Research Foundation)—project number: 392323616 and the Hamburg University of Technology (TUHH) in the funding program "Open Access Publishing".

Acknowledgments: The authors gratefully acknowledge the support of our industrial partner, the "weyer group".

Conflicts of Interest: The authors declare no conflict of interest.

Nomenclature

HEN	Heat exchanger network
HRL	Heat recovery loop
MER	Maximum energy recovery
MINLP	Mixed integer nonlinear programming
TAC	Total annual costs
Δp	Pressure drop (Pa)
ΔT_m	Logarithmic mean temperature difference (LMTD) (K)
η_{pump}	Pump efficiency
μ	Dynamic viscosity (Pa s)
ϱ	Density (kg m^{-3})
A	Heat transfer area of heat exchanger (m^2)
AF	Annualization factor (yr^{-1})
C_{CU}	Cold utility cost per unit duty (\$ kW^{-1} yr^{-1})
C_{el}	Electricity costs (\$ kW^{-1} h^{-1})
C_{HU}	Hot utility cost per unit duty (\$ kW^{-1} yr^{-1})
C_{HX}	Heat exchanger capital costs (\$)
c_p	Specific heat capacity flow rate (kJ kg^{-1} K^{-1})
C_{pipe}	Capital costs for piping (\$)
C_{pump}	Capital costs for pumps (\$)
$C_{pump,op}$	Pump operating costs (\$ yr^{-1})
C_{TAC}	Total annual costs (\$ yr^{-1})
D_i	Inner diameter (m)
D_{out}	Outer diameter (m)
f	Fanning friction factor
F	Relative fitness value
h	Heat transfer coefficient (kW m^{-2} K^{-1})
$\dot{m}$	mass flow rate (kg s^{-1})
N_c	Number of cold process streams
N_h	Number of hot process streams
N_S	Number of stages of a stage-wise superstructure
NTU	Number of transfer units
$Pcul$	Pipe capital costs (\$ m^{-1})
$\dot{Q}$	Heat load (kW)
R	Ratio of stream heat capacity flow rates
Re	Reynolds number
t	Plant operation duration (h yr^{-1})
T	Stream temperature (°C)

T_{out}^+	Upper bounds of target temperature (°C)	
T_{out}^-	Lower bounds of target temperature (°C)	
u	Velocity (m s^{-1})	
U	Overall heat transfer coefficient (kW m^{-2} K^{-1})	
$\dot{V}$	Volumetric flow rate (m^3 s^{-1})	
$\dot{W}$	Heat capacity flow rate (kW K^{-1})	
wt_{pipe}	Specific pipe weight (kg/m)	
X	Match-based costs ($ yr^{-1})	
z	Binary variable	
c	Cold stream	
CU	Cold utility	
h	Hot stream	
HU	Hot utility	
i	Stage index	
ijk	Index of heat exchanger in superstructure	
j	Hot stream index	
k	Cold stream index	
max	Maximum	
min	Minimum	
$\prime$	Inlet temperature	
$\prime\prime$	Outlet temperature	

Appendix A

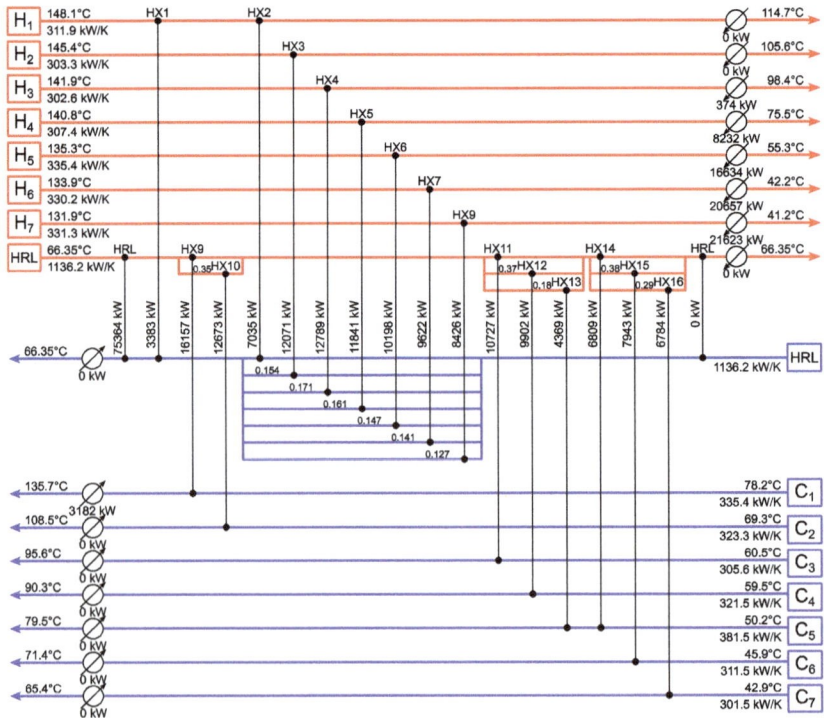

Figure A1. Grid-diagram representation of the optimal HEN configuration found of example 2 including a HRL in the superstructure layout (TAC: 1.54 M$/yr).

References

1. Rathjens, M.; Fieg, G. Design of cost-optimal heat exchanger networks considering individual, match-dependent cost functions. *Chem. Eng. Tran.* **2018**, *70*, 601–606. [CrossRef]
2. Linnhoff, B.; Hindmarsh, E. The pinch design method for heat exchanger networks. *Chem. Eng. Sci.* **1983**, *38*, 745–763. [CrossRef]
3. Papoulias, S.A.; Grossmann, I.E. A structural optimization approach in process synthesis—II. Heat recovery networks. *Comput. Chem. Eng.* **1983**, *7*, 707–721. [CrossRef]
4. Klemeš, J.J.; Kravanja, Z. Forty years of Heat Integration: Pinch Analysis (PA) and Mathematical Programming (MP). *Curr. Opin. Chem. Eng.* **2013**, *2*, 461–474. [CrossRef]
5. Dhole, V.R.; Linnhoff, B. Total site targets for fuel, co-generation, emissions, and cooling. *Comput. Chem. Eng.* **1993**, *17*, S101–S109. [CrossRef]
6. Chew, K.H.; Klemeš, J.J.; Wan Alwi, S.R.; Abdul Manan, Z. Industrial implementation issues of Total Site Heat Integration. *Appl. Therm. Eng.* **2013**, *61*, 17–25. [CrossRef]
7. Liu, X.; Klemeš, J.J.; Varbanov, P.S.; Qian, Y.; Yang, S.; Liu, X.; Varbanov, P.S.; Klemes, J.J.; Wan Alwi, S.R.; Yong, J.Y. Safety issues consideration for direct and indirect heat transfer on total sites. *Chem. Eng. Trans.* **2015**, *45*, 151–156. [CrossRef]
8. Nemet, A.; Klemeš, J.J.; Moon, I.; Kravanja, Z. Safety Analysis Embedded in Heat Exchanger Network Synthesis. *Comput. Chem. Eng.* **2017**, *107*, 357–380. [CrossRef]
9. Nemet, A.; Klemeš, J.J.; Kravanja, Z. Process synthesis with simultaneous consideration of inherent safety-inherent risk footprint. *Front. Chem. Sci. Eng.* **2018**, *12*, 745–762. [CrossRef]
10. Escobar, M.; Trierweiler, J.O.; Grossmann, I.E. Simultaneous synthesis of heat exchanger networks with operability considerations: Flexibility and controllability. *Comput. Chem. Eng.* **2013**, *55*, 158–180. [CrossRef]
11. Rathjens, M.; Bohnenstädt, T.; Fieg, G.; Engel, O. Synthesis of heat exchanger networks taking into account cost and dynamic considerations. *Procedia Eng.* **2016**, *157*, 341–348. [CrossRef]
12. Pavão, L.V.; Miranda, C.B.; Costa, C.B.B.; Ravagnani, M.A.S.S. Efficient multiperiod heat exchanger network synthesis using a meta-heuristic approach. *Energy* **2018**, *142*, 356–372. [CrossRef]
13. Chang, C.; Chen, X.; Wang, Y.; Feng, X. An efficient optimization algorithm for waste Heat Integration using a heat recovery loop between two plants. *Appl. Therm. Eng.* **2016**, *105*, 799–806. [CrossRef]
14. Wang, Y.; Wang, W.; Feng, X. Heat integration across plants considering distance factor. *Chem. Eng. Trans.* **2013**, *35*, 25–30. [CrossRef]
15. Wang, Y.; Chang, C.; Feng, X. A systematic framework for multi-plants Heat Integration combining Direct and Indirect Heat Integration methods. *Energy* **2015**, *90*, 56–67. [CrossRef]
16. Chang, C.; Chen, X.; Wang, Y.; Feng, X. Simultaneous optimization of multi-plant heat integration using intermediate fluid circles. *Energy* **2017**, *121*, 306–317. [CrossRef]
17. Liew, P.Y.; Wan Alwi, S.R.; Klemeš, J.J. Total Site Heat Integration Targeting Algorithm Incorporating Plant Layout Issues. In *24th European Symposium on Computer Aided Process Engineering*; Klemes, J., Varbanov, P.S., Liew, P.Y., Eds.; Elsevier Science: Burlington, NJ, USA, 2014; pp. 1801–1806.
18. Pouransari, N.; Maréchal, F. Heat exchanger network design of large-scale industrial site with layout inspired constraints. *Comput. Chem. Eng.* **2014**, *71*, 426–445. [CrossRef]
19. Souza, R.D.; Khanam, S.; Mohanty, B. Synthesis of heat exchanger network considering pressure drop and layout of equipment exchanging heat. *Energy* **2016**, *101*, 484–495. [CrossRef]
20. Luo, X.; Wen, Q.-Y.; Fieg, G. A hybrid genetic algorithm for synthesis of heat exchanger networks. *Comput. Chem. Eng.* **2009**, *33*, 1169–1181. [CrossRef]
21. Yee, T.F.; Grossmann, I.E.; Kravanja, Z. Simultaneous optimization models for heat integration—I. Area and energy targeting and modeling of multi-stream exchangers. *Comput. Chem. Eng.* **1990**, *14*, 1151–1164. [CrossRef]
22. Fieg, G.; Luo, X.; Jeżowski, J. A monogenetic algorithm for optimal design of large-scale heat exchanger networks. *Chem. Eng. Process. Process Intensif.* **2009**, *48*, 1506–1516. [CrossRef]
23. Brandt, C. *Entwicklung und Implementierung Eines Hybriden Genetischen Algorithmus für die Automatisierte Auslegung von Kostenoptimalen Wärmeübertragernetzwerken*; Shaker: Herzogenrath, Germany, 2018.
24. Lewin, D.R.; Wang, H.; Shalev, O. A generalized method for HEN synthesis using stochastic optimization—I. General framework and MER optimal synthesis. *Comput. Chem. Eng.* **1998**, *22*, 1503–1513. [CrossRef]

25. Stegner, C.; Brandt, C.; Fieg, G. EVHE—A new method for the synthesis of HEN. *Comput. Chem. Eng.* **2014**, *64*, 95–102. [CrossRef]
26. Brandt, C.; Fieg, G.; Luo, X. Efficient synthesis of heat exchanger networks combining heuristic approaches with a genetic algorithm. *Heat Mass Transf.* **2011**, *47*, 1019–1026. [CrossRef]
27. Pho, T.K.; Lapidus, L. Topics in computer-aided design: Part II. Synthesis of optimal heat exchanger networks by tree searching algorithms. *Aiche J.* **1973**, *19*, 1182–1189. [CrossRef]
28. Nishida, N.; Liu, Y.A.; Lapidus, L. Studies in chemical process design and synthesis: III. A Simple and practical approach to the optimal synthesis of heat exchanger networks. *Aiche J.* **1977**, *23*, 77–93. [CrossRef]
29. Linnhoff, B.; Flower, J.R. Synthesis of heat exchanger networks: II. Evolutionary generation of networks with various criteria of optimality. *Aiche J.* **1978**, *24*, 642–654. [CrossRef]
30. Lewin, D.R. A generalized method for HEN synthesis using stochastic optimization—II. The synthesis of cost-optimal networks. *Comput. Chem. Eng.* **1998**, *22*, 1387–1405. [CrossRef]
31. Lin, B.; Miller, D.C. Solving heat exchanger network synthesis problems with Tabu Search. *Comput. Chem. Eng.* **2004**, *28*, 1451–1464. [CrossRef]
32. Pariyani, A.; Gupta, A.; Ghosh, P. Design of heat exchanger networks using randomized algorithm. *Comput. Chem. Eng.* **2006**, *30*, 1046–1053. [CrossRef]
33. Chakraborty, S.; Ghosh, P. Heat exchanger network synthesis: The possibility of randomization. *Chem. Eng. J.* **1999**, *72*, 209–216. [CrossRef]
34. Yerramsetty, K.M.; Murty, C.V.S. Synthesis of cost-optimal heat exchanger networks using differential evolution. *Comput. Chem. Eng.* **2008**, *32*, 1861–1876. [CrossRef]
35. Peng, F.; Cui, G. Efficient simultaneous synthesis for heat exchanger network with simulated annealing algorithm. *Appl. Therm. Eng.* **2015**, *78*, 136–149. [CrossRef]
36. Aguitoni, M.C.; Pavão, L.V.; Siqueira, P.H.; Jiménez, L.; Ravagnani, M.A.S.S. Synthesis of a cost-optimal heat exchanger network using genetic algorithm and differential evolution. *Chem. Eng. Trans.* **2018**, *70*, 979–984. [CrossRef]
37. Peters, M.S.; Timmerhaus, K.D.; West, R.E. *Plant Design and Economics for Chemical Engineers*, 5th ed.; McGraw-Hill: Boston, MA, USA, 2003.
38. Hipólito-Valencia, B.J.; Rubio-Castro, E.; Ponce-Ortega, J.M.; Serna-González, M.; Nápoles-Rivera, F.; El-Halwagi, M.M. Optimal design of inter-plant waste energy integration. *App. Therm. Eng.* **2014**, *62*, 633–652. [CrossRef]
39. Stijepovic, M.Z.; Linke, P. Optimal waste heat recovery and reuse in industrial zones. *Energy* **2011**. [CrossRef]
40. Jabbari, B.; Tahouni, N.; Ataei, A.; Panjeshahi, M.H. Design and optimization of CCHP system incorporated into kraft process, using Pinch Analysis with pressure drop consideration. *App. Therm. Eng.* **2013**, *61*, 88–97. [CrossRef]

© 2019 by the authors. Licensee MDPI, Basel, Switzerland. This article is an open access article distributed under the terms and conditions of the Creative Commons Attribution (CC BY) license (http://creativecommons.org/licenses/by/4.0/).

Article

Optimal Operational Adjustment of a Community-Based Off-Grid Polygeneration Plant using a Fuzzy Mixed Integer Linear Programming Model

Aristotle T. Ubando [1,*], Isidro Antonio V. Marfori III [1], Kathleen B. Aviso [2] and Raymond R. Tan [2]

1. Mechanical Engineering Department, De La Salle University, 2401 Taft Avenue, Manila 0922, Philippines; isidro.marfori@dlsu.edu.ph
2. Chemical Engineering Department, De La Salle University, 2401 Taft Avenue, Manila 0922, Philippines; kathleen.aviso@dlsu.edu.ph (K.B.A.); raymond.tan@dlsu.edu.ph (R.R.T.)
* Correspondence: aristotle.ubando@dlsu.edu.ph; Tel.: +632-524-4611

Received: 31 December 2018; Accepted: 7 February 2019; Published: 16 February 2019

Abstract: Community-based off-grid polygeneration plants based on micro-hydropower are a practical solution to provide clean energy and other essential utilities for rural areas with access to suitable rivers. Such plants can deliver co-products such as purified water and ice for refrigeration, which can improve standards of living in such remote locations. Although polygeneration gives advantages with respect to system efficiency, the interdependencies of the integrated process units may come as a potential disadvantage, due to susceptibility to cascading failures when one of the system components is partially or completely inoperable. In the case of a micro-hydropower-based polygeneration plant, a drought may reduce electricity output, which can, in turn, reduce the level of utilities available for use by the community. The study proposes a fuzzy mixed-integer linear programming model for the optimal operational adjustment of an off-grid micro-hydropower-based polygeneration plant seeking to maximize the satisfaction levels of the community utility demands, which are represented as fuzzy constraints. Three case studies are considered to demonstrate the developed model. The use of a diesel generator for back-up power is considered as an option to mitigate inoperability during extreme drought conditions.

Keywords: off-grid polygeneration; micro-hydropower plant; fuzzy optimization; mixed-integer linear programming; dual-turbine; multi-objective

1. Introduction

In developing countries, high priority is given to providing access to clean electricity and other essential utilities in remote communities [1]. Access to such amenities is critical to enhancing the quality of life and achieving the Sustainable Development Goals (SDGs) [2]. The seventh SDG focuses on universal access to affordable and reliable energy [3], but as recent as 2014, more than 1 billion people (i.e., about 15% of world population) still had no access to electricity [4]. The lack of access is especially true for geographically isolated rural communities, for which grid connectivity may not be economically or physically viable.

Efforts in providing power to off-grid communities have been pursued to address the rural electrification issue. From 2008 to 2017, the global off-grid renewable energy capacity has grown by about 300%, providing about 6.5 GW in total [5]. The integration of various renewable energy technologies to locally integrated energy sectors (LIES) have been found to significantly reduce the resource consumption and carbon footprint associated with energy supply [6]. Despite the dependence to climatic conditions and the inherent variability of renewable energy sources, a hybrid power system was proposed to combine renewable energy technologies with conventional energy sources while minimizing the cost and energy losses [7], and accommodating uncertainties, especially in remote areas [8]. The development of off-grid renewable energy systems largely varies and depend on the locally available resources and requirements of the community. In the past decade, an increase in the demand for solar energy solutions has been observed as solar photovoltaic (PV) technology matures [5]. However, for areas where access to flowing bodies of water is possible, micro-hydropower plants are a practical choice for irrigation and to generate electricity for the community [9]. Micro-hydropower plants are a robust and inexpensive renewable energy option [10]. In the context of hybrid power systems, micro-hydropower plants present a viable renewable energy option, given the right geographic location with good power generation potential [11]. A typical off-grid community will also require other utilities, such as purified water for household use. Aside from access to electricity and clean water, such communities may also require ice to preserve their produce (e.g., fish or vegetables). Polygeneration systems have been designed previously to provide such basic requirements [12]. These systems provide an efficient, integrated means to produce multiple material and energy products, which leads to improved system-wide efficiency and reduced emissions, waste, and natural resource consumption [13]. Polygeneration is generally defined as the co-production of various outputs along with electricity; the co-products can include heat (i.e., steam or hot water), cold energy (e.g., ice or chilled water), purified water, and chemical products [14]. In a polygeneration system, the interdependencies among integrated process units make the whole system vulnerable to cascading failures [10]. An operational disruption of such nature arising from any portion of the system leads to the causal operational disturbance cascading through the entire process network. Such disruption significantly affects the overall performance of the system causing divergence from the operational production state of the system [15]. Process systems engineering (PSE) models can play an important role to minimize production loss while maximizing opportunities to operate at an abnormal state for a brief period of time. The optimization approach may then be utilized to find the optimal operational adjustment of a polygeneration system at a perturbed state. A disruptive event, which can be caused by an external (e.g., natural disaster) or internal (e.g., equipment malfunction) trigger, can be measured via a dimensionless index known as inoperability [16]. The inoperability concept may be applied in representing operational disruption using the inoperability input-output modelling (IIM) framework [17]. In the context of a process network such as the polygeneration system, the inoperability can be defined as the loss of capacity of a system to operate relative to its normal state, and is represented by a dimensionless number ranging from a value of 0 (fully operational) to 1 (fully non-operational). This index has been used extensively in the literature. A linear programming (LP) model has been utilized for the optimal allocation of resources aiming to minimize losses in the event of a disruption [18]. The IIM theory was further developed in detail presenting case studies of disruption such as malevolent attacks [19]. The IIM framework has later been used as a sequential decision support tool for bioenergy systems [20]. It has been extended to deal with energy-related problems such as the optimization of the energy supply chain [21] and the identification of process bottlenecks in energy systems [22].

This work focuses on presenting a novel application of the inoperability concept on the operational adjustment of a community-based off-grid micro-hydropower-based polygeneration plant. Micro-hydropower plants generate up to 100 kW of electricity [23], which is enough for a small community with 20 to 30 households. Methods for the design [24], feasibility assessment [25], and identification of near-optimal solutions [26] for such facilities have been conducted previously. However, few studies have been conducted for micro-hydropower-based polygeneration systems for remote communities. An off-grid polygeneration system has been designed by Ray et al. [27] for an off-grid community in India using an LP model incorporating economic and reliability analysis. A generic decision support software was developed by Khalilpour and Vassallo [28] for the optimal design of a polygeneration system for off-grid distributed generation and storage system. However, neither of these studies incorporated micro-hydropower plants as a technology option. Recent studies on the optimal design of a micro-hydropower based polygeneration plant have been conducted using fuzzy optimization models. Such formulations are characterized by non-sharp boundaries of feasible regions, such that constraints can be partially satisfied as defined by their membership functions [29]. A fuzzy LP (FLP) model was developed by Ubando et al. [30] for the optimal design of an off-grid polygeneration plant using a superstructure that includes a micro-hydropower plant, a biomass-based Stirling engine, an ice plant, a diesel engine, and a sewage treatment facility. The work was later extended to a fuzzy mixed-integer linear programming (FMILP) model by Ubando et al. [12] considering vapor compression and absorption chillers as refrigeration options. However, these models were developed for system planning and design, and lack the capability to be used for optimal operational adjustment under abnormal conditions. An off-grid micro-hydropower-based polygeneration system is susceptible to drought which can reduce water flowrate available for the turbines; also, mechanical problems with system components may take longer to repair due to the time delays in bringing spare parts to the site. Kasivisvanathan et al. [15] developed a mixed-integer linear programming (MILP) model which provided the solution for the optimal operational adjustment of a gas-fired polygeneration plant for commercial use. In this study, the objective function is to maximize the operating profit (or minimize economic losses) of the system while employing the inoperability in each process in cases when the system experiences a disruption. A P-graph approach to the same problem was later developed by Tan et al. [31]. In the case of a polygeneration plant for a remote community, it is necessary to consider the demand for each product separately, depending on the needs of the inhabitants. This aspect is addressed in this work using fuzzy optimization using the approach developed by Zimmerman [29]. This approach is known as a "symmetric" fuzzy optimization model, where both constraints and (multiple) objectives are represented by fuzzy membership functions. FLP has been previously used for the optimal design of systems such as combined heat and power plants [32], trigeneration plants [33], and integrated biorefineries [34]. It was extended later to fuzzy MILP (FMILP) for the optimal planning of bioenergy parks [35], multifunctional bioenergy systems [36], and biofuel supply chain networks [37]. However, these models again focused on system design rather than abnormal operations.

This study develops an FMILP model for the optimal operational adjustment of a micro-hydropower-based polygeneration system for off-grid communities. It is assumed that the likely disruption is in the form of drought with varying intensity, which can be introduced into the model as an inoperability parameter. Three case studies are used to demonstrate the efficacy of the developed model under different drought scenarios. The case studies consider different turbine configurations and the option of having a diesel generator back-up. The rest of the paper is organized as follows. The formal problem statement is given, followed by the model formulation. Three case studies then follow. Lastly, conclusions and future prospects are discussed.

2. Problem Statement

The statement of the problem is formally defined similar to that proposed by Kasivisvanathan et al. [15] and stated as follows:

- An off-grid polygeneration system is assumed to have M number of products and N number of installed process units.
- The process units are characterized by fixed input-output stream proportions described by either the yield, efficiency, or the coefficient of performance, depending on the appropriate factor for each unit.
- Each process unit is defined by a minimum part-load operating level below which unstable or uneconomical operation occurs. The input-output ratios of streams for each unit remain fixed for the entire feasible operating range bound by a lower limit (minimum part-load operating level) and an upper limit (the rated capacity with a safety factor). The operational flexibility of the off-grid polygeneration system is defined by this operational range. The off-grid polygeneration system is further assumed to operate at a new steady state mode where the inoperability transpires.

- For each product stream, a fuzzy membership function is defined to describe the limits on net output, as dictated by basic requirements of the community inhabitants and are assumed to be constant. The fuzzy membership functions are assumed to be linear, and can thus be defined by specifying upper and lower limits. The upper limits signify normal requirements, while the lower limits signify the bare minimum requirements.
- The problem is to determine the optimal adjustment of operating capacities and allocation of streams for each process unit given an inoperability on the availability of water (drought scenario).

3. Fuzzy Mixed Integer Linear Programming Model

The basic MILP model for optimizing abnormal operations was developed in a previous work, and is given in the Appendix A [15]. The weakness of such a formulation is that it places extreme preference for product streams of higher economic value, often at the expense of the other lower-value products. While applicable to many commercial or industrial applications, in the case of remote rural communities, the actual value of basic needs might not be fully reflected in their prices. Instead, a given community may have a set of subsistence-level requirements for essential supplies, all of which need to be balanced based on what are deemed as tolerable shortages. These limits can in turn be modelled as fuzzy membership functions. Thus, the basic MILP model is modified into a fuzzy model to allow for such balanced rationing of capacity during a crisis [38]. The fuzzy MILP model is composed of the objective function and the constraints. The objective function of the model is shown in Equation (1):

$$\text{maximize } \lambda \tag{1}$$

where λ is the degree of satisfaction which is represented in every membership function in the model. The λ-value maximizes the lowest degree of satisfaction achieved among all identified fuzzy objectives. The degrees of satisfaction range from 0 to 1, with 1 signifying complete satisfaction, 0 indicating undesirable result, and values between 0 to 1 illustrating partial satisfaction of the objective. In Figure 1, the degree of satisfaction allows the translation of a variable-value shown in the abscissa into values ranging from 0 to 1 in the ordinate following a linear relation known as the linear membership function.

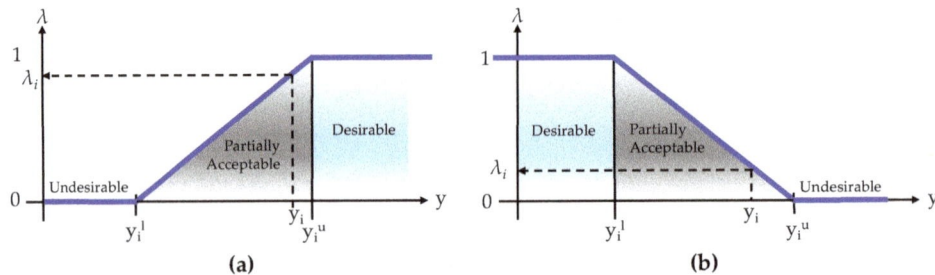

Figure 1. The fuzzy linear membership function for (**a**) maximization, and (**b**) minimization.

The optimal operational adjustment of the off-grid polygeneration system is subject to the following constraints. The technology matrix is introduced to define the optimal product output of the polygeneration system together with the process scaling vector as shown in Equation (2):

s.t.

$$\mathbf{A}\,\mathbf{x} = \mathbf{y} \tag{2}$$

where $\mathbf{A}$ is the technology matrix consisting of elements a_{ij} where i represents the product stream and j represents the technology considered. The technology matrix $\mathbf{A}$ is defined by a steady-state operational material and energy balance which are assumed to be scale invariant. In cases where the material or energy stream is identified as a raw material to a process, the value in matrix $\mathbf{A}$ would be represented by a negative value. Similarly, the positive values in matrix **A** represent the product or by-product streams generated from a process. The process scaling vector **x** with all elements equal to 1, indicates that all processes j in the system are operating in their optimal capacity. As the drought scenario sets in, the optimal operation adjustment in the off-grid polygeneration will be reflected in the change in values of **x**. The product output **y** of the off-grid polygeneration system represents the optimal production level of each product stream i. The product demand limits utilize membership functions which maximize the product output levels of the polygeneration system given a threshold demand range described in Equation (3). The application of the linear membership function for maximization was introduced by Zimmermann [39] using a fuzzy set framework:

$$\mathbf{y}_i \geq \mathbf{y}_i^l + \lambda(\mathbf{y}_i^u - \mathbf{y}_i^l) \qquad \forall\, i \qquad (3)$$

$$\mathbf{y}_i \leq \mathbf{y}_i^u + \lambda(\mathbf{y}_i^l - \mathbf{y}_i^u) \qquad \forall\, i = fuel \qquad (4)$$

where $\mathbf{y}_i^l$ is the lower threshold limit of each material or energy stream i, and $\mathbf{y}_i^u$ is the upper threshold limit of each material or energy stream i. Equation (3) represents the fuzzy linear membership function for maximizing a product stream. In Equation (3), any value less than $\mathbf{y}_i^l$ is considered undesirable while any value greater than $\mathbf{y}_i^u$ is highly desirable for a product stream $\mathbf{y}_i$. The values between $\mathbf{y}_i^l$ and $\mathbf{y}_i^u$ are considered as partially acceptable where degrees of satisfaction are described by a linear function with a slope of λ. Equation (4) represents the fuzzy linear membership function for minimizing a product stream which in this case is the fuel. For minimizing the fuel stream in Equation (4), $\mathbf{y}^l$ is the highly desired value, $\mathbf{y}^u$ is the undesirable value, and the values in between are treated as partially acceptable values. The diagram for the fuzzy linear membership function for maximization used in Equation (3) and for minimization used in Equation (4) are shown in Figure 1. The introduction of the drought scenario is described by Equation (5):

$$\mathbf{y}_i > (1 - D)\, \mathbf{y}_i^* \qquad \forall\, i \in R \qquad (5)$$

where D is the drought intensity factor with values ranging from 0% to 100%. A drought intensity factor of $D = 0$ represents a no drought scenario signifying full operational flow level of the river water during normal condition, while a drought intensity factor of $D = 1$ implies that there is no river water flowing. The $\mathbf{y}_i^*$ is the normal product output vector which represents the state of fully operational flow levels for all products i in the subset of R. The subset of R identifies all products which utilize the river water as raw material. The introduction of the binary variable vector is shown in Equations (6)–(8):

$$\mathbf{b}_j \mathbf{x}_j^l \leq \mathbf{x}_j \leq \mathbf{b}_j \mathbf{x}_j^u \qquad \forall\, j \qquad (6)$$

$$\mathbf{b}_j \in \{0,1\} \qquad \forall\, j \qquad (7)$$

$$f^l \leq \Sigma_j\, \mathbf{b}_j \leq f^u \qquad \forall\, j \in E \qquad (8)$$

The operational capacity limit of each technology j is described by the lower limit operational capacity $\mathbf{x}_j^l$ and the upper limit operational capacity $\mathbf{x}_j^u$. Equation (7) defines whether a technology j should operate ($\mathbf{b}_j = 1$) or not ($\mathbf{b}_j = 0$) during the chosen scenario. The topological binary constraint for the simultaneous selection of power generating technologies within a subset of E is shown in Equation (8) where the subset E represents all power generating technologies. The number of allowable operating technologies is limited by the minimum and maximum number defined by f^l and f^u, respectively. The introduction of the binary variable vector **b** in Equations (6)–(8), enables the developed model in becoming a mixed-integer linear programming which provided the selection of the appropriate technology for the off-grid polygeneration system. Lastly, Equation (9) ensures that the λ-value falls within the range of 0 to 1:

$$0 \leq \lambda \leq 1 \qquad (9)$$

It should be noted that equipment capital cost and stream prices are not included in the model. In the context of operational optimization, all equipment are assumed to have sunk costs, and there is no longer an option to modify them within any given scenario (Kasivisvanathan et al., [15]). Also, as the model deals with crisis conditions for remote rural communities, it is assumed that actual physical requirements for basic needs take precedence over cost considerations during the emergencies envisioned in this work. The model is solved utilizing Lingo 12.0, linked with MS Excel using the object linking and embedding (OLE) function of Lingo. The computer hardware used is powered by an Intel Core i7 processor with 8 GB of random access memory (RAM). Computational time for the examples in this work was negligible.

4. Case Study

Three case studies are presented to demonstrate the model developed. The first case study focuses on a basic off-grid community-based polygeneration system using a single turbine micro-hydropower plant. The second case study uses the same system in Case Study 1, but considers an additional back-up diesel generator. The third case study considers a dual turbine, instead of a single turbine, coupled with a back-up diesel generator.

4.1. Case Study 1

The case study considers a community of approximately 500 people requiring electricity, clean water, and ice. A single-turbine micro-hydropower plant produces 105 kW of electricity. Since this case study excludes the use of the diesel generator set as a back-up power generator, An ultra-filtration (UF) water treatment facility is considered to produce the required potable water for the community inhabitants, while a centralized ice plant provides for the cooling and refrigeration needs. The balanced process matrix of the case study is shown in Table 1. This matrix also shows the allocation of water flow to various uses within the community (including irrigation downstream of the micro-hydropower plant). The water flow allocation to the community shown in the table allows a dedicated water source for the production of clean water and ice. While a dedicated water flow allocation for the micro-hydropower plant is quantified in the table for power generation purposes. As per convention, positive and negative values in the process matrix signify outputs and inputs into any given process respectively.

Table 1. The balanced process matrix of case study 1.

Product Stream	Units	WTC [1]	WTM [2]	UFWT [3]	Ice Plant	MHP [4]	Product/Raw Materials
Clean Water	t/day	0	0	20	−5	0	15
Ice	t/day	0	0	0	5	0	5
Electricity	kW	0	0	−1	−4	105	100
Water to community supply	t/day	50	0	−50	0	0	0
Water to microhydro plant	t/day	0	52,500	0	0	−52,500	0
Rejected Water	t/day	0	0	30	0	0	30
River Water	t/day	−50	−52,500	0	0	0	−52,550
Diesel	t/day	0	0	0	0	0	0

[1] Water to Community; [2] Water to Microhydro; [3] Ultra-Filtration Water Treatment; [4] Microhydro Plant.

Under normal conditions, the community requires 100 kW of electricity, 15 t/day of purified water, and 5 t/day of ice. These values define the upper limits of the fuzzy membership functions of product requirements. To meet these demands, the polygeneration plant requires a river water flowrate of 52,550 t/day. Under crisis conditions, it is assumed that the community needs a bare minimum of 50 kW of electricity, 10 t/day of purified water, and 2 t/day of ice. These values define the lower limits of the fuzzy membership functions of the product demands. It is also assumed that the micro-hydropower plant has a minimum part-load operating level equivalent to 45% of its rated capacity. The drought index represents the fractional reduction of river water flowrate relative to normal conditions. The case study is then solved at increasing levels of drought, using increments of 10%. The polygeneration system schematic diagram is shown in Figure 2.

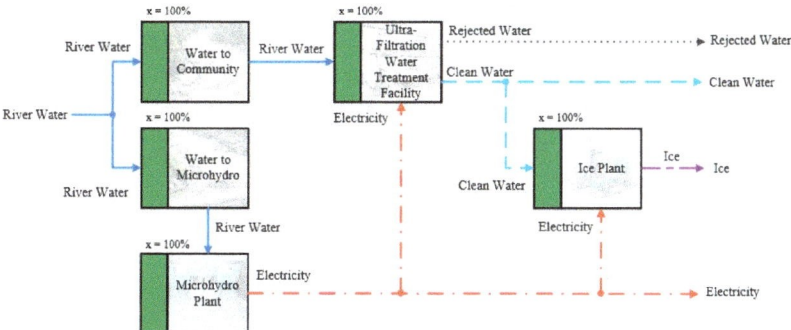

Figure 2. The process flowsheet of the polygeneration system of case study 1.

The optimal solutions to Case Study 1 for different drought intensities are shown in Table 2. The table shows the overall degree of satisfaction, λ, the optimal operating states of the process units, and the optimal net output levels of the products. As the drought level D increases progressively to 0.50 (which represents 50%), the value of λ, component operating states, and net output all decline. However, the increments of the λ-value are not evenly distributed due to the effect of the previously specified fuzzy membership functions of product requirements. For drought intensity levels of at least $D = 0.60$ (or 60%), it becomes impossible to find a feasible solution, which means that the bare minimum requirements of the community cannot be met. The optimal operating states of the polygeneration plant during a mild drought ($D = 0.10$) and moderate drought ($D = 0.50$) are shown in Figures 3 and 4, respectively.

Table 2. The resulting process capacity factor and the optimal production level for case study 1.

Drought Level, D	λ	Optimal Process Scaling Vector, x					Optimal Production Level, y		
		WTC [1]	WTM [2]	UFWT [3]	Ice Plant	MHP [4]	Electricity (kW)	Clean Water (t/day)	Ice (t/day)
0%	1.00	1.00	1.00	1.00	1.00	1.00	100.00	15.00	5.00
10%	0.80	0.92	0.90	0.92	0.88	0.90	90.05	14.01	4.40
20%	0.60	0.84	0.80	0.84	0.76	0.80	80.11	13.01	3.81
30%	0.40	0.76	0.70	0.76	0.64	0.70	70.16	12.02	3.21
40%	0.20	0.68	0.60	0.68	0.52	0.60	60.22	11.02	2.61
50%	0.01	0.60	0.50	0.60	0.40	0.50	50.27	10.03	2.02

[1] Water to Community; [2] Water to Microhydro; [3] Ultra-Filtration Water Treatment; [4] Microhydro Plant.

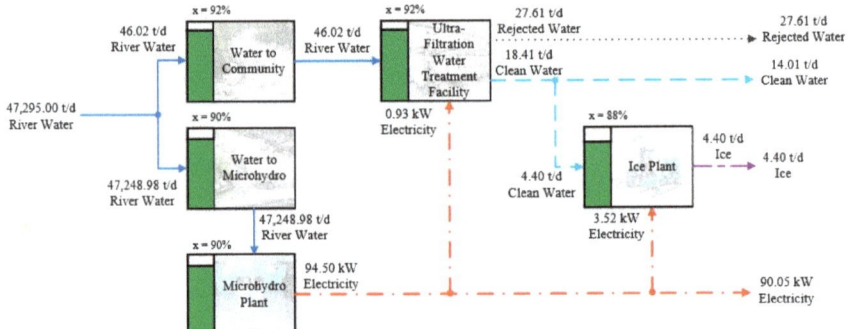

Figure 3. The optimal configuration of the polygeneration system of case study 1 under $D = 0.10$ (or 10%).

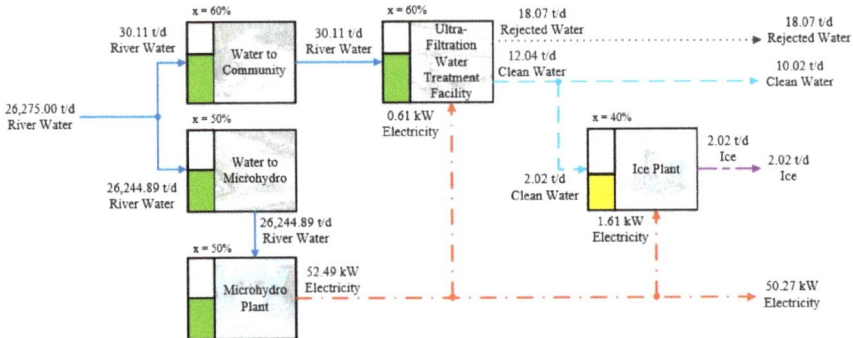

Figure 4. The optimal configuration of the polygeneration system of case study 1 under $D = 0.50$ (or 50%).

For $D = 0.10$, it can be seen that the ice plant process unit had the lowest operability of 88% while the water treatment facility can still operate at 92% of its normal operating capacity. This scenario was aggravated when D was up to 0.50 which resulted in the reduction in operating capacity of the ice plant to 40% and that of the water treatment facility to 60% of normal operating conditions. Furthermore, it can be noted that the ice plant is affected most since its disrupted state had an even higher disruption level (e.g., 12% and 60% disruption) than the experienced water reduction of 10% and 50%.

4.2. Case Study 2

Case study 2, on the other hand, considers the effect of the additional back-up diesel generator to the system in the previous example. This diesel generator can produce supplementary electricity as the drought level intensifies. The capacity of the diesel generator comes in standard capacity size which in this case is a 60 kW capacity. The balanced process matrix of the polygeneration system incorporating the diesel generator is shown in Table 3.

Table 3. The balanced process matrix of case study 2.

Product Stream	Units	WTC [1]	WTM [2]	UFWT [3]	Ice Plant	MHP [4]	DGS [5]	Product/Raw Materials
Clean Water	t/day	0	0	20	−5	0	0	15
Ice	t/day	0	0	0	5	0	0	5
Electricity	kW	0	0	−1	−4	105	60	160
Water to community supply	t/day	50	0	−50	0	0	0	0
Water to microhydro plant	t/day	0	52,500	0	0	−52,500	0	0
Rejected Water	t/day	0	0	30	0	0	0	30
River Water	t/day	−50	−52,500	0	0	0	0	−52,550
Diesel	t/day	0	0	0	0	0	−3.6	−3.6

[1] Water to Community; [2] Water to Microhydro; [3] Ultra-Filtration Water Treatment; [4] Microhydro Plant; [5] Diesel Generator Set.

The remote community is assumed to have access to an emergency stock of diesel fuel. This supply, as well as the ease of delivery of additional fuel, is the basis for the fuzzy membership function of the diesel fuel input (in the input-output framework used here, inputs can be treated simply as products with negative net flowrates). In the case study, the diesel generator can operate from 30% to 100% of its rated capacity. With the addition of the diesel back-up generator, a binary variable is added to the model to indicate whether the diesel generator is switched on or off as the drought level intensifies (see Equations (6)–(8)). This feature leads to an FMILP formulation. The system schematic diagram is shown in Figure 5.

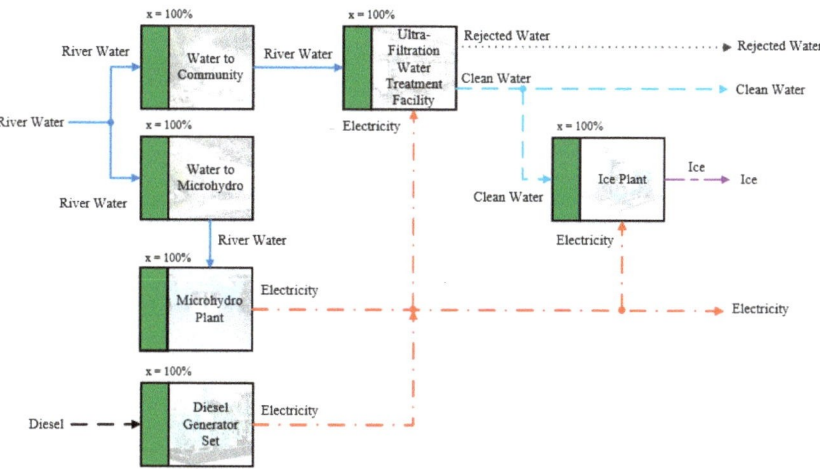

Figure 5. The process flowsheet of the polygeneration system of case study 2.

The optimal results of case study 2 at different levels of drought intensity are shown in Table 4. Note that the micro-hydropower plant ceases to operate at a drought level of $D = 0.60$ and beyond. The back-up diesel generator begins to operate at a drought level of $D = 0.20$ (or 20%) and progressively increases output under dryer conditions. The back-up generator is able to protect the community from the impacts of severe drought by extending the operational functionality of the polygeneration system even after the micro-hydropower plant becomes fully inoperable. This switch occurs at a drought level of $D = 0.60$ (or 60%), the electricity of the community is solely produced from the back-up diesel generator set at 90% capacity. This suggests that even at extreme drought conditions where the micro-hydropower plant fails to operate, a smaller capacity diesel generator set of about 54 kW may be considered if it is available in the market. The optimal operating states of the polygeneration plant at $D = 0.40$ and $D = 0.90$ are shown in Figures 6 and 7, respectively.

Table 4. The balanced process matrix of case study 2.

Drought Level, D	λ	Optimal Process Scaling Vector, x						Optimal Production Level, y		
		WTC [1]	WTM [2]	UFWT [3]	Ice Plant	MHP [4]	DGS [5]	Electricity (kW)	Clean Water (t/day)	Ice (t/day)
0%	1.00	1.00	1.00	1.00	1.00	1.00	0.00	100.00	15.00	5.00
10%	0.80	0.92	0.90	0.92	0.88	0.90	0.00	90.05	14.01	4.40
20%	0.68	0.87	0.67	0.87	0.81	0.67	0.30	83.88	13.39	4.03
30%	0.68	0.87	0.67	0.87	0.81	0.67	0.30	83.88	13.39	4.03
40%	0.61	0.85	0.60	0.85	0.77	0.60	0.36	80.66	13.07	3.84
50%	0.52	0.81	0.50	0.81	0.71	0.50	0.45	75.82	12.58	3.55
60%	0.03	0.61	0.00	0.61	0.42	0.00	0.90	51.67	10.17	2.10
70%	0.03	0.61	0.00	0.61	0.42	0.00	0.90	51.67	10.17	2.10
80%	0.03	0.61	0.00	0.61	0.42	0.00	0.90	51.67	10.17	2.10
90%	0.03	0.61	0.00	0.61	0.42	0.00	0.90	51.67	10.17	2.10

[1] Water to Community; [2] Water to Microhydro; [3] Ultra-Filtration Water Treatment; [4] Microhydro Plant; [5] Diesel Generator Set.

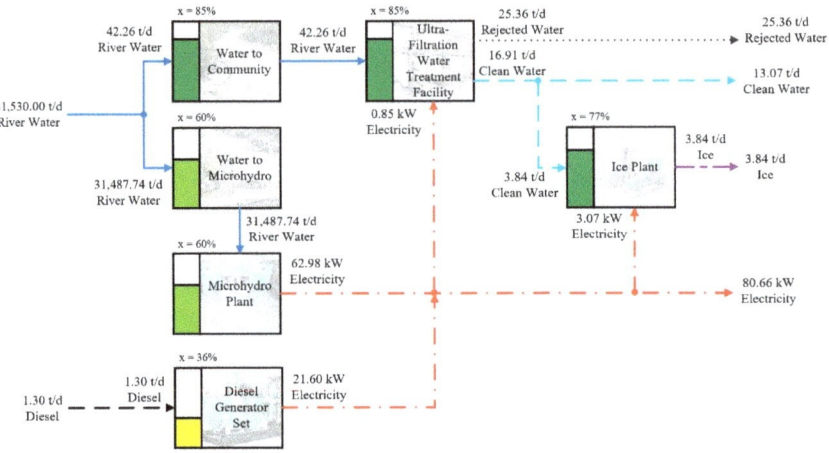

Figure 6. The optimal configuration of the polygeneration system of case study 2 under $D = 0.40$ (or 40%).

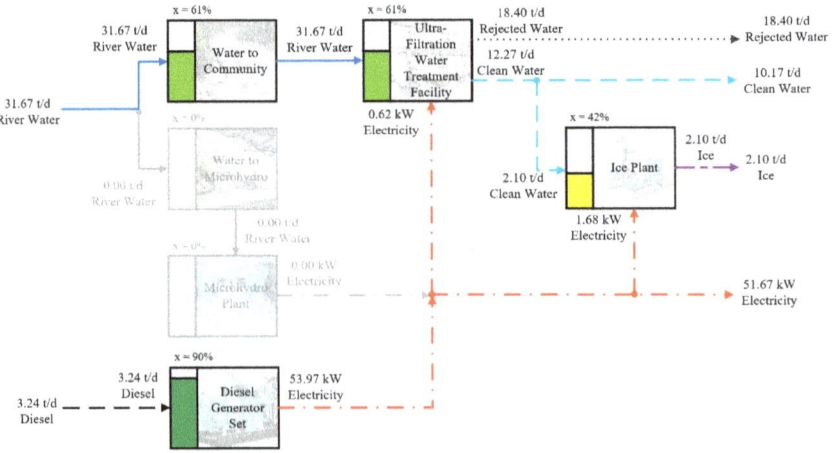

Figure 7. The optimal configuration of the polygeneration system of case study 2 under $D = 0.90$ (or 90%).

4.3. Case Study 3

This case study is similar to the previous one, except that a dual-turbine micro-hydropower plant is assumed. This dual-turbine micro-hydro plant consists of two smaller turbines rated at 70 kW and 35 kW, which together gives the system greater operational flexibility. The balanced process matrix for Case Study 3 is shown in Table 5 where the micro-hydropower turbines 1 and 2 are explicitly identified. With the reduction in the size of the turbine, this translates to a system-wide lower part-load operating capacity of 15 kW if only the 35 kW turbine is activated at 45% its rated capacity. The minimum part-load operating level for the 70 kW turbine and the 35 kW turbine of the micro-hydropower plant is defined as 45% [40]. Each turbine is operated individually allowing for a more flexible means to electricity generation during drought scenarios. Similar to case study 2, the back-up diesel generator set will be used as a support for the dual-turbine micro-hydropower plant if the need arises. The process flowsheet of case study 3 is shown in Figure 8.

Table 5. The balanced process matrix of case study 3.

Product Stream	Units	WTC [1]	WTM [2]	UFWT [3]	Ice Plant	MHT1 [4]	MHT2 [5]	DGS [6]	Product/Raw Materials
Clean Water	t/day	0	0	20	−5	0	0	0	15
Ice	t/day	0	0	0	5	0	0	0	5
Electricity	kW	0	0	−1	−4	70	35	60	160
Water to community supply	t/day	50	0	−50	0	0	0	0	0
Water to microhydro plant	t/day	0	52,500	0	0	−35,000	−17,500	0	0
Rejected Water	t/day	0	0	30	0	0	0	0	30
River Water	t/day	−50	−52,500	0	0	0	0	0	−52,550
Diesel	t/day	0	0	0	0	0	0	−3.6	−3.6

[1] Water to Community; [2] Water to Microhydro; [3] Ultra-Filtration Water Treatment; [4] Microhydro Turbine 1; [5] Microhydro Turbine 2; [6] Diesel Generator Set.

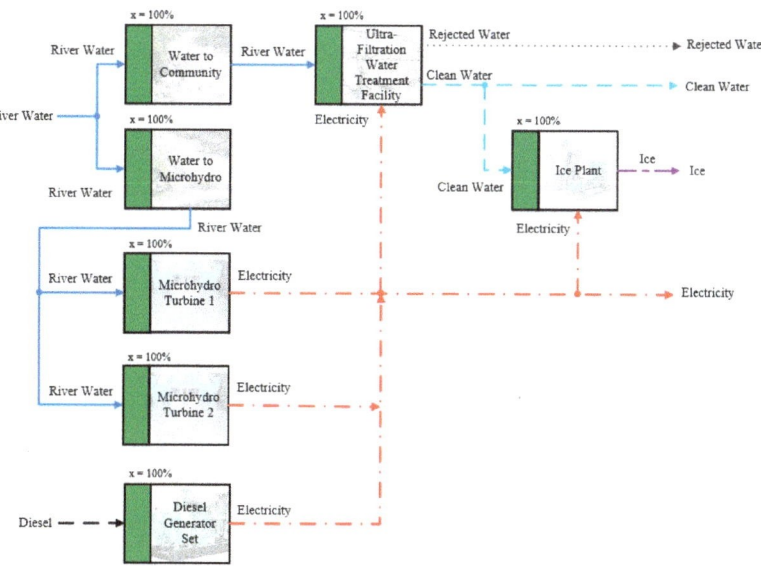

Figure 8. The process flowsheet of the polygeneration system of case study 3.

The optimal operational adjustment results for case study 3 using various levels of D is shown in Table 6. The table shows the dynamic operational capacity of the 2 micro-hydro turbines through varying drought intensity levels. It can be observed that turbine 2 will shut off at drought intensity levels of $D = 0.40$ and 0.50 and will operate again at $D = 0.70$ and 0.80.

Table 6. The resulting process capacity factor and the optimal production level for case study 3.

Drought Level, D	λ	Optimal Process Scaling Vector, x							Optimal Production Level, y		
		WTC[1]	WTM[2]	UFWT[3]	Ice Plant	MHT1[4]	MHT2[5]	DGS[6]	Electricity (kW)	Clean Water (t/day)	Ice (t/day)
0%	1.00	1.00	1.00	1.00	1.00	1.00	1.00	0.00	100.00	15.00	5.00
10%	0.80	0.92	0.90	0.92	0.88	1.00	0.70	0.00	90.05	14.01	4.40
20%	0.68	0.87	0.80	0.87	0.81	0.70	1.00	0.30	97.90	13.39	4.03
30%	0.68	0.87	0.70	0.87	0.81	0.55	1.00	0.30	87.39	13.39	4.03
40%	0.61	0.85	0.60	0.85	0.77	0.90	0.00	0.36	80.66	13.07	3.84
50%	0.52	0.81	0.50	0.81	0.71	0.75	0.00	0.45	75.82	12.58	3.55
60%	0.36	0.77	0.40	0.77	0.65	0.60	0.00	0.54	70.99	12.10	3.26
70%	0.32	0.73	0.30	0.73	0.59	0.00	0.90	0.63	66.15	11.62	2.97
80%	0.23	0.69	0.20	0.69	0.54	0.00	0.60	0.72	61.32	11.13	2.68
90%	0.13	0.61	0.00	0.61	0.42	0.00	0.00	0.90	51.67	10.17	2.10

[1] Water to Community; [2] Water to Microhydro; [3] Ultra-Filtration Water Treatment; [4] Microhydro Turbine 1; [5] Microhydro Turbine 2; [6] Diesel Generator Set.

Turbine 1, on the other hand, will continue to operate until it reaches a drought intensity of $D = 0.60$ and will shut off at $D = 0.70$ and beyond. At elevated drought intensity of $D = 0.70$ and 0.80, the smaller turbine provides better operational flexibility allowing appropriate operational adjustments. As a supplementary source of power, the back-up diesel generator begins to operate at a drought intensity of $D = 0.20$ and is partially operated until a drought intensity level of $D = 0.90$ at 90% part-load capacity. However, the dual-turbine remains operable over a wider range of conditions allowing operational flexibility on drought condition while the diesel generator partially operates from $D = 0.20$ to 0.90. The optimal configuration of the polygeneration system for case study 3 is shown in Figures 9 and 10 for $D = 0.30$ and $D = 0.90$, respectively.

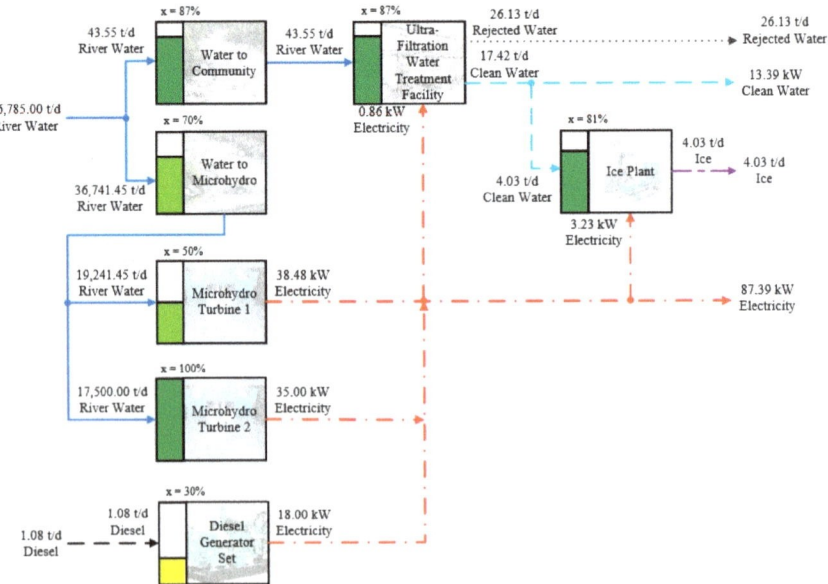

Figure 9. The optimal configuration of the polygeneration system of case study 3 under $D = 0.30$ (or 30%).

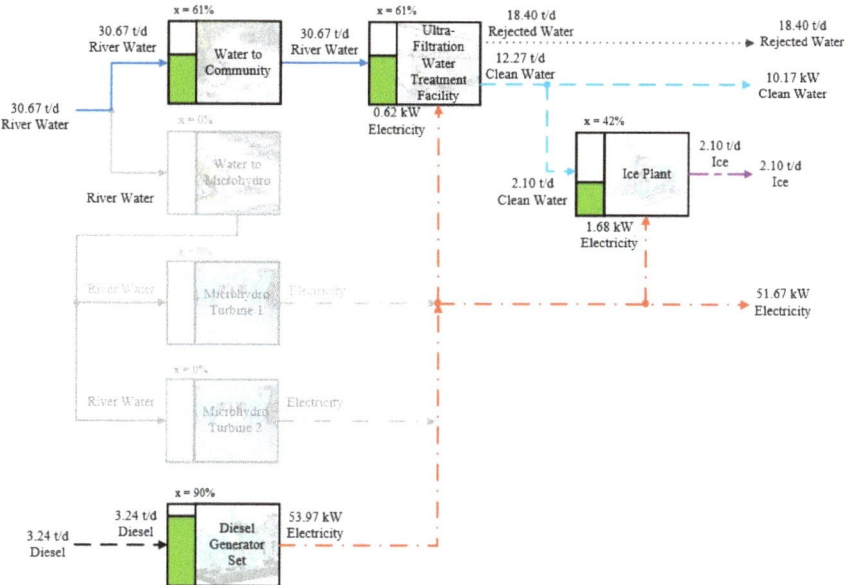

Figure 10. The optimal configuration of the polygeneration system of case study 3 under $D = 0.90$ (or 90%).

The comparison of the production level of electricity, water, and ice for the three case studies are shown in Figure 11a–c while the diesel fuel consumption for case studies 2 and 3 is shown in Figure 11d. The contribution of the power generating technologies to electricity production for each case study are shown in Figure 12. The back-up diesel generator set enabled an extended production of electricity, water, and ice for the community especially during low to severe drought intensity levels. The dual-turbine micro-hydro assembly with diesel generator set (Case Study 3) provides a more stable production of electricity, water, and ice and is found to be superior compared to the single-turbine micro-hydro plant with diesel generator (Case Study 2) especially for drought intensity levels from $D = 0.60$ to $D = 0.80$. With lower diesel fuel consumption and higher electricity generation, the dual-turbine micro-hydro plant with diesel generator is ideal for areas where drought risk is expected.

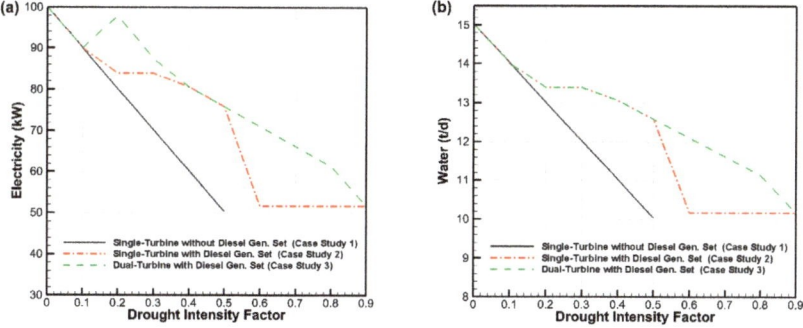

Figure 11. *Cont.*

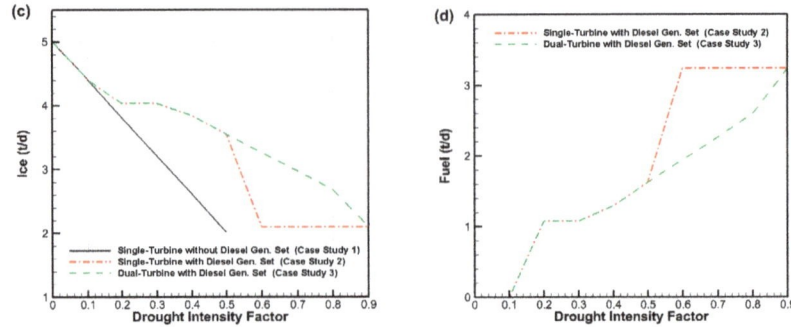

Figure 11. The optimal polygeneration product output of the polygeneration such as the (**a**) electricity, (**b**) water, and (**c**) ice, and the raw material such as the (**d**) diesel fuel.

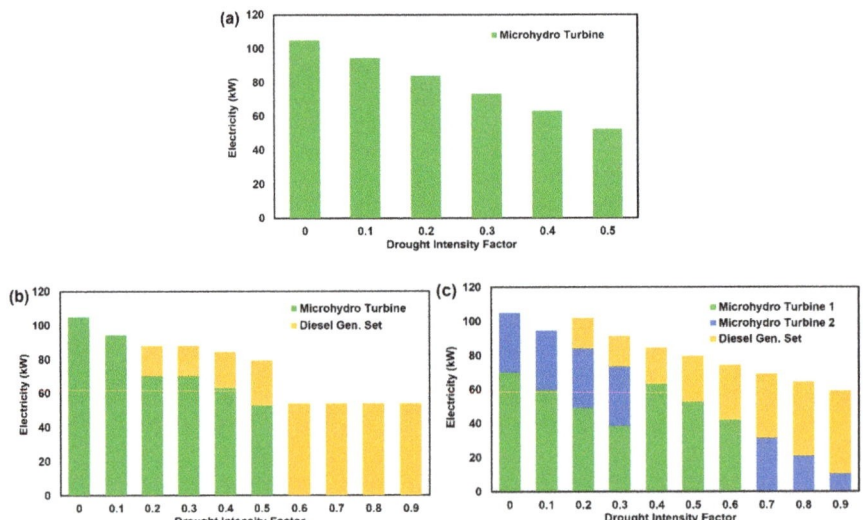

Figure 12. The resulting electricity generation contribution for (**a**) case study 1, (**b**) case study 2, (**c**) case study 3.

5. Conclusions

A fuzzy MILP model has been developed for the optimal operational adjustment of an off-grid micro-hydropower-based polygeneration system under varying drought intensity scenarios. The polygeneration system produces electricity, purified water, and ice, which are the basic needs of an off-grid community. The model accounts for the integration of the process units to give the optimal operational adjustments needed to partially satisfy the community demands for each product output, based on fuzzy constraints. Unlike conventional MILP formulations, the fuzzy bounds provide an optimal solution that gives a more balanced compromise on the reduction of outputs of the facility. A variant of the model also considers the minimization of the fuel consumption of the back-up diesel generator set. Case studies that consider different turbine configurations and diesel back-up options were solved to illustrate the model capabilities. This modelling framework is useful for off-grid electricity systems which have been established and have failed to take into consideration system disruptions resulting from the effects of climate change. The model provides decision-makers with a systematic procedure for identifying the optimal operational adjustment to ensure that the community product demands are satisfied during abnormal conditions. Furthermore, the model can provide insights on the management of external sources of energy.

Future work can apply this model formulation to a broader range of integrated polygeneration systems and consider multi-period models to accommodate variations in product demands. In addition, the potential for the use of alternative methods such as fuzzy P-graph [41] to deal with problems of this class should also be explored together with an examination of the availability and performance of alternative operating conditions.

Author Contributions: All authors contributed to the completion and publication of the study. R.R.T. conceptualized the general idea of the paper. I.A.V.M.III provided key technical information on off-grid microhydro plants. R.R.T., K.B.A., and A.T.U., contributed on the development of the model. K.B.A. thoroughly verified and validated the model. A.T.U. prepared and edited the manuscript, and responsible for drawing the figures. I.A.V.M.III, K.B.A., and R.R.T. revised and reviewed the manuscript.

Funding: This research was funded by Philippine Commission on Higher Education (CHED) PHERNet Project.

Acknowledgments: The research is supported by the Philippine Commission on Higher Education (CHED) PHERNet Sustainability Studies Program and the De La Salle Science Foundation (DLS-SFI) for the publication cost. The pioneering work of the late Godofredo Salazar is acknowledged in establishing the Center for Microhydro Research at De La Salle University.

Conflicts of Interest: The authors declare no conflict of interest.

Appendix A

The basic MILP model for optimizing the polygeneration system during abnormal operations is described by Equations (A1)–(A7) and is similar to the model introduced in [14].

$$\max = \mathbf{c}^T \mathbf{y}^* \tag{A1}$$

$$\mathbf{A} \cdot \mathbf{x} = \mathbf{y}^* \tag{A2}$$

$$\mathbf{b}^T \mathbf{x}^l \leq \mathbf{x} \leq \mathbf{b}^T \mathbf{x}^u \tag{A3}$$

$$\mathbf{b} \in \{0, 1\} \tag{A4}$$

$$y^*_{river\_water} = (1 - D)\, y_{river\_water} \tag{A5}$$

$$\mathbf{y}^* \geq \mathbf{y}^l \tag{A6}$$

$$f^l \leq \mathbf{b}^T \mathbf{e} \leq f^u \tag{A7}$$

The objective function (Equation (A1)) is to maximize the profit of the system, excluding capital cost recovery, where $\mathbf{c}^T$ represents the price vector for the material or energy streams and $\mathbf{y}^*$ is the net output vector during abnormal operations. The model is subject to material and energy balances (Equation (A2)) dependent on the process operational conditions where $\mathbf{A}$ is the balanced process matrix during normal operating conditions and $\mathbf{x}$ is the process scaling vector during abnormal conditions. The scaling vector is bound by the lower ($\mathbf{x}^l$) and upper ($\mathbf{x}^u$) operating capacity limits of each available technology (Equation (A3)) where $\mathbf{b}^T$ is a binary variable (Equation (A4)) indicating whether a technology is operational ($\mathbf{b}_i = 1$) or not ($\mathbf{b}_i = 0$) during abnormal operations. The abnormal conditions occur when there is insufficient river water (Equation (A5)) defined by the drought intensity factor, D. Furthermore, a minimum production level ($\mathbf{y}^l$) for identified product streams must be met (Equation (A6)). There is also a minimum (f^l) and maximum (f^u) number of operational technologies at any given time (Equation (A7)), vector $\mathbf{e}$ represents a column vector with all elements equal to 1.

References

1. The World Bank. *Designing Sustainable off-Grid Rural Electrification Projects: Principles and Practices*; Research Working Papers; The World Bank: Washington, DC, USA, 2008.

2. Banerjee, S.G.; Bhatia, M.; Portale, E.; Schers, J.; Dorner, D.; Azuela, G.E.; Jaques, I.S.; Ashok, P.E.; Bushueva, I.; Angelou, N.; et al. Global Tracking Framework. *Sustain. Energy All World Bank* **2013**, *3*, 289.
3. United Nations. *Affordable and Clean Energy: Energy Efficient*; United Nations: New York, NY, USA, 2018.
4. The World Bank. *State of Energy Access Report 2017*; The World Bank: Washington, DC, USA, 2017.
5. IRENA. *Off-Grid Renewable Energy Solutions: Global and Regional Status and Trends*; IRENA: Abu Dhabi, UAE, 2018; pp. 1–20.
6. Perry, S.; Klemeš, J.; Bulatov, I. Integrating waste and renewable energy to reduce the carbon footprint of locally integrated energy sectors. *Energy* **2008**, *33*, 1489–1497. [CrossRef]
7. Lee, J.Y.; Aviso, K.B.; Tan, R.R. Optimal Sizing and Design of Hybrid Power Systems. *ACS Sustain. Chem. Eng.* **2018**, *6*, 2482–2490. [CrossRef]
8. Lee, J.Y.; Lu, Y.C.; Aviso, K.B.; Tan, R.R. Mathematical programming for optimal design of hybrid power systems with uncertainties. *Chem. Eng. Trans.* **2018**, *70*, 67–72.
9. Geem, Z.W. Optimal Scheduling of Multiple Dam System Using Harmony Search Algorithm. In *Computational and Ambient Intelligence*; IWANN 2007. Lecture Notes in Computer Science; Sandoval, F., Prieto, A., Cabestany, J., Graña, M., Eds.; Springer: Berlin/Heidelberg, Germany, 2007; Volume 4507, pp. 316–323.
10. Agarkar, B.D.; Barve, S.B. A Review on Hybrid solar/wind/hydro power generation system. *Int. J. Curr. Eng. Technol.* **2011**, *4*, 188–191.
11. Walmsley, T.G.; Walmsley, M.R.W.; Varbanov, P.S.; Klemeš, J.J. Energy Ratio analysis and accounting for renewable and non-renewable electricity generation: A review. *Renew. Sustain. Energy Rev.* **2018**, *98*, 328–345. [CrossRef]
12. Ubando, A.T.; Marfori, I.A.V.; Aviso, K.B.; Tan, R.R. Optimal synthesis of a community-based off-grid polygeneration plant using fuzzy mixed integer linear programming model. *Chem. Eng. Trans.* **2018**, *70*, 955–960.
13. Serra, L.M.; Lozano, M.-A.; Ramos, J.; Ensinas, A.V.; Nebra, S.A. Polygeneration and efficient use of natural resources. *Energy* **2009**, *34*, 575–586. [CrossRef]
14. Adams, T.A.; Ghouse, J.H. Polygeneration of fuels and chemicals. *Curr. Opin. Chem. Eng.* **2015**, *10*, 87–93. [CrossRef]
15. Kasivisvanathan, H.; Barilea, I.D.U.; Ng, D.K.S.; Tan, R.R. Optimal operational adjustment in multi-functional energy systems in response to process inoperability. *Appl. Energy* **2013**, *102*, 492–500. [CrossRef]
16. Haimes, Y.Y.; Jiang, P. Leontief-based model of risk in complex interconnected infrastructures. *J. Infrastruct. Syst.* **2001**, *7*, 1–12. [CrossRef]
17. Leontief, W. *The Structure of American Economy, 1919–1929: An Empirical Application of Equilibrium Analysis*; Harvard University Press: Cambridge, MA, USA, 1941; p. 181.
18. Jiang, P.; Haimes, Y.Y. Risk management for Leontief-based interdependent systems. *Risk Anal.* **2004**, *24*, 1215–1229. [CrossRef] [PubMed]
19. Haimes, Y.Y.; Horowitz, B.M.; Lambert, J.H.; Santos, J.R.; Crowther, K.; Lian, C. Inoperability input-output model for interdependent infrastructure sectors. I: Theory and methodology. *J. Infrastruct. Syst.* **2005**, *11*, 80–92. [CrossRef]
20. Santos, J.R.; Barker, K.; Zelinke, P.J. Sequential decision-making in interdependent sectors with multiobjective inoperability decision trees: Application to biofuel subsidy analysis. *Econ. Syst. Res.* **2008**, *20*, 29–56. [CrossRef]
21. Tan, R.R. A general source-sink model with inoperability constraints for robust energy sector planning. *Appl. Energy* **2011**, *88*, 3759–3764. [CrossRef]
22. Tan, R.R.; Lam, H.L.; Kasivisvanathan, H.; Ng, D.K.S.; Foo, D.C.Y.; Kamal, M.; Hallaler, N.; Klemeš, J.J. An algebraic approach to identifying bottlenecks in linear process models of multifunctional energy systems. *Theoret. Found. Chem. Eng.* **2012**, *46*, 642–650. [CrossRef]
23. Hoq, T.; Nawshad, U.A.; Islam, N.; Syfullah, K.; Rahman, R. Micro Hydro Power: Promising Solution for Off-grid Renewable Energy Source. *Int. J. Sci. Eng. Res.* **2011**, *2*, 2–6.
24. Pasalli, Y.R.; Rehiara, A.B. Design Planning of Micro-hydro Power Plant in Hink River. *Procedia Environ. Sci.* **2014**, *20*, 55–63. [CrossRef]

25. Hongpeechar, B.; Krueasuk, W.; Poungching-Ngam, A.; Bhasaputra, P.; Pattaraprakorn, W. Feasibility study of micro hydro power plant for rural electrification in Thailand by using axial flux permanent magnet. In Proceedings of the 2011 International Conference and Utility Exhibition on Power and Energy Systems: Issues and Prospects for Asia, Pattaya City, Thailand, 28–30 September 2011; pp. 3–6.
26. Voll, P.; Jennings, M.; Hennen, M.; Shah, N.; Bardow, A. The optimum is not enough: A near-optimal solution paradigm for energy systems synthesis. *Energy* **2015**, *82*, 446–456. [CrossRef]
27. Ray, A.; Jana, K.; De, S. Polygeneration for an off-grid Indian village: Optimization by economic and reliability analysis. *Appl. Therm. Eng.* **2017**, *116*, 182–196. [CrossRef]
28. Khalilpour, K.R.; Vassallo, A. A generic framework for distributed multi-generation and multi-storage energy systems. *Energy* **2016**, *114*, 798–813. [CrossRef]
29. Zimmermann, H.J. Fuzzy programming and linear programming with several objective functions. *Fuzzy Sets Syst.* **1978**, *1*, 45–55. [CrossRef]
30. Ubando, A.T.; Marfori, I.A.; Culaba, A.B.; Dungca, J.R.; Promentilla, M.A.B.; Aviso, K.B.; Tan, R.R. A Systematic Approach for the Optimal Design of an Off-Grid Polygeneration System using Fuzzy Linear Programming Model. In *Computer Aided Chemical Engineering*; Espuña, A., Graells, M., Puigjaner, L., Eds.; Elsevier: Amsterdam, The Netherlands, 2017; Volume 40, pp. 2191–2196.
31. Tan, R.R.; Cayamanda, C.D.; Aviso, K.B. P-graph approach to optimal operational adjustment in polygeneration plants under conditions of process inoperability. *Appl. Energy* **2014**, *135*, 402–406. [CrossRef]
32. Chiu, G.M.K.; Aviso, K.B.; Ubando, A.T.; Tan, R.R. Fuzzy linear programming model for the optimal design of a combined cooling, heating, and power plant. In Proceedings of the 2017 IEEE 9th International Conference on Humanoid, Nanotechnology, Information Technology, Communication and Control, Environment and Management (HNICEM), Manila, Philippines, 1–3 December 2017; pp. 1–6.
33. Mayol, A.P.; Culaba, A.B.; Aviso, K.B.; Ng, D.K.S.; Tan, R.R.; Ubando, A.T. Fuzzy linear programming model for the optimal design of a trigeneration plant with product price variability. In Proceedings of the IEEE Region 10 Annual International Conference, Proceedings/TENCON, Singapore, 22–25 November 2017; pp. 3200–3205.
34. Ubando, A.T.; Culaba, A.B.; Tan, R.R.; Ng, D.K.S. A Systematic Approach for Optimization of an Algal Biorefinery Using Fuzzy Linear Programming. In *Computer Aided Chemical Engineering*; Karimi, I.A., Srinivasan, R., Eds.; Elsevier: Amsterdam, The Netherlands, 2012; Volume 31, pp. 805–809.
35. Ubando, A.T.; Culaba, A.B.; Aviso, K.B.; Tan, R.R.; Cuello, J.L.; Ng, D.K.S.; El-Halwagi, M.M. Fuzzy mixed integer non-linear programming model for the design of an algae-based eco-industrial park with prospective selection of support tenants under product price variability. *J. Clean. Prod.* **2016**, *136*, 183–196. [CrossRef]
36. Ubando, A.T.; Culaba, A.B.; Aviso, K.B.; Ng, D.K.S.; Tan, R.R. Fuzzy mixed-integer linear programming model for optimizing a multi-functional bioenergy system with biochar production for negative carbon emissions. *Clean Technol. Environ. Policy* **2014**, *16*, 1537–1549. [CrossRef]
37. Bairamzadeh, S.; Saidi-Mehrabad, M.; Pishvaee, M.S. Modelling different types of uncertainty in biofuel supply network design and planning: A robust optimization approach. *Renew. Energy* **2018**, *116*, 500–517. [CrossRef]
38. Tan, R.R.; Aviso, K.B.; Cayamanda, C.D.; Chiu, A.S.F.; Promentilla, M.A.B.; Ubando, A.T.; Yu, K.D.S. A fuzzy linear programming enterprise input–output model for optimal crisis operations in industrial complexes. *Int. J. Prod. Econ.* **2016**, *181*, 410–418. [CrossRef]
39. Zimmermann, H.-J. *Fuzzy Set Theory- and Its Applications*; Springer Seience+Business Media: New York, NY, USA, 1992.
40. Elbatran, A.H.; Yaakob, O.B.; Ahmed, Y.M.; Shabara, H.M. Operation, performance and economic analysis of low head micro-hydropower turbines for rural and remote areas: A review. *Renew. Sustain. Energy Rev.* **2015**, *43*, 40–50. [CrossRef]
41. Aviso, K.B.; Tan, R.R. Fuzzy P-graph for optimal synthesis of cogeneration and trigeneration systems. *Energy* **2018**, *154*, 258–268. [CrossRef]

 © 2019 by the authors. Licensee MDPI, Basel, Switzerland. This article is an open access article distributed under the terms and conditions of the Creative Commons Attribution (CC BY) license (http://creativecommons.org/licenses/by/4.0/).

Article

Performant and Simple Numerical Modeling of District Heating Pipes with Heat Accumulation

Libor Kudela *, Radomir Chylek and Jiri Pospisil

Energy Institute, Faculty of Mechanical Engineering, Brno University of Technology—VUT Brno, Technicka 2896/2, 61669 Brno, Czech Republic; Radomir.Chylek@vut.cz (R.C.); jiri.pospisil@vutbr.cz (J.P.)
* Correspondence: Libor.Kudela@vutbr.cz; Tel.: +420-54-114-2579

Received: 24 January 2019; Accepted: 13 February 2019; Published: 16 February 2019

Abstract: This paper compares approaches for accurate numerical modeling of transients in the pipe element of district heating systems. The distribution grid itself affects the heat flow dynamics of a district heating network, which subsequently governs the heat delays and entire efficiency of the distribution. For an efficient control of the network, a control system must be able to predict how "temperature waves" move through the network. This prediction must be sufficiently accurate for real-time computations of operational parameters. Future control systems may also benefit from the accumulation capabilities of pipes. In this article, the key physical phenomena affecting the transients in pipes were identified, and an efficient numerical model of aboveground district heating pipe with heat accumulation was developed. The model used analytical methods for the evaluation of source terms. Physics of heat transfer in the pipe shells was captured by one-dimensional finite element method that is based on the steady-state solution. Simple advection scheme was used for discretization of the fluid region. Method of lines and time integration was used for marching. The complexity of simulated physical phenomena was highly flexible and allowed to trade accuracy for computational time. In comparison with the very finely discretized model, highly comparable transients were obtained even for the thick accumulation wall.

Keywords: district heating; heat accumulation; pipe; numerical model; Modelica language; Julia language; performance

1. Introduction

In the next few decades, the district heating (DH) will undergo substantial changes. The major operational changes associated with the 4th generation of DH are lower temperatures in distribution grids (the heat-carrier will be liquid water), enhancement of accumulation, and use of heat pumps or other temperature boosters [1,2]. Utilization of renewable sources and waste heat (both often fluctuate) aim at the reduction of carbon dioxide emissions and fossil fuel consumption [3]. Consumers, suppliers, and network providers will become members of the multilevel heat market. The production and consumption of heat or cold can be shifted between seasons using large long-term thermal storage, which has a lower relative price than smaller ones [4]. In such a complicated arrangement, the network must be sufficiently flexible, predictive, and intelligent.

Advanced and holistic operational optimization of the control strategies will become crucial. Although the operational optimization is usually being performed on already existing grids with given topologies, the simultaneous optimization of both grid topology and operation may be more beneficial than a separate approach. In other words, some grid topologies can allow better control strategies while others do not have to. The ideal control system should take long-term consequences into account (delays are most often affected by thermal inertia) [5]. The controller has to find the control actions by minimizing the sum of all the penalties (energy losses, demand mismatches, temporary price, etc.)

defined inside the models of the individual components. This is called operational optimizations. It is crucial to fit as many simulations as possible to the control horizon (several minutes) to find the optimal values of the discrete control variables for the prediction horizon (several hours, days, weeks or even months). This depends on the models being fast and accurate. Regarding the pipes themselves, the correct prediction of step change propagation is most important as it determines the delay in temperature delivery. These delays are not only caused by pure advection through the system.

To ensure the sufficient performance of the model, the relative simulation time (ratio of physical time and computational time) is to be reduced. This might be achieved by a combination of several principles, such as lowering the number of used spatial derivatives (e.g., reducing to the one-dimensional description) or avoiding some of the discretizations altogether by using the analytic description of heat transfer [6] and friction [7] for all regimes in the pipes. When it comes to the hydraulic calculations, which are necessary to evaluate mass flows of heat carrier in the branches, it is usually considered satisfactory to use the static model, since the pressure waves spread significantly faster than thermal waves. It is also helpful to derive specific solutions that take advantage of known facts rather than those that cover a wide range of possible options (e.g., the presumption of incompressible fluid simplifies the dealing with the mass conservation laws significantly).

In previous work, a few main approaches that deal with simulations of DH grids are utilized. The first group is based on the presumption that the grid operates mostly in steady-state. Advanced techniques based on this can be found at [8], where authors develop a model and a method for parameter calibration that allows to better fit the measured data to simulation.

Another great example of steady-state estimation of DH grid is [9]. In this paper, the authors use automated customer meter system to evaluate mass flows, temperatures and consequently heat losses in pipes of a DH network. The optimization algorithm is utilized to estimate temperatures in the pipes, so simulated temperatures in the customer's location fit the measured data.

The second group of models, which include dynamic effects to some extent, is based on highly simplified models that directly link inlet temperature signal to the outlet signal. In the so-called node method [10], nodes that are connected by pipes of known geometrical and thermal properties represent the network. Based on mass flows in the pipes and the temperatures of water in individual nodes, the method steps through the time to evaluate temperature in all the nodes for all the time steps. It utilizes known time delays caused by advection and thermal inertias of the pipes.

Authors of [11] develop new so-called function method for the evaluation of thermal transients in DH grids. The method is based on the insertion of a Fourier series into the simplified Partial Differential Equation (PDE) of the fluid and solid regions of the circular pipe to derive the relationship between inlet and outlet temperature. Time delay and relative attenuations are incorporated into the model. The new method is reported to be about 37% faster than the standard node method. Model is validated in real DH system. No axial turbulent diffusion is considered.

Pure advection of inlet time of the water to the pipe is used in [12]. This value is then subtracted from the time when the liquid leaves the pipe. The resultant value represents the time the liquid spends inside the pipe which is then used for estimating the temperature drop due to heat loss. To incorporate thermal inertia, equivalent mixing volume is placed at the outlet. This results in a so-called plug-flow model. The model accurately simulated 168 h of operation of district heating grid in Pongau (Austria) in less than 2 s. No axial diffusion is incorporated.

Authors of [13] focus on transients in pipes with stable mass flows. They use steady-state axial temperature distributions and convolution with the axial diffusion kernel to arrive at an explicit mathematical function that describes the temporary transient behavior. Their solution is then compared with the numerical solution of the appropriate Partial Differential Equation (PDE) with positive results. No dynamic re-accumulation in the pipe walls is included.

A steady-state heat transfer model is combined with a variable transport delay model in [14] to obtain a fast model capable of simulations with variable flows. The model resulted in approximately 4000 times less intensive computation in comparison with one-dimensional PDE approach (both models were developed in Matlab/Simulink). This speedup factor rises with the utilization of larger time steps but at the cost of higher error. Axial diffusion and thermal inertia of the wall is considered.

Comparison of the node method and pseudo transient approach with experimental data is conducted in [15]. Authors of the article conclude that thermal diffusion caused by turbulent behavior of the fluid inside the pipe has a significant effect on the thermal wave propagation. The effect of turbulent diffusion is more significant with sharper inlet temperature changes and that the proper modeling should consider this. Further, they conclude that the pseudo transient approach predicts the thermal propagation more accurately than the node method.

The third group of models is based on the discretization of the PDE describing the pipe in one or more dimensions (numerical/physical models). Method of characteristics with the third-order numerical scheme is used in [16] to solve the energy Equation. It seems to preserve the shapes and amplitudes of the heat wave, which indicates that there is little or none numerical diffusion in the solution. Axial diffusion and thermal dynamics of the wall is not incorporated.

Improved third order numerical solution with total variation diminishing property (reduces oscillations near sudden temperature changes) is used in [17]. Simplified dynamic accumulation of heat into the pipe's wall is considered. An indicator that describes the influence of the thermal inertia of the wall onto the wave propagation is developed. The proposed model is validated in a real system. Axial turbulent diffusion is not incorporated into the model.

Although numerical models are usually regarded as slow, they have one significant advantage. They can capture most of the dynamic phenomena affecting the traveling thermal wave in considerable detail. According to the review of relevant literature given above, any of the states of the art models (of any group) does not incorporate all the mentioned effects at once. To speed up numerical models, it is desirable to limit the number of operations as well as the number of memory allocations (saving fewer data and creating fewer temporaries). To obtain a smooth solution, there is a need to choose minor steps unless an appropriate nonlinear interpolation is possible.

Usually, problems, such as advection, are numerically solved by finite volume methods (FVM). Either explicit or implicit schemes are used. The PDE is often discretized along all dimensions, including the time. The quadratic upstream interpolation with estimation terms (QUICKEST) is often considered to be the most efficient scheme [18–21]. A good FVM scheme should preserve sharp temperature changes while not introducing unphysical oscillation. Non-oscillatory schemes are nonlinear and usually involve a flux limiter, which makes them more expensive during run-time. The only linear non-oscillatory FVM is first order upwind scheme, but it suffers from severe unphysical (numerical) diffusion. Luckily, when real diffusion is involved, the oscillation of higher-order schemes fades out. An example of this is presented in the following section.

Heat transfer is very often solved by finite element method (FEM), which is based on a weak formulation of the PDEs. Nodal values of the stencil amplify local non-zero normalized field functions (shape functions). The global temperature field function is then a superposition of the magnified local fields. Therefore, a good FEM scheme is produced by shape functions that closely resemble the true local temperature distribution. In the following section, there is a detailed description of an approach that results in the stencil coefficients for the radial model of a solid wall. Steady-state biased shape functions are used to capture the less transient periods very accurately even through roughest meshes.

In order to facilitate the various models of individual components of a DH system, all components should be solved simultaneously to include their mutual effects. Method of lines can be utilized here. In this method, the PDEs are approximated by sets of ordinary differential equations (ODE), where only the spatial derivatives are discretized. Such models are easy to combine using an ODE solver.

Open-source environments with powerful ODE solvers can be used to solve these systems. Two interesting ones are the OpenModelica and Julia languages. The first uses a component-oriented, Equation-based language called Modelica. It specializes in the modeling of complex systems dealing with various physics backgrounds [22]. There are many libraries containing predefined models of components which can be easily combined together to create a complex system. The OpenModelica compiler compiles code that solves efficiently time-dependent variables based on given parameters. It uses solvers for Differential algebraic Equations (DAE) to solve implicit systems, which allows the user to specify the relationship between variables without the need for being explicit. Communication with MATLAB can be also established through scripting and file exchange [23] (this is advantageous because MATLAB's default ODE solvers are very slow).

Julia is a general-purpose, high-level, dynamic language that approaches the speeds of statically-compiled languages like C [24]. There are several tools for benchmarking and profiling of written algorithms, which helps to identify their potential speedups. There are many well-developed packages already, amongst which is a very performant (meaning that it is working well enough to be considered functional but may not yet be fully optimized) package dealing with Differential Equations [25]. This package contains many solvers and can even interface to advanced C libraries, such as Sundials [26]. The results use interpolation with predefined tolerance to return values at an arbitrary time. This together makes the language highly suitable for scientific computing.

The aim of this paper was to derive and test specialized and fast numerical model of a simple DH pipe that incorporates sharp interface preservation, turbulent diffusion, and detailed radial thermal inertia evaluation caused by the wall.

2. Materials and Methods

This chapter contains a very in-depth description of the mathematics behind the model. In order to ensure that the resultant model is fast, a one-dimensional PDE is used to avoid two spatial derivatives. Here, the incompressibility of the fluid is assumed. Individual variables depend on the position and time as denoted by variables x and t in the following equation:

$$\frac{dT(x,t)}{dt} = -u(t)\frac{dT(x,t)}{dx} + D\frac{d^2T(x,t)}{dx^2} + S(x,t), \tag{1}$$

where T is the local cross-sectional temperature of the fluid, u is the axial fluid velocity, D is the axial diffusion coefficient, and S is a source term (dealing with heat flow from inner wall and friction). Because the method of lines is used here, only the spatial derivatives must be discretized. The pipe is divided into n_a cylindrical volumes, where spatially dependent variables are localized (see Figure 1). Localized temperatures represent the volume-averaged temperature and, localized sources represent the incoming heat from the solid shell divided by the heat capacity of one volume element.

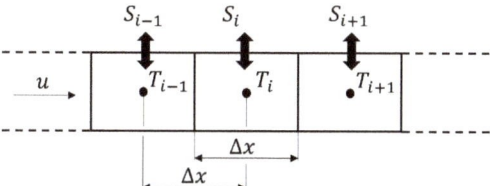

Figure 1. Division of pipe's fluid region into volumes, with localized spatial dependent variables.

Finite differences are used for discretization. There is a simple relationship between the finite difference schemes for any derivation order (including zero) and given n stencil points [27]:

$$c = \frac{1}{\Delta x^m} Z^{-1}\delta, \tag{2}$$

where c is a vector of the stencil coefficients, m is the derivation order, Z is an $n \times n$ matrix such as:

$$Z_{i,j} = \left(\frac{x_i - x_0}{\Delta x}\right)^{j-1}, \tag{3}$$

in which x_i is the axial position of a stencil point, x_0 is the axial position of the point in which the derivative is demanded, and δ represents a vector such as:

$$\delta_i = \begin{cases} 0 & i \neq m+1 \\ m! & i = m+1 \end{cases}. \tag{4}$$

The first term on the right-hand side of Equation (1) dealing with advection is discretized in two ways. The first way utilizes Equation (2) directly, which results in the finite difference method (FDM). The second way is based on FVM, where Equation (2) is used for the evaluation of the values at element faces (zero order derivatives). The discretization of the advection part should be good in shock capturing; therefore, 660 tests are conducted to compare this ability among the different schemes. By filling rows with the stencil coefficients, a script written in Julia constructs the matrix A that approximates the derivative operator. Among the input parameters, there is a number of used upwind and downwind stencil points. While filling the first few rows of matrix A, both these parameters are automatically decreased. When filling the last few rows, only the downwind parameter is decreased because of the missing nodes. For example, the maximum number of available upwind nodes at the inlet is one (the boundary temperature). No oscillation is caused by central schemes near the boundaries (unless a central scheme is used explicitly) because the local upwind bias is always preserved (first rows) or increased (last rows). The boundary vector A_{in} is constructed as well. Both are used to evaluate the pure advection problem in the following form:

$$\frac{dT}{dt} = -u(AT + T_{in}A_{in}), \tag{5}$$

where T is the vector of localized fluid volume temperatures and T_{in} is the time-dependent scalar representing the inlet (boundary) temperature signal. Problems are solved using normalized variables, and the behavior of the schemes is presented in a normalized variable diagram (NVD). Sundials library is used for solving the problem in time, namely the CVODE algorithm with backward differencing and with relative and absolute tolerances of 10^{-6}. It has an adaptive time step and produces a continuous temperature signal at the outlet by using Hermite interpolation. Examples of NVDs are shown in Figure 2. The diagrams contain hints about the schemes (U and D are the number of upwind and downwind nodes, respectively, and n is a number of volume elements).

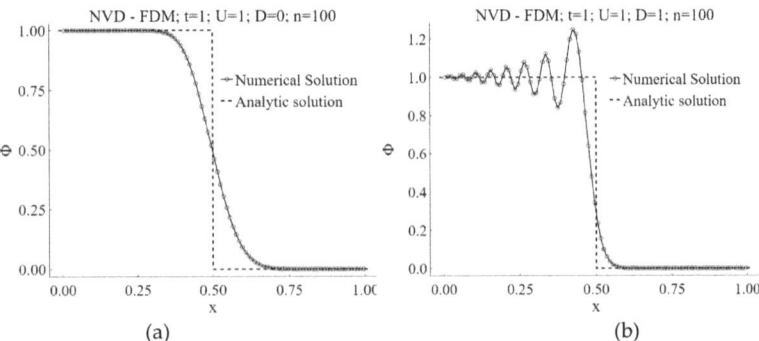

Figure 2. Examples of bad-behaving advection schemes. (**a**): Low order upwind biased scheme causes a lot of numerical diffusion; (**b**): Low order central scheme is being too oscillatory.

It is observed that central schemes have a stronger tendency to oscillate (Figure 2b) and are more computationally expensive than upwind-biased schemes (see Figure 3). Overall, higher order upwind schemes (see Figure 4) seem to show a good compromise between the degree of their oscillation, their ability for shock capturing, and computational expensiveness.

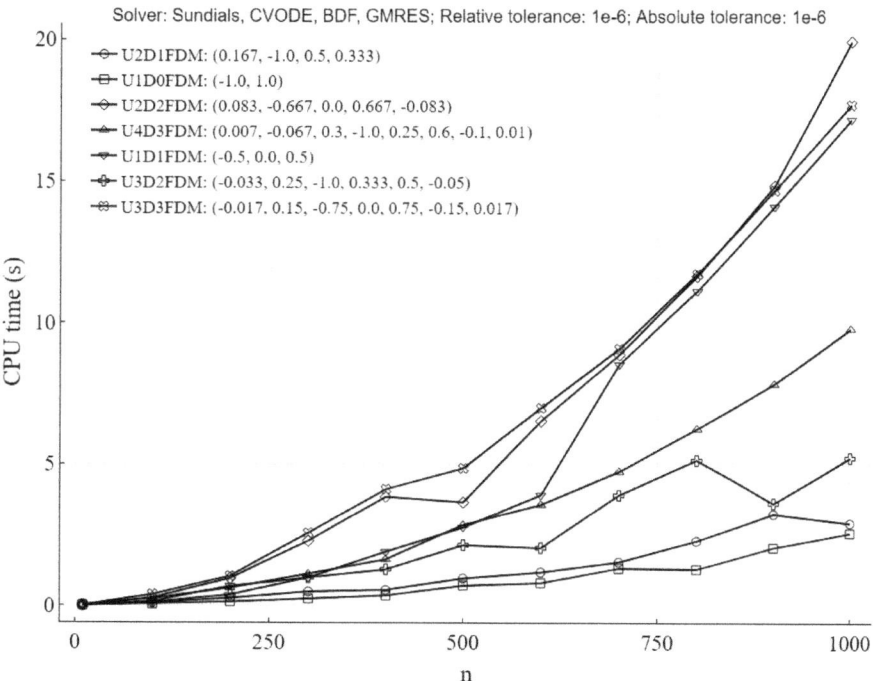

Figure 3. Comparison of computational time of the tests between upwind-biased and central schemes for FDM. Similar results can be observed for FVM. The values in brackets represent rounded row stencil coefficients in the typical row of the derivative operator A (when $\Delta x = 1$).

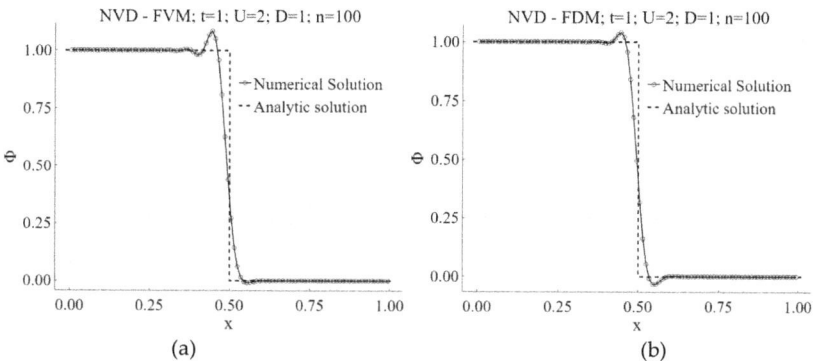

Figure 4. Examples of well-behaving advection schemes. (**a**): Higher order upwind biased FVM scheme with mild numerical diffusion and mild oscillations; (**b**): Higher order upwind biased FDM scheme with mild numerical diffusion and very mild oscillations.

The first order central scheme is used for the construction of B, which is the matrix that approximates the second derivative operator dealing with axial diffusion in Equation (1). Boundary vector B_{in} allows the calculation of the second derivative at the inlet node as well. Equation (5) is extended and results in Equation (6):

$$\frac{dT}{dt} = -u(AT + T_{in}A_{in}) + D(BT + T_{in}B_{in}), \qquad (6)$$

diffusion coefficient D may either be lumped or localized into the form:

$$D_i = w_{1,i} \cdot u \cdot d_1 + w_{2,i}, \tag{7}$$

The first term in Equation (7) deals with turbulent diffusion. The second term, which is strictly non-negative, corresponds to velocity-independent effects, such as conductive diffusion (which is shown later to be negligible) or spreading due to effects of gravity. If the second term in Equation (7) is neglected, the Equation (6) can be simplified:

$$\frac{dT}{dt} = u[(d_1 B - A)T + (d_1 B_{in} - A_{in})T_{in}] = u(MT + M_{in}T_{in}), \tag{8}$$

where M and M_{in} are the constant matrix operator and boundary vector of the fluid region, respectively. This step should improve the performance of the model. It also makes the Peclét number independent of flow velocity. This means that when proper axial discretization is chosen to ensure non-oscillatory behavior of the method, it is preserved for all velocities of the fluid. Automatic axial discretization, based on values $w_{1,i}$, is then possible. The weights in Equation (7) allow the tuning of the pipe to mimic a real installation more accurately (similar weighting is possible for the source term as well).

Using the analytic solution to step input in the NVD, the effect of D_i on oscillation of Equation (6) is studied. Any value of the numerical solution that is outside of the unit interval is considered to be an overshot caused by oscillation. This is a very simple way to estimate the residual oscillation of Equation (6). FDM is chosen for the model because of its lower oscillations (see Figure 5).

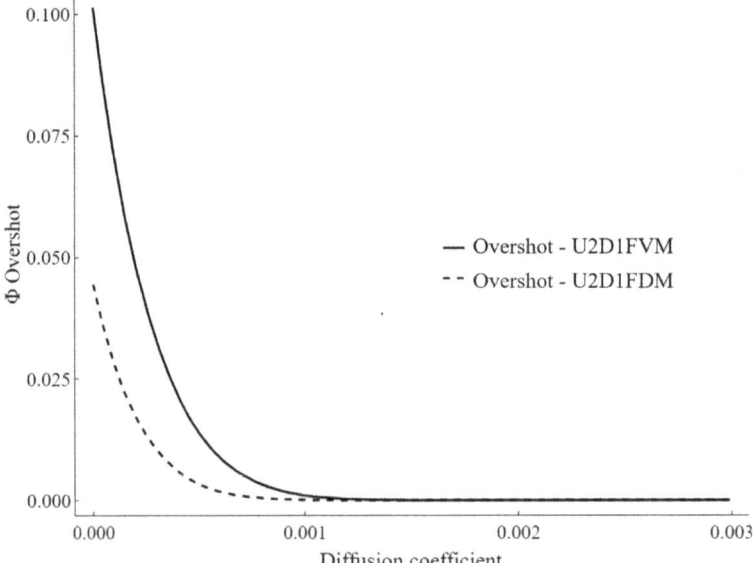

Figure 5. Effect of physical diffusion on unphysical oscillation on sharp interfaces for $n_a = 100$ (the critical Peclét number, which makes the residual oscillation less than 10^{-6}. It is approximately 6.47 for the U2D1FVM and 8.51 for the U2D1FDM).

The last term of Equation (1) is more complicated since it must deal with heat dynamics in the solid shells surrounding the pipe. From the fluid viewpoint, the only important value is the inner wall temperature of the first shell. Using publication [6], the Nusselt number for all flow regimes is:

$$Nu = \begin{cases} 3.66 \text{ if } Re \leq 2300 \\ 3.52 \cdot U^4 - 45.148 \cdot U^3 + 212.13 \cdot U^2 - 427.45 \cdot U^2 + 316.08 \text{ if } 2300 \leq Re \leq 3100 \\ \frac{f_r}{8} \cdot \frac{(Re-1000) \cdot Pr}{1+12.7 \cdot \left(\frac{f_r}{8}\right)^{1/2} \cdot \left(Pr^{2/3}-1\right)} \text{ if } Re < 3000 \end{cases}, \quad (9)$$

where Re is the Reynolds number, $U = Re/1000$, f_r is the Darcy-Weisbach friction factor, and Pr is the Prandtl Number. The friction factor might be evaluated using the Churchill Equation presented in [7] for all flow regimes:

$$f = 8 \cdot \left(\left(\frac{8}{Re}\right)^{12} + \frac{1}{(\theta_1 + \theta_2)^{3/2}}\right)^{1/12}, \quad (10)$$

where

$$\theta_1 = \left(-2.457 \cdot \log\left(\left(\frac{7}{Re}\right)^{0.9} + 0.27 \cdot \left(\frac{\varepsilon}{d_1}\right)\right)\right)^{16},$$
$$\theta_2 = \left(\frac{37530}{Re}\right)^{16} \quad (11)$$

in which ε is the roughness of the inner wall and d_1 is the inner diameter. It is a simple step to evaluate the local source term S in the form of a vector:

$$S_i = \frac{4 \cdot Nu \cdot \lambda}{d_1^2 \cdot \rho \cdot c_p} \cdot (T_{sn,i,1} - T_i). \quad (12)$$

where $T_{sn,i,1}$ represents the inner wall (later first node of the solid region) temperatures of the solid shell, c_p is the specific heat capacity of the fluid, ρ is the mass density of the fluid, and λ is the heat conductivity of the fluid.

The time-dependent distribution of temperature in the solid shells affects the axial transients of the heat waves (charging and discharging of the accumulation layers). Tangential and axial heat conductions in the shells are ignored because even when the flow velocity is small, the rate in which the heat is being transported by advection is often much higher than that by conduction. This might be expressed in the form of the Péclet number for conduction in both the fluid:

$$Pe_f = d_1 \cdot u \cdot \frac{\rho \cdot c_p}{\lambda} \approx d_1 \cdot u \cdot 6,29 \cdot 10^6 \gg 1, \quad (13)$$

and the solid (e.g., steel):

$$Pe_s = d_1 \cdot u \cdot \frac{\rho_s \cdot c_{p,s}}{\lambda_s} \approx d_1 \cdot u \cdot 8 \cdot 10^4 \gg 1. \quad (14)$$

As shown, the inner diameters of the pipes or the flow velocities would have to be very small for the conductive axial diffusion to be comparable with advection. This is not common with real applications in the DH systems which also reduces the number of necessary spatial derivatives used in the model. The fluid volumes can be mathematically coupled with one-dimensional models of purely radial heat transfer (later solid segments) described by PDE in the form:

$$\frac{dT_{s,i}}{dt} \cdot \rho_s \cdot c_{ps} = \frac{1}{r}\frac{d}{dr}\left(\lambda_s \cdot r \cdot \frac{dT_{s,i}}{dr}\right), \quad (15)$$

where r is the radial coordinate, $T_{s,i}$ is the continuous temperature in the solid segment, ρ_s is the mass density, c_{ps} is the specific heat capacity, and λ_s is the heat conductivity. Equation (15) might be rewritten into the weak form using the shape function V (notice that integrated term in the bracket affects the system only at the boundaries).

$$2 \cdot \pi \cdot \Delta x \cdot \int_{r_1}^{r_2} r \cdot V \cdot \frac{dT_{s,i}}{dt} \cdot \rho_s \cdot c_{ps} \cdot dr = 2 \cdot \pi \cdot \Delta x \cdot \left[-\int_{r_1}^{r_2} r \cdot \frac{dV}{dr} \cdot \frac{dT_{s,i}}{dr} \cdot \lambda_s \cdot dr + \left[r \cdot V \cdot \lambda_s \cdot \frac{dT_{s,i}}{dr}\right]_{r_1}^{r_2}\right]. \quad (16)$$

The steady state temperature distribution is used as the base for the nodal shape functions:

$$T_{se,i,j}(r) = \frac{\ln(r/R_j) \cdot (T_{sn,i,j+1} - T_{sn,i,j})}{\ln(R_{j+1}/R_j)} + T_{sn,i,j} = T_{sn,i,j+1} \cdot \frac{\ln(r/R_j)}{\ln(R_{j+1}/R_j)} + T_{sn,i,j}\left(1 - \frac{\ln(r/R_j)}{\ln(R_{j+1}/R_j)}\right), \quad (17)$$

where $T_{se,i,j}$ are the temperature distributions in the cylindrical elements, $T_{sn,i,j}$ are nodal temperatures, and R_j are radial coordinates of the nodes. Equation (17) is the solution to the following PDE with given boundary conditions (see also Figure 6):

$$\begin{aligned} \frac{1}{r}\frac{d}{dr}\left(\lambda_{s,j} \cdot r \cdot \frac{dT_{se,i,j}}{dr}\right) &= 0 \\ T_{se,i,j}(R_{j+1}) &= T_{sn,i,j+1} \\ T_{se,i,j}(R_j) &= T_{sn,i,j} \end{aligned} \quad (18)$$

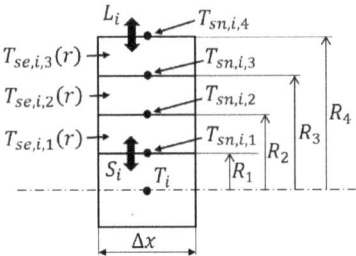

Figure 6. Graphical representation of the variables.

Nodal shape functions are derived from Equation (17) by considering the effect of one node to the global temperature distribution. Therefore, they take the following piecewise form (notice that boundary nodes are neighboring one element only, so the shape functions have only the appropriate half):

$$V_{sn,j}(r) = \begin{cases} \begin{cases} 1 - \frac{\ln(r/R_j)}{\ln(R_{j+1}/R_j)}, & R_j \leq r \leq R_{j+1} \\ 0, & R_j > r > R_{j+1} \end{cases}, j = 1 \\ \begin{cases} \frac{\ln(r/R_{j-1})}{\ln(R_j/R_{j-1})}, & R_{j-1} \leq r < R_j \\ 1 - \frac{\ln(r/R_j)}{\ln(R_{j+1}/R_j)}, & R_j \leq r \leq R_{j+1} \\ 0, & R_{j-1} > r > R_{j+1} \end{cases}, 2 \leq j \leq n-1, \\ \begin{cases} \frac{\ln(r/R_{j-1})}{\ln(R_j/R_{j-1})}, & R_{j-1} \leq r < R_j \\ 0, & R_{j-1} > r > R_j \end{cases}, j = n \end{cases} \quad (19)$$

When mass densities and specific heat capacities of the elements are constant, the following integration (according to the left-hand side of Equation (16)) results in discretized heat capacity:

$$C_{sn,j} = 2 \cdot \pi \cdot \Delta x \cdot \left[\rho_{s,j-1} \cdot c_{ps,j-1} \cdot \int_{R_{j-1}}^{R_j} r \cdot V_{sn,j} \cdot dr + \rho_{s,j} \cdot c_{ps,j} \cdot \int_{R_j}^{R_{j+1}} r \cdot V_{sn,j} \cdot dr\right], \quad (20)$$

where $C_{sn,j}$ are the nodal heat capacities, $\rho_{se,i,j}$ are the mass densities of the elements, and $c_{ps,j}$ is the specific heat capacity of the elements. For inside nodes, the integration of (20) results in:

$$C_{sn,j} = \pi \cdot \Delta x \cdot \left[\rho_{s,j-1} \cdot c_{ps,j-1} \cdot \left(\frac{R_{j-1}^2 - R_j^2}{2 \cdot \ln(R_j/R_{j-1})} + R_j^2\right) - \rho_{s,j} \cdot c_{ps,j} \cdot \left(\frac{R_j^2 - R_{j+1}^2}{2 \cdot \ln(R_{j+1}/R_j)} + R_j^2\right)\right], \quad (21)$$

For the boundary near the fluid region, there is only the second term in the bracket (there is no element below), while for the outer boundary node, there is only the first term (there is no element above).

The net heat flow incoming to the nodes in the solid region is (according to the right-hand side of Equation (16)):

$$H_{sn,i,j} = 2 \cdot \pi \cdot \Delta x \cdot [T_{sn,i,j-1} \cdot h_{sn,j-1,j} + T_{sn,i,j} \cdot h_{sn,j,j} + T_{sn,i,j+1} \cdot h_{sn,j+1,j}], \quad (22)$$

where

$$h_{sn,j-1,j} = -\int_{R_{j-1}}^{R_j} r \cdot \lambda_{s,j-1} \cdot \frac{dV_{sn,j-1}}{dr} \frac{dV_{sn,j}}{dr} dr = \frac{\lambda_{s,j-1}}{\ln(R_j/R_{j-1})}$$

$$h_{sn,j,j} = -\int_{R_{j-1}}^{R_j} r \cdot \lambda_{s,j-1} \cdot \left(\frac{dV_{sn,j}}{dr}\right)^2 dr - \int_{R_j}^{R_{j+1}} r \cdot \lambda_{s,j} \cdot \left(\frac{dV_{sn,j}}{dr}\right)^2 dr = \frac{-\lambda_{s,j-1}}{\ln(R_j/R_{j-1})} + \frac{-\lambda_{s,j}}{\ln(R_{j+1}/R_j)}. \quad (23)$$

$$h_{sn,j,j} = -\int_{R_j}^{R_{j+1}} r \cdot \lambda_{s,j} \cdot \frac{dV_{sn,j+1}}{dr} \frac{dV_{sn,j}}{dr} dr = \frac{\lambda_{s,j}}{\ln(R_{j+1}/R_j)}$$

The nodal net heat flow changes for boundary nodes. For the inner boundary node (here the integrated term in brackets of Equation (16) is replaced by heat flow from fluid to solid), the net heat flow is:

$$H_{sn,i,1} = 2 \cdot \pi \cdot \Delta x \cdot [T_{sn,i,1} \cdot h_{sn,1,1} + T_{sn,i,2} \cdot h_{sn,2,1}] + \pi \cdot \Delta x \cdot Nu \cdot \lambda \cdot (T_i - T_{sn,i,1}), \quad (24)$$

where

$$h_{sn,1,1} = -\int_{R_1}^{R_2} r \cdot \lambda_{s,1} \cdot \left(\frac{dV_{sn,1}}{dr}\right)^2 dr = \frac{-\lambda_{s,1}}{\ln(R_2/R_1)}$$

$$h_{sn,2,1} = -\int_{R_1}^{R_2} r \cdot \lambda_{s,1} \cdot \frac{dV_{sn,2}}{dr} \frac{dV_{sn,1}}{dr} dr = \frac{\lambda_{s,1}}{\ln(R_2/R_1)} \quad (25)$$

and similarly for the outer boundary node:

$$H_{sn,i,n} = 2 \cdot \pi \cdot \Delta x \cdot [T_{sn,i,n-1} \cdot h_{sn,n-1,n} + T_{sn,i,n} \cdot h_{sn,n,n}] + \alpha_{out} \cdot 2 \cdot \pi \cdot R_n \cdot \Delta x \cdot (T_{air} - T_{sn,i,n}), \quad (26)$$

where the last term represents true heat loss (heat transfer to the environment), α_{out} is the convective heat transfer coefficient at the outer wall, T_{air} is the ambient temperature and:

$$h_{sn,n-1,n} = -\int_{R_{n-1}}^{R_n} r \cdot \lambda_{s,n-1} \cdot \frac{dV_{sn,n-1}}{dr} \frac{dV_{sn,n}}{dr} dr = \frac{\lambda_{s,n-1}}{\ln(R_n/R_{n-1})}$$

$$h_{sn,n,n} = -\int_{R_{n-1}}^{R_n} r \cdot \lambda_{s,n-1} \cdot \left(\frac{dV_{sn,n}}{dr}\right)^2 dr = \frac{-\lambda_{s,n-1}}{\ln(R_n/R_{n-1})} \quad (27)$$

Using Equations (16), (21) and (22), the set of ODEs describing heat dynamics in the solid is:

$$\frac{dT_{sn,i,j}}{dt} = \frac{H_{sn,i,j}}{C_{sn,j}}, \quad (28)$$

Values in matrix $h_{sn,i,j}$ are nonzero only for values given by Equations (23), (25) and (27). Under such conditions, the following matrices might be constructed:

$$k_{i,j} = 2 \cdot \pi \cdot \Delta x \cdot \frac{h_{sn,i,j}}{C_{sn,j}}, \quad (29)$$

$$k_{in,i,j} = \begin{cases} j = 1, \ \pi \cdot \Delta x \cdot Nu \cdot \lambda \cdot (T_i - T_{sn,i,1})/C_{sn,1} \\ 2 \leq j \leq n-1, \ 0 \\ j = n, \ \alpha_{out} \cdot 2 \cdot \pi \cdot R_n \cdot \Delta x \cdot (T_{air} - T_{sn,i,n})/C_{sn,n} \end{cases} \quad (30)$$

Using Equations (29) and (30), the relation (28) might be expressed in a form similar to Equation (6):

$$\frac{dT_{sn,i}}{dt} = k \cdot T_{sn,i} + k_{in,i}. \quad (31)$$

where $T_{sn,i}$ are vectors of the nodal temperatures of the solid segments, k is a constant matrix (constructed by Equation (29)) capturing the heat dynamics of the nodes inside segments, and $k_{in,i}$ are the time-dependent vectors containing boundary conditions of the segments. Banded matrices can be used in order to express the dynamics fully in a single equation:

$$\frac{dT_{sn}}{dt} = K \cdot T_{sn} + K_{in}. \quad (32)$$

where T_{sn} is the column vector filled with column vectors $T_{sn,i}$, K is the banded square matrix filled with k on its diagonal, and K_{in} is the time-dependent boundary vector filed with column vectors $k_{sn,i}$.

This description of the solid region will satisfy the analytic solution completely at steady states but will have a tendency to deviate more with rapid changes of fluid temperature. Individual layers can be further divided into sublayers, which will increase the accuracy during nonstationary events.

3. Results and Discussion

The new model of radial heat transfer is tested for two different cases (see Table 1). The first case is not an actual installation but serves as an example of a case with very strong accumulation. A sudden change in the fluid temperature causes high temporal heat flux between the fluid and the wall. A quick reduction in the heat flux follows due to a rise of the innermost wall surface temperature. The radial elements increasing in size from the inner wall outward should capture this better. Smaller time-steps must be then used to ensure numerical stability. Element sizes are defined by a grow factor. Qualities of the solutions at the innermost piece of the wall are studied using a step change in fluid temperature from 0 to 10 K.

Table 1. The property material of the steel circular wall (inner pipe diameter is 0.6 m).

Property	Case 1 Accumulation	Case 2 No Accumulation
Layer thickness (mm)	100	4
Mass density (kg/m^3)	8030	8030
Heat conductivity (W/mK)	16.27	16.27
Specific heat capacity (J/kgK)	502.45	502.45

Since the temperature of the fluid steps up to 10 K and the correct inner wall temperature after 10 s is about 2.16 K, the error in heat flux caused by the use of 5 subdivisions (with grow factor 1) would be over 10%. Grow factor 1.0 with 10 subdivisions provides almost equivalent results to an advanced solution that uses grow factor 2.0 with 5 subdivisions, but it is slightly faster (see Figure 7). The thin wall in the second case behaves much better. It produces solutions accurate enough even with lower subdivisions (compare Figures 8–11 with Figures 12–15). This ensures good computational speed for such arrangements. To further estimate the accuracy of the new radial model for the case with strong accumulation, the maximal errors are estimated using temperature refinements between two consecutive levels of subdivision (see Figures 8–11). The whole event of heat recharging is used (7200 s). A tolerance of 10^{-12} is chosen to ensure that the time stepping method can always take advantage of a superior spatial discretization.

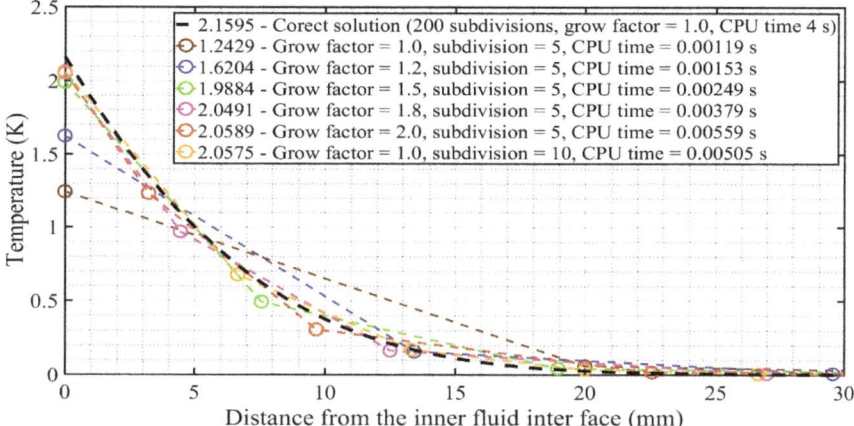

Figure 7. Demonstration of the solution quality (in the radial model) at 10 s for different subdivisions (Case 1).

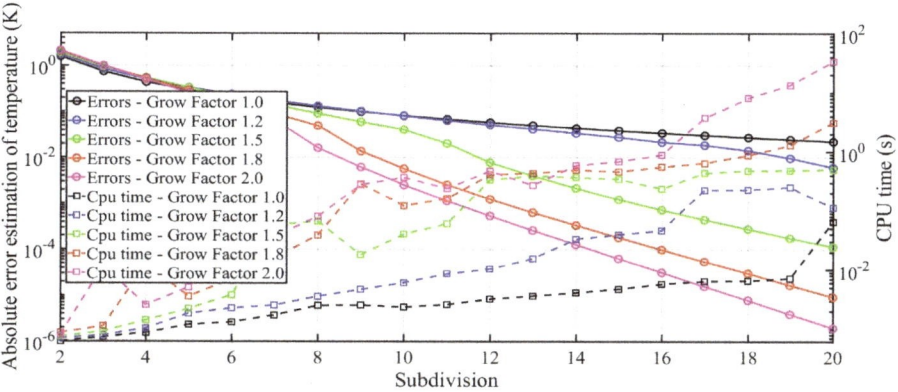

Figure 8. Absolute maximum temperature error estimations of the inner wall surface (Case 1).

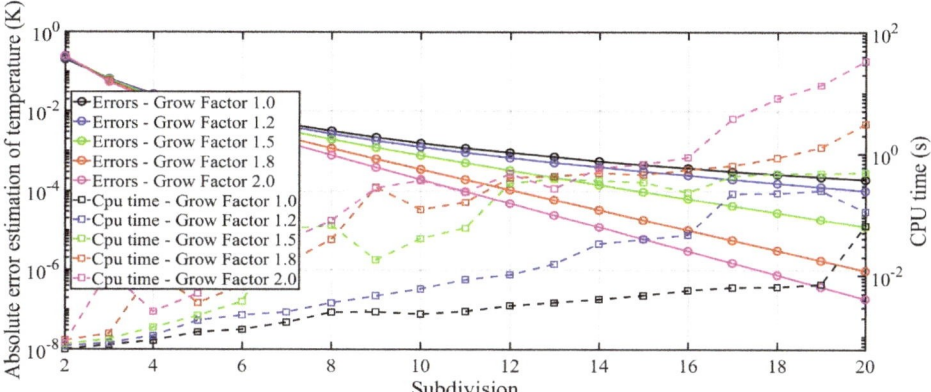

Figure 9. Absolute maximum temperature error estimations of the outer surface (Case 1).

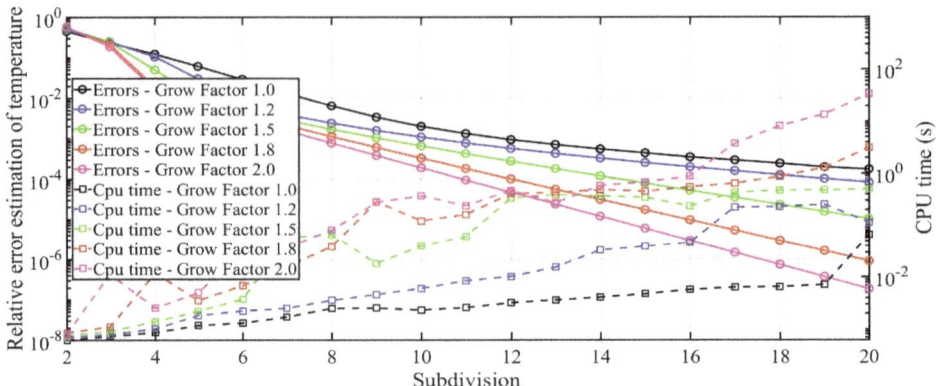

Figure 10. Relative maximum temperature error estimations of the inner surface (Case 1).

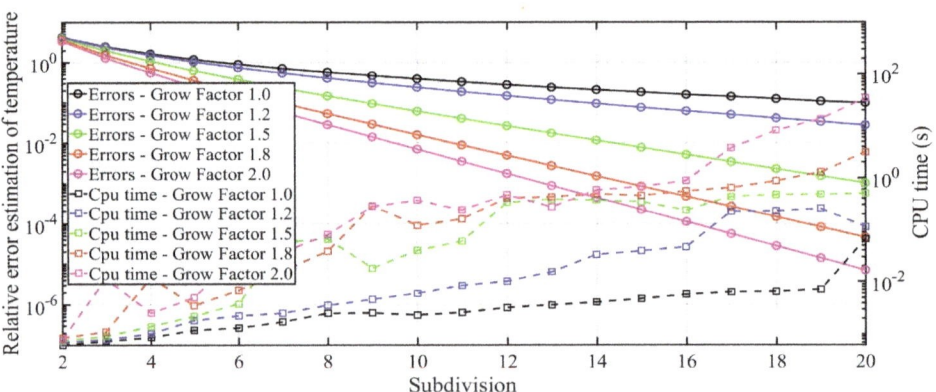

Figure 11. Relative maximum temperature error estimations of the outer surface (Case 1).

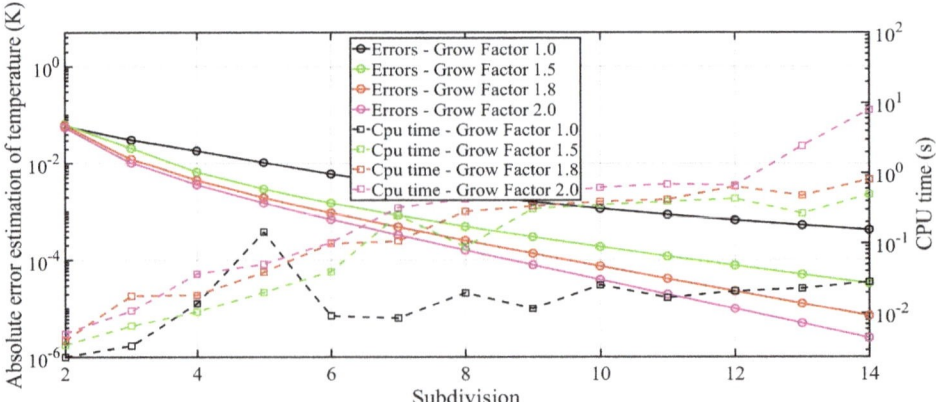

Figure 12. Absolute maximum temperature error estimations of the inner surface (Case 2).

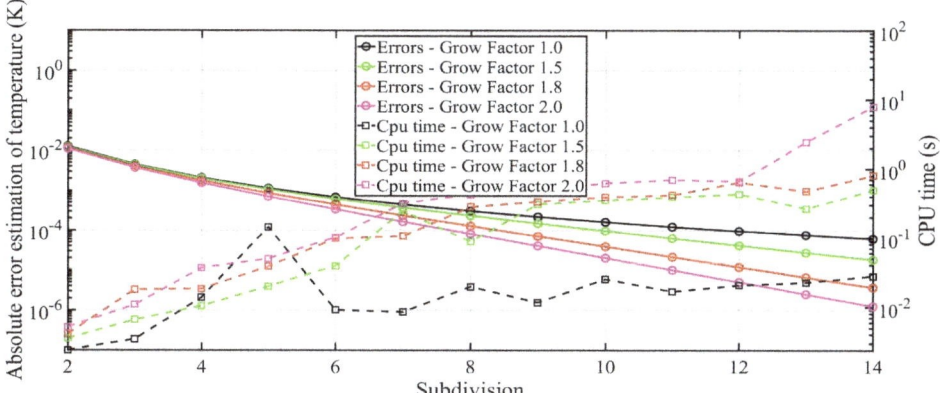

Figure 13. Absolute maximum temperature error estimations of the outer surface (Case 2).

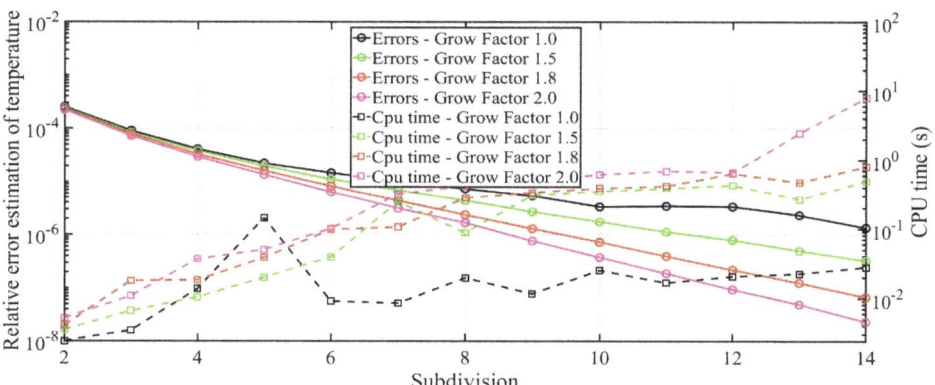

Figure 14. Relative maximum temperature error estimations of the inner surface (Case 2).

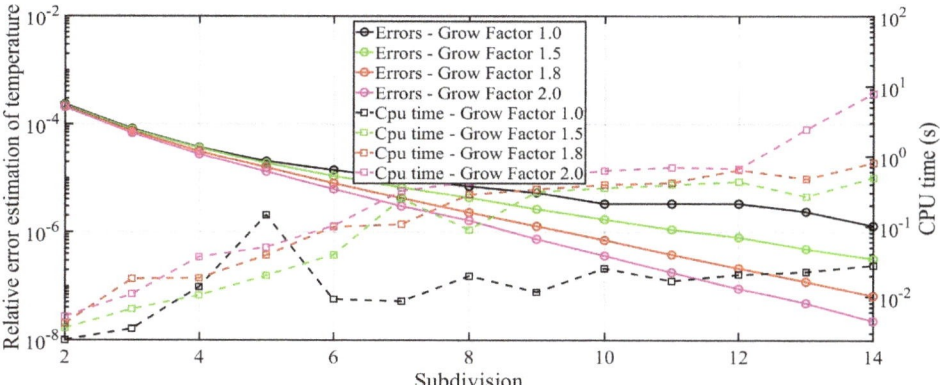

Figure 15. Relative maximum temperature error estimations of the outer surface (Case 2).

The errors in the second case for lower subdivisions are equivalent to errors in the first case for high subdivisions. This is because the inertia of the thin wall is not significant. The relative errors of the inner and outer walls are almost the same (Figures 13 and 14).

The composite model of the pipe written in OpenModelica is tested in a similar way (Julia was significantly slower here, possibly due to the use of numerical jacobians instead of symbolic ones). The 100 m long pipe with a cross-section corresponding to the first case of the previous test set is initialized with a temperature of 323.15 K. After steady-state is reached, a 30 K rise and drop in the inlet temperature generate the transient process. The duration of the tests was 16,000 s. The tolerance is set to 10^{-6}, and the values are saved every 3 s. Error estimations are evaluated using a discrete gradient (backward differencing) between the set of discretization. Using contours, the "willingness to refine maximal errors" of the model can be mapped (see Figures 16 and 17). The computational time can be mapped directly (see Figure 18).

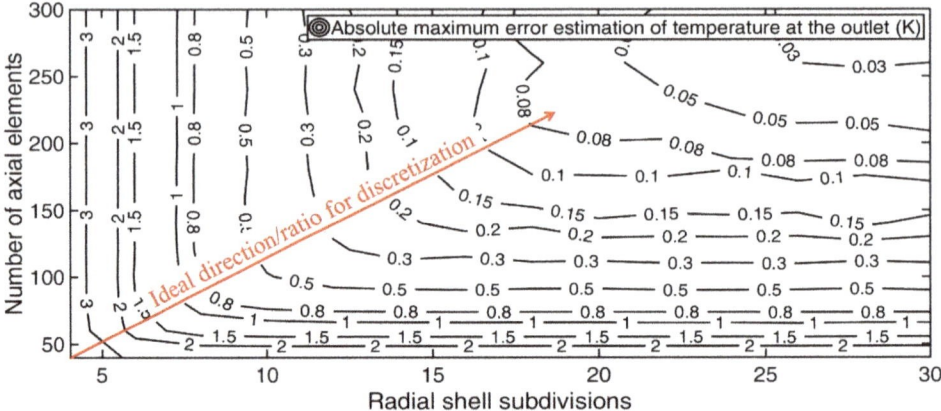

Figure 16. Estimations of maximal absolute errors at the outlet of the pipe (Case 1), where the red arrow signifies the ideal direction for discretization.

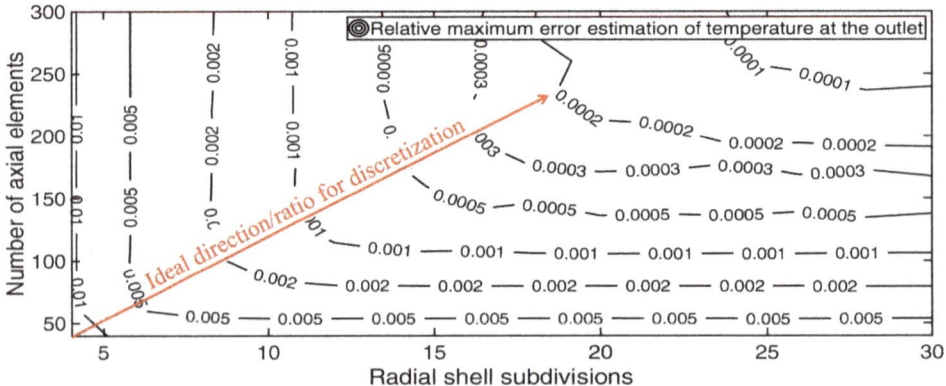

Figure 17. Estimations of maximal relative errors at the outlet of the pipe (Case 1), where the red arrow signifies the ideal direction for discretization.

Similar tests for the second case clearly demonstrate the independence on the radial number of subdivisions (see Figures 19 and 20). There is nothing to be refined in the radial direction. Further subdivisions only take more computational time (see Figure 21). The higher computational time in the right top corner of Figure 21 is caused by smaller time steps to ensure numerical stability.

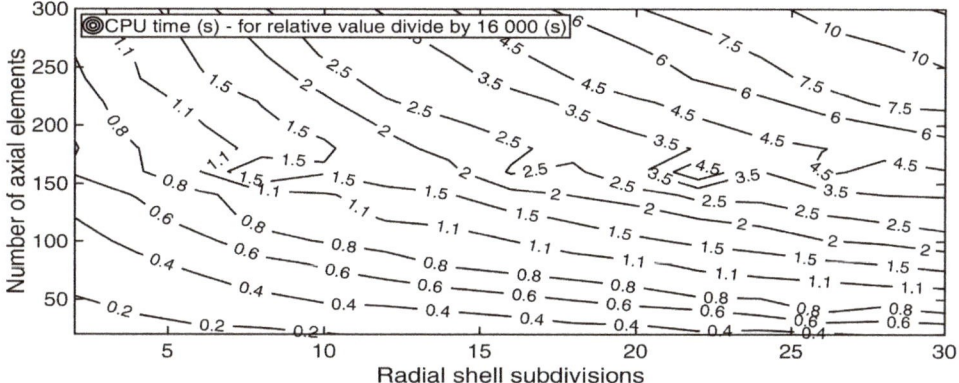

Figure 18. Computation times of the composite model tests (Case 1).

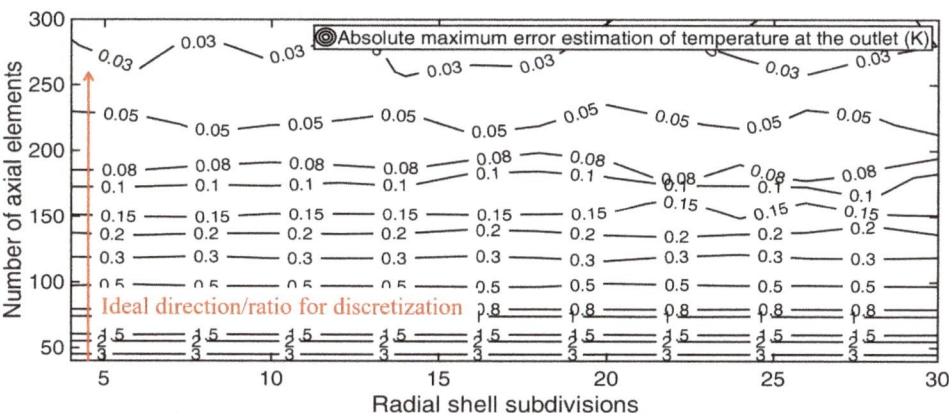

Figure 19. Estimations of maximal absolute errors at the outlet of the pipe (Case 2), where the red arrow signifies the ideal direction for discretization.

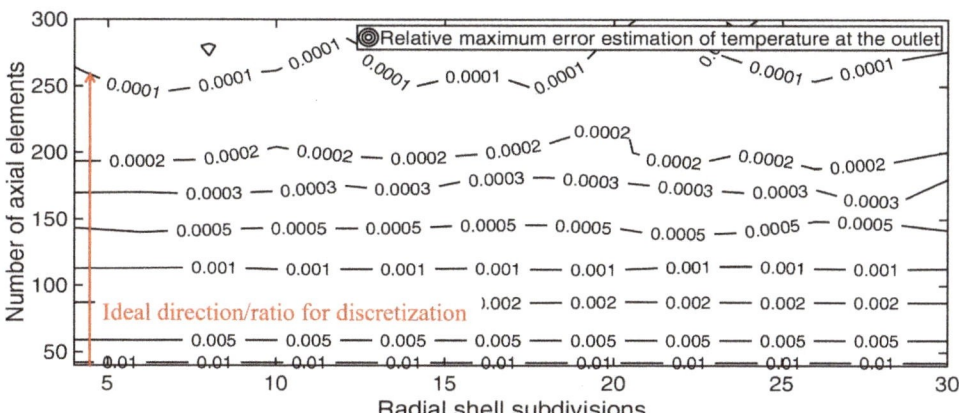

Figure 20. Estimations of maximal relative errors at the outlet of the pipe (Case 2), where the red arrow signifies the ideal direction for discretization.

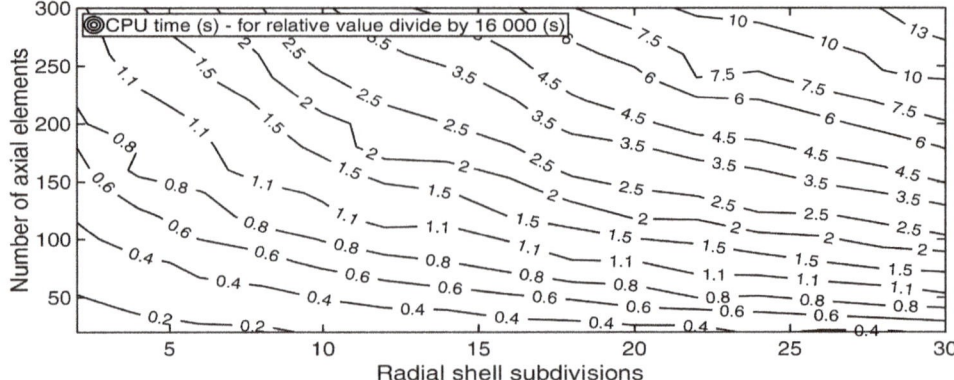

Figure 21. Computation times of the composite model tests (Case 2).

The Figure 22 demonstrates the effect of the strong accumulation using the outlet temperature perspective. The above results deal with the model alone. Unfortunately, real data were not available to the authors to validate the model properly. The model is at least compared (partially validated) to a very fine model developed in Fluent (axisymmetric 2D in Figures 23 and 24) and STAR-CCM+ (fully 3D in Figures 25 and 26). The tolerance of the new model was lowered to 10^{-5} and samples were saved every 10 s to demonstrate the effectiveness of the new model. The time span of the simulation was 16,000 s.

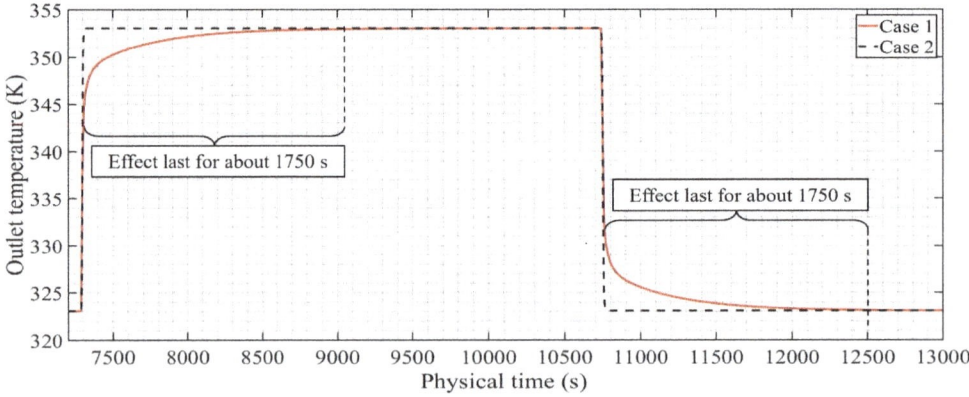

Figure 22. Comparison of Case 1 and Case 2 (with constant mass flow of 282.74 kg/s).

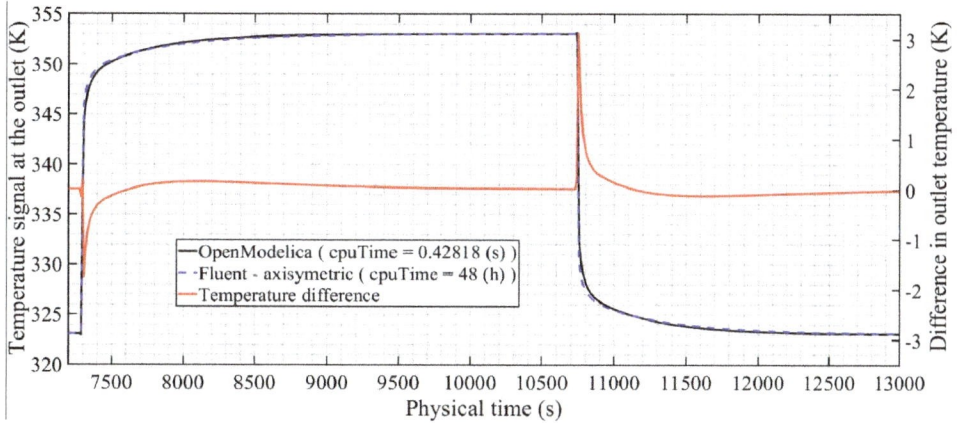

Figure 23. Comparison with a model from Fluent (the solid material property differ slightly here—mass density is 7850 kg/m^3, heat conductivity is 50 W/(mK), and specific heat capacity is 500 J/(kgK)).

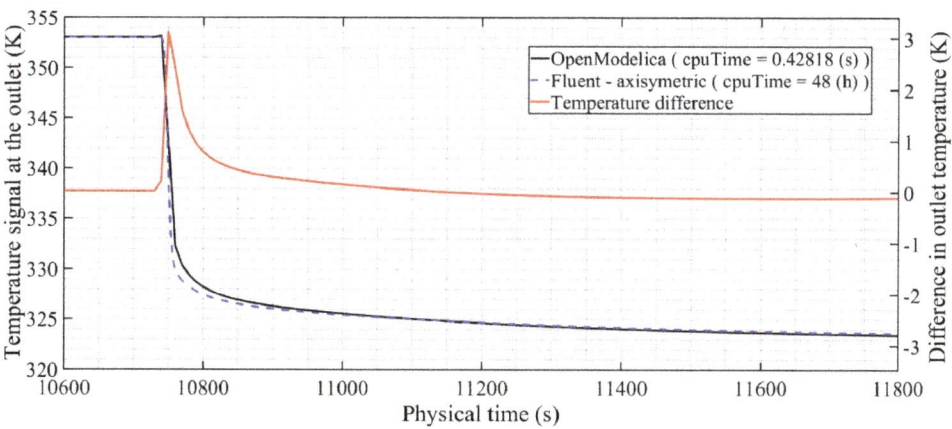

Figure 24. Detail of the difference in temperature on comparison with a model from Fluent.

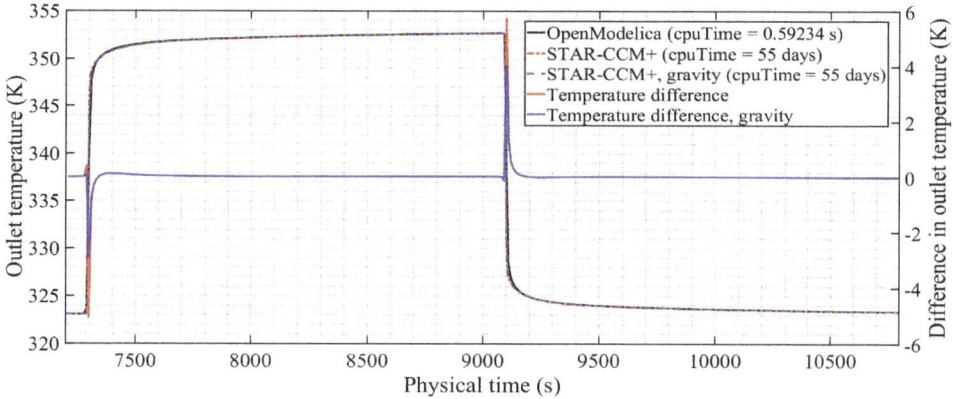

Figure 25. Comparison with a model from STAR-CCM+.

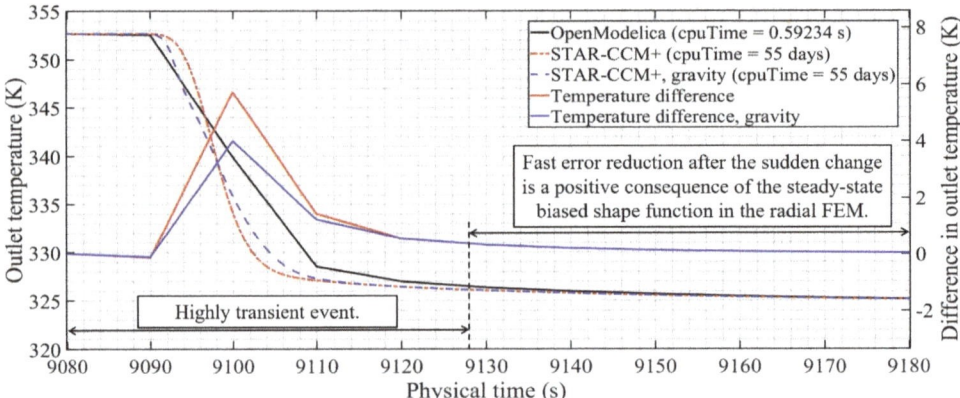

Figure 26. Detail of the difference in temperature on comparison with a model from STAR-CCM+.

Although the maximal temperature difference is higher than the model from STAR-CCM+, the temperature difference falls close to zero quicker than the model from Fluent. The effect of gravity on wave propagation is such that it behaves as additional axial diffusion, which can be tuned up by adjusting weights in Equation (7). Figure 26 shows this fact by the more time-spread rise in temperature at the outlet. Figures 27 and 28 show how the gravity spreads the temperature front in the STAR-CCM+ model 7 s after both hot and cold waves entered the pipe.

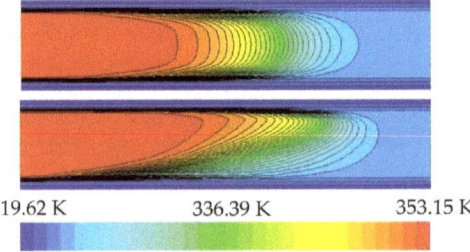

Figure 27. Effect of gravity on the temperature distribution of a hot wave (the upper part shows the temperature front without effects of gravity, the lower part shows the contrary).

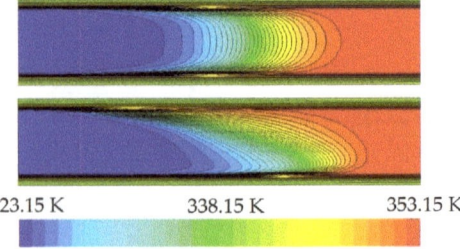

Figure 28. Effect of gravity on the temperature distribution of cold wave (the upper picture shows the temperature front without effects of gravity, the lower picture shows the contrary).

The comparison of the new model with the very fine numerical counterparts shows that the new model overestimates the wave to reach the outlet slightly later (asymmetric peaks in temperature difference in the Figures 23–26). This might be caused by the fact that the temperature and velocity profiles in the new model are considered to be fully developed right from the beginning, while with the very fine counterparts they are not.

4. Summary of the Results

- The shape function used in the radial model as well as the use of analytic description of heat exchange between the fluid and the inner wall surface ensures that the solution of the new model agree with the analytic solution completely even for the roughest radial discretization (also visible through fast stabilization in Figure 22 after the sudden temperature change).
- The new model allows simple modeling of circular pipes with an arbitrary number of shells and their material properties (e.g., a steel wall followed by insulation and a plastic shell). All shells then consider radial dynamic effects based on the newly derived radial model, and OpenModelica's marching algorithm automatically adapts the time step to ensure its numerical stability.
- Although elements growing in size in the radial direction better capture the sudden changes of temperature, the necessity of smaller time steps to ensure the stability of the marching algorithm completely negates this advantage (compare the CPU time in Figure 7) and makes the uniform mesh more preferable.
- With the preferred use of the uniform mesh in the radial model, the thick wall has to be modeled with higher number of elements, while the thin wall is captured well with very few or even single element (compare error estimations for grow factor 1 in Figures 8–15 or the error maps in Figures 16 and 19).
- The use of the method of lines and appropriate algorithm with adaptable time step makes the precision easily tradable for computation time, without the necessity to focus on the numerical stability conditions of the composite model (trading the capturing of the transients for CPU time, the steady-states will be captured accurately even with rough radial meshes).
- There are regions in which a sole improvement of the mesh in one direction will not result in superior solution (e.g., see the region of 100 axial and 20 radial elements in Figure 16, where sole improvement of radial discretization does not lower errors). There is, however, an ideal direction for the discretization in each case.
- Figure 22 shows how does strong thermal inertia of the wall affect the outlet temperature in idealized case (the effect lasts approximately for half an hour for the thick steel wall).
- Maximal deviation in outlet temperature in comparison with Fluent (2D) is about 3 K.
- Maximal deviation in outlet temperature in comparison with STAR-CCM+ (3D) is about 6 K for the case without gravity and 4 K for the case that includes gravity.
- During all simulations, positive weights in the definition of axial diffusion are deployed, which shows that turbulent axial mixing plays a role in thermal wave propagation.
- Presence of gravity in fully horizontal pipes acts as additional axial diffusion from the viewpoint of cross-sectional temperature average (comparison of the two curves from STAR-CCM+ in Figure 26).
- Relative simulation time of the new model for a single 100 m long pipe with strong accumulation is approximately in order of 3.75×10^{-5}.
- A simple quasi-static evaluation of the pressure losses (the friction coefficient is already being evaluated during thermal simulation) would allow the use of the new model for simulation of whole grids with automatic mass flows evaluation in each branch (e.g., OpenModelica connector definition includes Kirchhoff laws by default).

5. Conclusions

This paper deals with transients in circular pipes, where the accumulation of heat into surrounding material has been considered in detail. Model is a combination of the analytic description of a heat transfer between the fluid and the solid wall, simple advection scheme, turbulent diffusion, and a finite element scheme dealing with the circular wall in which the shape functions are based on the steady-state solution. Method of lines is used for time integration. The new model represents an effective numerical solution to the thermal wave propagation problem, which includes advanced evaluation of thermal inertia of the pipe's wall and the axial diffusion that is caused by turbulence. The diffusion coefficient evaluation is mainly proportional to flow velocity.

Overall, it can be said that strong accumulation affects the thermal wave propagation for a stretched period of time in comparison with the non-accumulation case. The effect of the re-accumulation is not only dependent on the pure amount of re-accumulated heat, but also on the particular way this event happens. In other words, to capture the event accurately is to focus on the proper evaluation of the innermost surface temperature of the pipe's wall because it directly affects the rate in which the temperature of the fluid changes. The discretization of both axial and radial direction must go hand in hand to improve the solution. During comparison with the fine models from Fluent and STAR-CCM+, deployment of positive weights in the diffusion coefficient evaluation are always necessary to match the simulated data. This means that turbulent diffusion plays a role in temperature wave propagation. Presence of physical diffusion also reduces unphysical oscillation on sharp interfaces. Gravity is shown to resemble the behavior of an additional axial diffusion as well. The new model resembles the transient behavior of the accumulating pipe wall with highly comparable results to the equivalent, very fine models from Fluent and STAR-CCM+, which should represent the ideal physics-based modeling. The validation with a branched real system needs to be done in the future.

The composite model had run faster in OpenModelica than in Julia possibly because of a non-optimized jacobian evaluation. It might be possible to speed up the Julia execution using function definition macros from Differential Equations package, which could be able to derive symbolic jacobian automatically. The model can be extended by pressure loss evaluation, which would allow simulation of whole DH grids based on Kirchhoff laws. Regularization of mass flows would be necessary here to avoid singularities (division by zero) in the nodes.

The future work should aim at a self-identification algorithm using inlet/outlet temperature signals from real installations to calibrate the appropriate weights in the model. It may be possible to generalize the approach for more complex configurations, such as twin-pipes, buried pipes, and so on. An optimization algorithm can be used to find the matrix coefficient describing the radial heat dynamics. A background from machine learning may be utilized for such an endeavor. The scaling of the new model, when used for simulation of whole grids with automatic mass flow evaluation in each branch based on the pressure drops between individual nodes, is also part of the future work.

Author Contributions: Conceptualization, L.K.; Data curation, L.K. and R.C.; Formal analysis, L.K.; Funding acquisition, J.P.; Investigation, L.K.; Methodology, L.K. and R.C.; Project administration, J.P.; Software, L.K. and R.C.; Supervision, J.P.; Validation, L.K. and R.C.; Visualization, L.K., R.C., and J.P.; Writing-original draft, L.K. and J.P.; Writing-review & editing, L.K.

Funding: This paper has been supported by the project "Computer Simulations for Effective Low-Emission Energy" funded as project No. CZ.02.1.01/0.0/0.0/16_026/0008392 by Operational Programme Research, Development and Education, Priority axis 1: Strengthening capacity for high-quality research.

Conflicts of Interest: The authors declare no conflict of interest.

References

1. Lund, H.; Werner, S.; Wiltshire, R.; Svendsen, S.; Thorsen, J.E.; Hvelplund, F.; Mathiesen, B.V. 4th Generation District Heating (4GDH). Integrating smart thermal grids into future sustainable energy systems. *Energy* **2014**, *68*, 1–11. [CrossRef]
2. Lund, R.; Østergaard, D.S.; Yang, X.; Mathiesen, B.V. Comparison of Low-temperature District Heating Concepts in a Long-Term Energy System Perspective. *Int. J. Sustain. Energy Plan. Manag.* **2017**, *12*, 5–18. [CrossRef]

3. Connolly, D.; Lund, H.; Mathiesen, B.V.; Werner, S.; Möller, B.; Persson, U.; Boermans, T.; Trier, D.; Østergaard, P.A.; Nielsen, S. Heat Roadmap Europe: Combining district heating with heat savings to decarbonise the EU energy system. *Energy Policy* **2014**, *65*, 475–489. [CrossRef]
4. Lund, H.; Østergaard, P.A.; Connolly, D.; Ridjan, I.; Mathiesen, B.V.; Hvelplund, F.; Thellufsen, J.Z.; Sorknæs, P. Energy Storage and Smart Energy Systems. *Int. J. Sustain. Energy Plan. Manag.* **2016**, *11*, 3–14. [CrossRef]
5. Vandermeulen, A.; van der Heijde, B.; Helsen, L. Controlling district heating and cooling networks to unlock flexibility: A review. *Energy* **2018**, *151*, 103–115. [CrossRef]
6. Abraham, J.P.; Sparrow, E.M.; Tong, J.C.K. Heat transfer in all pipe flow regimes: Laminar, transitional/intermittent, and turbulent. *Int. J. Heat Mass Transf.* **2009**, *52*, 557–563. [CrossRef]
7. Churchill, S.W. Friction factor Equations spans all fluid-flow regimes. *Chem. Eng.* **1977**, *84*, 91–92.
8. Wang, J.; Zhou, Z.; Zhao, J. A method for the steady-state thermal simulation of district heating systems and model parameters calibration. *Energy Convers. Manag.* **2016**, *120*, 294–305. [CrossRef]
9. Fang, T.; Lahdelma, R. State estimation of district heating network based on customer measurements. *Appl. Therm. Eng.* **2014**, *73*, 1211–1221. [CrossRef]
10. Benonysson, A. *Dynamic Modelling and Operational Optimization of District Heating Systems*; Lab. of Heating and Air Conditioning, Technical University of Denmark: Lyngby, Denmark, 1991.
11. Zheng, J.; Zhou, Z.; Zhao, J.; Wang, J. Function method for dynamic temperature simulation of district heating network. *Appl. Therm. Eng.* **2017**, *123*, 682–688. [CrossRef]
12. van der Heijde, B.; Fuchs, M.; Ribas Tugores, C.; Schweiger, G.; Sartor, K.; Basciotti, D.; Müller, D.; Nytsch-Geusen, C.; Wetter, M.; Helsen, L. Dynamic Equation-based thermo-hydraulic pipe model for district heating and cooling systems. *Energy Convers. Manag.* **2017**, *151*, 158–169. [CrossRef]
13. Chertkov, M.; Novitsky, N.N. Thermal Transients in District Heating Systems. *Energy* **2018**. [CrossRef]
14. Duquette, J.; Rowe, A.; Wild, P. Thermal performance of a steady state physical pipe model for simulating district heating grids with variable flow. *Appl. Energy* **2016**, *178*, 383–393. [CrossRef]
15. Benonysson, A.; Bohm, B.; Ravn3t, H.F. Operational optimization in a district heating system. *Energy Convers. Manag.* **1995**, *36*, 297–314. [CrossRef]
16. Stevanovic, V.D.; Zivkovic, B.; Prica, S.; Maslovaric, B.; Karamarkovic, V.; Trkulja, V. Prediction of thermal transients in district heating systems. *Energy Convers. Manag.* **2009**, *50*, 2167–2173. [CrossRef]
17. Wang, H.; Meng, H. Improved thermal transient modeling with new 3-order numerical solution for a district heating network with consideration of the pipe wall's thermal inertia. *Energy* **2018**, *160*, 171–183. [CrossRef]
18. Leonard, B.P. *Universal Limiter for Transient Interpolation Modeling of the Advective Transport Equations: The ULTIMATE Conservative Difference Scheme*; NASA Lewis Research Center: Cleveland, OH, USA, 1988.
19. Leonard, B.P. A stable and accurate convective modelling procedure based on quadratic upstream interpolation. *Comput. Methods Appl. Mech. Eng.* **1979**, *19*, 59–98. [CrossRef]
20. Grosswindhager, S.; Voigt, A.; Kozek, M. Linear Finite-Difference Schemes for Energy Transport in District Heating Networks. In Proceedings of the 2nd International Conference on Computer Modelling and Simulation, Mumbai, India, 7–9 January 2011; pp. 5–7.
21. Neumann, L.E.; Šimůnek, J.; Cook, F.J. Implementation of quadratic upstream interpolation schemes for solute transport into HYDRUS-1D. *Environ. Model. Softw.* **2011**, *26*, 1298–1308. [CrossRef]
22. Modelica Association Modelica Language Specification Documentation Release 3.3 Revision 1 (+ Sphinx conversion). Available online: https://media.readthedocs.org/pdf/modelica/numbered/modelica.pdf (accessed on 9 November 2018).
23. Kudela, L. OpenModelicaFromMatlab. Available online: https://github.com/LiborKudela/OpenModelicaFromMatlab (accessed on 9 November 2018).
24. Bezanson, J.; Edelman, A.; Karpinski, S.; Shah, V.B. Julia: A Fresh Approach to Numerical Computing. *SIAM Rev.* **2017**, *59*, 65–98. [CrossRef]
25. Rackauckas, C.; Nie, Q. DifferentialEquations.jl—A Performant and Feature-Rich Ecosystem for Solving Differential Equations in Julia. *J. Open Res. Softw.* **2017**, *5*. [CrossRef]

26. Hindmarsh, A.C.; Brown, P.N.; Grant, K.E.; Lee, S.L.; Serban, R.; Shumaker, D.E.; Woodward, C.S. SUNDIALS: Suite of Nonlinear and Differential/Algebraic Equation Solvers. *ACM Trans. Math. Softw.* **2005**, *31*, 363–396. [CrossRef]
27. Taylor, C. Finite Difference Coefficients Calculator. Available online: http://web.media.mit.edu/~{}crtaylor/calculator.html (accessed on 10 November 2018).

 © 2019 by the authors. Licensee MDPI, Basel, Switzerland. This article is an open access article distributed under the terms and conditions of the Creative Commons Attribution (CC BY) license (http://creativecommons.org/licenses/by/4.0/).

Article

Incorporating the Concept of Flexible Operation in the Design of Solar Collector Fields for Industrial Applications

Guillermo Martínez-Rodríguez, Amanda L. Fuentes-Silva, Juan R. Lizárraga-Morazán and Martín Picón-Núñez *

Department of Chemical Engineering, University of Guanajuato, Guanajuato 36050, Mexico; guimarod@ugto.mx (G.M.-R.); lucerofs@ugto.mx (A.L.F.-S.); lizarragamorazan@hotmail.com (J.R.L.-M.)
* Correspondence: picon@ugto.mx

Received: 30 December 2018; Accepted: 5 February 2019; Published: 12 February 2019

Abstract: This work introduces the concept of flexible operation in the design of solar thermal utility systems for low temperature processes. The design objectives are: (a) The supply of the thermal needs of the process (heat duty and minimum required temperature), and (b) the maximization of the operating time during the day. The approach shows how the network structure is defined by adjusting the mass flow rate and the inlet temperature of the working fluid to achieve the smallest collector surface area. This work emphasizes the need to specify the solar network structure, which is comprised of two main elements: The number of lines in parallel and the number of collectors in series in each line. The former of these two design specifications is related to the heat load that the system will supply, while the latter is directly related to the delivery temperature. A stepwise design approach is demonstrated using two case studies where it is shown that the detailed design of the solar collector network structure is fundamental for a successful thermal integration with minimum investment. In this paper, the design methodology is based on flat-plate solar collectors, but it can be extended to any other type of low temperature solar technology.

Keywords: solar collector network; minimum number of solar collectors; maximum operating time; flexible operation

1. Introduction

The structure and size of a solar collector field is determined by the number of collectors placed in series in a line and by the number of lines placed in parallel. The definition of the structure must take into consideration operational and economical aspects. From the operational point of view, the design must meet the background process heat duty and the required process temperature taking into consideration the driving force for heat transfer. These targets must be met for the longer operation time that the technology and the availability of solar radiation can provide considering the daily and seasonal variability. Additionally, the network surface area must be the smallest possible for the minimum investment. Meeting all these requirements is not a straightforward task since some of the design objectives oppose each other. For instance, maximizing the daylight working hours would mean larger surface areas, while higher process target temperatures can only be achieved at reduced working time in the day. Additionally, meeting the objectives in summer requires less surface area than that needed for operation over the winter.

The situations mentioned above call for a design approach that not only considers the duty targets, but also considers the combination of operating conditions that are most favorable to ensure operation throughout the year, while maximizing the operating hours.

In design, the inlet temperature to the first collector in a network is an important design variable that can be manipulated to achieve the operational objectives and maintain investment at low levels. To this end, this work introduces a design approach of solar collector fields for flexible operation and minimum surface area.

The design of solar collector fields for low temperature applications (up to 100 °C) has been the subject of several research works. One of the first attempts to systematize a design approach was introduced by Oonk et al. [1], who used the collector thermal efficiency curve to approximate the number of collectors in a series required to achieve the process duties. Picón-Núñez et al. [2–4] extended the concept of thermal length and hydraulic length to the design of solar collector networks, which emerge from the thermal and hydraulic considerations in the design of heat transfer equipment. In the case of evacuated-tube technology, Martínez-Rodríguez et al. [5] further developed the methodology extending the concept of maximum number of collectors in series for a given solar radiation.

In the design of heat transfer equipment, there is a design space region where many possible and feasible designs exist [6]. In the case of solar collectors, the same applies and the existence of a design space can be availed for by means of the manipulation of operating variables such as inlet temperature, delivery temperature, and mass flow rate to search for the design that minimizes the solar-collector area and maximizes the operating time considering the variability of the solar radiation. The final selection is the one that has the flexibility to meet the process duties while maximizing the operating time throughout the year. While a considerable amount of work has been done on the integration of solar systems into industrial processes, very few works emphasize the importance of the accurate design of the solar collector fields. Most of the work published on thermal integration has focused on the process side and the delivery of the hot utility. To this end, it is assumed that the solar collector area is fixed with no further description of its structure and its performance. The rationale behind it is that the specification of the total surface area is enough to guarantee the supply of the thermal needs of the process. However, the specific structure of the solar field must be specified, otherwise it risks that the thermal targets will not be met. As has been mentioned, the temperature delivery depends on the number of collectors placed in series in a line, while the thermal duty depends on the number of lines placed in parallel. For instance, Martínez-Rodríguez et al. [7] reported the use of solar thermal heat to run a corn-derivative production plant and determined that a total of 580 collectors were needed and the network structure was specified as: 20 lines, each containing 29 collectors in series. Cases where the collector surface area is fixed without definition to its structure are the works by El-Nashar [8], who reported a solar plant with a total of 1,064 evacuated-tube collectors; the work by Quijera et al. [9] who reported on a case study involving the integration of solar heat into a tuna fish production plant where it was assumed that only 10% of the heat load could be met with a 358.6 m^2 network. In a second work, Qujiera et al. [10] analyzed a dairy plant where a 1,939.2 m^2 solar field with 646 collectors was required to supply 50% of the total heat duty of the process. In the case of a dairy factory in New Zealand, Walmsley et al. [11] reported that part of the hot utility consumption was supplied using a total surface area of 1000 m^2.

Specific studies on the performance of solar collector networks have been published by Tiang et al. [12], who developed an experimentally validated quasi-dynamic simulation model for solar collector fields using TRNSYS. The solar network was composed of parabolic and flat-plate collectors to provide energy for district heating. The thermo-solar plant combined 5960 m^2 flat-plate collectors with 4039 m^2 of parabolic collectors. Ampuño et al. [13] proposed a dynamic model for the control and simulation of a solar network located in Almeria, Spain, to supply the thermal duty of a multi-effect desalination unit. The solar plant consisted of 60 flat-plate collectors with five parallel lines. The first line contains four collectors connected in cascade, while the other four lines contain seven collectors each. Shresthaa et al. [14] carried out the thermo-hydraulic study of two large solar fields in Chemnitz, Germany to provide district heating. The system has been operating since 2016 and is composed of two sets of fields totaling a surface of 2092.99 m^2 with 172 flat-plate collectors. Lauterbach et al. [15] evaluated the thermal performance of a thermo-solar system that supplies heat, hot water, and air conditioning to a 6200 m³/year throughput brewing. The solar field consists of 22 collectors with a surface area of 169 m². The solar network was monitored to get experimental data to be used in the validation of a thermal model. Their findings showed that the efficiency of the system is strongly affected by the geometry and by the operating variables such as mass flow rate and delivery temperature.

There are various industrial sectors that contain process operations whose operating temperatures are within the ranges of the operation of low temperature solar collectors [16]. For instance, the dairy industry where pasteurization takes place at 85 °C; the textile industry with temperatures below 100 °C except the fixing process that takes place between 160 °C and 180 °C [17]. In the pulp and paper industry, most temperatures are below 100 °C except by the bleaching process that operates between 130 °C and 150 °C [18]. Other examples are the beverage and meat industry, with temperatures below 80 °C [19,20], and the leather tanning industry with temperatures of around 45 °C [21] and the greenhouses industry with temperatures around 30 °C [22].

In design, the number of possible solutions to achieve a certain delivery temperature in a set of collectors arranged in series is large, especially, if the inlet temperature of the working fluid is changed. Likewise, there are many combinations of parallel arrangements able to meet the heat load if the working fluid mass flow rate is changed. The main features sought in a solar collector network are:

1. It exhibits the minimum surface area,
2. It provides the largest operating time in the day,
3. It fulfills the heat load and target temperature of the process.

Flexible operation of a solar system refers to the capability of the system to deliver the required thermal duties over the year as the ambient conditions vary. To this end, the operating variable that can be manipulated is the working fluid inlet temperature. Such manipulation can only be achieved by means of heat storage. The ideal scenario is that where the number of collectors in series and the number of lines in parallel are minimized, while the operating time is maximized. The reduction of the number of solar collectors works favorably in as much as the investment is concerned, since it reduces the pay-back time of the project.

This work introduces a stepwise design approach for solar collector networks. It incorporates the assessment of the effect that the design variables have upon the size of a solar installation. It also establishes a design strategy to obtain the network of solar collectors with the smallest surface that provides the longest operation during the day.

2. Thermal Performance

2.1. Assessment of the Design Variables

In design, the working fluid inlet temperature and mass flow rate are important design variables that can be manipulated to achieve the operational objectives and maintain investment at low levels. This section aims at presenting a close analysis of the effect of operating variables and ambient conditions upon the performance of networks of solar collectors and introduces the concept of minimum temperature rise (ΔT_{min}) to determine the maximum number of collectors in series for a given solar radiation intensity.

2.1.1. Effect of Solar Radiation Level

Figure 1 shows the total energy received per unit area in a day with ambient conditions typical of summer, spring, fall, and winter. The geographic coordinates where the measurements were made are: Latitude: 21°01′06″ N; longitude: 101°15′32″ O; and altitude: 2010 m above sea level. Collector tilt angle: 30°. During winter the total amount of energy received is 18.87 MJ/m²; over spring, the amount is 32.22% higher (27.84 MJ/m²). In summer, the increase is 34.90% compared to winter, while the fall is only 22.28% superior. From this information, it follows that the maximum outlet temperatures and the energy absorbed by a network structure will depend on the ambient conditions [23].

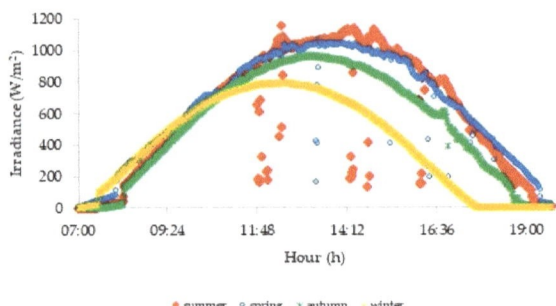

Figure 1. Solar irradiance measured at latitude: 21°01′06″ N; longitude: 101°15′32″ O; and altitude: 2010 m above sea level. Collector tilt angle: 30°.

The thermal efficiency of a solar collector reduces as the inlet temperature to the first collector of the network increases. Therefore, in a set of flat plate collectors in series, the efficiency of the downstream units is reduced as the operating temperature increases. Figure 2 depicts the irradiance and the energy absorbed by a network comprised of 28 collectors in series. From the data shown, only 40% of the total energy received is absorbed. This indicates that there are major challenges in technology development. An additional problem is that not all the energy absorbed is available at the required temperature.

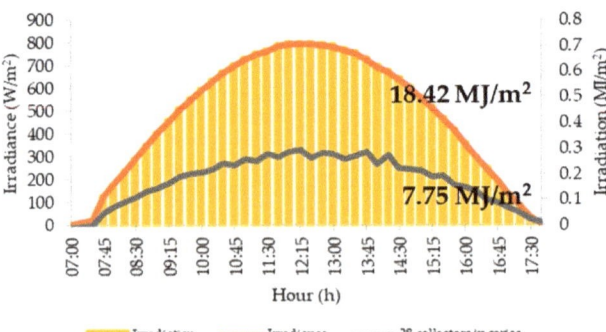

Figure 2. Plot of irradiance and total energy absorbed during a day for a 28-collector network in series.

The highest and lowest solar radiation magnitudes are found in the summer and the winter, respectively. A comparison between these two periods is shown in Figure 3.

Design and performance analysis of solar collector networks are fundamental for understanding the operation of solar collector networks. While design is carried out at fixed operating and ambient conditions, the analysis of the performance of the network must be carried out for a range of operating conditions. There is a minimum solar radiation intensity where the amount of energy received is enough for the fluid to reach a desired target temperature for the installed surface area. For instance, if an outlet temperature of 60 °C is desired, it can only be achieved when the solar radiation is around 450 W/m^2. The time of the day when this condition exists varies between seasons and latitudes, however, for such a radiation level, the target temperature can effectively be attained.

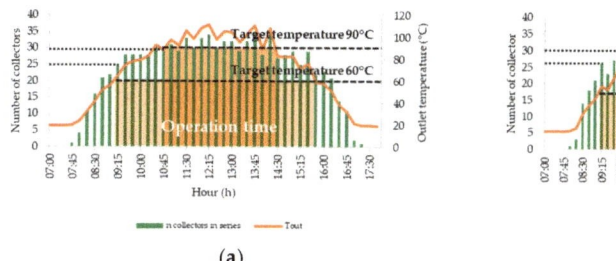 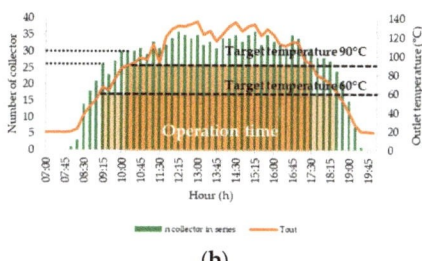

(a)   (b)

Figure 3. Maximum number of collectors in series and maximum practical outlet temperature during the day: (**a**) typical day in winter; (**b**) a typical day in summer.

In solar collector network design, the operating conditions and the amount of solar radiation determine the maximum thermodynamic outlet temperature. However, such value can only be achieved with an infinite surface area and it would not be economical to force a design to reach this value. Therefore, for practical purposes, with the aim of finding the required surface area, or the required number of solar collectors, a design consideration must be formulated. Thus, the term maximum practical attainable temperature is introduced. To describe the concept, consider a design approach where solar collectors are placed in series, as they are needed. The decision to add a new collector in series is made only if the temperature difference between the inlet and outlet of the last collector is greater or equal than a specified value (ΔT_{min}). The smaller this value, the closer to the maximum thermodynamic outlet temperature, but the larger the number of collectors required. In this work, it is assumed that a practical design criterion for this value is $\Delta T_{min} = 1\ °C$ (see Section 2.1.3).

To illustrate the concept described above, Figure 3a shows a case in a typical day in the winter. Starting from 9:15 h, 25 collectors in series can supply a temperature of 60 °C. As the day goes by, the outlet temperature increases and after midday reduces again until 15:45 h when the solar radiation is still enough for the outlet temperature to be 60 °C. After this time, the 25-collector network can no longer supply the target temperature. For a desired temperature of 90 °C, Figure 3a indicates that operation in the winter can start from 10:45 h and 30 collectors are required. The operation can go until 14:30, and after this time, the target temperature can no longer be attained. Figure 3b shows a similar analysis for summer conditions. Figure 4 is a schematic of a 25-collector network in series and Figure 5 shows the outlet temperature profile during the day for winter conditions.

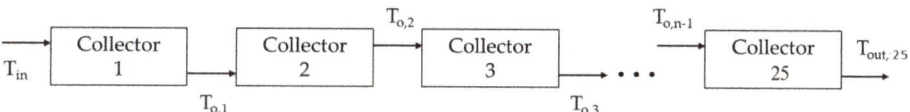

Figure 4. Solar collector network in series.

The type of working fluid is another variable that must be considered in design. For instance, for water, the maximum outlet temperature is limited to 100 °C when the pressure is 1 atm. In this case, not all the energy content of the irradiance curve will be used. Under such situations, it might be advisable to increase the operating pressure or shift to a different working fluid with a higher boiling point.

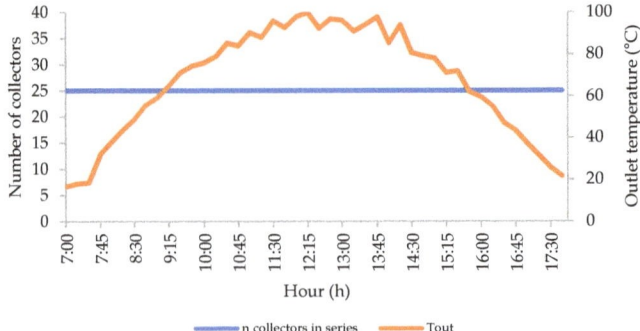

Figure 5. Outlet temperature profile of a 25-collector network during the day.

2.1.2. Effect of Inlet Temperature upon the Delivery Temperature

For a given set of collectors arranged in series, the outlet temperature strongly depends on the working fluid inlet temperature. For the purposes of design, this can be seen from a different perspective: For a fixed outlet temperature, the number of collectors in series is a function of the inlet temperature. Figure 6 shows the temperature profiles of a 28-series collector network with different inlet temperatures. Feed water temperatures are: 19 °C, 40 °C, and 60 °C. The mass flow rate is kept at a constant value of 0.05 kg/s. If the network is to deliver a target temperature of 95 °C, it can be seen that, for an inlet temperature of 19 °C, the operating period where the target temperature is achieved is 4 h and 30 min (270 min), with an average outlet temperature of 94.37 °C. For an inlet temperature of 40 °C, the operating period increases to 5 h (300 min) and the average outlet temperature is 98 °C. If the inlet temperature is increased to 60 °C, the system reaches an average outlet temperature of 99.03 °C, over an operating period of 6 h (360 min). Under this last condition, the system collects 39% more energy compared to the scenario when the inlet temperature corresponds to ambient conditions.

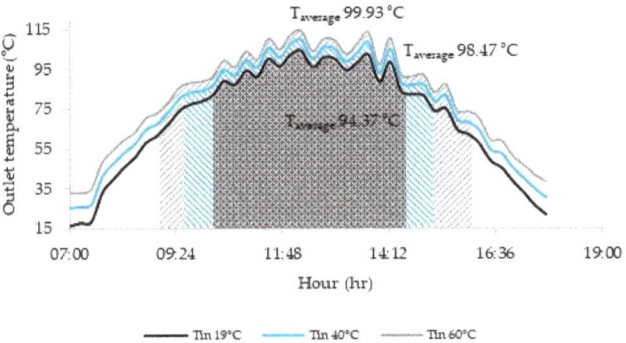

Figure 6. Outlet temperature profiles of a 28-series collector network for different inlet temperatures.

In the operation of a solar collector network, the flow rate per line determines the thermal performance of the set of collectors in series. For a given solar collector network, the modification of the mass flow rate influences the outlet temperature as well as the period the system provides the required heat load. For a scenario where the number of collectors in series is kept constant, increasing the mass flow rate reduces the temperature delivered by the network. Figure 7 shows the temperature profiles for different mass flow rates for a 28-series collector network that operates with an inlet temperature of 60 °C and a flow rate per line of 0.05 kg/s. The time interval over which the network delivers a temperature equal or larger than 95 °C is 6 h. Within this period, the average temperature is 99.03 °C. If the flow rate increases to 0.067 kg/s, the average outlet temperature is 98 °C and the working period reduces to 4 h and 45 min.

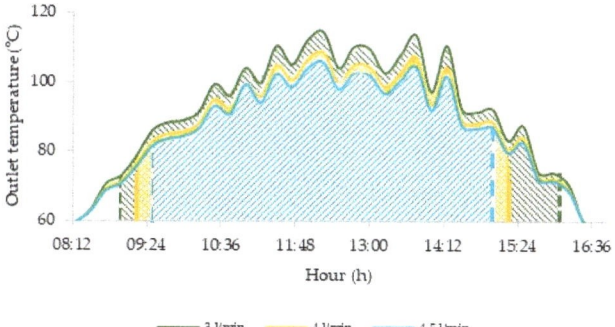

Figure 7. Delivery temperature profiles for different mass flow rates for a 28-series collector network.

Inlet temperature and mass flow rate can be manipulated strategically to reduce the number of solar collectors required to meet the process heat duty. Starting from the design of the network for the most critical ambient conditions, the inlet temperature is first increased. This will cause a rise in the delivery temperature and an increase in the available heat load for the process. Considering that the process heat duty is a fixed value, the flow rate per line can be increased to restore the delivery temperature of the solar plant. The new number of lines in parallel is determined from the heat balance and the flow rate per line. This approach is demonstrated in the case studies below.

2.1.3. Minimum Temperature Increment in a Network of Solar Collectors

The minimum number of solar collectors can be determined considering the ΔT_{min}. This is a design parameter defined as the difference between the outlet temperature $(T_{out})^n$ and inlet temperature $(T_{out})^{n-1}$ in a collector (Equation (1)).

$$\Delta T_{min} = (T_{out})^n - (T_{out})^{n-1} \qquad (1)$$

where n refers to the position of a collectors in a network of collectors in series.

The criterion to determine the maximum number of collectors for a given set of operating conditions is based on the value of ΔT_{min}. The choice of its value is such that the number of collectors remains at a reasonable level with the aim of maintaining the investment at the lowest. Therefore, an analysis is carried out considering a range of solar irradiation levels and the fixed operating conditions shown in Table 1.

Table 1. Operating conditions for the determination of the maximum number of solar collectors.

Variable	Range
Solar radiation, G (W/m²)	500–900
Mass flow rate, m (kg/s)	0.03
Wind velocity, v_v (m/s)	2
Inlet temperature, T_{in} (°C)	20
Ambient temperature, T_{amb} (°C)	17

The thermal model is used to calculate the number of collectors in series. The results are shown in Table 2. From the results in Table 2, it is seen that, for a value of $\Delta T_{min} = 1$, the maximum temperature attainable is 79.70 °C when the solar irradiation is 500 W/m². This temperature is achieved with 23 collectors in series. For a solar radiation of 700 W/m², a maximum temperature of 102.49 °C is reached with 26 collectors. Figure 8 shows the variation of the number of collectors with ΔT_{min} for various solar radiations.

Table 2. Maximum number of solar collectors for different values of ΔT_{min} and different levels of solar radiation.

ΔT_{min}	G = 500 W/m²		G = 700 W/m²		G = 900 W/m²	
	# collectors	T_{out}	# collectors	T_{out}	# collectors	T_{out}
4	5	42.25	9	68.22	12	93.03
3	9	55.35	13	81.12	15	102.46
2	15	69.03	18.00	92.25	19.00	111.60
1	23	79.70	26.00	102.49	27.00	122.12
0.7	28	83.60	29.00	104.79	30.00	124.37
0.5	32	85.75	33.00	107.03	34.00	126.51
0.3	38	87.90	39.00	109.17	39.00	128.21
0.2	43	89.02	43.00	110.06	43.00	129.08
0.1	51	90.07	51.00	111.10	50.00	129.96
0.05	59	90.60	58.00	111.55	56.00	130.35
0.0001	132	91.15	125.00	112.04	118.00	130.82

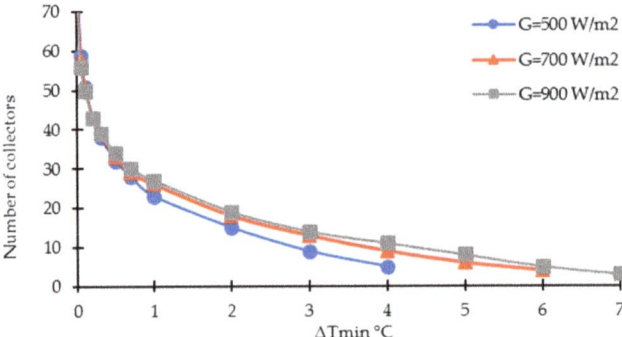

Figure 8. Maximum number of solar collectors with ΔT_{min} for various solar radiation intensities.

In principle, from a thermodynamic point of view, as the number of collectors tends to infinity, there is a limiting temperature that can be attained. For instance, for a solar radiation of 500 W/m² the limiting temperature is 91.15 °C; for a radiation of 700 W/m², the limiting temperature is 112.04 °C and for 900 W/m² is 130.82°C. The specification of a maximum ΔT_{min} in design allows the linking of three design parameters: Maximum delivery temperature, solar radiation, and operating time. For instance, if it is desirable to attain a delivery temperature of 79.70 °C, this can be achieved with 23 collectors in series from the moment of the day when the solar radiation is 500 W/m². Before this time, the temperature cannot be reached. After this time, the temperature delivered by the 23-collectors in series will increase to the point of maximum solar intensity, then, as the day goes by, it will reduce to the point when the solar radiation comes back to 500 W/m². After this point, the outlet temperature will no longer be met. The use of this concept in design is demonstrated in two case studies below.

3. Design of a Network of Solar Collectors

Thermal Model for Flat Plate Solar Collectors

An experimentally validated thermal model is used to design a solar collector network [2,3]. The model was validated according to the international norms for testing assays [24]. The aim of the model is to determine the number of collectors placed in series to achieve a desired target temperature. The methodology to design a collector network in series is to determine the outlet temperature of each unit using the approach described in this section. Starting from the first collector, the outlet temperature is calculated. Then, this outlet temperature becomes the inlet temperature of the next collector. If the difference between the inlet temperature and the outlet temperature is greater than the minimum acceptable value (Section 2.1.3), the collector is added to the network. The process continues until the difference is smaller than the minimum acceptable value. A counter is used to find the number of collectors and the outlet temperature for a given solar radiation. The main assumptions of the model are:

- Steady state conditions
- The heat flux is one-dimensional
- The flow distribution inside the tubes is identical
- The effect of dust on the optical properties of the transparent cover is negligible
- Constant physical properties
- Heat losses to ambient are considered
- The temperatures of the plate and the cover are uniform at any moment in time

To determine the outlet temperature of a single flat plate collector, various energy balance equations must be written: (a) The overall energy balance, (b) the useful heat, and (c) the heat lost to the ambient from the upper side of the collector. Unknown variables are: The temperature of the surface (T_s), temperature of the cover (T_c), and outlet temperature (T_{out}). Therefore, the three heat balance equations must be solved simultaneously.

The overall heat balance is:

$$mC_p(T_{out} - T_{in}) = G\tau\alpha \, A_s - U_L(T_s - T_{amb})A_s \qquad (2)$$

The useful heat is equal to the total energy received by the collector minus the energy lost to the surrounding and is expressed as:

$$mC_p(T_{out} - T_{in}) = h_{cw}A_s LMTD \qquad (3)$$

where:

$$LMTD = \frac{T_{in} - T_{out}}{\ln\left(\frac{T_s - T_{out}}{T_s - T_{in}}\right)} \qquad (4)$$

The heat lost to ambient from the upper section of the collector can be expressed as:

$$h_{csc+rsc}(T_s - T_c) = U_L(T_s - T_{amb}) \qquad (5)$$

For the solution of the system of equations, heat transfer coefficients must be determined. The specific terms are: h_{cw}, heat transfer coefficient between the water and the tubes; h_{csc} the convective heat transfer coefficient between the surface and the cover; h_{rsc}, is the heat transfer coefficient by radiation from the surface to the cover; h_{cca}, is the convective heat transfer coefficient between the cover and ambient; and h_{rca}, is the heat transfer coefficient by radiation from the cover to ambient. Expressions to determine these parameters are taken from Reference [3]. For the calculation of the convective heat transfer coefficient between the cover and ambient, a wind velocity of 2.5 m/s was used.

The overall heat transfer coefficient of losses is obtained from:

$$U_L = \frac{1}{\sum R} = \left(\frac{1}{h_{csc} + h_{rsc}} + \frac{1}{h_{cca} + h_{rca}}\right)^{-1} \qquad (6)$$

The collectors are orientated toward the south with a tilt angle of 30 degrees. The solution of the system starts by assuming a surface temperature (T_s) and a cover temperature (T_c). Then, the Prandtl

(Pr), Rayleigh number (Ra), and Nusselt number (Nu) between parallel plates using the Hollands equation [25] are calculated and the heat transfer coefficients are calculated: h_{csc}, h_{cca}, and h_{rsc} y h_{rca}.

The heat transfer coefficient h_{csc} is calculated from Nusselt number correlations.

$$Nu = \frac{h_{csc} L_e}{k} \qquad (7)$$

where L_e is the spacing between the plate and the transparent cover, and k is the thermal conductivity of the fluid. In the tube, the appropriate Nusselt number is determined according to the flow regime: Laminar, turbulent, or transitional. General correlations of Nusselt number have the form:

$$Nu = b Re^x Pr^y \qquad (8)$$

where x and y are constants; Nu is the Nusselt number; Re is the Reynolds number, and Pr is the Prandtl number.

$$Re = \frac{\rho v d}{\mu} \qquad (9)$$

$$Pr = \frac{C_p \mu}{k} \qquad (10)$$

Equations (2), (3) and (5) can be solved simultaneously to give T_{out}. The while model has been implemented in a Matlab platform using a Newton-Raphson numerical method. Figure 9 shows the solution algorithm in a block diagram. The physical properties of water and optical properties of the glass cover used in the calculations are presented in Tables 3 and 4 respectively.

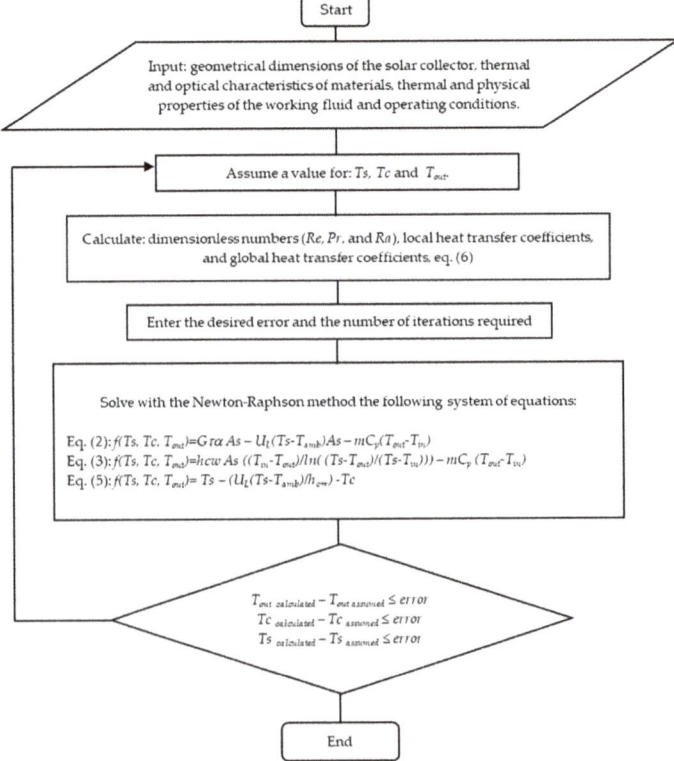

Figure 9. Algorithm to determine the outlet temperature of a flat-plate collector.

Table 3. Optical properties of the glass cover.

Description	Data
Transmittance of glass cover	0.92
Emissivity of the glass cover	0.88
Absorbance of the selective surface	0.96
Emissivity of the selective surface	0.05

Table 4. Physical properties of water.

Description	Data
Density	998.49 kg/m^3
Specific heat	4.182 kJ/kg K
Dynamic viscosity	0.001028 kg/m s
Thermal conductivity	0.58 W/m K

4. Case Studies

To exemplify the concept of flexible design of solar collectors for industrial applications, two case studies are analyzed. In the first case, a canned fish production factory is studied, where hot water is required at a temperature of 95 °C, while the second case refers to a dairy plant with temperature requirements below 100°C. The flat plate collectors used for the designs exhibit the geometrical features shown in Table 5.

Table 5. Geometrical features of the flat plate collector.

Total area (length × width) (1.995 × 0.995)	1.985 m^2
Number of tubes (copper)	8
Tube length	1.970 m
Tube outer diameter	0.0054 m
Header outer diameter	0.0210 m
Distance between tubes (from center to center)	0.1125 m
Fin dimensions (aluminum with a selective layer)	0.1125 m
Thickness of glass cover (tempered)	0.0045 m
Spacing between transparent cover and absorbing plate	0.025 m
Side insulation (glass fiber)	0.0125 m
Bottom insulation (glass fiber)	0.055 m

4.1. Canned Fish Plant

The flow diagram of the process is shown in Figure 10 [9]. The plant processes 50 t/day of raw fish to produce 48.5 t/day of canned product. The factory operates five days a week over winter and six days a week over summer with a total of 250 days a year, from 06:00 to 24:00 h. Three batches a day processed and hot water at a temperature of 95 °C is required.

The first operation stage begins at 06:00 and finishes at 11:00 h. It consists of the fish preparation. In this stage tap water is used for washing. Subsequently, the fish sections are treated in two pre-cookers that come into operation twice a day to process 9281 kg/batch. This stage takes place at a temperature of 102 °C for 180 min. Heating is provided by saturated steam at 200 °C. Next, cold water at 4 °C is employed to wash the tuna fish pieces for 30 min. When the temperature drops to 50 °C, the fish is taken out of the pre-cookers and is hand-cleaned. The different fish pieces are classified by size and weight to be canned. Cans are filled up with olive oil, salt, and other additives at 50 °C. Then are sealed and washed before being sent to the sterilization stage. This stage lasts 90 min and takes place at 118 °C. Three autoclaves operate in parallel three times a day.

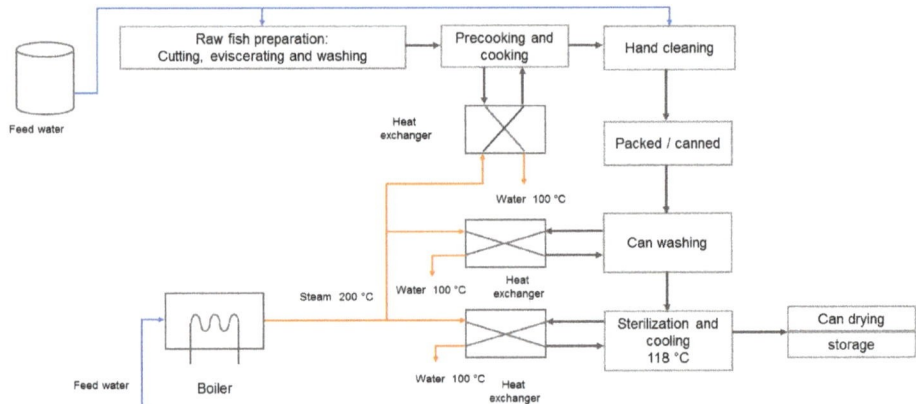

Figure 10. Simplified flow diagram of the canned-fish process. Adjusted from Reference [9].

The first step in the sizing of the solar collector network consists in the determination of the maximum number of collectors required for operation. Winter is the most critical season to reach the specified total heat load; therefore, the calculations are performed for the ambient conditions that prevail over this time.

4.1.1. Design of the Solar Collector Network

The total collector count in a network structure can be refined to maximize the operation and reduce the surface area in two different scenarios. In the first scenario, the number of collectors in series required to meet the target process temperature is recalculated considering different inlet temperatures. The second scenario is analyzed considering the increased inlet temperature and the manipulation of the mass flow rate per line. Such combination results in a reduction of the lines in parallel to restore the heat load. These scenarios are analyzed below.

Design for Increased Inlet Temperature

For a total process heat load of 5.91 GJ, an inlet temperature of 19 °C and a mass flow rate per line of 0.05 kg/s, 29 collectors in series are needed to reach a delivery temperature of 95 °C starting from 11:00 h in winter (Figure 11a). The heat load is met with 31 lines in parallel giving a total of 899 collectors. Figure 8a also shows the operating time where the network delivers a temperature equal to or above 95 °C. The number of collectors in series to reach 95 °C can be reduced if the inlet temperature is increased. Important to note, that under new inlet temperature conditions, the working fluid must be fed to the network at the time where the target temperature is reached. For instance, for an inlet temperature of 40 °C (Figure 11b), 23 collectors are required, and the water can be fed to the system at 11:00 h. Additionally, it is important to observe that operation time is maintained. With less collectors in series, the total number of collectors reduces to 713. With a further increase of the inlet temperature to 60 °C, Figure 11c shows that the system can still be fed at 11:00 h; in this case, only 18 collectors in series are required, thus reducing the total collector count to 558, with a 31 × 18 collector structure. This represents a reduction of 37% in the total surface area.

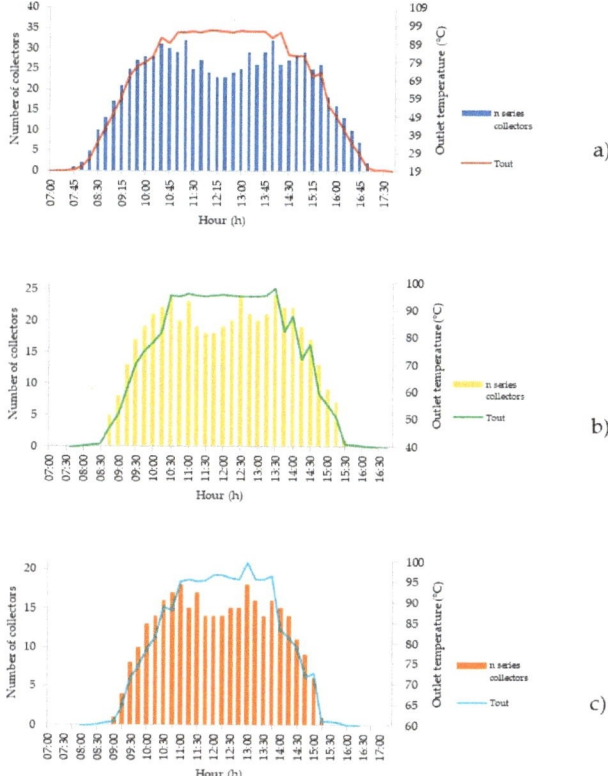

Figure 11. Solar collector design plot for a mass flow rate of 0.05 kg/s and (**a**) inlet temperature of 19 °C; (**b**) 40 °C; and (**c**) 60 °C. The projection of the bars on to the right-hand and the left-hand side axes indicate the maximum outlet temperature achieved at that time of the day and the maximum number of collectors to reach that temperature.

Design for Increased Mass Flow Rate

An alternative strategy for the reduction of the number of collectors can be achieved by means of the manipulation of the mass flow rate per line. Table 6 shows the performance of a 23 series collector network for different mass flow rates per line and an inlet temperature of 60 °C. An increase of flow rate to 0.067 kg/s allows the reduction of the number of parallel lines to 23, giving a total number of collectors of 529 in an arrangement of 23 parallel lines with 23 collectors each. The largest flow rate that ensures a delivery temperature of 95 °C is 0.075 kg/s. The total number of collectors reduces to 483 with 21 lines in parallel and 23 collectors per line. With respect to the original design, the overall reduction is 46%.

Table 6. Redesign of the solar network for 23 collectors in series and different flow rates.

Flow Rate (kg/s)	Start (h)	End (h)	Operating Time (min)	Average T (°C)	No. of Parallel Lines	Total No. of Collectors
0.05	11:00	14:15	195	103.18	29	667
0.067	11:00	14:15	195	97.85	23	529
0.075	11:00	14:15	195	95.72	21	483

Following the procedure described above, refined solar networks for the four seasons considering their corresponding average solar irradiance are presented in Table 7. In all cases, the inlet temperature and the mass flow rate per line are 60 °C and 0.075 kg/s. The number of collectors per line is kept at 23.

Table 7. Solar collector network design for the different seasons.

Season	Start (h)	End (h)	Operating Time (min)	Average T (°C)	No. of Parallel Lines	Total No. of Collectors
Spring	10:30	15:00	270	100.95	14	345
Summer	11:15	17:00	345	107.90	11	253
Fall	11:15	15:15	240	105.04	16	368
Winter	11:00	14:15	195	95.72	21	483

The largest solar field is required during the winter with a structure of 21 parallel lines with 23 collectors per line (21 × 23). The operation of this network during the year is shown in Table 8. The results indicate that for most part of the year the network delivers and excess of heat, except for winter where it meets the specified heat load.

Table 8. Heat duty and delivery temperature for a network designed for winter conditions and operated in a different season. Solar network operation with 0.075 kg/s and a feed temperature of 60 °C.

Season	Start (h)	End (h)	Operating Time (min)	Average T (°C)	Heat Duty Excess (%)
Spring	10:30	15:00	270	100.95	47.94
Summer	11:15	17:00	345	107.90	105.07
Fall	11:15	15:15	240	105.04	38.08
Winter	11:00	14:15	195	95.72	0

4.2. Dairy Process

This plant produces yoghurt, cheese, and non-fermented milk drinks from caw milk as raw material [10]. From the 20,000 L a day pasteurized milk, 15% is sent to the production of yoghurt, 80% to produce cheese, and the other 5% to produce milk drinks. The plant operates seven days a week and 360 days a year. The whole process consists of four stages: Pretreatment and milk pasteurization, production of yoghurt, non-fermented milk drinks, and cheese production. Figure 12 shows a simplified block diagram of the production process.

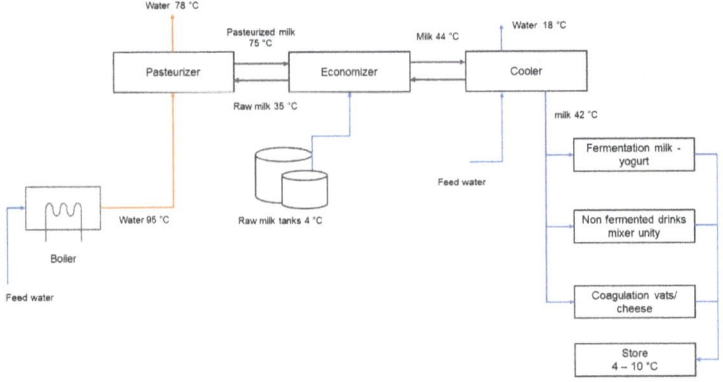

Figure 12. Simplified flow diagram of the dairy plant. Adjusted from Reference [10].

Operation takes 12 h a day, beginning at 06:00 h and finishing at 18:00. For milk pasteurization a boiler operating with natural gas produces hot water at 95 °C. The thermal efficiency of the boiler is assumed to be 92%. To provide a heat load of 880.20 kW, 45.70 m^3 of water at a temperature of 95 °C are needed every day. The total energy consumption in this period is 15.85 GJ.

For this plant, the whole thermal duty can be fully satisfied using a low temperature solar collector network. The pasteurization process takes place at 85 °C.

Design of the Solar Collector Network

The network of solar collectors was designed considering winter conditions. The daily heat duty is 4401.01 kWh (15.85 GJ). A 2233-collector network was designed with 77 parallel lines and 29 collectors per line (77 × 29), operating with a water feed temperature of 19°C and a flow rate of 0.05 kg/s. The network delivers the required temperature from 11:00 h until 14:15 h (195 min). The average delivery temperature is 98.04°C. As in the case study above, the total number of collectors can be reduced through the increase of the inlet temperature. Thus, for an inlet temperature of 60 °C, 23 collectors in series are required giving a collector count of 1771 (77 × 23), resulting in a reduction of 20% in collector area.

To explore other design options, a flow rate of 0.067 kg/s and a feed temperature of 60 °C are analyzed. The resulting network contains 1426 collectors in an arrangement of 62 lines in parallel and 23 collectors per line (62 × 23). Further increase of the flow rate to 0.075 kg/s with the same 60 °C inlet temperature results in a network with 1311 collectors containing 57 parallel lines and 23 collectors in series per line (57 × 23). With respect to the initial design, the total surface area is reduced in 41%. The results are summarized in Table 9.

Table 9. Redesign of the solar network for 23 collectors in series and different flow rates.

Flow Rate (kg/s)	Start (h)	End (h)	Operating Time (min)	Average T (°C)	No. of Parallel Lines	Total No. of Collectors
0.05	11:00	14:15	195	103.18	77	1771
0.067	11:00	14:15	195	97.85	62	1426
0.075	11:00	14:15	195	95.72	57	1311

Designs for the other seasons are shown in Table 10. Inlet temperature is maintained at 60°C, the flow rate at 0.075 kg/s and 23 collectors in series are used in all cases.

Table 10. Solar collector network design for the different seasons.

Season	Start (h)	End (h)	Operating Time (min)	Average T (°C)	No. of Parallel Lines	Total No. of Collectors
Spring	10:30	15:00	270	100.95	39	897
Summer	11:15	'17:00	345	107.90	28	644
Fall	11:15	15:15	240	105.04	41	943
Winter	11:00	14:15	195	95.72	57	1311

The operation of a network designed under the operating condition that prevail in winter will deliver a higher heat load in the other months of the year. The performance of a 57 × 23 network in other seasons in terms of operating time, delivery temperature, and heat load are shown in Table 11.

Table 11. Heat duty and delivery temperature for a network designed in winter and operated in a different season. Operation with 0.075 kg/s and an inlet temperature of 60 °C.

Season	Start (h)	End (h)	Operating Time (min)	Average T (°C)	Heat Duty Excess (%)
Spring	10:30	15:00	270	100.95	49.73
Summer	11:15	17:00	345	107.90	107.54
Fall	11:15	15:15	240	105.04	39.74
Winter	11:00	14:15	195	95.72	0

5. Analysis of Results

There is a design space where different solar networks that can supply the heat duty of a process coexist. Out of the many design options, there are few that minimize the investment. A design space is determined by: (a) Process conditions such as the specified target temperature; (b) ambient conditions such as the solar irradiation, ambient temperature, and wind velocity; and (c) solar plant operating conditions such as the fluid inlet temperature and the mass flow rate. The maximum delivery temperature of a solar plant is a function of the number of collectors placed in series. For a solar plant to exhibit a flexible operation, its design must be able to supply the process thermal needs throughout the year and at the same time, maximize the operating time. Fluid inlet temperature and fluid mass flow rate are two design variables that can be manipulated to achieve the design objectives. Increased inlet temperatures tend to reduce the number of solar collectors. Mass flow rate is an operating parameter that can be used to control the outlet temperature in periods of higher solar radiation. For a flexible operation, it is assumed that the manipulation of inlet temperature and mass flow rate is possible using a heat storage system (not discussed in this work). The results obtained in the case studies are further analyzed below.

5.1. Case Study 1

In the first case study, the solar plant was considered to provide a fraction of 0.115 of the total process heat duty in the winter with a hot temperature of 95 °C. With a water flow rate of 0.05 kg/s and an inlet temperature of 19 °C, 899 collectors arranged in 31 lines of 29 collectors per line were needed in the first approach. If the temperature of the feed is increased to 60 °C, the number of collectors in series to attain a target temperature of 95 °C is reduced to 18 with 31 lines in parallel resulting in a total number of collectors of 558. Over the summer, with 0.05 kg/s and a feed temperature of 60 °C, the outlet temperature increases to 103.18 °C. For the purposes of avoiding phase change in the summer, the water flow rate can be raised to 0.075 kg/s. With this flow rate, the system requires more units in series. Therefore, 23 collectors in series are analyzed. Under these conditions, the number of lines in parallel reduces from 31 to 21, giving a total of 483 collectors keeping the operating time in 195 min.

5.1.1. Flexible Operation of the Solar Network

The solar network with a structure of 21 × 23 supplies a daily heat load of 5.91 GJ. For an inlet temperature of 60 °C and a flow rate of 0.075 kg/s, the average delivery temperatures and excess heat load with respect to winter in the different seasons are:

- $T_{average,\ winter}$ = 95.72 °C
- $T_{average,\ fall}$ = 105.04 °C; excess heat load: 38.08 %
- $T_{average,\ spring}$ = 100.95 °C, excess heat load: 47.94 %
- $T_{average,\ summer}$ = 107.90 °C, excess heat load: 105.07 %

5.1.2. Solar Network

Figure 13 shows the network structure (21 × 23) with the process thermal requirements (heat load and temperature). This structure exhibits the minimum number of collectors and maximizes the operating time.

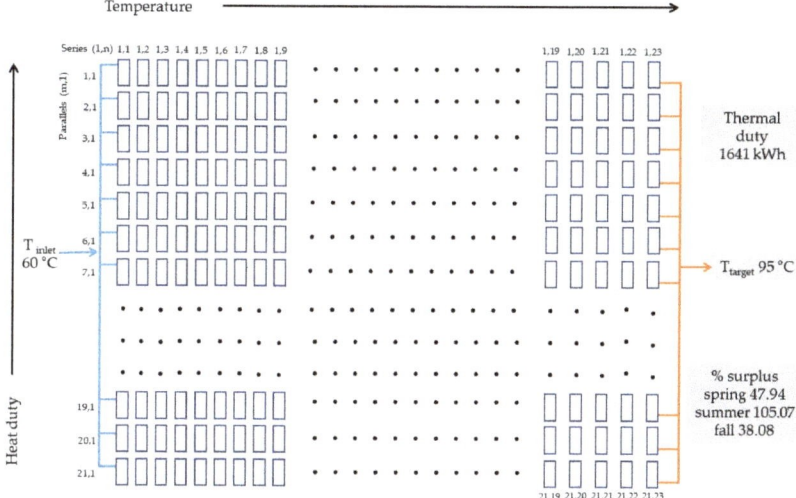

Figure 13. Solar network structure for the canned-fish process. Case study 1.

5.2. Case Study 2

The target temperature of case study 2 is 95 °C, and the number of collectors in series per line required to achieve this target is 29. The total heat load is 15,843.63 MJ. In this case, the whole heat duty can be supplied by the solar plant with 0.05 kg/s and an inlet temperature of 19 °C. The design approach reveals that for these conditions, 77 lines in parallel each with 29 collectors in series are required. If the temperature of the feed is increased to 60 °C, the number of collectors in series to attain a target temperature of 95 °C is 23. Thus, with 77 lines in parallel, the total number of collectors reduces to 1771. In the summer, the outlet temperature increases to 113.54 °C. For the purposes of avoiding phase change, the water flow rate can be raised to 0.075 kg/s. Under these conditions, the number of lines in parallel reduces from 77 to 57, giving a total of 1311 collectors.

5.2.1. Flexible Operation of the Solar Network

The network structure of 57 × 23 solar collectors supplies a daily heat load of 15.85 GJ. For an inlet temperature of 60 °C and a flow rate of 0.075 kg/s, the average delivery temperatures and excess heat load with respect to winter in the different seasons are:

- $T_{average,\ winter}$ = 95.72 °C
- $T_{average,\ fall}$ = 105.04 °C, excess heat load: 39.74 %
- $T_{average,\ spring}$ = 100.95 °C, excess heat load: 49.73 %
- $T_{average,\ summer}$ = 107.90 °C, excess heat load: 107.54 %

5.2.2. Solar network

Figure 14 shows the network structure (57 × 23) with the process thermal requirements (heat load and temperature). This structure exhibits the minimum number of collectors and maximizing the operating time.

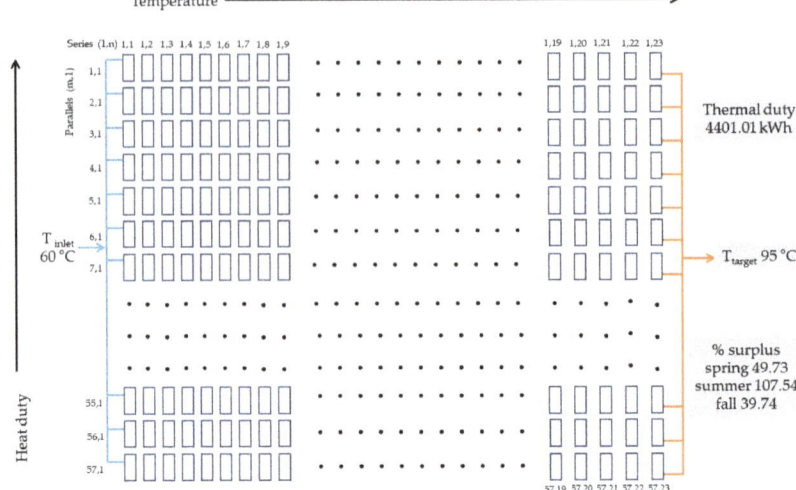

Figure 14. Solar network structure for the dairy factory. Case study 2.

6. Conclusions

A solar collector network is a thermal structure that absorbs solar energy, transforms it into heat, and delivers it to a user. When the user is an industrial process, the operation of the solar plant must be capable of supplying the process thermal needs under the most critical ambient conditions. The thermal design under these conditions fixes the maximum working hours wherein the process thermal requirements are achieved. For operation under more favorable atmospheric conditions, the excess heat collected can be taken advantage of to increase the plant working time. Heat storage can be used to provide a flexible operation. The features of a flexible solar collector network are:

- It is designed for the less favorable ambient conditions to meet the process thermal needs
- It contains a heat storage system to aid in the conditioning of the inlet temperature
- It allows to maximize the working hours by regulating the inlet temperature

A complete approach to the integration of solar heat into process industries must involve, on the one hand, the detailed design and analysis of the operation of the solar collector plant, and on the other, the specification of the way heat is transmitted to the process. This work covers the design for flexible operation of a solar collector plant.

The design of a thermal plant must specify the total number of solar collectors and its arrangement. The target temperature is achieved by placing collectors in series, while the heat load is achieved by adding lines in parallel. The manipulation of the inlet temperature and the mass flow rate are fundamental to reduce the size of the solar plant. It was shown that significant reduction in collector surface area can be achieved by proper design and this has a direct impact on plant investment.

Author Contributions: Conceptualization, G.M.-R., J.R.L.-M. and M.P.-Z.; methodology, G.M.-R. and J.R.L.-M.; software, J.R.L.-M. and A.L.F.-S.; validation, G.M.-R. and J.R.L.-M.; formal analysis, M.P.-Z. and G.M.-R.; investigation, G.M.-R., J.R.L.-M., M.P.-Z. and A.L.F.-S.; resources, G.M.-R.; data curation, G.M.-R. and A.L.F.-S.; writing-original draft preparation, G.M.-R., J.R.L.-M. and M.P.-Z.; writing-review and editing, A.L.F.-S. and M.P.-Z.; visualization, A.L.F.-S.; supervision, G.M.-R.; project administration, G.M.-R.; funding acquisition, G.M.-R. and M.P.-Z.

Funding: This research received no external funding.

Conflicts of Interest: The authors declare no conflict of interest.

Nomenclature

A_S	Heat transfer surface area (m^2)
C_p	Heat capacity, J/kg °C
d	Tube inner diameter, m
G	Solar irradiance, W/m^2
h_{csc}	Coefficient of heat transfer by convection from surface to cover, W/(m^2 K)
h_{rsc}	Radiative coefficient between the surface and the transparent cover, W/(m^2 K)
h_{cca}	Heat transfer coefficient between the cover and ambient, W/(m^2 K)
h_{rca}	Radiative coefficient between the cover and ambient, W/(m^2 K)
h_{cw}	Heat transfer coefficient between the water and the tubes, W/(m^2 K)
k	Thermal conductivity of the fluid, W/(m^2 K)
LMTD	Logarithmic mean temperature difference, °C
L_e	Spacing between the plate and the cover, m
$\dot{m}$	Mass flow rate, kg/s
Nu	Nusselt number
Pr	Prandtl number
R	Thermal resistance
Re	Reynolds number
T_{amb}	Ambient temperature, °C
$T_{average}$	Fluid average temperature, °C
T_c	Cover temperature, °C
T_{in}	Inlet temperature to the collector, °C
ΔT_{min}	Temperature difference between the outlet and inlet temperature of the last collector, °C
T_{out}	Outlet temperature from tube, °C
T_s	Surface temperature, °C
U_L	Overall heat transfer coefficient of losses to ambient, W/(m^2 °C)
v	Fluid velocity, m/s

Greek letters

μ	Fluid viscosity, kg/(m s)
ρ	Fluid density, kg/m^3
$\tau\alpha$	normal transmittance-absorptance product

References

1. Oonk, R.L.; Jones, D.E.; Cole-Appel, B.E. Calculation of performance of N collectors in series from test data on a single collector. *Sol. Energy* **1979**, *23*, 535–536. [CrossRef]
2. Picón-Núñez, M.; Martínez-Rodríguez, G.; Fuentes-Silva, A.L. Thermo-hydraulic Design of Solar Collector Networks for Industrial Applications. *Chem. Eng. Trans.* **2013**, *35*, 457–462. [CrossRef]
3. Picón-Núñez, M.; Martínez-Rodríguez, G.; Fuentes-Silva, A.L. Design of solar Collector Networks for Industrial Applications. *Appl. Therm. Eng.* **2014**, *70*, 1238–1245. [CrossRef]
4. Picón-Núñez, M.; Martínez-Rodríguez, G.; Fuentes-Silva, A.L. Targeting and design of evacuated-tube Solar Collector Networks. *Chem. Eng. Trans.* **2016**, *52*, 859–864. [CrossRef]
5. Martínez-Rodríguez, G.; Fuentes-Silva, A.L.; Picón-Núñez, M. Solar Thermal Networks operating with Evacuated-tube Collectors. *Energy* **2018**, *146*, 26–33. [CrossRef]
6. Picón-Núñez, M.; Polley, G.T.; Riesco-Ávila, J.M. Design Space for the Sizing and Selection of Heat Exchangers of the Compact Type. *Chem. Eng. Trans.* **2012**, *29*, 217–222. [CrossRef]
7. Martínez-Rodríguez, G.; Fuentes-Silva, A.L.; Picón-Núñez, M. Targeting the maximum outlet temperature of solar collectors. *Chem. Eng. Trans.* **2018**, *70*, 1567–1572. [CrossRef]
8. El-Nashar, A.M. Seasonal effect of dust deposition on a field of evacuated tube collectors on the performance of a solar desalination plant. *Desalination* **2009**, *239*, 66–81. [CrossRef]
9. Quijera, J.A.; González-Alriols, M.; Labidi, J. Integration of a solar thermal system in canned fish factory. *Appl. Therm. Eng.* **2014**, *70*, 1062–1072. [CrossRef]

10. Quijera, J.A.; González-Alriols, M.; Labidi, J. Integration of a solar thermal system in a dairy process. *Renew. Energy* **2011**, *36*, 1843–1853. [CrossRef]
11. Walmsley, M.R.W.; Walmsley, T.G.; Atkins, M.J.; Neale, J.R. Options for Solar Thermal and Heat Recovery Loop Hybrid System Design. *Chem. Eng. Trans.* **2014**, *39*, 361–366. [CrossRef]
12. Tiang, Z.; Perers, B.; Furbo, S.; Fan, J. Analysis and validation of a quasi-dynamic model for a solar collector field with flat plate collectors and parabolic trough collectors in series for district heating. *Energy* **2018**, *142*, 130–138. [CrossRef]
13. Ampuño, G.; Roca, L.; Berenguel, M.; Gil, J.D.; Pérez, M.; Normey-Ricod, J.E. Modeling and simulation of a solar field based on flat-plate collectors. *Sol. Energy* **2018**, *170*, 369–378. [CrossRef]
14. Shresthaa, N.L.; Frotscher, O.; Urbaneck, T.; Oppelt, T.; Göschel, T.; Uhlig, U.; Frey, H. Thermal and hydraulic investigation of large-scale solar collector field. *Energy Procedia* **2018**, *149*, 605–614. [CrossRef]
15. Lauterbach, C.; Schmitt, B.; Vajen, K. System analysis of a low-temperature solar process heat system. *Sol. Energy* **2014**, *101*, 117–130. [CrossRef]
16. Kalogirou, S. The potential of solar industrial process heat applications. *Appl. Energy* **2003**, *76*, 337–361. [CrossRef]
17. Jia, H.; Cheng, X.; Zhu, J.; Li, Z.; Guo, J. Mathematical and experimental analysis on solar thermal energy harvesting performance of the textile-based solar thermal energy collector. *Renew. Energy* **2018**, *129*, 553–560. [CrossRef]
18. Sharma, A.K.; Sharma, C.; Mullick, S.C.; Kandpal, T.C. Carbon mitigation potential of solar industrial process heating: Paper industry in India. *J. Clean. Prod.* **2016**, *112*, 1683–1691. [CrossRef]
19. Pietruschka, D.; Fedrizzi, R.; Orioli, F.; Söll, R.; Staus, R. Demonstration of three large scale solar process heat applications with different solar thermal collector technologies. *Energy Procedia* **2012**, *30*, 755–764. [CrossRef]
20. Mekhilef, S.; Saidur, R.; Safari, A. A review on solar energy use in industries. *Renew. Sustain. Energy Rev.* **2011**, *15*, 1777–1790. [CrossRef]
21. Farjana, S.H.; Huda, N.; Parvez Mahmud, M.A.; Saidur, R. Solar Process heat in industrial systems—A global review. *Renew. Sustain. Energy Rev.* **2017**, *82*, 2270–2286. [CrossRef]
22. Norton, B. Industrial and agricultural applications of solar heat. *Compr. Renew. Energy* **2012**, *3*, 567–594. [CrossRef]
23. Private Database. Solar Collectors Testing Laboratory from University of Guanajuato. Unpublished work. 2018.
24. Martínez-Rodríguez, G.; Fuentes-Silva, A.L.; Picón-Núñez, M. A parameter design tool for solar collectors. In *Solar Collectors, Applications and Performance*, 1st ed.; Picón-Núñez, M., Ed.; Nova Science Publishers Inc.: New York, NY, USA, 2018; pp. 51–84.
25. Hollands, K.G.T.; Shewen, E.C. Optimization of flow passage geometry for air heating plate-type solar collectors. *ASME J. Sol. Energy Eng.* **1981**, *103*, 323–330. [CrossRef]

© 2019 by the authors. Licensee MDPI, Basel, Switzerland. This article is an open access article distributed under the terms and conditions of the Creative Commons Attribution (CC BY) license (http://creativecommons.org/licenses/by/4.0/).

Article

Analysis of Fired Equipment within the Framework of Low-Cost Modelling Systems

Dominika Fialová * and Zdeněk Jegla

Institute of Process Engineering, Faculty of Mechanical Engineering, Brno University of Technology, Technická 2, 616 69 Brno, Czech Republic; jegla@fme.vutbr.cz
* Correspondence: dominika.fialova@vutbr.cz; Tel.: +42-054-114-4917

Received: 31 December 2018; Accepted: 5 February 2019; Published: 7 February 2019

Abstract: Fired equipment suffers from local overloading and fouling of heat transfer surfaces, products are not of the required quality, and operating costs are increased due to the high pressure drop of process fluids. Such operational issues are affected by the non-uniform distribution of fluid flow and heat flux variability. Detailed numerical analyses are often applied to troubleshoot these problems. However, is this common practice effective? Is it not better to prevent problems from occurring by using quality equipment design? It is, according to the general consensus. Still, the experience of designing fired apparatuses reveals that the established standards do not reflect the real maldistribution sufficiently. In addition, as found from the given overview of modelling approaches, the radiant chamber and the convection section are usually analysed separately without significant continuity. A comprehensive framework is hence introduced. The proposed procedure clearly defines the interconnection of traditional thermal-hydraulic calculations and low-cost modelling systems for radiant and convection sections. A suitable combination of simplified methods allows for the reliable design of complex equipment and fast identification of problematic areas. The utilisation of selected low-cost models, i.e., the second phase of the systematic framework, is presented regarding the example of a steam boiler.

Keywords: simplified methods; design procedure; convection section; radiant section; flow distribution; heat flux distribution; boiler

1. Introduction

Hot utilities are the most energy-consuming and therefore the costlier apparatuses in any industrial plant. Tubular fired heaters (cylindrical and cabin type) are distinctive components of petrochemical plants, continuous furnaces are used in metal processing, and boilers produce superheated steam for process and power purposes. Recently, waste-to-energy applications have become increasingly important and the waste incinerator furnace is their key component. All these processes and power apparatuses share two main features, i.e., the arrangement and the insufficient reliability of equipment design or operation, especially in the context of strict emission limits and pressure to achieve a greater efficiency.

Despite many similarities, boilers and furnaces tend to be described by different computation models, which are employed to design the major parts: the radiant and convection sections. A limiting factor of design procedures is the lack of necessary information. Since experiments can be carried out on the fired equipment in only a very limited scale, simplified (low-cost) modelling methods and numerical analyses based on computational fluid dynamics (CFD) have an irreplaceable role in the designing process. Furthermore, the apparatus is usually not equipped with a sufficient number of measuring devices for the purpose of examining the thermal and flow behaviour [1]. While basic measurement and control devices ensure safe operation, the collected data serves only as support information for the validation of simplified or numerical models.

In this article, we want to show that it is possible to improve the efficiency of the design procedure and increase the reliability of furnaces and boilers via a comprehensive computational methodology using up-to-date, low-cost models. Of course, this approach relies not only on accurate and fast individual models but also on their precise interconnection. Hence, the main objective of this work is to introduce the unifying framework that will link calculations of the radiant and convection sections. The chief motivation, i.e., the operational problems of fired equipment and current unsystematic (inefficient) approaches to modelling, is outlined in the first two parts of this paper. The description of the novel, low-cost modelling framework follows. The last part is devoted to a case study of a steam boiler. Through this practical application of the proposed framework, the selected low-cost modelling systems are discussed in detail.

Fired Equipment and Maldistribution Issues

Industrial boilers and fired heaters (see Figure 1) consist of two parts, a radiant section (also called radiant or combustion chamber) and a convection section. The shape of the radiant chamber and the configuration of heat transfer surfaces are designed depending on the required heat duty and the properties of the process fluids and fuel (problematic ones are mainly fouling propensity and heat sensitivity). The entire apparatus is then tailor-made to the specific application [2].

In view of transferred heat, the dominant part of the equipment is the combustion chamber where the prevailing mechanism of heat transfer is radiation. Additionally, the heat transfer is the most intense in this part of the apparatus. Convective heat transfer predominates in the heat exchanger zone that follows. The heat transfer in this section may be enhanced to improve operational efficiency. The enhanced heat exchanger is even more sensitive to any non-uniformity of fluid flow than conventional equipment with plain tubes. Therefore, enhancements, e.g., tube fins, must be utilised only with regard to the flue gas temperature level and tendency to foul [3].

The exchanger section of a water-tube steam boiler possesses three parts:

- an economizer—serves to heat feed water, it is usually situated in areas with a lower flue gas temperature;
- an evaporator—tube-side fluid changes phase in this heat exchanger, which is often integrated into the membrane walls of the radiant and convection sections;
- a superheater—production of superheated steam that is used for a particular process or power application is finalised in this heat exchanger. It usually forms the first heat transfer surface in the convection section, but it is frequently installed above the radiant chamber.

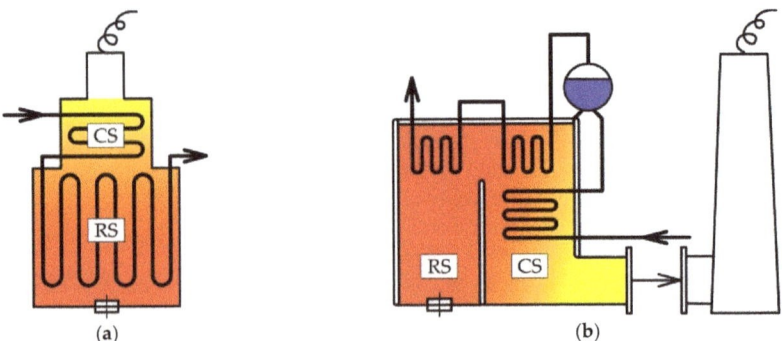

Figure 1. Simplified schematic (**a**) of a cylindrical fired heater, and (**b**) of a two-pass boiler. Abbreviations CS and RS stand for convection and radiant sections, respectively.

For the reliable design of both parts of fired equipment (radiant and convection sections), it is necessary to identify the distribution of heat flux and process fluids. Recognition of heat flux variability crucially affects the faultless performance and service life of a tubular system in a radiant chamber. Design of the radiant chamber in fired heaters is based on established procedures, such as American Petroleum Institute (API) Standard 530 [4] and API Standard 560 [5]. Nevertheless, their calculation methods inadequately predict the longitudinal and circumferential maldistribution of heat flux. An assumption of average radiant heat flux and the neglecting of the burner's actual thermal profile contributes to a decrease in efficiency and increased wear of the whole equipment or of some of its parts. Underestimation of this issue often causes serious operational problems of fired heaters: increased fouling of the tubular system (e.g., deposition of coke), overheating of the tube material with subsequent deformation and a dramatically increased pressure drop. In the worst-case scenario, an accident may happen [6], and in a better case, only unplanned shutdown procedure is activated, followed by cleaning and minor repairs [7].

Significant inaccuracies of the common design practice are particularly related to the recent use of low-NO_x and ultralow-NO_x burners. If these burners are compared with conventional burner types, thermal behaviour (flame length and width) is noticeably different [8]. Consequently, the peak thermal loading of the radiant heat transfer surfaces greatly varies [9].

The distribution of the flue gas flow, as the hot stream, is also essential in the convection section of the fired equipment. Typically, the heat exchangers contain dense tube bundles, which on the one hand increase the heat transfer, and on the other hand, complex geometry is more prone to the non-uniformity of the tube-side fluid flow [10]. The problematic flow behaviour in the tubes causes uneven loading of the heat transfer surfaces, especially when it is negatively influenced by flue gas maldistribution. An example of such a problematic area may be the turning position between the first and second pass of the steam boiler illustrated in Figure 1b.

The consequences of the described underestimation or negligence in the design procedure are addressed by equipment troubleshooting [11]. Of course, operational issues cannot be completely eliminated, but the goal of reliable design is to reduce the number and severity of problems as much as possible. An important feature of the suggested framework is the minimization of the potential operational difficulties thanks to taking the heat and fluid flow distribution into account as early as the initial phase of the design procedure. The use of the low-cost models for initial design calculations allows for fast identification of problem areas. In the next phase, a detailed but time-consuming CFD simulation can effectively analyse these bottlenecks.

2. Overview of the Modelling Approaches

In general, an increasingly prevalent trend in modelling different parts of an apparatus is the use of CFD simulations. It is the most flexible yet at the same time the most demanding approach. If the final stage of the new equipment design is to be solved or if problems of the already operating units are being investigated, the use of numerical models is undoubtedly beneficial. According to the broad overview provided by Aslam Bhutta et al. [12], the CFD thermal analyses of different heat exchangers can achieve an agreement with measured data within 5 %. The high accuracy of the results compensates for the long process of preparation and computation itself. The long time required to obtain results, of course, increases the cost of these analyses.

The basic challenges of CFD modelling involve the creation of a high-quality mesh whose size should be as small as possible [13]. In order to determine a suitable number of elements, a grid independence test is carried out. As an example, analyses of distribution systems [14] can be mentioned. Gandhi et al. [14] tested meshes consisting of hexahedral and tetrahedral elements. A similar flow geometry (with a slightly higher number of tubes) was also investigated by Zhou et al. [15] who used relatively new polyhedral cells in combination with prismatic elements. Regarding the flow system including only 14 tubes and 2 headers, the grid independence test [15] indicated that the optimal mesh should contain almost three million cells. It should be noted that these two studies did not examine heat transfer, and moreover, the maximum number of tubes was only 50 [14] and 70 [15]. In practice, heat transfer apparatuses with much larger tube bundles are employed. Typical representatives of such equipment are, for example, heat exchangers in boilers and heat recovery steam generators (HRSGs). It is apparent that the thermal analyses of large distribution systems tend to be computationally intensive and highly time-consuming. Various approaches that balance accurate data and the computing costs of CFD models will be described as follows.

Poursaeidi and Arablu [16] combined two-dimensional (2D) CFD models with three-dimensional (3D) models of one pass of a steam boiler to address both combustion and non-uniform steam distribution. First, the 2D models evaluate the division of flow in manifolds. One tube in a simplified 2D geometry stands for a row with ten tubes in real conditions [16]. Then, a 3D combustion model yields temperature fields in the combustion chamber. Finally, the boiler is divided into simpler heat exchangers. In these parts, heat transfer is analysed using 3D CFD models, which employ the results obtained previously. The combination of 3D and 2D CFD simulations was also used for analysing heat transfer in a vertical fired heater [17]. In this case, the 3D model provides an insight into the longitudinal distribution of the heat flux in the combustion chamber. The 2D models then specify the effect of geometrical imperfections on the circumferential heat flux variability.

The model described by Gómez et al. [1] calculates thermal fields on the sides of both fluids (flue gas and steam), as well as the tube wall temperature, and the shell-side flow-field in the convection section of a boiler in the power industry. This model employs the special types of elements that substitute tubes (so-called sub-grid features) and headers (virtual elements). Via modelling the entire structures that are similar in their geometry, physical properties and topology, rather than meshing the individual tubes, it is possible to significantly reduce the size of the mesh. The major advantage of this model is the interconnection of the tube banks in the convection section, which improves the accuracy of the obtained results.

Models with individual HRSG tubes were compared by Galindo-García et al. [18] with models that replaced tube bundles with porous layers. In order to decrease the number of cells, Galindo-García et al. [18] relied on a non-conformal mesh, with individual parts of HRSG being linked by an interface. Additional shortening of evaluation time brings a steady state simulation. Another example of the use of the porous layer is the work of Nad' et al. [19], who focused on flue gas flow in an industrial boiler. Replacement of tube bundles with larger (porous) zones is advantageous for the evaluation of flow in the shell, but this technique causes loss of information about the local thermal loading of heat transfer surfaces. From this point of view, Gómez et al. [1] offered the most comprehensive approach, which can be modified so that, according to its authors, it can assume the non-uniform distribution of the fluid in tubes.

Most authors use various versions of a commercial software Ansys Fluent, except for, e.g., Zhou et al. [15] (Siemens' STAR-CCM+) and Manickam et al. [20]. The last-mentioned authors [20] solved a waste heat recovery boiler problem using the software CFD-FLOW3D (now provided by Ansys as a software tool CFX).

Taking into account the before-mentioned cases, it is clear that the use of CFD in an initial design phase is rather ineffective. Contrarily, in the initial phase, it is necessary to quickly exclude completely inappropriate configurations, adjust the main dimensions and select the most suitable configuration for further detailed calculations. For the purpose of this sizing procedure, the simplified (i.e., low-cost) models and methods that will be described as follows fit much better.

Considering the convection section, the models of heat exchangers can be classified into three categories. The first category of models focuses on the distribution of the working fluids while heat transfer (convection) is neglected. For the second group, the calculations concentrate on heat transfer whilst flow distribution is always supposed to be known and usually also to be uniform (see, e.g., Lan et al. [21]). If needed, the potentially non-uniform distribution of fluid flow must be estimated by another calculation tool. This is the case, for example, for the cell method [22] or a cross flow calculation presented by Shah and Sekulić [23]. Models that capture both phenomena in a simplified manner represent the last category.

The first group includes fast isothermal models [24,25], which predict the flow rates and pressures in dividing flow manifolds. The model [24] describes the distribution systems in differential form. These equations are elegant but depend on a knowledge of actual velocity profiles in tube inlet ducts. More practical is a formulation of correction coefficients that was used in the work [25]. Here, the coefficients (the coefficient of static pressure regain and discharge coefficient) are calculated using upstream and downstream values of respective variables.

The non-uniform heating of the water–steam mixture at different levels of the distribution system was analysed by Ngoma and Godard [26]. Unlike the previous models, this parallel system contained only one row of tubes. The results revealed that in the case of non-uniform thermal loading, the process fluid flow was slightly more uniformly distributed. A similar effect of (non-uniform) heating on the flow of refrigerant was observed by Cho et al. [27] when the fluid was heated in the last third of the microchannels. The opposite effect had local heating applied in only one of the nine sections of the distribution system [27]. It caused intense evaporation of the refrigerant, which resulted in a significant reduction in the mass flow rates in the respective channels.

If maldistribution in microchannels is taken into consideration, Baek et al. [28] reported an interesting comparison of effectiveness-NTU (i.e., number of transfer units) expressions. The obtained results showed that the linear maldistribution of a hot stream in the vertical direction had a particularly negative effect on the heat exchanger effectiveness. This one-dimensional (1D) model [28] analyses the pure countercurrent configuration of the heat exchanger, which in most industrial applications is just the ideal we are trying to converge toward.

A more accurate description is offered by models and relationships that deal with crossflow exchangers. This type of flow arrangement can be further sub-divided into pure crossflow, multi-pass parallel crossflow, and multi-pass counter-crossflow as shown in Figure 2. Multi-pass crossflow exchangers with different configurations are the objectives of, e.g., Cabezas-Gómez et al. [29], who discussed the applicability of the effectiveness-NTU simplified equations. Cabezas-Gómez et al. [29] point out that heat exchangers with more than six rows (or passes in terms of our work) should not be evaluated on effectiveness by simple expressions for the purely concurrent or purely countercurrent arrangement. There is a grey area where the theoretical relationships considering the number of tubes is no longer appropriate, yet the before-mentioned substitution cannot be made either. This situation arises if the number of passes is higher than four and lower than approx. 20 and 50 in the cases of parallel and countercurrent cross exchangers, respectively.

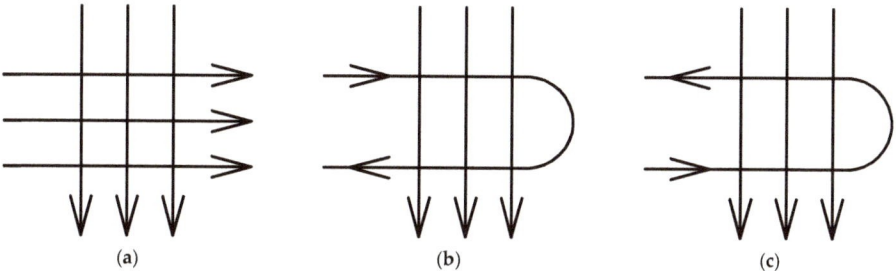

Figure 2. Schematics illustrating different types of a cross flow arrangement: (**a**) pure crossflow, (**b**) two-pass parallel crossflow, and (**c**) two-pass counter-crossflow

Another unacceptable simplification that occurs in the models of multi-pass distribution systems is the creation of virtual mixing chambers at the turning points. This interconnection of the discretised parts of the heat exchanger (different types of flow geometries are sketched out, e.g., by Pignotti and Shah [30]) is frequently adopted by commercial software. The computational tools average the values of fluid properties in these virtual mixing chambers, unrealistically affecting the obtained temperature field, as well as the calculated thermal loading of the heat transfer surfaces.

As previously mentioned, calculations of fired equipment must adhere to the design standards. The design calculations of the furnace and boiler radiant section are based on the well-stirred furnace (WSF) model [31]. This model for determining radiant heat transferred to a radiant tubular system divides the furnace (or radiant chamber in general) into three zones representing the radiating flue gas, tubular system, and refractory walls. The 1D model [31] has a large number of simplifications (see, for instance, the comparison of this model with numerical simulations [32]); nevertheless it forms the basis for 2D thermal analyses in radiant section. The WSF model is still used in computational tools, for example, to optimize operating conditions [33] or to predict the fouling rate of crude oil [34].

Weaknesses of the WSF model (especially the strong underestimation of real heat flux distribution) are particularly evident in apparatuses with a significant temperature gradient due to the large difference between the two main dimensions. An example of such constructions is a cylindrical radiant chamber with its large height (or length) and relatively small diameter. The (long) radiant section of the total length L can be discretized into a series of segments of length dx. Heat transfer is then sequentially solved in each segment using a 1D plug-flow (PF) model (see, e.g., Hewitt et al. [35]). This model is sometimes called a long furnace model and can be applied in, e.g., the technical-economic analysis of possible energy savings [36]. The PF model requires the local volumetric heat release rate as a function of furnace length. If the flame is parallel to the furnace axis, the local volumetric heat release rate can be determined using an experiment or via a CFD simulation. Both ways are far from trivial. The disadvantages of the PF model were summarized by Jegla in his work [37]. Instead, he introduced a modified plug-flow (MPF) model. One of the improvements—the inclusion of the burner test results not only in the validation of the MPF model [38], but also in the calculation algorithm itself [8]—allows this model to accurately predict the distribution of heat flux in the radiant chamber.

The basic MPF model provides excellent results for cylindrical equipment with one burner, i.e., the equipment that is similar to the test combustion chamber [8]. Industrial fired equipment, however, usually contains more burners and the radiant section can have a rectangular cross-section (such as cabin furnaces), as well as the standard circular cross-section. The presence of multiple burners affects the flue gas flow (its mixing) and the overall character of the heat transfer in the radiant section. The cylindrical shape of a furnace is more favourable for the (uniformly) distributed heat and fluid flow than the cabin type [17]. These parameters are considered by the adapted modified plug-flow (hereafter AMPF) model, which was introduced by Jegla [7]. Of the models describing the design of the radiant chambers, the AMPF model is the most complex but also the most versatile low-cost technique.

On the one hand, simplified methods have, by their very nature, a good deal of limitations that the user must pay attention to. However, on the other hand, this modelling approach has a significant advantage in low computational requirements. This is the main reason for their use in design calculations. Moreover, some restrictions can be eliminated by using a suitable combination of models, as shown, for example, in the before-mentioned works [7,10]. If several simplified models (or different ways of estimation) are systematically linked, it is even possible to analyse the entire complex equipment possessing radiant and convection sections.

3. Proposed Calculation Framework

The common practice of designing fired equipment suffers from the inaccuracy of the employed empirical approximation. Therefore, the basic calculations of the established design standards are variously supplemented by the low-cost tools for modelling radiant and convection sections or these parts are being analysed in detail by CFD. In spite of the apparent interconnection of both parts, modelling activities are predominantly conducted separately. This may not be an issue if only one part of the apparatus is of interest. However, in the case of designing or rating the entire equipment, such a procedure is not systematic, and it also reduces the quality of the complete equipment design. Based on long-term experience with the design of fired equipment for process and power industries, there has been a need for a comprehensive methodology that would improve the reliability of the current design practice.

The aim of this work is to introduce such a framework that aggregates the basic thermal-hydraulic calculations of the radiant and convection section and fast and accurate low-cost computational models. Because of this systematic combination, the novel framework significantly increases the reliability of the fired equipment design. The calculation framework primarily serves the initial design of combustion equipment; however, it can be utilised for the thermal rating analysis of existing equipment.

The outline of the proposed approach is in Figure 3. In essence, this calculation procedure has two phases. In the first phase, standard thermal-hydraulic calculations are employed to obtain initial data on:

1. a radiant section—i.e., to evaluate the amount of radiant heat duty, the main dimensions of the radiant chamber and the tubular system;
2. a convection section—among other things, to design the main parameters of tubular heat exchangers (number of tubes, tube geometry, etc.), average input, output temperatures and heat duty.

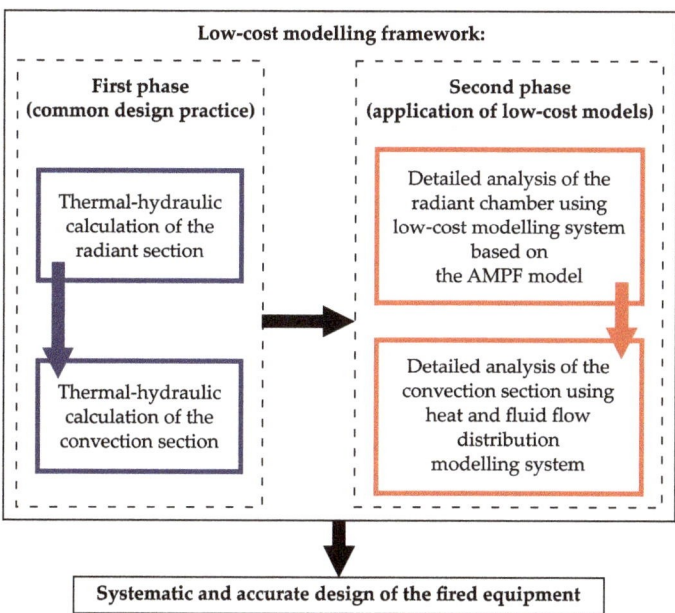

Figure 3. Proposed systematic framework of the novel design procedure showing the interconnection between all calculations

The second phase of the calculation procedure serves as the thorough investigation into the thermal-hydraulic behaviour of the respective equipment. It includes the analyses of heat and fluid flow maldistribution in the radiant and convection sections. The idea is to avoid CFD simulations as much as possible, despite the fact that they provide a very good insight into the heat transfer system. The considerable disadvantage of the CFD approach is its inability to flexibly respond to any modifications of the equipment dimensions that are often in the initial phase of the design procedure [38].

For the design of the radiant chamber, a low-cost modelling system using the AMPF model [7] is chosen. This universal three-step method takes into account the real local thermal load of the radiant tubular system. At first, the burner heat flux profile is experimentally determined. Then, the MPF model, which uses the results of the previous tests, is applied to identify the fuel burnt profile. In the last step, the design of the combustion chamber is completed using the AMPF model.

Information about flue gas distribution is subsequently taken over by another simplified modelling system for the exposed parts of the convection section. The heat and fluid flow distribution modelling system, which was presented by Jegla and Fialová [10], identifies locations that are potentially risky in view of thermal loading (overheating but also subcooling).

Both separate modelling systems are situated in the context of designing the complete fired equipment in order that the framework allows a continuous segue between the individual parts of the calculation. Such a homogeneous design of the complex fired equipment thus maintains a high degree of accuracy.

In the following part, the application of the calculation framework will be demonstrated, as well as a more detailed description of the utilised modelling systems, using an industrial example.

4. Case Study

The purpose of this section is to discuss the principles of the selected low-cost models in detail and to apply them to a case study of the steam boiler. In other words, the case study focuses on the second part of the suggested framework. First of all, the equipment in question will be described. The other two parts will be devoted to the modelling methods for the calculation of the heat flux distribution in the boiler radiant chamber and for the calculation of the temperature profile in the steam superheater.

At this point, it is necessary to point out that the case study did not deal with grassroots design, but with the equipment being operated. However, the use of low-cost models was the same as in the case of new designs. Since this was the investigation into the operated apparatus, the conclusions of both low-cost modelling systems were only of a recommending nature. Model results only indicated problematic locations or suggest their more appropriate configuration. Appropriate design modifications that would lead to the elimination of operational problems were not included in the case study. Also, the proposed framework did not contain a technical-economic analysis that would certainly precede the possible retrofit.

4.1. Industrial Boiler

The objective of the case study was a three-pass steam boiler with natural circulation (a schematic of the respective boiler will be shown in following subsection). The combustion chamber had a rectangular cross-section, measuring 5×7.2 m, and a total height of 18.4 m. At the heights of 7.5 m and 10 m, there were four burners, which enabled the combustion of liquid fuels (heavy fuel oil mixed with liquid tar waste) and natural gas. The boiler employed membrane walls with tubes of diameter 57 mm and steel strips 75 mm in width. This inbuilt tubular system served as an evaporator and cooled all walls of the combustion chamber and the second pass of the boiler (already a convection section). A total heat transfer area that ensured the production of saturated steam was 550 m^2. The first heat exchanger area in the convection section was a final steam superheater. At the output of this heat exchanger, superheated steam had the following parameters: a temperature of approx. 370 °C and a mass flow of 16.7 kg/s. Downstream of the final superheater was a primary superheater. Economizers were located at the bottom of the second boiler pass and in the third pass.

Due to the previous operational problems, the whole boiler had undergone a non-destructive examination. The photograph in Figure 4 was taken during this inspection. The main overloaded heat transfer surface was identified in tubes of the final steam superheater, especially in the area before entering the collector. This was manifested as an increased material loss of these tubes. The general CFD simulation [19], which was utilised for the necessary troubleshooting, revealed that wall-thinning was induced by the non-uniform distribution of flue gas. The critical location was between the radiant chamber and the convection section where flue gas flow changed its direction. Therefore, the low-cost modelling system [10] was applied to a detailed analysis of the final steam superheater. The main topological and geometrical parameters of the heat exchanger, as well as the relevant properties of both fluids, are arranged in Table 1.

Figure 4. Tube bundle of the final steam superheater; an orange colouration is a distinctive feature of the accumulated corrosion layer. Grey and black deposits of flue gas components are highly visible between tubes.

Table 1. General parameters of the final steam superheater.

Parameter		Value	
Flow arrangement		Counter-crossflow	
Number of tube passes		6	
Tube bundle:	Number of tubes	198	
	Number of tube columns	66	
	Number of tube rows	3	
		Hot Stream	Cold Stream
Flow data:	Fluid	Flue gas	Superheated steam
	Mass flow rate (kg/s)	17.8	16.7
	Inlet temperature (°C)	734.0	248.2
	Outlet temperature (°C)	495.5	369.0

4.2. Low-Cost Modelling of the Radiant Section

As mentioned earlier in this paper, the MPF model [8] is sufficient only for single-burner cylindrical equipment without an inbuilt tubular heat transfer system. In contrast, the AMPF model [7] considers: (i) tube coils or membrane walls; and (ii) the rectangular, as well as circular, cross-section of a combustion chamber. Both models demonstrated extremely good accuracy regarding the modelled heat flux distribution in the radiant section. On the one hand, the low-cost AMPF model can design the industrial equipment respecting the real heat flux variability. On the other hand, the AMPF model requires information on the thermal characteristic of a burner in a suitable form, which can be provided by the MPF model. Therefore, the complete modelling procedure contained three steps, as shown in Figure 5.

The first step was determining the thermal characteristic of the burner, either the intended one (in case of brand-new designs or retrofit applications), or the one already installed (if rating calculations are needed). Of course, in situ measurements or the usage of a test facility was the most expensive approach as it necessitated not only specialised measurement equipment but also trained and experienced staff. The need for operating or experimental data may, therefore, appear to be a significant disadvantage of the whole modelling system. However, one cannot rely on results obtained by exact calculations with inaccurate inputs. Comparing the measured data with the results of numerical simulations indicates that the CFD approach is not able to fully capture the thermal and flow behaviour in the combustion chamber [38].

Once reliable primary data was available (from the previous measurements of the burner), the MPF model was applied to identify the real heat flux distribution. Another possibility (which is often omitted) is to ask a burner manufacturer for the heat flux profile of the respective burner (for example, as a part of tender documentation). If the thermal behaviour of the burner including the fuel burn profile is known, the first two steps (the experiment and the application of MPF model) can be skipped, passing straight to designing and evaluating the radiant chamber by the AMPF model. The AMPF model formed the final step of the low-cost modelling system for designing and analysing radiant chambers.

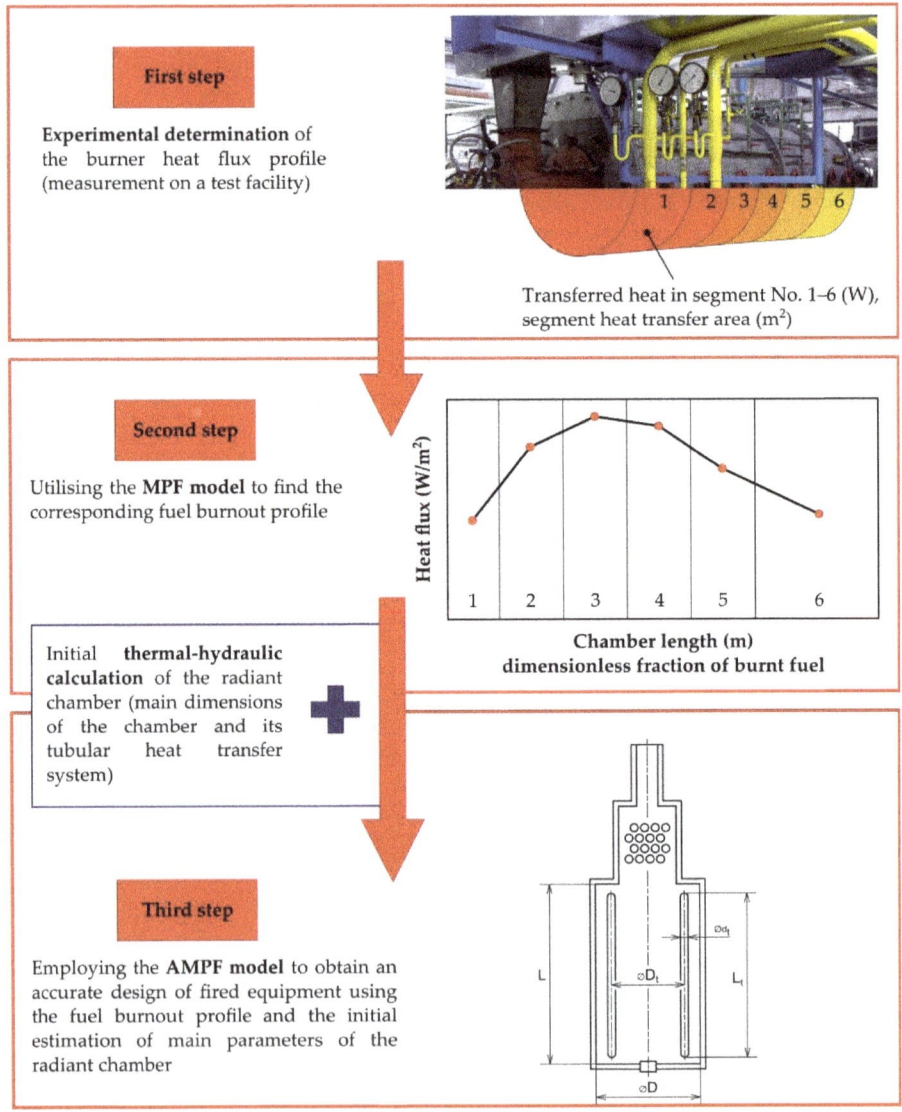

Figure 5. Schematic flow chart of the three-step modelling system for an accurate and reliable design of fired equipment

One of the main features of the AMPF model is the optimal 1D discretization, i.e., the size of one element, $L_{segment}$. According to the sensitivity study [7], the choice of the suitable division into computational segments must take into account the position of burners and the maximum difference of a flue gas temperature of 100 K in each segment. This criterion can be expressed as:

$$L_{segment} \leq 0.15 \times L \tag{1}$$

where L denotes length (or height) of the radiant section.

To satisfy this Condition (1), the discussed radiant chamber was divided into eight parts so that the length of one segment was 2.3 m. As shown in Figure 6, the burners were located in the second and third segments. Then the AMPF model iteratively evaluated the profile of the heat flux along the radiant chamber. It is important to note that the described boiler had the specific construction that allowed, at the very beginning of the calculation, the setting of 50% burnout of the combustion mixture in the two segments with burners. The experimental step and the utilisation of the MPF model (the first and second step of the modelling system) can be avoided by this initialisation in all cases where the burners are oriented orthogonally to the direction of the flue gas flow in the radiant chamber.

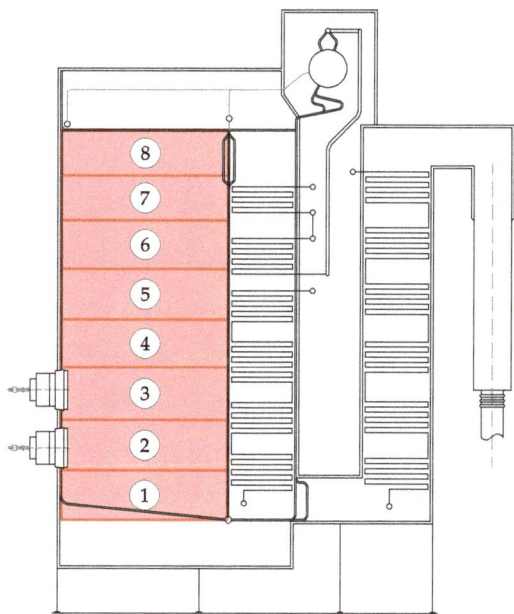

Figure 6. Radiant chamber of the discussed steam boiler discretised via the AMPF model

In addition to the heat loading of the radiant tubes, it is possible to estimate the real state of the water–steam mixture if the real heat flux distribution is utilised for a following thermal-hydraulic calculation. By adding a standard AMPF result calculation, a critical location can be identified, e.g., where there is a risk of the so-called dry out and overheating of tubes.

4.3. Low-Cost Modelling of the Convection Tube Bank

Input data on the temperatures and flows (or speeds) of process fluids can be obtained by operational measurement or some previous simulations of the respective heat transfer equipment. However, experience of troubleshooting shows that the measured operating data is not as detailed as is necessary for equipment analysis. In the case of a new apparatus, of course, any data is completely missing and simplified or CFD models are irreplaceable. Another feasible approach is also utilisation of some of the established commercial software that is specialised for the heat exchanger modelling.

Although the CFD analysis [19] provided valuable information about the flue gas flow in the boiler space, it was not possible to model such complex fired equipment in detail. Thus, the tubular heat exchangers in the second part of the respective boiler were replaced by porous layers with the same pressure losses. This general CFD model [19] was able to investigate the flow field characteristics just above the discussed superheater. For the purpose of 2D calculations [10], the obtained velocity and temperature fields were divided by the number of tube columns. The flue gas temperatures and velocities in each of these sections (columns) serve as input data for the 2D calculation of the temperature distribution across the tube bundle. If such a CFD analysis is not available, it is possible to make use of data on flue gas thermal behaviour from the previous evaluation of the radiant chamber.

The inadequate simplification of commercial software, i.e., virtual mixing headers already mentioned in the second part of this work, was also noticed in the troubleshooting analysis of the superheater discussed here (see Figure 7). This simplification showed its impact on the steam temperature, which was averaged at the entrance of each tube pass. As an example, temperature data from the last two (the fifth and sixth) tube passes are listed in Table 2. The values at the end of the fifth pass and at the beginning of the sixth tube pass are written in italics to highlight the difference between the steam temperatures before and after averaging by the described software simplification.

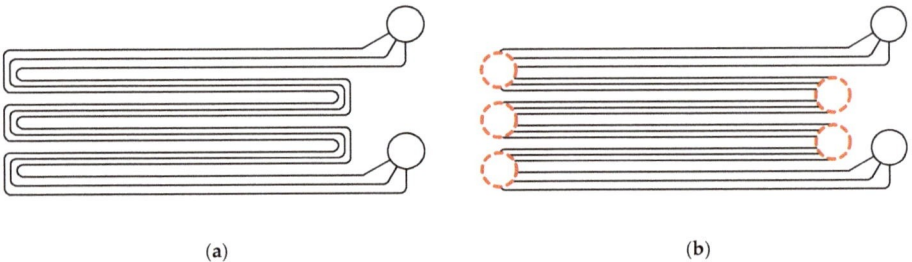

(a) (b)

Figure 7. Channel system of the steam superheater discussed here: (**a**) schematic of the real flow geometry, and (**b**) representation of the heat exchanger using virtual mixing headers.

Table 2. Steam temperatures in the inlet and outlet zones of the tubes in the last two tube passes.

		*j*th Tube Element						
		1	2	3	...	8	9	10
6th pass	Tube 1	370.5	397.6	364.7	...	347.1	344.1	341.1
	Tube 2	369.0	366.2	363.5	...	346.8	343.9	341.1
	Tube 3	367.6	365.0	362.4	...	346.5	343.8	341.1
5th pass	Tube 3	316.5	319.1	321.7	...	337.3	339.8	*342.4*
	Tube 2	316.5	319.0	321.5	...	336.2	338.7	*341.1*
	Tube 1	316.5	318.9	321.2	...	335.3	337.6	*339.9*

Information about the distribution of the fluid in the tubes was obtained using the isothermal model [25], which was fast because of its simplicity, and still sufficiently precise for calculation purposes. With respect to the location of the dividing and collecting manifold in the boiler (see Figure 8), heat transfer in these headers was neglected. Consequently, steam in the distributor (before dividing into the bundle) was assumed to be completely mixed at a constant temperature of 248.2 °C.

Figure 8. The 2D method applied to a model of the tubular steam superheater. Hot fluid (flue gas) is shown in red, cold streams (steam) are outlined as blue lines.

The core of the modelling system is a 2D analysis of the heat distribution across the tube bundle. Due to the change in the direction of the steam flow, the overall calculation is divided into solutions of the so-called subexchangers, which represent the individual tube passes (also shown in Figure 8). Based on the cell method [22], the temperature field is obtained using dimensionless temperature formulae. Then, considering a general cell [i, j], the dimensionless outlet temperatures θ_{out} of both process fluids can be calculated with the following equations:

$$\theta_{out,1}[i,j] = (1 - P_1[i,j]) \times \theta_{out,1}[i-1,j] + P_1[i,j] \times \theta_{out,2}[i,j-1]$$
$$\theta_{out,2}[i,j] = (1 - R_1[i,j] \cdot P_1[i,j]) \times \theta_{out,2}[i,j-1] + R_1[i,j] \cdot P_1[i,j] \times \theta_{out,1}[i-1,j] \quad (2)$$

where P is temperature effectiveness and R the ratio of heat capacity rates of two fluids. Subscripts 1 and 2 denote variables related to the fluid with lower and higher heat capacity rate, respectively. In this case the hot stream (flue gas) had the lower heat capacity rate.

The interconnection of the described models is represented schematically in Figure 9. As can be seen, the individual streams were kept separate across the whole tube bundle. Thus, the flow data avoided undesirable averaging. The presented technique yielded more accurate data on temperatures and mass flow rates of process fluid, thereby improving the accuracy of the predicted local overloading of heat transfer surface.

Not only Hewitt [22], but also other authors (e.g., Ptáčník [39] or the already-mentioned Cabezas-Gómez et al. [29]) point out that the thermal efficiency in Equation (2) is influenced by the flow arrangement (concurrent, countercurrent, and crossflow), mixing of each fluid, number of transfer units, and the ratio of heat capacity rates. Furthermore, Cabezas-Gómez et al. [29] argue that an effectiveness error of 0.01% may cause inaccuracies in the following calculations on the order of units of percent. The future research will, therefore, address the possibilities of P–NTU relationships, i.e., whether it is sufficient to use standard P–NTU formulas (a broad overview was given, e.g., by Shah and Sekulić [23]) without great inaccuracies, or it is appropriate and necessary to employ the so-called countercurrent coefficient described by Ptáčník [39]. In the current calculation, the overall heat transfer coefficient is assumed to be constant across the entire tube bundle. For that reason, the future work will also include refinements of this aspect of the calculation model, i.e., to specify the overall heat transfer coefficient in individual cells with respect to the local values of physical properties of process fluids.

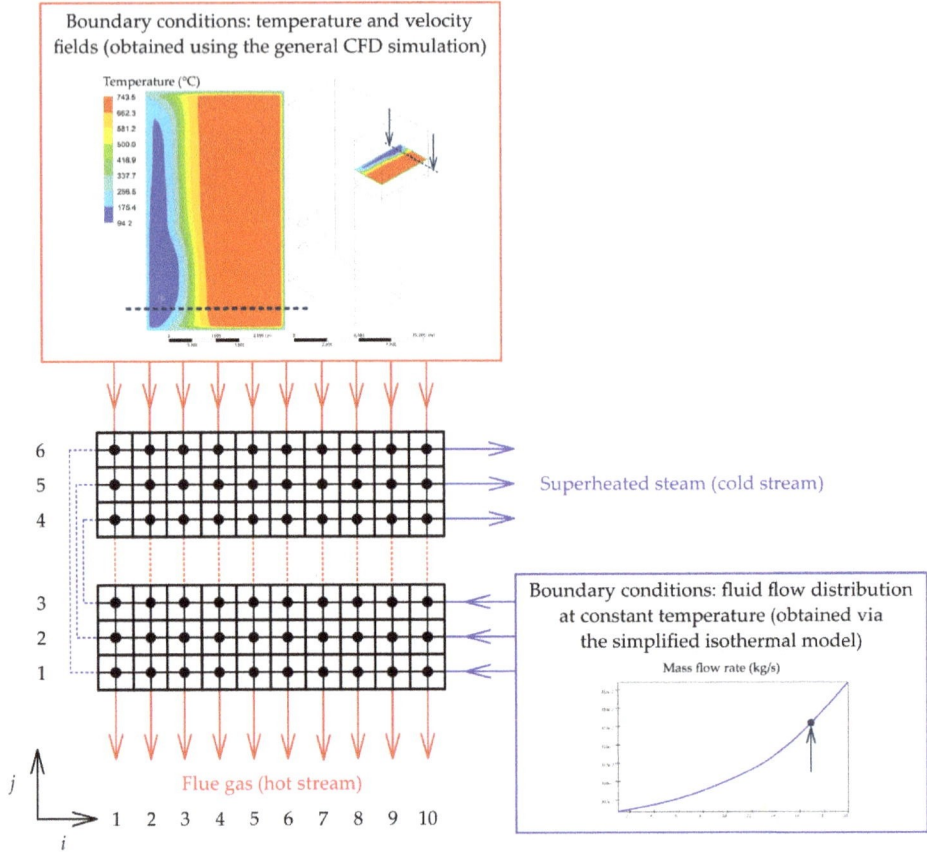

Figure 9. Concept of the heat and fluid flow distribution modelling system illustrated on two passes of the heat exchanger in question. Flow data of the flue gas are provided by the previous general CFD analysis [19] while distribution of steam was pre-solved using a simplified model based on Bailey's approach [25].

5. Conclusions

In this paper, a novel framework for the efficient design of fired equipment has been introduced. The traditional thermal-hydraulic calculation of the radiant and convection section (i.e., the first phase of the calculation procedure) was supplemented by using low-cost modelling systems that take into account the real (non-uniform) distribution of heat flux and process fluids. Designing the combustion chamber was done using the AMPF model [7], the problematic heat exchanger in the convection section was analysed using the heat and fluid flow distribution modelling system [10] presented at the 21st Conference on Process Integration, Modelling and Optimisation for Energy Saving and Pollution Reduction PRES 2018. The application of these low-cost models (the second phase of calculations in the proposed framework) has been demonstrated in the industrial steam boiler case.

In the context of simplified methods, this framework is a logical outcome of the long-term development that has taken place at the Institute of Process Engineering of the Faculty of Mechanical Engineering, Brno University of Technology. First, the individual simplified models were unified for the analysis of thermal behaviour and flow distribution in combustion chambers and heat exchangers. Second, the presented framework links calculations of radiant and convection sections and offers a systematic approach to designing and fast rating calculations of complex fired equipment. It should be emphasized that the low-cost modelling procedure does not replace CFD analyses neither for purposes of troubleshooting nor for the final detailed design of the equipment. On the contrary, the described procedure enables detailed CFD analyses to concentrate effectively on critical locations identified using low-cost models.

Future work will focus on further refinement of 2D heat distribution calculations in the convection section.

Author Contributions: Conceptualization, Z.J.; methodology, Z.J. and D.F.; writing—original draft, D.F.; writing—review and editing, Z.J.; visualization, D.F.; supervision, Z.J.

Funding: This research was supported by the EU project Strategic Partnership for Environmental Technologies and Energy Production, funded as project No. CZ.02.1.01/0.0/0.0/16_026/0008413 by Czech Republic Operational Programme Research, Development and Education, Priority Axis 1: Strengthening capacity for high-quality research.

Acknowledgments: Thanks to M. Na' who provided the authors with CDF data. Also, the authors wish to acknowledge the consultations offered by V. Turek.

Conflicts of Interest: The authors declare no conflict of interest.

References

1. Gómez, A.; Fueyo, N.; Díez, L.I. Modelling and simulation of fluid flow and heat transfer in the convective zone of a power-generation boiler. *Appl. Therm. Eng.* **2008**, *28*, 532–546. [CrossRef]
2. Hájek, J.; Jegla, Z. Standards for fired heater design: Analysis of two dominant heat flux variation factors. *Appl. Therm. Eng.* **2017**, *125*, 702–713. [CrossRef]
3. Stehlík, P. Conventional versus specific types of heat exchangers in the case of polluted flue gas as the process fluid—A review. *Appl. Therm. Eng.* **2011**, *31*, 1–13. [CrossRef]
4. API, Standard 530. *Calculation of Heater-tube Thickness in Petroleum Refineries*, 6th ed.; American Petroleum Institute: Washington, DC, USA, 2008.
5. API, Standard 560. *Fired Heaters for General Refinery Service*, 4th ed.; American Petroleum Institute: Washington, DC, USA, 2007.
6. Jegla, Z.; Kohoutek, J.; Stehlík, P. Design and operating aspects influencing fouling inside radiant coils of fired heaters operated in crude oil distillation plants. In Proceedings of the Heat Exchanger Fouling and Cleaning IX, Crete Island, Greece, 5–10 June 2011; Malayeri, M.R., Müller-Steinhagen, H., Watkinson, A.P., Eds.; pp. 7–14.
7. Jegla, Z. Innovative adaptation of MPF model to recognition of thermal behaviour of operated industrial low emission burner system. *Chem. Eng. Trans.* **2016**, *52*, 667–672. [CrossRef]
8. Jegla, Z.; Kilkovský, S.; Turek, V. Novel approach to proper design of combustion and radiant chambers. *Appl. Therm. Eng.* **2016**, *105*, 876–886. [CrossRef]
9. Baukal, C.E. *The John Zink Hamworthy Combustion Handbook: Volume 1—Fundamentals*, 2nd ed.; CRC Press: Boca Raton, FL, USA, 2013; ISBN 9781439839621.
10. Jegla, Z.; Fialová, D. Development of heat and fluid flow distribution modelling system for analysing multiple-distributed designs of process and power equipment. *Chem. Eng. Trans.* **2018**, *70*, 1471–1476. [CrossRef]
11. Nekvasil, R.; Jegla, Z. Boiler reheater chamber cracking analysis. *All Power* **2014**, *5*, 11–13.
12. Aslam Bhutta, M.M.; Hayat, N.; Bashir, M.H.; Khan, A.R.; Ahmad, K.N.; Khan, S. CFD applications in various heat exchangers design: A review. *Appl. Therm. Eng.* **2012**, *32*, 1–12. [CrossRef]
13. Turek, V.; Fialová, D.; Jegla, Z. Efficient flow modelling in equipment containing porous elements. *Chem. Eng. Trans.* **2016**, *52*, 487–492. [CrossRef]
14. Gandhi, M.S.; Ganguli, A.A.; Joshi, J.B.; Vijayan, P.K. CFD simulation for steam distribution in header and tube assemblies. *Chem. Eng. Res. Des.* **2012**, *90*, 487–506. [CrossRef]

15. Zhou, J.; Sun, Z.; Ding, M.; Bian, H.; Zhang, N.; Meng, Z. CFD simulation for flow distribution in manifolds of central-type compact parallel flow heat exchangers. *Appl. Therm. Eng.* **2017**, *126*, 670–677. [CrossRef]
16. Poursaeidi, E.; Arablu, M. Using CFD to study combustion and steam flow distribution effects on reheater tubes operation. *J. Fluids Eng.* **2011**, *133*, 051303:1–051303:11. [CrossRef]
17. Jegla, Z.; Vondál, J.; Hájek, J. Standards for fired heater design: An assessment based on computational modelling. *Appl. Therm. Eng.* **2015**, *89*, 1068–1078. [CrossRef]
18. Galindo-García, I.F.; Vázquez-Barragán, A.K.; Rossano-Román, M. CFD simulations of heat recovery steam generators including tube banks. In Proceedings of the ASME 2014 Power Conference (POWER 2014), Baltimore, MD, USA, 28–31 July 2014; POWER2014-32261. pp. 1–9. [CrossRef]
19. Naď, M.; Jegla, Z.; Létal, T.; Lošák, P.; Buzík, J. Thermal load non-uniformity estimation for superheater tube bundle damage evaluation. *MATEC Web Conf.* **2018**, *157*, 02033:1–02033:10. [CrossRef]
20. Manickam, M.; Schwarz, M.P.; Perry, J. CFD modelling of waste heat recovery boiler. *Appl. Math. Model.* **1998**, *22*, 823–840. [CrossRef]
21. Lan, J.; Zhu, L.; Zhao, J. Modeling and Analysis of Cross-Flow Heat Exchanger Based on the Distributed Parameter Method. In Proceedings of the ASME 2012 International Mechanical Engineering Congress & Exposition (IMECE 2012), Houston, TX, USA, 9–15 November 2012; IMECE2012-86235. pp. 417–425. [CrossRef]
22. Gaddis, E.S. Effectiveness of multipass shell-and-tube heat exchangers with segmental baffles (cell method). In *Heat Exchanger Design Handbook*; Hewitt, G.F., Ed.; Begell House: New York, NY, USA, 1998; ISBN 1-56700-094-0.
23. Shah, R.K.; Sekulić, D.P. *Fundamentals of Heat Exchanger Design*; John Wiley & Sons: Hoboken, NJ, USA, 2003; pp. 256–258. ISBN 0-471-32171-0.
24. Bajura, R.A.; Jones, E.H. Flow Distribution Manifolds. *J. Fluids Eng.* **1976**, *98*, 654–665. [CrossRef]
25. Bailey, B.J. Fluid flow in perforated pipes. *J. Mech. Eng. Sci.* **1975**, *17*, 338–347. [CrossRef]
26. Ngoma, G.D.; Godard, F. Flow distribution in an eight level channel system. *Appl. Therm. Eng.* **2005**, *25*, 831–849. [CrossRef]
27. Cho, E.S.; Choi, J.W.; Yoon, J.S.; Kim, M.S. Modeling and simulation on the mass flow distribution in microchannel heat sinks with non-uniform heat flux conditions. *Int. J. Heat Mass Transf.* **2010**, *53*, 1341–1348. [CrossRef]
28. Baek, S.; Lee, C.; Jeong, S. Effect of flow maldistribution and axial conduction on compact microchannel heat exchanger. *Cryogenics* **2014**, *60*, 49–61. [CrossRef]
29. Cabezas-Gómez, L.; Navarro, H.A.; Saiz-Jabardo, J.M. Thermal performance of multipass parallel and counter-cross-flow heat exchangers. *J. Heat Transf.* **2007**, *129*, 282–290. [CrossRef]
30. Pignotti, A.; Shah, R.K. Effectiveness-number of transfer units relationships for heat exchanger complex flow arrangements. *Int. J. Heat Mass Transf.* **1992**, *35*, 1275–1291. [CrossRef]
31. Lobo, W.E.; Evans, J.E. Heat transfer in radiant section of petroleum heaters. *Trans. Am. Inst. Chem. Eng.* **1939**, *35*, 743–751.
32. Jegla, Z.; Hájek, J.; Vondál, J. Numerical analysis of heat transfer in radiant section of fired heater with realistic imperfect geometry of tube coil. *Chem. Eng. Trans.* **2014**, *39*, 889–894. [CrossRef]
33. Li, C.; Hu, G.; Zhong, W.; Cheng, H.; Du, W.; Qian, F. Comprehensive simulation and optimization of an ethylene dichloride cracker based on the one-dimensional Lobo–Evans method and computational fluid dynamics. *Ind. Eng. Chem. Res.* **2013**, *52*, 645–657. [CrossRef]
34. Morales-Fuentes, A.; Polley, G.T.; Picón-Núñez, M.; Martínez-Martínez, S. Modeling the thermo-hydraulic performance of direct fired heaters for crude processing. *Appl. Therm. Eng.* **2012**, *39*, 157–162. [CrossRef]
35. Hewitt, G.F.; Shires, G.L.; Bott, T.R. *Process Heat Transfer*, 1st ed.; CRC Press: Boca Raton, FL, USA, 1994; ISBN 9780849399183.
36. Tucker, R.; Ward, J. Identifying and quantifying energy savings on fired plant using low cost modelling techniques. *Appl. Energy* **2012**, *89*, 127–132. [CrossRef]
37. Jegla, Z. Development of modified plug-flow furnace model for identification of burner thermal behavior. *Chem. Eng. Trans.* **2013**, *35*, 1195–1200. [CrossRef]

38. Jegla, Z.; Horsák, J.; Turek, V.; Kilkovský, B.; Tichý, J. Validation of developed modified plug-flow furnace model for identification of burner thermal behaviour. *Chem. Eng. Trans.* **2015**, *45*, 1189–1194. [CrossRef]
39. Ptáčník, R. Analysis, Synthesis, and Retrofit Design of Heat Exchanger Networks Containing Multipass Crossflow Heat Exchangers. Ph.D. Thesis, Brno University of Technology, Brno, Czech Republic, 1991.

© 2019 by the authors. Licensee MDPI, Basel, Switzerland. This article is an open access article distributed under the terms and conditions of the Creative Commons Attribution (CC BY) license (http://creativecommons.org/licenses/by/4.0/).

Article

Evaluation of the Complexity, Controllability and Observability of Heat Exchanger Networks Based on Structural Analysis of Network Representations

Daniel Leitold [1,2], Agnes Vathy-Fogarassy [1,2] and Janos Abonyi [2,*]

1. Department of Computer Science and Systems Technology, University of Pannonia, Egyetem u. 10, H-8200 Veszprém, Hungary; leitold@dcs.uni-pannon.hu (D.L.); vathy@dcs.uni-pannon.hu (A.V.-F.)
2. MTA-PE Lendület Complex Systems Monitoring Research Group, University of Pannonia, Egyetem u. 10., POB. 158, H-8200 Veszprém, Hungary
* Correspondence: janos@abonyilab.com; Tel.: +36-88-624000 (ext. 6078)

Received: 25 December 2018; Accepted: 2 February 2019; Published: 6 February 2019

Abstract: The design and retrofit of Heat Exchanger Networks (HENs) can be based on several objectives and optimisation algorithms. As each method results in an individual network topology that has a significant effect on the operability of the system, control-relevant HEN design and analysis are becoming more and more essential tasks. This work proposes a network science-based analysis tool for the qualification of controllability and observability of HENs. With the proposed methodology, the main characteristics of HEN design methods are determined, the effect of structural properties of HENs on their dynamical behaviour revealed, and the potentials of the network-based HEN representations discussed. Our findings are based on the systematic analysis of almost 50 benchmark problems related to 20 different design methodologies.

Keywords: heat exchanger network; structural controllability; structural observability; operability; network science; sensor and actuator placement

PACS: 02.30.Yy; 02.40.Pc; 02.40.Re; 02.50.Sk; 89.20.Ff; 89.75.-k; 89.75.Da; 89.75.Fb; 89.75.Hc; 89.75.Kd

MSC: 93B07; 93B51

1. Introduction

More now than ever, industrial processes are integrated to increase efficiency [1]. Process integration (PI) dates back to 1970 when PI was the response to the oil crisis, and, since then, this field has been a hot topic as it can be utilised to minimise energy and water usage [2], waste as well as emissions [3]. Since then, tools developed for Heat and Energy Integration have become essential elements of Process Design [4], including Energy Storage Systems [5]. Meanwhile, PI increases the efficiency of the technologies, and the increased complexity also complicates operations. Heat Integration also often results in more complex but less operable technologies [6]. Thus, a methodology to support the integrated system design must have a good qualitative and quantitative model that highlights the characteristics of the processes [7].

Although many works deal with the controllability [8] and observability [9] of complex systems, the connection between the complexity of the system and the difficulty of operations has not been examined in details. However, the analysis of the structural motifs of the building elements of complex systems and their models can highlight potentially useful information that can support the design and operation. The structure-relevant analysis of Heat Exchanger Networks (HENs) has already been proven to be beneficial, e.g., the resilience index (RI) quantifies the ability of a HEN to deal with disturbances [10], the controllability index measures the controllability of the HENs [11], and structural analysis can be applied to determine the locations of additional sensors that can reveal otherwise indistinguishable faults [12].

The connection between the structural properties of HENs designed by different methodologies and the operability of the system has yet to be examined in details. The optimisation of HENs can be formalised in several ways. Algorithms can focus on the minimum number of matches [13], the Maximum Energy Recovery (MER) [14], the Minimum Energy-Capital cost [15], the minimum Total Annualised Cost (TAC) [16], the minimum number of exchangers [17], or, in the case of retrofit design, the minimum number of additional exchangers and the additional area of the exchangers or piping costs [18]. The goals listed above can be achieved by different algorithms, such as the Pinch methodology [19], dual-temperature approach method [20], pseudo-pinch [21], Supertargeting [14], State-Space approach [16], branch- and bound-based algorithms [22], or the application of Genetic Algorithm and Simulated Annealing (GA and SA, respectively) [23].

Although there are no systematic studies related to how optimisation algorithms affect the structural properties of HENs with regard to operability, there are some well-known relations. The pinch methodology determines the minimum temperature difference, ΔT_{min}, which divides the HEN into two sub-networks: above the Pinch and below the pinch [19]. The method enables an MER design to be created, i.e., it minimises the energy required. The price of the MER design is the increased number/area and installation costs of heat exchangers. In contrast, optimisation based on the minimum number of matches decreases the number of heat exchangers in addition to the installation cost [24], and makes the operation easier [17]. As this approach has a negative impact on TAC and utility costs, other approaches aim to minimise them [25]. The tradeoff between the targets is continuously changing from one study to the next, and it is unequivocal that the tradeoff has a significant effect on the structure of the designed HEN, along with the controllability and observability of the system.

The dynamical characteristics of the HENs have increasingly become the focus of attention, such as controllability [26,27], observability [28], flexibility [29] or operability [30,31]. The trends mentioned above highlight that it is more critical to automatically qualify the operability-related dynamical properties of HENs based on structural information. For this purpose, HENs can be transformed into the networks of state variables [32], streams and matches [33], or networks of state-space representations [34].

Over the last five years, the system analysis based on the network has spread quickly, as the number and location of necessary actuators [8] and sensors [9] can be determined efficiently by utilising the structure-based maximum matching algorithm. The maximum matching algorithm is a hot topic in other fields such as pattern recognition [35] or machine vision [36] and method to determine maximum matching in large graphs also proposed [37]. The maximum matching algorithm is widely applied also in the fields of pattern recognition [35], machine vision [36], and it can scale up to handle large graphs [37]. With this methodology, the systems are analysed in terms of how does the correlation degree affect the controllability [38], robustness [39] and energy demand [40]. How structural motifs in the network influence the controllability of a transcription network [41], a human protein-protein interaction network [42] or cancer metabolic networks [43] has also been studied.

These promising applications suggest that it could be beneficial to utilise this methodology to reveal the effect of the complexity of HENs in terms of their operability and to address the question concerning what kind of measures of network science can be applied to compare the complexity of HENs obtained by different design methodologies.

Section 2 presents three different network-based representations of HENs to explore motifs that have a significant effect on structural controllability and observability. To evaluate the structural controllability and observability of HENs, the minimum sets of driver and sensor nodes were generated. Additionally, an extended and more exhaustive version of the maximum matching-based approach was used [8] since it can underestimate the number of controllers required [44]. The generated set of driver nodes Was evaluated structurally through the number and location of driver nodes to determine control-relevant installation costs. The relative degree of the resulting dynamical system was also analysed to assess the sluggishness [45] and difficulty of the operation. A new methodology to decrease the relative degree is introduced in Section 2. Section 3 presents how the proposed approach can be used in the systematic analysis of almost 50 HEN design problems and how the developed measures can be used to compare different design methodologies.

2. Network-Based Evaluation of the Complexity, Controllability and Observability of HENs

The workflow of the proposed method is depicted in Figure 1. The method handles several goal-oriented representations of HENs to analyse their specific properties. As this work focuses on graph-based approaches, approaches based on temperature-enthalpy, heat-content or temperature-interval diagrams [46] are not discussed. As our goal was to study dynamic structural characteristics, this section presents the three most relevant models that can be used for such a purpose. After the introduction of the network-based representations, the systematic analysis of the extracted network is presented (see Figure 1). For this purpose, a MATLAB toolbox was developed, which is available at www.abonyilab.com.

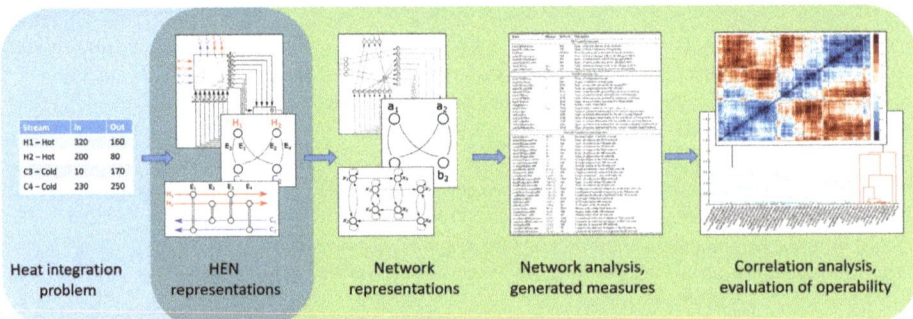

Figure 1. Workflow of the proposed methodology.

2.1. Network Representations of HENs

Besides the widely applied Process Flow Diagrams (PFDs), HENs of hot and cold streams are represented in three different ways, as illustrated in Figure 2.

The first classical representation is the state-space (SS) approach (Figure 2a), which consists of the Distribution Network (DN) and the superstructure operator [34]. The DN determines how units are located on the streams, while the superstructure operator shows the interactions between the streams. The SS-based network representation (Figure 2b) can be defined as $G_{SS}(V_{SS}, E_{SS})$ based on the V_{SS} set of vertices that contain the inlet and outlet of the streams and the E_{SS} edges that present the connections between the heat exchangers.

The second classical representation shows how streams are matched according to the units, namely heat exchangers, utility heaters and utility coolers (Figure 2c) [33]. The second SM-based network representation is almost the same as the classical one (Figure 2d), and it can be defined as $G_{SM}(V_{SM}, E_{SM})$, where the nodes ($V_{SM} = \{V_{hs}, V_{cs}, V_{hp}, V_{cw}\}$) denote the sets of hot streams (V_{hs})d an cold streams (V_{cs}), the high pressure steam (V_{hp}) and the cold water (V_{cw}). The edges ($E_{SM} = \{E_{he}, E_{uh}, E_{uc}\} \subseteq V_{SM} \times V_{SM}$) represent the heat exchangers ($E_{he} \subseteq V_{hs} \times V_{cs}$), the utility heaters ($E_{uh} \subseteq V_{hp} \times V_{cs}$) and the utility coolers ($E_{uc} \subseteq V_{hs} \times V_{cw}$). The two partitions of the nodes are the hot streams (Process and utility, $\{V_{hs} \cup V_{hp}\}$) and the cold streams (Process and utility, $\{V_{cs} \cup V_{cw}\}$). The weights of the nodes represent the heat capacity of the streams, while the weights of the edges represent the heat load on the units.

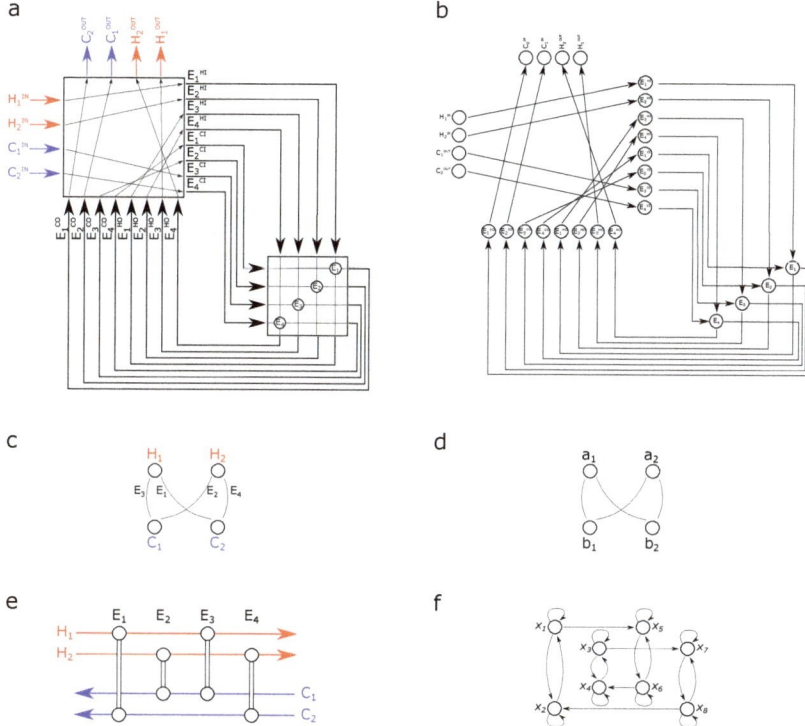

Figure 2. Classical representations of HENs and their graph-based models. The representations are as follows: (**a**) state-space approach (SS); (**b**) network of state-space approach; (**c**) streams and matches (SM); (**d**) bipartite network representation of streams and matches; (**e**) grid diagram; and (**f**) network representation of state variables.

The third and most common classical representation of a HEN is the Grid diagram (GD) [6,47], where the streams are presented as they are connected through the heat exchangers and influenced by utility units (Figure 2e). The dynamics of the units, i.e., the heat exchangers, utility coolers and utility heaters can be described by a differential-algebraic system of equations (DAE) as [48]:

$$\frac{dT_{ho}}{dt} = \frac{v_h}{V_h}(T_{hi} - T_{ho}) + \frac{UA}{c_{ph}\rho_h V_h}(T_{co} - T_{ho}), \tag{1}$$

$$\frac{dT_{co}}{dt} = \frac{v_c}{V_c}(T_{ci} - T_{co}) + \frac{UA}{c_{pc}\rho_c V_c}(T_{ho} - T_{co}), \tag{2}$$

where T_{hi}, T_{ci}, T_{ho} and T_{co} denote the temperatures of the hot input, cold input, hot output and cold output streams, respectively; v_h and v_c represent the flow rates of the cold and hot streams; V_h and V_c stand for the volumes of the hot and cold side tanks of the heat exchanger; U is the heat transfer coefficient; A denotes the heat transfer area of the heat exchanger; and c_p and ρ are the specific heat and density of the streams.

In the above representation, the state variables are the temperatures of the outlet streams of the heat exchanger $x(t) = [T_{ho}, T_{co}]^T$, the temperature of the inlet streams are regarded as disturbances $d(t) = [T_{hi}, T_{ci}]^T$, and $v_c(t)$ and $v_h(t)$ are time-varying parameters [48]. The dynamics of the utility units are similar. Utility coolers can be described by Equation (1) by considering the temperature of the cold water stream T_{co} as a controlled input variable. Analogously, utility heaters are modeled by

Equation (2) where T_{ho} denotes the controlled inputs. HENs based on these building blocks can be represented by linear state-space models in the general form of Equations (3) and (4).

$$\dot{x} = Ax + Bu + \Gamma d \qquad (3)$$

$$y = Cx + Du \qquad (4)$$

Recently, Liu et al. introduced a network science-based representation of dynamical systems and a methodology to determine the minimal input configuration that ensures the controllability of the system [8]. The third, state-space model-based network representation (Figure 2f) easily lends itself to the application of this methodology. The resultant network can be described as a graph $G_{DAE} = (V_{DAE}, E_{DAE})$, where vertices represent the state variables x, while the edges are derived from the structure matrix of the state-transition matrix A as if $A_{ij} \neq 0$, then an edge from x_j to x_i exists. As a result of Equations (1) and (2), the simplest building blocks of HENs can be defined as the heat exchanger cells, the utility coolers and the utility heaters. Their grid diagram and state-space-based network representations can be seen in Figure 3.

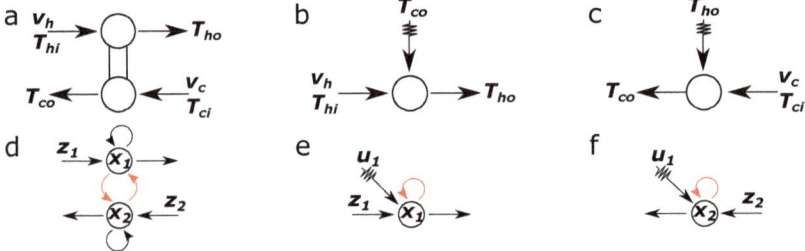

Figure 3. The grid diagram and state-space network representations of: a heat exchanger cell (**a,d**); a utility cooler (**b,e**); and a utility heater (**c,f**). Red edges denote loops in the network representation, which create a strongly connected component in each elementary building block of HENs. In the case of the heat exchanger cell, the hot and cold sides of the exchanger belong to the same component even though they also have their own intrinsic dynamics, as in the case of utility units.

The SS-based network representation can be easily applied to HENs even when the streams are mixed and split. In the other two network representations, the handling of mixers and bypasses is less straightforward.

The number of vertices and edges of the SM-based network representation directly represents the streams and units, respectively. Loops and paths can be identified more easily in this network representation than in any other representations. This property is essential when the goal is not the MER design but to minimise the number of units that can be readily determined by the equation $U_{un} = N_s + L - S$, where U_{un} denotes the number of units, N_s the number of streams ($|V_{GD}|$), L the number of independent loops and S the number of separate components in the network [33]. Loops and paths should be excluded from the designs, where the target is to minimise the number of units. The presence of splitters and mixers does not influence the property $U_{un} = N_s + L - S$ when Pinch Partitioning is excluded [49]. Furthermore, with this representation the degree of freedom of the heat loads can also be determined as $H - N_s + S$, where H denotes the number of heat exchangers ($|E_{he}|$) in the network.

The DAE-based network representation is suitable to analyse the operability of HENs. A system is controllable if it can be derived from an initial state to any desired final state over a finite period of time, while it is observable if any of the internal states can be reproduced by the knowledge of the initial state in addition to all the inputs and outputs [50]. The system is structurally controllable (observable) if the controllability (or observability) matrix, $\mathcal{C} = [B, AB, \ldots, A^{n-1}B]$ (or $\mathcal{O} = [C^T, (CA)^T, \ldots, (CA^{n-1})^T]^T$) is of full rank, $rank(\mathcal{C}) = n$ (or $rank(\mathcal{O}) = n$) [51].

Liu et al. utilised the maximum matching algorithm on the DAE-based network representation to determine the minimum number of actuators (or sensors) and create a controllable (or observable) input (or output) configuration [8]. Nevertheless, how the network is represented is critical. If the adjacency matrix of the network is identical to the structure matrix of the state-transition matrix of the linear dynamical equation seen in Equation (3), then observability can be analysed, while in the reversed direction the controllability of the system can be investigated.

As each motif in the DAE-based network representation possesses a self-loop (coloured in red in Figure 3), i.e., the diagonal elements are non-zero values in the state-transition matrix, maximum matching selects these edges and leaves no unmatched node that can be a driver (or sensor) node. Since each HEN is constructed from these motifs, the approach is unusable in the case of HENs. Nevertheless, this method can be extended to be suitable for dynamical systems that exhibit this kind of behaviour. This method is referred to as the path-finding method [44] as it is also based on maximum matching, but following the maximum matching it searches for circles which can be cut off into Hamiltonian paths.

Since only the DAE network representation is detailed enough to determine the location of actuators and sensors, the relative degree can only be interpreted by this representation. To evaluate the "physical closeness" or "direct effect" of the control configuration of a nonlinear system with state equation $\dot{x} = f(x) + \sum_{j=1}^{m} g_j(x)u_j + \sum_{\kappa=1}^{p} w_\kappa(x)d_\kappa$ and output equation $y_i = h_i(x)$, relative order is introduced as a structural measure of the initial sluggishness of the response [45]. The relative degree can be determined by the standard Lie derivative. Therefore, for scalar field $h_i(x)$ and vector field $f(x)$ it is defined as $\mathcal{L}_f h_i(x) = \sum_{l=1}^{n} (\partial h(x)/\partial x_l) f_l(x))$, where $f_l(x)$ denotes the row element l of $f(x)$. Higher order Lie derivatives are defined as $\mathcal{L}_f^k h_i(x) = \mathcal{L}_f \mathcal{L}_f^{k-1} h_i(x)$, while mixed Lie derivatives $\mathcal{L}_{g_j} \mathcal{L}_f^{k-1} h_i(x)$ in an obvious way [45]. Then, relative degree r_i is defined for output y_i as the smallest integer for which $[\mathcal{L}_{g_1} \mathcal{L}_f^{r_i-1} h_i(x) \cdots \mathcal{L}_{g_m} \mathcal{L}_f^{r_i-1} h_i(x)] \neq [0 \cdots 0]$ with respect to input vector $\mathbf{u}$ if exists, otherwise $r_i = \infty$. The relative order r_{ij} of output y_i with respect to input u_j is the smallest integer for which $\mathcal{L}_{g_j} \mathcal{L}_f^{r_{ij}-1} h_i(x) \neq 0$ if exists, otherwise $r_{ij} = \infty$. The relationship between r_i and r_{ij} can be defined as $r_i = min(r_{i1}, r_{i2}, \ldots, r_{im})$. Analogously to r_{ij}, the relative order $\rho_{i\kappa}$ of output y_i with respect to disturbance d_κ can be defined as the smallest integer $\mathcal{L}_{w_\kappa} \mathcal{L}_f^{\rho_{i\kappa}} h_i(x) \neq 0$ if exists, otherwise $\rho_{i\kappa} = \infty$. For linear systems that are defined above, r_{ij} can be defined as the smallest integer for which $c_i \mathbf{A}^{r_{ij}-1} b_j \neq 0$, where c_i is row i of $\mathbf{C}$ and b_j is column j of $\mathbf{B}$. The relative order $\rho_{i\kappa}$ with respect to the disturbance d is the smallest integer for which $c_i \mathbf{A}^{\rho_{i\kappa}-1} \gamma_\kappa \neq 0$, where γ_κ denotes column κ of Γ. By utilizing the DAE network representation, the relative degree r_{ij} can be defined as the shortest path ℓ_{ij} from input u_j to output y_i minus one, $r_{ij} = \ell_{ij} - 1$. Analogously, the relative degree $\rho_{i\kappa}$ can be defined as the geodesic path $\ell_{i\kappa}$ from disturbance d_κ to output y_i minus one ($\rho_{i\kappa} = \ell_{i\kappa} - 1$). The visual representation of the relative degree can be seen in Figure 4.

As can be seen in Figure 4, the shaded area represents clusters of state variables that can be achieved by a relative degree that is smaller than r_i. The operability of HENs can be improved by minimising the highest relative degree with the addition of actuators or sensors [52] as will be presented in the following subsection.

The analysis of the previously introduced network representations can highlight the structural properties of HENs. The developed network-based measures are summarised in Tables 1 and 2.

Table 1. Building elements of the studied network-based representations of HENs.

Property	SS	SM	DAE
node (V_i)	Connection of streams.	Stream.	State variable.
edge (E_i)	Stream.	Match: a unit.	Dynamical effect between two state variables.
node weight (n_i)	The nodes are not weighted.	The heat load of the stream.	The output temperature of the exchanger.
edge weight (w_i)	Edges are not weighted.	Heat load exchanged between the streams.	Dynamical effect between state variables determined by the state-transition matrix.
direction	The network is directed, edge directions are based on the direction of the streams [34]. The network is more transparent when—similarly to Grid diagrams—the hot streams run from the left to the right and cold vice versa to follow the direction of Composite Curves.	The network is undirected.	The network is directed, the direction of edges is based on matrix **A** [8] that reflects the direction of the streams and the heat transfer.
loops	Loops can reflect superfluous units.	Loops appear when there are more units than necessary.	Loops create strongly connected components that can be controlled (or observed) by only one driver (or sensor) node.
maximum matching [8]	n/a	n/a	Unmatched nodes provide driver and sensor nodes.
relative order [45]	n/a	n/a	Difficulty in operation.
number of components (S)		Components are the disconnected sub-networks in the HEN.	

Table 2. Structural properties of the SS, SM and DAE-based network representations of HENs. The operator $|X|$ yields the cardinality of set X, and the operator $\langle X \rangle$ yields the average of the values in X.

Property	Calculation	SS	SM	DAE
number of nodes	$N = |V|$	Starting- and endpoints of the streams.	Number of streams (Process and utility).	Number of state variables necessary to describe the dynamics of HEN.
number of edges	$M = |E|$	Number of stream intervals.	Number of units.	Nonzero elements of the state-transition matrix.
degree	$k_i^{out} = \sum_{j \in V} A_{ij}$, $k_i^{in} = \sum_{j \in V} A_{ji}$, $k_i = k_i^{out} + k_i^{in}$	Fixed for the nodes, $k_i = 1$ for stream source and drains, $k_i = 2$ for endpoints of stream intervals and $k_i = 4$ for heat exchanger units.	k_i is the number of units on stream.	Fixed for the nodes, k_i moves from 2 to 4 in case of heat exchanger, and from 1 to 2 in case of utility units (loops are excluded).
shortest path	$\ell_{i,j}$	The minimum of length of paths from node i to node j.		
distance matrix	D	The distance matrix contains the shortest distance between node i and j, $D_{i,j}$. The distance of node j is $D_i = \sum_j D_{ij}$.		
cycle rank [53]	L	The number of independent loops in the network.		
total walk count (TWC) [54]	$TWC(G) = \sum_{l=1}^{N-1} \sum_i \sum_j A^l$	The number of walks is proportional to the complexity, as the walks can represent the effect of an input signal in the system. More walks means more effect in the HEN.		
coefficient of network complexity (CNC) [55]	$CNC(G) = M^2/N$	Due to the fixed degree, the CNC is constrained.	The proportion of units to streams.	Due to the fixed degree, the CNC is constrained.
eccentricity [54]	$Ecc(i) = \max_{j \in V}(\ell_{ij})$ $Ecc(G) = \max_{j \in V} Ecc(i)$	Maximum length of heat loads.	n/a.	The relative order of the system.
Wiener index [56]	$W(G) = \sum_{(v,w) \in E} \ell_{v,w}$	Reflects the size of the HEN, and the complexity as long paths increase $W(G)$.	In undirected case, $W(G)$ should be divided by two. Since the network is undirected, $W(G)$ approximates the size of the system.	Longer geodesic paths represent higher order dynamics.
A/D index [54]	$AD(G) = \langle k_i \rangle / \langle d_i \rangle$	Since the degree is fixed, the index approximates the closeness of the units.	As the order of heat exchanger units is not given, the A/D index does not necessarily characterise heat loads or HEN.	Since degree is fixed, the index approximates the closeness of units.
Balaban-J index [57]	$BJ(G) = \frac{M}{L+1} \sum_{(v,w) \in E} \sqrt{D_v D_w}$	Balaban-J index can only be calculated to undirected graphs.	It approximates the size of the HEN.	Balaban-J index can only be calculated to undirected graphs.
connectivity index [56]	$CI(G) = \sum_{(v,w) \in E} 1/\sqrt{k_v k_w}$	Since the degree is fixed, the index cannot characterise the HEN.	Shows how the streams are interconnected.	Since the degree is fixed, the index cannot characterise the HEN.

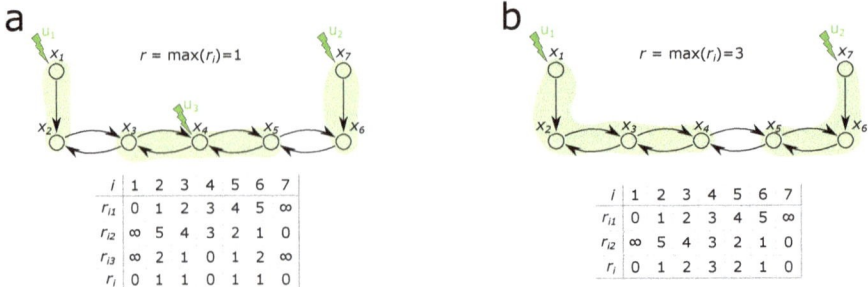

Figure 4. Visual representation of relative degree in the DAE-based network representation. (**a**) The relative degree is equal to one if three actuators were placed in the network, while (**b**) it is three if only two actuators were assigned to the system.

2.2. Operability-Focused Sensor and Controller Placement in HENs

Two methods are used to extend the minimal input and output configurations [58]. The first approach utilises the set-covering method. Firstly, the allowed maximal relative degree r_{max} is defined. Secondly, the set of nodes W_i reachable from node i over a maximum of r_{max} steps was determined. In the formulation of the algorithm, $\mathcal{U}$ denotes the set of all the state variables, C the set of the actuators necessary to ensure structural controllability, and O the set of sensors necessary for structural observability. In the case of input (or output) configuration, J represents the set of necessary driver (or sensor) nodes, such that P is the set of state variables that is covered by J, namely $P = \cup_{j \in J} W_j$. The goal is to minimise J such that $P = \mathcal{U}$ and $C \subset J$ (or $O \subset J$); moreover, $r_u \leq r_{max}$, and $\forall u \in \mathcal{U}$. When initial input or output configurations are not given, then the method yields a global optimum for the problem. A greedy algorithm was applied to solve the set-covering problem [59].

The second approach uses two network-specific measures, namely the closeness and the node betweenness centrality measures. The closeness centrality of node i can be calculated using Equation (5).

$$Cc(i) = \frac{N-1}{\sum_{j \neq i} \ell_{i,j}} \qquad (5)$$

Betweenness centrality calculates the number of geodesic paths that intercept node i ($\sigma_{st}(i)$) and divides this by the number of all the geodesic paths (σ_{st}) for each start s and target t node, such that $s \neq i \neq t$. The betweenness centrality of node i is shown in Equation (6).

$$Bc(i) = \sum_{s \neq i \neq t} \frac{\sigma_{st}(i)}{\sigma_{st}} \qquad (6)$$

For each component that exceeds r_{max}, a node with the highest centrality measure is selected as an additional actuator (or sensor), i.e., $C = C \cup \{i : \max(Cc(i)Bc(i)), i \notin C\}$ in the case of the input configuration and $O = O \cup \{i : \max(Cc(i)Bc(i)), i \notin O\}$ in the case of the output configuration. The steps are repeated iteratively until all the components exhibit an order not in excessive of r_{max}.

For all HENs presented, the number of additional actuators and sensors required to manage the system with r_{max} was calculated by the set-covering method (global optimum), the retrofit set-covering method (using existing configurations) and the retrofit centrality-based method (using current configurations). In this analysis, the maximal order was determined as $r_{max} = \lfloor d_{max}/4 \rfloor$, where d_{max} denotes the diameter of the network.

3. Systematic Analysis of HENs

3.1. The Studied Benchmark Problems

The well-known benchmark sets of Furman and Sahinidis (2004) [60], Chen et al. (2015) [61,62] and Grossmann (2017) [63] were used to study how the proposed measures can be used to evaluate HENs and compare different design methodologies of HENs. The benchmark problems and the applied methods are summarised in Table 3, while the notation of the utilised methods and their objective functions are shown in Table 4. These problem sets contain 48 problems and 23 methods, as well as 639 different HENs, of which 539 are unique. Fifty-three different measures, as summarised in Table 5, were generated for all the HENs.

Table 3. Benchmark HENs and their optimisation methods that are used during the analysis. As the heuristic methods, CP, CRR, CSH, FLPR, GP, GSH, LFM, LHM, LHM-LP, LRR, SS, WFG and WFM were applied to each problem. This set of methods is denoted as HEU referring to heuristic methods. The methods are introduced in Table 4, where abbreviations are also presented.

Problem	Hot Streams	Cold Streams	Methods	Source
Furman and Sahinidis (2004)				
4sp1	2	2	HEU, BB	[24]
6sp-cf1	3	3	HEU, RET	[18]
6sp-gg1	3	3	HEU	[64]
6sp1	3	3	HEU, BB	[24]
7sp-cm1	3	4	HEU	[65]
7sp-s1	6	1	HEU	[66]
7sp-torw1	4	3	HEU	[67]
7sp1	3	4	HEU, SYN	[25]
7sp2	3	3	HEU, SYN	[25]
7sp4	6	1	HEU	[68]
8sp-fs1	5	3	HEU, E	[69]
8sp1	4	4	HEU	[70]
9sp-al1	4	5	HEU, E, MER, EC, ECR, ST	[15]
9sp-has1	5	4	HEU	[71]
10sp-la1	4	5	HEU	[14]
10sp-ol1	4	6	HEU	[66]
10sp1	5	5	HEU, BB	[22]
12sp1	9	3	HEU	[70]
14sp1	7	7	HEU	[70]
15sp-tkm	9	6	HEU, E	[72]
20sp1	10	10	HEU	[70]
22sp-ph	11	11	HEU	[73]
22sp1	11	11	HEU	[16]
23sp1	11	12	HEU, DS	[17]
28sp-as1	16	12	HEU	[74]
37sp-yfyv	21	16	HEU, GASA	[23]
Chen et al. (2015a,b)				
balanced5	5	5	HEU	[61,62]
balanced8	8	8	HEU	[61,62]
balanced10	10	10	HEU	[61,62]
balanced12	12	12	HEU	[61,62]
balanced15	15	15	HEU	[61,62]
unbalanced5	5	5	HEU	[61,62]
unbalanced10	10	10	HEU	[61,62]
unbalanced15	15	15	HEU	[61,62]
unbalanced17	17	17	HEU	[61,62]
unbalanced20	20	20	HEU	[61,62]
Grossmann (2017)				
balanced12_random0	12	12	HEU	[63]
balanced12_random1	12	12	HEU	[63]
balanced12_random2	12	12	HEU	[63]
balanced15_random0	15	15	HEU	[63]
balanced15_random1	15	15	HEU	[63]
balanced15_random2	15	15	HEU	[63]
unbalanced17_random0	17	17	HEU	[63]
unbalanced17_random1	17	17	HEU	[63]
unbalanced17_random2	17	17	HEU	[63]
unbalanced20_random0	20	20	HEU	[63]
unbalanced20_random1	20	20	HEU	[63]
unbalanced20_random2	20	20	HEU	[63]

Table 4. Utilised methods and their objective functions.

Name	Method	Objective Function	Source
BB	Branch and bound	Minimises the heat exchanger and utility costs.	[24]
CP	CPLEX-solved transportation model.	Minimum number of matches.	[13]
CRR	Covering Relaxation Rounding	Minimum number of matches.	[13]
CSH	CPLEX-solved transshipment model.	Minimum number of matches.	[13]
DS	Decomposition Strategy	Minimises the number of exchangers.	[17]
E	No method declared	Existing network, not optimised.	
EC	Pinch methodology	Energy-Capital cost minimisation.	[14,15]
ECR	Pinch methodology	Energy-Capital Retrofit cost minimisation.	[14,15]
FLPR	Fractional LP Rounding	Minimum number of matches.	[13]
GASA	Genetic Algorithm with Simulated Annealing	Minimises the annual cost.	[23]
GP	Gurobi-solved transportation model.	Minimum number of matches.	[13]
GSH	Gurobi-solved transshipment model.	Minimum number of matches.	[13]
LFM	Largest Fraction Match	Minimum number of matches.	[13]
LHM	Largest Heat Match Greedy	Minimum number of matches.	[13]
LHM-LP	Largest Heat Match LP-based	Minimum number of matches.	[13]
LRR	Lagrangian Relaxation Rounding	Minimum number of matches.	[13]
MER	Pinch methodology	Maximum Energy Recovery, Minimum Energy Requirement.	[14,15]
RET	Structural modifications by categories	Minimises the cost of new exchangers, exchanger areas and piping.	[18]
SS	Shortest Stream	Minimum number of matches.	[13]
ST	Pinch methodology	Supertargeting.	[14,15]
SYN	Heuristic stage-by-stage structuring synthesis	Minimises the annual cost.	[25]
WFG	Water Filling Greedy	Minimum number of matches.	[13]
WFM	Water Filling MILP	Minimum number of matches.	[13]

Table 5. Calculated measures and their short descriptions.

Name	Measure	Network	Description
HEN-specific measures			
numOfHotStream		SM	Num. of the hot streams in the problem.
numOfColdStream		SM	Num. of the hot streams in the problem.
hasPinch		SS,DAE	True, if Pinch point is determined for the problem.
numOfExchangers		SM	Num. of heat exchanger cells in the designed HEN.
numOfUtilityHeater		SM	Num. of utility heater cells in the designed HEN.
numOfUtilityCooler		SM	Num. of utility cooler cells in the designed HEN.
numOfUnits	U_{un}	SM	Num. of heat exchanger units in the designed HEN.
numOfMinUnits	U_{min}	SM	Num. of minimum units necessary for the problem.
Dynamical properties			
numOfIndLoop	L	SM	Num. of independent loops.
degOfFreedom		SM	Degree of freedom of heat loads.
numOfCompDAE		DAE	Num. of sub-networks in the designed HEN.
numOfCompSM	S	SM	Num. of components in the SM network.
numOfUMCon		DAE	Num. of driver nodes granted by maximum matching.
numOfDriver		DAE	Num. of actuator nodes necessary for controllability.
numOfUMObs		DAE	Num. of sensor nodes granted by maximum matching.
numOfSensor		DAE	Num. of sensor nodes necessary for observability.
sluggishness	r	DAE	Relative order of the HEN.
targetOrder	r_{max}	DAE	Target relative order (in our case: $\lfloor d_{max}/4 \rfloor$).
actCovSize		DAE	Num. of actuators determined by the set-covering method.
senCovSize		DAE	Num. of sensors determined by the set-covering method.
actCovRetSize		DAE	Num. of actuators determined by the retrofit set-covering method.
senCovRetSize		DAE	Num. of sensors determined by the retrofit set-covering method.
actNetMesSizeRet		DAE	Num. of actuators determined by the retrofit centrality-based method.
senNetMesSizeRet		DAE	Num. of sensors determined by the retrofit centrality-based method.
Network-based structural properties			
balabanJIndex	$BJ(G)$	SM	Balaban-J index of the SM network.
numOfNodes.ofDAE	N	DAE	Num. of nodes in the DAE network.
numOfNodes.ofSM	N	SM	Num. of nodes in the SM network.
numOfNodes.ofSS	N	SS	Num. of nodes in the SS network.
numOfEdges.ofDAE	M	DAE	Num. of edges in the DAE network.
numOfEdges.ofSM	M	SM	Num. of edges in the SM network.
numOfEdges.ofSS	M	SS	Num. of edges in the SS network.
averageDegree.ofDAE	$\langle k_i \rangle$	DAE	Average degree in the DAE network.
averageDegree.ofSM	$\langle k_i \rangle$	SM	Average degree in the SM network.
averageDegree.ofSS	$\langle k_i \rangle$	SS	Average degree in the SS network.
eccentricity.ofDAE	$Ecc(G)$	DAE	Graph eccentricity value of DAE network.
eccentricity.ofSM	$Ecc(G)$	SM	Graph eccentricity value of SM network.
eccentricity.ofSS	$Ecc(G)$	SS	Graph eccentricity value of SS network.
totalWalkCount.ofDAE	$TWC(G)$	DAE	Num. of walks in the DAE network.
totalWalkCount.ofSM	$TWC(G)$	SM	Num. of walks in the SM network.
totalWalkCount.ofSS	$TWC(G)$	SS	Num. of walks in the SS network.
coeffNetwCompl.ofDAE	$CNC(G)$	DAE	Coefficient of network complexity in the DAE network.
coeffNetwCompl.ofSM	$CNC(G)$	SM	Coefficient of network complexity in the SM network.
coeffNetwCompl.ofSS	$CNC(G)$	SS	Coefficient of network complexity in the SS network.
adIndex.ofDAE	$AD(G)$	DAE	A/D index of the DAE network.
adIndex.ofSM	$AD(G)$	SM	A/D index of the SM network.
adIndex.ofSS	$AD(G)$	SS	A/D index of the SS network.
wienerIndex.ofDAE	$W(G)$	DAE	Wiener index of the DAE network.
wienerIndex.ofSM	$W(G)$	SM	Wiener index of the SM network.
wienerIndex.ofSS	$W(G)$	SS	Wiener index of the SS network.
connInd.ofDAE.inConn	$CJ(G)$	DAE	Connectivity index for in-degree in DAE network.
connInd.ofDAE.outConn	$CJ(G)$	DAE	Connectivity index for out-degree in DAE network.
connInd.ofSM	$CJ(G)$	SM	Connectivity index for SM network.
connInd.ofSS.inConn	$CJ(G)$	SS	Connectivity index for in-degree in the SS network.
connInd.ofSS.inConn	$CJ(G)$	SS	Connectivity index for out-degree in the SS network.

3.2. Analysis of the 9sp-al1 Problem

The 9sp-al1 benchmark problem was studied in detail to demonstrate the applicability of the proposed methodology. As can be seen in Figure 5, the solutions to the problem generated by the E, MER, EC and ECR methods significantly differed [14]. The first two solutions were based on the pinch point analysis to ensure Minimal Energy Requirement design, while the second two solutions were obtained by minimising energy and capital costs in the case of a new (EC) and retrofit (ECR) design.

The networks extracted from the HEN 9sp-al1-E solution (Figure 5a) can be seen in Figure 6. In DAE-based network representation (Figure 6b), the driver and sensor nodes are denoted by green- and orange-coloured symbols, respectively, while in SM-based network representation, a critical path is denoted by the colour green (Figure 6c).

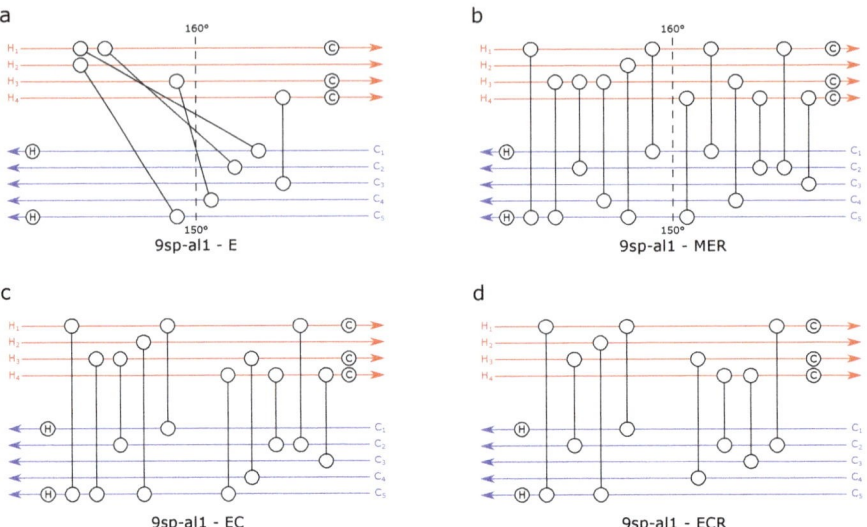

Figure 5. Four designed HEN for the 9sp-al1 problem [14]: (**a**) existing network of an aromatic complexes in Europe; (**b**) maximum Energy Recovery design of the aromatic complexes; (**c**) grassroot (new) design with optimisation for the Energy-Capital tradeoff; and (**d**) retrofit design with optimisation for the Energy-Capital tradeoff.

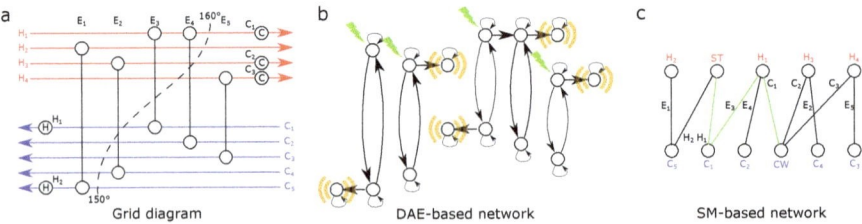

Figure 6. Networks extracted from the HEN 9sp-al1-E: (**a**) grid diagram of the HEN seen in Figure 5a; (**b**) DAE-based network representation, in which the green symbols represent the nodes where the input signal should be shared, and the orange symbols stand for the outputs to grant structural controllability and observability; and (**c**) SM-based network representation, in which green edges show paths between high-pressure steam and cold water, i.e., between utility units which should be broken in order to reach the minimal number of units. In this HEN, no loop appears.

The clusters (communities) in the DAE-based network representation were also analysed by community detection algorithms [75]. The clusters of the HEN 9sp-al1-MER can be seen in Figure 7. Such clustering of the network can highlight useful information, e.g., it can show how the HEN is integrated and how the pinch isolates the clusters.

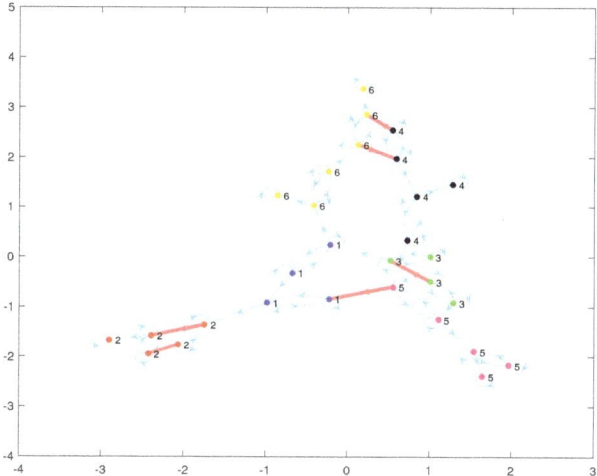

Figure 7. Communities of the DAE-based network representation of HEN 9sp-al1-MER were also identified. The colours of the nodes and their labels denote the community IDs. The highlighted edges are intercept Pinch.

All measures introduced in Table 5 were calculated for the presented HENs, and the results of the analysis are presented in Table 6.

As our analysis also aimed to identify structural parameters that have an impact on the dynamical properties of the HEN, firstly, the correlation between the proposed measures in these four examples was studied (see Figure 8). The results are discussed in the following section together with the analysis of all 639 HENs.

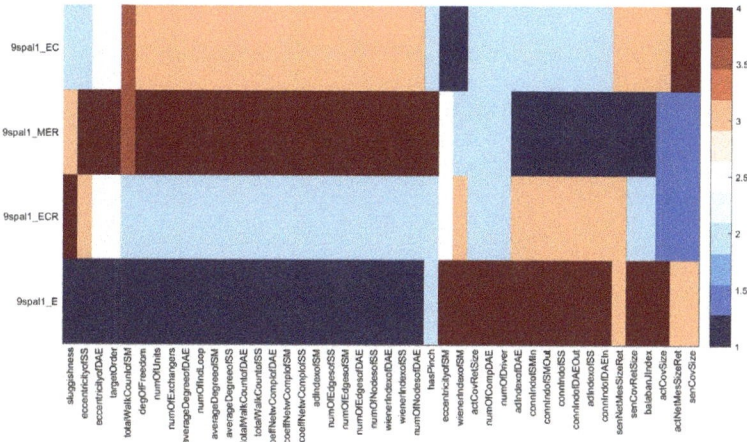

Figure 8. Correlations between the proposed network measures according to four HENs of the 9sp-al1 problem. The rows and columns were ordered based on similarities calculated based on the Spearman's rank correlation. The colours of the map represent the rank values.

Table 6. Calculated measures and their short descriptions.

Name	9sp-al1 - E	9sp-al1 - MER	9sp-al1 - EC	9sp-al1 - ECR
HEN-specific measures				
numOfHotStream	4	4	4	4
numOfColdStream	5	5	5	5
hasPinch	0	1	0	0
numOfExchangers	5	12	10	8
numOfUtilityHeater	2	2	2	2
numOfUtilityCooler	3	3	3	3
numOfUnits	10	17	15	13
numOfMinUnits	10	10	10	10
Dynamical properties				
numOfIndLoop	0	7	5	3
degOfFreedom	−5	2	0	−2
numOfCompDAE	4	1	1	1
numOfCompSM	1	1	1	1
numOfUMCon	0	0	0	0
numOfDriver	4	1	1	1
numOfUMObs	0	0	0	0
numOfSensor	5	5	5	5
sluggishness	2	9	8	11
targetOrder	1	3	2	2
actCovSize	9	7	8	7
senCovSize	5	4	6	4
actCovRetSize	9	7	7	7
senCovRetSize	7	4	6	5
actNetMesSizeRet	9	8	10	8
senNetMesSizeRet	7	4	7	7
Network-based structural properties				
balabanJIndex	9.9212	4.3990	5.6234	5.2789
numOfNodesofDAE	15	29	25	21
numOfNodesofSM	11	11	11	11
numOfNodesofSS	72	107	97	87
numOfEdgesofDAE	31	73	61	49
numOfEdgesofSM	10	17	15	13
numOfEdgesofSS	71	113	101	89
averageDegreeofDAE	2.1333	3.0345	2.8800	2.6667
averageDegreeofSM	1.8182	3.0909	2.7273	2.3636
averageDegreeofSS	1.9722	2.1121	2.0825	2.0460
eccentricityofDAE	3	12	11	11
eccentricityofSM	7	5	4	5
eccentricityofSS	13	28	22	25
totalWalkCountofDAE	802,788	1.02127E+13	92,080,948,186	1,364,370,909
totalWalkCountofSM	34,258	2,114,490	2,114,490	410,032
totalWalkCountofSS	646	9138	3274	1977
coeffNetwComplofDAE	64.0667	183.7586	148.8400	114.3333
coeffNetwComplofSM	9.0909	26.2727	20.4545	15.3636
coeffNetwComplofSS	70.0139	119.3364	105.1649	91.0460
adIndexofDAE	0.8421	0.0522	0.0661	0.0787
adIndexofSM	0.1136	0.2636	0.2586	0.1857
adIndexofSS	0.0446	0.0067	0.0089	0.0107
wienerIndexofDAE	38	1687	1090	712
wienerIndexofSM	176	129	116	140
wienerIndexofSS	3,181	33,524	22,586	16,582
connIndofDAEIn	0.0154	0.0053	0.0065	0.0085
connIndofDAEOut	0.0147	0.0052	0.0063	0.0082
connIndofSS	0.0246	0.0082	0.0103	0.0143
connIndofSMIn	0.0131	0.0077	0.0087	0.0100
connIndofSMOut	0.0131	0.0077	0.0087	0.0100

3.3. Results and Discussion of Systematic Correlation Analysis

Figure 9 shows the similarity-based ordering of the 539 unique networks (on the rows) and the measures (on the columns). The colours of the map represent rankings of the networks based on the given measures. As can be seen, the map is clustered and on the right-hand side measures that are negatively correlated to the measures on the left-hand side are presented. This phenomenon originates from the calculation of the measures (some of these are calculated mostly as the reciprocal of the measures on the left-hand side). The similarities between the most important measures are visualised by a dendrogram in Figure 10.

Based on the analysis of the results, the following conclusions can be made:

Firstly, it is visible that the studied HENs were clustered and the members of such clusters exhibited similar dynamical properties. To confirm this, the Spearman distance for each network based on the measures was generated and visualised on a heat map, as can be seen in Figure 11. On the heat map, three groups can be easily distinguished, and these groups consisted mostly of problems similar in size.

Secondly, the number of unmatched nodes that were generated by maximum matching was equal to zero for all HENs. This was caused by the SCCs and the intrinsic dynamics introduced in Figure 3. Thus, additional driver and sensor nodes should be determined by methods such as path-finding [44].

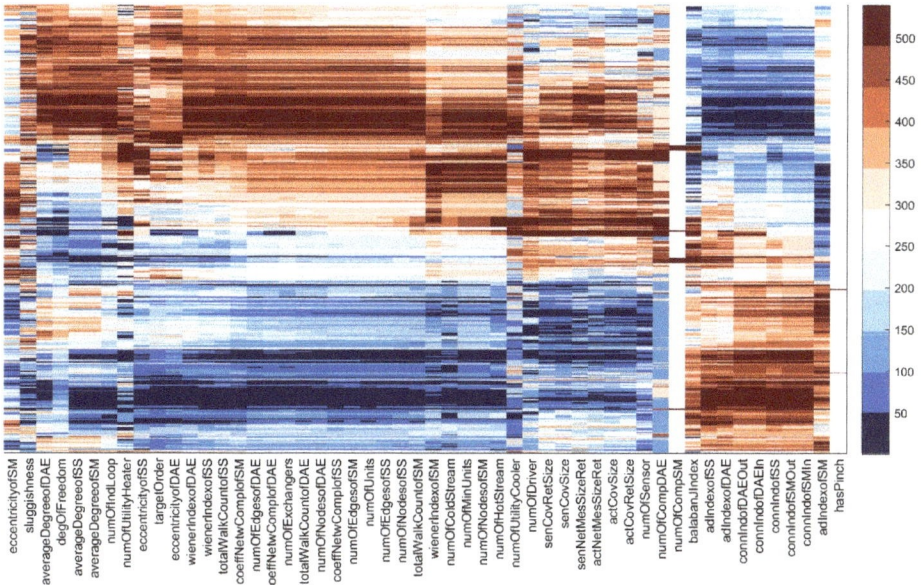

Figure 9. Similarity-based ordering of the 539 HENs (rows) and developed measures (columns). The networks are ranked according to each measure, the similarities were calculated by Spearman's rank correlation and the colours of the map represent the rank values.

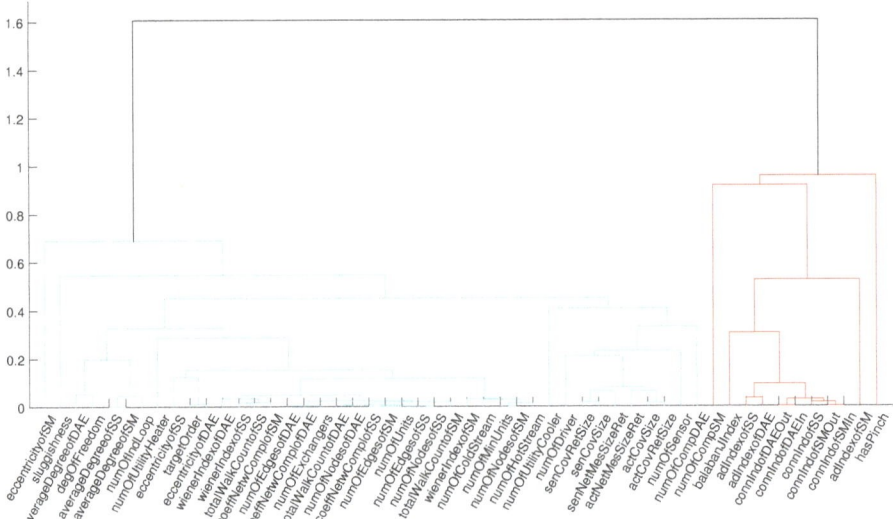

Figure 10. Correlations between the measures. Dendrogram presenting the distances between the measures for all unique HENs.

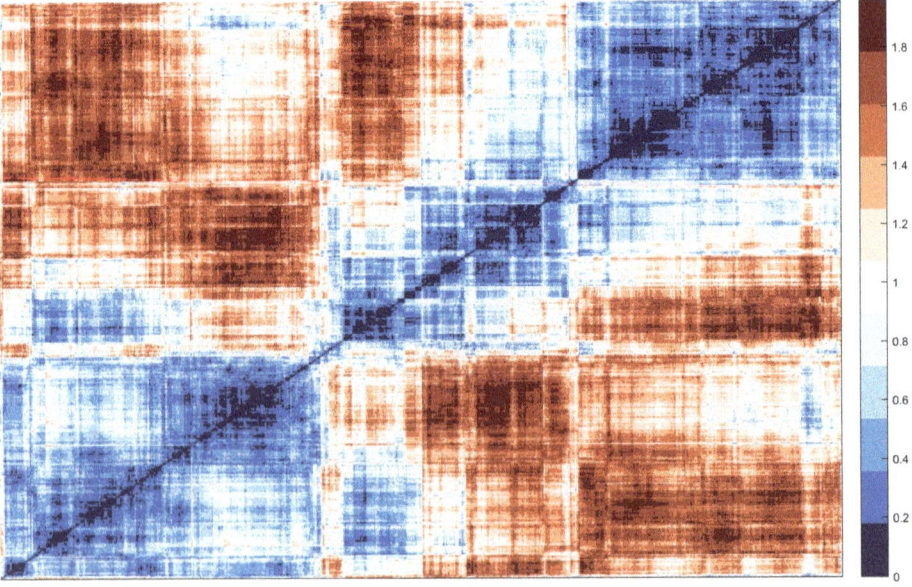

Figure 11. Spearman's rank correlation-based clustering of 539 unique HENs. The clusters of HENs can be easily detected. The method ordered the networks according to their similarities which were mainly determined by the complexity and the applied design methodology.

Thirdly, the number of the additional driver and sensor nodes correlated with the number of components in DAE-based representation, i.e., the number of sub-networks in the HEN. Thus, more interconnected streams demanded fewer driver and sensor nodes.

Fourthly, a smaller number of driver and sensor nodes resulted in an increased relative order, thus more complex operability.

Fifthly, the number of independent loops in the SM-based network representation correlated with the number of units in the HEN, and the average degree in SS- and SM-based network representations. The number of loops also correlated negatively with connectivity indexes. As the number of independent loops resulted in enhanced heat recovery, it attracted an increase in the number of heat exchangers too. Thus, a larger average degree meant that energy integration was greater, and more heat exchangers were utilised.

Finally, the comparison of the proposed methods showed that the results significantly varied from problem to problem. Only one correlation could be determined: the CRR, LHM and LRR algorithms solved the problem using more exchangers than other methods.

4. Conclusions

A network science-based analytical approach is proposed for the structural analysis of heat exchanger networks.

The proposed methodology utilises three different network-based representations that highlight various aspects of the dynamics of the HEN. The extracted networks were analysed by the toolbox of network science to create structural operability and complexity measures.

The analysis of more than 600 HENs confirmed that the proposed approach can be efficiently applied to the fast screening of HENs based on their structural properties.

A significant result of the analysis is that the popular maximum matching-based method used to determine the sensor and driver nodes in dynamical systems was not suitable for heat exchanger networks due to the intrinsic dynamics. It was highlighted that, with the more interconnected the HENs, fewer actuators and sensors are needed to ensure structural controllability and observability, which results in a higher relative order and the increased difficulty of the operability.

It was also highlighted that the degree-based structural measures can refer to the level of Heat Integration. The methodology was suitable for classifying the different methods that grant the HEN, as some techniques tended to determine more units in the network than others.

Although the proposed methodology has been introduced for the analysis of HENs, it can be applied to a broader class of dynamical systems. In our further research, we will generalise the results by examining the integrated and multi-objective design of processes and their control systems and studying the network topology-based properties of dynamical systems.

Author Contributions: D.L. reviewed the literature on network science and heat exchanger networks, developed the algorithms, designed and performed the experiments, and wrote the related sections. A.V.-F. participated in the formalisation of the methodology. J.A. conceived and designed the core concept, developed the algorithms and proofread the paper.

Funding: This research was supported by the National Research, Development and Innovation Office NKFIH, through the project OTKA-116674 (Process mining and deep learning in the natural sciences and process development) and the EFOP-3.6.1- 16-2016- 00015 Smart Specialization Strategy (S3) Comprehensive Institutional Development Program. Daniel Leitold was supported by the ÚNKP-18-3 New National Excellence Program of the Ministry of Human Capacities.

Conflicts of Interest: The authors declare no conflicts of interest. The founding sponsors had no role in the design of the study; in the collection, analyses, or interpretation of data; in the writing of the manuscript; nor in the decision to publish the results.

Abbreviations

CNC	Coefficient of Network Complexity
DAE	Differential Algebraic Equations
DN	Distribution Network
GA	Genetic Algorithm
GD	Grid Diagram
HEN	Heat Exchanger Network
MER	Maximum Energy Recovery or Minimum Energy Requirement
PI	Process Integration
PFD	Process Flow Diagram
RI	Resilience Index
SA	Simulated Annealing
SCC	Strongly Connected Component
SM	Streams and Matches
SS	State Space
TAC	Total Annualised Cost
TWC	Total Walk Count

Nomenclature

G_{SS}	SS-based graph
V_{SS}	Set of nodes of SS-based representation
E_{SS}	Set of edges of SS-based representation
G_{SM}	SM-based graph
V_{SM}	Set of nodes of SM-based representation
E_{SM}	Set of edges of SM-based representation
V_{hs}	Set of nodes of hot streams in SM-based representation
V_{cs}	Set of nodes of cold streams in SM-based representation
V_{hp}	Node of high pressure steam in SM-based representation
V_{cw}	Node of cold water in SM-based representation
E_{he}	Set of edges of heat exchangers in SM-based representation
E_{uh}	Set of edges of utility heaters in SM-based representation
E_{uc}	Set of edges of utility coolers in SM-based representation
G_{DAE}	DAE-based graph
V_{DAE}	Set of nodes of DAE-based representation
E_{DAE}	Set of edges of DAE-based representation
T_{hi}	Hot input temperature of a heat exchanger
T_{ci}	Cold input temperature of a heat exchanger
T_{ho}	Hot output temperature of a heat exchanger
T_{co}	Cold output temperature of a heat exchanger
t	Time
v_h	Flow rate of hot stream
v_c	Flow rate of cold stream
V_h	Volume of hot side of tank
V_c	Volume of cold side of tank
U	Heat transfer coefficient
A	Heat transfer area
c_{ph}	Heat capacity of hot stream
c_{pc}	Heat capacity of cold stream
ρ_h	Density of hot stream
ρ_c	Density of cold stream
x	Vector of state variables
n	Number of state variables
u	Vector of inputs
d	Vector of disturbances
y	Vector of outputs
$\mathbf{A}$	State-transition matrix
$\mathbf{B}$	Input matrix
b_j	Column j of matrix $\mathbf{B}$
Γ	Disturbance matrix
$\mathbf{C}$	Output matrix
c_j	Row j of matrix $\mathbf{C}$
$\mathbf{D}$	Feedthrough (or feedforward) matrix

$\mathcal{C}$	Controllability matrix
$\mathcal{O}$	Observability matrix
U_{un}	Number of units
N_s	Number of streams
L	Number of independent loops
S	Number of separate components
H	Number of heat exchangers
$\mathcal{L}$	Lie derivative
r_{ij}	Relative degree of input j and output i
$r_i = \min_j r_{ij}$	Relative degree of output i
$r = \max_i r_i$	Relative degree of the system
r_{max}	Upper bound for r
V_i	Node i from node set V
E_i	Edge i from edge set E
n_i	Weight of node i
w_i	Weight of edge i
N	Number of nodes in the network
M	Number of edges in the network
k_i^{out}, k_i^{in}, k_i	Out-, in-, and simple degree of node i
$\langle k_i \rangle$	Average node degree
ℓ_{ij}	Shortest path between nodes i and j
D	Distance matrix
$TWC(G)$	Total walk count measure of graph G
$CNC(G)$	Coefficient of network complexity of graph G
$Ecc(i), Ecc(G)$	Eccentricity of node i, or graph G
$W(G)$	Wiener index of graph G
$AD(G)$	A/D index of graph G
$BJ(G)$	Balaban-J index of graph G
$CI(G)$	Connectivity index of graph G
W_i	Set of reachable nodes from node i in r_{max}
$\mathcal{U}$	Set of all state variables
C	Set of necessary driver nodes
O	Set of necessary sensor nodes
J	Set of driver/sensor nodes of the input/output configuration
P	Set of state variables covered by set J
$Cc(i)$	Closeness centrality of node i
$Bc(i)$	Betweenness centrality of node i
σ_{st}	Number of shortest paths from node s to node t
$\sigma_{st}(i)$	Number of shortest paths from node s to node t that intercept node i
d_{max}	Diameter of the network

References

1. Chrissis, M.B.; Konrad, M.; Shrum, S. *CMMI Guidlines for Process Integration and Product Improvement*; Addison-Wesley Longman Publishing Co., Inc.: Boston, MA, USA, 2003.
2. Varbanov, P.S.; Walmsley, T.G.; Klemes, J.J.; Wang, Y.; Jia, X.X. Footprint Reduction Strategy for Industrial Site Operation. *Chem. Eng. Trans.* **2018**, *67*, 607–612, doi:10.3303/CET1867102. [CrossRef]
3. Klemes, J.J. *Handbook of Process Integration (PI): Minimisation of Energy and Water Use, Waste and Emissions*; Elsevier: Cambridge, UK, 2013.
4. Klemeš, J.J.; Varbanov, P.S.; Walmsley, T.G.; Jia, X. New directions in the implementation of Pinch Methodology (PM). *Renew. Sustain. Energy Rev.* **2018**, *98*, 439–468, doi:10.1016/j.rser.2018.09.030. [CrossRef]
5. Jamaluddin, K.; Wan Alwi, S.R.; Manan, Z.A.; Klemes, J.J. Pinch Analysis Methodology for Trigeneration with Energy Storage System Design. *Chem. Eng. Trans.* **2018**, *70*, 1885–1890, doi:10.3303/CET1870315. [CrossRef]
6. Kemp, I.C. *Pinch Analysis and Process Integration: A User Guide on Process Integration for the Efficient Use of Energy*; Elsevier: Oxford, UK, 2011.
7. Zafiriou, E. *The Integration of Process Design and Control*; Pergamon: Oxford, UK, 1994.
8. Liu, Y.Y.; Slotine, J.J.; Barabási, A.L. Controllability of complex networks. *Nature* **2011**, *473*, 167. [CrossRef] [PubMed]
9. Liu, Y.Y.; Slotine, J.J.; Barabási, A.L. Observability of complex systems. *Proc. Natl. Acad. Sci. USA* **2013**, *110*, 2460–2465. [CrossRef] [PubMed]

10. Saboo, A.K.; Morari, M.; Woodcock, D.C. Design of resilient processing plants—VIII. A resilience index for heat exchanger networks. *Chem. Eng. Sci.* **1985**, *40*, 1553–1565. [CrossRef]
11. Westphalen, D.L.; Young, B.R.; Svrcek, W.Y. A controllability index for heat exchanger networks. *Ind. Eng. Chem. Res.* **2003**, *42*, 4659–4667. [CrossRef]
12. Düştegör, D.; Frisk, E.; Cocquempot, V.; Krysander, M.; Staroswiecki, M. Structural analysis of fault isolability in the DAMADICS benchmark. *Control Eng. Pract.* **2006**, *14*, 597–608. [CrossRef]
13. Letsios, D.; Kouyialis, G.; Misener, R. Heuristics with performance guarantees for the minimum number of matches problem in heat recovery network design. *Comput. Chem. Eng.* **2018**, *113*, 57–85. [CrossRef]
14. Linnhoff, B.; Ahmad, S. Supertargeting: Optimum synthesis of energy management systems. *J. Energy Resour. Technol.* **1989**, *111*, 121–130. [CrossRef]
15. Ahmad, S.; Linnhoff, B. Supertargeting: Different process structures for different economics. *J. Energy Resour. Technol.* **1989**, *111*, 131–136. [CrossRef]
16. Bagajewicz, M.J.; Pham, R.; Manousiouthakis, V. On the state space approach to mass/heat exchanger network design. *Chem. Eng. Sci.* **1998**, *53*, 2595–2621. [CrossRef]
17. Mocsny, D.; Govind, R. Decomposition strategy for the synthesis of minimum-unit heat exchanger networks. *AIChE J.* **1984**, *30*, 853–856. [CrossRef]
18. Ciric, A.; Floudas, C. A retrofit approach for heat exchanger networks. *Comput. Chem. Eng.* **1989**, *13*, 703–715. [CrossRef]
19. Linnhoff, B.; Hindmarsh, E. The pinch design method for heat exchanger networks. *Chem. Eng. Sci.* **1983**, *38*, 745–763. [CrossRef]
20. Trivedi, K.; O'Neill, B.; Roach, J. Synthesis of heat exchanger networks featuring multiple pinch points. *Comput. Chem. Eng.* **1989**, *13*, 291–294. [CrossRef]
21. Wood, R.; Suaysompol, K.; O'Neill, B.; Roach, J.; Trivedi, K. A new option for heat exchanger network design. *Chem. Eng. Prog.* **1991**, *87*, 38–43.
22. Pho, T.; Lapidus, L. Topics in computer-aided design: Part II. Synthesis of optimal heat exchanger networks by tree searching algorithms. *AIChE J.* **1973**, *19*, 1182–1189. [CrossRef]
23. Yu, H.; Fang, H.; Yao, P.; Yuan, Y. A combined genetic algorithm/simulated annealing algorithm for large scale system energy integration. *Comput. Chem. Eng.* **2000**, *24*, 2023–2035. [CrossRef]
24. Lee, K.F.; Masso, A.; Rudd, D. Branch and bound synthesis of integrated process designs. *Ind. Eng. Chem. Fund.* **1970**, *9*, 48–58. [CrossRef]
25. Masso, A.; Rudd, D. The synthesis of system designs. II. Heuristic structuring. *AIChE J.* **1969**, *15*, 10–17. [CrossRef]
26. Miranda, C.B.; Costa, C.B.B.C.; Andrade, C.M.G.; Ravagnani, M.A.S.S. Controllability and Resiliency Analysis in Heat Exchanger Networks. *Chem. Eng. Trans.* **2017**, *61*, 1609–1614.
27. Svensson, E.; Eriksson, K.; Bengtsson, F.; Wik, T. *Design of Heat Exchanger Networks with Good Controllability*; Technical Report; CIT Industriell Energi AB: Göteborg, Sweden, 2018.
28. Varga, E.; Hangos, K. The effect of the heat exchanger network topology on the network control properties. *Control Eng. Pract.* **1993**, *1*, 375–380. [CrossRef]
29. Escobar, M.; Trierweiler, J.O.; Grossmann, I.E. Simultaneous synthesis of heat exchanger networks with operability considerations: Flexibility and controllability. *Comput. Chem. Eng.* **2013**, *55*, 158–180. [CrossRef]
30. Calandranis, J.; Stephanopoulos, G. Structural operability analysis of heat exchanger networks. *Chem. Eng. Res. Des.* **1986**, *64*, 347–364.
31. Zhelev, T.; Varbanov, P.; Seikova, I. HEN's operability analysis for better process integrated retrofit. *Hung. J. Ind. Chem.* **1998**, *26*, 81–88.
32. Leitold, D.; Vathy-Fogarassy, A.; Abonyi, J. Network Distance-Based Simulated Annealing and Fuzzy Clustering for Sensor Placement Ensuring Observability and Minimal Relative Degree. *Sensors* **2018**, *18*, 3096. [CrossRef]
33. Linnhoff, B.; Mason, D.R.; Wardle, I. Understanding heat exchanger networks. *Comput. Chem. Eng.* **1979**, *3*, 295–302. [CrossRef]
34. Bagajewicz, M.J.; Manousiouthakis, V. Mass/heat-exchange network representation of distillation networks. *AIChE J.* **1992**, *38*, 1769–1800. [CrossRef]
35. Foggia, P.; Percannella, G.; Vento, M. Graph matching and learning in pattern recognition in the last 10 years. *Int. J. Pattern Recognit. Artif. Intell.* **2014**, *28*, 1450001. [CrossRef]

36. Conte, D.; Foggia, P.; Sansone, C.; Vento, M. Thirty years of graph matching in pattern recognition. *Int. J. Pattern Recognit. Artif. Intell.* **2004**, *18*, 265–298. [CrossRef]
37. Carletti, V.; Foggia, P.; Saggese, A.; Vento, M. Challenging the time complexity of exact subgraph isomorphism for huge and dense graphs with VF3. *IEEE Trans. Pattern Anal. Mach. Intell.* **2018**, *40*, 804–818. [CrossRef] [PubMed]
38. Pósfai, M.; Liu, Y.Y.; Slotine, J.J.; Barabási, A.L. Effect of correlations on network controllability. *Sci. Rep.* **2013**, *3*, 1067. [CrossRef] [PubMed]
39. Liu, X.; Mo, Y.; Pequito, S.; Sinopoli, B.; Kar, S.; Aguiar, A.P. Minimum robust sensor placement for large scale linear time-invariant systems: A structured systems approach. *IFAC Proc. Vol.* **2013**, *46*, 417–424. [CrossRef]
40. Yan, G.; Tsekenis, G.; Barzel, B.; Slotine, J.J.; Liu, Y.Y.; Barabási, A.L. Spectrum of controlling and observing complex networks. *Nat. Phys.* **2015**, *11*, 779–786. [CrossRef]
41. Mangan, S.; Alon, U. Structure and function of the feed-forward loop network motif. *Proc. Natl. Acad. Sci. USA* **2003**, *100*, 11980–11985. [CrossRef] [PubMed]
42. Vinayagam, A.; Gibson, T.E.; Lee, H.J.; Yilmazel, B.; Roesel, C.; Hu, Y.; Kwon, Y.; Sharma, A.; Liu, Y.Y.; Perrimon, N.; et al. Controllability analysis of the directed human protein interaction network identifies disease genes and drug targets. *Proc. Natl. Acad. Sci. USA* **2016**, *113*, 4976–4981. [CrossRef]
43. Asgari, Y.; Salehzadeh-Yazdi, A.; Schreiber, F.; Masoudi-Nejad, A. Controllability in cancer metabolic networks according to drug targets as driver nodes. *PLoS ONE* **2013**, *8*, e79397. [CrossRef]
44. Leitold, D.; Vathy-Fogarassy, Á.; Abonyi, J. Controllability and observability in complex networks–the effect of connection types. *Sci. Rep.* **2017**, *7*, 151. [CrossRef]
45. Daoutidis, P.; Kravaris, C. Structural evaluation of control configurations for multivariable nonlinear processes. *Chem. Eng. Sci.* **1992**, *47*, 1091–1107. [CrossRef]
46. Nishida, N.; Stephanopoulos, G.; Westerberg, A.W. A review of process synthesis. *AIChE J.* **1981**, *27*, 321–351. [CrossRef]
47. Linnhoff, B.; Flower, J.R. Synthesis of heat exchanger networks: I. Systematic generation of energy optimal networks. *AIChE J.* **1978**, *24*, 633–642. [CrossRef]
48. Varga, E.; Hangos, K.; Szigeti, F. Controllability and observability of heat exchanger networks in the time-varying parameter case. *Control Eng. Pract.* **1995**, *3*, 1409–1419. [CrossRef]
49. Wood, R.; Wilcox, R.; Grossmann, I.E. A note on the minimum number of units for heat exchanger network synthesis. *Chem. Eng. Commun.* **1985**, *39*, 371–380. [CrossRef]
50. Reinschke, K.J. *Multivariable Control: A Graph Theoretic Approach*; Springer: Berlin/Heidelberg, Germany, 1988.
51. Kalman, R.E. Mathematical description of linear dynamical systems. *J. Soc. Ind. Appl. Math. Ser. A Control* **1963**, *1*, 152–192. [CrossRef]
52. Letellier, C.; Sendiña-Nadal, I.; Aguirre, L.A. A nonlinear graph-based theory for dynamical network observability. *Phys. Rev. E* **2018**. [CrossRef] [PubMed]
53. Volkmann, L. Estimations for the number of cycles in a graph. *Periodica Math. Hung.* **1996**, *33*, 153–161. [CrossRef]
54. Bonchev, D.; Buck, G.A. Quantitative measures of network complexity. In *Complexity in Chemistry, Biology, and Ecology*; Springer: Richmond, VA, USA, 2005; pp. 191–235.
55. Latva-Koivisto, A.M. *Finding a Complexity Measure for Business Process Models*; Helsinki University of Technology, Systems Analysis Laboratory: Helsinki, Finland, 2001.
56. Dehmer, M.; Kraus, V.; Emmert-Streib, F.; Pickl, S. *Quantitative Graph Theory*; CRC Press: Danvers, MA, USA, 2014.
57. Balaban, A.T. Highly discriminating distance-based topological index. *Chem. Phys. Lett.* **1982**, *89*, 399–404. [CrossRef]
58. Leitold, D.; Vathy-Fogarassy, A.; Abonyi, J. Design-Oriented Structural Controllability and Observability Analysis of Heat Exchanger Networks. *Chem. Eng. Trans.* **2018**, *70*, 595–600.
59. Gori, F.; Folino, G.; Jetten, M.S.; Marchiori, E. MTR: Taxonomic annotation of short metagenomic reads using clustering at multiple taxonomic ranks. *Bioinformatics* **2010**, *27*, 196–203. [CrossRef]
60. Furman, K.C.; Sahinidis, N.V. Approximation algorithms for the minimum number of matches problem in heat exchanger network synthesis. *Ind. Eng. Chem. Res.* **2004**, *43*, 3554–3565. [CrossRef]

61. Chen, Y.; Grossmann, I.E.; Miller, D.C. Computational strategies for large-scale MILP transshipment models for heat exchanger network synthesis. *Comput. Chem. Eng.* **2015**, *82*, 68–83. [CrossRef]
62. Chen, Y.; Grossmann, I.E.; Miller, D.C. Large-Scale MILP Transshipment Models for Heat Exchanger Network Synthesis. 2015. A Collaboration of Carnegie Mellon University and IBM Research. Available online: https://www.minlp.org/library/problem/index.php?i=191&lib=MINLP (accessed on 4 February 2019).
63. Grossmann, I.E. *Personal Communication*; Carnegie Mellon University: Pittsburgh, PA, USA, 2017.
64. Gundersen, T.; Grossmann, I.E. Improved optimization strategies for automated heat exchanger network synthesis through physical insights. *Comput. Chem. Eng.* **1990**, *14*, 925–944. [CrossRef]
65. Colberg, R.; Morari, M. Area and capital cost targets for heat exchanger network synthesis with constrained matches and unequal heat transfer coefficients. *Comput. Chem. Eng.* **1990**, *14*, 1–22. [CrossRef]
66. Shenoy, U.V. *Heat Exchanger Network Synthesis: Process Optimization by Energy and Resource Analysis*; Gulf Professional Publishing: Houston, TX, USA, 1995.
67. Trivedi, K.; O'Neill, B.; Roach, J.; Wood, R. Systematic energy relaxation in MER heat exchanger networks. *Comput. Chem. Eng.* **1990**, *14*, 601–611. [CrossRef]
68. Dolan, W.; Cummings, P.; Le Van, M. Algorithmic efficiency of simulated annealing for heat exchanger network design. *Comput. Chem. Eng.* **1990**, *14*, 1039–1050. [CrossRef]
69. Farhanieh, B.; Sunden, B. Analysis of an existing heat exchanger network and effects of heat pump installations. *Heat Recov. Syst. CHP* **1990**, *10*, 285–296. [CrossRef]
70. Grossmann, I.E.; Sargent, R. Optimum design of heat exchanger networks. *Comput. Chem. Eng.* **1978**, *2*, 1–7. [CrossRef]
71. Hall, S.; Ahmad, S.; Smith, R. Capital cost targets for heat exchanger networks comprising mixed materials of construction, pressure ratings and exchanger types. *Comput. Chem. Eng.* **1990**, *14*, 319–335. [CrossRef]
72. Tantimuratha, L.; Kokossis, A.; Müller, F. The heat exchanger network design as a paradigm of technology integration. *Appl. Therm. Eng.* **2000**, *20*, 1589–1605. [CrossRef]
73. Polley, G.; Heggs, P. Don't let the pinch pinch you. *Chem. Eng. Prog.* **1999**, *95*, 27–36.
74. Ahmad, S.; Smith, R. Targets and design for minimum number of shells in heat exchanger networks. *Chem. Eng. Res. Des.* **1989**, *67*, 481–494.
75. Arenas, A.; Fernandez, A.; Gomez, S. Analysis of the structure of complex networks at different resolution levels. *New J. Phys.* **2008**, *10*, 053039. [CrossRef]

© 2019 by the authors. Licensee MDPI, Basel, Switzerland. This article is an open access article distributed under the terms and conditions of the Creative Commons Attribution (CC BY) license (http://creativecommons.org/licenses/by/4.0/).

Article

Exergy-Based and Economic Evaluation of Liquefaction Processes for Cryogenics Energy Storage

Sarah Hamdy [1], Francisco Moser [2], Tatiana Morosuk [2,*] and George Tsatsaronis [2,*]

[1] Energy Engineering Department, Technische Universität Berlin, 10587 Berlin, Germany; sarah.hamdy@tu-berlin.de
[2] Institute for Energy Engineering, Technische Universität Berlin, 10587 Berlin, Germany; f.moser.k@googlemail.com
* Correspondence: tetyana.morozyuk@tu-berlin.de (T.M.); georgios.tsatsaronis@tu-berlin.de (G.T.)

Received: 29 December 2018; Accepted: 31 January 2019; Published: 4 February 2019

Abstract: Cryogenics-based energy storage (CES) is a thermo-electric bulk-energy storage technology, which stores electricity in the form of a liquefied gas at cryogenic temperatures. The charging process is an energy-intensive gas liquefaction process and the limiting factor to CES round trip efficiency (RTE). During discharge, the liquefied gas is pressurized, evaporated and then super-heated to drive a gas turbine. The cold released during evaporation can be stored and supplied to the subsequent charging process. In this research, exergy-based methods are applied to quantify the effect of cold storage on the thermodynamic performance of six liquefaction processes and to identify the most cost-efficient process. For all liquefaction processes assessed, the integration of cold storage was shown to multiply the liquid yield, reduce the specific power requirement by 50–70% and increase the exergetic efficiency by 30–100%. The Claude-based liquefaction processes reached the highest exergetic efficiencies (76–82%). The processes reached their maximum efficiency at different liquefaction pressures. The Heylandt process reaches the highest RTE (50%) and the lowest specific power requirement (1021 kJ/kg). The lowest production cost of liquid air (18.4 €/ton) and the lowest specific investment cost (<700 €/kW$_{char}$) were achieved by the Kapitza process.

Keywords: cryogenic energy storage; air liquefaction; exergy analysis; economic analysis; exergoeconomic analysis

1. Introduction

The interest in electricity storage has significantly increased with higher shares of intermittent renewable energy sources in the grid. In particular, grid-scale electricity storage with low costs are considered suitable to integrate renewable electricity generation and introduce flexibility to the power grid. Cryogenics-based energy storage (CES), frequently referred to as liquid air energy storage (LAES), is the only energy storage technology so far, which is capable to store large quantities of electricity without geographical limitations or a substantial negative environmental impact.

The thermo-electric energy storage technology stores electricity in the form of a liquefied gas (air) at a cryogenic temperature. The integrated methods of operation (charge, storage, discharge) are displayed in Figure 1. An energy-intensive liquefaction process forms the charging process of CES. The liquefied gas (cryogen) is stored in a site-independent insulated storage tank at approximately ambient pressure and a cryogenic temperature (e.g., −194 °C). The compression process of the liquefaction is presented separately, as in the adiabatic CES the heat of compression is recovered and stored to be used in the discharge process. In the discharge process, the liquefied gas is pumped to supercritical pressure in a cryogenic pump, evaporated and superheated, with thermal energy provided by the heat storage, and supplied to a series of expanders regaining a part of the electricity charged to the system.

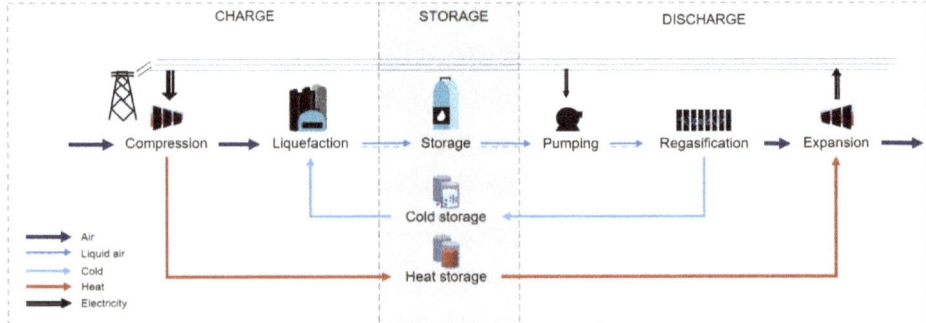

Figure 1. Illustration of cryogenic energy storage steps of operation (charge, storage, discharge), heat and cold recovery and storage.

The cold exergy rejected during the evaporation process is stored in order to increase the efficiency of the liquefaction process (charge). The CES system is composed of well-known components from the industrial gas and liquefied natural gas (LNG) supply chain.

As CES systems are based on mature technology, developers expect comparatively fast progress towards commercialization, competitive costs and efficiency enhancement. CES exergy densities are by approximately two orders of magnitudes higher than of competing technologies such as pumped hydro and compressed air energy storage reaching values higher than 430 kJ/kg. A detailed comparison of CES characteristics to other energy storages can be found in [1]. Moreover, long cycle life, low storage costs, the economy of scale and the independent sizing of charge and discharge unit speak for economic viability. Yet, the adiabatic CES systems upper limit to efficiencies is 45–50%. The thermal integration at the system level is crucial to its performance, which is the reason why the integration of cold storage into the liquefaction process is the subject of this paper.

State of the Art

Both, cryogenic energy storage and air liquefaction, are no new concepts. Large-scale air liquefaction for industrial purposes became commercial in the 1940s [2] and the first conception of storing electricity in liquid air dates back to the year 1977 [3].

Nowadays CES is rated as a pre-commercial technology being evaluated with a technology readiness level (TRL) of about 8 [4]. The CES concept was confirmed viable in testing, after Mitsubishi extended an existing air liquefier with the first pilot cryogenic power recovery unit (2.6 MW) [5]. The second pilot plant was the first integrated CES plant (350 kW/2.5 MWh) which was the result of joint research between the University of Leeds and Highview Power Storage Ltd. (London, UK) in the year 2011. The results were published in 2015: the CES economic viability was confirmed and a positive outlook on performance and costs was given [6]. A demonstration plant of 5 MW/15 MWh started operation in 2018, demonstrating a number of balancing services [7].

The significance of the liquefaction process to the CES's performance was addressed by [8] as "the key part" of the system as the discharge unit is relatively simple and efficient. The liquefaction was found to account for more than 70% of the overall exergy destruction (MW) of the CES system by the authors in a comparative exergy analysis of two 10 MW CES systems [9].

One of the key findings from the testing of the first pilot plant was the significant increase of system efficiency by cold recovery and storage [6]. The effect of cold recycle was firstly quantified by the authors as the introduction of cold storage doubled the liquid yield of the liquefaction process of the analyzed system [9].

Large-scale air liquefaction has been commercial for several decades. A number of processes exist. The simplest (no moving parts) and first-industrialized configuration is the Linde process, where purified compressed air is cooled and undergoes isenthalpic (free) expansion in a throttling valve, thus brought to its due point by the Joule-Thomson effect [10]. Gas liquefaction is nowadays performed in more complex configurations [11].

Recently, a number of publications have discussed the thermodynamic performance of CES. In the reviewed literature, CES systems with different liquefaction processes, pressures and cold storage configurations are presented in Table 1. Two kinds of a cold storage configuration are presented: (1) quartzite gravel based packed bed store with dry air as secondary working fluid, and (2) a two-tank fluid storage with methanol and propane (or R218) as secondary working fluids and storage media on two different temperature levels.

The liquid yield γ, the ratio between the mass flow of the air liquefied in the liquefaction process and the mass flow of the compressed air, is an indication of the charging-unit performance. The liquid yield varies strongly from one publication to the other. The liquid yield increases with liquefaction pressure. Yet, with increased pressures, the power consumption of the compression process increases as well. This is why the liquid yield cannot be considered as the sole indicator for the performance of the liquefaction process.

In general, different assumptions are made in the different references, e.g., ideal dry air was assumed, heat and pressure losses in most components as well as heat losses in the cold box were neglected [12], or assumed lower than 8% [6] which is why comparing the various configurations is problematic.

Three comparative evaluations of air liquefaction processes in CES systems were presented in [8,13,14]. Borri et al. [13] compared three air liquefaction processes (Linde-Hampson, Claude, Collins) for application in a micro-scale CES. The Claude process was identified as the most suitable air liquefaction process. The Linde-Hampson process (with a Joule-Thomson valve only) was found to be inferior and the second cold expander used in the Collins process was claimed to be economically not feasible. Yet, the integration of cold recovery and storage was not considered. Li [8] came to the same conclusion, that the throttle-valve-based Linde-Hampson system is not applicable for CES. Therefore, only the integration of a cold expander instead of a throttling valve in the Linde process and an expander process, employing a refrigeration process with Helium as working fluid, are compared in [8].

Table 1. Parameters of CES systems presented in [5,8,9,12,14–19]: Liquefaction processes, liquefaction pressure p_{char}, liquid yield γ, cold storage configuration, discharge pressure p_{dis} and round trip efficiency η_{RTE}.

Source	Process	p_{char}, bar	γ, -	Cold Storage Configuration	p_{dis}, bar	η_{RTE}, %
[12]	Linde-Hampson	120	0.83	fluid tanks (CH$_4$O, C$_3$H$_8$)	50	50–60
[15]	Integr. Linde-Hampson	90	0.60	fluid tanks (CH$_4$O, C$_3$H$_8$)	120	60
[9]	Heylandt	180	0.61	fluid tanks (CH$_4$O, C$_3$H$_8$)	150	41
[16]	Modified Claude	180	0.86	packed bed gravel (air)	75	48.5
[17]	2 Turbine Claude/Collins	54	NA	packed bed gravel (air)	150	47
[6]	4 Turbine Claude	56.8	0.551 [1]	packed bed gravel (air)	190	>50
[18]	Linde-Hampson	180	0.842	fluid tanks (CH$_4$O, C$_3$H$_8$)	65	50
[19]	Linde-Hampson	140	NA	NA	70	47.2
[11]	Linde-Hampson	20	0.70	direct integration (ideal)	100	20–50
[8]	Linde-Hampson [2]	~130	0.44–0.74	fluid tanks (CH$_4$O, R218)	112–120	28–37
[8]	Expander cycle	NA	NA	fluid tanks (CH$_4$O, R218)	NA	40–46
[20]	Single expander	135	0.84	fluid tanks (CH$_4$O, C$_3$H$_8$)	80	50–58

[1] calculated from: 12 h charging, 3:1 (charge-to-discharge ratio), $\dot{m}_{char}$ = 34.1 kg/s. [2] with cold expander/throttling valve.

Abdo et al. [14] compared the by Chen et al [21] patented CES system design based on a simple Linde-Hamson liquefaction process to two alternative systems based on the Claude and the Collins process. The heat of compression was taken into account but cold storage was not comprised. The Claude and Collins process showed similar thermodynamic performance with greater RTE that the Linde based system. Despite the Linde-Hampson having the lowest specific costs, the Claude-based system was evaluated the best option. The present paper aims to compare a number of air liquefaction process configurations with integrated cold storage in order to identify the most suitable process for implementation in CES systems.

2. Methods

For a comparative analysis, six liquefaction processes were simulated with and without integration of cold storage under similar conditions. Results from energetic, exergetic, economic and exergoeconomic analyses were used to identify the most cost-effective liquefaction process for CES with cold storage.

2.1. Design and Simulation

Aspen Plus® (Version 9, Aspen Technology Inc., Bedford, MA, USA) was chosen as a suitable software for process simulation. With the aid of the simulation software, all mass and energy balances are fulfilled and the specific enthalpy and entropy values of all streams and substances are calculated. The Peng-Robinson equation of state was employed and the simulation was performed under steady-state conditions. Fortran routines are integrated to calculate exergy values for the exergetic analysis. Six liquefaction processes were simulated: the simple Linde, the precooled Linde, the dual pressure Linde, the simple Claude, the Kapitza and the Heylandt process. At first, the liquefaction processes were manually optimized and later modified to accommodate the cold storage. The assumptions made in simulation are given in Table 2.

Table 2. Assumptions made in simulation.

Parameter	Value, Unit
Isentropic efficiencies (compressors, expanders)	$\eta_{is,CM}$ = 87% [8], $\eta_{is,EX}$ = 80% [17]
Intercooler exit temperature and pinch	$T_{exit,\,IC}$ = 25 °C, $\Delta T_{pinch,\,IC}$ = 5 K [15]
Main heat exchanger pinch temperature difference	$\Delta T_{pinch,\,MHE}$ = 1–3 K [8,17]
Maximal pressure of compression	$p_{max,\,CM}$ = 200 bar [22]
Ambient conditions	T_{amb} = 15 °C, p_{amb} = 1.013 bar

The overall system configuration is shown in Figure 2. The pretreated air enters the analyzed system at 15 °C, 1.013 bar and a molar composition of 79% N_2 and 21% O_2 (a1). The compression block is the same for all systems. The air exits the last intercooler of the three-stage compression at a temperature of 25 °C and a pressure of $p_{max,CM}$ of 200 bar (a2). The largest part of the thermal energy increase during compression is recovered in a heat storage. The heat storage is realized with pressurized water tanks (5 bar, 205 °C). The design of the liquefaction block is different for each system. Two types of liquefaction processes can be distinguished: Linde-based (Figure 3) and Claude-based (Figure 4) liquefaction processes. The liquefied air exits the flasher and is stored at a temperature of −192 °C and slightly elevated pressure 1.3 bar. The liquid is stored in an insulated storage tank with boil-off losses of 0.2 %$_{Vol}$.

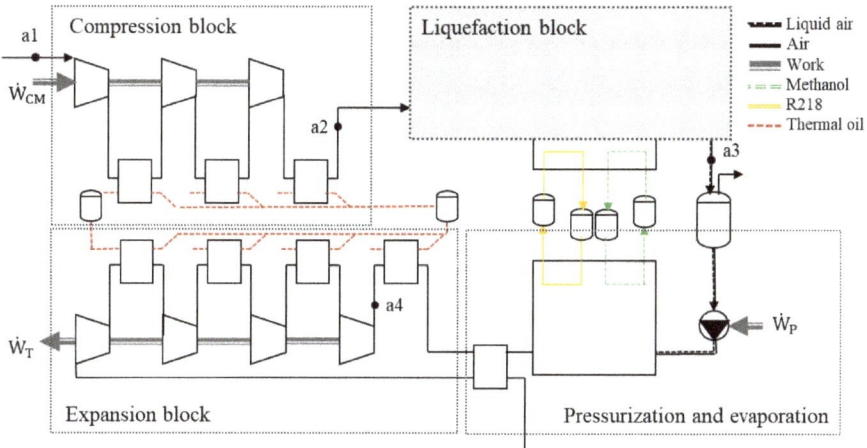

Figure 2. Flowsheet of the adiabatic CES system with "black box" air liquefaction block.

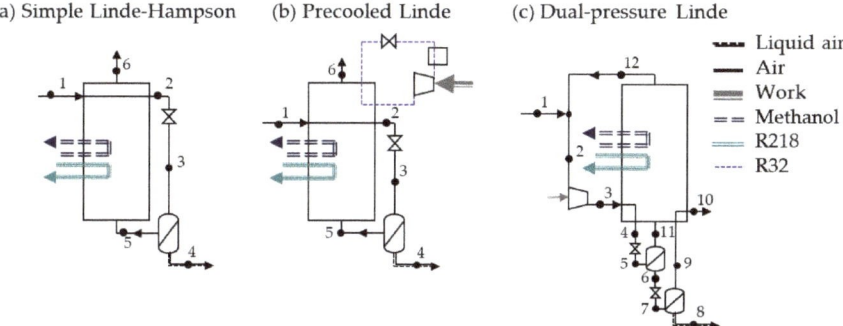

Figure 3. Flowsheets of Linde-based air liquefaction processes with cold recycle.

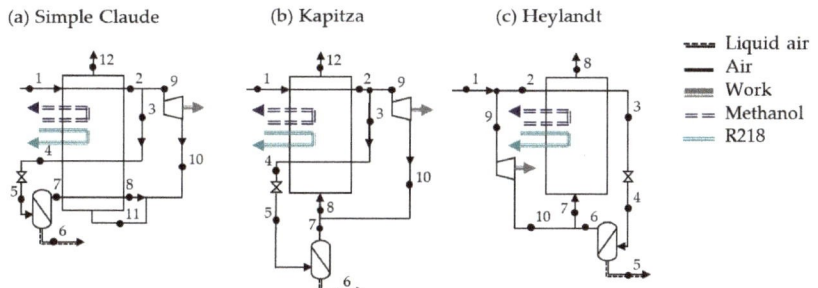

Figure 4. Flowsheets of Claude-based air liquefaction processes with cold recycle.

During discharge, the liquid air is pressurized to 150 bar, evaporated in heat exchange to the cold storage media, superheated (T_{a4} = 195 °C) and fed to the four-stage expander with reheat. The specific power output of the discharge unit is constant for all systems (w_{dis} = 470 kJ/kg of liquid air).

The assumed method of cold storage uses two fluid tanks and two circulating working fluids that recover the high-grade and low-grade cold rejected in the evaporation process. Reviewing a number of refrigerants, *R218* and *methanol* are shown to be advantageous with respect to toxicity, flammability, boiling and freezing temperatures [9]. The cold in the temperature interval −180 to

−61 °C is recovered by *R218*, while the cold at higher temperatures (−19 to −59 °C) is captured and stored using *methanol*. The amount of cold recovered is determined by the amount of air liquefied in the liquefaction process. The mass flow rates of the cold storage media are therefore determined by a ratio of the mass flow rate of the liquefied air $\dot{m}_{a3}$:

$$\dot{m}_{R218} = 2.29 \cdot \dot{m}_{a3} \tag{1}$$

$$\dot{m}_{methanol} = 0.49 \cdot \dot{m}_{a3} \tag{2}$$

The ratio is adjusted to the optimal heat transfer between the evaporating liquid air and the cold storage media. Thermal losses in the cold storage were accounted for and are equivalent to 4 K/cycle.

The liquefaction processes are shown in Figures 3 and 4. A detailed description of the liquefaction processes and the fundamental concept can be found in fundamental publications e.g., [23]. The stream values (mass flow $\dot{m}$, temperature T and pressure p) can be found in Tables 3 and 4.

The Linde-Hampson process, Figure 3a, is the most straightforward of all liquefaction processes. The process consists of only four sets of components: the compressor(s), the main heat exchanger (MHE), the throttling valve and the flash tank. After compression, the temperature of the air is reduced (below −100 °C) in the MHE. The low-temperature high-pressure air is throttled reducing the temperature close to the dew point resulting in partial condensation. In the flash tank, the liquid air is separated and stored. The gaseous air is supplied back to the MHE to precool the compressed air. The efficiency of the simple Linde-Hampson process strongly depends on the temperature of the high-pressure gas at the inlet of the MHE.

The precooled Linde-Hampson process, shown in Figure 3b, intends to achieve a better performance and a higher liquid yield by lowering the temperature of the air with the addition of a compression refrigeration process. Working fluids such as ammonia, carbon dioxide or Freon compounds are commonly used for the secondary refrigeration cycle.

In the dual-pressure Linde process (Figure 3c) the heat transfer in the MHE is improved by introducing a second pressure level. The air enters the liquefaction process at an intermediate-pressure (1). Together with the recycled stream, the pressure of the air is elevated further to the high-pressure level (3). The gas is cooled and throttled to the intermediate-pressure level (5). The gaseous and the liquid air are separated in the intermediate-pressure flash tank. The gaseous part is fed back to the MHE to precool the entering air stream (3) to (4) and is mixed to the entering intermediate-pressure air stream (1). The liquefied air is fed to the second pressure-stage. This modification reduces the specific work required to liquefy the air at the expense of the share of air liquefied.

The Claude process and its modifications are the most commonly employed process in commercial air liquefaction plants, as its efficiency is higher than that of the Linde process [23]. In the Claude process the cooling of compressed air is provided by a cold recycle stream—a part of the pressurized air that underwent an isentropic expansion in cold expanders [6]. The application of a cold expander avoids part of the exergy destruction in the throttling process and reduces the required power for liquefaction by the power output of the expander ($\dot{W}_{char} = \sum \dot{W}_{CM} - \dot{W}_{EX}$). The stream exiting the expander ($\dot{m}_{10}$) is used to cool the air stream entering the MHE. The expander does not replace the throttling valve before the flash tank.

The Kapitza process is analogous to the Claude process but with the difference that the third partition of the MHE (or low-temperature heat exchanger) is eliminated. In other words, while using a multi-stream heat exchanger, stream 7 is not fed to the MHE before mixing. Streams 7 and 10 tend to have only a small temperature difference, which is why the difference in heat exchanger area and performance is little. The Heyland process is also adopted from the Claude process. Nevertheless, it can also be seen as a variation of the precooled Linde-Hampson process using air as a refrigerant. The precooling process—the splitting of the stream before entering the MHE—improves the heat transfer process in the MHE [23].

Table 3. Stream values for the states indicated in the flowsheets in Figure 3.

Stream	Variable, Unit		Simple Linde With Storage		Precooled Linde With Storage		Dual-Pressure Linde With Storage	
1	$\dot{m}$	kg/h	100.0	100.0	100.0	100.0	100.0	100.0
	T	°C	25.0	25.0	25.0	25.0	25.0	25.0
	p	bar	200.0	200.0	200.0	200.0	33.4	33.4
2	$\dot{m}$	kg/h	100.0	100.0	100.0	100.0	547.3	137.3
	T	°C	−102.4	−125.3	−113.6	−138.7	24.1	24.2
	p	bar	200.0	1.03	200.0	200.0	30.4	30.4
3	$\dot{m}$	kg/h	100.0	100	100	100	547.3	137.3
	T	°C	−191.8	−191.7	−192.3	−193.1	25.0	25.0
	p	bar	1.03	1.03	1.03	1.03	200.0	200.0
4	$\dot{m}$	kg/h	9.0	31.2	19.8	44.1	547.3	137.3
	T	°C	−191.8	−192.7	−192.3	−193.1	−105.0	−124.5
	p	bar	1.03	1.03	1.03	1.03	200.0	200.0
5	$\dot{m}$	kg/h	91.0	68.8	80.2	55.9	547.3	137.3
	T	°C	−191.8	−192.7	−192.3	−193.1	−146.2	−146.2
	p	bar	1.03	1.03	1.03	1.03	30.4	30.4
6	$\dot{m}$	kg/h	91.0	68.8	80.2	55.9	100.0	100.0
	T	°C	24.0	24.0	24.0	−95.8	−146.2	−146.2
	p	bar	1.03	1.03	1.03	1.03	30.4	30.4
7	$\dot{m}$	kg/h	-	-	-	-	100.0	100.0
	T	°C	-	-	-	-	−192.9	−192.9
	p	bar	-	-	-	-	1.03	1.03
8	$\dot{m}$	kg/h	-	-	-	-	40.0	40.0
	T	°C	-	-	-	-	−193.0	−193.0
	p	bar	-	-	-	-	1.03	1.03
9	$\dot{m}$	kg/h	-	-	-	-	60.0	60.0
	T	°C	-	-	-	-	−193.0	−193.0
	p	bar	-	-	-	-	1.03	1.03
10	$\dot{m}$	kg/h	-	-	-	-	60.0	60.0
	T	°C	-	-	-	-	24.0	24.0
	p	bar	-	-	-	-	1.03	1.03
11	$\dot{m}$	kg/h	-	-	-	-	447.3	37.3
	T	°C	-	-	-	-	−146.2	−146.2
	p	bar	-	-	-	-	30.4	30.4
12	$\dot{m}$	kg/h	-	-	-	-	447.3	37.3
	T	°C	-	-	-	-	24.0	24.0
	p	bar	-	-	-	-	30.4	30.4

Table 4. Stream values for the states indicated in the flowsheets in Figure 4.

Stream	Variable,	Unit	Claude	With Storage	Kapitza	With Storage	Heylandt	With Storage
1	$\dot{m}$	kg/h	100.0	100.0	100.0	100.0	100.0	100.0
	T	°C	25.0	25.0	25.0	25.0	25.0	25.0
	p	bar	200.0	200.0	200.0	200.0	200.0	200.0
2	$\dot{m}$	kg/h	100.0	100.0	100.0	100.0	36.0	76.1
	T	°C	−4.0	−2.0	−4.0	−2.0	25.0	25.0
	p	bar	200.0	200.0	200.0	200.0	200.0	200.0
3	$\dot{m}$	kg/h	34.3	73.2	34.3	73.3	36.0	76.1
	T	°C	−4.0	−2.0	−4.0	−2.0	−177.6	−180.5
	p	bar	200.0	200.0	200.0	200.0	200.0	200.0
4	$\dot{m}$	kg/h	34.3	73.2	34.3	73.3	36.0	76.1
	T	°C	−190.8	−182.8	−190.6	−182.8	−193.9	−194.0
	p	bar	200.0	200.0	200.0	200.0	1.03	1.03
5	$\dot{m}$	kg/h	34.3	73.2	34.3	73.3	28.6	62.3
	T	°C	−194.1	−194.0	−194.1	−194.0	−193.9	−194.0
	p	bar	1.03	200.0	1.03	1.03	1.03	1.03
6	$\dot{m}$	kg/h	31.2	61.5	31.1	61.5	7.4	13.7
	T	°C	−194.1	−194.0	−194.1	−194.0	−176.5	−179.0
	p	bar	1.03	1.03	1.03	1.03	1.03	1.03
7	$\dot{m}$	kg/h	3.1	11.8	3.2	11.8	71.4	37.7
	T	°C	−194.1	−194.0	−194.1	−194.0	−176.4	−177.4
	p	bar	1.03	1.03	1.03	1.03	1.03	1.03
8	$\dot{m}$	kg/h	3.1	11.8	68.9	38.5	71.4	37.7
	T	°C	−192.0	−191.0	−191.7	−192.1	−7.3	24.0
	p	bar	1.03	1.03	1.03	1.03	1.03	1.03
9	$\dot{m}$	kg/h	65.7	26.8	65.7	26.7	64.0	24.0
	T	°C	−4.0	−2.0	−4.0	−2.0	25.0	25.0
	p	bar	200.0	200.0	200.0	200.0	200.0	200.0
10	$\dot{m}$	kg/h	65.7	26.8	65.7	26.7	64.0	24.0
	T	°C	−191.6	−191.2	−191.6	−191.2	−176.4	−176.4
	p	bar	1.03	1.03	1.03	1.03	1.03	1.03
11	$\dot{m}$	kg/h	68.8	73.0	-	-	-	-
	T	°C	−191.7	−182.8	-	-	-	-
	p	bar	1.03	200.0	-	-	-	-
12	$\dot{m}$	kg/h	68.8	38.6	68.9	73.3	-	-
	T	°C	24.0	23.4	24.0	24.0	-	-
	p	bar	1.03	1.03	1.03	1.03	-	-

The performance of the Claude-based processes is dependent on the splitting ratio r. The splitting ratio is defined as the mass flow through the expander $\dot{m}_{EX}$ over the mass flow through the last compression step $\dot{m}_{CM}$:

$$r = \frac{\dot{m}_{EX}}{\dot{m}_{CM}} \qquad (3)$$

The Kapitza process dates back to 1939 when the inventor suggested the use of centrifugal expansion turbines in the Claude process [10]. Most modern liquefiers utilize expansion turbines proposed by Kapitza [10,24] and most high-pressure air liquefaction plants operate with the Heylandt process. Highview Power Storage Ltd. base their charging unit on the Claude process relying on the maturity of the process and the trouble-free scale-up [25]. The pilot plant operates with a Claude-based liquefaction process similar to the Kapitza configuration [26]. The operation pressures for the different liquefaction processes differ [23]. For a better comparison, the liquefaction pressure is kept to 200 bar [23] first and later varied in sensitivity analysis.

2.2. Energetic and Exergetic Analyses

The six liquefaction processes were compared with and without cold storage in energetic and exergetic analyses at the system level. For the three most efficient processes, sensitivity analyses and exergetic analyses at the component level were further undertaken. The exergetic analysis is adopted from [27]. The exergetic efficiency ε, the liquid yield γ and the specific power requirement w of the systems were used as a basis for comparing the process' performance with and without cold recovery. The parameters are defined below:

$$\varepsilon = \frac{\dot{E}_{liquid\ air} + \dot{E}_{q,hot}}{\dot{W}_{char} + \dot{E}_{q,cold}} [-] \quad (4)$$

$$\gamma = \frac{\dot{m}_{liquid\ air}}{\dot{m}_{CM}} [-] \quad (5)$$

$$w = \frac{\dot{W}_{char}}{\dot{m}_{liquid\ air}} \left[kJ/kg_{liquid\ air} \right] \quad (6)$$

The general definition of the exergetic efficiency ε is the ratio of the exergy of the product $\dot{E}_P$ and the exergy of the fuel $\dot{E}_F$. The fuel supplied to the liquefaction system is the charging power $\dot{W}_{char}$ and the exergy of the low-temperature exergy supplied by the cold storage $\dot{E}_{q,cold}$:

$$\dot{W}_{char} = \sum \dot{W}_{CM} - \dot{W}_{EX} \quad (7)$$

$$\dot{E}_{q,cold} = \left| (1 - T_0/T_{cold}) \cdot \dot{Q}_{cold} \right| = \dot{m}_{liquid} \cdot \Delta e_{R218} + \dot{m}_{methanol} \cdot \Delta e_{methanol} \quad (8)$$

T_{cold} (or T_{hot}) denote the thermodynamic mean temperatures at which the low-temperature energy (or the heat) is supplied. Both the exergy of the liquefied air $\dot{E}_{liquid\ air}$ and the exergy of the heat supplied to the heat storage $\dot{E}_{q,hot}$ are products of the liquefaction process:

$$\dot{E}_{liquid\ air} = \dot{m}_{liquid} \cdot e_{liquid\ air} \quad (9)$$

$$\dot{E}_{Q,hot} = (1 - T_0/T_{hot}) \cdot \dot{Q}_{heat} \quad (10)$$

The definitions of fuel and product for CES system components can be found in [9]. As the systems partially operate below the ambient temperature, the physical exergy is split into its mechanical and thermal parts, according to [28].

The liquefaction processes with the best performance with cold storage were identified (200 bar) and a sensitivity analysis was performed. In sensitivity analyses the splitting ratio r and liquefaction pressure $p_{max,CM}$ were varied. For the optimal liquefaction pressure and splitting ratio, the three systems were compared using economic and exergoeconomic analyses.

The round-trip efficiency (RTE) of the systems was calculated as base for comparison. The RTE is defined as the ratio between the electricity charged and the electricity discharged:

$$\eta_{RTE} = \frac{\dot{W}_{dis}}{\dot{W}_{char} \cdot \frac{\tau_{char}}{\tau_{dis}}} \quad (11)$$

In contrast to evaluating the charging system only, for the overall system the charging duration τ_{char} and the discharge duration τ_{dis} need to be accounted for. Reason for this is that the charge-to-discharge ratio ($\frac{\tau_{char}}{\tau_{dis}}$) may be unequal to one. For calculation of the RTE an exergy density of approx. 445–465 kJ/kg and a charge-to-discharge ratio of two was accounted for.

2.3. Economic Analysis

The economic analysis was performed on the optimal system configuration ($p_{max,CM}, r \to \varepsilon_{max}$) of the best performing processes. The processes were sized to 20 MW charging power $\dot{W}_{char}$. The total revenue requirement (TRR) method was applied [22]. The bare module costs (BMC) of the components were estimated with a number of methods. Cost estimating charts [29,30], cost estimating equations [8] and past purchase orders [27,31,32] were considered. Pressure and temperate ranges were also taken into account. The costs were adjusted to €2017 with the chemical engineering cost indexes of the reference years (CEPCI$_{2017}$ = 567.5 [33]). The derived cost equations of the BMC for each type of component can be found in [34].

The assumptions made in the economic analyses are summarized in Table 5. The operation and maintenance costs (OMC) are assumed as a percentage of the fixed capital investment (FCI) which ranges from 1.5% to 3% of the plant purchase price per year [35]. The system is assumed to operate at low electricity prices.

Table 5. Assumptions made in economic analysis.

Assumption	Value
Service facilities, architectural work	30% of BMC
Contingencies	15% of BMC
Effective interest rate	8%
Average inflation rate	3%
Plant economic life	30 years
Annual full load operation	2882 h/a
Annual OMC	1.5% of FCI
Mean cost of charged electricity	17.2 €/MWh

For better comparability the specific investment costs are determined. The total capital investment (TCI) of the charging unit is levelized to the charging capacity of the storage (€/kW$_{char}$).

For the exergoeconomic analysis, the levelized cost rate $\dot{Z}_k$ of each component k needs to be determined. The component cost rate considers the costs associated with the capital investment $\dot{Z}_k^{CI}$ and the operation and maintenance costs $\dot{Z}_k^{OM}$ of the respective component. The component cost rate is calculated over the levelized carrying charges CC_L, the levelized operation and maintenance costs OMC_L, the annual operation time of the component τ and the share of the investment costs BMC_k associated with the k-th component in the total bare-module costs BMC_{tot} of the overall system:

$$\dot{Z}_k = \dot{Z}_k^{CI} + \dot{Z}_k^{OM} = \frac{BMC_k}{BMC_{tot}} \cdot \frac{(CC_L + OMC_L)}{\tau} \quad (12)$$

2.4. Exergoeconomic Analysis

The exergoeconomic analysis was applied to the best performing liquefaction processes. Aim is to identify the cost-effectiveness of the processes, the costs associated with the thermodynamic inefficiencies and the potential for cost reduction in the processes. This is achieved by "exergy costing" [27], where the average cost per unit of exergy of each stream in the process is calculated with the aid of cost balances and auxiliary equations. The cost balance for the k-th component of the process is expressed by:

$$\sum \dot{C}_{out,k} + \dot{C}_{k,W} = \dot{C}_{k,Q} + \sum \dot{C}_{in,k} + \dot{Z}_k \quad (13)$$

The cost balance needs to be fulfilled for each component in the system to determine the costs of the exiting streams. The sum of the costs associated with the n entering streams of matter $\sum \dot{C}_{in,k}$, the cost rate of the respective component $\dot{Z}_k$ and the cost of heat supplied to the component $\dot{C}_{k,Q}$ are equal to the sum of costs associated with the m exiting streams of matter $\sum \dot{C}_{out,k}$ and the work done by the system. Each stream of matter, heat or work with associated exergy transfer rate has an average cost per unit of exergy c_n (€/GJ):

$$\dot{C}_n = c_n \cdot \dot{E}_n \tag{14}$$

$$\dot{C}_W = c_W \cdot \dot{W} \tag{15}$$

$$\dot{C}_Q = c_Q \cdot \dot{E}_Q \tag{16}$$

All costs associated with the streams entering the overall system need to be known. The specific cost of the incoming air is set to c_1 = 0 €/MWh while the specific costs of the electricity is c_W = 17.5 €/MWh. The specific exergy costs of the entering cold storage media streams are assumed equal to the cost per unit of exergy of the liquid air:

$$c_{R218,\,in} = c_{methanol,in} = c_{liquid\;air} \tag{17}$$

If more than one stream exits the component, auxiliary equations based on the "fuel and product" approach are necessary [24]. The cost balance at the component level can also be formulated as:

$$\dot{C}_{P,k} = \dot{C}_{F,k} + \dot{Z}_k \tag{18}$$

The cost associated with the thermodynamic inefficiencies—the exergy destruction—is calculated by the average cost per unit of exergy of the fuel to the component $c_{F,k}$ and the exergy destruction $\dot{E}_{D,k}$ of the respective component:

$$\dot{C}_{D,k} = c_{F,k} \cdot \dot{E}_{D,k} \tag{19}$$

The components which are of high importance to the system's cost-effectiveness are determined by the sum of cost associated with the initial investment of the component $\dot{Z}_k$ and the cost associated with the exergy destruction $\dot{C}_{D,k}$. The exergoeconomic factor can be used to determine the type of changes required to improve the cost effectiveness of the respective component:

$$f = \frac{\dot{Z}_k}{\dot{Z}_k + \dot{C}_{D,k}} \tag{20}$$

In the performed exergoeconomic analysis the major contributors to the overall costs are identified and their potential for cost reduction is compared. Moreover, the results facilitate a subsequent iterative optimization.

3. Results and Discussions

3.1. Energetic and Exergetic Analyses

The results of the energetic and exergetic analysis of each liquefaction configuration before and after the integration of cold storage are shown in Figure 5. The integration of cold storage significantly increases the liquid yield. The exergy of the product increases correspondingly:

$$\dot{E}_P \uparrow = \dot{E}_{liquid\;air} \uparrow + \dot{E}_{Q,hot}. \tag{21}$$

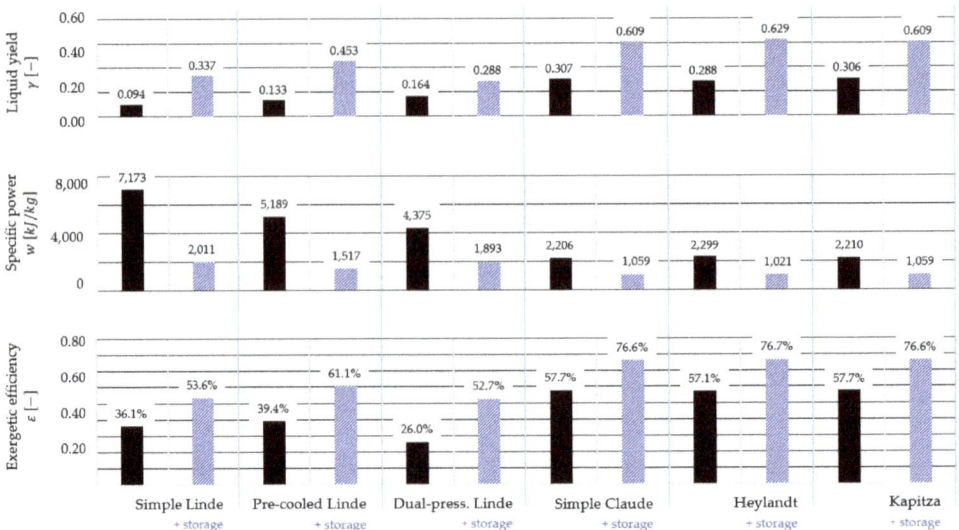

Figure 5. Results of exergy analysis of the liquefaction processes with/without integrated cold storage.

With a higher share of air liquefied the cold supplied by the cold storage increases ($\dot{E}_{cold}\uparrow$). A substantial reduction of the specific power required to produce one kg of liquid air is observed for all processes ($\dot{W}_{char}\downarrow$). Thus, the exergetic efficiency is considerably augmented with the addition of cold storage:

$$\varepsilon \uparrow = \frac{\dot{E}_{liquid\ air}\uparrow + \dot{E}_{Q,hot}}{\dot{E}_{cold}\uparrow + \dot{W}_{char}\downarrow} \tag{22}$$

The improvements were most significant in the simple Linde and the precooled Linde configuration where the exergetic efficiency increased significantly. Despite the liquid yield of the precooled Linde reaching a compatible value (0.453), its specific power requirement and exergetic efficiency cannot level with the Claude-based configurations. The simple Claude process, the Heylandt process and the Kapitza process reach the highest exergetic efficiencies (76.6%, 76.7% and 76.6), and have the lowest specific power requirement (1059, 1021 and 1059 kJ/kg$_{liquid\ air}$) and the highest liquid yields (0.609, 0.629, 0.609).

For the most efficient liquefaction configurations, the Claude-based processes, a sensitivity analysis was conducted. The compression pressure was varied (80–200 bar) and the splitting ratio r was reduced to its absolute minimum value. The effect of these variations on the exergetic efficiency ε can be seen in Figure 6. The share of air liquefied increases with a reduction in the value of the splitting ratio ($r = \dot{m}_{EX}/\dot{m}_{CM}$), as a greater mass flow enters the MHE and throttling process. The temperature difference in the MHE decreases with a reduction in "cold feed" ($\dot{m}_{EX}$) and a simultaneous increase in "hot feed" ($\dot{m}_{CM} - \dot{m}_{EX}$). The minimum splitting ratio is therefore restricted by the minimum pinch temperature ($\Delta T_{MHE1,\ min} \to 1$ K).

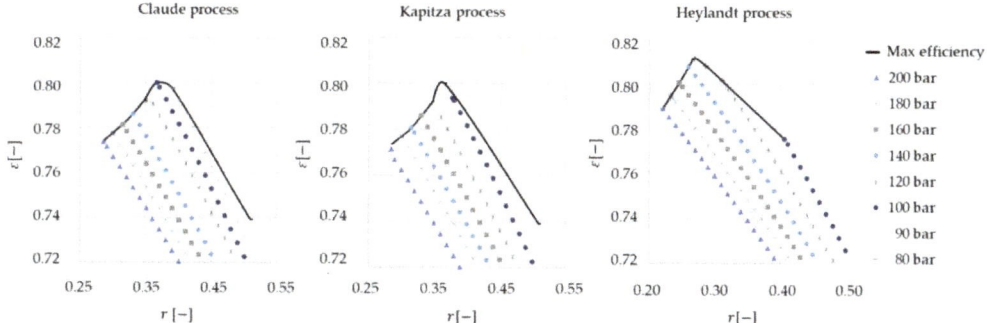

Figure 6. Sensitivity analysis results of the Claude, the Kapitza and the Heylandt process: exergetic efficiency ε over splitting ratio r, for various values of the liquefaction pressure. The maximum efficiency line is indicated with a solid black line.

By minimizing the splitting ratio for the respective compression pressure, a maximum efficiency line can be obtained. In Figure 7, the maximum exergetic efficiency curve of the Claude process, the Kapitza process and the Heylandt process are compared. The maximum liquid yield and the minimum specific power consumption graphs are also compared in Figures 8 and 9 respectively.

The thermodynamic performances of the Claude and Kapitza processes are almost the same. Reason for this is the temperature difference of only 3.3 K of the two mixing streams. The three processes reach their maximum efficiency at different pressures (Figure 7). This confirms that comparing the systems at a single pressure level is not sufficient. For liquefaction pressures of 120 bar and above the Heylandt process performs better reaching its optimum of approximately 81.2% (at 130 bar). The optimal configuration of the Claude and the Kapitza process is at about 100 bar reaching 80% exergetic efficiency.

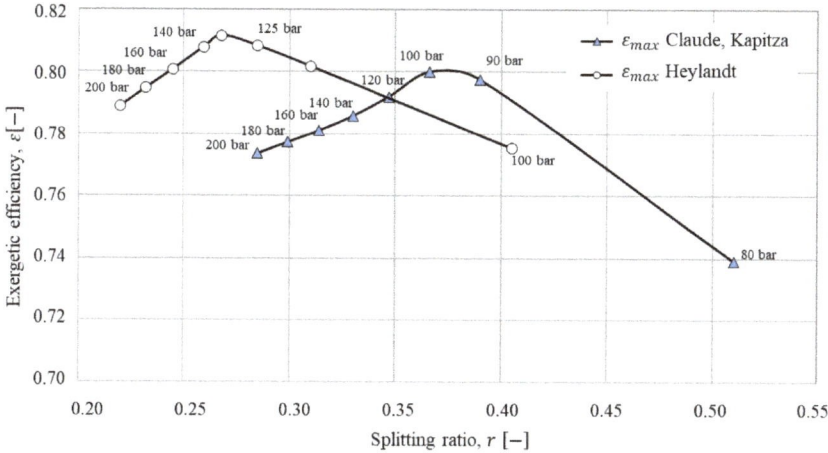

Figure 7. Maximum exergetic efficiency graphs as a function of the splitting ratio r for the Claude, Kapitza and Heylandt processes.

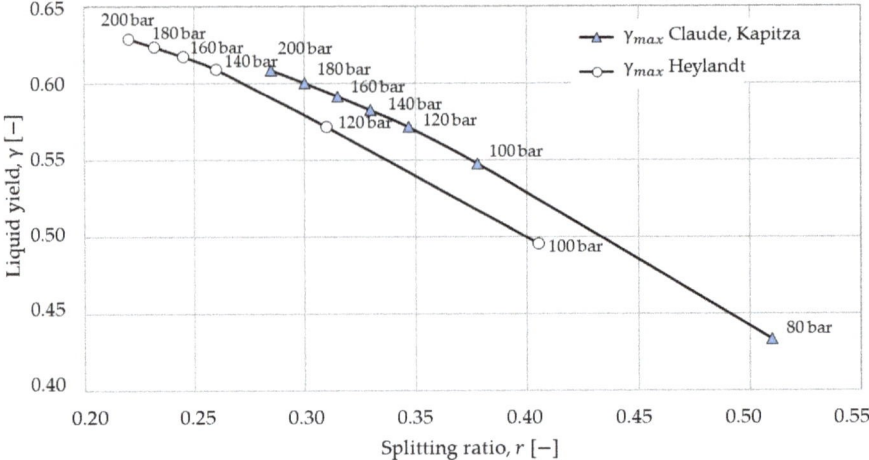

Figure 8. Maximum liquid yield graphs of the three Claude-based processes for different pressures and splitting ratios.

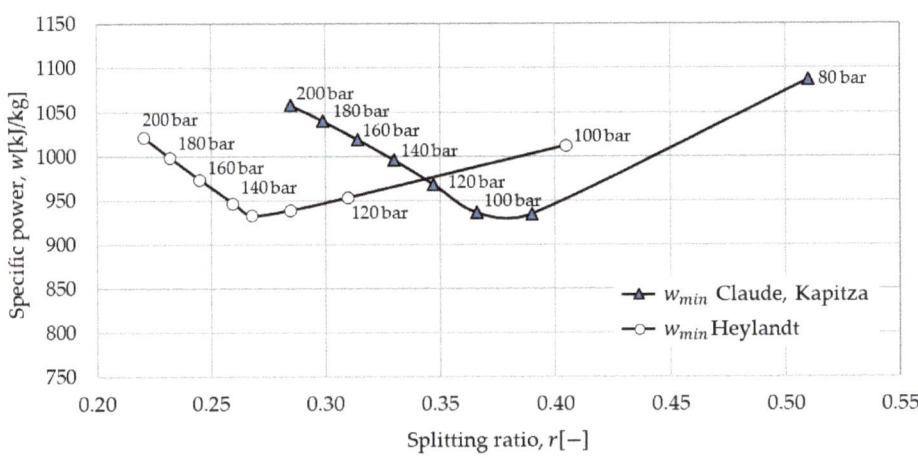

Figure 9. Minimum specific power graphs of the three Claude-based processes for different pressures and splitting ratios.

3.2. Economic Analysis

The economic analysis was conducted for the optimal system configuration for each of the three Claude-based systems. The system design parameters are given in Table 6. The charging power $\dot{W}_{char}$, the liquefaction capacity $\dot{m}_{liquid\ air}$, and the storage capacity ($\dot{W}_{dis} \cdot \tau_{dis}$) are similar for all systems. The liquid yield γ and the charging pressure p_{CM} of the Heylandt process is slightly higher.

Table 6. Design parameters for liquefaction systems evaluated in economic analysis.

Parameter	Unit	Claude	Heylandt	Kapitza
Liquefaction pressure	bar	95	130	95
Charging capacity	MW	20	20	20
Liquefaction capacity	tons/day	606	608	606
Storage capacity	MWh	76.6	78.4	76.6
Liquid yield	-	0.54	0.59	0.54

Figure 10 shows the BMC broken down to the component groups: expander, compressors, intercoolers, main heat exchanger and other components. The heat exchangers are responsible for 70–80% of the investment costs for all processes. The results of economic analysis of the Claude and the Kapitza process differ despite similar performance in energetic and exergetic analysis. The small difference in size of the MHE results in a noteworthy difference in costs. The total revenue requirements for the Claude, Heylandt and Kapitza systems amount to 2770 €/a, 2915 €/a and 2670 €/a respectively.

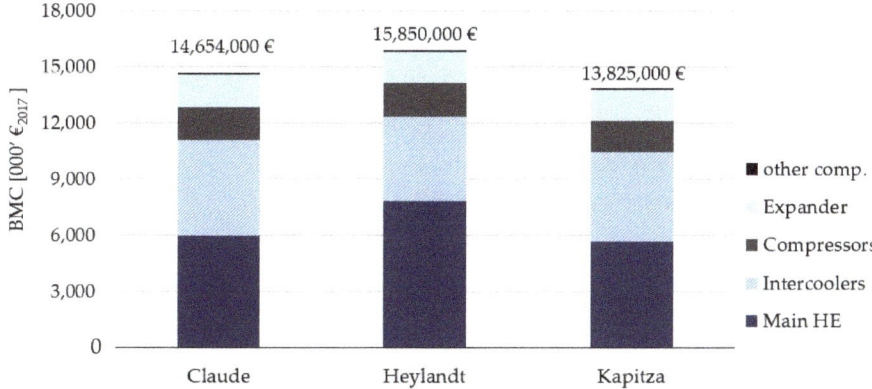

Figure 10. Bare module costs of the evaluated Claude-based systems with indicated cost shares of the contributing component groups.

The Heylandt system is not competitive in regards to its specific investment per unit of exergy stored despite the slightly higher energy output of the process, see Table 6. The specific investment per kW installed capacity for the Claude, Heylandt and Kapitza systems amount to 733 €/kW$_{char}$, 792 €/kW$_{char}$ and 691 €/kW$_{char}$, respectively.

3.3. Exergoeconomic Analysis

The results of the exergoeconomic analysis at the component level are shown in Figure 11. All analyzed systems show an elevated exergoeconomic factor ($f \gg 0.5$) which indicates that the costs associated with the purchase and maintenance of the components $\dot{Z}_k$ dominates the cost picture ($\dot{Z}_k \gg \dot{C}_{D,k}$). The cost of exergy destruction in the components $\dot{C}_{D,k}$ is a minor contributor to the costs of the final product. When investment costs dominate, a reduction in investment costs while accepting lower efficiencies is recommended to lower the total costs.

The exergoeconomic factor of several components in the Heylandt system is higher than in the other two systems. This indicates that the Heylandt system leaves more room for improvement of the cost-effectiveness of the system. Yet, regarding the significantly higher average cost per unit of exergy of the product $c_{P,tot}$ (Table 7), the reduction in costs may not be substantial enough to surpass the other configurations.

The average cost of exergy of the fuel $c_{F,tot}$ is relatively high in comparison to the low average cost of the electricity $c_{electricity}$ = 17.5 €/MWh. Reason for this is the average cost of low-temperature exergy supplied by the cold storage $c_{q,cold}$ which is relatively high and amounts to the average cost of exergy of the liquid air $c_{liquid\ air}$. The average cost of exergy of the heat supplied to heat storage has the lowest value for the Heylandt process while the average cost of exergy of liquid air is the most expensive. Regarding the average cost of the exergy of the final product $c_{P,tot}$, the Kapitza process performs best.

This conclusion is not expected to be changed with increase in the system size. The reason is that the heat exchangers are the major contributors to the costs of the liquefaction systems, and the cost of heat exchanger increase linearly with scale—for all systems equally.

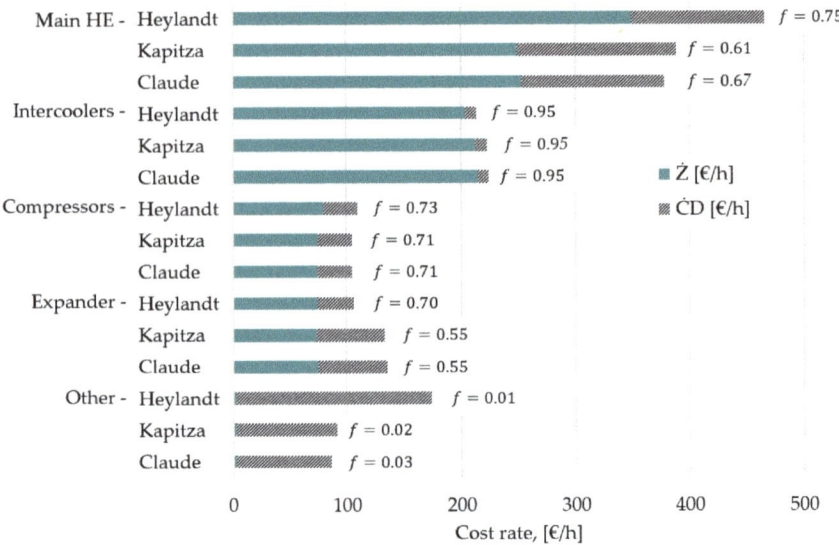

Figure 11. Sum of the cost rates associated with the initial investment of the component $\dot{Z}_k$ and the exergy destruction $\dot{C}_{D,k}$ and exergoeconomic factor f of the respective component(s).

Table 7. Results of exergoeconomic analysis for the three evaluated systems.

Parameter	Claude	Heylandt	Kapitza	Unit
Average cost of exergy of the fuel, $c_{F,tot}$	44.3	61.2	44.0	€/MWh
Average cost of exergy of the losses, $c_{L,tot}$	39.3	43.7	39.2	€/MWh
Average cost of exergy of the product, $c_{P,tot}$	88.8	114.8	88.0	€/MWh
Average cost of exergy of the liquid air, $c_{liquid\ air}$	95.4	133.4	94.6	€/MWh
Average cost of exergy of the heat, $c_{q,hot}$	67.9	63.1	67.5	€/MWh

No previous publications considered the effect of integrating cold storage on the selection of the liquefaction process. Thus, the results are validated by drawing comparison to values given in literature for air liquefaction processes without cold storage (Table 8) and values reported in previous publications for CES system characteristics (Table 9).

Table 8. Final results of the three Claude-based systems compared to air liquefaction processes.

Parameter	Unit	Claude	Heylandt	Kapitza	Reference
Specific power consumption	kWh/ton	264.0	263.3	264.0	520–760 [13], 439 [35]
Production cost of liquid air	€/ton	18.6	25.9	18.4	37–48 [35]

Table 9. Final results of the evaluation of the three Claude-based systems compared to CES system.

Parameter	Unit	Claude	Heylandt	Kapitza	Reference
CES specific investment	€/kW$_{dis}$	939	923	911	500–3000 [6,7,36]
CES round-trip efficiency	%	46.9	49.0	46.9	40–60 [6,9,36]

The specific power consumption of air liquefaction processes reported in [13] and [35] is twice as large than in the presented systems. The integration of cold storage thus not only decreases the specific power consumption to half but also reduces the production cost of liquid air from 37–48 €/ton [35] to 18.4–25.9 €/ton (Table 8).

Assuming a TCI for the 40 MW discharge unit of 17.1 Mio €, the specific investment of the CES systems based on the Claude processes reach values lower than 1000 €/kW$_{dis}$, see Table 9. The specific investment costs of the total CES system is approximated from 500–3,000 €/kW [6,7,36] in literature. The levelized cost of discharged electricity (LCOE$_{dis}$) of the CES systems based on the Claude, the Heylandt, and the Kapitza process are expected to reach 175.6 €/MWh$_{el}$, 175.3 €/MWh$_{el}$ and 172.0 €/MWh$_{el}$, respectively. For industrial application 120-200 €/MWh are set as goal. A sensitivity analysis of the LCOE and comparison to other technologies was reported in [34]. The final RTE of 47–49% are also in line with the expected 40–60%, which confirms the presented results.

In Table 10, the specific investment costs and RTE of other bulk-energy storage technologies are given. The competing bulk-energy storage technology are also capital intense which makes CES competitive with compressed air energy storage (CAES), pumped hydro storage (PHS) and hydrogen-based energy storage (H$_2$). Regarding the RTE of PHS and CAES, CES efficiency is still the greatest obstacle. The high exergy density of CES (120–200 kWh/m^3 [36])—the absence of geographical constraints—remains the technologies greatest advantage.

Table 10. Specific investment cost and RTE of competing bulk-energy technologies.

Parameter	Unit	CAES	PHS	H2
Specific investment cost	€/kWdis	500–2200 [37]	350–1,500 [37]	> 2000+ [38]
Round-trip efficiency	%	40–5 [37]	75–85 [39,40]	30–50 [38,41]
Exergy density	kWh/m^3	0.5–1.5 [36]	3–6 [36]	133–785 [41], 500+ [36]

4. Conclusions

This paper presents the state-of-the-art of cryogenic energy storage with regards to air liquefaction processes, thermodynamic parameters and cold storage configurations. Six air liquefaction processes within the charge unit of CES were investigated and results obtained from the exergy-based analysis were compared. The effect of cold storage integration on different liquefaction processes was firstly quantified.

- The integration of the charging unit with cold exergy recovery was shown to substantially augment the liquid yield γ, significantly reduce the specific power requirement w_{char} and significantly improve the exergetic efficiency ε of all liquefaction processes assessed.
- The simple Claude, the Heylandt and the Kapitza processes were found to reach the highest exergetic efficiencies and liquid yields, as well as the lowest specific power requirements for liquefaction.
- The sensitivity analysis showed that for liquefaction pressures of 125 bar and higher, the Heylandt process reaches the highest exergetic efficiencies, at lower pressures the Claude and the Kapitza process are superior.
- The economic analysis revealed that the Kapitza process-based system has the lowest specific investment cost and total revenue requirement.
- The exergoeconomic analysis demonstrated that the Kapitza process is the most cost-effective liquefaction process to be considered for CES with cold storage. The average cost of the exergy of the final product was the lowest in the Kapitza process.
- The results were compared to values from literature. The specific power consumption of the presented air liquefaction processes with cold storage ($\leq$264 kWh/ton) was found to be approximately half the values reported in literature. The production cost of liquid air was found to be significantly reduced with the integrating cold storage (18–26 €/ton).
- The final results on system level were found to be in line with the values reported for CES specific investment cost and RTE. Finally, CES was evaluated cost-competitive with other bulk-energy storage technologies.

Author Contributions: Conceptualization, investigation and data curation, S.H.; methodology, G.T.; software and validation, S.H. and F.M.; formal analysis, S.H. and T.M.; writing—original draft preparation, S.H.; writing—review and editing, T.M. and G.T.; supervision, T.M.; project administration, T.M.

Funding: Sarah Hamdy acknowledges the financial support of the Federal Ministry of Education and Research (BMBF—Bundesministerium für Bildung und Forschung) under the Transnational Education project (ID 57128418) of the German Academic Exchange Service (DAAD).

Conflicts of Interest: The authors declare no conflict of interest.

References

1. Hamdy, S.; Morosuk, T.; Tsatsaronis, G. Cryogenic Energy Storage: Characteristics, Potential Applications and Economic Benefit. In *Recent Developments in Cryogenics Research*; Putselyk, S., Ed.; Nova Science, Inc.: New York, NY, USA, 2019; pp. 277–310.
2. Li, Y.; Chen, H.; Ding, Y. Fundamentals and applications of cryogen as a thermal energy carrier: A critical assessment. *Int. J. Therm. Sci.* **2010**, *49*, 941–949. [CrossRef]
3. Smith, E. Storage of Electrical Energy using Supercritical Liquid Air. *Proc. Inst. Mech. Eng.* **1977**, *191*, 289–298. [CrossRef]
4. European Energy Research Alliance. *Liquid Air Energy Storage, EERA Joint Program SP4—Mechanical Storage Fact Sheet 3*; European Energy Research Alliance: Brussels, Belgium, 2016.
5. Kishimoto, K.; Hasegawa, K.; Asano, T. *Development of Generator of Liquid Air Storage Energy System*; Mitsubishi Heavy Industries, Ltd.: Tokyo, Japan, 1998.
6. Morgan, R.; Nelmes, S.; Gibson, E.; Brett, G. An analysis of a large-scale liquid air energy storage system. In Proceedings of the Institution of Civil Engineers, London, UK, 1 September 2015.
7. Highview Enterprises Ltd. Highview Power Storage. 10 May 2018. Available online: http://www.highview-power.com/ (accessed on 15 December 2018).
8. Li, Y. Cryogen Based Energy Storage: Process Modelling and Optimization. Ph.D. Thesis, University of Leeds, Leeds, UK, 2011.
9. Hamdy, S.; Morosuk, T.; Tsatsaronis, G. Cryogenics-based energy storage: Evaluation of cold exergy recovery cycles. *Energy* **2017**, *138*, 1069–1080. [CrossRef]
10. Kerry, F.G. *Industrial Gas Handbook: Gas Separation and Purification*; Taylor & Francis Group, LLC: Boca Raton, FL, USA, 2007.
11. Ameel, B.; T'Joen, C.; De Kerpel, K.; De Jaeger, P.; Huisseune, H.; Van Belleghem, M. Thermodynamic analysis of energy storage with a liquid air Rankine cycle. *Appl. Therm. Eng.* **2013**, *52*, 130–140. [CrossRef]
12. Peng, H.; Shan, X.; Yang, Y.; Ling, X. A study on performance of a liquid air energy storage system with packed bed units. *Appl. Energy* **2018**, *211*, 126–135. [CrossRef]
13. Borri, E.; Tafone, A.; Romagnoli, A.; Comodi, G. A preliminary study on the optimal configuration and operating range of a "microgrid scale" air liquefaction plant for Liquid Air Energy Storage. *Energy Convers. Manag.* **2017**, *143*, 275–285. [CrossRef]
14. Abdo, R.F.; Pedro, H.T.; Koury, R.N.; Machado, L.; Coimbra, C.F.; Porto, M.P. Performance Evaluation of various cryogenic energy storage systems. *Energy* **2015**, *90*, 1024–1032. [CrossRef]
15. She, X.; Peng, X.; Nie, B.; Leng, G.; Zhang, X.; Weng, L.; Tong, L.; Zheng, L.; Wang, L.; Ding, Y. Enhancement of round trip efficiency of liquid air energy storage through effective utilization of heat of compression. *Appl. Energy* **2017**, *206*, 1632–1642. [CrossRef]
16. Sciacovelli, A.; Vecchi, A.; Ding, Y. Liquid air energy storage (LAES) with packed bed cold thermal storage—From component to system level performance through dynamic modelling. *Appl. Energy* **2017**, *190*, 84–98. [CrossRef]
17. Morgan, R.; Nelmes, S.; Gibson, E.; Brett, G. Liquid air energy storage—Analysis and first results from pilot scale demonstration plant. *Appl. Energy* **2015**, *137*, 845–853. [CrossRef]
18. Guizzi, G.L.; Manno, M.; Tolomei, L.M.; Vitali, R.M. Thermodynamic analysis of a liquid air energy storage system. *Energy* **2015**, *93*, 1639–1647. [CrossRef]
19. Xue, X.; Wang, S.X.; Zhang, X.L.; Cui, C.; Chen, L.; Zhou, Y.; Wang, J. Thermodynamic analysis of a novel liquid air energy storage system. *Phys. Procedia* **2015**, *67*, 733–738. [CrossRef]

20. Peng, X.; She, X.; Cong, L.; Zhang, T.; Li, C.; Li, Y.; Wang, L.; Tong, L.; Ding, Y. Thermodynamic study on the effect of cold and heat recovery on performance of liquid air energy storage. *Appl. Energy* **2018**, *221*, 86–99. [CrossRef]
21. Chen, H.; Ding, Y.; Peters, T.; Berger, F. A method of storing energy and a cryogenic energy storage system. US Patent EP1989400A1, 27 February 2006.
22. Li, Y.; Cao, H.; Wang, S.; Jin, Y.; Li, D.; Wang, X.; Ding, Y. Load shifting of nuclear power plants using cryogenic energy storage technology. *Appl. Energy* **2014**, *113*, 1710–1716. [CrossRef]
23. Barron, R.F. *Cryogenic Systems*, 2nd ed.; Oxford University Press, Clarendon Press: New York, NY, USA; Oxford, UK, 1985.
24. Herron, D.M.; Agrawal, R. *Air Liquefaction: Distillation*; Air Products and Chemicals, Inc.: Allentown, PA, USA, 2000.
25. Brett, G.; Barnett, M. The application of liquid air energy storage for large scale long duration solutions to grid balancing. *EPJ Web Conf.* **2014**, *79*, 03002. [CrossRef]
26. Highview Power Storage Ltd. *Performance and Technology Review*. March 2012. Available online: https://businessdocbox.com/Green_Solutions/77092496-Highview-power-storage-technology-and-performance-review.html (accessed on 1 January 2019).
27. Bejan, A.; Tsatsaronis, G.; Moran, M. *Thermal Design and Optimization*; John Wiley & Sons Inc.: New York, NY, USA, 1996.
28. Morosuk, T.; Tsatsaronis, G. Splitting physical exergy: Theory and application. *Energy* **2019**, *167*, 698–707. [CrossRef]
29. Ulrich, G.D.; Vasudevan, P.T. Chapter 5 Capital Cost Estimation. In *Chemical Engineering—Process Design and Economics—A Practical Guide*; Process Publishing: New Hampshire, NH, USA, 2004; pp. 352–419.
30. Peters, M.S.; Timmerhaus, K.D.; West, R.E. Chapter 12 Materials-Handling Equipment—Design and Costs. In *Plant Design and Economics of Chemical Engineers*; McGraw-Hill Companies, Inc.: New York, NY, USA, 2003; pp. 485–589.
31. Xu, G.; Liang, F.; Yang, Y.; Hu, Y.; Zhang, K.; Liu, W. An Improved CO_2 Separation and Purification System Based on Cryogenic Separation and Distillation Theory. *Energies* **2014**, *7*, 3484–3502. [CrossRef]
32. Smith, R. *Chemical Process Design and Integration*; John Wiley & Sons, Ltd.: Manchester, UK, 2014.
33. Chemical Engineering. Available online: https://www.chemengonline.com/cepci-updates-january-2018-prelim-and-december-2017-final/ (accessed on 12 September 2018).
34. Hamdy, S.; Morosuk, T.; Tsatsaronis, G. Exergetic and economic assessment of integrated cryogenic energy storage systems. *Cryogenics* **2019**. accepted manuscript.
35. Strahan, D.; Akhurst, M.; Atkins, D.A.; Arbon, P.I.; Ayres, M. *Liquid Air in the Energy and Transport Systems*; Center for Low Carbon Futures: York, UK, 2013.
36. Chen, H.; Cong, T.N.; Yang, W.; Tan, C.; Li, Y.; Ding, Y. Progress in electrical energy storage system: A critical review. *Prog. Nat. Sci.* **2009**, *19*, 291–312. [CrossRef]
37. EERA/EASE. European Association for Storage of Energy. Available online: http://ease-storage.eu/wp-content/uploads/2015/10/EASE-EERA-recommendations-Roadmap-LR.pdf (accessed on 16 January 2019).
38. Welder, L.; Stenzel, P.; Ebersbach, N.; Markewitz, P.; Robinius, M.; Emonts, B.; Stolten, D. Design and Evaluation of Hydrogen Electricity Reconversion Pathways in National Energy Systems Using Spatially and Temporally Resolved Energy System Optimization. *Int. J. Hydrog. Energy* **2018**, in press. [CrossRef]
39. Ibrahim, H.; Ilinca, A.; Perron, J. Energy storage systems—Characteristics and comparisons. *Renew. Sustain. Energy Rev.* **2008**, *12*, 1221–1250. [CrossRef]
40. Kaldellis, J.K.; Zafirakis, D. Optimum energy storage techniques for the improvement of renewable energy sources-based electricity generation economic efficiency. *Energy* **2007**, *32*, 2295–2305. [CrossRef]
41. Li, Y.; Chen, H.; Zhang, X.; Tan, C.; Ding, Y. Renewable energy carriers: Hydrogen or liquid air/nitrogen? *Int. J. Therm. Sci.* **2010**, *49*, 941–949. [CrossRef]

© 2019 by the authors. Licensee MDPI, Basel, Switzerland. This article is an open access article distributed under the terms and conditions of the Creative Commons Attribution (CC BY) license (http://creativecommons.org/licenses/by/4.0/).

Article

Analyzing the Energy Consumption, GHG Emission, and Cost of Seawater Desalination in China

Xuexiu Jia [1,*], Jiří Jaromír Klemeš [1], Petar Sabev Varbanov [1] and Sharifah Rafidah Wan Alwi [2]

1   Sustainable Process Integration Laboratory—SPIL, NETME Centre, Faculty of Mechanical Engineering, Brno University of Technology, Technická 2896/2,616 00, Brno, Czech Republic; jiri.klemes@vutbr.cz (J.J.K.); varbanov@fme.vutbr.cz (P.S.V.)
2   Process Systems Engineering Centre (PROSPECT), Research Institute for Sustainable Environment and School of Chemical and Energy Engineering, Universiti Teknologi Malaysia (UTM), 81310 UTM Johor Bahru, Johor, Malaysia; syarifah@utm.my
*   Correspondence: jia@fme.vutbr.cz; Tel.: +420-777-365-648

Received: 17 December 2018; Accepted: 30 January 2019; Published: 31 January 2019

Abstract: Seawater desalination is considered a technique with high water supply potential and has become an emerging alternative for freshwater supply in China. The increase of the capacity also increases energy consumption and greenhouse gases (GHG) emissions, which has not been well investigated in studies. This study has analyzed the current development of seawater desalination in China, including the capacity, distribution, processes, as well as the desalted water use. Energy consumption and GHG emissions of overall desalination in China, as well as for the provinces, are calculated covering the period of 2006–2016. The unit product cost of seawater desalination plants specifying processes is also estimated. The results showed that 1) The installed capacity maintained increased from 2006 to 2016, and reverse osmosis is the major process used for seawater desalination in China. 2) The energy consumption increased from 81 MWh/y to 1,561 MWh/y during the 11 years. The overall GHG emission increase from 85 Mt CO_{2eq}/y to 1,628 Mt CO_{2eq}/y. Tianjin had the largest GHG emissions, following are Hebei and Shandong, with emissions of 4.1 Mt CO_{2eq}/y, 2.2 Mt CO_{2eq}/y. and 1.0 Mt CO_{2eq}/y. 3) The unit product cost of seawater desalination is higher than other water supply alternatives, and it differentiates the desalination processes. The average unit product cost of the reverse osmosis process is 0.96 USD and 2.5 USD for the multiple-effect distillation process. The potential for future works should specify different energy forms, e.g. heat and power. Alternatives of process integration should be investigated—e.g. efficiency of using the energy, heat integration, and renewables in water desalination, as well as the utilization of total site heat integration.

Keywords: water desalination; water supply; water shortage; energy demand; environmental impacts; specific energy consumption

1. Introduction

Increasing water scarcity has become a global issue. Freshwater supply is limited and has been remarkably affected by the degradation of water quality in natural water bodies, while the demand for freshwater has continued to increase. Besides water consumption minimization by improving water use efficiency, conventional water treatment and desalination are employed to reclaim the polluted water and freshwater to increase the supply. Especially in water-scarce regions, where the water source is mainly from precipitation, the water supply has been unreliable due to the influence of global climate change [1]. Water desalination has been widely applied in the world. A report from the Water Desalination Report by the International Desalination Association presented the current installed capacities of the world desalination by countries [2]. Figure 1 shows the water desalination capacities of the world by countries from 2010 to 2016 [2]. In the last six years, the world total water desalination capacity, including brackish water and seawater desalination, increased steadily with an annual rate of about 9%. A large proportion came from the Gulf region, and not surprisingly, the Kingdom of Saudi Arabia (KSA) and the United Arab Emirates (UAE), which take a proportion of 15% and 11% of the world's total desalination capacities in 2016. Next is the USA, which takes 10% of the world total installed desalination capacity. China has the fourth largest water desalination capacity, with a share of 5% of the world total installed capacity.

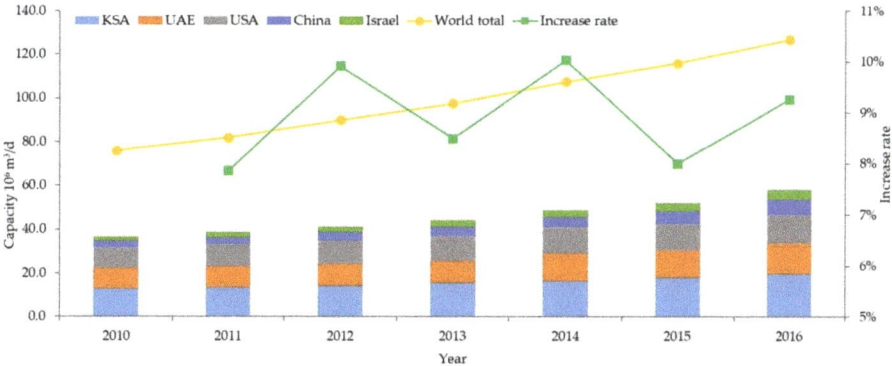

Figure 1. Water desalination capacities of the world and selected countries from 2010 to 2016 (derived from [2]). KSA: Kingdom of Saudi Arabia; UAE: United Arab Emirates.

Water desalination is an energy intensive approach for freshwater production [3], and the rapid increase of installed capacity has resulted in increasing resource (mainly energy) consumption and environmental impacts. Based on the water desalination capacities and the energy consumption factor provided by [4], the energy consumption of the world overall water desalination is estimated and as shown in Figure 2.

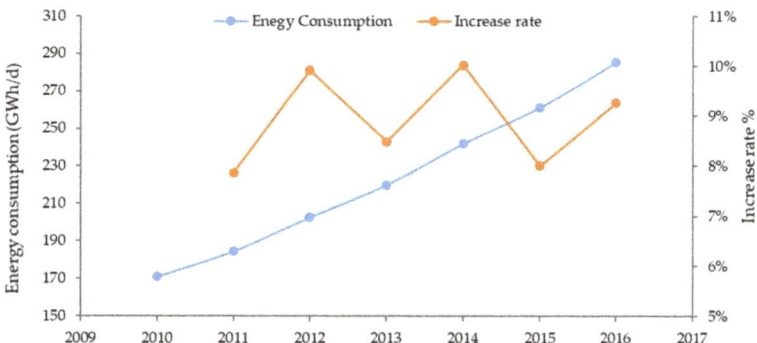

Figure 2. Water desalination energy consumption (electricity) from 2010 to 2016.

The environmental impact of water desalination has been focused on theoretical and scenario analyses [5]. Cornejo et al. [6] found that reverse osmosis (RO) technologies have lower GHG emissions than thermal desalination technologies. The estimated GHG emissions footprint of seawater RO desalination (0.4–6.7 kg CO_{2eq}/m^3) is generally larger than brackish water RO desalination (0.4–2.5 kg CO_{2eq}/m^3) and water reuse systems (0.1–2.4 kg CO_{2eq}/m^3). Shrestha et al. [7] determined that the associated CO_2 emissions for seawater desalination (0.25 Mt/y) are 47.5 % higher than that for water conveyance (0.17 Mt/y). The GHG footprint values vary due to the variability of location, technologies, life cycle stages, parameters considered, etc.

For producing freshwater, seawater desalination has been strongly implemented in the Gulf region and is emerging in East Asia, where are facing serious water stress issues. China has the fourth largest capacity of seawater desalination in the world, and water desalination is still an emerging industry. The major driving force is increasing water shortage. China is becoming one of the countries with a severe water shortage especially in the most developed northeast region [8]. For example, in 2016, the average water resource per capita was 2,355 m^3 [9], which is about 40% of the world average value [10]. For the capital city of China, Beijing, the amount is 162 m^3 [11], which is less than 3% of the world average value. The population and urban land in these water stress areas are still increasing [8], which indicates an increasing demand for freshwater. One fact is that about 71% of the Earth's surface is covered by water, and the oceans hold about 96.5% of all Earth's water [12]. Consequently, the water shortage is a shortage of clean freshwater, which will lead to an increase in economic cost and resource consumption, as well as potential environmental impacts.

Facing the water shortage issue, China is making a major effort with increasing the water use efficiency and eliminating water waste. On the other hand, for the regions with severe water shortage, there are mainly two possibilities to increase the amount of available freshwater. One solution is water transfer projects, including the South–North Water Transfer Project and the Water Transfer from Yellow River to Qingdao Project [13]. These projects are carried out by constructing water channels to transfer freshwater from water-rich areas to water-scarce regions, mainly Beijing, Hebei, Henan, and Shandong. Another action is the promotion of seawater desalination techniques and projects. From 2006 to 2016, the installed seawater desalination has increased from 20×10^6 m^3/y to 390×10^6 m^3/y [14]. Until 2016, there have been 15 newly released standards by the government to facilitate the promotion and management of water desalination projects [14]. Seawater desalination has been considered as one of the most promising techniques due to the abundance of seawater and the improving operating efficiency [15]. With an increased capacity, the potential of resource consumption and environmental impacts are also concerned.

The majority of studies have focused on either the advancement of the desalination process or specific case plants. Sores et al. [16] proposed and tested a novel supercharger which can be applied to a seawater desalination RO system and found that the efficiency of the tidal supercharger is currently lower than 20%, although the efficiency increased from 12% to 14%, with the seawater flow rate increasing from 290 m^3/h to 440 m^3/h. The application of renewable energies is also discussed in the current studies. For instance, Zuo et al. [17] proposed a model of a wind supercharged solar chimney power plant combined with seawater desalination and claimed that the utilization of solar energy can be raised by 70% with integration. Li et al. developed a high-efficiency membrane for seawater desalination using solar energy [18], and the results showed a 90% efficiency of converting solar energy. Desalination plants in China have seldom been discussed. Liu et al. [19] carried out the systems process analysis of the freshwater cost of seawater desalination, with a case study of a 25,000 m^3/d seawater desalination plant in Huanghua Port, Hebei Province, China. The study determined that the freshwater consumption of the plant is 4.5×10^5 m^3/y, which is 5% of the annual freshwater production (9.2×10^6 m^3/y). The World Resources Institute [15] investigated the carbon footprint of different scenarios in Qingdao. It showed that in 2020 with a water desalination capacity of 400×10^3 m^3/d, the carbon footprint of the water desalination will be 541.31 kt CO_{2eq}/y (with a cost of 8 CNY/m^3), which is 1.81 times more expensive than water supplies from surface and groundwater (with a cost of 1.17 CNY/m^3). Other studies also reported that water desalination with various techniques has other atmospheric emissions, e.g. dust, NOx, and SOx [20].

Most of the literature have focused on either the advancement of the desalination process or the specific case plants. The overall picture of the development and environmental performance, as well as the cost of seawater desalination in China, has not been thoroughly discussed. There is an urgent need to analyze the current development of the seawater desalination in China and to benchmark the energy consumption, emissions, as well as the cost. This can provide an overall picture of the environmental and economic performance, and facilitate energy consumption minimization, GHG reduction, and efficiency improvement in seawater desalination implementations. The aim of this paper is to provide fundamental remarks for the further studies of the water-energy nexus of seawater desalination. In order to do this, the paper first overviewed the current development and processes of seawater desalination projects in China. Then, the energy demand, GHG emissions, as well as the unit product cost of the seawater desalination plants were estimated. Based on the calculation, the future development and promising directions of water desalination studies are discussed.

2. Seawater Desalination in China

With the promotion and development of water desalination projects and more advanced technology, the total capacity of seawater desalination plants increased from 20×10^6 m^3/y to 390×10^6 m^3/y from 2006 to 2016, which can be seen from Figure 3 [14]. Due to the introduction of relevant policies and standards by the government, seawater desalination has been one of the promising solutions to the water shortage in China.

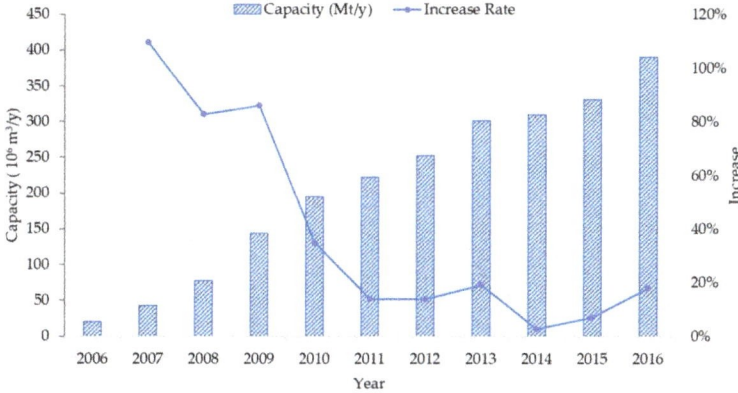

Figure 3. Total seawater desalination plant capacities in China from 2006 to 2016 (Based on [14]).

According to the State Oceanic Administration of China [14], there have been 131 seawater desalination plants/projects up until 2016 in mainly coastal cities in China, and the total capacity has reached to 1.19×10^6 m^3/d. The plants with different capacities are evenly distributed.

As shown in Table 1, there are 36 plants/projects with a capacity larger than 10 km^3/d, with a total capacity of 10^6 m^3/d. 38 plants have a capacity of 10^3 m^3/d to 10^4 m^3/d, with a total capacity of 1.2×10^5 m^3/d. Another 57 smaller plants/projects have a total capacity of 1.1×10^4 m^3/d. The largest water desalination plant in China is in Tianjin, with a total capacity of 2×10^5 m^3/d.

Table 1. Water desalination projects development in China until 2016 (Based on [14]).

Item	Description	Value
Quantity	Number of plants	131
	Total Project capacity	1.19×10^6 m^3/d
Capacity	Plants capacity > 10^4 m^3/d	Number of plants: 36, Total capacity 10^6 m^3/d
	Plants capacity (1–10^4 m^3)	Number of plants: 38, Total capacity 118×10^3 m^3/d
	Plants capacity < 10^3 km^3	Number of plants: 57, Total capacity 11×10^3 m^3/d
Benefits	The added value of seawater desalination plants	1.5×10^9 CNY/y

Regionally, large-scale seawater desalination plants are located on the northeastern coast of China with relatively severe water scarcity issues, e.g. Tianjin, Shandong, and Hebei. Figure 4 shows the seawater desalination capacities of the plants by desalination process, as well as the estimated water stress index of the provinces with seawater desalination plants. Tianjin, as a coastal city next to Beijing (the capital of China), has the largest total capacity of 317 × 10^3 m^3/d, with RO and multiple-effect distillation (MED) being the main processes, multi-stage flash (MSF) and electrodialysis (ED) are less applied in China. The Tianjin Beijing seawater desalination plant, with a capacity of 200 × 10^3 m^3/d, is the largest seawater desalination plant in China. Tianjin also has the largest installed seawater desalination plants with MED process. Shandong has the second largest seawater desalination capacity (282 × 10^3 m^3/d). The largest water desalination plant in Shandong is located in Qingdao, an international city facing severe water shortage issues, with a capacity of 10^5 m^3/d. The total capacity of water desalination plants in Qingdao reached 235× 10^3 m^3/d in 2016, which takes 83% of the total capacity of Shandong province [21]. One large-scale plant in southern China is in Zhejiang Province, with a capacity of 228 × 10^3 m^3/d. Other seawater desalination plants in southern China (e.g. Guangdong, Fujian, Jiangsu, and Hainan) are mainly small plants. Most of the plants are built after 2009, and the main processes applied are reverse osmosis (RO, with a proportion of 86%) and MED (with a proportion of 12%). Other technologies such as MSF and ED are also applied in few plants, with a total ratio of 2%.

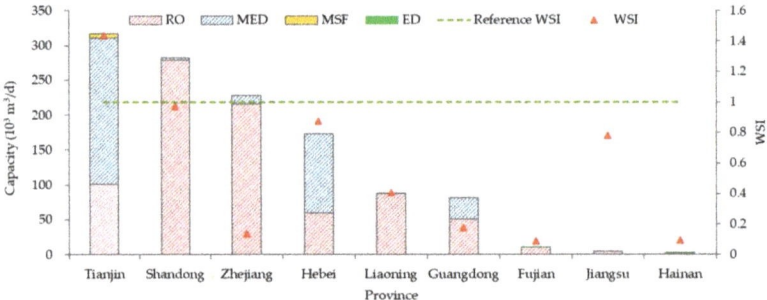

Figure 4. Water desalination capacity by provinces (derived from [14]). WSI: water stress index, RO: reverse osmosis; MED: multiple-effect distillation; MSF: multi-stage flash; ED: electrodialysis.

Figure 4 also indicates that the provinces/cities with large seawater desalination plants also have a higher water stress index (WSI), which is the ratio of the annual water consumption and the available natural water resources [22]. A higher WSI indicates that consumption is closer to the available water resources, thus higher water stress. When the value is higher than 1, it means the water consumption of the region has exceeded the available freshwater resources, and the region has a high water stress level. The WSI of the selected provinces are estimated with the method of [22] based on the data from [9], and the values are shown in Figure 4. The green dash line is the reference WSI with a value of 1. For most of the selected provinces, provinces with higher water stress have higher seawater desalination capacity, which indicates the driving factors of the development of seawater desalination. Two exceptions are Zhejiang and Jiangsu. Zhejiang has the third largest seawater desalination capacity, but the water stress is rather small. Jiangsu has a very small seawater desalination capacity but with relatively high water stress.

In terms of water use, as shown in Figure 5, a major proportion (66.6%) of desalted water is used in industries, and followed by domestic use with a ratio of 33.1%, and a small fraction (0.3%) is used for other purposes, e.g. watering in parks and greenbelts. Within industry, main users of desalinated water are fossil fuel power plants (31.6% of total), steel making industries (13.1%), and petrol chemical industries (12.3% of total). The major use in industries indicates that the quality of the desalted seawater needs to meet the quality requirement for industrial use.

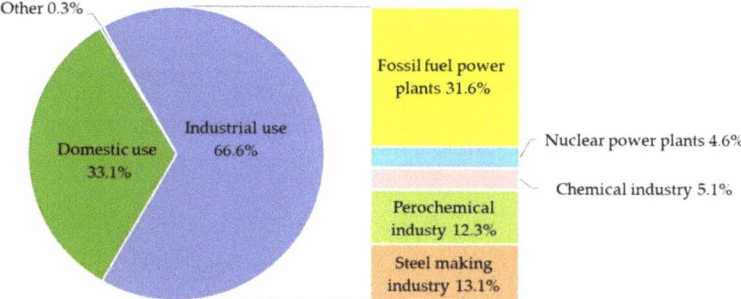

Figure 5. Distribution of desalinated water use in 2016 (Based on [14]).

Water treatment, seawater desalination, and water resource transfer are the main solutions for the freshwater shortage in China. Wastewater treatment, as the most conventional approach, has been well developed and implemented. The wastewater treatment approach can only offset part of the consumed water by removing the contaminants from the polluted water and return the treated water back to the available natural resource, but usually, the water quality of the discharged water is not as high as previous. When there is not enough available natural water resource, the cycle of supply- use-treatment-return cycle would be difficult to maintain. Water transfer is usually a huge project that has considerable economic and ecological cost, with large-scale changes to the inhabitants along the channel. Due to this reason, the existing water transfer projects in China are still under controversial discussion [23]. Seawater desalination is considered as one of the most promising approaches to produce freshwater.

Although seawater desalination is a reliable water supply and is not vulnerable to climate change, it consumes a lot of energy and it requires a lot of investment and public acceptance. The increasing capacity and wide distribution in different regions also result in the issue of increasing energy consumption, GHG emissions, as well as the economic cost. It is important to analyze and benchmark the environmental and economic performance, in order to provide insightful data for the further planning and optimization of energy use and emission reduction in seawater desalination.

3. Methodology

The energy consumption, GHG emission, and the unit product cost of the seawater desalination plants in China are assessed for the year of 2006–2016. The seawater desalination capacity data of China overall and selected provinces in 2016 are mainly derived from [14], and other data sources are explained where mentioned.

3.1. Energy Consumption

The energy consumption of seawater desalination plants in China can be calculated using mass balance equations along with the specific energy consumption (SEC) and the capacity of the plants. SEC in kWh/m^3 desalinated water, is one of the most critical factors characterizing the performance of the water supply [24].

$$EC_a = \sum SEC_i \times C_i \times 365 \times P_a \tag{1}$$

where EC_a is the annual energy consumption of the plant, kWh/y; SEC_i is the specific energy consumption of seawater desalination plants with process i, kWh/m^3; the energy consumption of all desalination processes involved in this study (RO, MED, MSF, and ED) are converted to the form of electricity, kWh/m^3; C_i is the capacity of the desalination plants with process i, m^3/d. P_a is the availability of the plant, when specific data is not available, P_a is set as 90% based on the study of [25].

The specific energy consumption of various desalination processes is listed in Table 2.

Table 2. Specific energy consumption (SEC) of different processes (based on [4]).

Process	Specific Energy Consumption kWh/m³
RO (seawater)	5.0
MED	17.9
MSF	23.4
ED	4.1

3.2. GHG Emissions

Since there are no major direct GHG emissions in water desalination, the estimation of GHG emissions mainly considers the emissions from energy consumption. The estimation method is based on the calculation method of GHG emission from processing proposed by ISCC [26], which is shown as follows:

$$G_e = EM_e + EM_{in} + EM_{waste} \quad (2)$$

where G_e is the annual GHG emissions of the desalination plants, t CO_{2eq}/y; EM_e is the emission of energy consumption, t CO_{2eq}/y; EM_{in} is the emission of material inputs, t CO_{2eq}/y; and EM_{waste}, is the indirect emission from treating the waste generated from the desalination processes, t CO_{2eq}/y. In this study, the GHG emission from energy consumption is estimated, and the emissions of material input (seawater) and waste (brine) are not considered due to the limit of data availability.

$$EM_e = EC_a \times E_f \quad (3)$$

where EC_e is the annual energy consumption of process i, kWh/m³; and Ef is the emission factors, t CO_{2eq}/kWh. In this study, the Ef is set as 1.04 t CO_{2eq}/kWh according to the reference [27].

3.3. Unit Product Cost

The cost of water desalination mainly includes capital cost and operating cost, with the later mainly consisting of energy cost for plant operation and the cost for maintenance. In this study, the unit product cost, which is the cost per m³ desalted water, is calculated based on the method proposed by [25].

Estimation of the unit product cost is calculated as follows:

$$UPC = \frac{\frac{CC}{Pl} + OP_a}{Ca \times Pa} \quad (4)$$

where UPC is the unit product cost, USD/m³; CC is the capital cost of the plant over the lifespan, USD; Pl is the plant life, y; OP_a is the annual operating cost; USD; Ca is the capacity of the plant, USD; Pa is the plant availability, %.

Capital Cost

The capital cost is calculated according to the power law rule:

$$\frac{CC_x}{CC_{rf}} = \left[\frac{Ca_x}{Ca_{rf}}\right]^m \quad (5)$$

where CC_x and Ca_x are the capacity (m³/d) and capital cost (MUSD) of the studied plant; CC_{rf} and Ca_{rf} are the capacity (m³/d) and capital cost (MUSD) of the reference plant, m is the power value.

Consequently, the capital cost of plant x can be calculated as:

$$CC_x = e^{m \times \ln(Ca_x) - m \times \ln(Ca_{rf}) + \ln(CC_{rf})} \quad (6)$$

According to [25], m is set as 0.8 for seawater desalinate plants. A dataset of the year 2016 from the Carlsbad Desalination Plant, in San Diego County, USA, is selected as the reference plant to estimate the overall capital cost of the desalination plants (shown in Table 3).

Table 3. Basic data of the Carlsbad Desalination Plant, in San Diego County [25].

Parameters	Value	Units
Total capacity	204,390	m^3/d
Feedwater TDS	34,500	mg/L
Process	RO 4 stages	-
Capital cost of the plant	537 M	USD

Operating Cost

The annual operating cost includes the energy consumption (electrical power), maintenance, labor, membrane replacement, as well as the cost for the chemicals. The operating cost is dependent on the operating process of the desalination plants, but in general, energy cost is the major component. Zhou et al. [28] studied the cost of thermal processes and found that energy cost is 87% of the total operating cost.

Wittholz et al. [25] investigated the cost of water desalination, and analyzed the breakdowns of the cost, including fixed cost (capital cost) and operating cost (maintenance, material and energy cost, etc). The contribution of the energy cost to operating cost is also estimated (Table 4).

Table 4. Average cost breakdowns of different desalination processes (based on [25]).

Process	Fixed Cost Contribution	Operating Contribution	Energy Cost Contribution in Operating Cost	EOP
RO (seawater)	35%	65%	35%	54%
RO (brackish water)	35%	65%	30%	46%
MED	40%	60%	45%	75%
MSF	40%	60%	45%	75%

The operating cost can be calculated with energy cost and the energy/operating cost ratio as in Equation (7):

$$OP_a = Eco/EOP \tag{7}$$

where OPa is the annual operating cost, MUSD, Eco is the energy cost, MUSD, and EOP is the ratio of energy cost and operating cost, %, which is shown in Table 4.

The cost of energy consumption can be calculated based on the specific energy consumption of different processes and the capacity as well as the price of the electric power, as shown in Equation (8):

$$Eco = SEC_i \times C_i \times P_e \tag{8}$$

where SECi is the specific energy consumption, kWh/m^3; the energy consumption of all desalination processes involved in this study (RO, MED, MSF, and ED) is converted to the form of electricity, kWh/m^3; P_e is the price of the electricity supplied for desalination plant, USD, the price is estimated for the year of 2016.

The electricity price for water desalination plant is referred from reference [29]. The cost is estimated from water input to the gate of the plant; water conveyance and distribution, as well as the brine disposal are yet considered.

4. Results

4.1. Energy Consumption and GHG Emissions

The energy consumption and GHG emissions of the Chinese seawater desalination are determined for the period of 2006–2016 (Figure 6). The annual energy consumption increased in the 11 y from 81 MWh/y to 1,561 MWh/y, with an increasing rate of 182%. The GHG emissions increased from 85 MtCO$_{2eq}$/y to 1,628 MtCO$_{2eq}$/y.

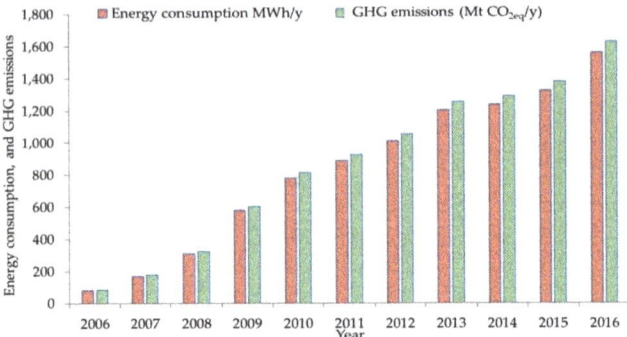

Figure 6. Overall seawater desalination energy consumption and GHG emissions from 2006–2016.

The breakdowns of GHG emissions by province and desalination processes are estimated for the year of 2016, as shown in Figure 7.

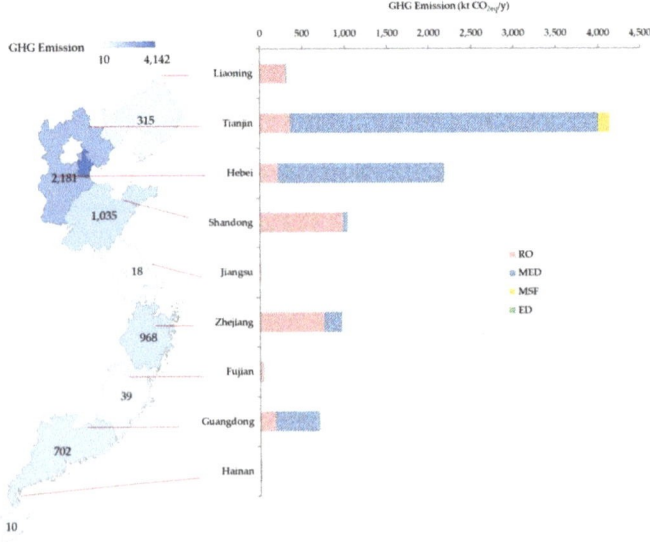

Figure 7. GHG emissions of seawater desalination: Regional distribution and breakdowns by processes.

The map in Figure 7 showed the east costal line in China and marks the provinces with seawater desalination plants. The deeper color indicates higher provincial seawater desalination GHG emissions. The total GHG emissions of the seawater desalination plants in China in 2016 are 9,409 MtCO$_{2eq}$. The provinces of Tianjin, Hebei, and Shandong are the middle three contributors. The GHG emissions of these three provinces are 7,359 MtCO$_{2eq}$, which is 78.2% of the total seawater desalination GHG emission in China. Regarding the desalination process, MED plants contribute the most to the GHG emissions. Tianjin has the highest GHG emissions of 4,142 MtCO$_{2eq}$ in China, and the MED plants contributed more than 88.0% (3,645 MtCO$_{2eq}$) of the total seawater desalination GHG emissions of Tianjin. This is a major contributor to the overall emissions of all seawater desalination plants in China in 2016. Shandong Province has the second largest seawater desalination capacity, but the GHG emissions are much lower (1,035 MtCO$_{2eq}$) compared to Hebei, which has larger emissions (2,183 MtCO$_{2eq}$) with smaller desalination capacity. The main reason is that the seawater desalination plants in Shandong are all RO plants. For Hebei Province, about 65% of the total desalination capacity is MED, which is more energy and GHG emission intensive. On the other hand, the south part of China has relatively lower GHG emissions, due to a smaller capacity and less energy intensive processes.

4.2. Unit Product Cost

The Unit Product Cost (UPC) is correlated with the desalination processes and the capacity of the plant, the type of energy used for the plant, etc [30]. In this study, the energy consumptions of all processes are converted to electricity, therefore the impact of energy source is not analyzed. The UPC of seawater desalination plants in different provinces in 2016 is determined as shown in Figure 8, and the desalination process is specified.

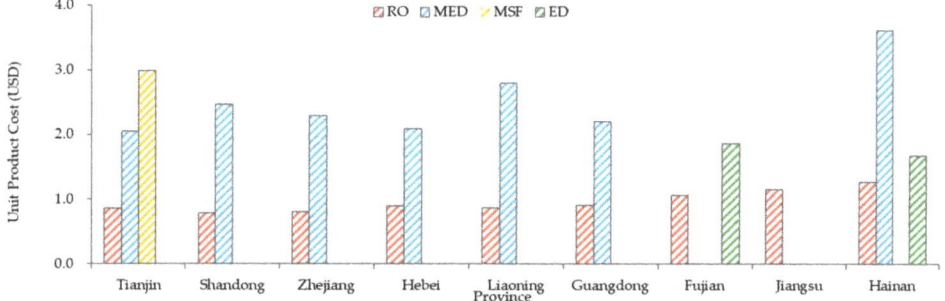

Figure 8. Unit product cost of water desalination by province in 2016.

The UPC of RO desalination plants shows a slight difference. Firstly, the UPC of MED, MSF, and ED is much higher than RO, which is inconsistent with the conclusion of other studies. For example, a case study of Qingdao [14], Shandong Province, China showed the average economic cost of seawater desalination process is 8 CNY/ m^3 (approx. 1.16 USD based on the current exchange rate 0.15), with an RO plant capacity of 3 × 10^3 m^3/d. Hainan has the highest UPC for RO seawater desalination of 1.3 USD, Shandong and Zhejiang have the lowest UPC for RO seawater desalination of 0.8 USD. For MED seawater desalination, Hainan has the highest UPC of 3.6 USD, while Tianjin has the lowest UPC of 2.0 USD. The only province with MSF process, Tianjin, has a UPC for MSF desalination of 3.0 USD. The UPC for ED process desalination of Fujian and Hainan are 1.9 and 1.7 USD. It can be seen that for different processes, the UPC in increasing order is RO < MED < ED < MSF, with RO as the cheapest and most applied desalination process. For the same process, the price varies within a reasonable range, e.g. the UPC of RO process desalination plants in the selected provinces in increasing order are: Shandong = Zhejiang < Tianjin = Hebei = Liaoning = Guangdong < Fujian < Jiangsu < Hainan.

5. Discussion and Future Directions

5.1. Discussion

Seawater desalination has high water supply potential, but high energy consumption, GHG emissions, and cost. Water desalination has a considerable water supply potential due to the abundance of its water source, and the desalination technology is improving. On the other hand, this type of water supply is still supported by higher cost and intensive energy consumption and GHG emission. Even though the cost and energy use are decreasing [7], seawater desalination is still an energy intensive and costly approach compared to other freshwater alternatives. The World Resources Institute investigated the energy consumption of water desalination plants in Qingdao, Shandong province in China. The results showed that electricity is the main energy used for water desalination [15], and the SEC of different water supplies are shown in Table 5.

Table 5. Energy demand and water supply potentials of water supply alternatives (from [15]).

Water Supplies	Energy Demand (kWh/m^3)	Water Supply Potential (Mm^3)
Surface water	0.43	980
Water transfer from Yellow River	0.70	1090
Groundwater	0.78	1700
Reclaimed water	0.82	2100
Water transfer from Yangtze river	1.14	2410
Brackish water desalination	1.40	2470
Seawater desalination (RO) [24]	4.34	2650

It showed that for the various water supply alternatives, surface water has the lowest energy demand for per unit water (about 10% of seawater desalination), and at the same time with the lowest water supply potential (37% of seawater desalination). On the other hand, seawater desalination (RO process) has the highest energy demand with the highest water supply potential. For other desalination processes, the energy demand is even higher.

This indicates that with the current situation of techniques, water desalination is still an option with higher cost. Considering the continuously increasing population and the demand for freshwater, seawater desalination has the potential of providing a stable amount of fresh water and should be viewed as a crucial component in the future development of non-conventional water supplies [31]. The energy demand and thus GHG emissions are highly related to the capacity and techniques applied in the plant [30]. The study of [4] showed that for a water desalination plant with capacity from 5 to 15 m^3/d, the total electricity consumption would vary from 14.45 to 21.35 kWh/m^3, and the MED process has the highest energy demand, which is inconsistent with the results of this study (Figure 8). This indicates the potential for energy consumption and GHG emission reduction can be optimized with improving the combination of capacities and processes. According to the results of [14], a case study of Qingdao, the overall cost of different water supplies from the lowest to the highest are local surface water < groundwater < reclaimed water < brackish water desalination < seawater desalination. Seawater desalination is still currently a more expensive approach for producing freshwater compared to other water supply alternatives.

Determining the energy consumption, GHG emissions, as well as the cost of seawater desalination would be helpful to identify the potential of improving the energy efficiency and productivity of seawater desalination. New and advanced technologies, e.g. low energy reverse osmosis membranes, improved energy recovery devices, highly energy efficient pumps, and optimized pre-treatment systems, can enhance the energy efficiency of seawater desalination. The application of energy recovery units in the desalination processes can also highly increase the energy use efficiency [3]. Increasing the efficiency of the process and the application of energy recovery system will reduce its energy consumption and thereby its CO_2 emissions.

The energy consumption, GHG emissions, and the cost of seawater desalination in China are higher than the average values of the major desalination contributors in East Asia, e.g. Singapore. Singapore has the second largest seawater desalination capacity of 0.45 × 10^6 m^3/d [32], which is 36% of the seawater desalination capacity of China. The average UPC of the two large scale RO seawater desalination plants is 3.5 kWh/m^3 and the cost is estimated as 0.75 USD/m^3 [33], and the price is estimated to increase due to the higher price of energy. The energy consumption and cost of per m^3

desalinated water in Singapore are lower than in China. However, as these indicators are affected by the specific process, type, quality, and price of the energy, as well as the location of the plants, etc. The comparison between different countries would be limited to provide insightful information, but a comprehensive analysis of different water supply alternatives, or different desalination processes in the same region would facilitate regional water use management.

5.2. Future Works

The correlation between the UPC and the capacity needs further investigation. In this study, the UPC and capacity of seawater desalination plants showed a very obvious correlation. Identifying this correlation would be helpful for the optimization of seawater desalination. Figure 9 showed the trend line of the UPC-availability plots. As MSF and ED are rarely applied, there is not enough sample data, and only RO and MED are discussed. For RO plants, the correlation between the UPC and the capacity fits the logarithmic correlation with a R^2 of 0.9847 and 0.9359 for MED plants.

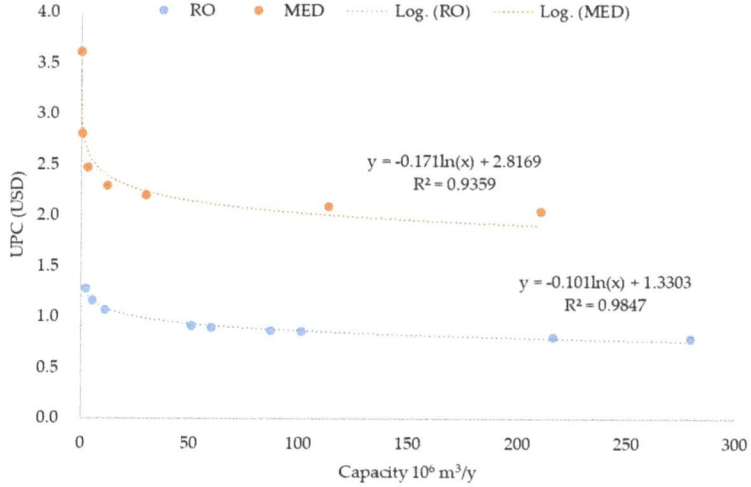

Figure 9. The correlation between unit product cost (UPC) and the seawater desalination capacity.

The figure showed a decreasing UPC with the capacity increasing. According to a statistic of China seawater desalination [14], in 2016, large plants (capacity > 10^4 m^3/d), the average cost is 6.22 CNY/m^3 (0.90 USD), and the average of desalination cost in medium plants (10^4 < capacity < 10^3) is 7.20 CNY/m^3 (1.05 USD). In this study, Shandong has the lowest UPC (0.8 USD) with the highest RO capacity of 0.28 × 10^6 m^3/d. Following is Zhejiang, with the same UPC and a RO capacity of 0.22 × 10^6 m^3/d Tianjin. For MED the trend is not clear. The correlation analysis between the capacity and the UPC should be further investigated with more sufficient data.

Potentials for energy consumption and GHG emission reduction in seawater desalination. The water-energy integration in seawater desalination, e.g. seawater desalination plants utilizing renewable energy [34], low potential heat integrated with seawater multiple-effect thermal desalination [35], which are worth further investigation. The implementation of the integration of energy and seawater desalination still needs a comprehensive review.

In addition to energy consumption and GHG emission, there are other potential environmental issues of seawater desalination. Along with the desalted seawater, a remarkable amount of brine is also produced from the process. A common approach is currently to dispose the brine back to the sea, which might cause harm to the regional aquatic life due to high salinity [36]. The utilization and treatment technologies of the brine is still. Similarly, chemical and thermal pollutions in the near-plant areas, should be further investigated in the future works.

A limitation of this work, which should form the potential future works, is less investigation of the energy consumption by different forms of energy, e.g. heat and electricity in different processes. The usage of renewable energies should also be further reviewed. To cover the whole life cycle, the waste generated during the desalination process, including brine, wastewater, and waste heat should be investigated.

6. Conclusions

Seawater desalination is an emerging approach for producing freshwater in China, and the number of plants, as well as the installed capacity, are increasing. Simultaneously, the energy consumption, and environmental impacts are also increasing. This study initially investigated the energy consumption, GHG emission of the seawater desalination in China from 2006 to 2016, and the unit product cost in 2016. The key findings and conclusions are:

(1) With the increasing installed capacity of seawater desalination from 2006 to 2016, the energy consumption and GHG emission increased from 81 MWh/y to 1,561 MWh/y during the 11 years. The overall GHG emission increase from 85 Mt CO_{2eq}/y to 1,628 Mt CO_{2eq}/y, with an increasing rate of 180%. Tianjin has the largest GHG emissions, followed by Hebei and Shandong, with emissions of 4.1 $MtCO_{2eq}$/y, 2.2 $MtCO_{2eq}$/y. and 1.0 $MtCO_{2eq}$/y.

(2) The unit product cost (UPC) of seawater desalination is higher than other water supply alternatives, and it differentiates the desalination processes. The UPC of the RO process varies from 0.8 USD to 1.3 USD in 2016, and the UPC of MED, MSF, and ED are 2.0 USD–3.6 USD, 3.0 USD, and 1.7 USD to 1.9 USD. Tianjin which has the largest overall seawater desalination capacity has the relatively lowest UPC for RO and MED.

(3) Seawater desalination is now being highly encouraged and developed in China and is becoming a critical water supply alternative for cities with serious water scarcity. The cost, energy demand and GHG emissions are still considerably higher than surface water supply. There is potential for energy consumption, GHG emission and cost reduction with the application of energy recovery units, the integration of desalination plants and renewable energies or low potential heat, as well as the development of new technologies.

Limitation of this work, and also the potential for future works are, (1) Energy consumption should specify different energy forms, e.g. heat and power. In this study, due to the limit of data, the energy used in different processes are converted into electricity. But it is necessary to investigate different energy forms for a more detailed analysis, and the cost analysis would be more accurate. (2) When data is available, the capital cost of the plants should be calculated based on the type of process. (3) Alternatives for process integration should be investigated—e.g. efficiency of using the energy, heat integration, and renewables in water desalination, as well as the utilization of total site heat integration.

Author Contributions: All the authors contributed to publishing this paper and developed the research idea and provided the supervision. X.J. prepared the manuscript and was responsible for the data processing. J.J.K., P.S.V., and S.R.W.A. reviewed and edited the manuscript. X.J., J.J.K., P.S.V., and S.R.W.A. revised the paper.

Funding: The EU supported project Sustainable Process Integration Laboratory—SPIL funded as project No. CZ 02.1.01/0.0/0.0/15_003/0000456, by Czech Republic Operational Programme Research and Development, Education, Priority 1: Strengthening capacity for quality research, in the collaboration agreement with the Universiti Teknologi Malaysia (UTM) based on the SPIL project have been gratefully acknowledged.

Conflicts of Interest: The authors declare no conflict of interest.

Nomenclature

C_a	The capacity of the plant, MUSD. Subscript x and rf represents the capacity of the studied plant and the reference plant
CC	Capital cost of the plant over the lifespan, USD; Subscript x and rf represents the capital cost of the studied plant and the reference plant
C_i	Capacity of the desalination plants with process i, m^3/d
CO_2	Carbon Dioxide
EC_a	Annual energy consumption, kWh/y
EC_e	Annual energy consumption of process i, kWh/m^3
Ec_o	The cost due to energy consumption, MUSD
ED	Electrodialysis
Ef	Emission factors of per unit electricity, t CO$_{2eq}$/kWh
EM_e	Emissions of energy consumption, tCO$_{2eq}$/y
EM_{in}	Emissions of material inputs, tCO$_{2eq}$/y
EOP	Ratio of energy cost and operating cost, %
G_e	Annual GHG emissions, tCO$_{2eq}$/y
KSA	Kingdom of Saudi Arabia
MED	Multiple-effect distillation
MSF	Multi-stage flash
OP_a	The annual operating cost; MUSD
P_a	Availability of the plant, %
P_e	The price of the electricity supplied for seawater desalination plant, USD
Pl	The plant life, y
RO	Reverse osmosis
SEC	Specific energy consumption, kWh/m^3 desalinated water
UAE	United Arab Emirates
UPC	Unit Product Cost, USD/m^3
WSI	Water Stress Index

References

1. Boulay, A.-M.; Bare, J.; Benini, L.; Berger, M.; Lathuillière, M.J.; Manzardo, A.; Margni, M.; Motoshita, M.; Núñez, M.; Pastor, A.V.; et al. The WULCA consensus characterization model for water scarcity footprints: Assessing impacts of water consumption based on available water remaining (AWARE). *Int. J. Life Cycle Assess.* **2018**, *23*, 368–378. [CrossRef]
2. International Desalination Association. *IDA Desalination Yearbook 2016–2017*; International Desalination Association: Topsfield, MA, USA, 2017.
3. Attarde, D.; Jain, M.; Singh, P.K.; Gupta, S.K. Energy-efficient seawater desalination and wastewater treatment using osmotically driven membrane processes. *Desalination* **2017**, *413*, 86–100. [CrossRef]
4. Al-Karaghouli, A.; Kazmerski, L.L. Energy consumption and water production cost of conventional and renewable-energy-powered desalination processes. *Renew. Sustain. Energy Rev.* **2013**, *24*, 343–356. [CrossRef]
5. Čuček, L.; Klemeš, J.J.; Kravanja, Z. A Review of Footprint analysis tools for monitoring impacts on sustainability. *J. Clean. Prod.* **2012**, *34*, 9–20. [CrossRef]
6. Cornejo, P.; Santana, M.; Hokanson, D.; Mihelcic, J.R.; Zhang, Q. Carbon footprint of water reuse and desalination: A review of greenhouse gas emissions and estimation tools. *J. Water Reuse Desalin.* **2014**, *4*, 238–252. [CrossRef]
7. Shrestha, E.; Ahmad, S.; Johnson, W.; Shrestha, P.; Batista, J.R. Carbon footprint of water conveyance versus desalination as alternatives to expand water supply. *Desalination* **2011**, *280*, 33–43. [CrossRef]
8. Li, J.; Liu, Z.; He, C.; Yue, H.; Gou, S. Water shortages raised a legitimate concern over the sustainable development of the drylands of northern China: Evidence from the water stress index. *Sci. Total Environ.* **2017**, *590–591*, 739–750. [CrossRef]
9. National Data: Resource and Environment. Available online: Data.stats.gov.cn/easyquery.htm?cn=C01 (accessed on 3 December 2018).

10. Renewable Internal Freshwater Resources Per Capita (Cubic Meters) | Data. Available online: https://data.worldbank.org/indicator/ER.H2O.INTR.PC?view=chart (accessed on 3 December 2018).
11. Regional Data by Provinces-Beijing-Resource and Environment. Available online: Data.stats.gov.cn/english/easyquery.htm?cn=E0103 (accessed on 3 December 2018).
12. Where Is Earth's Water? USGS Water-Science School. Available online: https://water.usgs.gov/edu/earthwherewater.html (accessed on 3 December 2018).
13. Sheng, J.; Webber, M. Incentive-compatible payments for watershed services along the Eastern Route of China's South-North Water Transfer Project. *Ecosyst. Serv.* **2017**, *25*, 213–226. [CrossRef]
14. Annual Report for China Desalination. Available online: www.soa.gov.cn/zwgk/hygb/hykjnb_2186/201707/t20170719_57029.html (accessed on 3 December 2018).
15. Hua, W.; Lijin, Z.; Xiaotian, F.; Spooner, S. *Water-Energy Nexus in the Urban Water Source Selection: A Case Study from Qingdao*; World Resources Institute Report; Water Resources Institute: Beijing, China, 2014; pp. 49–78.
16. Soares, C.G. *Advances in Renewable Energies Offshore: Proceedings of the 3rd International Conference on Renewable Energies Offshore (RENEW 2018), October 8–10, 2018, Lisbon, Portugal*; CRC Press: Boca Raton, FL, USA, 2018; ISBN 978-0-429-99955-0.
17. Zuo, L.; Ding, L.; Chen, J.; Zhou, X.; Xu, B.; Liu, Z. Comprehensive study of wind supercharged solar chimney power plant combined with seawater desalination. *Sol. Energy* **2018**, *166*, 59–70. [CrossRef]
18. Li, G.; Law, W.-C.; Cheung Chan, K. Floating, highly efficient, and scalable graphene membranes for seawater desalination using solar energy. *Green Chem.* **2018**, *20*, 3689–3695. [CrossRef]
19. Liu, S.Y.; Zhang, G.X.; Han, M.Y.; Wu, X.D.; Li, Y.L.; Chen, K.; Meng, J.; Shao, L.; Wei, W.D.; Chen, G.Q. Freshwater costs of seawater desalination: Systems process analysis for the case plant in China. *J. Clean. Prod.* **2018**, *212*, 677–686. [CrossRef]
20. Raluy, G.; Serra, L.; Uche, J. Life cycle assessment of MSF, MED and RO desalination technologies. *Energy* **2006**, *31*, 2361–2372. [CrossRef]
21. Karabelas, A.J.; Koutsou, C.P.; Kostoglou, M.; Sioumiddleoulos, D.C. Analysis of specific energy consumption in reverse osmosis desalination processes. *Desalination* **2018**, *431*, 15–21. [CrossRef]
22. Pfister, S.; Koehler, A.; Hellweg, S. Assessing the Environmental Impacts of Freshwater Consumption in LCA. *Environ. Sci. Technol.* **2009**, *43*, 4098–4104. [CrossRef] [PubMed]
23. Wilson, M.C.; Li, X.-Y.; Ma, Y.-J.; Smith, A.T.; Wu, J. A Review of the Economic, Social, and Environmental Impacts of China's South-North Water Transfer Project: A Sustainability Perspective. *Sustainability* **2017**, *9*, 1489. [CrossRef]
24. Caldera, U.; Bogdanov, D.; Breyer, C. Local cost of seawater RO desalination based on solar PV and wind energy: A global estimate. *Desalination* **2016**, *385*, 207–216. [CrossRef]
25. Wittholz, M.K.; O'Neill, B.K.; Colby, C.B.; Lewis, D. Estimating the cost of desalination plants using a cost database. *Desalination* **2008**, *229*, 10–20. [CrossRef]
26. Riedel, S. ISCC 205 Greenhouse Gas Emissions. In *Greenhouse Gas Emissions*; ISCC: Köln, Germany, 2017.
27. Brander, M.; Sood, A.; Wylie, C.; Haughton, A.; Lovell, J. Technical Paper | Electricity-Specific Emission Factors for Grid Electricity. *Ecometrica* **2011**, *22*, 1–22.
28. Zhou, Y.; Tol, R.S.J. Implications of desalination for water resources in China—An economic perspective. *Desalination* **2004**, *164*, 225–240. [CrossRef]
29. Electricity Price for Water Desalination Plants in Shandong. Available online: www.sdwj.gov.cn/ggfw/jggl/zls/11/171295.shtml (accessed on 6 December 2018).
30. Pinto, F.S.; Marques, R.C. Desalination projects economic feasibility: A standardization of cost determinants. *Renew. Sustain. Energy Rev.* **2017**, *78*, 904–915. [CrossRef]
31. Zarzo, D.; Prats, D. Desalination and energy consumption. What can we expect in the near future? *Desalination* **2018**, *427*, 1–9. [CrossRef]
32. Water Technology—Tuaspring Desalination and Integrated Power Plant. Available online: www.water-technology.net/projects/tuaspring-desalination-and-integrated-power-plant/ (accessed on 9 January 2019).
33. Water Technology—Tuas Seawater Desalination Plant. Available online: www.water-technology.net/projects/tuas-seawater-desalination/ (accessed on 9 January 2019).
34. Grubert, E.A.; Stillwell, A.S.; Webber, M.E. Where does solar-aided seawater desalination make sense? A method for identifying sustainable sites. *Desalination* **2014**, *339*, 10–17. [CrossRef]

35. Wang, Y.; Lior, N. Thermoeconomic analysis of a low-temperature multi-effect thermal desalination system coupled with an absorption heat pump. *Energy* **2011**, *36*, 3878–3887. [CrossRef]
36. Dai, J.Y.; Wu, L.Y.; Zhang, Y.G.; Tang, Z.X. Brief analysis on environmental influence and comprehensive utilization of brine from thermal desalination. *Guangdong Chem. Ind.* **2018**, *365*, 48–51. (In Chinese)

© 2019 by the authors. Licensee MDPI, Basel, Switzerland. This article is an open access article distributed under the terms and conditions of the Creative Commons Attribution (CC BY) license (http://creativecommons.org/licenses/by/4.0/).

Article

Numerical Analysis of Longitudinal Residual Stresses and Deflections in a T-joint Welded Structure Using a Local Preheating Technique

Mato Perić [1], Ivica Garašić [2], Sandro Nižetić [3,*] and Hrvoje Dedić-Jandrek [3]

1. Bestprojekt, Bureau of Energetics and Mechanical Engineering Ltd., Petrovaradinska 7, 10000 Zagreb, Croatia; mato.peric@fsb.hr
2. Faculty of Mechanical Engineering and Naval Architecture, University of Zagreb, I. Lučića 5, 10000 Zagreb, Croatia; ivica.garasic@fsb.hr
3. LTEF—Laboratory for Thermodynamics and Energy Efficiency, Faculty of Electrical Engineering, Mechanical Engineering and Naval Architecture, University of Split, Rudjera Boskovića 32, 21 000 Split, Croatia; hdedicja@fesb.hr
* Correspondence: snizetic@fesb.hr; Tel.: +385-21-305-954

Received: 16 November 2018; Accepted: 11 December 2018; Published: 14 December 2018

Abstract: In this paper a numerical analysis of a T-joint fillet weld is performed to investigate the influences of different preheat temperatures and the interpass time on the longitudinal residual stress fields and structure deflections. In the frame of the numerical investigations, two thermo-mechanical finite element models, denoted M2 and M3, were analyzed and the results obtained were then compared with the model M1, where the preheating technique was not applied. It is concluded that by applying the preheat temperature prior to the start of welding the post-welding deformations of welded structures can be significantly reduced. The increase of the preheat temperature increased the longitudinal residual stress field at the ends of the plates. The influence of the interpass time between two weld passes on the longitudinal residual stress state and plate deflection was investigated on two preheated numerical models, M4 and M5, with an interpass time of 60 s and 120 s, respectively. The results obtained were then compared with the preheated model M3, where there was no time gap between the two weld passes. It can be concluded that with the increase of interpass time, the plate deflections significantly increase, while the influence of the interpass time on the longitudinal residual stress field can be neglected.

Keywords: welding residual stress; welding deflection; T-joint fillet weld; preheat temperature; interpass time; finite element analysis

1. Introduction

The welding technique is one of the most frequently used engineering methods of joining structural components in many industrial fields. The large localized heat generation during welding and subsequent very fast cooling of the melted material to ambient temperature have, as a consequence, the occurrence of permanent residual stresses and dimensional imperfections in the welded structure. Such imperfections can cause large inconveniences during the structure assembly, while high residual stresses can have a detrimental impact on its integrity and durability [1–4]. The elimination of these consequences using conventional post-weld thermal or mechanical treatments requires an extended production time and incurs additional financial expenses. For these reasons, it is highly desirable to carry out measures that will lessen these consequences prior to, or during, the welding process. To mitigate the residual stress and plate deflections in a single pass T-joint fillet weld Gannon et al. [5] and Fu et al. [6] numerically and experimentally studied the influence of various welding sequences on the residual stress field. It was concluded that the welding sequence has a negligible impact on the residual stress distribution pattern, but has an influence on the longitudinal stress peaks. Jiang and Yahiaoui [7] presented a three-dimensional thermomechanical model to investigate the effect of welding sequences on the residual stress field in a multipass welded piping branch junction. Li et al. [8] dealt with water cooling effects on welding residual stresses in a core shroud. In their contribution, Moat et al. [9] concluded that martensitic filler metals with low transformation temperatures can efficiently reduce the welding residual stresses. Cozzolino et al. [10] investigated the mitigation of residual stresses and distortions using a post-weld rolling technique. Schenk et al. [11] studied the influence of different clamping conditions in a T-joint fillet welded structure. Chuvas et al. [12] investigated welding residual stress relief with mechanical vibrations using the X-ray diffraction technique and Monte Carlo method on a butt-welded model of two plates. A significant longitudinal residual stress reduction of 40% was observed after the application of mechanical vibration procedures, while in the transversal direction, the residual stress was reduced by 20%. A finite element model with a clearer insight into the influential parameters of the vibration stress treatment was presented by Yang [13]. In this study, the impact of frequency, vibration amplitude and welding parameters on the residual stress were investigated for resonant and non-resonant cases. Samardžić et al. [14] experimentally elaborated the influence of vibrations on the residual stress in a T-joint fillet weld structure and concluded that this method can reduce the residual stresses in specific areas of the structure, depending on the position of the force inductor. Dong et al. [15] suggested a simple engineering scheme for estimating residual stress reduction that is based on post-weld heat treatment temperature, material type and component wall thickness. Fu et al. [16] investigated the structure deflections under a constant preheat temperature during the welding process in a T-joint welded structure. They concluded that the preheating during the welding has a significant effect on structure deformations, while its influence on the longitudinal and tensile residual stresses is much smaller. In their numerical study, Kala et al. [17] investigated the influence of the interpass time between various sequences in multipass butt-welded steel plates. They concluded that the interpass time has a significant effect on the structure deformations. Moreover, they concluded that the interpass time has a large influence on the phase-transformations in weldment. Ilman et al. [18] carried out an investigation to mitigate the residual stresses in butt-welded plates using static thermal tensioning to improve the fatigue performance of the welded structure. Zhang et al. [19] introduced a multi-beam preheating method to reduce structure deformation. They concluded that a multi-beam preheating method can successfully reduce the compressive stress in the welding area. In their study, Okano and Mochizuki [20] applied a trailing heat sink using a water-cooling device. They reduced the longitudinal tensile zone width, longitudinal bending and angular deformations by up to 70%. Meanwhile, the tensile longitudinal stresses in the weld area were negligibly reduced by the heat sink usage in this case. Zubairuddin et al. [21] numerically and experimentally investigated butt-welded plates during a GTA welding process and concluded that the preheating procedure before welding can significantly reduce welding distortions.

As can be seen above, there are numerous mitigation techniques that can be undertaken before or during the welding process to reduce residual stresses and structure deformations. It is important to note that the mitigation techniques mentioned above are primarily suitable for indoor factory fabrication when structures are of smaller dimensions. In this work, a study of the influence of local preheat temperature and the interpass time on the longitudinal residual stress field and deformations was carried out on a T-joint fillet welded structure, in the case where the preheat temperature was kept constant only at the beginning of the welding process. This technique is more appropriate for outdoor on-site welding of large structures.

The paper consists of six sections: In Section 1 a survey of the literature is given; in Section 2, the geometry of a T-joint fillet welded model, the welding parameters and the preheat temperature volume are defined; a detailed description of the numerical model is provided in Section 3; in Section 4 the data about the application of the preheat temperature and interpass time to the numerical models are provided; Section 5 contains an analysis of the preheat temperature influence and interpass time on the residual stress fields and deflections of the T-joint fillet welded model; in the last section, the conclusions of the investigations are summarized.

2. T-joint Fillet Weld Geometry and Welding Conditions

To analyze the influence of the preheat temperature and the interpass time between two weld passes on the longitudinal residual stress (residual stress in welding direction) field and plate deflection, a T-joint fillet weld model taken from Deng et al. [22] was considered. The plates were joined with two single pass welds using the MAG procedure without any time gap in between and their geometry is shown in Figure 1. The welding conditions are given in Table 1. The plate material was SM400A carbon steel, for which the temperature-dependent thermal and mechanical properties are shown in Figures 2 and 3, and the elemental composition is given in Table 2.

Table 1. Welding conditions [22].

Welding Current (A)	Welding Voltage (V)	Welding Speed (mm/min)	Angle of Torch (°)
270	29	400	45

Table 2. Elemental composition of SM400A steel (mass%) [22].

C	Si	Mn	P	S
0.23	-	0.56	<0.035	<0.035

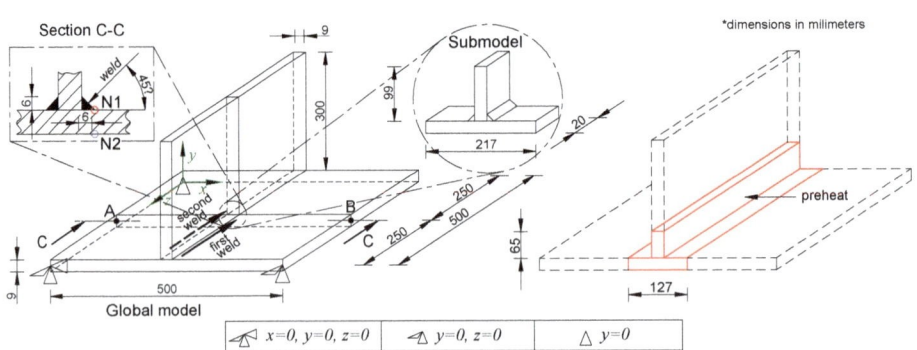

Figure 1. T-joint geometry and volume of preheating.

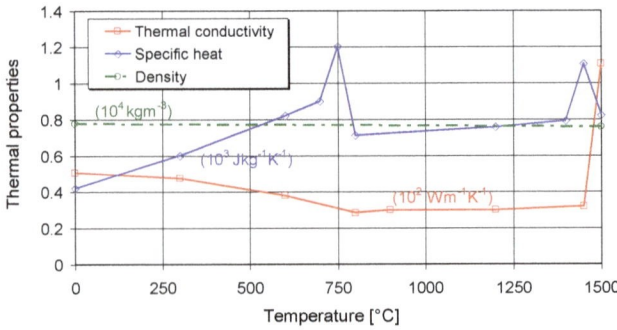

Figure 2. Thermal properties of SM400A steel [22].

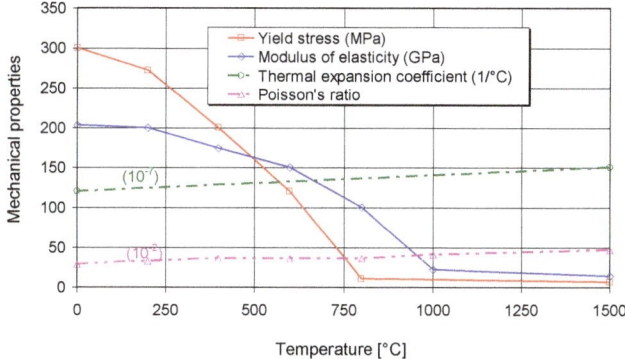

Figure 3. Mechanical properties of SM400A steel [22].

Since only the experimental measurements of horizontal plate deflections, using a Vernier caliper, are provided in Reference [22], the validation of longitudinal residual stresses was performed using two idealized solutions from the literature [23], which could provide acceptable engineering results for single-pass T-joint structures welded with MAG technology [5,24,25], and are appropriate for the evaluation of the reference model M1 (Table 3). In these two simple solutions it is assumed that the maximum residual stress along the line A-B in the middle plane of the horizontal plate (Figure 4) is tensile and that it reaches the yield stress of the material. The compressive residual stress along the same line governs the rest of the horizontal plate and can be calculated as follows:

$$\sigma_c = \frac{2b_t}{b - 2b_t} \text{ [MPa]} \tag{1}$$

or

$$\sigma_c = \frac{2b_t + t_p \sigma_y + b_s t_w \sigma_y}{(b - 2b_t)t_p + A_s + b_s t_w} \text{ [MPa]} \tag{2}$$

where

$$b_t = \frac{t_w}{2} + \frac{0.26 \Delta Q}{t_w + 2t_p} \text{ [mm]} \tag{3}$$

$$b_s = \frac{t_w}{t_p}\left(b_t - \frac{t_w}{2}\right) \text{ [mm]} \tag{4}$$

$$\Delta Q = 78.8 l^2 \tag{5}$$

$$l = 0.7 t_w \text{ when } t < 10 \text{ mm} \tag{6}$$

$$l = 7.0 \text{ when } t \geq 10 \text{ mm} \tag{7}$$

Table 3. List of the simulated numerical models.

Model Name	Preheating Application	Interpass Time
M1	No	t = 0 s
M2	Yes, T = 100 °C	t = 0 s
M3	Yes, T = 150 °C	t = 0 s
M4	Yes, T = 150 °C	t = 60 s
M5	Yes, T = 150 °C	t = 120 s

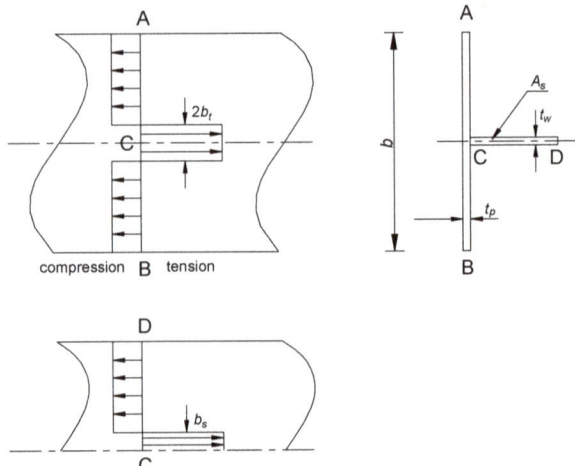

Figure 4. Idealised longitudinal residual stress distribution [23].

In the equations given above, b is the width of the horizontal plate in mm, t_w and t_p denote the thicknesses of the horizontal and vertical plates in mm, $2b_t$ is the tensile zone width of the horizontal plate, b_s is the tensile zone width of the vertical plate, σ_y is the yield stress of the material at room temperature, while A_s is the cross section of the vertical plate in mm^2.

3. Numerical Model

A numerical simulation was performed employing a sequentially coupled thermal-elastic-plastic model [26–29]. In this case, the welding simulation process consisted of two separate numerical analyses, i.e., one thermal and one mechanical. In the thermal analysis, the key equation for nonlinear transient heat transfer can be written in the form of:

$$\frac{\partial}{\partial x}\left(k_x \frac{\partial T}{\partial x}\right) + \frac{\partial}{\partial y}\left(k_y \frac{\partial T}{\partial y}\right) + \frac{\partial}{\partial z}\left(k_y \frac{\partial T}{\partial z}\right) + Q = \rho C \frac{\partial T}{\partial t} \tag{8}$$

In Equation (8) k_x, k_y, and k_z are the thermal conductivity components in the x, y and z directions; T is the body temperature; Q is the generated heat input; ρ is the material density; C is the specific heat capacity of the material; and t is time, respectively. A general solution for Equation (8) can be obtained when the following initial boundary conditions on the outer model surfaces are taken into account:

$$T(x,y,z,0) = T_0(x,y,z) \tag{9}$$

$$\left(k_x \frac{\partial T}{\partial x} N_x + k_y \frac{\partial T}{\partial y} N_y + k_z \frac{\partial T}{\partial z} N_z\right) + q_s + h_c(T - T_\infty) + h_r(T - T_r) = 0 \tag{10}$$

where N_x, N_y, and N_z are the direction cosine of the normal to the boundary; h_c denotes the convective heat transfer coefficient; h_r is the radiation heat transfer coefficient; q_s represents the heat flux on the outer body boundaries; T_r denotes the radiation temperature; and T_∞ is the ambient temperature. Heat loss due to radiation can be expressed by the following expression:

$$h_r = \sigma_{Bolt}\varepsilon_{surf}F(T^2 + T_r^2)(T + T_r) \tag{11}$$

where $\sigma_{Bolt} = 5.67 \times 10^{-8}$ Wm^{-2}K^{-4} denotes the Stefan–Boltzmann constant; ε_{surf} is the surface emissivity factor; and F is the configuration factor. The generated heat input applied to the weld volume can be expressed as follows:

$$Q = \frac{\eta UI}{V_H} \quad (12)$$

In Equation (12) η represents the efficiency of the welding process, I is the welding current, U denotes the arc voltage, and V_H is the weld volume. Although it is usual in the literature for the temperature distribution calculation in the MAG welding process to be performed as a combination of Gaussian and a uniformly distributed volumetric heat flux model [5,22], in this study, a pure volumetric heat flux with uniformly distributed heat input was used to speed up the simulation process, $Q = 5.22 \times 10^{10}$ Jm^{-3}s^{-1} per weld volume was applied and its value was obtained from Equation (12). The MAG welding process efficiency $\eta = 80\%$ was taken according to the EN 1011-1 [30]. On the model boundaries, the convection heat transfer coefficient $h_c = 10$ Wm^{-2}K^{-1} and the surface emissivity $\varepsilon_{surf} = 0.9$ were assumed. During the thermal analysis, the element birth and date method [31–33] was employed for the simulation of weld filler addition.

To cut down calculation time, the mechanical analysis was performed simultaneously in one step, without the application of the element and birth technique [34] and elastic-perfectly plastic behavior of the material was assumed here. Because the impact of metallurgical phase transformations on the residual stress field in low-carbon steel is relatively small [35], it was neglected in this study. Furthermore, creep material behavior was also neglected because the exposure period of the material to high temperatures due to welding is very short. Keeping in the mind that phase transformations and creep material behavior are neglected, the total strain increment $d\varepsilon_{total}$ can finally be written as follows:

$$d\varepsilon_{total} = \{d\varepsilon_e\} + \{d\varepsilon_p\} + \{d\varepsilon_{th}\} \quad (13)$$

where $\{d\varepsilon_e\}$, $\{d\varepsilon_p\}$ and $\{d\varepsilon_{th}\}$ are elastic, plastic and thermal strain increments, respectively.

It is important to note that the plates of the T-joint sample are free welded, without any mechanical constraints, but they are added in the mechanical numerical simulation only to prevent plate motions as a rigid body. The applied mechanical constraints are shown in Figure 1. Because data about thermal and mechanical properties of the weld filler material are not provided in Reference [22], it was assumed that they are the same as the base metal ones.

A finite element mesh containing 19,188 elements is shown in Figure 5. Its sensitivity was checked on a small part of the T-joint with a very high density using the submodeling technique [36]. The dimensions of the submodel and volume of submodeling are given in Figure 1.

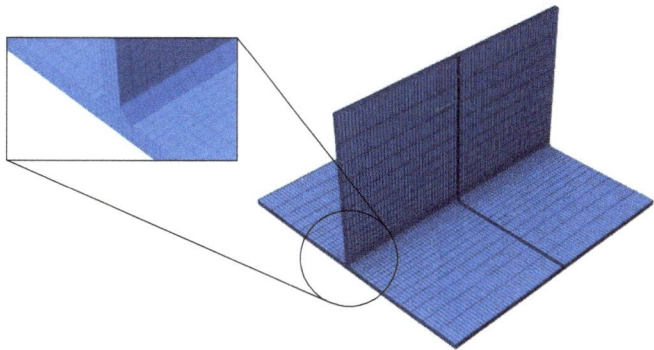

Figure 5. T-joint finite elements mesh.

The same mesh was employed in both the thermal and mechanical analysis, except that the three-dimensional 8-node brick DC3D8 finite elements from the thermal analysis were converted into 8-node brick C3D8I finite elements with incompatible modes in the mechanical analysis. The numerical simulation was performed using Abaqus/Standard software.

4. Application of Preheat Temperature and Interpass Time in the Numerical Models

In the frame of the numerical preheat temperature investigations on the longitudinal residual stress (stress in welding direction) fields and horizontal plate deflections, two thermomechanical finite element models, denoted M2 and M3 were analyzed. These models were locally preheated (Figure 1) at 100 °C and 150 °C, respectively. The obtained results were then compared with the reference model M1, where no preheat was applied. In the numerical simulations the preheat temperature was accomplished with the *INITIAL TEMPERATURE option in Abaqus/Standard software before the heat flux application.

To analyze the influence of interpass time between the two weld passes on the longitudinal residual stress fields and deflections, the numerical models M4 and M5 with an interpass time of 60 s and 120 s, respectively, were considered. In both the M4 and M5 numerical models, a preheat temperature of 150 °C was assumed. The results of the two investigated interpass time models obtained were then compared with model M3 where the interpass time was 0 s. In all the preheated models, it was assumed that a structure volume of 500 × 127 × 65 mm^3 dimensions (Figure 1) was preheated before the start of welding. The selected dimensions of the preheated volume satisfied the minimum prescribed requirements according to the ISO 13916 [37] norm. Therefore, it is important to point out that the local preheating technique was applied in this work, where preheating is only applied in areas close to the weld, while the rest of the structure is not preheated. Since the preheated part of the structure attempts to expand, the non-preheated area resists it, which introduces residual stresses into the structure before the start of the welding. This approach differs from Reference [16] where the entire structure was preheated at the same temperature and there was no additional introduction of residual stresses before welding. Also, unlike Reference [20], where the structure was continuously preheated during the welding process, in this study the preheating was applied once at the beginning of the welding process. All the numerical models considered in this study are given in Table 3.

5. Results and Discussion

5.1. Residual Stress and Deflection Distributions—Reference Model

Since the obtained thermal field from the thermal analysis is a burden on the mechanical analysis, special attention is devoted to its accuracy. For this purpose, the numerically calculated peak temperatures of model M1 were compared with a more appropriate model from the literature [5], where the heat flux is defined as a combination of volumetric and Gaussian surface heat flux. Figure 6 shows the temperature histories of model M1 at nodes N1 and N2 for the first 250 s after the beginning of the welding process. The comparison of the peak temperatures between model M1 and the model from the literature is given in Table 4, where it can be seen that the differences are negligible. It can be concluded that the impact of the heat flux simplification has little effect on temperature distribution and that the model presented can be applied in the mechanical analysis.

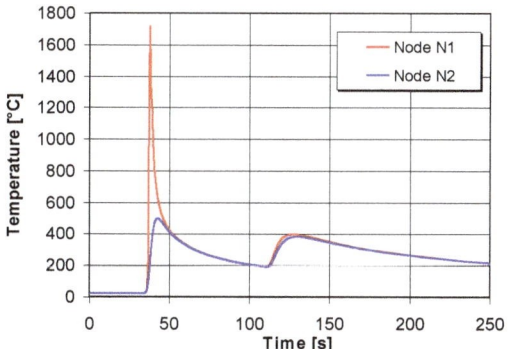

Figure 6. Temperature histories at nodes N1 and N2 (Figure 1).

Table 4. Peak temperatures at nodes N1 and N2.

Peak Temperatures (°C)	Node N1 (1st Pass)	Node N2 (1st Pass)	Node N1 (2nd Pass)	Node N1 (2nd Pass)
Current study	1712	496	398	381
Gannon et. al. [5]	1730	500	374	356

Figure 7 shows the numerically calculated deflection distribution of the horizontal plate (deflection in y-direction) at the middle plane along the line A-B (Figure 1) for the reference model M1, after the completion of welding and cooling process to room temperature. The obtained numerical deflection corresponds very well with the experimental measurement. Figure 8 shows the longitudinal residual stress profile (σ_z, stress in z-direction) at the middle plane of the horizontal plate along the A-B line (Figure 1) compared with two analytical solutions from Equations (1) and (2). It can be seen that tensile residual stresses are in the weld area, while in the rest of the model they are compressive. The maximum numerically calculated tensile stresses are approximately 5% higher than the analytically calculated ones. The numerically calculated compressive stresses are very close to the analytical ones calculated according to Equations (1) and (2) of 48.2 MPa and 42.5 MPa, respectively. Comparing the numerically calculated width of the tensile zone with the analytically calculated values it can be seen that they corresponded well. The numerically calculated tensile zone width is approximately 4% lower than the analytically calculated ones. Based on these results, it can be concluded that the numerical model presented is sufficiently accurate and it can be applied to the other four numerical models given in Table 3.

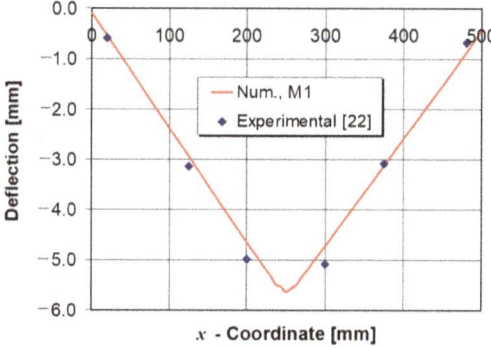

Figure 7. Middle-plane deflection profile along A-B line (Figure 1), model M1.

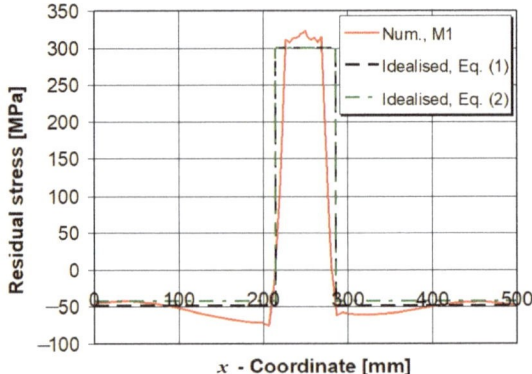

Figure 8. Middle-plane longitudinal residual stress profile along A-B line (Figure 1), model M1.

5.2. Influence of Preheat Temperature on the Longitudinal Residual Stress and Horizontal Plate Deflection

Figure 9 shows a comparison of the numerically calculated horizontal plate deflections in the preheated models M2 and M3 compared with the reference model M1. Here it is evident that with an increase of the preheat temperature, the deflections of the horizontal plate significantly decrease. Unfortunately, the increase of the preheat temperature level is limited by several factors such as: welding technology, heat input, steel group, chemical composition, diffusible hydrogen, required microstructure, and plate thickness. As for the T-joint welded model discussed in this study, it should not exceed 150 °C according to the EN-1011-2 norm [38]. In the case of the preheat temperature being higher than prescribed in Reference [38], certain issues can be expected regarding the mechanical properties in the weld joint, especially in the weld metal, such as metal softening in the heat-affected zone. For the preheated models M2 and M3, the horizontal plate deflections are approximately 12% and 22% lower, respectively, in comparison with the reference model M1. If the entire volume of the T-joint model was hypothetically preheated at 150 °C prior to the start of welding, the peak deflection would be 4.5 mm which is very close to the value of the locally preheated model M3 used in this study, of 4.6 mm, i.e., the deflection differences are within a range of 2%. Furthermore, it should be pointed out that from an energy point of view, the use of local preheat leads to 76% in energy savings compared to preheating an entire model of the same dimensions. Keeping these two conclusions in mind, as well as the fact that the recommendations in the norm [38] are conservative, the model presented offers the possibility for structure volume optimization which is preheated.

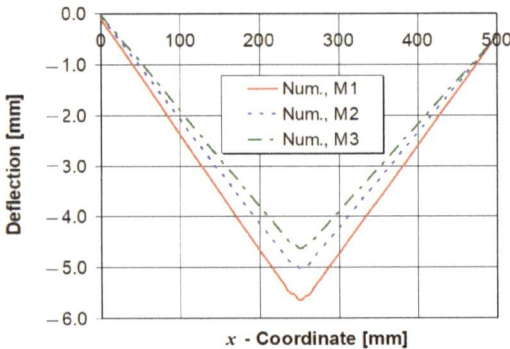

Figure 9. Middle-plane deflection profile along A-B line (Figure 1), models M1, M2 and M3.

Furthermore, Figure 10 shows that the preheat temperature increase minimally affects the longitudinal tensile stress and the tensile zone widths. However, the increase of preheat temperature significantly increases the longitudinal compressive stresses at the ends of the plates. Since the increase of temperature in the preheated part of the welded model causes its extension in the longitudinal direction, the non-preheated part resists it. It can be concluded that the application of the local preheating technique prior to welding causes higher pressure compressive residual stresses outside the weld area in comparison with the models where the entire volume is preheated before the start of welding [16]. This phenomenon will be far more visible in welded models with higher preheat temperatures than in the model analyzed in this study.

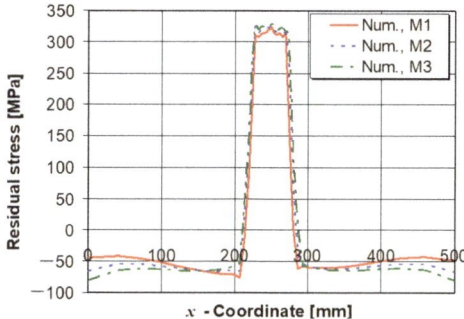

Figure 10. Middle-plane longitudinal residual stress profile along A-B line (Figure 1), models M1, M2 and M3.

5.3. Influence of Interpass Time on the Longitudinal Residual Stress and Horizontal Plate Deflection

The influence of the interpass time between the two weld passes was investigated on the preheated models M4 and M5 and the obtained values were then compared with the M3 model where the interpass time was 0 s. Figure 11 shows that with the increase of interpass time, the horizontal plate deflection increased as well, because the positive effect of the preheat procedure vanishes during the cooling process. For example, when the interpass time is hypothetically extended up to approximately 70 min, the full model completely cools down to room temperature before the start of the second weld pass. In such a case the maximum horizontal plate deflection becomes only about 5% lower than in the reference model M1, without applying preheat and without interpass time application. It can be stated that when the interpass time is too long, the preheating procedure is almost useless. In Figure 12 it is shown that the influence of the interpass time on the longitudinal tensile residual stress and the width of the tensile stress zone can be neglected.

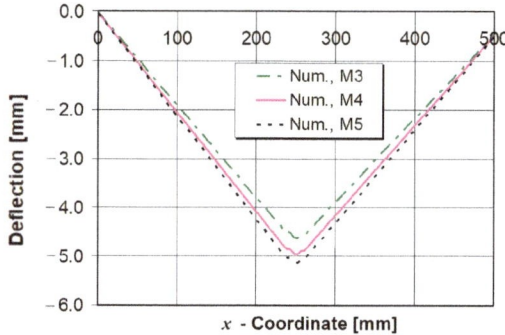

Figure 11. Middle-plane deflection profile along A-B line (Figure 1), models M3, M4 and M5.

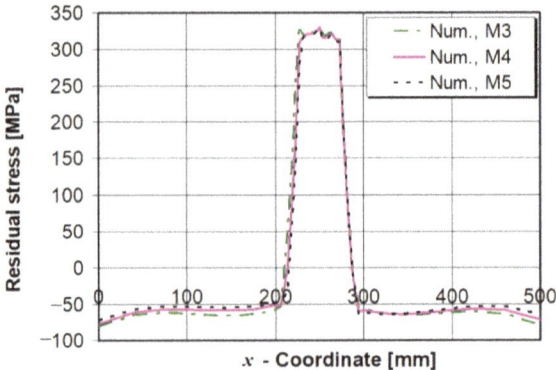

Figure 12. Middle-plane residual stress profile along A-B line (Figure 1), models M3, M4 and M5.

This is due to a smaller temperature gradient difference between the preheated and non-preheated parts of the model in comparison with models M1, M2 and M3 without the application of interpass time.

6. Conclusions

In this study, the influences of preheat temperature and interpass time on the longitudinal residual stress and vertical deflection of a T-joint fillet model were investigated. The local preheating technique, where the preheat temperature was kept constant only at the beginning of the welding process was simulated in the numerical calculations. The conclusions are as follows:

- The increase of the preheat temperature decreases the horizontal plate deflection of a T-joint very quickly.
- The influence of preheat temperature on the longitudinal tensile residual stress and the tensile stress zone width is negligible.
- The application of local preheating increases the compressive longitudinal stresses due to the increased temperature gradients between the preheated and non-preheated parts of the model. This occurrence is much more pronounced than in the models where the entire volume is preheated before the start of welding.
- The increase of interpass time increases the plate deflections. In cases when interpass time is prolonged, the positive effects of preheating vanish.
- The increase of interpass time minimally affects the longitudinal tensile residual stress and its tensile zone width.
- The effect of interpass time increase on compressive longitudinal stresses in preheated models can be neglected.

Generally speaking, the application of a local preheating procedure is very useful when the reduction of plate deflections is of primary concern. In this case, a preheat temperature should be imposed as high as practically possible, without interpass time, as was demonstrated with model M3. When the compressive longitudinal residual stress at the ends of the plates is minimized, the optimal solution is model M1 without any localized preheating and with no time gap between the two weld passes. In order to reduce the energy consumption that is needed for the preheat procedure, in the next phase of the investigations, the emphasis will be on optimization of the preheat zone size and temperature level in the high-productive buried arc welding of thick steel plates, when the heat inputs are very high.

Author Contributions: Conceptualization, methodology, data curation, investigation, writing—original draft preparation, formal analysis, funding acquisition, supervision and project administration, M.P.; data curation and supervision, I.G. and S.N.; data curation, H.D-J.

Funding: This research received no external funding.

Acknowledgments: The authors wish to express their deep gratitude to ISO Plus Croatia for financial support and to its engineers Ivica Cvrlje and Bruno Repar for their valuable comments.

Conflicts of Interest: The authors declare no conflicts of interest.

References

1. Boljanović, S.; Maksimović, S. Computational mixed mode failure analysis under fatigue loadings with constant amplitude and overload. *Eng. Fract. Mech.* **2017**, *174*, 168–179. [CrossRef]
2. Chen, Z.; Xiong, Y.; Qiu, H.; Lin, G.; Li, Z. Stress intensity factor-based prediction of solidification crack growth during welding of high strength steel. *J. Mater. Process. Technol.* **2018**, *252*, 270–278. [CrossRef]
3. Oh, S.H.; Ryu, T.Y.; Park, S.H.; Won, M.G.; Kang, S.J.; Lee, K.S.; Lee, S.H.; Kim, M.K.; Choi, J.B. Evaluation of J-groove weld residual stress and crack growth rate of PWSCC in reactor pressure vessel closure head. *J. Mech. Sci. Technol.* **2015**, *29*, 1225–1230. [CrossRef]
4. Božić, Ž.; Schmauder, S.; Wolf, H. The effect of residual stresses on fatigue crack propagation in welded stiffened panels. *Eng. Fail. Anal.* **2018**, *84*, 346–357. [CrossRef]
5. Gannon, L.; Liu, Y.; Pegg, M.; Smith, M. Effect of welding sequence on residual stress and distortion in flat-bar stiffened plates. *Mar. Struct.* **2010**, *23*, 385–404. [CrossRef]
6. Fu, G.M.; Lourenço, I.L.; Duan, M.; Estefen, S.E. Influence of the welding sequence on residual stress and distortion of fillet welded structures. *Mar. Struct.* **2016**, *46*, 30–55. [CrossRef]
7. Jiang, W.; Yahiaoui, K. Effect of welding sequence on residual stress distribution in a multipass welded piping branch junction. *Int. J. Press. Vessels Pip.* **2012**, *95*, 39–47. [CrossRef]
8. Li, Y.; Yoshiyuki, K.; Igarashi, K. Effects of thermal load and cooling condition on weld residual stress in a core shroud with numerical simulation. *Nucl. Eng. Des.* **2012**, *242*, 100–107. [CrossRef]
9. Moat, R.J.; Stone, H.J.; Shirzadi, A.A.; Francis, J.A.; Kundu, S.; Mark, A.F.; Bhadesia, H.K.D.H.; Karlsson, L.; Withers, P.J. Design of weld fillers for mitigation of residual stresses in ferritic and austenitic steel welds. *Sci. Technol. Weld. Join.* **2011**, *16*, 279–284. [CrossRef]
10. Cozzolino, L.D.; Coules, E.; Colegrove, A.P.; Wen, S. Investigation of post-weld rolling methods to reduce residual stress and distortion. *J. Mater. Process. Technol.* **2017**, *247*, 243–256. [CrossRef]
11. Schenk, T.; Richardson, I.M.; Kraska, M.; Ohnimus, S. Influence of clamping on distortion of welded S355 T-joints. *Sci. Technol. Weld. Join.* **2009**, *14*, 369–375. [CrossRef]
12. Chuvas, T.C.; Castello, D.A.; Fonseca, M.P.C. Residual stress relief of welded joints by mechanical vibrations. *J. Braz. Soc. Mech. Sci. Eng.* **2016**, *38*, 2449–2457. [CrossRef]
13. Yang, Y.P. Understanding of vibration stress relief with computation modeling. *J. Mater. Eng. Perform.* **2009**, *18*, 856–862. [CrossRef]
14. Samardžić, I.; Vuherer, T.; Marić, D.; Konjatić, P. Influence of vibrations on residual stresses distribution in welded joints. *Metalurgija* **2015**, *54*, 527–530.
15. Dong, P.; Song, S.; Zhang, J. Analysis of residual stress relief mechanisms in post-weld heat treatment. *Int. J. Press. Vessels Pip.* **2014**, *122*, 6–14. [CrossRef]
16. Fu, G.M.; Lourenço, I.L.; Duan, M.; Estefen, S.E. Effects of preheat and interpass temperature on the residual stress and distortion on the T-joint weld. In Proceedings of the ASME 2014 33rd International Conference on Ocean, Offshore and Arctic Engineering OMAE2014, San Francisco, CA, USA, 8–13 June 2014.
17. Kala, S.R.; Prasad, N.S.; Phanikumar, G. Numerical studies on effect of interpass time on distortion and residual stresses in multipass welding. *Adv. Mater. Res.* **2013**, *601*, 31–36. [CrossRef]
18. Ilman, M.N.; Kusmono; Muslih, M.R.; Subeki, N.; Wibovo, H. Mitigating distortion and residual stress by static thermal tensioning to improve fatigue crack growth performance of MIG AA5083 welds. *Mater. Des.* **2016**, *99*, 273–283. [CrossRef]
19. Zhang, W.; Fu, H.; Fan, J.; Li, R.; Xu, H.; Liu, F.; Qi, B. Influence of multi-beam preheating temperature and stress on the buckling distortion in electron beam welding. *Mater. Des.* **2018**, *139*, 439–446. [CrossRef]
20. Okano, S.; Mochizuki, M. Experimental and numerical investigation of trailing heat sink effect on weld residual stress and distortion of austenitic stainless steel. *ISIJ Int.* **2016**, *56*, 647–653. [CrossRef]

21. Zubairuddin, M.; Albert, S.K.; Vasudevan, M.; Mahadevan, S.; Chaudri, V.; Suri, V.K. Thermo-mechanical analysis of preheat effect on grade P91 steel during GTA welding. *Mater. Manuf. Process.* **2016**, *31*, 366–371. [CrossRef]
22. Deng, D.; Liang, W.; Murakawa, H. Determination of welding deformation in fillet-welded joint by means of numerical simulation and comparison with experimental measurements. *J. Mater. Process. Technol.* **2007**, *183*, 219–225. [CrossRef]
23. Yao, T.; Astrup, O.C.; Caridis, P.; Chen, Y.N.; Cho, S.R.; Dow, R.S.; Niho, O.; Rigo, P. Ultimate hull girder strength. In Proceedings of the 14th International Ship and Offshore Structure Congress, Nagasaki, Japan, 2–6 October 2000; Volume 2, pp. 323–333.
24. Chen, B.K.; Soares, G.C. Effects of plate configurations on the weld induced deformations and strength of fillet-welded plates. *Mar. Struct.* **2016**, *50*, 243–259. [CrossRef]
25. Gannon, L.; Liu, Y.; Peg, N.; Smith, M.J. Effect of welding-induced residual stress and distortion on ship hull girder ultimate strength. *Mar. Struct.* **2012**, *25*, 25–49. [CrossRef]
26. Perić, M.; Tonković, Z.; Rodić, A.; Surjak, M.; Garašić, I.; Boras, I.; Švaić, S. Numerical analysis and experimental investigation in a T-joint fillet weld. *Mater. Des.* **2014**, *53*, 1052–1063. [CrossRef]
27. Perić, M.; Stamenković, D.; Milković, V. Comparison of residual stresses in butt-welded plates using software packages Abaqus and Ansys. *Sci. Tech. Rev.* **2010**, *60*, 22–26.
28. Chen, Z.; Gongrong, L. A study on the spring-back deflections and constraint forces of a bottom grillage cling welding. *Ships Offshore Struct.* **2017**, *12*, 1077–1085. [CrossRef]
29. Lostado, R.L.; Garcia, R.E.; Martinez, R.F.; Martinez Calvo, M.A. Using genetic algorithms with multi-objective optimization to adjust finite element models of welded joints. *Metals* **2018**, *8*, 230. [CrossRef]
30. EN 1011-1. Welding—Recomendations for welding of metallic materials—Part 1. General guidance for arc welding. *Change* **2009**, *30*, 9.
31. Seleš, K.; Perić, M.; Tonković, Z. Numerical simulation of a welding process using a prescribed temperature approach. *J. Constr. Steel Res.* **2018**, *145*, 49–57. [CrossRef]
32. Perić, M.; Tonković, Z.; Garašić, I.; Vuherer, T. An engineering approach for a T-joint fillet welding simulation using simplified material properties. *Ocean Eng.* **2016**, *128*, 13–21. [CrossRef]
33. Menéndez, M.C.; Rodríguez, E.; Ottolini, M.; Caixas, J.; Guirao, J. Analysis of the effect of the electron-beam welding sequence for a fixed manufacturing route using finite element simulations applied to ITER vacuum vessel manufacture. *Fusion Eng. Des.* **2016**, *104*, 84–92. [CrossRef]
34. Perić, M.; Tonković, Z.; Karšaj, I.; Stamenković, D. A simplified engineering method for a T-joint welding simulation. *Therm. Sci.* **2018**, *22*, S867–S873. [CrossRef]
35. Deng, D. FEM prediction of welding residual stress and distortion in carbon steel considering phase transformation effects. *Mater. Des.* **2009**, *30*, 359–366. [CrossRef]
36. Marenić, E.; Skozrit, I.; Tonković, Z. On the calculation of stress intensity factors and J-integrals using the submodeling technique. *J. Press. Vessel Technol.* **2010**, *132*, 041203–041212. [CrossRef]
37. ISO (International Organization for Standardization). *Welding—Measurement of Preheating Temperature, Interpass Temperature and Preheat Maintenance Temperature*; ISO: Geneva, Switzerland, 2017.
38. EN 1011-2. Welding—Recommendations for welding of metallic materials—Part 2: Arc welding of ferritic steels. *Change* **2001**, *30*, 15–23.

© 2018 by the authors. Licensee MDPI, Basel, Switzerland. This article is an open access article distributed under the terms and conditions of the Creative Commons Attribution (CC BY) license (http://creativecommons.org/licenses/by/4.0/).

Article

Assessing Energy and Environmental Efficiency of the Spanish Agri-Food System Using the LCA/DEA Methodology

Jara Laso [1,*], Daniel Hoehn [1], María Margallo [1], Isabel García-Herrero [1], Laura Batlle-Bayer [2], Alba Bala [2], Pere Fullana-i-Palmer [2], Ian Vázquez-Rowe [3], Angel Irabien [1] and Rubén Aldaco [1]

1. Department of Chemical and Biomolecular Engineering, University of Cantabria, Avda. de los Castros s/n, 39005 Santander, Spain; daniel.hoehn@unican.es (D.H.); margallom@unican.es (M.M.); isabel.garciaherrero@unican.es (I.G.-H.); irabiena@unican.es (A.I.); aldacor@unican.es (R.A.)
2. UNESCO Chair in Life Cycle and Climate Change ESCI-UPF, Universitat Pompeu Fabra, Pg. Pujades 1, 08003 Barcelona, Spain; laura.batlle@esci.upf.edu (L.B.-B.); alba.bala@esci.upf.edu (A.B.); pere.fullana@esci.upf.edu (P.F.-i-P.)
3. Peruvian LCA Network, Department of Engineering, Pontificia Universidad Católica del Perú, Av. Universitaria 1801, San Miguel, Lima 15088, Peru; ian.vazquez@pucp.pe
* Correspondence: jara.laso@unican.es; Tel.: +34-942-846-531

Received: 31 October 2018; Accepted: 30 November 2018; Published: 4 December 2018

Abstract: Feeding the world's population sustainably is a major challenge of our society, and was stated as one of the key priorities for development cooperation by the European Union (EU) policy framework on food security. However, with the current trend of natural resource exploitation, food systems consume around 30% of final energy use, generating up to 30% of greenhouse gas (GHG) emissions. Given the expected increase of global population (nine billion people by 2050) and the amount of food losses and waste generated (one-third of global food production), improving the efficiency of food systems along the supply chain is essential to ensure food security. This study combines life-cycle assessment (LCA) and data envelopment analysis (DEA) to assess the efficiency of Spanish agri-food system and to propose improvement actions in order to reduce energy usage and GHG emissions. An average energy saving of approximately 70% is estimated for the Spanish agri-food system in order to be efficient. This study highlights the importance of the DEA method as a tool for energy optimization, identifying efficient and inefficient food systems. This approach could be adopted by administrations, policy-makers, and producers as a helpful instrument to support decision-making and improve the sustainability of agri-food systems.

Keywords: data envelopment analysis; energy efficiency; food loss and waste; life-cycle assessment

1. Introduction

The food industry is one of the major manufacturing sectors, representing 15% of the sales in the European Union and more than 1,120,000 million euros [1]. However, the rapid growth of global population and its corresponding consequences, such as the increasing demand of food production, caused an intensive use of energy resources [2]. In this sense, a considerable use of energy, estimated at 30% of the total energy consumption, is attributed to agri-food systems [1]. Given the strong dependency on fossil fuels of our current energy system, such energy consumption is responsible for 20–30% of total anthropogenic greenhouse gases (GHG) [3]. This fact results in the depletion of fossil resources and GHG emissions being of major concern. Energy is required at every stage in food production, including the cultivation and harvesting of crops, animal husbandry, transportation and distribution, and food processing for consumption, with the agricultural stage as the most critical [4]. This causes a low efficiency for food systems, with around 10 kcal of fossil fuel energy to produce 1 kcal of food [1]. Some foods are very efficient in their use of resources to produce a nutritious meal, such as fruits and vegetables, while others are very inefficient, such as animal-based products. In general, animal foods require eight times more energy per calorie than plant-based commodities [5].

Concerns regarding inefficiency are wider when food losses and waste (FLW) are taken into account. According to the Organization for Economic Co-operation and Development (OECD) [6], more than one-third of the food produced worldwide is lost along the supply chain, involving around 38% of the energy consumed in its production. In fact, FLW was identified as a major global concern putting at risk environmental, economic, and health security [7]. This food produced is, therefore, lost or wasted from initial agricultural production down to final household consumption. Since the embodied amount of energy builds up along the chain, the later the waste occurs, the higher the energy waste and the related GHG emissions will be. In addition, due to the expected global population increase, a 60% rise in food production was forecasted for the year 2050, entailing a 50% rise in global energy consumption [8]. Therefore, the efficient use of energy is a necessary step toward reducing environmental hazards, preventing destruction of natural resources, and ensuring food sustainability [9].

With the aim of attaining sustainability, many analytical tools are applied individually or combined to determine an appropriate productive performance. Life-cycle assessment (LCA) is a powerful tool that assesses the environmental impacts of products, processes, and services; it gained in acceptance since it first appeared in the 1990s and is today well established [10]. Many authors used LCA to assess the energy efficiency of their processes. For instance, Marique and Rossi [11], Ingrao et al. [12], and Berg and Fuglseth [13] applied LCA to energy uses in the building sector, while Li and Feng [14] used it to improve the energy recovery from sewage sludge comparing different pathways. Sundaram et al. [15] assessed the energy efficiency of the production of gasoline from biogas and pyrolysis oil using an LCA approach. Regarding the agri-food sector, Carrasquer et al. [16] created a new indicator to estimate the water and energy efficiency in agro-industries based on life-cycle thinking methodologies. Skunca et al. [17] performed the LCA of the chicken meat chain to estimate the cumulative energy demand (CED), while Pires-Gaspar et al. [18] carried out an energy-LCA of Portuguese peach production using energy efficiency indicators.

On the other hand, additional methods to compute which resource/technology combinations provide the highest amounts of net energy to society are reported as energy return ratios (ERRs) [19]. The most well-known indicator is the energy return on investment (EROI). In fact, its use is widespread within the energy sector to determine the energy that is returned from an energy-collecting process as compared to the embodied energy to provide this energy [20]. However, in recent years, some authors highlighted the benefits of calculating EROI ratios to monitor energy return in food systems [21]. For instance, Laso et al. [22,23], Vázquez-Rowe et al. [20], Tyedmers [24], and Ramos et al. [25] applied the EROI ratio to fishery systems. Cancino-Espinoza et al. [26] computed the EROI of organic quinoa, and Pérez Neira et al. [27] estimated the EROI of tomato production.

In contrast to traditional EERs, the use of the data envelopment analysis (DEA) enables the identification of energy-saving targets for energy-inefficient processes [28]. DEA is a non-parametric multi-input/output linear approach for the calculation of the relative efficiency of a set of comparable decision-making units (DMUs) [29]. In recent years, the application of DEA increased in the fields of environmental and energy. For instance, DEA was applied in the building sector [30–32], the power industry [33,34], the food production sector [1,9,35], and for agricultural production [36–38]. Moreover, it was also used with economic and social variables [3]. The combination of LCA and DEA for the identification and quantification of potential environmental consequences of operational inefficiencies experienced a pronounced increase, and enriched the interpretation of results [39,40]. Several previous studies used the joint LCA/DEA method in different agricultural production systems: fisheries [41,42], grapes for winemaking [39], wheat [43], rice [44], and dairy farms [45,46].

In this contribution, a hybrid approach which combines EROI, LCA, and DEA is proposed to assess the energy and environmental efficiency of agri-food systems. The capabilities of this combined approach are illustrated through its application to the Spanish agri-food system in order to explore the operational inefficiencies and improvement actions.

2. Materials and Methods

Figure 1 depicts the methodology described in this analysis, which attempted to assess the energy and environmental efficiency of Spanish agri-food system using a life-cycle approach. Firstly, an LCA was conducted to estimate the primary energy demand (PED) and related global warming potential (GWP). Secondly, the EROI was estimated to describe the energy efficiency of food categories under study. Finally, DEA was synergistically combined with LCA results to assess the efficiency of the system.

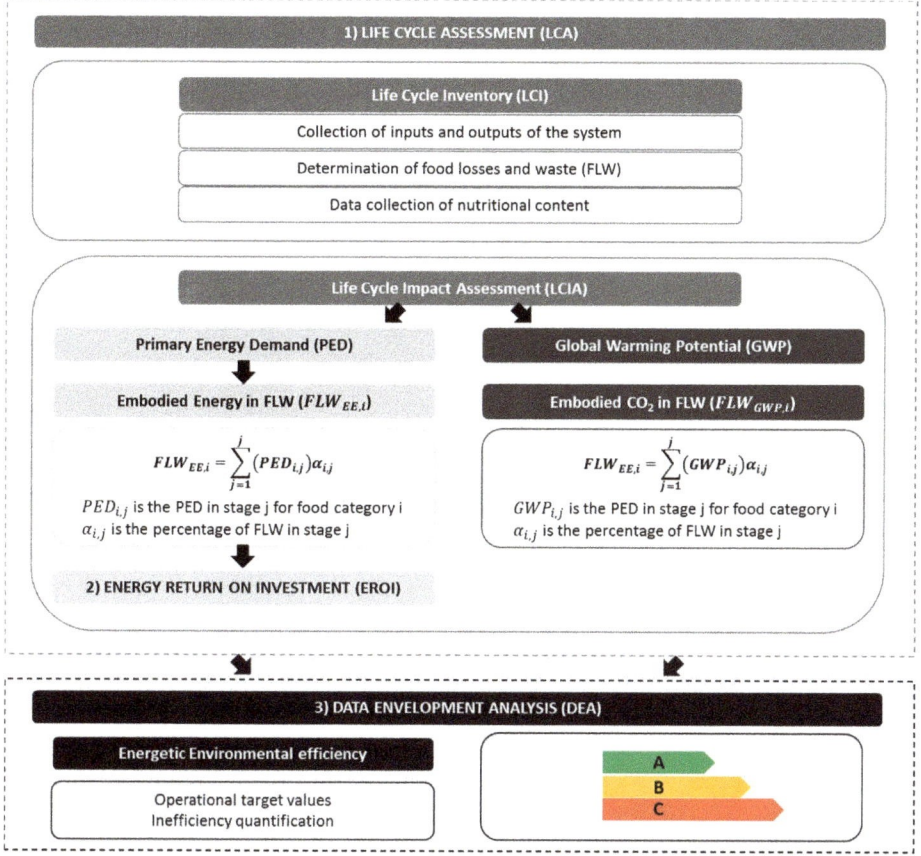

Figure 1. Schematic representation of the methodology applied.

2.1. Life-Cycle Assessment (LCA)

LCA is a powerful tool for performing the potential environmental assessment of impacts and resource consumption throughout a product's life cycle [47]. In this regard, LCA became one of the most relevant methodologies to help organizations perform their activities in the most environmental friendly way along the whole value chain.

In the current study, LCA was conducted following the recommendations of International Organization for Standardization (ISO) 14040 and 14044 international standards [48,49], and was divided into four stages.

(i). Goal and Scope Definition

This stage consisted of a detailed definition of the intended application of the study in terms of the system boundaries and functional unit (FU) [50,51]. Moreover, allocation procedures, cut-off rules, and assumptions were also defined in this step.

(ii). Life-Cycle Inventory (LCI)

In this stage, all relevant energy, water, and material consumptions, and emissions, effluents, and residues of the process in a specific temporality and geography were collected [52]. Therefore, in this stage, it was necessary to determine the FLW flows. Moreover, in this study, the LCI also collected data about the nutritional content of the different food categories.

(iii). Life-Cycle Impact Assessment (LCIA)

The LCIA step transformed the inputs and outputs of the LCI into environmental impact over all of the stages involved in the supply chain. In this study, the impact assessment method selected followed the International Reference Life Cycle Data Product Environmental Footprint (ILCD/PEF) recommendations v1.09 for determining the global warming potential (GWP). The consumption of primary energy resources (net calorific value, PED) was determined according to PE International's [53] life-cycle inventory. Once the PED and GWP embodied along the agri-food supply chain (FSC) were determined, the embodied energy and GHG emissions in FLW could be determined using FLW rates from Gustvasson et al. [54], as indicated Equations (1) and (2).

$$FLW_{EE,i} = \sum_{j=1}^{j} (PED_{i,j}) \alpha_{i,j}, \tag{1}$$

$$FLW_{GWP,i} = \sum_{j=1}^{j} (GWP_{i,j}) \alpha_{i,j}, \tag{2}$$

where $PED_{i,j}$ is the PED in stage j for food category i, $\alpha_{i,j}$ is the percentage of FLW in stage j for food category i, and $GWP_{i,j}$ is the GWP of stage j for food category i (see the example in Section S4 of the Supplementary Materials).

(iv). Interpretation of the Results

2.2. Energy Return on Investment (EROI)

The concept of EROI is part of the field of net energy analysis (NEA), and is one way of measuring and comparing the net energy availability from different energy sources and processes. In general, EROI can be defined as "the ratio between the energy returned from an energy-gathering activity compared to the embodied energy in that process" [55]. Although this concept was used initially to develop an energy-focused approach to the economy [56], the concept was adapted to calculate ratios between food energy output and food production energy inputs. In this sense, the most commonly used EROI perspective is the human-edible food energy return on industrially energy investment, as indicated in Equation (3). This provides an anthropocentric perspective on the non-renewable resource efficiencies of competing food production technologies [5]. Therefore, the higher the EROI of a food system is, the more "valuable" it is in terms of producing (nutritionally) useful energy output.

$$EROI = \frac{\sum_{j=1}^{j} Food_{i,j} \times NC_{i,j}}{\sum_{j=1}^{j} PED_{i,j}}, \tag{3}$$

where $Food_{i,j}$ represents the amount of the food category i available in stage j after withdrawing the FLW, $NC_{i,j}$ is the nutritional content of food category i in supply chain j expressed in kcal, and $PED_{i,j}$ is the PED in stage j for food category i calculated using LCA.

2.3. Data Envelopment Analysis (DEA)

DEA is a non-parametric method widely used to compare the inputs and outputs of a set of homogenous DMUs. This method focuses on evaluating the performance of DMUs based on the evaluation of relative efficiency of comparable DMUs by estimating an empirical efficient boundary [9].

Charnes, Cooper, and Rhodes (CCR) introduced the DEA method for the first time. The original CCR model was applicable to the assumption of constant returns to scale (CRS) [57]. In this sense, the CCR model considers that the efficiency frontier is a straight line intersecting the point of origin and the best performer(s), as shown in Figure 2a. The best performer is determined by the highest ratio of output to input; thus, as DMU_2 has this condition, it is considered as the reference DMU for all other units [38]. The remaining four DMUs in Figure 2a are inefficient since they use more amounts of input than DMU P_2 to produce one unit of output.

Banker et al. [58] modified the CCR model by introducing the so-called "convexity constraint", which changed the efficiency frontier from being a straight line to a convex hull (see Figure 2b). The new model, referred to as BCC, was built based on variable returns to scale (VRS). This model presents two advantages as compared to the CCR model: (i) more units can potentially be considered efficient, and (ii) inefficient units can be compared to more appropriate peers [38]. As shown in Figure 2b, using the BCC model, a higher number of units are efficient (DMU_1, DMU_2 and DMU_3).

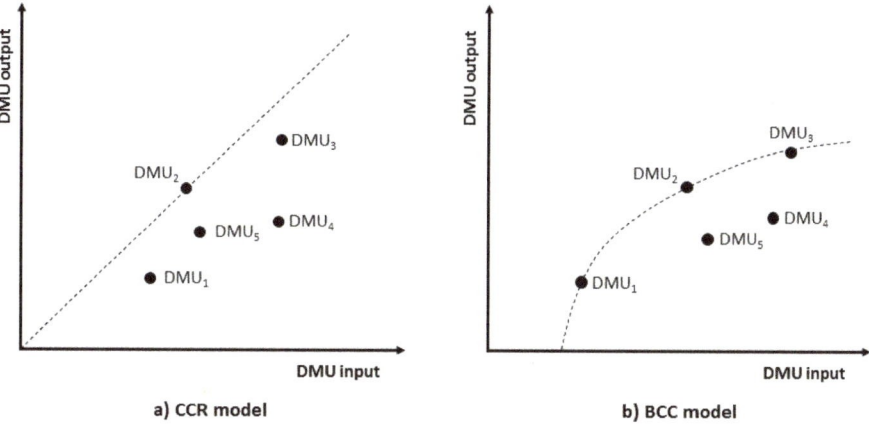

Figure 2. Graphical representation of the efficiency frontier of the (**a**) Charnes, Cooper, and Rhodes (CCR) model, and (**b**) the CCR model modified by Banker et al. (BCC model). Adapted from Hosseinzadeh-Bandbafha et al. [38].

The slacks-based measure (SBM) model is a typical extension of DEA. Although CCR and BCC approaches calculate the efficiencies of the DMUs based on the proportional decrease/increase of inputs/outputs, the SBM model uses the input excess and output shortfall of each DMU to measure its efficiency [59]. Moreover, it is possible to create and estimate models that provide input-oriented or output-oriented approaches for both CRS and VRS envelopments. An input-oriented model attempts to reduce input variables while remaining within the envelopment space. On the other hand, an output-oriented model increases output variables while remaining within the envelopment space [37]. In this study, an input-oriented approach was selected because multiple inputs were used, while there was only one output.

The DEA linear programming (LP) model (M1) deployed to generate CCR efficiency factors of DMUs is as follows [54]:

M1 (CCR model input-oriented to be solved for each DMU k_0):

$$\min \theta_{CCR}(k_0) = \sum_{j=1}^{n} u_j y_{jk_0};$$

s.t.

$$\sum_{i=1}^{m} v_i x_{ik_0} = 1;$$

$$\sum_{j=1}^{n} u_j y_{jk_0} - \sum_{i=1}^{m} v_i x_{ik_0} \leq 0, \quad k = 1, \ldots, K; \; j = 1, \ldots, n; \; i = 1, \ldots, m;$$

$$u_j \geq 0,$$
$$v_i \geq 0,$$

where u_j is the weight for output j, v_i is the weight for input i, m is the number of inputs, n is the number of outputs, k is the number of DMUs, y_{jk} is the amount of output j of DMU k, and x_{jk} is the amount of input i of DMU k.

On the other hand, the DEA LP model deployed to generate BCC efficiency factors of DMUs is as follows [58]:

M2 (BCC model input-oriented to be solved for each DMU k_0):

$$\min \theta_{BCC}(k_0) = \sum_{j=1}^{n} u_j y_{jk_0} - u(k_0);$$

s.t.

$$\sum_{i=1}^{m} v_i x_{ik_0} = 1;$$

$$\sum_{j=1}^{n} u_j y_{jk_0} - \sum_{i=1}^{m} v_i x_{ik_0} - u_{(k_0)} \leq 0,$$

$$k = 1, \ldots, K; \; j = 1, \ldots, n; \; i = 1, \ldots, m;$$

$$u_j \geq 0,$$
$$v_i \geq 0,$$

where u_j is the weight for output j, v_i is the weight for input i, m is the number of inputs, n is the number of outputs, k is the number of DMUs, y_{jk} is the amount of output j of DMU k, and x_{jk} is the amount of input i of DMU k.

To combine the LCA and DEA methods, a five-step procedure was followed [60,61]: (i) individual LCI for each of the DMUs; (ii) LCIA for every DMU included in the inventory (in this case, using the PED and GWP indicators, as explained in Section 2.1); (iii) determination of the operational efficiency for each DMU; (iv) LCIA of the target DMUs; (v) quantification of the environmental consequences of operational inefficiencies [42].

3. Case Study: The Spanish Food Basket

3.1. Goal and Scope

The Spanish Agency of Food Security and Nutrition (AECOSAN), along with the Spanish Society of Community Nutrition (SENC), is the institution that sets the nutritional recommendations for the Spanish population. These recommendations take into account the main diet-related public health challenges, as well as cultural habits of the Spanish population, to promote healthy eating and regular physical activity [62].

In this study, a basket of products was selected based on the consumption data reported by the Ministry of Agriculture, Fisheries, and Food (MAPAMA) [63]. These food commodities were classified according to eleven categories following the Food and Agriculture Organization of the United Nations (FAOSTAT) classification: eggs, meat and animal fat, fish and seafood, dairy, cereals, sweets, pulses, vegetable oils, vegetables, fruits, and roots and tubers. The food commodities considered in each category are collected in Table 1.

The goal and scope of this study was to assess the energetic and environmental efficiency of the Spanish food basket by means of the combination of LCA and DEA approaches. An additional goal of this analysis was to determine whether the use of DEA methodology is a viable tool for assessing the energy efficiency of food systems with FLW. The results are expected to provide an overview regarding the energy efficiency of the different food categories under study with the aim of identifying the most inefficient and the stages where improvement measures should be applied.

Table 1. Food commodities included in the study.

Food Category	Commodities Included
Cereals	Wheat, rice, maize, and others
Roots and tubers	Potatoes
Sugar	Sugar
Vegetable oils	Sunflower seed oil, palm oil, olive oil, and others
Vegetables	Tomatoes, onions, and others
Fruit	Oranges and mandarins, grapes (excluding grapes for winemaking), apples, and others
Pulses	Beans, peas, and others
Meat and animal fat	Beef, pork, lamb, and poultry
Fish and seafood	Fish and seafood
Dairy	Milk, cheese, and butter
Eggs	Eggs

3.2. Function, Functional Unit, and System Boundaries

The methodology proposed included three main parts: LCA, EROI, and the combination of LCA and DEA. Therefore, the function of the study was the estimation of the environmental impacts of the Spanish food basket and the determination of its energy efficiency. To quantify this function, it was necessary to define an FU to which inputs and outputs would be referred. In this case, the FU was described as the food basket with the representative food products consumed by a Spanish citizen in a year, covering the daily energy requirement of 2000 kcal of an adult.

The system boundaries comprise the entire supply chain of a food system following Garcia-Herrero et al. [4] (see Figure 3), i.e., agricultural production, postharvest and storage, industrial processing, distribution (i.e., retail/wholesale), and consumption. The consumption stage was divided into household consumption and related extradomestic consumption.

3.3. Life-Cycle Inventory (LCI)

An extensive LCI was built up using data from the literature (for more details, see Batlle-Bayer et al. [64]) and the PE GaBi database [53]. The food balance sheet and the unavoidable FLW percentages of each food category were collected from García-Herrero et al. [3], where production values were mainly sourced from Eurostat [65–68]. Imported food products were taken into account based on trade statistics compiled by the Spanish Tax Agency (AEAT) [69]. The food balance sheet for Spain in 2015 and the FLW percentages are available in Sections S1 and S2 in the Supplementary Materials (SM).

For transportation, 400-km and 100-km distances to wholesale and retailers were assumed, respectively. Electricity consumption for retail storage was considered, assuming two days of storage for products requiring cooling conditions, and 15 days under freezing conditions. Transport to home was estimated based on Milà i Canals et al. [70]. Regarding home storage, data from the LCA Food

Database was used [71]. For cooking, energy factors of Foster et al. [72] were taken (for more details, see Batlle-Bayer et al. [64]).

On the other hand, nutritional data for the EROI estimation were obtained from the food composition tables of the Institute for Education in Nutrition and Dietetics from Spain (CESNID) [73]. Such tables are registered in the FAO's International Network of Food Data Systems [74]. The nutritional content for each food category under study is collected in Section S3 in the Supplementary Materials.

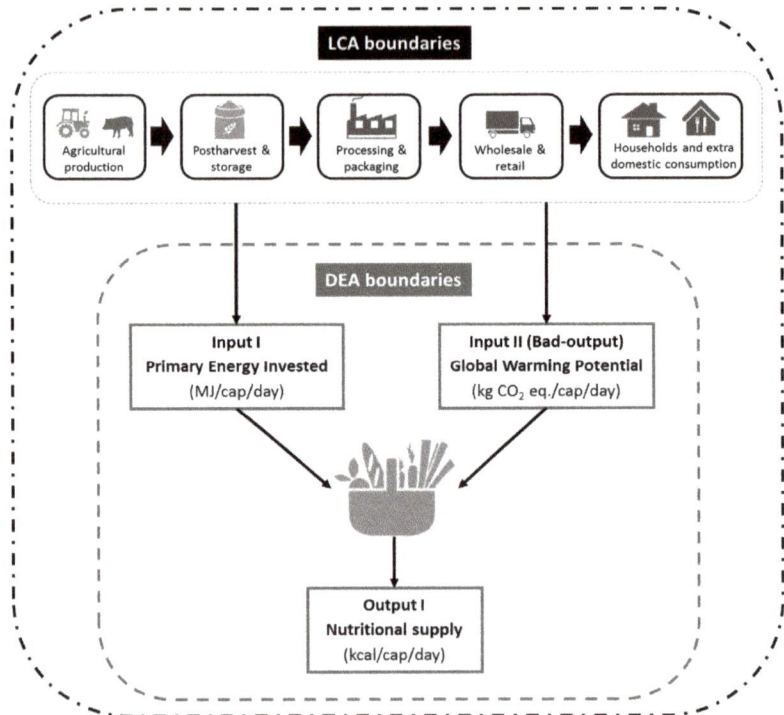

Figure 3. System boundaries for the Spanish food basket using a life-cycle assessment (LCA)/data envelopment analysis (DEA) approach.

3.4. Selection of the DEA Model

The selected DEA model was the slacks-based measure of efficiency (SBM). The choice of model was based on its flexibility concerning the computation of the DMUs irrespective of the units of measure used for the different inputs/outputs [42]. Moreover, for the sake of further discussion, the CRS and VRS approaches were computed in parallel in the current study under an input-oriented approach.

3.5. Inputs and Output Selection for the DEA Matrix

One DEA matrix composed of 11 DMUs was computed in this study. The number of DMUs under study must satisfy Equation (4). Each DMU included two different inputs (see Figure 3). As input 1, the primary energy invested in food production was computed in MJ/cap/day, whereas the GWP was considered as input 2. As mentioned previously, an input-oriented approach was based on the assumption that inputs have to be minimized and outputs have to be maximized. However, in some situations, undesirable (bad) inputs and outputs may be presented in the production process [75]. In this case, the GWP was a bad output that was computed as an input to be minimized. Finally, the nutritional energy supply was computed in kcal/cap/day as output.

$$n \geq \max \{m \times s, 3 \times (m+s)\}, \tag{4}$$

where m is the number of inputs used in the DEA study, and s is the number of outputs involved.

4. Results and Discussion

Following the methodology described in Section 2, the LCA of the 11 food categories selected was performed, and the PED and GWP were calculated. These values were obtained from a previous study developed by Batlle-Bayer et al. [64]. As shown in Table 2, the total energy invested in food production daily amounts to around 19,500 kcal for an average Spanish citizen. This involves an emission of close to 4 kg of CO_2 equivalent per day, with meat production as the most responsible: 28% for PED and 42% in terms of CO_2 equivalent emissions. Dairy, and fish and seafood are the second main categories contributing to GWP, each representing 12% of total emissions. Results suggest a high correlation between the investment of energy along the FSC and the generation of GHG emissions, owing to the high dependency of the energy matrix on fossil fuels. Conversely, this is not directly related to the nutritional energy provided to consumers. As displayed in Table 1, the major contributors to the Spanish average diet are vegetable oils and cereals, responsible for nearly half of the energy supply. However, the contribution of meat to nutritional energy supply is relegated to less than 12%, with eggs as the food category providing the lowest amount of nutritional energy to consumers (2%). These results are in agreement with data reported by the European Commission [76], which confirm that livestock products, such as meat, fish and seafood, and dairy products, incorporate a substantial amount of energy; in addition, the consumption of dairy and meat products has a major role in GHG emissions.

Table 2. Energy invested from cradle to plate for the food categories under study, related CO_2 eq. emissions, and nutritional energy provided to consumer. PED—primary energy demand; GWP—global warming potential; FLW—food losses and waste.

Food Category	PED (kcal/cap/d)	Energy Provided to Consumer (kcal)	GWP (g CO_2 eq./cap/d)	Embodied Energy in FLW (kcal/cap/d)	Embodied GWP in FLW (g CO_2 eq./cap/d)
Eggs	1059	41	221	163	35
Meat	5465	261	1673	1162	378
Fish and seafood	3170	99	468	852	128
Dairy	1411	289	496	137	55
Cereals	2717	456	372	1042	143
Sweets	156	103	28	35	8
Pulses	490	58	85	142	27
Vegetable Oils	717	461	158	150	39
Vegetables	3297	72	261	745	70
Fruits	690	159	159	171	52
Roots	330	53	51	88	16
Total	19,501	2000	3971	4685	951

When considering FLW, it was observed that embodied energy waste duplicates the daily energy supplied to consumers (4685 kcal/cap/d). As shown in Figure 4a, the largest contributor to this fact is the meat category, accounting for 25%. This is directly related to the unnecessary emission of CO_2 eq., which also contributes 40% to the total GWP (Figure 4b). It is followed by cereals and fish and seafood categories, for which a 22% and 18% embodied energy waste was estimated, respectively, whereas a 15% and 12% of embodied GWP waste was calculated.

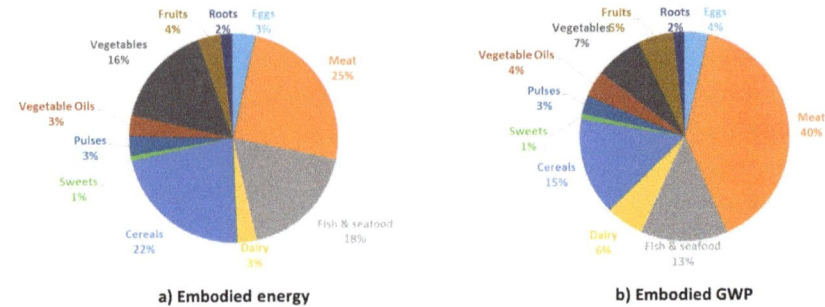

Figure 4. Contribution of each food category in the average Spanish diet to the (**a**) embodied energy, and (**b**) embodied global warming potential (GWP).

These findings suggest the need for estimating the efficiency of the Spanish agri-food system. As a first approach, the EROI indicator was used to estimate the energetic efficiency. As displayed in Table 3, the maximum EROI was observed for sweets and vegetable oils, which means that such food categories are the most efficient from an energy perspective. These results also suggest the need of studying more in-depth nutrients that should be either encouraged or limited, in order to provide a nutritional efficiency perspective. Results also show that animal-based products such as eggs, meat, and fish present the lowest EROI values (3.1–4.0%), which agree with the study of Pelletier et al. [5], where an eightfold difference between animal- and plant-based products was determined. An exception was observed in our results for vegetables. This is mainly due to the inclusion of processed commodities in this category, particularly tomato sauce, which exhibits an EROI value of 1.2%.

Table 3. Energy return on investment (EROI) values for food categories in the Spanish diet.

Food Category	Eggs	Meat	Fish and Seafood	Dairy	Cereals	Sweets
EROI (%)	3.90	3.96	3.14	20.52	16.79	66.00
Food category	Pulses	Vegetable Oils	Vegetables	Fruits		Roots
EROI (%)	11.75	64.27	2.18	23.05		16.10

However, these efficiency results do not consider the environmental impacts of food production along the supply chain. For this reason, DEA was combined with LCA, allowing a wider scope of the efficiency of the Spanish agri-food system to be provided. To perform an LCA/DEA analysis, it is firstly necessary to elaborate the DEA matrix. Table 4 represents the DEA matrix, composed of 11 DMUs, with each one represented by two inputs and one output, as explained in Section 3.4.

Table 4. Data envelopment analysis (DEA) matrix under study. DMU—decision-making unit.

Food Category	DMU	Input 1	Bad Output (Input 2)	Output 1
		Primary Energy Invested (MJ/cap/day)	GWP (g CO_2 eq/cap/day)	Nutritional Supply (kcal/cap/day)
Eggs	1	4.43	221	41.3
Meat	2	22.84	1672	216
Fish and seafood	3	13.25	468	99.7
Dairy	4	5.90	496	289
Cereals	5	11.36	371	456
Sweets	6	0.53	28.5	103
Pulses	7	2.05	85.0	57.6
Vegetable Oils	8	3.00	158	461
Vegetables	9	13.78	261	71.7
Fruits	10	2.84	154	151
Roots	11	1.38	50.5	53.1

DEA Frontier was the software used for the computation of the DEA matrix [77]. Results using CRS and VRS perspectives were compared and are shown in Figure 5, which includes the efficiency score for each DMU. A food category was considered inefficient with an efficiency score ϕ < 1, whereas φ = 1 represented an efficient food category. According to that, when the CRS perspectives were considered, one food category was efficient (sweets), while a 0.97 efficiency was observed for vegetable oils (see Figure 5a). The CRS approach presented a wide range of efficiencies, between 0.06 and 0.97. Egg production was the least efficient food category, followed by fish, meat, and vegetables, with similar efficiencies ranging from 0.06 to 0.08. On the other hand, when the VRS model was applied, a total of two food categories were considered efficient (sweets and vegetable fats; see Figure 5b). Despite the fact that the VRS approach presented an additional efficient DMU, the range of efficiencies was lower than when using a CRS perspective, ranging from 0.06 to 0.56. Finally, the average efficiencies (including efficient DMUs) were similar: 0.33 ± 0.34 for CRS and 0.40 ± 0.34 for VRS. Some authors state that using a VRS approach rather than a CRS formulation leads to a higher number of efficient DMUs since the constraint set for CRS is less restrictive and, consequently, lower efficiency scores are possible [78]. In addition, the convexity constraint added in the VRS simply guarantees that each DMU is only compared to others of similar size [43]. Therefore, considering that the unit of reference (i.e., DMU) is, in each case, a different food category and that the food categories inventoried in this study present different characteristics, we considered that the use of the VRS approach should prevail.

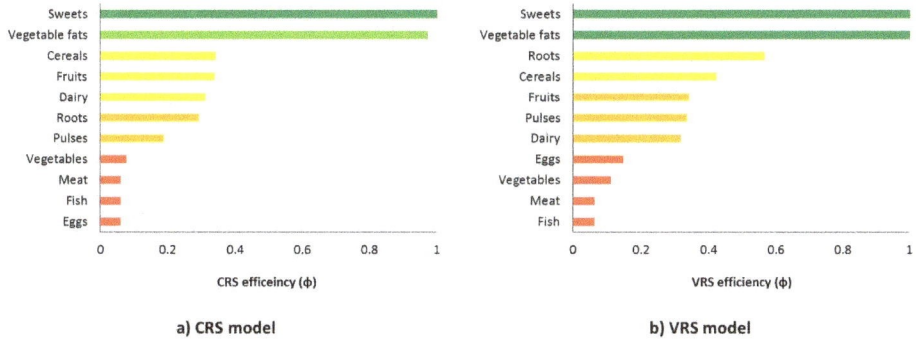

Figure 5. Efficiency (ϕ) score of the selected food categories (i.e., decision-making units (DMUs)) using the (**a**) constant return to scale (CRS) model, and (**b**) variable return to scale (VRS) model.

The previous findings agree with the pattern observed for the EROI assessment. While this can be taken as a verification of both approaches, DEA also provides a further exploration of the operational inefficiencies and the improvement actions of the system under study. Hence, the DEA model was also used to formulate new virtual and efficient values for the inputs of the inefficient DMUs, by projecting the inefficient scores on the efficient targets established. Both PED and GWP were subjected to minimization while maintaining the same nutritional supply.

Figures 6 and 7 compare the current environmental impacts of the different DMUs with the target PED and GWP calculated with the LCA/DEA methodology. In other words, it presents the environmental savings that would be reached in these DMUs if they were to operate under the efficient conditions projected in the computed matrix. In this regard, when the CRS perspective was used, primary energy invested savings ranged from 2.83% to 96.7%, while, when the VRS approach was used, reductions ranged from 52.9% to 95.3%. The highest reductions were obtained by the most inefficient food categories (i.e., vegetables, meat, fish, and eggs). On the other hand, GWP savings followed the same trend, ranging from 19% to 96.4% for CRS and from 43.6% to 95.8% for VRS. Similarly, the highest reductions were also reached by vegetables, meat, fish, and eggs.

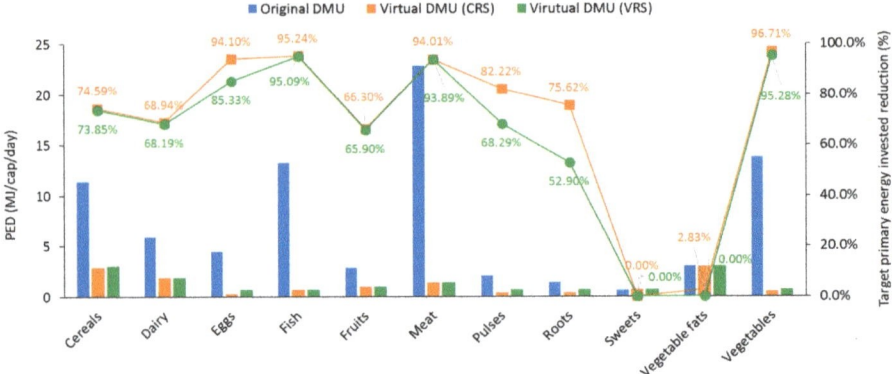

Figure 6. Primary energy invested (PED) for original DMUs (blue bar) and virtual targets for the CRS model (orange bar) and VRS model (green bar). The lines represent the overall environmental improvements for the CRS model (orange line) and VRS model (green line).

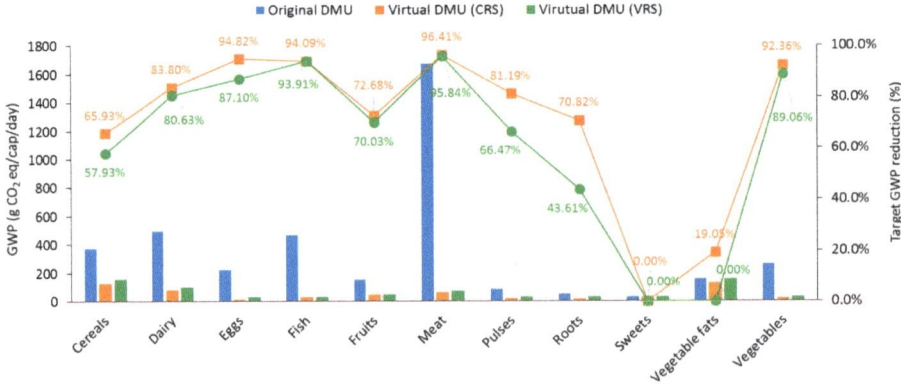

Figure 7. Global warming potential (GWP) for original DMUs (blue bar) and virtual targets for the CRS model (orange bar) and VRS model (green bar). The lines represent the overall environmental improvements for the CRS model (orange line) and VRS model (green line).

The results suggest that the planification of citizens' diet may affect the quantity of embodied energy waste and related environmental impacts. For this reason, specific strategies should be addressed to the categories revealing larger inefficiency scores, such as meat, eggs, fish, and vegetables, which all have better nutritional reputation than sweets or processed cereals. Figure 8 displays the contribution of each stage of the FSC to the PED for eggs, meat, fish and seafood, and vegetable categories. As stated by Batlle-Bayer et al. [64], in terms of the life-cycle processes, the primary production phases were the major contributors to the PED for eggs (85%), meat (65%), and fish and seafood (59%), while the household stage was the main contributor for vegetables (52%). These results are related to the production of feed in the case of eggs and meat production [60]. According to Ghasempour and Ahmadi [79], the production of feed (corn, soybean, and wheat), as well as the poultry equipment, used in the production of eggs is associated with the most energy consumption, whereas Skunca et al. [17] suggested the use of grain legumes as a protein source in feed instead of soybean-based ingredients. This fact could decrease the resource consumption since the cultivation of grain legumes does not require mineral fertilizer application. Moreover, in the case of beef, the use of electricity and diesel during the suckling cow–calf stage was also significant in the primary production phase [64].

Other authors, such as Avadí et al. [80,81], Laso et al. [22], and Vázquez-Rowe et al. [82] assessed the life cycle of processed fish products identifying the fishery stage as one of the most energy-intensive due to the use of fossil fuels. Batlle-Bayer et al. [64] analyzed eight fish and seafood species (i.e., mussels, shrimps and prawns, Atlantic mackerel, European hake, sardines, salmon, tuna, and octopus), identifying the use of diesel as the main contributor (around 98%) in the fishery stage, and the use of electricity (around 97%) in the culture stage (mussels farming).

On the other hand, for vegetables, several foods were considered (tomato, lettuce, and veggies). The fact that household consumption was the stage with the highest contribution to PED was due to the use of energy when cooking veggies (i.e., boiling and frying) and cold storage for preserving their shelf life. However, the agricultural production stage for vegetables was also important, representing 26% of total PED, due to the cultivation of lettuce, in particular, the use of diesel for machinery. In fact, although the amount of PED required for meat, fish and seafood, and vegetable consumption is similar (Table 4), it was distributed in different stages. Finally, the processing stage represented 19% of the total impact due to the inclusion of processed products, such as tomato processed products, which use fuel oil and electricity in their processing.

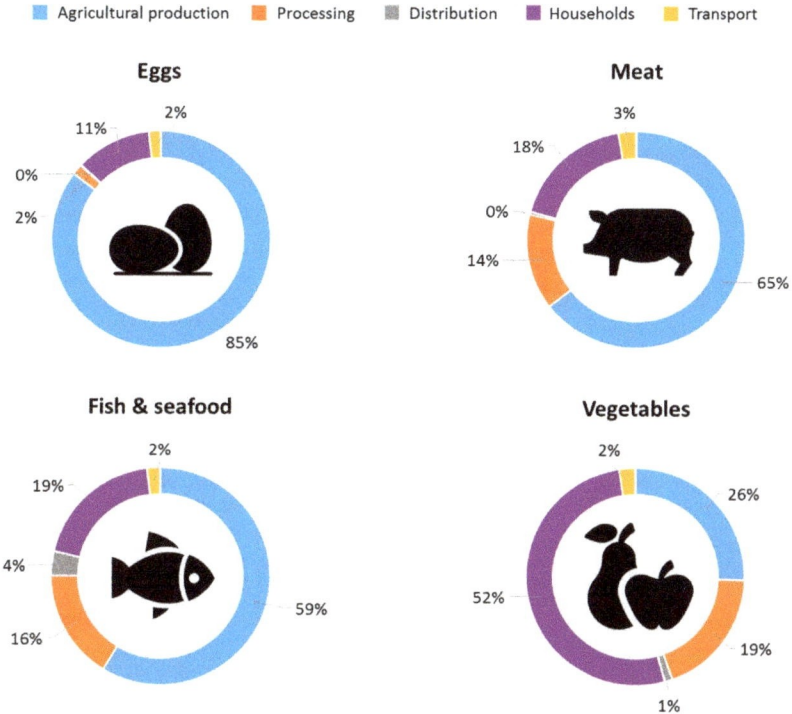

Figure 8. Distribution of the primary energy demand (PED) between the different stages of the food supply chain for the most inefficient food categories: eggs, meat, fish and seafood, and vegetables.

5. Conclusions

Food systems are heavily reliant on energy resources, especially non-renewable resources. This causes significant amounts of GHG emissions. In this study, the efficiency of the Spanish agri-food system was addressed from an energy and environmental perspective. Firstly, an LCA was performed to determine the PED and GWP impact results of the food basket with the representative food products consumed in and out of home by a Spanish adult in a year. Thereafter, the EROI indicator was used to perform an energy-based efficiency assessment. Finally, DEA was coupled with LCA to include environmental aspects in the efficiency assessment. Both approaches provided similar results, suggesting a high correlation between PED and CO_2 eq. emissions. Best results were obtained for the categories of sweets and vegetable oils ($\phi = 1$), while vegetables, fish, eggs, and meat exhibited the lowest efficiency, below 0.1. As expected, animal-based products required more energy resources in their production than vegetable-based products. However, it is important to mention that these types of food (i.e., sweets and vegetable oils) are considered "empty kcals", and their nutritional value is very low, whereas eggs, meat, and fish are most valuable in terms of nutritional value. Therefore, it would be interesting to considering other nutrients, such as proteins, in future assessments. On the other hand, the unexpected results for vegetables are due to the inclusion of processed products in this category. On average, approximately 70% energy-saving potential is estimated for the Spanish agri-food system if it were to be efficient, with a similar reduction in related GHG emissions. These results suggest the need for improving the efficiency of the FSC by introducing circular economy strategies, such as establishing appropriate food waste management measures and the consequent reduction of FLW. Therefore, the methodology proposed is a useful tool for promoting the circular economy of food. The introduction of nutritional- and energy-based criteria, in addition to the environmental pillar, provides an integrated framework for proposing integrated reduction targets. However, as ongoing research, other criteria, such as economic and social aspects, could be considered. From the DEA results, it is possible to define specific strategies for the categories revealing larger inefficiency scores, such as meat, eggs, fish, and vegetables. In terms of the life-cycle processes, the primary production phases were the major contributors to PED for eggs (85%), meat (65%), and fish and seafood (59%), due to the production of feed (corn, soybean, and wheat) in the case of eggs and meat, and the use of diesel and electricity in the fishery and cultivation stages, respectively. On the other hand, the household stage was the main contributor for vegetables (52%) due to the use of electricity for cooking and cooling.

Finally, we can establish that DEA is a useful tool for energy optimization, identifying efficient and inefficient food systems. This approach may be adopted by institutions, policy-makers, and producers as a helpful instrument to support decision-making and improve the sustainability of agri-food systems.

Supplementary Materials: The following are available online at http://www.mdpi.com/1996-1073/11/12/3395/s1. Section S1: Food Balance Sheet Construction; Section S2: Avoidable and Unavoidable Food Loss Calculation; Section S3: Nutritional Food Loss Calculation; Section S4: Example of Calculations According to the Methodologies Presented in Section 2.

Author Contributions: Conceptualization, R.A.; investigation, I.G.-H., J.L., and D.H.; methodology, L.B.-B., and I.V.-R.; formal and technical analysis, I.V.-R., J.L., P.F., A.B., A.I., L.B.-B., and R.A.; supervision, M.M., J.L., and R.A.; writing and editing of manuscript, I.G.-H. and J.L.

Funding: This work was carried out under the financial support of the Project Ceres-Procom: food production and consumption strategies for climate change mitigation (CTM2016-76176-C2-1-R) (AEI/FEDER, UE) financed by the Ministry of Economy and Competitiveness of the Government of Spain.

Acknowledgments: The authors want to acknowledge the Industrial PhD Program 2017 (BOC N°200 18/10/2017) and to the UNESCO Chair in Life Cycle and Climate Change.

Conflicts of Interest: The authors declare no conflict of interest.

References

1. Pimentel, D.; Pimentel, M.H. (Eds.) *Food, Energy and Society*; CRC press: Boston, MA, USA, 2008.
2. Bolandnazar, E.; Keyhani, A.; Omid, M. Determination of efficient and inefficient greenhouse cucumber producers using Data Envelopment Analysis approach, a case study: Jiroft city in Iran. *J. Clean. Prod.* **2014**, *79*, 108–115. [CrossRef]
3. Cucchiella, F.; D'Adamo, I.; Gastaldi, M.; Miliacca, M. Efficiency and allocation of emission allowances and energy consumption over more sustainable European economies. *J. Clean. Prod.* **2018**, *182*, 805–817. [CrossRef]

4. Garcia-Herrero, I.; Hoeh, D.; Margallo, M.; Laso, J.; Bala, A.; Batlle-Bayer, I.; Fullana, P.; Vázquez-Rowe, I.; Gonzalez, M.J.; Durá, M.J.; et al. On the estimation of potential food waste reduction to support sustainable production and consumption policies. *Food Policy* **2018**, *80*, 24–38. [CrossRef]
5. Pelletier, N.; Audsley, E.; Brodt, S.; Garnett, T.; Henriksson, P.; Kendall, A.; Troell, M. Energy intensity of agriculture and food systems. *Annu. Rev. Environ. Resour.* **2011**, *36*, 233–246. [CrossRef]
6. OECD. Improving energy efficiency in the Agro-Food chain. In *OECD Green Growth Studies*; OECD Publishing: Paris, France, 2017.
7. Morone, P.; Falcone, P.M.; Lopolito, A. How to promote a new and sustainable food consumption model: A fuzzy cognitive map study. *J. Clean. Prod.* **2019**, *208*, 563–574. [CrossRef]
8. Vora, N.; Shah, A.; Bilec, M.M.; Khanna, V. Food-energy-water nexus: Quantifying embodied energy and GHG emissions from irrigation through virtual water transfers in food trade. *ACS Sustain. Chem. Eng.* **2017**, *5*, 2119–2128. [CrossRef]
9. Khoshroo, A.; Mulwa, R.; Emrouznejad, A.; Arabi, B. A non-parametric Data Envelopment Analysis approach for improving energy efficiency of grape production. *Energy* **2013**, *63*, 189–194. [CrossRef]
10. Arushanyan, Y.; Björklund, A.; Eriksson, O.; Finnveden, G.; Söderman, M.L.; Sundqvist, J.; Stenmarck, A. Environmental Assessment of Possible Future Waste Management Scenarios. *Energies* **2017**, *10*, 247. [CrossRef]
11. Marique, A.; Rossi, B. Cradle-to-grave life-cycle assessment within the built environment: Comparison between the refurbishment and the complete reconstruction of an office building in Belgium. *J. Environ. Manag.* **2018**, *224*, 396–405. [CrossRef]
12. Ingrao, C.; Messineo, A.; Beltramo, R.; Yigitcanlar, T.; Iioppolo, G. How can life cycle thinking support sustainability of buidings? Investigating life cycle assessment applications for energy efficiency and environmental performance. *J. Clean. Prod.* **2018**, *201*, 556–569. [CrossRef]
13. Berg, F.; Fuglseth, M. Life cycle assessment and historic buidings: Energy-efficiency refurbishment versus new construction in Norway. *J. Arch. Conserv.* **2018**, *24*, 152–167.
14. Li, H.; Feng, K. Life cycle assessment of the environmental impacts and energy efficiency of an integration of sludge anaerobic digestion and pyrolysis. *J. Clean. Prod.* **2018**, *195*, 476–485. [CrossRef]
15. Sundaram, S.; Kolb, G.; Hessel, V.; Wang, Q. Energy-Efficient Routes for the Production of Gasoline from Biogas and Pyrolysis Oil—Process Design and Life-Cycle Assessment. *Ind. Eng. Chem. Res.* **2017**, *56*, 3373–3387. [CrossRef] [PubMed]
16. Carrasquer, B.; Uche, J.; Martinez-Gracia, A. A new indicator to estimate the efficiency of water and energy use in agro-industries. *J. Clean. Prod.* **2017**, *143*, 462–473. [CrossRef]
17. Skunca, D.; Tomasevic, I.; Nastasijevic, I.; Tomovic, V.; Djekic, I. Life cycle assessment of the chicken meat chain. *J. Clean. Prod.* **2018**, *184*, 440–450. [CrossRef]
18. Pires-Gaspar, J.; Dinis-Gaspar, P.; Dinho da Silva, P.; Simoes, M.P.; Espirito-Santo, C. Energy Life-Cycle Assessment of Fruit Products—Case Study of Beira Interior's Peach (Portugal). *Sustainability* **2018**, *10*, 3530. [CrossRef]
19. Murphy, D.J.; Carbajales-Dale, M.; Moeller, D. Comparing Apples to Apples: Why the Net Energy Analysis Community Needs to Adopt the Life-Cycle Analysis Framework. *Energies* **2016**, *9*, 917. [CrossRef]
20. Vázquez-Rowe, I.; Villanueva-Rey, P.; Moreira, M.T.; Feijoo, G. Edible Protein Energy on Investment Ratio (ep-EROI) for Spanish Seafood Products. *AMBIO* **2014**, *43*, 381–394. [CrossRef]
21. Pelletier, N.; Tyedmers, P. An ecological economic critique of the use of market information in life cycle assessment research. *J. Ind. Ecol.* **2011**, *15*, 342–354. [CrossRef]
22. Laso, J.; Vázquez-Rowe, I.; Margallo, M.; Crujeiras, R.M.; Irabien, A.; Aldaco, R. Life cycle assessment of European anchovy (*Engraulis encrasicolus*) landed by purse seine vessels in northern Spain. *Int. J. Life Cycle Assess.* **2018**, *23*, 1107–1125. [CrossRef]
23. Laso, J.; Margallo, M.; García-Herrero, I.; Fullana, P.; Bala, A.; Gazulla, C.; Polettini, A.; Kahhat, R.; Vázquez-Rowe, I.; Irabien, A.; et al. Combined application of life cycle assessment and linear programming to evaluate food waste-to-food strategies: Seeking for answers in the nexus approach. *Waste Manag.* **2018**, *80*, 186–197. [CrossRef] [PubMed]
24. Tyedmers, P. Energy consumed by North Atlantic fisheries. In *Fisheries Impacts on North Atlantic Ecosystems: Catch, Effort and National/Regional Datasets*; Zeller, D., Watson, R., Pauly, D., Eds.; Fisheries Centre Research Reports; Fisheries Centre, University of British Columbia: Kelowna, BC, Canada, 2001; Volume 9.

25. Ramos, S.; Vázquez-Rowe, I.; Artetxe, I.; Moreira, M.T.; Feijoo, G.; Zufia, J. Environmental assessment of the Atlantic mackerel (*Scomber scombrus*) season in the Basque Country. Increasing the timeline delimitation in fishery LCA studies. *Int. J. Life Cycle Assess.* **2011**, *16*, 599–610. [CrossRef]
26. Cancino-Espinoza, E.; Vázquez-Rowe, I.; Quispe, I. Organic quinoa (*Chenopodium quinoa* L.) production in Peru: Environmental hotspots and food security considerations using Life Cycle Assessment. *Sci. Total Environ.* **2018**, *637–638*, 221–232. [CrossRef]
27. Pérez Neira, D.; Soler Montiel, M.; Delgado Cabeza, M.; Reigada, A. Energy use and carbon footprint of the tomato production in heated multi-tunnel greenhouses in Almeria within an exporting agri-food system context. *Sci. Total Environ.* **2018**, *628–629*, 1627–1636. [CrossRef] [PubMed]
28. Masuda, K. Energy Efficiency of Intensive Rice Production in Japan: An Application of Data Envelopment Analysis. *Sustainability* **2018**, *10*, 120. [CrossRef]
29. Mardani, A.; Streimikiene, D.; Balezentis, T.; Zameri Mat Saman, M.; Nor, K.; Khoshnava, S. Data Envelopment Analysis in Energy and Environmental Economics: An Overview of the State-of-the-Art and Recent Development Trends. *Energies* **2018**, *11*, 2002. [CrossRef]
30. Lee, W.S.; Lee, K.P. Benchmarking the performance of building energy management using data envelopment analysis. *Appl. Therm. Eng.* **2009**, *29*, 3269–3273. [CrossRef]
31. Hu, X.; Liu, C. Slacks-based data envelopment analysis for eco-efficiency assessment in the Australian construction industry. *Constr. Manag. Econ.* **2017**, *35*, 693–706. [CrossRef]
32. Chen, Y.; Liu, B.; Shen, Y.; Wang, X. The energy efficiency of China's regional construction industry based on the three-stage DEA model and the DEA-DA model. *KSCE. J. Civ. Eng.* **2016**, *20*, 34–47. [CrossRef]
33. Duan, N.; Guo, J.P.; Xie, B.C. Is there a difference between the energy and CO_2 emission performance for China's thermal power industry? A bootstrapped directional distance function approach. *Appl. Energy* **2016**, *162*, 1552–1563. [CrossRef]
34. Sözen, A.; Alp, I.; Özdemir, A. Assessment of operational and environmental performance of the thermal power plants in Turkey by using data envelopment analysis. *Energy Policy* **2010**, *38*, 6194–6203. [CrossRef]
35. Khoshnevisan, B.; Rafiee, S.; Omid, M.; Mousazadeh, H. Applying data envelopment analysis approach to improve energy efficiency and reduce GHG (greenhouse gas) emissions of wheat production. *Energy* **2013**, *58*, 588–593. [CrossRef]
36. Yang, Z.; Wang, D.; Du, T.; Zhang, A.; Zhou, Y. Total-Factor Energy Efficiency in China's Agricultural Sector: Trends, Disparities and Potentials. *Energies* **2018**, *11*, 853. [CrossRef]
37. Hosseinzadeh-Bandbafha, H.; Safarzadeh, D.; Ahmadi, E.; Nabavi-Pelesaraei, A.; Hosseinzadeh-Bandbafha, E. Applying data envelopment analysis to evaluation of energy efficiency and decreasing of greenhouse gas emissions of fattening farms. *Energy* **2017**, *120*, 652–662. [CrossRef]
38. Hosseinzadeh-Bandbafha, H.; Safarzadeh, D.; Ahmadi, E.; Nabavi-Pelesaraei, A. Optimization of energy consumption of dairy farms using data envelopment analysis—A case study: Qazvin city of Iran. *J. Saudi Soc. Agric. Sci.* **2018**, *17*, 217–228. [CrossRef]
39. Mohseni, P.; Borghei, A.M.; Khanali, M. Coupled life cycle assessment and data envelopment analysis for mitigation of environmental impacts and enhancement of energy efficiency in grape production. *J. Clean. Prod.* **2018**, *197*, 937–947. [CrossRef]
40. Paramesh, V.; Arunachalam, V.; Nikkhah, A.; Das, B.; Ghnimi, S. Optimization of energy consumption and environmental impacts of arecanut production through coupled data envelopment analysis and life cycle assessment. *J. Clean. Prod.* **2018**, *203*, 674–684. [CrossRef]
41. Vázquez-Rowe, I.; Iribarren, D.; Moreira, M.T.; Feijoo, G. Combined application of life cycle assessment and data envelopment analysis as a methodological approach for the assessment of fisheries. *Int. J. Life Cycle Assess.* **2010**, *15*, 272–283. [CrossRef]
42. Laso, J.; Vázquez-Rowe, I.; Margallo, M.; Irabien, A.; Aldaco, R. Revisiting the LCA+DEA method in fishing fleets. How should we be measuring efficiency? *Mar. Pol.* **2018**, *91*, 34–40. [CrossRef]
43. Masuda, K. Measuring eco-efficiency of wheat production in Japan: A combined application of life cycle assessment and data envelopment analysis. *J. Clean. Prod.* **2016**, 373–381. [CrossRef]
44. Nabavi-Pelesaraei, A.; Rafiee, S.; Mohtasebi, S.S.; Hosseinzadeh-Bandbafha, H.; Chau, K. Energy consumption enhancement and environmental life cycle assessment in paddy production using optimization techniques. *J. Clean. Prod.* **2017**, *162*, 571–586. [CrossRef]

45. Cecchini, L.; Venanzi, S.; Pierri, A.; Chiorri, M. Environmental efficiency analysis and estimation of CO_2 abatement cost in dairy cattle farms in Umbria (Italy): A SBM-DEA model with undesirable output. *J. Clean. Prod.* **2018**, *197*, 894–907. [CrossRef]
46. Iribarren, D.; Hospido, A.; Moreira, M.T.; Feijoo, G. Benchmarking environmental and operational parameters through eco-efficiency criteria for dairy farms. *Sci. Total Environ.* **2011**, *409*, 1786–1798. [CrossRef] [PubMed]
47. Margallo, M.; Onandía, R.; Aldaco, R.; Irabien, A. When life cycle thinking is necessary for decision making: Emerging cleaner technologies in the chlor-alkali industry. *Chem. Eng. Trans.* **2016**, *52*, 475–480.
48. *ISO 14040: Environmental Management—Life Cycle Assessment—Principles and Framework*; International Organization for Standardization: London, UK, 2006.
49. *ISO 14044: Environmental Management—Life Cycle Assessment—Requirements and Guidelines*; International Organization for Standardization: London, UK, 2006.
50. Rebitzer, G.; Ekvall, T.; Frischknecht, R.; Hunkeler, D.; Norris, G.; Rydberg, T.; Schmidt, W.P.; Suh, S.; Weidema, B.P.; Pennington, D.W. Life cycle assessment—Part 1: Framework, goal and scope definition, inventory analysis, and applications. *Environ. Int.* **2004**, *30*, 701–720. [CrossRef] [PubMed]
51. Guinée, J.B.; Udo de Haes, H.A.; Huppes, G. Quantitative life cycle assessment of products. 1: Goal definition and inventory. *J. Clean. Prod.* **1993**, *1*, 3–13. [CrossRef]
52. Margallo, M.; Dominguez-Ramos, A.; Aldaco, R.; Bala, A.; Fullana, P.; Irabien, A. Environmental sustainability assessment in the process industry: A case study of waste-to-energy plants in Spain. *Resour. Conserv. Recycl.* **2014**, *93*, 144–155. [CrossRef]
53. PE International. *Gabi 6 Software and Database on Life Cycle Assessment*; PE International: Leinfelden-Echterdingen, Germany, 2014.
54. Gustavsson, J.; Cederberg, C.; Sonesson, U.; Emanuelsson, A. *The Methodology of the FAO Study "Global Food Losses and Food Waste—Extent, Causes and Prevention"*; FAO, The Swedish Institute for Food and Biotechnology (SIK): Göteborg, Sweden, 2013.
55. Brand-Correa, L.; Brockway, P.; Copeland, C.; Foxon, T.; Owen, A.; Taylor, P. Developing an Input-Output Based Method to Estimate a National-Level Energy Return on Investment (EROI). *Energies* **2017**, *10*, 534. [CrossRef]
56. Hall, C.A.S.; Kiltgaard, K.A. *Energy and the Wealth of Nations. Understanding the Biophysical Economy*; Springer: New York, NY, USA, 2012.
57. Charnes, A.; Cooper, W.W.; Rhodes, E. Measuring the efficiency of decision making units. *Eur. J. Oper. Res.* **1978**, *2*, 429–444. [CrossRef]
58. Banker, R.D.; Charnes, A.; Cooper, W.W. Some models for estimating technical and scale inefficiencies in data envelopment analysis. *Manag. Sci.* **1984**, *30*, 1078–1092. [CrossRef]
59. Chu, J.; Wu, J.; Song, M. An SBM-DEA model with parallel computing design for environmental efficiency evaluation in the big data context: A transportation system application. *Ann. Oper. Res.* **2018**, *270*, 105–124. [CrossRef]
60. Iribarren, D.; Vázquez-Rowe, I.; Moreira, M.T.; Feijoo, G. Further potentials in the joint implementation of life cycle assessment and data envelopment analysis. *Sci. Total Environ.* **2010**, *408*, 5265–5272. [CrossRef] [PubMed]
61. Avadí, A.; Vázquez-Rowe, I.; Fréon, P. Eco-efficiency assessment of the Peruvian anchoveta steel and wooden fleets using LCA + DEA framework. *J. Clean. Prod.* **2014**, *70*, 118–131. [CrossRef]
62. Carrillo-Álvarez, E.; Pintó-Domingo, G.; Cussó-Parcerisas, I.; Riera-Romani, J. *The Spanish Healthy Food Basket Complete Report. Pilot Project for the Development of a Common Methodology on Reference Budgets in Europe*; Grup de Recerca en Pedagogia, Societat i Innovació amb el suport de les TIC (PSITIC): Barcelona, Spain, 2016.
63. MAPAMA, Spanish Ministry of Agriculture, Fishery, Food and Environment. Household Consumption Database. 2017. Available online: www.mapama.gob.es/es/alimentacion/temas/consumo-y-comercializacion-ydistribucion-alimentaria/panel-de-consumo-alimentario/base-de-datos-de-consumo-en-hogares/ (accessed on 19 September 2018). (In Spanish)
64. Batlle-Bayer, L.; Bala, A.; García-Herrero, I.; Lemaire, E.; Song, G.; Aldaco, R.; Fullana, P. National Dietary Guidelines: A potential tool to reduce greenhouse gas emissions of current dietary patterns. The case of Spain. *J. Clean. Prod.*. Under review.
65. Eurostat. Fishery Production in All Fishing Regions (tag00117). 2015. Available online: http://ec.europa.eu/eurostat/data/database (accessed on 19 September 2018).

66. Eurostat. Production and Utilization of Milk on the Farm—Annual Data (apro_mk_ farm). 2015. Available online: http://ec.europa.eu/eurostat/web/agriculture/data/database (accessed on 19 September 2018).
67. Eurostat. Slaughtering in Slaughterhouses—Annual Data (apro_mt_pann). 2015. Available online: http://ec.europa.eu/eurostat/web/agriculture/data/database (accessed on 19 September 2018).
68. Eurostat. Crop Products—Annual Data (apro_cpp_crop). 2015. Available online: https://ec.europa.eu/eurostat/web/agriculture/data/database (accessed on 19 September 2018).
69. Datacomex. Estadísticas del comercio exterior español, Agencia Española de Administración Tributaria. 2008. Available online: http://datacomex.comercio.es/ (accessed on 19 September 2018).
70. Milà i Canals, L.; Munoz, I.; Mclaren, S.J.; Brandão, M. *LCA Methodology and Modelling Considerations for Vegetable Production and Consumption*; Centre for Environmental Strategy, University of Surrey: Surrey, UK, 2007.
71. Nielsen, P.H.; Nielsen, A.M.; Weidema, B.P.; Dalgaard, R.; Halberg, N. LCA Food Database. 2003. Available online: http://www.lcafood.dk/ (accessed on 24 September 2018).
72. Foster, C.; Green, K.; Bleda, M.; Dewick, P.; Evans, B.; Flynn, M.J. Production and consumption: A research report completed for the department for environment. In *Food and Rural Affairs by Manchester Business School*; DEFRA: London, UK, 2006.
73. Farran, A.; Zamora, R.; Cervera, P. Nutrition and Dietetics Institute. In *Food Composition Tables from CESNID*; Barcelona University: Barcelona, Spain, 2004.
74. FAO-INFOODS. International Network of Food Data Systems. 2018. Available online: http://www.fao.org/infoods/infoods/tablas-y-bases-de-datos/es/ (accessed on 24 September 2018).
75. Jahanshahloo, G.R.; Hosseinzadeh, F.; Shoja, N.; Tohidi, G.; Razavyan, S. Undesirable inputs and outputs in DEA models. *Appl. Math. Comp.* **2005**, *169*, 917–925. [CrossRef]
76. Monforti-Ferrario, R.; Pinedo Pascua, I. *Energy Use in the EU Food Sector: State of Play and Opportunities for Improvement*; JRC Science and Policy Report; European Commission: Brussels, Belgium, 2015, ISBN 978-92-79-48299-1.
77. DEA Frontier. Joe Zhu's Research on Data Envelopment Analysis. 2018. Available online: http://www.deafrontier.net/index.html (accessed on 4 October 2018).
78. Murillo-Zamorano, L.R. Economic efficiency and frontier techniques. *J. Econ. Surv.* **2004**, *18*, 33–78. [CrossRef]
79. Ghasempour, A.; Ebrahim, A. Assessment of the environment impacts of egg production chain using life cycle assessment. *J. Environ. Manag.* **2016**, *183*, 980–987. [CrossRef]
80. Avadí, A.; Fréon, P.; Quispe, I. Environmental assessment of Peruvian anchoveta food production: Is less refined better? *Int. J. Life Cycle Assess.* **2014**, *19*, 1276–1293. [CrossRef]
81. Avadí, A.; Bolaños, C.; Sandoval, I.; Ycaza, C. Life cycle assessment of Ecuadorian processed tuna. *Int. J. Life Cycle Assess.* **2015**, *20*, 1415–1428. [CrossRef]
82. Vázquez-Rowe, I.; Villanueva-Rey, P.; Hospido, A.; Moreira, M.T.; Feijoo, G. Life cycle assessment of European pilchard (*Sardina pilchardus*) consumption. A case study for Galicia (NW Spain). *Sci. Total Environ.* **2014**, *475*, 48–60. [CrossRef]

© 2018 by the authors. Licensee MDPI, Basel, Switzerland. This article is an open access article distributed under the terms and conditions of the Creative Commons Attribution (CC BY) license (http://creativecommons.org/licenses/by/4.0/).

Article

Particulate Matter Produced by Micro-Scale Biomass Combustion in an Oxygen-Lean Atmosphere

Jan Poláčik [1], Ladislav Šnajdárek [1], Michal Špiláček [2,*], Jiří Pospíšil [1] and Tomáš Sitek [1]

1. Energy Institute, NETME Centre, Brno University of Technology, Technická 2896/2, 61669 Brno, Czech Republic; jan.polacik@vutbr.cz (J.P.); snajdarek@fme.vutbr.cz (L.Š.); pospisil.j@fme.vutbr.cz (J.P.); tomas.sitek@vutbr.cz (T.S.)
2. Sustainable Process Integration Laboratory—SPIL, NETME Centre, Brno University of Technology, Technická 2896/2, 61669 Brno, Czech Republic
* Correspondence: michal.spilacek@vutbr.cz; Tel.: +420-541-142-573

Received: 30 October 2018; Accepted: 21 November 2018; Published: 1 December 2018

Abstract: This article extends earlier research by the authors that was devoted to the experimental evaluation of ultra-fine particles produced by the laboratory combustion of beechwood samples. These particles can have severe influence on human health. The current paper presents a parametrical study carried out to assess the influence of the composition of the atmosphere and the temperature on the production of ultra-fine particles during the micro-scale combustion process. The paper presents a laboratory procedure that incorporate the thermogravimetric analysis (TGA) and detailed monitoring of the size distribution of the produced fine particles. The study utilises the laboratory scale identification of the formation and growth of the fine particles during the temperature increase of beech wood samples. It also compares the particle emissions produced by beech heartwood and beech bark. The size of the emitted particles is very strongly influenced by the concentration of light volatiles released from the heated wood sample. From the experimental study, decreasing oxygen content in the atmosphere generally results in higher particulate matter (PM) production.

Keywords: particulate matter; fine particles; combustion particles; nucleation; particle growth

1. Introduction

In recent years, the monitoring of air pollution has been heavily focused on particulate matter (PM) [1,2]. The negative impact of fine and ultra-fine particles has been studied by many authors. Ultra-fine particles are dangerous to human health [3] and have much greater active surface than larger particles with comparable weight and, therefore, have a greater ability to bind to other harmful substances [4]. Particles smaller than 1 micrometre (PM_1) are less than the minimum size for the lung's self-cleaning ability. The smallest particles can even diffuse through the wall of alveolar sacs and into the bloodstream [3]. The ultra-fine particles are produced during combustion processes due to a rise in temperature. Internal combustion engines and small furnaces are the main producers of ultra-fine particles in urban areas [5] and most of the particles produced by these combustion processes are PM_1 [6], which becomes increasingly problematic in urban areas with a high density of residents using cars and small furnaces. The production of ultra-fine particles from these sources is significant and the final emission concentrations are dependent on the conditions of dispersion. Unfortunately, city buildings are a great obstacle for air flow at the canopy layer of air and the low air velocity creates conditions for a long residence time of particulates in air [7]. Ultra-fine particles do not sediment and can remain in the environment for days and weeks. They are separated from the air by touching a solid surface like walls, roads, vegetation, or a liquid surface like water bodies or raindrops. The ultra-fine particles can grow to fine particles by (i) coagulation mechanisms, that aggregates colloidal and macromolecular organic particles into larger clusters, (ii) by agglomeration, where particles bonding is based on the adhesion of the surfaces, by (iii) oxidation reactions or by (iv) condensation of condensable vapours on the particle surface [8].

This paper focuses on the emissions of ultra-fine particles released by the combustion process occurring in a small biomass furnace. There is a vast variety of those furnaces and in each of them are different combusting conditions. Some parts of the furnaces have intense air flow that causes the flying of ash. Sides of a combustion chamber are intensively heated by exothermic reactions as there is a high concentration of volatiles in the combustion chamber. These volatiles undergo incomplete combustion due to a lack of intake air and then cool down and condense – a mechanism resulting in the nucleation of ultra-fine particles.

The measurement of the ultra-fine particles dispersed in the flue gas is important for identifying the effects of furnace operating parameters (heat power output, a surplus of combustion air, temperature) on the resulting emissions of the particles. Such measurements are also used for comparison of different fuels in one furnace [9], but these experimental measurements do not provide detailed insight at the sub-stages of the process of nucleation and subsequent agglomeration of particles. Numerous studies published by different authors deal with this topic, but the process of particle nucleation is still not sufficiently explained and clarified [10–12]. It is necessary to continue with the experimental investigation of numerous samples and analyse the result for a better understanding of the nucleation and the growth process. These processes are studied in laboratory conditions with ideal combustion of fuel samples. This experimental research is the main target of this paper and uses thermogravimetric analysis (TGA) for the identification of particles emitted from a small sample of wood. The measuring instrument provides the particle size distribution of particulate matter. It is very important to understand the nucleation and creation of these particles in the combustion process to create viable strategies for their elimination.

An earlier study by the authors [13] dealt with a laboratory investigation of fine particulate matter production from the controlled heating of beechwood samples in the atmosphere with 21% oxygen. This paper extends this previous experimental investigation by using lower oxygen content in the atmosphere during laboratory combustion. The limiting test case uses an inert atmosphere with 100% nitrogen.

2. Particulate Formation

When heat flux is applied to the solid phase of woods it is divided into three distinctive layers: (i) the char layer, (ii) the pyrolysis layer and (iii) raw wood. Between the char and pyrolysis layer is identified the char front in which occurs transition from raw wood into char by thermal decomposition (pyrolysis). This transition is usually considered to take place at the 300 °C isotherm, called the char-line [14].

When external heat flux affects the wood surface, part of the heat is reflected from the surface. Convective heat transfer between surrounding gases and the wood surface also occurs. In the pyrolysis layer, water evaporation occurs first and is later accompanied by pyrolysis reactions and the production of gas volatiles. Vapour and gas volatiles penetrate through pores and leave the wood.

Released volatile components have different values of partial pressure. When they reach their respective saturation point, the formation of a new phase begins – this is the nucleation process. The molecules are clustered into ultra-fine particles up to 0.1 µm in size. These particles can further grow by mechanisms mentioned in the introduction section. Particles that are formed in the combustion chamber are called primary particles. Particles formed in the flue gas duct and in the atmosphere are called secondary particles. Organic particles have different levels of volatility that divides them into volatile organic compounds (VOC) and semi-volatile organic compounds (SVOC) [15].

2.1. Soot

Soot is a spherical particle of impure carbon coated in polyaromatic hydrocarbons (PAHs) [16]. These pollutants are usually the result of a fuel-rich combustion at high temperatures (over 900 °C).

Combustion at low temperatures (under 700 °C) results in CO, VOC, smoke species, and soot coated in oxidized PAHs [17]. Production of soot can also occur in fuel-lean conditions if some parts of the fuel cannot be accessed by the combustion air both in the fuel bed and in the freeboard flue gas.

Modelling of soot formation is based on experimental work on laminar flames of simple and clean fuels like ethylene and methylene, and to date, no model describes the formation precisely. But there is an agreement on the basic mechanism of soot formation, and a mechanism known as HACA —hydrogen abstraction–C_2H_2 addition, that also describes the evolution of PAHs, manages to describe the nucleation of soot [18]. For the description of this mechanism, a basic chemical and physical frame was created. In the chemical frame, the main precursor of soot formation is considered to be acetylene C_2H_2 that is reacting to create higher hydrocarbons (aromatics with one or two rings) that are the basis of PAH [19]. Once the PAH is formed, the physical frame can be used to describe the rest of the formation. This consists of four steps: formation (nucleation), coagulation, condensation, and surface growth [19]. Coagulation and condensation are shown in Figure 1. These two steps are creating agglomerates that, after further reactions, turn into impure carbon nuclei. Oxidation by O_2 and OH appears at all steps [18]. The main source of these pollutants is incomplete combustion and is controllable to some degree, because when biomass is combusted on grate pyrolysis always takes place, by the quality of the combustion process.

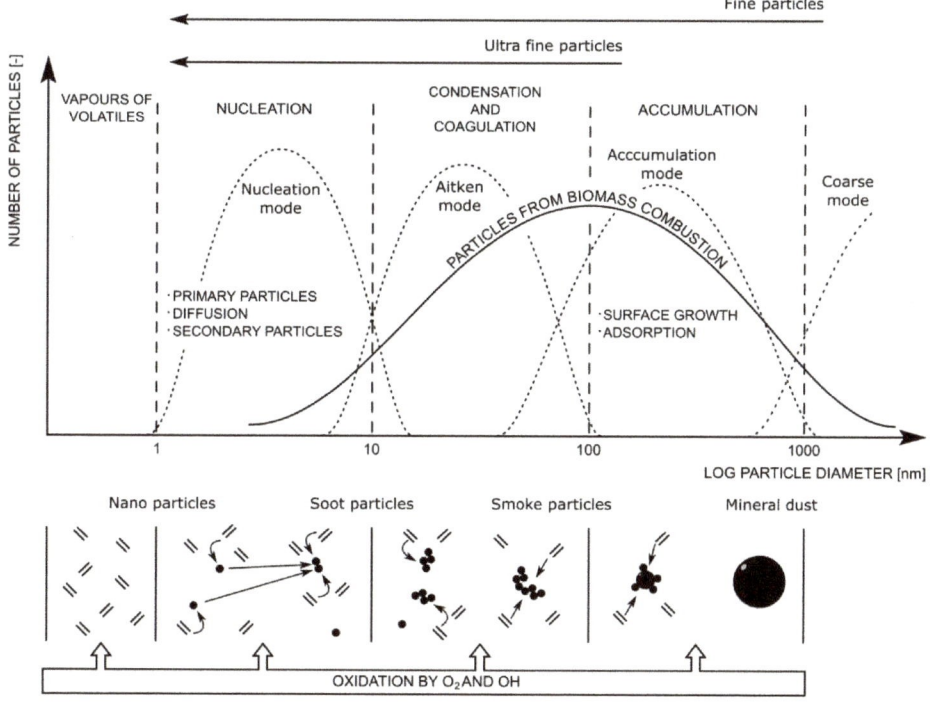

Figure 1. Nucleation and growth of combustion particles.

2.2. Fly Ash

Biomass generally contains trace amounts of metals and the most important of them is potassium [17]. These metal elements are released during combustion and create the compounds responsible for fouling the equipment after condensing on its walls. But some of these compounds leave the combustion equipment in the form of aerosols that are dangerous for human health by themselves or when they create particles of fly ash in a similar fashion to the formation of soot particles. The reason for this is that these mechanisms are interconnected.

The inception and growth of ultra-fine particles are very complex processes in which it is difficult to quantify the intensity of individual incremental changes. A certain help in the quantification is numerical modelling, but despite the great advances in the development of models and the greater computational resources, it is not yet possible to account for every process and create an exact model of the particulate's inception and growth. Additionally, models of biomass combustion do not usually account for particulates [20].

3. Experimental Procedure and Measurement Device

From the previously stated reasons, the focus of this paper is on the laboratory measurements of the size distribution of ultra-fine particles emitted from the combustion of beech wood. The developed laboratory procedure incorporates the advantages of thermogravimetric analysis and the detailed monitoring of the size distribution of the produced fine particles. Thermogravimetric analysis (TGA) allows to monitor the exact influence temperature has on a small fuel sample according to the desired heating schedule by monitoring the weight of the heat affected sample and identifying the weight loss. TGA can also change the composition of the atmosphere flowing around the sample. By analysing the development of temperature change, it further identifies the presence of endothermic and exothermic reactions.

The measurement was carried out by utilising a STA 449 F3 Jupiter TGA device (NETZSCH, Selb, Germany). The base component of the STA-449 analyser is a very precise digital weighing system with a vertical design. The analysed samples are placed into a shielded ceramic module (TG-module). This module is linked to the weighting system itself. For the whole duration of the measurement, the entire module is located in a gas-tight laboratory furnace with a controlled heating rate. For each measurement the device provides a TGA curve showing the relationship between weight change and the temperature of the sample.

The gas phase volatiles released by pyrolysis from the fuel sample during the measurement are scattered in the test atmosphere and removed from the device. Subsequently, the resulting stream is cooled down to the ambient temperature by flowing through the connected pipeline, creating an aerosol stream that enters a Scanning Mobility Particle Sizer (SMPS). In SMPS are separated fractions of different particle size. The production of fine particles during the process of combustion of the wood sample was measured with a TSI-SMPS device (Model 3080-Series Electrostatic Classifiers including CPC 3775, TSI Inc., Minneapolis, MN, USA). The SMPS consists of an aerosol neutralizer and differential mobility analyser (DMA).

Almost every particle has some level of electric charge. The DMA requires the aerosol to achieve a uniform and steady charge distribution. This is achieved with an aerosol neutralizer device, which provides a process that neutralises the charge of particles. After neutralisation, the particles pass through a bipolar charger and are all charged to a unified level. Then, the aerosol flows into the DMA where sizing occurs, see Figure 2.

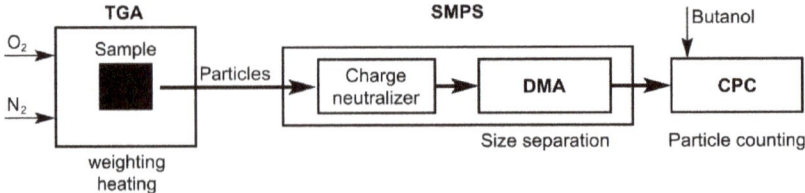

Figure 2. Scheme of the measuring equipment and the particles' flow.

As a particle sizing system, the DMA separates particles by size for high-resolution measurements of the particle size distribution and allows particles in the range of 10 to 1000 nanometers in diameter to be classified. The size of the particles in the polydisperse aerosol flowing though the DMA is determined by their electrical mobility and as only particles of specific sizes are selected, a highly monodisperse aerosol leaves the DMA. This aerosol continues to a condensation particle counter (CPC). The purpose of this device is to measure the particle number concentration and identify particles in the particle-size fractions.

The CPC function is based on condensation of butanol vapours on the monodisperse aerosol's particles that act as condensation nuclei. The condensation occurs in a cooled condenser and the particles grow into larger droplets of such size that they can be counted by an optical detector when passing through it [9].

4. Experimental Results and Discussion

Many parameters vary during the combustion process. Temperature and combustion air surplus being the most variable depending on the location in the combustion chamber. The measuring apparatus described in the previous section was used for the research of nucleation and growth of the fine particles in the flue gas from beechwood combustion.

Plants and trees are mainly composed by cellulose fibrils [21]. Next to the cellulose cell walls of wood are composed by lignin and hemicellulose. Cellulose and hemicellulose together are called holocellulose. These three together principally represents the total carbohydrate fraction and are considered as the bulk of wood. The proportions of holocellulose and lignin are different in various parts of woody plants, and, therefore, their residues (sawdust, bark) have different compositions. The beech is a genus of trees and the beechwood have the following composition: extractables (2%), lignin (20%) and holocellulose (78%) [22], as shown in the Table 1.

Table 1. Chemical composition of beechwood. Adapted with permission from Elsevier [22].

Description	Value
Density	730 kg/m^3
Humidity	6.1%
Lignin	20%
Holocellulose	78%
α-cellulose	50%
Extractives	2%

Blocks of beechwood were used as testing samples with each block weighing 80 mg. The samples were heated up in the thermogravimetric analyser from 20 °C to 620 °C over a period of 120 min. The constant temperature increase was set at +5 °C/min. The experiments were carried out for beech heartwood and beech bark. To simulate the wide spectrum of possible conditions of the combustion process the samples were heated up in atmospheres with different oxygen concentrations, namely with the concentrations of 0%, 5%, 10% and 15% of oxygen. Each atmosphere was prepared by mixing pure oxygen and pure nitrogen from pressurised vessels connected to the TGA device.

Mass loss occurs during the controlled heating up the samples as shown in Figure 3. This figure presents the temperature pattern of wood samples (increasing curve) and the associated weight loss of the samples (decreasing curves). Water evaporates from the tested samples during the temperature increase up to 120 °C, then light volatile compounds are released in the interval from 120 °C to 250 °C. In the interval from 250 °C to 300 °C occurs a rapid loss of a sample's weight induced by the intensive release of remaining volatile compounds. The release of these volatile compounds is intensified by the heat generated from exothermic reactions that commence within these temperatures. The interval from 300 °C to 500 °C is associated with the gradual decomposition of carbon in charcoal. The relations presented at the Figure 3 show the different intensity of incineration of samples in different atmospheres. Higher concentration of the oxygen causes more intensive combustion of charcoal. In the atmosphere formed only by pure nitrogen, 30% of the sample remained unburned in form of charcoal. Influence of the oxygen content in the atmosphere is similar for beech heartwood and beech bark.

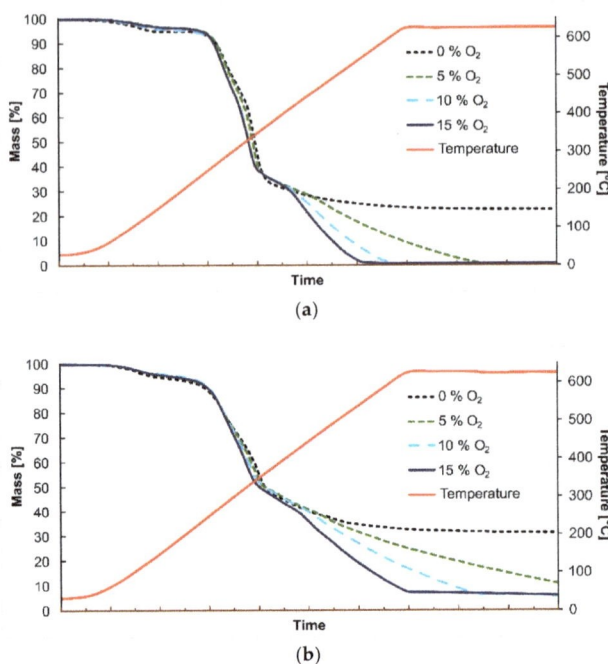

Figure 3. Results from the thermogravimetric analysis of beech samples in different atmospheres. (a) Beech heartwood; (b) Beech bark.

The thermal decomposition of the beechwood samples is influenced by the intensity of the oxidation reactions. The result of the thermal decomposition is the production of ultra-fine particles scattered in the gaseous products that leave the TGA apparatus. Figure 4. presents the obtained distribution of the generated particle concentration for all tested concentrations of oxygen in the atmosphere. The presented results represent the sum of all particles generated during the increase of temperature from 20 °C to 620 °C over the testing period of 120 min. Significant concentrations of the ultra-fine particles occurred within the particle mobility diameter of 50 nm–340 nm and the maximum concentration corresponds to a particle size around 140 nm. This corresponds to the typical concentration peak of combustion particles formed by the coagulation process. At 300 °C, the highest concentrations of emitted particles were measured. This temperature is close to the ignition temperature of the samples and causes an intense production of volatile compounds that subsequently lead to the nucleation of a great number of particles that in turn, coagulate into large particles.

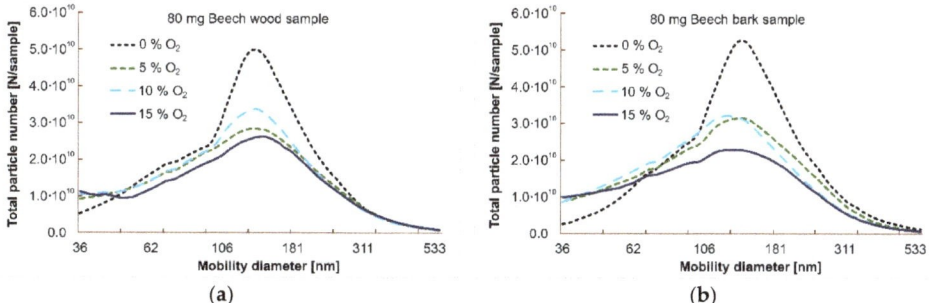

Figure 4. The distribution curves of the total particle number generated over the testing period. (**a**) Beech heartwood; (**b**) Beech bark.

Low concentrations of oxygen in the atmosphere generally caused a higher number of produced particles. This relationship is valid for particle size bigger than 50 nm. In the case of a pure nitrogen atmosphere, significant concentrations of particulate matter were obtained. This is due to the absence of most oxidation reactions, also leading to a decrease in the local temperature affecting other reactions. The dependence of the production of fine parts on the oxygen concentration in the atmosphere is more pronounced when burning beech bark compared to the beech heartwood. Beech bark produces a higher number of particles in the atmosphere of pure nitrogen. The same sample produces a lower particle number in the atmosphere with 15% oxygen and more.

Figure 5 shows the total mass of all particles generated during the increase of temperature from 20 °C to 620 °C over the testing period of 120 min. The maximum total particle mass was identified for particle sizes close to 200 nm. As oxygen concentration in the atmosphere increases, the total particle mass generally decreases. Threshold oxygen concentrations in the atmosphere were identified to be between 5% and 10% of oxygen. For oxygen concentrations up to 5%, the higher total particle mass was achieved by burning beech bark; however, for oxygen concentrations of 10% and above, the higher total particle mass was achieved by burning beech heartwood.

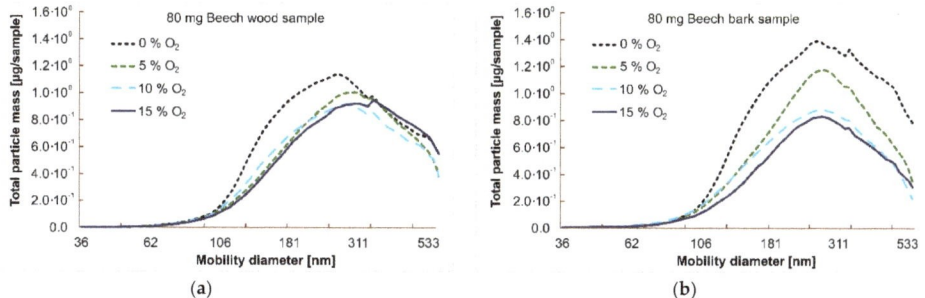

Figure 5. The relationship between the total particle mass generated over the testing period and the mobility diameter of the particles. (**a**) Beech heartwood; (**b**) Beech bark.

5. Conclusions

The fine and ultra-fine particles produced by combustion processes represent a significant health risk. The entire process of inception and growth of these particles is very complex and still not sufficiently described. The study of this issue in actual combustion devices is very difficult due to the instability and diversity of the combustion process due to the wide ranges of devices and fuels. The research presented in this paper focused on the emission of ultra-fine particles from the controlled heating of small samples of beechwood (heartwood and bark). A connection between thermogravimetric analysis and the mobility diameter measurement of particulates was realised for this purpose. Other studies looking into the topic of particulate emissions from biomass combustion were interested in: (i) the chemical composition of released volatile compounds from which particulate matter can form [23–25], (ii) mass amount of particulate matter released by different types of biomass [23,26–28] and (iii) number of released particles [23,26,28]. These studies were conducted on small (domestic) biomass burners for different types of wood (beech included) and under various conditions specified by % of the nominal output of the device under observation. But wider range of atmospheric conditions is required for modern approaches for biomass combustion [29], pyrolysis (0% O_2) included. The study carried out in this paper was a parametric sweep tracking the influence of different concentrations of oxygen in the atmosphere on the production of particulates during a controlled heat up.

From the experimental measurements it can be concluded that increasing the temperature leads to a higher emission of particles produced by the combustion process. This applies until the ignition temperature is reached. This is most likely due to the increasing concentration of volatile vapours. Just before the ignition temperature the highest amount of volatile vapours is evaporating from the sample leading to the most intensive nucleation of particles resulting in their high concentrations, coagulation and growth. When the ignition temperature of the volatile vapours is reached, which can only happen if oxygen is present, the concentration of these vapours diminishes and so does the production of the particulates. This diminishing process occurs until the maximum experimental temperature of 620 °C.

The utilisation of the thermogravimetric analysis is suitable for the experimental measurement of particulates produced by combustion and is a suitable tool for detailed insight into the sub-processes of the particulate's inception and growth during the combustion process, which can be further expanded by performing the same study for other types of woods. With the information from [30] about the effective density of the particulate matter a deeper insight into the researched problem can be obtained.

Author Contributions: Conceptualization, J.P. (Jan Poláčik) and J.P. (Jiří Pospíšil); Data curation, J.P. (Jan Poláčik) and T.S. (Tomáš Sitek); Methodology, L.Š. (Ladislav Šnajdárek); Supervision, J.P. (Jiří Pospíšil); Visualization, M.Š.; Writing—original draft, M.Š. and J.P. (Jiří Pospíšil).

Funding: This paper has been supported by the EU projects: Sustainable Process Integration Laboratory – SPIL funded as project No. CZ.02.1.01/0.0/0.0/15_003/0000456 and Computer Simulations for Effective Low-Emission Energy funded as project No. CZ.02.1.01/0.0/0.0/16_026/0008392 by Czech Republic Operational Programme Research, Development and Education, Priority 1: Strengthening capacity for high-quality research and the collaboration.

Conflicts of Interest: The authors declare no conflict of interest.

References

1. Chen, D.; Liu, X.; Han, J.; Jiang, M.; Xu, Y.; Xu, M. Measurements of particulate matter concentration by the light scattering method: Optimization of the detection angle. *Fuel Process. Technol.* **2018**, *179*, 124–134. [CrossRef]
2. Carminati, M.; Ferrari, G.; Sampietro, M. Emerging miniaturized technologies for airborne particulate matter pervasive monitoring. *Measurement* **2017**, *101*, 250–256. [CrossRef]

3. Santibáñez-Andrade, M.; Quezada-Maldonado, E.M.; Osornio-Vargas, Á.; Sánchez-Pérez, Y.; García-Cuellar, C.M. Air pollution and genomic instability: The role of particulate matter in lung carcinogenesis. *Environ. Pollut.* **2017**, *229*, 412–422. [CrossRef] [PubMed]
4. Baldock, J.A.; Smernik, R.J. Chemical composition and bioavailability of thermally altered Pinus resinosa (Red pine) wood. *Org. Geochem.* **2002**, *33*, 1093–1109. [CrossRef]
5. Karagulian, F.; Belis, C.A.; Dora, C.F.C.; Prüss-Ustün, A.M.; Bonjour, S.; Adair-Rohani, H.; Amann, M. Contributions to cities' ambient particulate matter (PM): A systematic review of local source contributions at global level. *Atmos. Environ.* **2015**, *120*, 475–483. [CrossRef]
6. Cheng, Y.; Zou, S.C.; Lee, S.C.; Chow, J.C.; Ho, K.F.; Watson, J.G.; Han, Y.M.; Zhang, R.J.; Zhang, F.; Yau, P.S.; et al. Characteristics and source apportionment of PM1 emissions at a roadside station. *J. Hazard. Mater.* **2011**, *195*, 82–91. [CrossRef] [PubMed]
7. Kukkonen, J.; Karl, M.; Keuken, M.P.; Denier van der Gon, H.A.C.; Denby, B.R.; Singh, V.; Douros, J.; Manders, A.; Samaras, Z.; Moussiopoulos, N.; et al. Modelling the dispersion of particle numbers in five European cities. *Geosci. Model Dev.* **2016**, *9*, 451–478. [CrossRef]
8. Obaidullah, M.; Bram, S.; De Ruyck, J.; Verma, V.K. A Review on Particle Emissions from Small Scale Biomass Combustion. *Int. J. Renew. Energy Res.* **2012**, *2*, 147–159.
9. Kantová, N.; Nosek, R.; Holubčík, M.; Jandačka, J. The Formation of Particulate Matter during the Combustion of Different Fuels and Air Temperatures. *Renew. Energy Sources Eng. Technol. Innov.* **2018**, 47–52. [CrossRef]
10. Kwiatkowski, K.; Zuk, P.J.; Dudyński, M.; Bajer, K. Pyrolysis and gasification of single biomass particle—New openFoam solver. *J. Phys. Conf. Ser.* **2014**, *530*. [CrossRef]
11. Lupascu, A.; Easter, R.; Zaveri, R.; Shrivastava, M.; Pekour, M.; Tomlinson, J.; Yang, Q.; Matsui, H.; Hodzic, A.; Zhang, Q.; et al. Modeling particle nucleation and growth over northern California during the 2010 CARES campaign. *Atmos. Chem. Phys.* **2015**, *15*, 12283–12313. [CrossRef]
12. Appel, J.; Bockhorn, H.; Frenklach, M. Kinetic modeling of soot formation with detailed chemistry and physics: Laminar premixed flames of C2 hydrocarbons. *Combust. Flame* **2000**, *121*, 122–136. [CrossRef]
13. Poláčik, J.; Pospíšil, J.; Šnajdárek, L.; Sitek, T. Influence of temperature on the production and size distribution of fine particles released from beech wood samples. *Matec Web Conf.* **2018**, *168*, 1–9. [CrossRef]
14. Cachim, P.B.; Franssen, J.M. Comparison between the charring rate model and the conductive model of Eurocode 5. *Fire Mater.* **2009**, *33*, 129–143. [CrossRef]
15. Sippula, O. Fine Particle Formation and Emissions in Biomass Combustion Title. Ph.D. Thesis, University of Eastern Finland, Kuopio City, Finland, 2010.
16. Battin-Leclerc, F.; Simmie, J.M.; Blurock, E. *Cleaner Combustion: Developing Detailed Chemical Kinetic Models*, 1st ed.; Springer: New York, NY, USA, 2013;, ISBN 978-1-4471-6909-3.
17. Williams, A.; Jones, J.M.; Ma, L.; Pourkashanian, M. Pollutants from the combustion of solid biomass fuels. *Prog. Energy Combust. Sci.* **2012**, *38*, 113–137. [CrossRef]
18. Desgroux, P.; Faccinetto, A.; Mercier, X.; Mouton, T.; Aubagnac Karkar, D.; El Bakali, A. Comparative study of the soot formation process in a "nucleation" and a "sooting" low pressure premixed methane flame. *Combust. Flame* **2017**, *184*, 153–166. [CrossRef]
19. Saini, R.; De, A. Assessment of soot formation models in lifted ethylene/air turbulent diffusion flame. *Therm. Sci. Eng. Prog.* **2017**, *3*, 49–61. [CrossRef]
20. Abricka, M.; Barmina, I.; Valdmanis, R.; Zake, M.; Kalis, H. Experimental and Numerical Studies on Integrated Gasification and Combustion of Biomass. *Chem. Eng. Trans.* **2016**, *50*, 127–132. [CrossRef]
21. Poletto, M.; Pistor, V.; Zatera, J.A. Structural Characteristics and Thermal Properties of Native Cellulose. In *Cellulose—Fundamental Aspects*; IntechOpen Limited: London, UK, 2013; pp. 45–68, ISBN 978-953-51-1183-2.
22. Di Blasi, C.; Branca, C.; Santoro, A.; Gonzalez Hernandez, E. Pyrolytic behavior and products of some wood varieties. *Combust. Flame* **2001**, *124*, 165–177. [CrossRef]
23. Weimer, S.; Alfarra, M.R.; Schreiber, D.; Mohr, M.; Prévôt, A.S.H.; Baltensperger, U. Organic aerosol mass spectral signatures from wood-burning emissions: Influence of burning conditions and wood type. *J. Geophys. Res.* **2008**, *113*. [CrossRef]

24. Hilbers, T.J.; Wang, Z.; Pecha, B.; Westerhof, R.J.M.; Kersten, S.R.A.; Pelaez-Samaniego, M.R.; Garcia-Perez, M. Cellulose-Lignin interactions during slow and fast pyrolysis. *J. Anal. Appl. Pyrolysis* **2015**, *114*, 197–207. [CrossRef]
25. Quan, C.; Gao, N.; Song, Q. Pyrolysis of biomass components in a TGA and a fixed-bed reactor: Thermochemical behaviors, kinetics, and product characterization. *J. Anal. Appl. Pyrolysis* **2016**, *121*, 84–92. [CrossRef]
26. Johansson, L.S.; Tullin, C.; Leckner, B.; Sjövall, P. Particle emissions from biomass combustion in small combustors. *Biomass Bioenergy* **2003**, *25*, 435–446. [CrossRef]
27. Zosima, A.T.; Ochsenkühn-Petropoulou, M. Characterization of the Particulate Matter and Carbonaceous Particles Produced by Biomass Combustion. *Anal. Lett.* **2015**, *49*, 1102–1113. [CrossRef]
28. Torvela, T.; Tissari, J.; Sippula, O.; Kaivosoja, T.; Leskinen, J.; Virén, A.; Lähde, A.; Jokiniemi, J. Effect of wood combustion conditions on the morphology of freshly emitted fine particles. *Atmos. Environ.* **2014**, *87*, 65–76. [CrossRef]
29. Rodilla, I.; Contreras, M.L.; Bahillo, A. Thermogravimetric and mass spectrometric (TG-MS) analysis of sub-bituminous coal-energy crops blends in N_2, air and CO_2/O_2 atmospheres. *Fuel* **2018**, *215*, 506–514. [CrossRef]
30. Zhai, J.; Lu, X.; Li, L.; Zhang, Q.; Zhang, C.; Chen, H.; Yang, X.; Chen, J. Size-resolved chemical composition, effective density, and optical properties of biomass burning particles. *Atmos. Chem. Phys.* **2017**, *17*, 7481–7493. [CrossRef]

© 2018 by the authors. Licensee MDPI, Basel, Switzerland. This article is an open access article distributed under the terms and conditions of the Creative Commons Attribution (CC BY) license (http://creativecommons.org/licenses/by/4.0/).

MDPI
St. Alban-Anlage 66
4052 Basel
Switzerland
Tel. +41 61 683 77 34
Fax +41 61 302 89 18
www.mdpi.com

Energies Editorial Office
E-mail: energies@mdpi.com
www.mdpi.com/journal/energies

www.ingramcontent.com/pod-product-compliance
Lightning Source LLC
LaVergne TN
LVHW071934080526
838202LV00064B/6606